Fundamentals of RF and Microwave Transistor Amplifiers

Fundamentals of RF and Microwave Transistor Amplifiers

Inder J. Bahl

WILEY

A John Wiley & Sons, Inc., Publication

Published by John Wiley & Sons, Inc., Hoboken, New Jersey
Published simultaneously in Canada.

Library of Congress Cataloging-in-Publication Data:

Bahl, I. J.
Fundamentals of RF and Microwave Transistor Amplifiers / Inder Bahl.
p. cm.
Includes bibliographical references and index.
ISBN 978-0-470-39166-2 (cloth)
1. Amplifiers, Radio frequency. 2. Microwave amplifiers. 3. Transistor amplifiers. I. Title.
TK6565.A55B26 2009
621.3815′35–dc22

2008049895

10 9 8 7 6 5 4 3 2 1

Contents in Brief

Contents

Foreword

It has been a pleasure to review a book manuscript called *Fundamentals of RF and Microwave Transistor Amplifiers* and to reflect on the career of the author, my friend and colleague of more than 25 years, Dr. Inder Bahl.

Dr. Bahl is a man who is passionate about microwaves. Microwaves are his work and his play. I recall a time when those of us working with Dr. Bahl could tell when he was on vacation because as he sat at his desk working his latest microwave project, he wasn't wearing a tie! Dr. Bahl has been a prodigious contributor to the microwave art, authoring or coauthoring over 150 papers, 12 books, serving as the editor of *International Journal of RF and Microwave Computer-Aided Engineering*, and successfully completing literally hundreds of low noise, power, and control MMIC designs. Beyond the science, engineering, and the math, Dr. Bahl has a canny intuitive feel for how microwave circuits behave. It is almost as though Dr. Bahl can surf the microwaves through the circuit, feeling the gains and losses, experiencing the discontinuities, and uncovering the hidden gremlins!

The writing of this text is a gift from Dr. Bahl to the microwave community. Authoring a text of the scope and magnitude of *Fundamentals of RF and Microwave Transistor Amplifiers* is a monumental task the success of which is a tribute to Dr. Bahl's broad and extensive experience in the field, and to his dedication. His goal is to support and encourage others to participate in the microwave art to which Dr. Bahl has dedicated his life, and to share his excitement!

In this book, Dr. Bahl has outlined for his readers the keys to successful transistor amplifier design. In the following text, you will get the opportunity to see the world of solid-state RF and microwave amplifiers through the eyes of a master. What does he think about? What's important? How does he proceed with a design? Sit back, read, enjoy, and get prepared for what the future will bring. The excitement is just beginning!

Dr. Edward L. Griffin

Roanoke, Virginia
December 2008

Preface

Amplifiers have played a vital role in the development of high-performance and low-cost solutions for front-end RF and microwave systems. Numerous articles scattered in a wide array of technical journals and conference proceedings, book chapters, and even books have been published on amplifiers. However, no comprehensive text dedicated to this topic covering both theory and practical aspects exists. Thus there is an urgent need to bring out a book on this subject to fill the void.

This book evolved basically with my transistor amplifier design experience over the past 28 years. I have been actively involved with numerous amplifier designs from the concept level to the end products. The present book provides a comprehensive treatment of RF and microwave low-noise and power amplifier circuits including, low noise, narrowband, broadband, linear, high power, high efficiency, and high voltage. The topics discussed include modeling, analysis, design, packaging, and thermal and fabrication considerations. The elements of the book are self-contained and cover practical aspects in detail. The book also includes extensive design information in the form of equations, tables, graphs, and examples and has a unique integration of theory and practical aspects of amplifier circuits. Amplifier related design problems range from matching networks to biasing and stability. Practical examples (over 80 fully solved) make it simple to understand the concepts of amplifier design.

Simple design equations are also included to help the reader understand design concepts. In addition to the solved examples, over 160 problems are provided to help readers test their basic amplifier and circuit design skills. With its emphasis on theory, design, and practical aspects geared toward day-to-day applications, this book is intended for students, teachers, scientists, and practicing engineers. Students are required to have prior knowledge of topics such as solid state device basics, theory of transmission lines, basic circuit theory, and electromagnetics taught at the undergraduate level. It is hoped that through this book, readers will benefit in their quest to understand RF and microwave transistor amplifier circuit design.

The unique features of this book include in-depth study of transistor amplifiers, extensive design equations and figures, treatment of the practical aspects of amplifier circuits, and description of fabrication technologies. It provides a broad view of solid state transistor amplifiers. It has dedicated chapters on topics such as stability analysis, high-efficiency amplifiers, broadband amplifiers, monolithic amplifiers, high-voltage design, biasing of amplifiers, thermal design, power combining, integrated function amplifiers, and amplifier measurements. This book is not intended to cover any specific application, however; its purpose is to present essential background material in the fundamentals of amplifier design including both theoretical and practical aspects. RF and microwave circuits using Si bipolar and CMOS technologies have made tremendous progress and an enormous number of papers have been published in recent years.

This topic is covered in a limited scope because there are several excellent books available on this subject and amplifier designs are largely based on analog design concepts.

The book is divided into 22 chapters, with the material treated precisely and thoroughly and covering various aspects of amplifiers in each chapter. These chapters present the basic principles, analysis, techniques, and designs used in transistor amplifiers and provide the foundation for the analysis and design of RF and microwave transistor amplifiers. Design procedures and examples are provided in each chapter. The step-by-step procedures help to eliminate any doubts and help the student sharpen his/her design skills. In addition, technical information and remarks on various components, devices, and circuits update the reader on the most widely used microwave techniques. It is hoped that the selection of topics and their presentation will meet the expectations of the readers. Like any other comprehensive book, the work of other researchers is included or cited for further reading. This book also includes a comprehensive list of references. Finally, most chapters have a set of problems.

Chapter 1 provides an introduction to transistor amplifiers and their applications in both commercial and military systems. Chapter 2 establishes the basic amplifier analysis parameters and representation of RF and microwave networks. Fundamental network analysis tools such as impedance, admittance, *ABCD*, and scattering matrix techniques are introduced together with the properties of multiport networks. Relationships between the commonly used matrix representation forms are established to permit the researcher or designer to work in the system of his/her preference.

Chapter 3 deals with the definition of amplifier terms and characterization of amplifiers. Fundamental amplifier parameters are defined, together with a brief introduction to reliability. The intent of this chapter is to define amplifier characteristics at one place for quick reference. Chapter 4 deals with transistors, including Si bipolar, GaAs FETs, GaAs pHEMTs, GaAs HBTs, Si MOSFETs, and SiGe HBTs. The treatment is limited to emphasize characteristics that will be of interest to students and design engineers. Chapter 5 deals with linear and nonlinear transistor models. These models are the backbone of amplifier designs and are based on equivalent circuit formulations. The device models included are for low-noise and low-power applications. The devices included are MESFETs, pHEMTs, HBTs, and MOSFETs. The EC model, model scaling, and source-pull and load-pull characterizations are also detailed in this chapter.

In Chapter 6, the fundamentals of transmission lines and lumped elements are considered, including their characteristics. Since the book is primarily devoted to planar circuits, the characteristics of commonly used planar transmission media such as microstrip line and coplanar waveguide are described. Discontinuities and coupling aspects in the microstrip line are also covered. The design of lumped elements such as capacitors, inductors, and resistors is treated at the end of the chapter. All of this leads naturally into Chapter 7 on impedance matching networks that are fundamental to any microwave circuit or system. Impedance matching circuits for narrowband and wideband applications and their design techniques are discussed. Finally, their practical realization aspects are considered.

Analyses of most commonly used amplifier classes are discussed in Chapter 8. It also provides a comparison of the various amplifier classes used for high-efficiency applications. High-efficiency operation of the power amplifiers can be obtained by operating the transistor in class B or C as well as using load impedances for class E or F. The switched-mode class-E tuned power amplifiers are widely used at low RF frequencies while class-F amplifiers are realized up to microwave frequencies. Practical aspects of high-efficiency amplifiers are provided in greater detail in later chapters.

The next five chapters describe amplifier designs, beginning with Chapter 9 on amplifier design methods. The fundamental design methods for amplifiers, including linear, nonlinear, and statistical, are described. Nonlinear circuit analysis, employing time and frequency domain simulations, is a powerful CAD tool for design and optimization of power amplifiers developed for different applications. The advantages of such tools include accurate design predicting nonlinear behavior, first-pass success for MMICs, and reduced product development time and cost. Design procedures and typical design examples are presented for GaAs FETs, pHEMTs, and HBTs, and for Si CMOS and SiGe HBT amplifiers. Using the material in Chapter 8 as background, high-efficiency amplifier design techniques are discussed in Chapter 10. This critical component is treated in depth, starting with an overdriven class-A amplifier and concluding with harmonic tuning techniques for high PAE. Design examples are provided and important design considerations for high PAE are discussed.

The design of broadband amplifiers is described in Chapter 11. Bandwidth for amplifiers is always an important consideration, and several methods used in broadbanding amplifiers are presented. This includes reactive/resistive, feedback, balanced, and distributed amplifiers. Critical design considerations for the broadband amplifiers are discussed. These circuits are seeing widespread use, particularly in electronic warfare, countermeasures, and support systems. The design of linear amplifiers is presented in Chapter 12. As in previous chapters, the emphasis is on the design of these components and thus different circuit possibilities for each component, design considerations, limitations of design, and so on are presented. Linearization techniques are discussed together with design techniques, realization aspects, and special design considerations.

Chapter 13 covers the fast evolution of high-voltage transistors, which are gaining widespread acceptance and application. Pros and cons of high-voltage transistor based amplifies are described. Si bipolar and LDMOS, GaAs MESFET, pHEMT, and HBT and SiC MESFET and GaN HEMT devices are considered, and their design techniques are presented. Both the hybrid and monolithic high-voltage amplifiers are described. High-voltage operation employing several low-voltage transistors in series is also discussed.

Using the material presented in the previous chapters, it is naturally hoped that the reader will embark on the design and fabrication of microwave integrated circuit (MIC) amplifiers. While mastering the complexity of the manufacture of these circuits can only come from years of practical experience, Chapters 14 and 15 expose one to the types of hybrid and monolithic MIC amplifiers, along with their design considerations, fabrication procedures, and design criteria. Adequate material is presented to allow the designer to make a good choice of substrates and materials for design and fabrication, for both hybrid MICs and monolithic MICs. Examples of hybrid MIC amplifiers are illustrated. Chapter 15 deals with monolithic microwave integrated circuit (MMIC) amplifiers. Several types of amplifiers are discussed and MMIC examples are presented.

The next four chapters are devoted to the practical design aspect of various types of amplifiers. These include thermal design, stability analyses, biasing networks, and power combining. Chapter 16 deals with the thermal design of power amplifiers. Thermal models for channel temperature calculation for both transistors and amplifiers are described. Practical methods for thermal resistance determination are also discussed. Amplifier stability is the topic of Chapter 17. A comprehensive treatment is given of theoretical and practical aspects of amplifier stability with several examples.

Biasing is another important aspect for successful amplifier design, which is treated in Chapter 18. The biasing of transistors is discussed first, followed by a detailed description of biasing networks. Biasing of multistage amplifiers as well as biasing for low-frequency stabilization of power amplifiers are discussed. Chapter 19 gives an overview of power combining techniques. After the basics of power combining are described, the fundamental differences between the device and circuit combining techniques are presented. Power combiners are also covered in this chapter together with the methods for design and analysis for both hybrids and couplers. Practical examples of multichip MMIC based combined HPAs are included.

Applications of amplifiers in modern commercial and military systems require cost-effective solutions. A popular technique to achieve product cost goals is to integrate more functions into a single MMIC amplifier chip or into a package or module. For example, a high level of integration at the MMIC chip level reduces the number of chips and interconnects and results in low test and assembly costs, which in turn increases the reliability and reduces the subsystem cost. Chapter 20 includes examples of this type of integration such as a limiter/LNA, transmitter chains with several stages, amplifiers with variable gain and output power, amplifiers with built-in power monitors, temperature compensation for gain, and output mismatch protection.

Chapter 21 deals with amplifier packaging issues. Both plastic and ceramic packages are described. Design of the feed and cavity for ceramic packages is treated in detail. Brief descriptions of die attach and wire bonding techniques are also included. Measurements of transistor and amplifiers are covered in the last chapter. The characterization of transistors is described first, followed by amplifier testing including S-parameters, noise figure, source-pull and load-pull characterization, VSWR, output power versus input power, PAE, harmonics, distortion, phase noise, and recovery time. At the end, several appendix are included to facilitate the designs of readers.

This book contains enough material for a one-year course at the senior or graduate level. With judicious selection of specific topics, one can use the book for one-semester, two-semester, or two-quarter courses. Problems are given at the end of most chapters. They have been tested to ensure that their level of difficulty and complexity is suitable for the student.

This book is dedicated to all my colleagues who have done pioneer work in the advancement of microwave engineering. I am also indebted to Dr. Edward L. Griffin who introduced me to the wonderful field of microwave amplifiers. He reviewed the manuscript and made excellent suggestions. Many friends and colleagues, at Tyco Electronics and elsewhere, have significantly contributed to improvements in this book. I particularly want to thank George Studtmann for critically reviewing and editing the complete manuscript, making numerous suggestions to greatly enhance the text, and also providing HBT amplifier examples. I wish to thank Mark Dayton, Andy Peake, Tom Winslow, Jain Zhao, James Perdue, and Gordon Tracy for providing critical reviews of chapters and for their support. The preparation of this book has depended on my organization and a number of very supportive individuals, including David Conway, Michael Rachlin, Janice Blackwood, and Neil Alls. I owe a special note of thanks to Linda Blankenship for expertly transforming some of my handwritten text into word-processing documents. I would like to thank Tyco Electronics management for its support and encouragement. This book became a reality only because of the great support I received from George Telecki and his staff including Lucy Hitz and Lisa Morano Van Horn at John Wiley & Sons.

Finally, I want to express my deep appreciation to my wife, Subhash Bahl, for her love, encouragement, enduring unselfishness, and support. Her patience allowed me to work during many evenings, holidays, and weekends to complete this gigantic task. Especially, I wish to thank my daughter, Preeti, my son-in-law, Ashutosh, my son, Puneet, and my grandsons, Karan and Rohan, for their love, support, and patience. They all truly deserve much of the credit.

INDER J. BAHL

Roanoke, Virginia
March 2009

Chapter 1

Introduction

Among electronic circuits, signal amplification is one of the most important radiofrequency (RF) and microwave circuit functions. The introduction of radar during World War II provided the first significant application requiring amplification of microwave signals. In recent times, the wireless communication revolution has provided an explosion of RF and microwave amplification applications. During the last two decades, amplifier technology has made tremendous progress in terms of devices (low noise and power), circuit computer-aided design (CAD) tools, fabrication, packaging, and applications. Low-cost power amplifiers for wireless applications are a testament to this explosion.

Early microwave amplifiers were the exclusive province of vacuum tube devices such as Klystrons [1–3], traveling-wave tube (TWT) amplifiers [2–4], and magnetrons [2, 3]. Today, microwave amplification is dominated by solid state amplifiers except for applications at high output powers ($\gtrsim$100 watts). Today, the most common vacuum tube application is the 900-watt microwave oven using a 2.45-GHz magnetron. The power levels achievable for tube amplifiers are on the order of 10^3 higher than achievable for solid state amplifiers. The microwave oven magnetron, with a manufacturing cost of about \$10 ($\sim$\$0.01/watt), has no solid state competition in sight. Likewise, today's \$0.50/watt 900-MHz to 2-GHz cell phone solid state transistor amplifier and \$0.30/watt 200–500-W L/S-band base station transistor power amplifiers have no tube competition.

Solid state amplifiers are of two general classes: those based on two-terminal negative resistance diode devices, and those based on three-terminal devices known as transistors. Early solid state amplifiers were dominated by two-terminal devices because diodes are typically much easier to fabricate than transistors. Quite an array of two-terminal amplifier designs have been introduced, including parametric amplification (varactor diodes) [5–8], tunneling diodes [7–9], transferred electron diodes (Gunn and LSA diodes) [8, 10, 11], and avalanche transit-time diodes (IMPATT, TRAPATT, and BARITT) [8, 12]. Such diodes are used only for special amplifier functions.

1.1 TRANSISTOR AMPLIFIER

Today, solid state amplification is dominated by use of three-terminal transistors [13–36]. Using a small voltage applied at the input terminal of the device, one can control, in an efficient manner, a large current at the output terminal when the

Fundamentals of RF and Microwave Transistor Amplifiers. By Inder J. Bahl

common terminal is grounded. This is the source for the name *transistor*, which is a unification of the words *transfer resistor*.

Solid state transistors may be grouped into two categories: bipolar and unipolar devices. The bipolar devices are comprised of silicon (Si) bipolar junction transistors (BJTs) and silicon germanium (SiGe) and gallium arsenide (GaAs) heterojunction bipolar transistors (HBTs). The unipolar devices include Si metal oxide semiconductor field effect transistors (MOSFETs), GaAs metal semiconductor field effect transistors (MESFETs), and pseudomorphic high electron mobility transistors (pHEMTs). The switchover to three-terminal devices was largely due to cost. Diodes are typically less expensive to manufacture than transistors but the associated circuitry to achieve gain from a two-terminal device is much more expensive than that for a three-terminal device. For example, a transistor (without any matching network) connected between 50-ohm input and output terminals can provide 15–20 dB gain at radiofrequencies and 6–8 dB at 20 GHz. In addition, design of three-terminal amplifiers for stable operation and routine high-yield manufacturing is exceedingly simple.

Signal amplification is a fundamental function in all RF and microwave systems. When the strength of a weak signal is increased by a device using a direct current (DC) power supply, the device along with its matching and biasing circuitry is known as an amplifier. Here the DC power from the power supply is converted into RF power to enhance the incoming signal strength. If a device is a transistor, the signal is applied to the input terminal (gate/base) and the amplified signal appears at the output (drain/collector) and the common terminal (source/emitter) is usually grounded. The matching networks help in exciting the device and collecting the output signal more efficiently. Figure 1.1 shows a schematic representation of a single-stage transistor amplifier. Basic constituents are a transistor, input and output matching networks, bias circuitry, and input and output RF connections. The DC bias and RF connections may be made to connectors if housed in a fixture or to lead frame if assembled in a package depending on the amplifier fabrication scheme.

There are various types of amplifiers used at RF and microwave frequencies. Basic types consist of low-noise, buffer, variable gain, linear power, saturated high-power, high-efficiency, narrowband, and broadband amplifiers. The design of

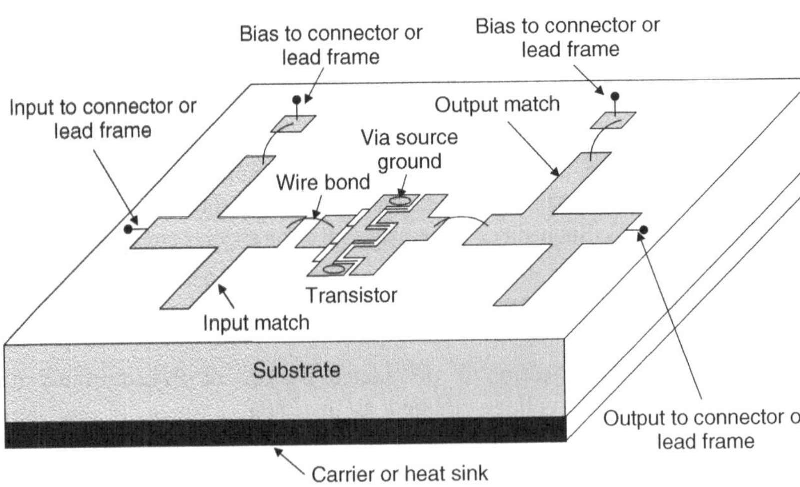

Figure 1.1 Schematic representation of a transistor amplifier.

amplifiers requires essentially device models/S-parameters, CAD tools, matching and biasing networks, and fabrication technology. Each type mandates additional insights to meet required amplifier specifications. For example, a low-noise amplifier (LNA) needs a low-noise device and a low-loss input matching network while a power amplifier (PA) requires a power device and low-loss output matching network.

RF and microwave amplifiers have the following characteristics:

- Band-limited RF response
- Less than 100% DC to RF conversion efficiency
- Nonlinearity that generates mixing products between multiple signals
- RF coupled and no DC response
- Power-dependent amplitude and phase difference between the output and input
- Temperature-dependent gain, higher gain at lower temperatures and vice versa

1.2 EARLY HISTORY OF TRANSISTOR AMPLIFIERS

The use of Si based bipolar transistors and GaAs based MESFET for amplifiers have been reported since the mid-1960s and early 1970s, respectively. Most of the initial work on Si based bipolar transistor amplifiers was below C-band frequencies, whereas GaAs based MESFET amplifiers were designed above L-band frequencies (see Appendix C for frequency band designations). Low-noise HEMTs were reported in the early 1980s. Internally matched narrowband MESFET power amplifiers working from S- through X-band were available during the 1980s and Ku-band amplifiers were introduced in the early 1990s.

The GaAs monolithic microwave integrated circuit (MMIC) amplifier was reported in 1976 and since then there has been tremendous progress in both LNAs and PAs. Some of the early development milestones in MMIC amplifiers are as follows:

- X-band low-power GaAs MESFET amplifier in 1976
- X-band GaAs MESFET power amplifier in 1979
- K-band GaAs MESFET LNA in 1979
- Q-band GaAs MESFET power amplifier in 1986
- V-band GaAs HEMT LNA in 1988
- X-band GaAs HEMT power amplifier in 1989
- W-band HEMT LNA/power amplifier in 1992

1.3 BENEFITS OF TRANSISTOR AMPLIFIERS

Major benefits of transistor amplifiers versus tube amplifiers are smaller size, lighter weight, higher reliability, high level of integration capability, high-volume and high-yield production capability, greater design flexibility, lower supply voltages, reduced maintenance, and unlimited application diversity. Transistors have much longer operating life (on the order of millions of hours) and require much lower warming time. Solid state amplifiers also do not require adjustment in the bias or the circuit, as required in tubes, over long periods of operation.

In comparison to solid state diode amplifiers, transistor amplifiers have greater flexibility in terms of designing matching networks, realizing high-stability circuits, and cascading amplifier stages in series for high gain. The outstanding progress made in monolithic amplifiers is attributed to three-terminal transistors, especially on GaAs substrates. Monolithic amplifiers are fabricated on wafers in batches, and hundreds or thousands can be manufactured at the same time. For example, over 15,000 amplifiers, each having a chip size of 1 mm^2, can be obtained on a single 6-inch diameter GaAs wafer. Thus monolithic amplifiers have a great advantage in terms of the manufacturing cost per unit. In general, monolithic amplifiers will have advantage in terms of size and weight over hybrid integrated techniques. It is worth mentioning that the weight of an individual or discrete chip resistor or a chip capacitor or an inductor is typically more than an entire monolithic amplifier chip. Many of today's high-volume applications using amplifiers are in hand-held gadgets. Both hybrid and monolithic MIC technologies are used and considered reliable. However, a well-qualified MMIC process can be more reliable because of the much lower part counts and far fewer wire bonds.

1.4 TRANSISTORS

During the past two decades outstanding progress has been made in microwave and millimeter-wave transistors. The low-noise and power performance as well as the operating voltages have significantly been advanced. Among low-noise devices, the pHEMT is the most popular due to its low noise figure and high gain characteristics. Other devices for small-signal applications are MESFETs, MOSFETs, and SiGe HBTs. Today, a designer has several different types of power transistors available as discrete devices (in chip or packaged form) or as part of a foundry service to design power amplifier MMICs. Several solid state devices are being used to develop power amplifier (PA) circuits including BJTs, laterally diffused metal oxide semiconductor (LDMOS) transistors, MESFETs, or simply FETs, both GaAs and indium phosphide (InP) based HEMTs, GaAs based HBTs and silicon carbide (SiC) based FETs, and gallium nitride (GaN) HEMTs. Each device technology has its own merits, and an optimum technology choice for a particular application depends not only on technical issues but also on economic issues such as cost, power supply requirements, time to develop a product, time to market a product, and existing or new markets.

HEMTs have the highest frequency of operation, lowest noise figure, and high power and PAE capability. Due to the semi-insulating property of GaAs substrates, the matching networks and passive components fabricated on GaAs have lower loss than on Si. The GaAs FET as a single discrete transistor has been widely used in hybrid microwave integrated circuit (MIC) amplifiers for broadband, medium-power, high-power, and high-efficiency applications. This wide utilization of GaAs FETs can be attributed to their high frequency of operation and versatility. However, increasing emphasis is being placed on new devices for better performance and higher frequency operation. HEMT and HBT devices offer potential advantages in microwave and millimeter-wave IC applications, arising from the use of heterojunctions to improve charge transport properties (as in HEMTs) or pn-junction injection characteristics (as in HBTs). HEMTs have a performance edge in ultra low-noise, high-linearity, and high-frequency applications. The MMICs produced using novel structures such as pseudomorphic and lattice matched HEMTs have significantly improved power and power added efficiency (PAE) performance and high-frequency (up to 280 GHz) operation. The pHEMTs that utilize multiple epitaxial III–V compound layers have shown excellent millimeter-wave power performance from Ku- through W-bands. HBTs are vertically oriented heterostructure devices and are very popular as low-cost power devices when operated using a single power supply. They offer better linearity and lower phase noise than FETs and HEMTs.

On the other hand, bipolar transistors require only a single power supply, have low leakage, low $1/f$ noise or phase noise, and are produced much cheaper on Si. The SiGe HBTs have the low-cost potential of Si BJTs and electrical performance similar to GaAs HBTs. Thus discrete silicon BJTs, SiGe HBTs, and MOSFETs have an edge over GaAs FETs, HEMTs, and HBTs in terms of cost at low microwave frequencies. For highly integrated RF front ends, GaAs FETs and HEMTs are superior to bipolar transistors and Si substrate based devices due to high performance multifunction devices and lower capacitive loss, respectively. The electrical performance and cost trade-offs between Si and GaAs generally favor silicon devices due to single power supply operation and lower cost, whereas GaAs based devices are preferred due to superior low-noise and power (high breakdown voltage) performance and high-frequency operation.

Many of these transistors are available as discrete devices as well as a foundry to design monolithic amplifiers. Discrete transistors are available in die form and in plastic and/or ceramic packages. The ceramic package devices are for high-frequency and high-power applications. Plastic packaged transistors are for low-cost and high-volume applications.

1.5 DESIGN OF AMPLIFIERS

The design of RF and microwave amplifiers has several facets impacting their performance. The most important factors include the selection of semiconductor technology, device models, circuit architecture and design methodology, matching networks, packaging, and thermal management. Thus amplifier design becomes an art, to meet several often conflicting requirements, and an experienced designer will outperform beginners.

The design of an amplifier for a particular application and frequency range is quite complicated in the sense that it has to meet physical, electrical, thermal, and cost requirements. Salient features of an amplifier design are given in Figure 1.2. The amplifier performance requirements in terms of frequency band, gain, noise figure, power output, PAE, linearity, and input and output VSWR are determined by the device sizes, the circuit design topology, matching networks, the number of gain stages, the aspect ratio for the devices between the stages, design methodology, fabrication technology, and packaging. More often it involves trade-offs in terms of size, electrical performance, reliability, and cost. The amplifier designs are normally performed using device S-parameters, linear and nonlinear models, and matching component models.

Design of amplifiers, in a broad sense, falls into two categories: low noise and power. In a low-noise amplifier, the transistor's input is matched for optimum noise figure and the transistor's output is conjugately matched to 50-Ω system impedance for maximum gain and return loss (RL). In a power amplifier, the 50-Ω system impedance is matched to a required load at the transistor's output for maximum power and the transistor's input is conjugately matched for maximum gain and RL. In linear amplifiers the input and output are matched for better linearity. Thus, in an amplifier, the device's input is either matched for minimum noise or maximum gain or linearity and output is matched for maximum gain or optimum power and PAE or linearity. The matching networks are comprised of distributed and lumped elements. In an amplifier, the supply

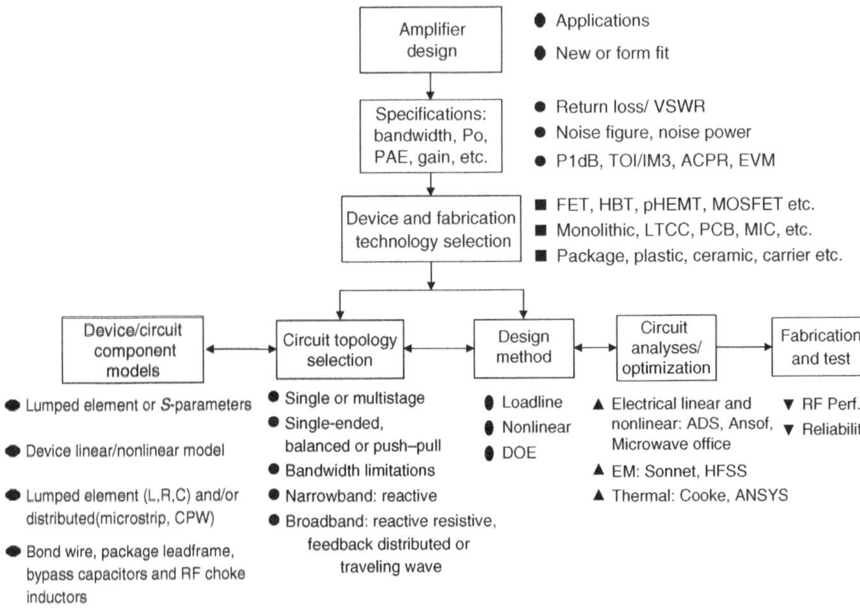

Figure 1.2 The salient features of amplifier design.

voltage (drain or collector) is applied through an RF choke or through the biasing circuit, which is usually an integral part of the matching network.

A low-noise or small-signal amplifier is designed using a device's noise model or noise parameters and *S*-parameters. The amplifier design must be conditionally stable. Narrowband power amplifier design can be carried out using a device's source-pull and load-pull data to design the amplifier's input and output matching circuits. This technique provides approximate electrical performance such as gain, power, and power added efficiency calculated by using measured small-signal *S*-parameters of the active device. However, for broadband applications, the aforementioned technique is very involved. Accurate nonlinear models for active devices provide a more suitable technique by using nonlinear computer-aided design (CAD) tools. These models help in determining matching networks over the desired bandwidth of the amplifier and also assist in simulating large-signal performance such as gain, VSWR, P_{1dB}, PAE, P_{sat}, TOI or ACPR or EVM, and harmonic levels. All these terms are defined in Chapter 3. These models also provide accurate solutions for multistage power amplifiers. The amplifier design must be conditionally stable, and also odd-mode, parametric, and low-frequency oscillation conditions must be prevented. However, in power amplifiers, unconditional stability is generally desired.

Power amplifiers are nonlinear circuits. Thus linearization of such circuits for multiple-carrier communication applications is required to minimize distortion. The design of such circuits can be obtained either by using measured source-pull and load-pull data, or accurate nonlinear models, or by using some sort of distortion cancellation technique.

The initial cost of developing MMIC based amplifiers is far greater than in hybrid based technology. Also, the tuning of the fabricated MMICs is difficult. Therefore the design phase of MMIC amplifiers becomes very critical to achieve first-pass success to minimize expensive design iterations. In the design of such products, an extensive accurate modeling library of active devices, passive circuit components, and other parasitic reactances including discontinuities, cross-coupling, bonding pads, connecting wires, and package lead frame becomes an integral part of an amplifier design.

1.6 AMPLIFIER MANUFACTURING TECHNOLOGIES

Several RF/microwave amplifier manufacturing technologies are being used to reduce component counts, size, and cost. These are printed circuit board, thin-film and thick-film integrated circuit (IC) hybrid, low-and high-temperature cofired ceramic, monolithic IC, and multichip module. Over the past decade, the trends in amplifier fabrication technology have shifted from hybrid IC to monolithic IC. The majority of power amplifiers are fabricated by using some sort of hybrid technology. At radiofrequencies, discrete matching components such as inductors, capacitors, and resistors are added on a printed circuit board to build power amplifiers. At microwave frequencies, thin-film technology is used to design hybrid amplifiers and internally matched amplifiers, to assemble MMIC amplifiers, and for high power combining. An MMIC technique for amplifiers is preferred especially for broadband, high-frequency, and large-volume applications.

The choice of suitable semiconductor technology depends on its performance capability and cost. For example, at S-band GaAs pHEMT and MESFET devices have superior performance compared to Si LDMOS transistors, however, Si LDMOS technology is mostly used to develop high-power amplifiers (on the order of hundreds of watts) for base station transmitters for cellular networks. Because the LDMOS is based on well-established and low-cost Si technology, meeting cost targets and providing desired gain, linearity, and reliability are not a problem. On the other hand, the GaAs pHEMT meets low-noise and power performance needs for millimeter applications.

1.7 APPLICATIONS OF AMPLIFIERS

In general, a microwave system requires a group of amplifiers. Low-noise amplifiers are integral parts of receivers while transmitters are based on several stages of power amplifiers. RF/microwave power amplifiers are important circuit components used in every system including cordless and cellular telephones, base station equipment, spaceborne, airborne, and ground based (fixed/mobile) satellite communications, wireless local area networks, terrestrial broadcast and telecommunications, point-to-point radio (PPR), very small aperture terminal (VSAT) wideband satellite communications, air traffic systems, global positioning system (GPS), phased array radar (PAR), electronic warfare (EW), and smart weapons. Most of these systems require low-cost (high-volume) and more reliable solid state power amplifiers. Cordless and cellular telephones require low-bias operation (2–5 V), single power supply, and very high-efficiency (analog versions) or high-linearity (digital versions) amplifiers. Cellular telephones may also require dual, triple or quad-mode operation including multiple frequencies in both digital and analog versions. Power amplifiers for point-to-point radio and very small aperture terminal applications are operated typically at 8 V. The output power requirements are in the range of 0.2–4 W. On the other hand, for a phased-array antenna (PAA), the amplifiers are typically operated at 10 V and the output power requirements are in the range of 20–40 W per element.

The power level of amplifiers is dictated by the intended application. For example, for cellular base stations and EW, the power levels are in tens to hundreds of watts, while for satellite and radar systems, their levels may be a magnitude higher. For portable wireless handsets and wireless LANs, the required power levels are an order of magnitude lower, usually less than 1 watt. Based on the modulation schemes, the handset requirements can be grouped into two categories: constant envelope and nonconstant envelope. In the former scheme, there is no information contained in the amplitude of the transmitted signal. In this case, the amplifiers are operated in high-efficiency mode. Common applications are groupe special mobile (GSM) and digital European cordless telecommunication (DECT). The latter scheme enhances the spectral efficiency of the signal by incorporating the intended information in the amplitude of the transmitted signal. Most popular applications are code division multiple access (CDMA), wideband code division multiple access (WCDMA), and local area network (LAN). Usually, the amplifiers are operated in linear mode at the cost of amplifier efficiencies. For wireless base stations, high linearity is of paramount importance in power amplifiers. For example, personal communication service (PCS) (1.8–2.0 GHz) requires power levels in the range of 5–200 W.

Modern active-aperture antenna subsystems for phased-array applications require hundreds or even thousands of transmit/receive (T/R) modules, each delivering tens of watts of output power. These phased-array antenna subsystems are employed in airborne communication and radar systems, ground based and ship based tactical radars, as well as space based radar and communication systems. Typical T/R module requirements for these systems include (a) small size, dictated by required antenna element-to-element spacing, (b) low weight, especially in airborne and spaced based systems, (c) precise control of insertion phase and amplitude for good beam pointing accuracy and low side lobe levels, (d) high reliability, (e) high power added efficiency (PAE) to reduce prime power and cooling requirements, and (f) low cost, since thousands of modules may be required for a single system. Thus 5–10% improvement in PAE can greatly affect the DC power requirements and thermal design techniques. The MMIC amplifier technique looks very attractive in realizing a couple of tens of watts of power. These chips are further combined using standard hybrid MIC techniques to obtain much higher power levels. Higher efficiency operation of these devices is becoming one of the most important factors in reducing prime power and cooling requirements for advanced systems. These characteristics are particularly useful for space and military applications where weight, size, and power added efficiency can impose severe limitations on the choice of components and systems.

Figures 1.3 through 1.7 show examples of typical microwave amplifiers. These include a small-signal MMIC amplifier, plastic packaged driver and high-voltage power amplifiers, hybrid power amplifiers, ceramic packaged power amplifiers, and MMIC power amplifiers. Figure 1.8 shows an example of a printed circuit board used for plastic amplifier testing.

The phase-array antenna, which uses a large number of elements, benefits significantly from low-cost transceiver (T/R) modules using several amplifiers. Figure 1.9a shows a simplified block diagram of a T/R module whose size and cost can be reduced drastically by monolithic integration of amplifiers or all microwave functions except the circulator and the antenna. Figure 1.9b shows an X-band radar GaAs MMIC chipset employing three amplifier chips: limiter/LNA, driver, and HPA and several buffer amplifiers used in the control chip for loss compensation.

Requirements for power amplifiers vary drastically from one application to another. Usually communication applications require linear operation, while for radar applications high PAE is of prime importance. Personal communication systems working in

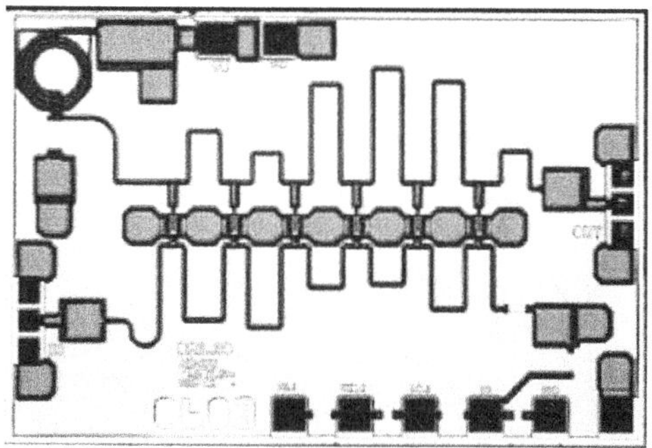

Figure 1.3 Single-stage broadband low-noise MMIC amplifier. Chip size is $3 \times 2\ mm^2$.

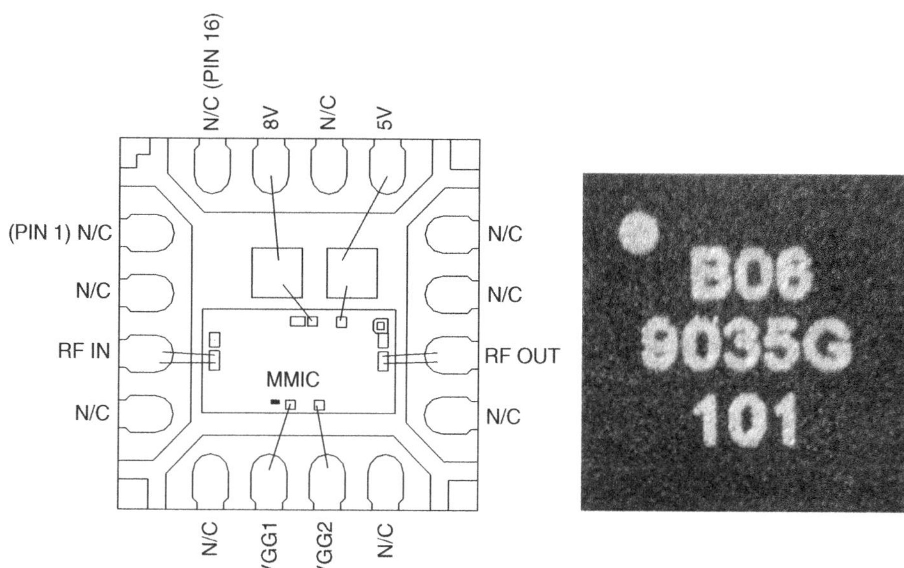

(a) (b)

Figure 1.4 Example of a plastic packaged driver amplifier: (a) MMIC wiring to the lead frame and (b) top view of the package. Small shunt objects on the supply lines are decoupling capacitors.

(a) (b)

Figure 1.5 Examples of hybrid MIC amplifiers: (a) LNA using pHEMT and (b) amplifier in a metal housing with connectors and single supply.

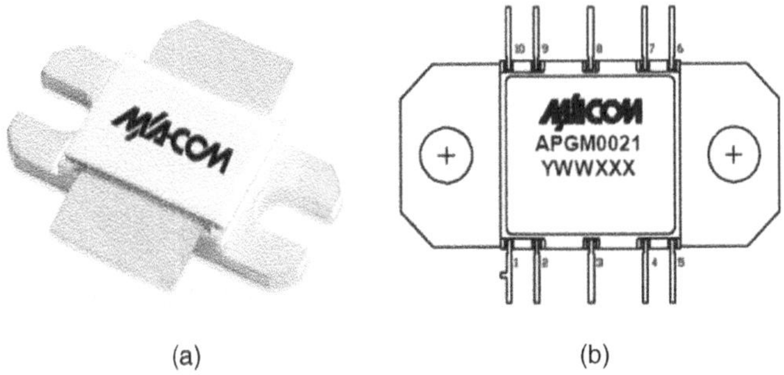

Figure 1.6 Examples of ceramic packaged devices: (a) discrete transistor and (b) MMIC amplifier.

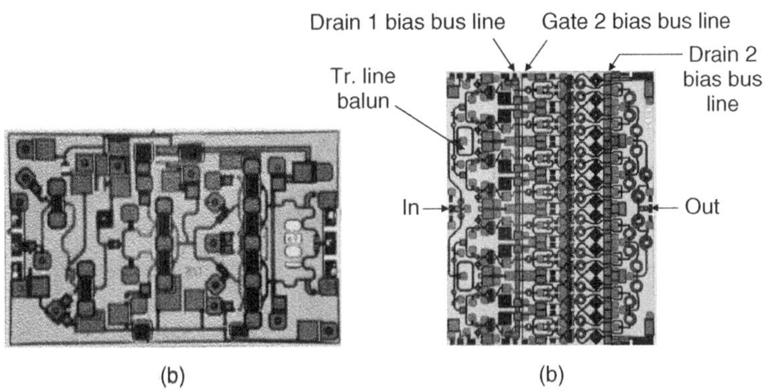

Figure 1.7 Examples of MMIC amplifiers: (a) K-band four-stage 1-W driver, chip size is 3.1×2.0 mm^2 and (b) 2–8 GHz 8-W power amplifier, chip size is 5.0×6.3 mm^2.

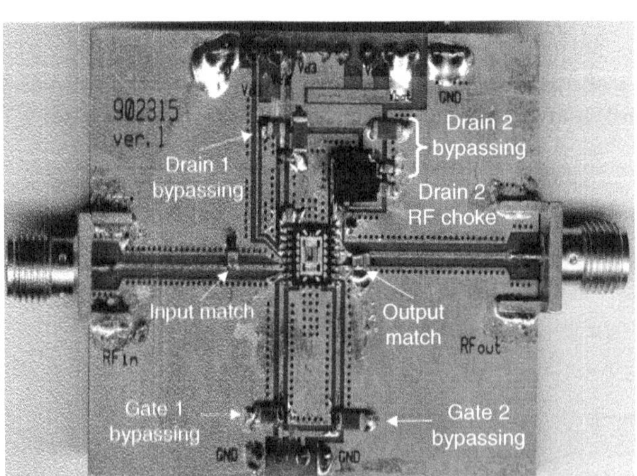

Figure 1.8 Example of a prototype PCB, with a power amplifier in a TSSOP 16-pin plastic package, for RF testing.

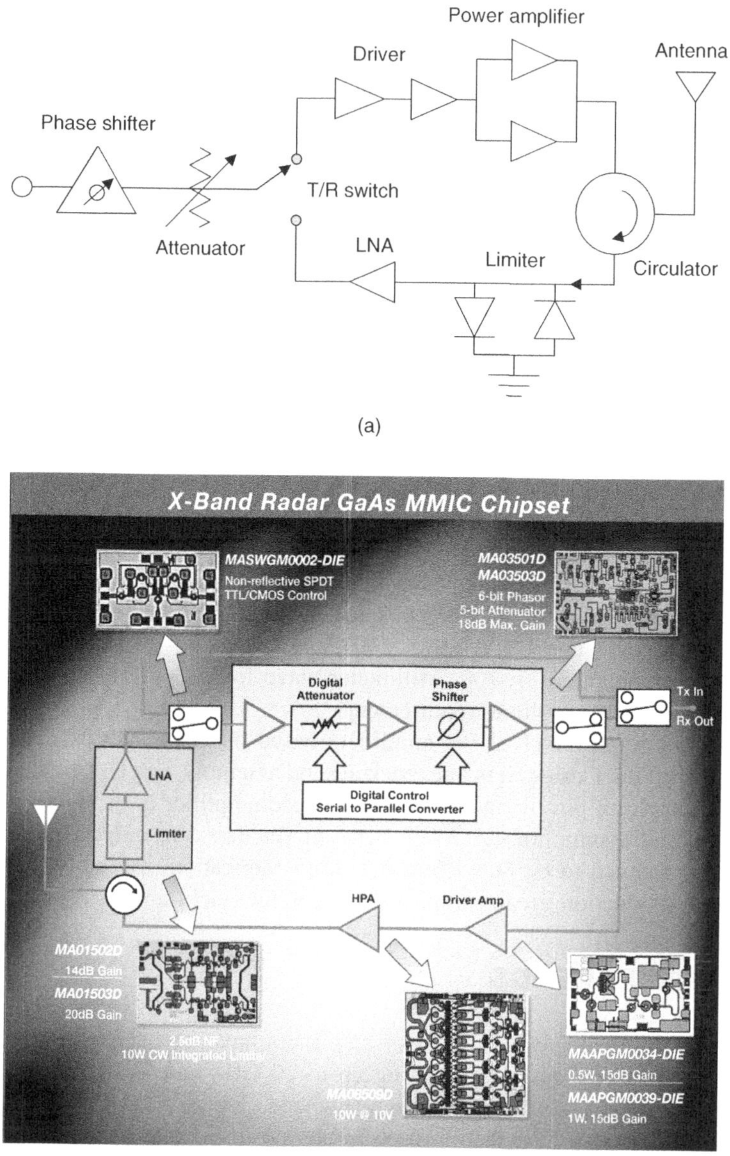

(a)

(b)

Figure 1.9 (a) A block diagram that illustrates the amplifier functions of a T/R module. (b) X-band radar GaAs MMIC chipset showing three amplifier chips: limiter/LNA, driver, and HPA and several buffer amplifiers used in the control chip for loss compensation.

the 800-MHz to 2.5-GHz range use different digital modulation and access schemes. They require high-efficiency and linear power amplifiers for hand-held as well as for base station applications.

There are several emerging commercial and military applications that require broadband and high-power amplifiers. These include broadband wireless access systems, communication, and electronic warfare.

Table 1.1 Percentage of Cost Split for Plastic and Ceramic Packaged Amplifiers

Item	Plastic	Ceramic
MMIC die	85	33
Package and assembly	5	58
Test	10	9

1.8 AMPLIFIER COST

The cost of an amplifier depends on its capability (power, PAE, noise figure, frequency, etc.), fabrication techniques, complexity (die, package, mechanical structure for support, etc.), and applications (low or high volume). Cost is first and foremost driven by manufacturing volume. Complex amplifiers in very high-volume commercial use cost less than relatively simple amplifiers with very little volume. One can buy low-noise and low-power amplifiers in plastic packages for as low as \$0.25–2.00, while high-power modules cost \$2000–5000. Microwave driver amplifiers (0.50–2 W) in die form or packaged sell for \$5–30. High-power (10–20 W) X-band MMIC amplifiers are available in the \$100–200 range in moderate quantities. RF and low microwave frequency range HPAs (100–200 W), with a mostly internally matched configuration, cost about \$100–200. Millimeter-wave frequency HPAs are still quite expensive because of low volume requirements.

The cost of low-power (1–2 W) amplifiers based on GaAs MMIC can approximately be split into three categories: die, package and assembly, and test yield. Table 1.1 provides an example of plastic and ceramic packaged amplifier costs. The cost model is based on several assumptions and represents production cost only. The end-to-end yield has been assumed to be 80%. Automation in fabrication, assembly, and testing in high-volume production greatly improves the product yield and final cost.

1.9 CURRENT TRENDS

Microwave and millimeter-wave transistor amplifiers have advanced dramatically. Si based CMOS technology circuits operating up to 70 GHz and GaAs/InP based technology operating up to 280 GHz have been realized. The Si LDMOS transistor is a primary power device for base station transmitters up to S-band frequencies. Devices such as pHEMTs and HBTs made on InP, SiC, and GaN substrate materials have performed beyond 100 GHz. SiC based GaN pHEMT technology has also advanced rapidly and is finding special application where high-power, high-efficiency, low-noise, broadband, and millimeter-wave operation are required. Extremely high-frequency circuits enable a wide range of new applications to be developed in communications, security, medicine, sensing, and imaging. Power amplifiers are vital components in evolving broadband wireless applications including TV broadcasting, voice over Internet protocol (VoIP), video on demand (VOD), online gaming, mobile streaming, and mobile video telephony.

Recently, SiC based transistors operating at 30–50 V have advanced rapidly. SiC MESFETs have increasing applications in high-power wideband at low microwave frequencies and GaN HEMTs on SiC are finding numerous applications where high-power and high-frequency operation is required. Much higher power densities for such devices meet the current need in reducing the cost of solid state power amplifiers. Another emerging technology is GaN HEMT on SiC. This technology has the potential of meeting cost targets for numerous applications, including base station and radar transmitters. These devices are capable of generating hundreds of watts at C/X-band, tens of watts in the millimeter-wave region, and 1–2 W at 100 GHz. Since this technology has an order of magnitude higher breakdown voltage and power density potentials along with an outstanding thermal dissipation substrate, it has all the ingredients required for high-power amplifiers.

The RF and microwave industry is still growing and there is strong evidence that it is being fully supported to meet current trends. New high-volume applications demand low-cost solutions for transmitters based on transistor amplifiers. Current trends are in the areas of improved device models and integrated CAD tools. In a new competitive business environment, it is essential to have accurate device models and suitable circuit design tools to develop state-of-the-art circuits to meet system requirements, including cost and production schedule. It becomes essentially important for amplifier design engineers to develop amplifier products for specific applications on time. For emerging wideband applications that require very high-power (50–200 W) amplifiers with PAE as high as 50%, new circuit topologies to meet these challenging performance goals will be required.

There are continuous trends for improving the performance of LNAs and PAs in order to make them cheaper for high-volume applications. Thus advancements in RFIC and MMIC technologies and packaging will continue with the pace set in the past decade. For high-volume applications, a package (plastic or ceramic) has become an integral part of RFICs and MMICs for power amplifiers. Achieving the smallest size and cheapest product cost requires inexpensive and high-performance leadless surface-mount and ball-grid array packages. Plastic packaging is a preferred technique for small-signal amplifiers and more and more power devices are being housed in plastic packages. Low thermal resistance is another important requirement for such packages.

1.10 BOOK ORGANIZATION

The book primarily deals with the operating principles and design of RF and microwave transistor amplifiers. It is organized into four parts consisting of 22 chapters using a top–down approach. An overview of the chapters is shown in Figure 1.10. In the first part there are seven chapters dealing with the fundamentals of amplifier design. It includes network theory, amplifier definitions, transistor basics and their models, impedance matching elements, and matching techniques. The second part focuses on the design of amplifiers in six chapters. Each of the six chapters deals with important features of various amplifier types. Topics such as amplifier analysis, design methods, high-efficiency, broadband, and linearization techniques, and high voltage and high power are discussed in this part. The other amplifier topics such as manufacturing techniques, biasing, thermal and stability analyses, and power combining constitute the third part, which has six chapters. In the fourth part there are three chapters: two chapters deal with integrated function amplifiers and packages, and the final chapter is devoted exclusively to transistor and amplifier measurements. These chapters will also help in understanding the basic requirements of a successful amplifier design. In addition, several appendixes are included at the end that will be useful for RF and microwave designers.

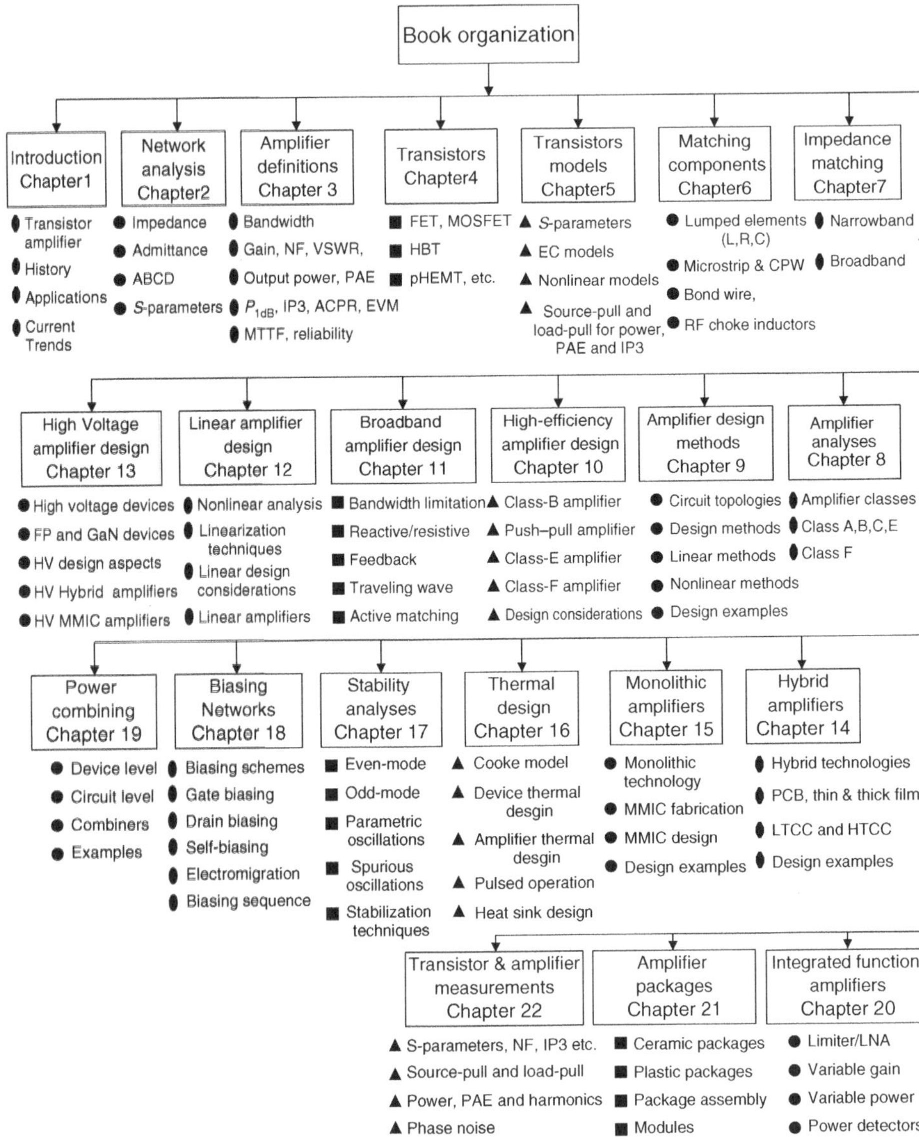

Figure 1.10 Book organization overview showing the connectivity between chapters.

REFERENCES

1. A. E. Harrison, *Klystron Tubes*, McGraw-Hill, New York, 1947.
2. S. Y. Liao, *Microwave Devices and Circuits*, Prentice Hall, Englewood Cliffs, NJ, 1980.
3. O. P. Gandhi, *Microwave Engineering and Applications*, Pergamon, New York, 1981.
4. J. R. Pierce, *Traveling Wave Tubes*, Van Nostrand, Princeton, NJ, 1950.
5. L. A. Blackwell and K. L. Kotzebue, *Semiconductor-Diode Parametric Amplifiers*, Prentice Hall, Englewood Cliffs, NJ, 1964.
6. P. Penfield and R. Rafuse, *Varactor Applications*, MIT Press, Cambridge, MA, 1962.
7. K. K. N. Chang, *Parametric and Tunnel Diodes*, Prentice Hall, Englewood Cliffs, NJ, 1964.

8. H. A. Watson, *Microwave Semiconductor Devices and Their Circuit Applications*, McGraw-Hill, New York, 1969.
9. F. Sterzer, Tunnel diode devices, in L. Young (Ed.), *Advances in Microwaves*, Vol. 2, Academic, New York, 1967.
10. F. Sterzer, Transferred electron (Gunn) amplifiers and oscillators for microwave applications, *Proc. IEEE*, Vol. 59, pp. 1155–1163, August 1971.
11. L. F. Eastman, *Gallium Arsenide Microwaves Bulk and Transit-Time Devices*, Artech House, Norwood, MA, 1973.
12. G. Haddad, *Avalanche Transit-Time Devices*, Artech House, Norwood, MA, 1973.
13. R. S. Engelbrecht and K. Kurokawa, A wideband low noise L-band balanced transistor amplifier, *Proc. IEEE*, Vol. 53, pp. 237–247, March 1965.
14. K. Kurokawa, Design theory of balanced transistor amplifiers, *Bell Syst. Tech. J.*, Vol. 44, pp. 1675–1698, October 1965.
15. H. F. Cooke, Microwave transistors: theory and design, *Proc. IEEE*, Vol. 59, pp. 1163–1181, August 1971.
16. H. Krauss, C. Bostian, and F. Raab, *Solid-State Radio Engineering*, John Wiley & Sons, Hoboken, NJ, 1980.
17. T. T. Ha, *Solid-State Microwave Amplifier Design*, John Wiley & Sons, Hoboken, NJ, 1981.
18. H. Fukui (Ed.), *Low-Noise Transistors and Amplifiers*, IEEE Press, New York, 1981.
19. R. S. Carson, *High-Frequency Amplifiers*, 2nd edition, John Wiley & Sons, Hoboken, NJ, 1982.
20. J. V. Dilorenzo and D. D. Khandelwal (Eds.), *GaAs FET Principles and Technology*, Artech House, Norwood, MA, 1982.
21. R. S. Pengally, *Microwave Field-Effect Transistors—Theory, Design and Applications*, John Wiley & Sons, Hoboken, NJ, 1982.
22. G. Gonzalez, *Microwave Transistor Amplifiers Analysis and Design*, Prentice Hall, Englewood Cliffs, NJ, 1984.
23. I. J. Bahl and P. Bhartia, *Microwave Solid State Circuit Design*, John Wiley & Sons, Hoboken, NJ, 1988.
24. R Soares (Ed.), *GaAs MESFET Circuit Design*, Artech House, Norwood, MA, 1989.
25. P. H. Ladbrooke, *MMIC Design: GaAs FETs and HEMTs*, Artech House, Norwood, MA, 1989.
26. R. Goyal (Ed.), *Monolithic Microwave Integrated Circuits: Technology and Design*, Artech House, Norwood, MA, 1989.
27. K. Chang (Ed.), *Handbook of Microwave and Optical Components*, Vol. 2, John Wiley & Sons, Hoboken, NJ, 1990.
28. G. D. Vendelin, A. M. Pavio, and U. O. Rohde, *Microwave Circuit Design Using Linear and Nonlinear Techniques*, John Wiley & Sons, Hoboken, NJ, 1990.
29. F. Ali and A. Gupta (Eds.), *HEMTs and HBTs: Devices, Fabrication and Circuits*, Artech House, Norwood, MA, 1991.
30. J. L. B. Walker (Ed.), *High-Power GaAs FET Amplifiers*, Artech House, Norwood, MA, 1993.
31. K. Chang, *Microwave Solid-State Circuits and Applications*, John Wiley & Sons, Hoboken, NJ, 1994.
32. R. J. Weber, *Introduction to Microwave Circuits*, IEEE Press, New York, 2001.
33. K. Chang, I. Bahl, and V. Nair, *RF and Microwave Circuit and Component Design for Wireless Systems*, John Wiley & Sons, Hoboken, NJ, 2002.
34. A. Grebennikov, *RF and Microwave Power Amplifier Design*, McGraw-Hill, New York, 2004.
35. I. A. Glover, S. R. Pennock, and P. R. Shepherd (Eds.), *Microwave Devices, Circuits and Subsystems*, John Wiley & Sons, Hoboken, NJ, 2005.
36. A. Raghavan, N. Srirattana, and J. Laskar, *Modeling and Design Techniques for RF Power Amplifiers*, John Wiley & Sons, Hoboken, NJ, 2008.

Chapter 2

Linear Network Analysis

Linear network analysis methods are the fundamental topics in RF and microwave circuit design. Basically it consists of mathematical relationships between the voltages and currents at various ports of a component, which may be a transistor or a passive element (bond wire, lumped element, microstrip section, coupler, etc.) or an amplifier. Linear characterization is performed at low power levels, hence the origin of "small-signal" network parameters. In this chapter features of various linear network analysis techniques are briefly discussed.

Microwave passive and active networks can be classified as multiport or N-port networks. These networks as stand-alone or their combinations are used frequently in the analysis, synthesis, and optimization of multiport and multistage circuits. In most cases, passive networks are treated as linear, and power- and temperature-independent components. However, for power amplifiers, a nonlinear analysis is required to measure their large signal performance, as a function of input power, including output power, PAE, linearity, and harmonic levels.

The determination of network characteristics at microwave frequencies is substantially different from those used for low-frequency circuits. For low-frequency networks, one can define (and measure) unique voltages and currents at various locations in the circuit. Unfortunately, the same is not true at microwave frequencies and beyond. At microwave frequencies, the networks are characterized in terms of equivalent voltage and current. Various types of matrix representations are used to analyze a microwave circuit. Popular network techniques are impedance (Z), admittance (Y), circuit constants/transmission parameters $ABCD$, and scattering (S) parameters. At microwave frequencies, like low frequencies, it is difficult to measure Z, Y, and $ABCD$ parameters. The only measurable quantities are scattering parameters or S-parameters. The reasons one cannot measure Z, Y, and $ABCD$ parameters at microwave frequencies are as follows.

- It is not easy to measure accurately voltages and currents on which these parameters are defined. High-frequency probes are available but associated parasitic reactance limits the frequency of operation.
- It is difficult to realize reference open and short circuits over a wide frequency range.
- Active devices such as BJTs, FETs, HEMTs, and HBTs exhibit instability when an open or short circuit is connected at the terminals of the device.

Fundamentals of RF and Microwave Transistor Amplifiers. By Inder J. Bahl

- Multiport devices (more than two ports) cannot be tested accurately.

At RF and microwave frequencies, both passive components and active devices are characterized using S-parameter measurements even up to 100 GHz. These measurements use wave propagation on a transmission line concept and measure incident, reflected, and transmitted waves generally between two ports. In a cascaded network of multiple components, the measured waves at the input and output ports of the network represent composite waves. These waves have both magnitude and phase information. In most microwave measurement systems, the measurement impedance for S-parameters is 50 Ω. The S-parameters can be modified for impedances other than 50 Ω and converted into Z, Y, or $ABCD$ parameters. An overview of the properties of these parameters and their conversions from one type to another [1–5] are discussed next.

2.1 IMPEDANCE MATRIX

Consider the N-port network shown in Figure 2.1. In the impedance matrix representation, the voltage at each port is related to the currents at the different ports as follows:

$$\begin{aligned} V_1 &= Z_{11}I_1 + Z_{12}I_2 + \cdots + Z_{1N}I_N \\ V_2 &= Z_{21}I_1 + Z_{22}I_2 + \cdots + Z_{2N}I_N \\ &\vdots \\ V_N &= Z_{N1}I_1 + Z_{N2}I_2 + \cdots + Z_{NN}I_N \end{aligned} \tag{2.1}$$

Here

$$Z_{11} = V_1/I_1, \quad I_2, I_3, \ldots, I_N = 0 \tag{2.2a}$$
$$Z_{12} = V_1/I_2, \quad I_1, I_3, \ldots, I_N = 0 \tag{2.2b}$$
$$Z_{21} = V_2/I_1, \quad I_2, I_3, \ldots, I_N = 0 \tag{2.2c}$$
$$Z_{22} = V_2/I_2, \quad I_1, I_3, \ldots, I_N = 0 \tag{2.2d}$$

By applying appropriate conditions, all Z-parameters can be determined.

In matrix notation, the above set of equations can be expressed as

$$[V] = [Z][I] \tag{2.3}$$

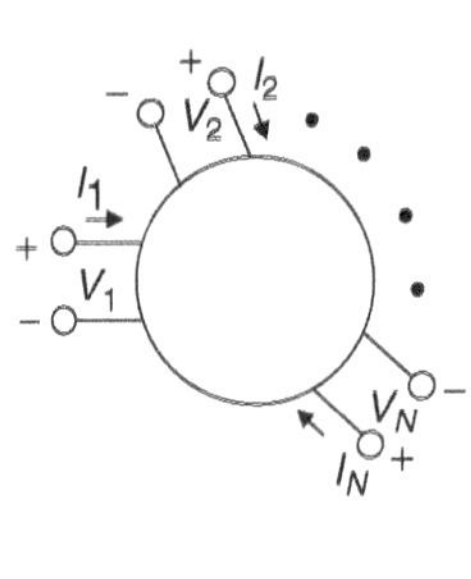

Figure 2.1 N-port network configuration.

where

$$[V] = \begin{bmatrix} V_1 \\ \vdots \\ V_N \end{bmatrix} \tag{2.4a}$$

$$[I] = \begin{bmatrix} I_1 \\ \vdots \\ I_N \end{bmatrix} \tag{2.4b}$$

and

$$[Z] = \begin{bmatrix} Z_{11} & Z_{12} & \cdots & Z_{1N} \\ Z_{21} & Z_{22} & \cdots & Z_{2N} \\ \vdots & \vdots & \vdots & \vdots \\ Z_{N1} & Z_{N2} & \cdots & Z_{NN} \end{bmatrix} \tag{2.5}$$

Figure 2.2 shows three typical impedance connection configurations. For a series case the impedance matrix cannot be defined. For a shunt case the impedance matrix is

$$[Z]_{\text{Sh}} = \begin{bmatrix} Z & Z \\ Z & Z \end{bmatrix} \tag{2.6}$$

EXAMPLE 2.1

Consider a T-equivalent impedance network as shown in Figure 2.2c. Determine its Z matrix, and Z_1, Z_2, and Z_3 in terms of Z matrix parameters.

SOLUTION In this case,

$$V_1 = Z_1\, I_1 + Z_3\, I_1 + Z_3\, I_2 = (Z_1 + Z_3)I_1 + Z_3 I_2 \tag{2.7}$$

$$V_2 = Z_2\, I_2 + Z_3\, I_2 + Z_3\, I_1 = Z_3\, I_1 + (Z_2 + Z_3)\, I_2 \tag{2.8}$$

or

$$[Z] = \begin{bmatrix} Z_1 + Z_3 & Z_3 \\ Z_3 & Z_2 + Z_3 \end{bmatrix} \tag{2.9}$$

Also, $Z_1 = Z_{11} - Z_{21}$, $Z_2 = Z_{22} - Z_{21}$, and $Z_3 = Z_{21}$. The Z matrix terms can also be determined using the definitions (2.2a–d). ■

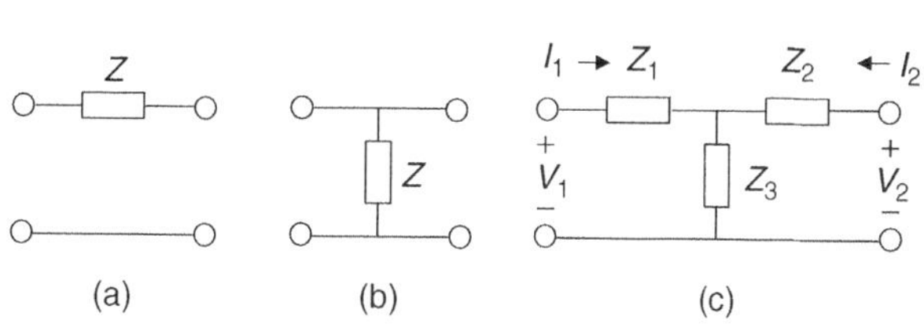

Figure 2.2 Impedance configurations: (a) series, (b) shunt, and (c) T-equivalent impedance network.

2.2 ADMITTANCE MATRIX

In the admittance matrix representation, the current at each port of the network as shown in Figure 2.1 is related to the voltages at the different ports as follows:

$$[I] = [Y][V] \tag{2.10}$$

where $[V]$ and $[I]$ are column vectors as defined by (2.4a) and (2.4b), respectively, and

$$[Y] = \begin{bmatrix} Y_{11} & Y_{12} & \cdots & Y_{1N} \\ Y_{21} & Y_{22} & \cdots & Y_{2N} \\ \vdots & \vdots & \vdots & \vdots \\ Y_{N1} & Y_{N2} & \cdots & Y_{NN} \end{bmatrix} \tag{2.11}$$

Figure 2.3 shows three typical admittance connection configurations. For a series case the admittance matrix is

$$[Y]_S = \begin{bmatrix} Y & -Y \\ -Y & Y \end{bmatrix} \tag{2.12}$$

For a shunt case the admittance matrix cannot be defined.

EXAMPLE 2.2

Consider a two-port π-equivalent admittance network as shown in Figure 2.3c. In this case an additional current source $g_m V_1$ (where g_m is the device transconductance defined in Chapter 4) at the output has been added. Determine its Y matrix.

SOLUTION The input and output currents may be written

$$I_1 = Y_1 V_1 + Y_3 V_1 - Y_3 V_2 = (Y_1 + Y_3)V_1 - Y_3 Y_2 \tag{2.13}$$

$$I_2 = g_m V_1 + Y_2 V_2 + Y_3 V_2 - Y_3 V_1 = (g_m - Y_3)V_1 + (Y_2 + Y_3) V_2 \tag{2.14}$$

or

$$[Y] = \begin{bmatrix} Y_1 + Y_3 & -Y_3 \\ g_m - Y_3 & Y_2 + Y_3 \end{bmatrix} \tag{2.15}$$

When g_m, Y_1, and Y_2 are zero, and $Y_3 = Y$, (2.15) reduces to (2.12). Table 2.1 provides the relationships between π- and T-equivalent circuit parameters. ■

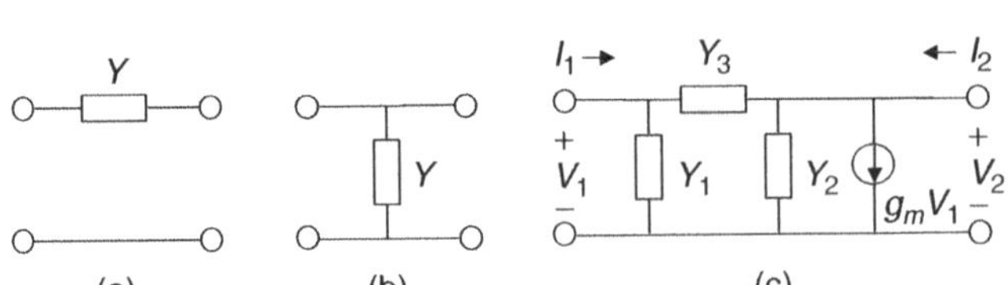

Figure 2.3 Admittance configurations: (a) series, (b) shunt, and (c) π-equivalent admittance network.

Table 2.1 Relationships Between π- and T-Equivalent Circuit Parameters

T-to-π Transformation	π- to-T Transformation
$Y_1 = \dfrac{Z_2}{Z_1Z_2 + Z_2Z_3 + Z_1Z_3}$	$Z_1 = \dfrac{Y_2}{Y_1Y_2 + Y_2Y_3 + Y_1Y_3}$
$Y_2 = \dfrac{Z_1}{Z_1Z_2 + Z_2Z_3 + Z_1Z_3}$	$Z_2 = \dfrac{Y_1}{Y_1Y_2 + Y_2Y_3 + Y_1Y_3}$
$Y_3 = \dfrac{Z_3}{Z_1Z_2 + Z_2Z_3 + Z_1Z_3}$	$Z_3 = \dfrac{Y_3}{Y_1Y_2 + Y_2Y_3 + Y_1Y_3}$

2.3 *ABCD* PARAMETERS

At microwave frequencies, the generalized circuit constants matrix ($ABCD$) and scattering matrix (S) methods of circuit analysis are used exclusively. Calculations are most easily made using *ABCD* parameters because (a) lumped elements and transmission line elements are related to the matrix elements by simple expressions, and (b) elements are cascaded simply by multiplying their matrices. The *ABCD* matrix method is just for two-port networks. On the other hand, the scattering matrix formulation is a more general method of representing microwave networks and can handle three or more ports.

ABCD parameters for a two-port network, such as shown in Figure 2.4, are defined as

$$\begin{bmatrix} V_1 \\ I_1 \end{bmatrix} = \begin{bmatrix} A & B \\ C & D \end{bmatrix} \begin{bmatrix} V_2 \\ I_2 \end{bmatrix} \tag{2.16}$$

The *ABCD* matrices for the T-network shown in Figure 2.2c and the π-network shown in Figure 2.3c, when $g_m V_1$ is zero, are

$$\begin{bmatrix} A & B \\ C & D \end{bmatrix}_{\mathrm{T}} = \begin{bmatrix} 1 + \dfrac{Z_1}{Z_3} & Z_1 + Z_2 + \dfrac{Z_1 Z_2}{Z_3} \\ \dfrac{1}{Z_3} & 1 + \dfrac{Z_2}{Z_3} \end{bmatrix} \tag{2.17a}$$

$$\begin{bmatrix} A & B \\ C & D \end{bmatrix}_{\pi} = \begin{bmatrix} 1 + \dfrac{Y_2}{Y_3} & \dfrac{1}{Y_3} \\ Y_1 + Y_2 + \dfrac{Y_1 Y_2}{Y_3} & 1 + \dfrac{Y_1}{Y_3} \end{bmatrix} \tag{2.17b}$$

In Figure 2.4 note that I_2 is shown to flow outward and becomes I_1 of the next two-port network in a cascaded chain. The *ABCD* matrices for commonly used microwave circuit

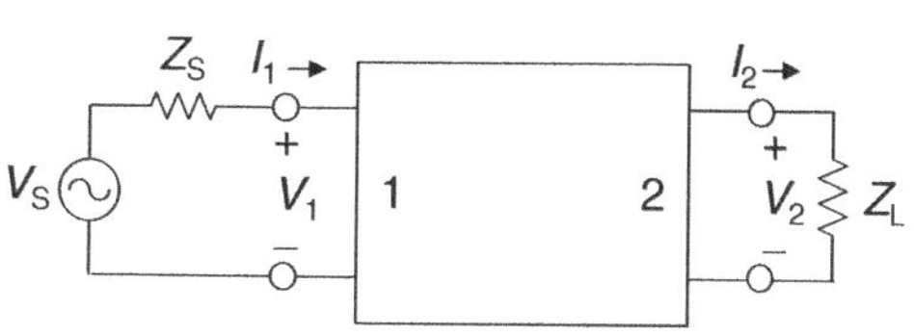

Figure 2.4 Two-port network representation.

Table 2.2 *ABCD* Matrices of Commonly Used Two-Port Networks

Network	ABCD Matrix
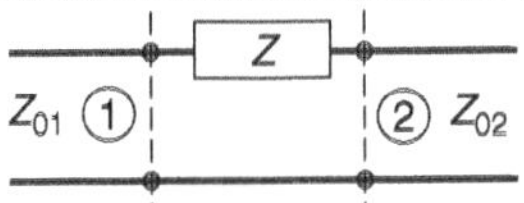	$\begin{bmatrix} 1 & Z \\ 0 & 1 \end{bmatrix}$
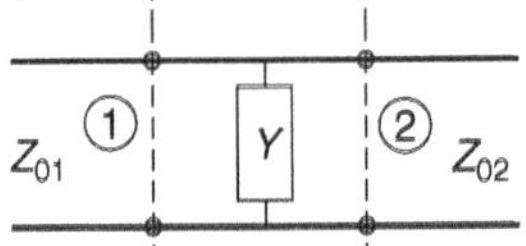	$\begin{bmatrix} 1 & 0 \\ Y & 1 \end{bmatrix}$
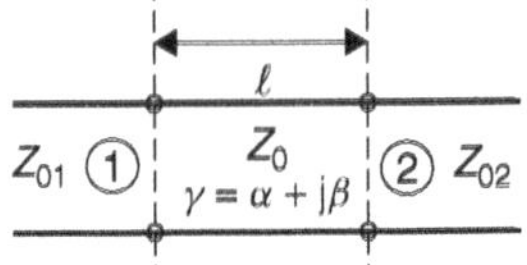	$\begin{bmatrix} \cosh \gamma\ell & Z_0 \sinh \gamma\ell \\ \dfrac{\sinh \gamma\ell}{Z_0} & \cosh \gamma\ell \end{bmatrix}$
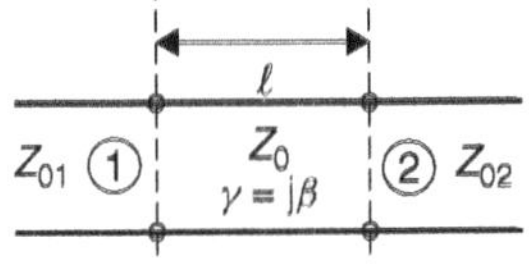	$\begin{bmatrix} \cos \beta\ell & jZ_0 \sin \beta\ell \\ \dfrac{j \sin \beta\ell}{Z_0} & \cos \beta\ell \end{bmatrix}$
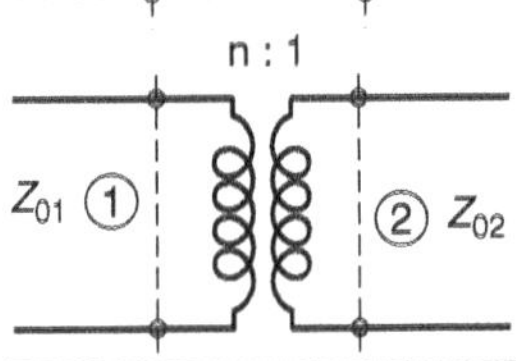	$\begin{bmatrix} n & 0 \\ 0 & 1/n \end{bmatrix}$

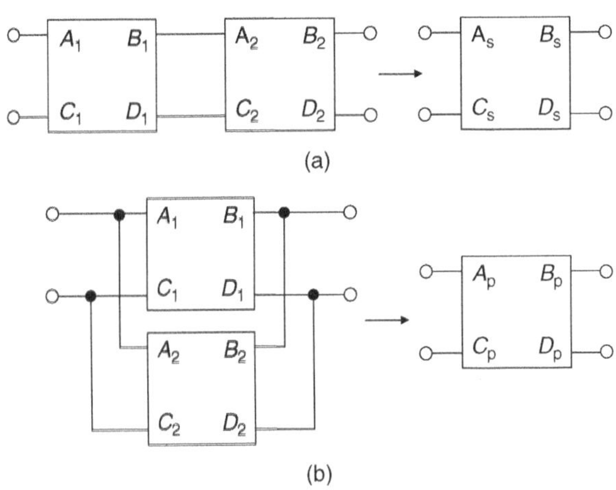

Figure 2.5 *ABCD* matrix operations: (a) series and (b) parallel.

elements are listed in Table 2.2. The connections for combining *ABCD* matrices in series and in parallel are given in Figure 2.5. The operations for such connections are as follows:

$$\text{Series}: \quad \begin{bmatrix} A_s & B_s \\ C_s & D_s \end{bmatrix} = \begin{bmatrix} A_1 & B_1 \\ C_1 & D_1 \end{bmatrix} \begin{bmatrix} A_2 & B_2 \\ C_2 & D_2 \end{bmatrix} = \begin{bmatrix} A_1A_2 + B_1C_2 & A_1B_2 + B_1D_2 \\ C_1A_2 + D_1C_2 & C_1B_2 + D_1D_2 \end{bmatrix} \tag{2.18a}$$

$$\text{Parallel:} \quad \begin{bmatrix} A_p & B_p \\ C_p & D_p \end{bmatrix} = \begin{bmatrix} \dfrac{A_1B_2 + A_2B_1}{B_1 + B_2} & \dfrac{B_1B_2}{B_1 + B_2} \\ C_1 + C_2 + \dfrac{(A_2 - A_1)(D_1 - D_2)}{B_1 + B_2} & \dfrac{D_1B_2 + D_2B_1}{B_1 + B_2} \end{bmatrix} \tag{2.18b}$$

ABCD matrices exhibit the following characteristics.

1. For reciprocal networks, the forward and backward characteristics are the same,

$$AD - BC = 1 \tag{2.19}$$

2. For symmetrical networks, which remain unaltered when the two ports are interchanged, we have $A = D$.

2.4 *S*-PARAMETERS

The use of *ABCD* parameters at microwave frequencies is not very convenient from the measurements point of view. Also, its main advantage of cascading network components does not hold when the network consists of components with three or more ports. The scattering matrix formulation is a more general method for representing microwave networks and the device under test is usually terminated in 50 Ω. A scattering matrix represents the relationship between variables a_n (proportional to the incoming wave at the nth port) and variables b_n (proportional to the outgoing wave at the nth port) defined in the following manner.

$$a_n = \frac{v_n^+}{\sqrt{Z_{0n}}} \tag{2.20a}$$

$$b_n = \frac{v_n^-}{\sqrt{Z_{0n}}} \tag{2.20b}$$

where v_n^+ and v_n^- represent voltages corresponding to the incoming and the outgoing waves in the transmission line (or the waveguide) connected to the nth port and Z_{0n} is the characteristic impedance of the line (or waveguide). Knowledge of v_n^+ and v_n^- is not required to evaluate coefficients of the scattering matrix. Relationships between b_n and a_n for the two-port network shown in Figure 2.6 may be written

$$b_1 = S_{11}a_1 + S_{12}a_2 \tag{2.21a}$$

$$b_2 = S_{21}a_1 + S_{22}a_2 \tag{2.21b}$$

Here

$$S_{11} = b_1/a_1, \qquad S_{21} = b_2/a_1, \qquad a_2 = 0 \tag{2.22a}$$

$$S_{12} = b_1/a_2, \qquad S_{22} = b_2/a_2, \qquad a_1 = 0 \tag{2.22b}$$

In general, for an n-port network, we have

$$[b] = [S][a] \tag{2.23}$$

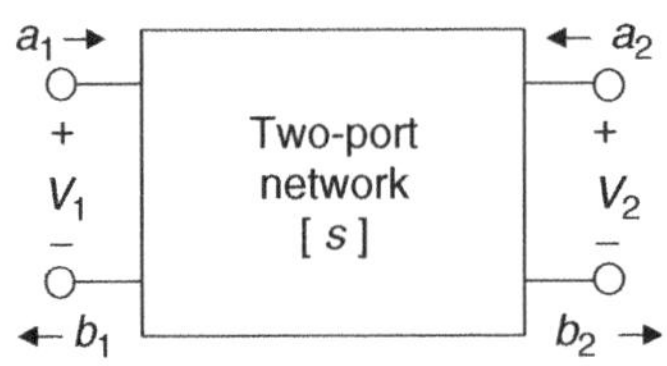

Figure 2.6 S-parameter representation of a two-port network.

Some of the important characteristics of S matrices are listed below.

1. For a reciprocal network, the S matrix is symmetrical, that is,

$$S = S^{\mathrm{t}}$$

where the superscript t indicates the transpose of a matrix.

2. In a lossless passive network,

$$\sum_{n=1}^{N} |S_{ni}|^2 = \sum_{n=1}^{N} S_{ni} S_{ni}^* = 1 \tag{2.24a}$$

for all $i = 1, 2, \ldots, N$. The asterisk designates the complex conjugate quantity.

3. Again, for lossless passive networks, the power conservation condition yields an orthogonality constraint given by

$$\sum_{n=1}^{N} S_{ns} S_{nr}^* = 0 \tag{2.24b}$$

for all $s, r = 1, 2, \ldots, N, s \neq r$.

At the nth port, the total voltage V_n and current I_n are given by

$$V_n = V_n^+ + V_n^- = \sqrt{Z_{0n}}(a_n + b_n) \tag{2.25a}$$

$$I_n = I_n^+ - I_n^- = (V_n^+ - V_n^-)/Z_{0n} = \frac{1}{\sqrt{Z_{0m}}}(a_n - b_n) \tag{2.25b}$$

Because the current flows in the axial direction, the total (net) current is given by the difference of the currents flowing in the positive and negative directions.

The net average power flow into the network is given by the usual low-frequency relation; that is,

$$P_n = \tfrac{1}{2}\,\mathrm{Re}(V_n \times I_n^*) \tag{2.26}$$

where V_n and I_n denote the total peak voltage and current, respectively. Using (2.25) in (2.26), we find

$$P_n = \tfrac{1}{2}(|a_n|^2 - |b_n|^2) + \text{imaginary term} = \tfrac{1}{2}(|a_n|^2 - |b_n|^2) \tag{2.27}$$

This equation shows that the power delivered to a port is equal to the power in the incident wave minus the power in the reflected wave at that port. This is an important term that defines the delivered, incident, and reflected powers. The terms S_{11}, S_{22} and S_{12}, S_{21} are defined as the reflection and transmission coefficients, respectively.

Attenuation in a circuit, α, expressed in decibels (dB), is given by

$$\alpha = -20 \log \left| \frac{b_2}{a_1} \right| \text{ dB} = -20 \log |S_{21}| \text{ dB} \tag{2.28}$$

For a passive circuit, $b_2 < a_1$ and value of α is a positive quantity. However, for an amplifier with gain, $b_2 > a_1$ and α becomes a negative quantity. In other words, the negative attenuation is nothing but gain. The transmission phase is given by

$$\phi = \text{angle of } S_{21} \tag{2.29}$$

EXAMPLE 2.3

Consider a transmission line section of characteristic impedance Z_0, length ℓ, and phase constant β as shown in Figure 2.7. Determine its S matrix.

SOLUTION This transmission line section can be regarded as a two-port network with incident waves a_1, a_2 and reflected waves b_1, b_2 at ports 1 and 2, respectively. Since the transmission line is matched at the input and output there are no reflections at the two ports; that is, S_{11} and S_{22} are zero. For a matched lossless line, we have

$$S_{12} = b_1/a_2 = a_2e^{-j\beta\ell}/a_2 = e^{-j\beta\ell}$$

$$S_{21} = b_2/a_1 = a_1e^{-j\beta\ell}/a_1 = e^{-j\beta\ell}$$

$$b_1 = S_{11}a_1 + S_{12}a_2 = 0.a_1 + a_2e^{-j\beta\ell} \tag{2.30a}$$

$$b_2 = S_{21}a_1 + S_{22}a_2 = a_1e^{-j\beta\ell} + 0 \cdot a_2 \tag{2.30b}$$

and we may obtain the scattering matrix for this network as

$$[S] = \begin{bmatrix} 0 & e^{-j\beta\ell} \\ e^{-j\beta\ell} & 0 \end{bmatrix} \tag{2.31}$$

Equation (2.31) may be used to determine the modified S matrix for any network when the reference plane for one of its ports is shifted away along the transmission line connected to that port. ■

EXAMPLE 2.4

Properties of the S matrix of a three-port network shown in Figure 2.8 may be used to show that a lossless, matched, and nonreciprocal three-port microwave circuit is only possible in the form of a three-port circulator. A circulator is a three-port matched network, in which the signal travels only in one direction from left to right.

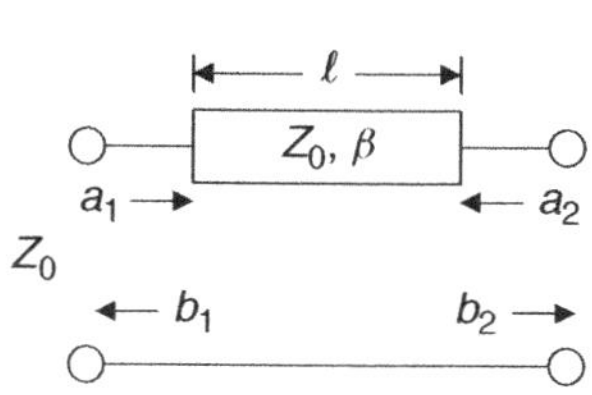

Figure 2.7 Network representation of a transmission line section.

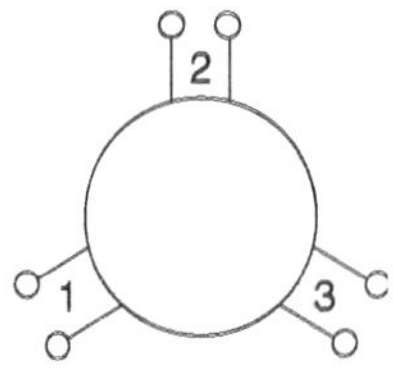

Figure 2.8 Three-port network representation of a component.

SOLUTION A perfectly matched three-port microwave circuit has an S matrix with $S_{ii} = 0$ (where $i = 1$, 2, and 3) and the S matrix is given by

$$[S] = \begin{bmatrix} 0 & S_{12} & S_{13} \\ S_{21} & 0 & S_{23} \\ S_{31} & S_{32} & 0 \end{bmatrix} \tag{2.32}$$

For a nonreciprocal three-port microwave circuit, the S matrix is no longer symmetrical; that is, $S_{ij} \neq S_{ji}$. However, if the circuit is lossless, conservation of power still requires that the S matrix satisfy (2.24a) and (2.24b). Applying these conditions, we have

$$S_{21}S_{21}^* + S_{31}S_{31}^* = 1 \tag{2.33a}$$
$$S_{12}S_{12}^* + S_{32}S_{32}^* = 1 \tag{2.33b}$$
$$S_{13}S_{13}^* + S_{23}S_{23}^* = 1 \tag{2.33c}$$
$$S_{31}S_{32}^* = S_{21}S_{23}^* = S_{12}S_{13}^* = 0 \tag{2.33d}$$

Let us assume that $S_{12} \neq 0$. Equation (2.33d) then yields $S_{13} = 0$. Equation (2.33c) now requires that $|S_{23}| = 1$. From (2.33d) $S_{21} = 0$, from (2.33a) $|S_{31}| = 1$, and from (2.33d) $S_{32} = 0$. Thus, if $S_{12} \neq 0$,

$$S_{13} = S_{21} = S_{32} = 0 \tag{2.34a}$$
$$|S_{23}| = |S_{31}| = |S_{12}| = 1 \tag{2.34b}$$

Consequently, there is a perfect transmission from port 3 into port 2, from port 2 into port 1, and from port 1 into port 3. Also, there is no transmission in the reverse directions, and the device therefore acts as a perfect circulator. Thus the S matrix of a lossless, matched, and nonreciprocal three-port microwave circuit may be written

$$[S] = \begin{bmatrix} 0 & S_{12} & 0 \\ 0 & 0 & S_{23} \\ S_{31} & 0 & 0 \end{bmatrix} \tag{2.35}$$

■

EXAMPLE 2.5

Now consider a network, shown in Figure 2.8, working as a 3-dB power divider. The input port 1 is matched and the input power is divided equally at the output ports 2 and 3. Show that ports 2 and 3 have a reflection coefficient of 0.5.

SOLUTION The scattering matrix coefficients for a 3-dB power divider are given as

$$S_{11} = 0 \tag{2.36a}$$
$$S_{21} = 1/\sqrt{2} \tag{2.36b}$$
$$S_{31} = 1/\sqrt{2} \tag{2.36c}$$

Equal power division implies $|S_{21}| = |S_{31}| = 1/\sqrt{2}$. Using this information, S matrix (2.32) becomes

$$[S] = \begin{bmatrix} 0 & 1/\sqrt{2} & 1/\sqrt{2} \\ 1/\sqrt{2} & S_{22} & S_{23} \\ 1/\sqrt{2} & S_{23} & S_{33} \end{bmatrix} \tag{2.37}$$

Using (2.24a), we have

$$\tfrac{1}{2} + S_{22}S_{22}^* + S_{23}S_{23}^* = 1 \tag{2.38a}$$
$$\tfrac{1}{2} + S_{23}S_{23}^* + S_{33}S_{33}^* = 1 \tag{2.38b}$$

These two relations lead to

$$|S_{22}|^2 = |S_{33}|^2 \tag{2.39}$$

Also, by using (2.24b), we have

$$S_{22}^*/\sqrt{2} + S_{23}^*/\sqrt{2} = 0 \tag{2.40}$$

which yields

$$S_{22} = -S_{23} \tag{2.41}$$

Substituting this in (2.38a), we have

$$\tfrac{1}{2} + S_{22}S_{22}^* + S_{22}S_{22}^* = 1 \tag{2.42a}$$
$$|S_{22}|^2 = \tfrac{1}{4} \tag{2.42b}$$
$$|S_{22}| = |S_{23}| = |S_{33}| = \tfrac{1}{2} \tag{2.42c}$$

Relation (2.42c) indicates clearly that ports 2 and 3 are not matched. The reflection coefficient at these ports will be 0.5. Figures 2.9 and 2.10 validate this by using a circuit schematic approach. In Figure 2.9, the output port 50-Ω impedance is transformed to 100 Ω (see Chapter 7) at port 1. Two 100-Ω impedances connected in parallel result in 50 Ω. Thus the input port is perfectly matched. Similarly, in Figure 2.10, the impedance at output port 3 is transformed to 100 Ω at port 1. A parallel combination of 50-Ω and 100-Ω impedances results in 33.33 Ω. The 33.33-Ω impedance is transformed to 150 Ω at output port 2, which when looking into the 50 Ω has a reflection coefficient of 0.5. In order to obtain a good match at the output ports, one has to use isolation resistors between the output ports, as discussed in Chapter 19. ■

The three above described examples demonstrate that the properties of scattering matrices may be used to study the characteristics of RF and microwave networks or circuits.

S-parameters for commonly used networks shown in Table 2.2 and terminated in input and output impedance of Z_0 are given below.

1. Series line section of characteristic impedance Z and electrical length $\gamma\ell$, where the propagation constant $\gamma = \alpha + j\beta$ and where α and β are the attenuation and phase constants, respectively:

$$[S]_L = \frac{1}{D_L}\begin{bmatrix} (Z^2 - Z_0^2)\sinh\gamma\ell & 2ZZ_0 \\ 2ZZ_0 & (Z^2 - Z_0^2)\sinh\gamma\ell \end{bmatrix} \tag{2.43}$$

where $D_L = 2ZZ_0\cosh\gamma\ell + (Z^2 + Z_0^2)\sinh\gamma\ell$.

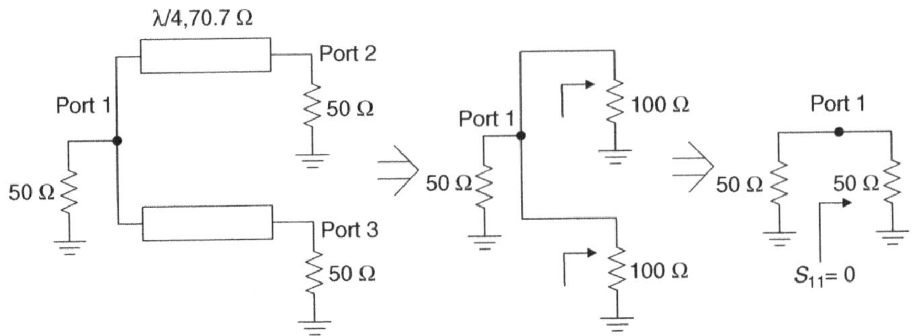

Figure 2.9 Input match representation of a 3-dB power divider usingλ/4 transmission line sections.

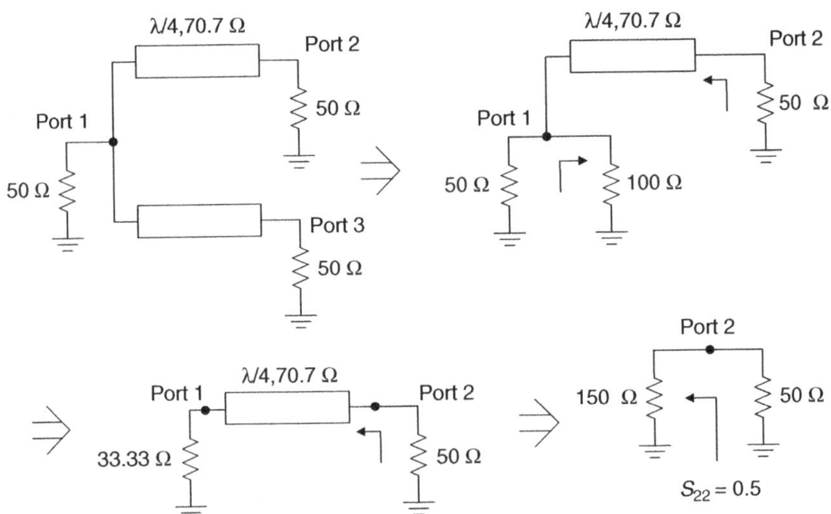

Figure 2.10 Output port match representation of a 3-dB power divider.

2. Series impedance, Z:

$$[S]_Z = \frac{1}{Z + 2Z_0}\begin{bmatrix} Z & 2Z_0 \\ 2Z_0 & Z \end{bmatrix} \tag{2.44}$$

3. Shunt admittance, Y:

$$[S]_Y = \frac{1}{Y + 2Y_0}\begin{bmatrix} -Y & 2Y_0 \\ 2Y_0 & -Y \end{bmatrix} \tag{2.45}$$

where $Y_0 = 1/Z_0$.

4. An ideal transformer with impedance transformation ratio n:1:

$$[S]_T = \frac{1}{1 + n^2}\begin{bmatrix} n^2 - 1 & 2n \\ 2n & 1 - n^2 \end{bmatrix} \tag{2.46}$$

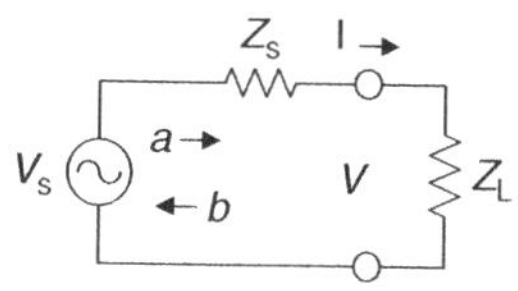

Figure 2.11 One-port network representation.

2.4.1 *S*-Parameters for a One-Port Network

Consider a one-port network as shown in Figure 2.11, where Z_S and Z_L are the source and load impedances and voltage V and current I have peak values. If the source and load have complex values, that is,

$$Z_S = R_S + jX_S, \quad Z_L = R_L + jX_L \tag{2.47}$$

there exists real power (dissipated or radiated) and imaginary power (stored in reactive component). Maximum power is delivered to the load when imaginary power is zero, that is, $X_L = X_S$.

The average power delivered to the load is given by

$$P_{av} = \frac{VI}{2} = \frac{1}{2}I^2 R_L = \frac{1}{2}\frac{V^2 R_L}{(R_S + R_L)^2} \tag{2.48}$$

A condition for maximum delivered power is obtained by $\partial P_{av}/\partial R_L = 0$. This gives $R_L = R_S$; that is, for maximum power deliver $Z_L = Z_S^*$ or

$$P_{av} = \frac{1}{8}\frac{V^2}{R_L} \tag{2.49}$$

For a one-port network, the forward wave, a, and the reflected wave, b, may be written

$$a = \frac{V + IZ_S}{2\sqrt{Z_S}} \tag{2.50a}$$

$$b = \frac{V - IZ_S}{2\sqrt{Z_S}} \tag{2.50b}$$

The one-port S-parameter (S_{11}) or the reflection coefficient is given by

$$S_{11} = \frac{b}{a} = \frac{V - IZ_S}{V + IZ_S} = \frac{Z_L - Z_S}{Z_L + Z_S} \tag{2.51}$$

The average power delivered to the load may also be calculated using (2.50) in (2.27) and leads to (2.48).

2.5 RELATIONSHIPS BETWEEN VARIOUS TWO-PORT PARAMETERS

The various two-port parameters (Z, Y, $ABCD$, and S) described earlier can be related to each other by applying their definitions and relating voltage and currents. These relations are given in Table 2.3 for unnormalized parameters for arbitrary input and output impedances.

Table 2.3 Conversion Relationships Between Various Representative Matrices of Two-Port Networks with Arbitrary Source and Load Impedances

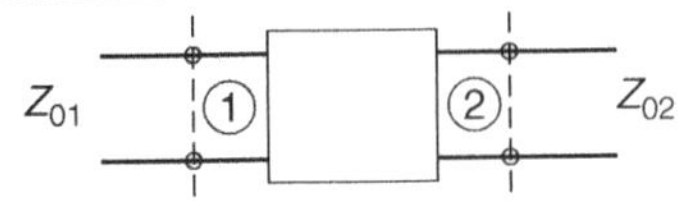

$$\begin{bmatrix} Z_{11} & Z_{12} \\ Z_{21} & Z_{22} \end{bmatrix} = \frac{1}{Y_{11}Y_{22} - Y_{12}Y_{21}} \begin{bmatrix} Y_{22} & -Y_{12} \\ -Y_{21} & Y_{11} \end{bmatrix}$$

$$\begin{bmatrix} Y_{11} & Y_{12} \\ Y_{21} & Y_{22} \end{bmatrix} = \frac{1}{Z_{11}Z_{22} - Z_{12}Z_{21}} \begin{bmatrix} Z_{22} & -Z_{12} \\ -Z_{21} & Z_{11} \end{bmatrix}$$

$$\begin{bmatrix} Z_{11} & Z_{12} \\ Z_{21} & Z_{22} \end{bmatrix} = \frac{1}{C} \begin{bmatrix} A & AD - BC \\ 1 & D \end{bmatrix}$$

$$\begin{bmatrix} Y_{11} & Y_{12} \\ Y_{21} & Y_{22} \end{bmatrix} = \frac{1}{B} \begin{bmatrix} D & -(AD - BC) \\ -1 & A \end{bmatrix}$$

$$\begin{bmatrix} A & B \\ C & D \end{bmatrix} = \frac{1}{Z_{21}} \begin{bmatrix} Z_{11} & (Z_{11}Z_{22} - Z_{12}Z_{21}) \\ 1 & Z_{22} \end{bmatrix}$$

$$\begin{bmatrix} A & B \\ C & D \end{bmatrix} = \frac{1}{-Y_{21}} \begin{bmatrix} Y_{22} & 1 \\ Y_{11}Y_{22} - Y_{12}Y_{21} & Y_{11} \end{bmatrix}$$

$$\begin{bmatrix} S_{11} & S_{12} \\ S_{21} & S_{22} \end{bmatrix} = \frac{1}{\left(\dfrac{Z_{11}}{Z_{01}} + 1\right)\left(\dfrac{Z_{22}}{Z_{02}} + 1\right) - \dfrac{Z_{12}Z_{21}}{Z_{01}Z_{02}}} \times \begin{bmatrix} \left(\dfrac{Z_{11}}{Z_{01}} - 1\right)\left(\dfrac{Z_{22}}{Z_{02}} + 1\right) - \dfrac{Z_{12}Z_{21}}{Z_{01}Z_{02}} & 2\dfrac{Z_{12}}{Z_{02}} \\ 2\dfrac{Z_{21}}{Z_{01}} & \left(\dfrac{Z_{11}}{Z_{01}} + 1\right)\left(\dfrac{Z_{22}}{Z_{02}} - 1\right) - \dfrac{Z_{12}Z_{21}}{Z_{01}Z_{02}} \end{bmatrix}$$

$$\begin{bmatrix} S_{11} & S_{12} \\ S_{21} & S_{22} \end{bmatrix} = \frac{1}{\left(1 + \dfrac{Y_{11}}{Y_{01}}\right)\left(1 + \dfrac{Y_{22}}{Y_{02}}\right) - \dfrac{Y_{12}Y_{21}}{Y_{01}Y_{02}}} \times \begin{bmatrix} \left(1 - \dfrac{Y_{11}}{Y_{01}}\right)\left(1 + \dfrac{Y_{22}}{Y_{02}}\right) + \dfrac{Y_{12}Y_{21}}{Y_{01}Y_{02}} & -2\dfrac{Y_{12}}{Y_{01}} \\ -2\dfrac{Y_{21}}{Y_{02}} & \left(1 + \dfrac{Y_{11}}{Y_{01}}\right)\left(1 - \dfrac{Y_{22}}{Y_{02}}\right) + \dfrac{Y_{12}Y_{21}}{Y_{01}Y_{02}} \end{bmatrix}$$

$$\begin{bmatrix} Y_{11} & Y_{12} \\ Y_{21} & Y_{22} \end{bmatrix} = \frac{1}{(1 + S_{11})(1 + S_{22}) - (S_{12}S_{21})} \times \begin{bmatrix} Y_{01}[(1 - S_{11})(1 + S_{22}) + S_{12}S_{21}] & -2Y_{01}S_{12} \\ -2Y_{02}S_{21} & Y_{02}[(1 + S_{11})(1 - S_{22}) + S_{12}S_{21}] \end{bmatrix}$$

Table 2.3 (*Continued*)

$$\begin{bmatrix} A & B \\ C & D \end{bmatrix} = \frac{1}{(2S_{21})} \begin{bmatrix} [(1+S_{11})(1-S_{22})+S_{12}S_{21}] & Z_{02}[(1+S_{11})(1+S_{22})-S_{12}S_{21}] \\ \dfrac{1}{Z_{01}}[(1-S_{11})(1-S_{22})-S_{12}S_{21}] & \dfrac{Z_{01}}{Z_{02}}[(1-S_{11})(1+S_{22})+S_{12}S_{21}] \end{bmatrix}$$

$$\begin{bmatrix} S_{11} & S_{12} \\ S_{21} & S_{22} \end{bmatrix} = \frac{1}{(B+CZ_{01}Z_{02})+(AZ_{02}+DZ_{02})} \times \begin{bmatrix} (B-CZ_{01}Z_{02})+(AZ_{02}-DZ_{01}) & 2Z_{01}(AD-BC) \\ 2Z_{02} & (B-CZ_{01}Z_{02})-(AZ_{02}-DZ_{01}) \end{bmatrix}$$

$$\begin{bmatrix} Z_{11} & Z_{12} \\ Z_{21} & Z_{22} \end{bmatrix} = \frac{1}{(1-S_{11})(1-S_{22})-S_{12}S_{21}} \times \begin{bmatrix} Z_{01}[(1+S_{11})(1-S_{22})+S_{12}S_{21}] & 2Z_{02}S_{12} \\ 2Z_{01}S_{21} & Z_{02}[(1-S_{11})(1+S_{22})+S_{12}S_{21}] \end{bmatrix}$$

REFERENCES

1. G. Gonzalez, *Microwave Transistor Amplifiers Analysis and Design*, Prentice Hall, Englewood Cliffs, NJ, 1984.
2. R. E. Collin, *Foundations of Microwave Engineering*, McGraw-Hill, New York, 1992.
3. D. M. Pozar, *Microwave Engineering*, John Wiley & Sons, Hoboken, NJ, 1992.
4. I. J. Bahl and P. Bhartia, *Microwave Solid State Circuit Design*, 2nd edition, John Wiley & Sons, Hoboken, NJ, 2003.
5. A. Grebennikov, *RF and Microwave Power Amplifier Design*, McGraw-Hill, New York, 2004.

PROBLEMS

2.1 Determine the Z-parameters for the network shown below.

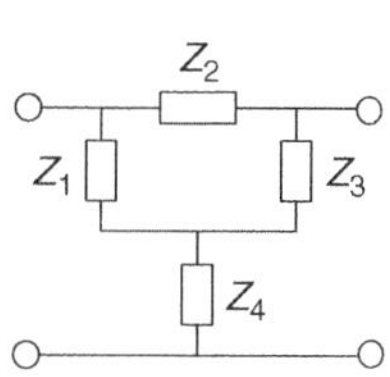

2.2 Determine the Y-parameters for the network shown below.

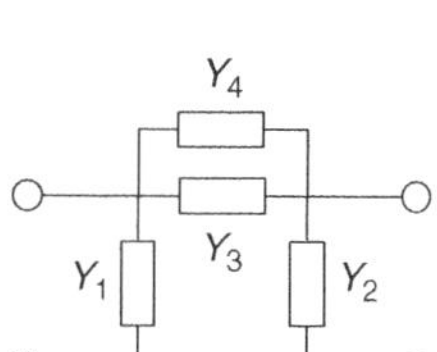

2.3 The output of a two-port network with S-parameters S_{11}, S_{12}, S_{21}, and S_{22} is terminated with an impedance Z_L. Determine the input return loss if the terminating impedance is Z_0.

2.4 A lossless line of characteristic impedance Z and length ℓ is terminated on both sides with an impedance Z_0. Derive an expression for the reflection coefficient S_{11}.

Calculate its value for a quarter-wave long line when Z_0 and Z values are 50 Ω and 25 Ω, respectively.

2.5 Prove that a lossless and reciprocal three-port network cannot be designed with all ports matched.

2.6 Show that for a lossless and reciprocal two-port network, $|S_{21}|^2 = 1 - |S_{11}|^2$.

2.7 Drive the popular expression for input impedance of a transmission line terminated in a load impedance Z_L as given below:

$$Z_{\text{in}} = Z_0 \frac{Z_L + jZ_0 \tan \beta\ell}{Z_0 + jZ_L \tan \beta\ell}$$

The line is lossless and its characteristic impedance, phase constant, and physical length are Z_0, β, and ℓ, respectively.

Chapter 3

Amplifier Characteristics and Definitions

Although many characteristics must be considered when designing an amplifier, the most important of these are frequency range or bandwidth, power gain, noise figure, power output, 1-dB gain compression point, input and output VSWR, power added efficiency (PAE), intermodulation distortion, adjacent channel power ratio (ACPR), dynamic range, phase noise, stability [1–7], and reliability [8–13]. These parameters are briefly described in this chapter. The stability is discussed in detail in Chapter 17 and the power added efficiency, which is most important for power amplifiers, is treated in Chapter 10. The measurement of phase noise in amplifiers is discussed in Chapter 22.

3.1 BANDWIDTH

The bandwidth definition of an amplifier is very complex, when one has to meet several conflicting specifications in terms of gain, noise figure, output power, gain flatness, PAE, TOI, VSWR, and so on. The output power specification may be at P_{1dB} or saturated power or 3–10 dB back-off from P_{1dB} for linear operation. Generally, the bandwidth of an amplifier is defined as the frequency range over which the circuit meets the specified minimum/maximum or typical aforementioned requirements.

Since the transistor S-parameters vary with frequency, broad bandwidth of an amplifier cannot be obtained easily by using conventional or ladder type matching networks. Such limitations are discussed in detail in Chapter 11. Let us briefly look into inherent bandwidth limitations of a transistor.

Consider input impedance and output admittance of a transistor as shown in Figure 3.1. In this case input and output 3-dB normalized bandwidths are given by

$$\frac{\Delta f_{\text{in}}}{f_0} = \frac{1}{Q_{\text{in}}} = 2\pi f_0 R_{\text{in}} C_{\text{in}} \tag{3.1a}$$

$$\frac{\Delta f_{\text{out}}}{f_0} = \frac{1}{Q_{\text{out}}} = \frac{1}{2\pi f_0 R_{\text{o}} C_{\text{o}}} \tag{3.1b}$$

Fundamentals of RF and Microwave Transistor Amplifiers. By Inder J. Bahl

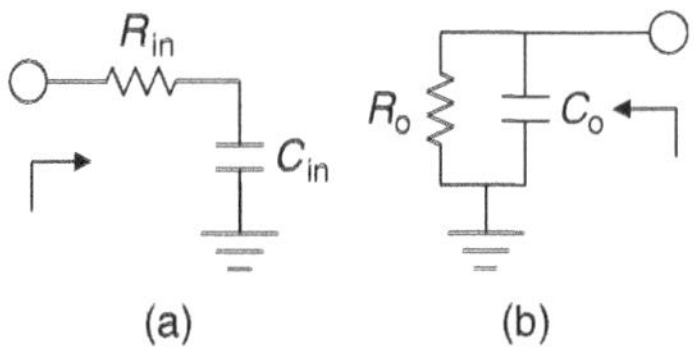

Figure 3.1 One-port equivalent circuit of a transistor: (a) input and (b) output.

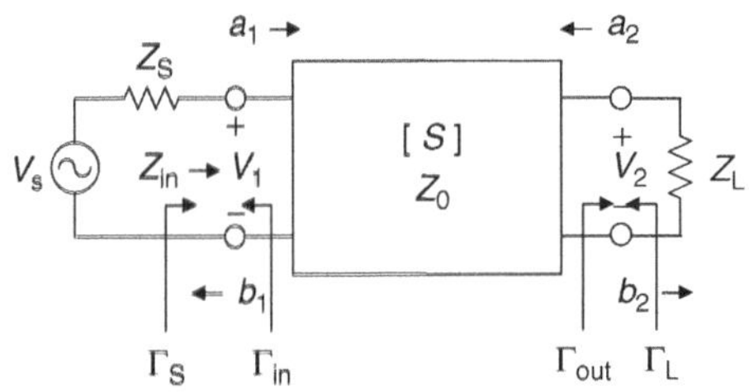

Figure 3.2 Two-port network for a transistor.

where f_0 is the center frequency. Consider a transistor (0.6-mm FET) having $R_{in} = 4\ \Omega$, $C_{in} = 1$ pF, $R_o = 200\ \Omega$, and $C_o = 0.16$ pF. The calculated values of Δf_{in} and Δf_{out} at 10 GHz are

$$\Delta f_{in} \cong 25\% \quad \text{and} \quad \Delta f_{out} = 50\%$$

This shows that the bandwidth of the transistor is limited by its input impedance, which can be increased by adding more resistance in series with the device. However, it reduces the transistor gain but improves its stability as discussed in Chapter 17.

3.2 POWER GAIN

The power gain of a two-port network, amplifier in this case, is defined as the ratio of the output power to input power. For a two-port network (shown in Fig. 3.2), the power gain can be defined in several ways. The three commonly used definitions are transducer power gain (G_T) or simply power gain, maximum available power gain (G_A), and available power gain (G_P). Their definitions are given below:

$$G_T = \frac{\text{power delivered to the load}}{\text{power available from the source}} = \frac{P_L}{P_{avs}} \tag{3.2}$$

$$G_A = \frac{\text{power available from the network}}{\text{power available from the source}} = \frac{P_N}{P_{avs}} \tag{3.3}$$

$$G_P = \frac{\text{power delivered to the load}}{\text{power delivered to the network}} = \frac{P_L}{P_{in}} \tag{3.4}$$

Here $G_A \geq G_T$ and $G_P \geq G_T$.

Next, we derive these gain terms expressed in the linear network S-parameters and arbitrary source and load reflection coefficients, Γ_S and Γ_L, respectively.

From S-parameter definitions,

$$b_1 = S_{11}a_1 + S_{12}a_2 = S_{11}a_1 + S_{12}\Gamma_L b_2 \tag{3.5}$$

$$b_2 = S_{21}a_1 + S_{22}a_2 = S_{21}a_1 + S_{22}\Gamma_L b_2 \tag{3.6}$$

where $\Gamma_L = a_2/b_2$. Eliminating b_2 from (3.5), we have

$$b_1 = a_1 \left[S_{11} + \frac{S_{12}S_{21}\Gamma_L}{1 - S_{22}\Gamma_L} \right]$$

or

$$\frac{b_1}{a_1} = \Gamma_{in} = S_{11} + \frac{S_{12}S_{21}\Gamma_L}{1 - S_{22}\Gamma_L} \tag{3.7}$$

Similarly,

$$\frac{b_2}{a_2} = \Gamma_{out} = S_{22} + \frac{S_{12}S_{21}\Gamma_S}{1 - S_{11}\Gamma_S} \tag{3.8}$$

The source and load reflection coefficients are given by

$$\Gamma_{S,L} = \frac{Z_{S,L} - Z_0}{Z_{S,L} + Z_0} \tag{3.9}$$

where Z_0 is the source and load impedance used for the measurement of device S-parameters. Also,

$$Z_{S,L} = Z_0 \frac{1 + \Gamma_{S,L}}{1 - \Gamma_{S,L}} = R_{S,L} + jX_{S,L} \tag{3.10}$$

$$Z_{in,out} = Z_0 \frac{1 + \Gamma_{in,out}}{1 - \Gamma_{in,out}} \tag{3.11}$$

The average power delivered to the network is

$$P_{in} = \tfrac{1}{2}[|a_1|^2 - |b_1|^2] \tag{3.12a}$$

or

$$P_{in} = \tfrac{1}{2}|a_1|^2[1 - |\Gamma_{in}|^2] \tag{3.12b}$$

Using (2.25a), the voltage V_1 may be expressed as

$$V_1 = V_S \frac{Z_{in}}{Z_S + Z_{in}} = \sqrt{Z_0}(a_1 + b_1) = \sqrt{Z_0}a_1 \left(1 + \frac{b_1}{a_1}\right)$$

or

$$V_S = \sqrt{Z_0}a_1(1 + \Gamma_{in})(Z_S + Z_{in})/Z_{in} \tag{3.12c}$$

Here V_S has the peak value. From (3.12c)

$$a_1 = \frac{V_S}{\sqrt{Z_0}} \frac{Z_{in}}{(1 + \Gamma_{in})(Z_S + Z_{in})} \tag{3.13}$$

From (3.10) and (3.11), we find

$$\frac{Z_{\text{in}}}{(1+\Gamma_{\text{in}})(Z_S+Z_{\text{in}})} = \frac{1-\Gamma_S}{2(1-\Gamma_S\Gamma_{\text{in}})} \tag{3.14}$$

From (3.12b), (3.13), and (3.14), we have

$$P_{\text{in}} = \frac{|V_S|^2}{8Z_0}\frac{|1-\Gamma_S|^2(1-|\Gamma_{\text{in}}|^2)}{|1-\Gamma_S\Gamma_{\text{in}}|^2} \tag{3.15a}$$

When the network is matched, $Z_S = Z_0$, $\Gamma_S = 0$, and $\Gamma_{\text{in}} = 0$,

$$P_{\text{in}} = \frac{|V_S|^2}{8Z_0} \tag{3.15b}$$

This represents the maximum available power from the source. The maximum deliverable power to the network occurs when

$$\Gamma_{\text{in}} = \Gamma_S^*$$

Under this condition, (3.15a) becomes

$$P_{\text{avs}} = \frac{|V_S|^2}{8Z_0}\frac{|1-\Gamma_S|^2}{1-|\Gamma_S|^2} \tag{3.16}$$

Next, we calculate the power delivered to the load, which is given by

$$P_L = \tfrac{1}{2}(|b_2|^2 - |a_2|^2) = \tfrac{1}{2}|b_2|^2(1-|\Gamma_L|^2) \tag{3.17}$$

From (3.6), we find

$$b_2 = \frac{S_{21}a_1}{1-S_{22}\Gamma_L} \tag{3.18}$$

From (3.13), (3.14), (3.17), and (3.18), we have

$$P_L = \frac{|V_S|^2}{8Z_0}\frac{|S_{21}|^2(1-|\Gamma_L|^2)|1-\Gamma_S|^2}{|1-S_{22}\Gamma_L|^2|1-\Gamma_S\Gamma_{\text{in}}|^2} \tag{3.19}$$

Maximum available power from the network or deliverable power to the load occurs when

$$\Gamma_L = \Gamma_{\text{out}}^* \tag{3.20}$$

Under this condition, (3.19) becomes

$$P_N = \frac{|V_S|^2}{8Z_0}\frac{|S_{21}|^2(1-|\Gamma_{\text{out}}|^2)|1-\Gamma_S|^2}{|1-S_{22}\Gamma_{\text{out}}^*|^2|1-\Gamma_S\Gamma_{\text{in}}|^2} \tag{3.21}$$

and represents the power available from the network.

From (3.2), (3.16), and (3.19), we find

$$G_T = \frac{|S_{21}|^2(1-|\Gamma_S|^2)(1-|\Gamma_L|^2)}{|1-\Gamma_S\Gamma_{in}|^2|1-S_{22}\Gamma_L|^2} \tag{3.22a}$$

Substituting Γ_{in} using (3.7), (3.22a) becomes

$$G_T = \frac{|S_{21}|^2(1-|\Gamma_S|^2)(1-|\Gamma_L|^2)}{|(1-S_{11}\Gamma_S)(1-S_{22}\Gamma_L)-S_{12}S_{21}\Gamma_S\Gamma_L|^2} \tag{3.22b}$$

The S-parameters are measured with usually 50 Ω as the input and output impedance, and arbitrary source impedance Z_S and load impedance Z_L are connected to the network. For unilateral transducer power gain, the reverse power gain is set to zero (i.e., $|S_{12}| = 0$), and (3.22b) becomes

$$G_{TU} = \frac{|S_{21}|^2(1-|\Gamma_S|^2)(1-|\Gamma_L|^2)}{|(1-S_{11}\Gamma_S)(1-S_{22}\Gamma_L)|^2} = G_S|S_{21}|^2 G_L \tag{3.23a}$$

where

$$G_S = \frac{(1-|\Gamma_S|^2)}{|1-S_{11}\Gamma_S|^2} \tag{3.23b}$$

and

$$G_L = \frac{(1-|\Gamma_L|^2)}{|1-S_{22}\Gamma_L|^2} \tag{3.23c}$$

The terms G_S and G_L represent the gain or loss of the input and output matching circuits, respectively.

The maximum unilateral power gain is attained when $\Gamma_S = S_{11}^*$ and $\Gamma_L = S_{22}^*$, that is, when the network is matched conjugately at the input and output ports. Then the maximum unilateral power gain, also called the maximum available gain from a unilateral device, is given by

$$G_A = G_{TUm} = \frac{|S_{21}|^2}{(1-|S_{11}|^2)(1-|S_{22}|^2)} \tag{3.24}$$

Thus the maximum available gain is the product of the transistor transducer power gain $|S_{21}|^2$ between source and load impedances (usually 50 Ω) and the increase in gain due to matching the input port $(1-|S_{11}|^2)^{-1}$ and matching the output port $(1-|S_{22}|^2)^{-1}$. In other words, a single-stage amplifier design consists of (a) designing an input matching network to give $\Gamma_S \cong 0$ and (b) designing an output matching network that simultaneously gives $\Gamma_L \cong 0$. As we will show in Chapter 17, the above mentioned conditions fail if the solid state devices are unstable.

Other definitions of gain are derived as follows. From (3.3), (3.16), and (3.21), we have

$$G_A = \frac{|S_{21}|^2(1-|\Gamma_S|^2)(1-|\Gamma_{out}|^2)}{|1-\Gamma_S\Gamma_{in}|^2|1-S_{22}\Gamma_{out}^*|^2} \tag{3.25a}$$

From (3.4), (3.15a), and (3.19), we have

$$G_P = \frac{|S_{21}|^2(1 - |\Gamma_L|^2)}{(1 - |\Gamma_{in}|^2)|1 - S_{22}\Gamma_L|^2} \tag{3.25b}$$

The most commonly used gain definition in amplifiers is transducer power gain, G_T; other gain definitions are normally used to characterize a transistor. The power gain G is usually expressed in decibels, that is,

$$G(\text{dB}) = 10 \log \text{G(power ratio)} \tag{3.26}$$

3.3 INPUT AND OUTPUT VSWR

The input and output VSWR values are commonly used to characterize an amplifier's circuit match to source and load impedance (usually 50 Ω). Mismatch between 50 Ω and the amplifier's input, and 50 Ω and the amplifier's output are measured as input and output reflection coefficients. The voltage reflection coefficient is related to VSWR using a standard definition as given in (3.27). When the voltage reflection coefficient $|\rho| = 0.333$ (VSWR = 2), the power reflection coefficient $|\rho|^2 = 0.11$ means 11% power is reflected. In most applications, a VSWR value of 2:1 is acceptable.

The reflection coefficients are measured at small-signal conditions as well as at large-signal conditions. The measurement of output reflection coefficient under high-power conditions is not trivial (see Chapter 22). The output of power amplifier circuits is designed to provide optimum power performance, not necessarily good VSWR. Normally, good input and output VSWR values in power amplifiers are achieved by using a balanced configuration or by using a traveling-wave combining technique.

Three related parameters—the return loss (RL), VSWR, and reflection coefficient (ρ)—are commonly used to characterize amplifier reflections. The return loss is the ratio of the input power (P_{in}) to reflected power (P_R):

$$RL = -10 \log \frac{P_R}{P_{in}} = -10 \log \left(\frac{VSWR - 1}{VSWR + 1} \right)^2 = -10 \log(|\rho|^2) \tag{3.27}$$

Some typical values for RL, ρ, and VSWR are given in Table 3.1. The return loss expressed in decibel is a positive quantity.

3.4 OUTPUT POWER

Power delivered to the load (P_L) is known as the output power, P_o or P_{out}, which is a strong function of the input power. When the gain is reduced at higher input power levels (as the device gets saturated or compressed) by 1 dB, the output power is defined as $P_{1\text{ dB}}$, which is normally used to characterize nonlinearity in amplifiers. At 3–5-dB gain compression, the output power is saturated and known as P_{sat}. Figure 3.3 shows a typical variation of gain with output power of an amplifier. Output power points at 1- and 3-dB gain compression are also shown.

Table 3.1 Return Loss and Related Parameters

RL (dB)	ρ (V)	VSWR
0	1	∞
0.5	0.9441	34.75
1	0.8913	17.39
2	0.7943	8.72
3	0.7079	5.85
4	0.6310	4.42
6	0.5012	3.01
8	0.3981	2.32
10	0.3162	1.93
12	0.2512	1.67
15	0.1778	1.43
20	0.1000	1.22
25	0.0560	1.12
30	0.0316	1.07

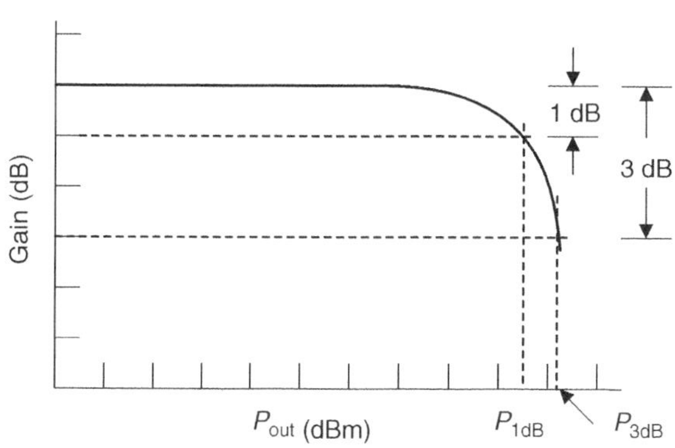

Figure 3.3 Gain versus output power of an amplifier.

3.5 POWER ADDED EFFICIENCY

For power amplifiers, power added efficiency is defined as

$$PAE = \frac{\text{output signal power} - \text{input signal power}}{\text{DC power}} = \frac{P_o - P_{in}}{P_{DC}}$$

$$= \frac{P_o}{P_{DC}}\left(1 - \frac{1}{G}\right) = \eta_D\left(1 - \frac{1}{G}\right) \quad (3.28a)$$

$$\eta_D = \frac{P_o}{P_{DC}} \quad (3.28b)$$

where η_D is known as the drain efficiency and frequently used for RF amplifiers, and $G(= P_o/P_{in})$ is the amplifier gain. For high-efficiency amplifiers, single-stage gain is required to be on the order of 10 dB or higher. For a 10-dB gain value, $PAE = 0.9\eta_D$, whereas for a 6-dB gain value, $PAE = 0.75\eta_D$. Thus a 10-W power amplifier with 10-dB gain requires 1-W input power and with 6-dB gain it requires 2.5-W input power, and reduces the PAE by about 17%.

Note that in (3.28a) the output power P_o excludes the harmonic power content. An accurate definition of *PAE* is given below:

$$PAE = 1 - P_{diss}/P_{DC} \tag{3.28c}$$

where

$$P_{diss} = P_{DC} + P_{in} - P_T \tag{3.28d}$$

and

$$P_T = P_o + \text{harmonic output power}$$

Since $P_T > P_o$, (328a) underestimates the PAE calculation.

3.6 INTERMODULATION DISTORTION

An amplifier is called linear when the output power increases linearly with the input power or power gain is constant with input power. As input power increases, the device current starts reaching its maximum value depending on its size. Near the limiting region the amplifier transfer function becomes nonlinear and a point is reached where the output power does not increase with the input power. One of the measures of nonlinearity of amplifiers is *intermodulation distortion*. When more than one carrier frequency is present in a nonlinear amplifier, due to mixing, multiple sidebands will be generated as intermodulation products. Intermodulation distortion can also arise from the combined effects of amplitude modulation (AM) introduced by a previous stage and AM to PM (phase modulation) conversion. Intermodulation distortion can be introduced by any nonlinear devices or amplifier stage. Since the intermodulation distortion is an important amplifier performance parameter, it is described below.

There are a number of different ways to measure the nonlinearity or distortion behavior of an amplifier. The simplest method is the measurement of the 1-dB gain compression power level, $P_{1\ dB}$. This provides a crude estimation of distortion but is not adequate. For a single-carrier system, third-order intermodulation distortion measurement known as IP3 or TOI is performed, whereas for a multicarrier system ACPR, EVM, and NPR measurements are commonly used. The nonlinearity in wireless power amplifiers gives rise to spectral regrowth and distortion in signals. There is a specified limit, depending on the modulation format, to spectral regrowth that is quantified by ACPR specification or NPR characterization or EVM measurement. The ACPR measures adjacent channel leakage. The NPR characterization is used in satellite systems employing different modulation formats at the same time. EVM is a measure of the quality of the received digitally modulated signal. It is defined as the magnitude of the difference between the ideal and actual received signal vectors.

3.6.1 IP3

The IP3 method that is very popular uses two closely spaced frequencies. When two signals at frequencies f_1 and f_2 are incident on an amplifier, the output of the amplifier contains these two signals, as well as intermodulation products at frequencies $mf_1 \pm nf_2$, where $m + n$ is known as the order of the intermodulation (IM) product. For a bandwidth of an octave or less, only the third-order IM products that appear at $2f_1 - f_2$ and $2f_2 - f_1$ are the major sources of intermodulation distortion. However, second-order IM products appear in ultra broadband amplifiers.The third-order intercept level and 1-dB compression point level for various amplifiers will be discussed in Chapter 12.

The ratio of third-order intermodulation products and the carrier is known as IM3, given by

$$IM3(\text{dBc}) = 10\log\frac{P_{2f_2-f_1}}{P_{f_2}} = 10\log\frac{P_{2f_1-f_2}}{P_{f_1}} \tag{3.29a}$$

where P_{f1}, P_{f2}, P_{2f1-f2}, and P_{2f2-f1} are the power outputs at frequencies f_1, f_2, $2f_1 - f_2$, and $2f_2 - f_1$, respectively. IM3 is described in units of dBc, because it is specified relative to the fundamental tone power output. Similarly, the second-order intermodulation product, IM2, is given by

$$IM2(\text{dBc}) = 10\log\frac{P_{f_1+f_2}}{P_{f_1}} \tag{3.29b}$$

The output third-order intermodulation product, OIP3 or simply IP3, and second-order intermodulation product, IP2, are then calculated using the following relationships:

$$IP3 = 0.5(3P_{f1} - IM3) \tag{3.29c}$$

$$IP2 = 2P_{f1} - IM2 \tag{3.29d}$$

3.6.2 ACPR

Adjacent channel power ratio (ACPR) is a commonly used figure of merit to evaluate the intermodulation distortion performance of RF power amplifiers designed for CDMA wireless communication systems. ACPR is a measure of spectral regrowth and appears in the signal sidebands and is analogous to IM3/IM5 for an analog RF amplifier.

The ACPR is defined as

$$ACPR = \frac{\text{power spectral density in the main channel 1}}{\text{power spectral density in the offset channel 2 or 3}} \tag{3.30}$$

Here offset frequencies and measurement bandwidths vary with system application. For the example shown in Figure 3.4a for CDMA power amplifiers, the two offsets are at 885 kHz and 1.98 MHz away from channel 1 center frequency and the measurement bandwidth specified is 30 kHz. The ACPR requirements are −42 dBc and −54 dBc at channel offset 2 and offset 3, respectively.

3.6.3 EVM

Wireless local area network (WLAN) power amplifiers, operating under IEEE802.11a,g standards, use a maximum 64 quadrature amplitude modulation (QAM) signal with an orthogonal frequency division multliplexing (OFDM) scheme. These wideband digital modulations demand PAs with high linearity, due to the inherent high peak-to-average ratio (PAR) of their signals. The nonlinear response of such amplifiers is characterized using error vector magnitude (EVM), because signal information is carried out on both magnitude and phase.

EVM is a measure of the deviation of the demodulated received symbol (I, Q), from the original transmitted data symbol (I_o, Q_o). The ratio of the error magnitude to the original symbol magnitude, shown in Figure 3.4b, is defined as EVM [1] and is given by

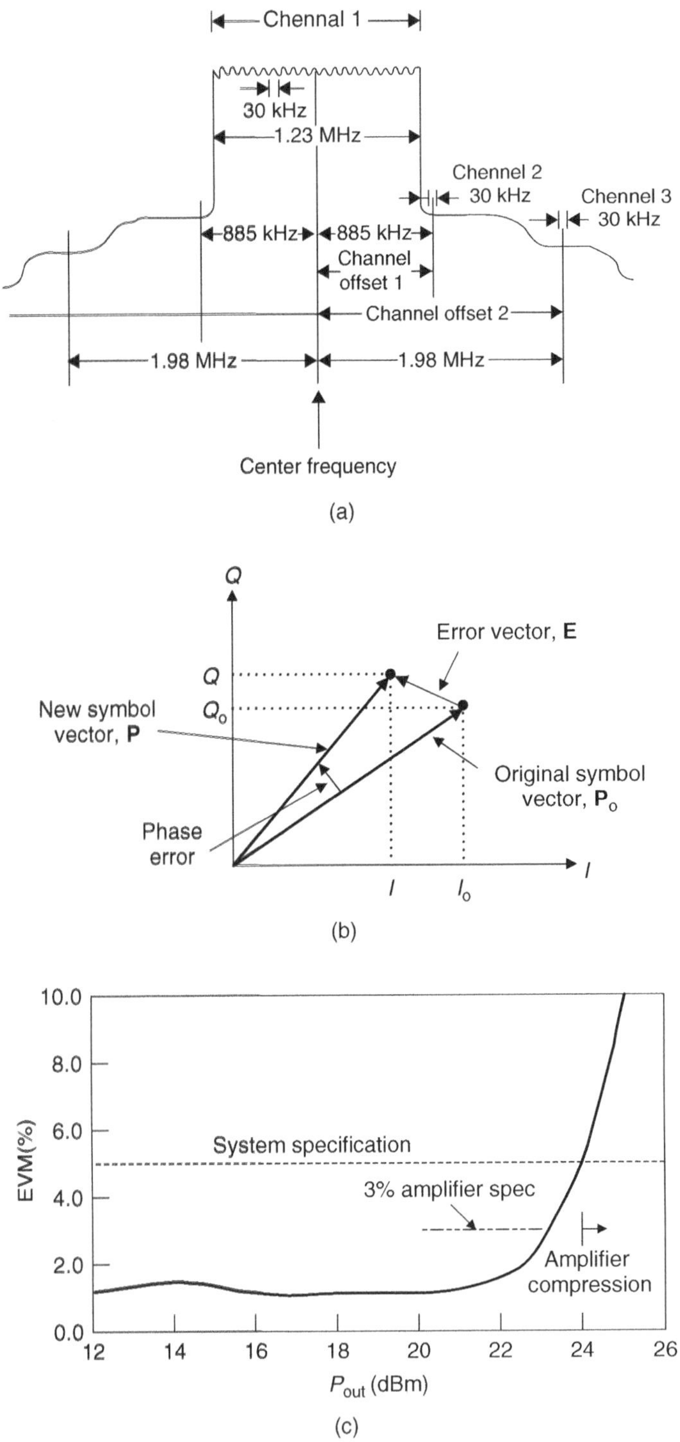

Figure 3.4 (a) CDMA adjacent channel power measurement frequency spectrum. (b) Plot of $I-Q$ constellation diagram showing the error vector magnitude. (c) Plot of amplifier EVM versus output power.

$$EVM = \frac{|E|}{|P_o|} \tag{3.31}$$

where $|E|$ is the magnitude of the error vector and $|P_o|$ is the magnitude of the average power of the constellation. EVM is always expressed as a percentage quantity and is generally required to be 2% or lower.

Figure 3.4c shows a typical measured amplifier EVM versus output power for a 5.8-GHz IEEE802.11a signal, OFDM, QAM-64, and 54 Mbps.

The noise power ratio (NPR) measurement used to evaluate nonlinearity in amplifiers is described in Chapter 22.

3.7 HARMONIC POWER

Distortion in power amplifiers is also evaluated in terms of harmonic power levels. In most applications, the measurements of the second harmonic power HP_2 at $2f_0$ and third harmonic power HP_3 at $3f_0$ are adequate. They are defined as

$$HP_2 = \frac{P_o(2f_0)}{P_o(f_0)} \tag{3.32a}$$

$$HP_3 = \frac{P_o(3f_0)}{P_o(f_0)} \tag{3.32b}$$

Or $HP_{2,\text{dBc}} = 10\log(HP_2)$, $HP_{3,\text{dBc}} = 10\log(HP_3)$. The harmonics contents are usually expressed in dBc, that is, their power levels are that many decibels down with respect to the fundamental power level at a given input power level. For example, for an HPA at $P_{in} = 10$ dBm, the output power levels are 40 dBm, 27 dBm, and 18 dBm at the fundamental, second, and third harmonic frequencies. In this case, the second and third harmonic levels are -13 dBc and -22 dBc, respectively.

3.8 PEAK-TO-AVERAGE RATIO

The ratio between the peak power (P_p) and the average power (P_a) of a signal is called the peak-to-average ratio and is given by [2]

$$\chi = \frac{P_p}{P_a} \quad \text{or} \quad 10\log\frac{P_p}{P_a} \quad \text{in dB} \tag{3.33a}$$

The peak-to-average ratio, ΔP_s, of an input signal consisting of N carriers, each having an average power P_i, is defined as

$$\Delta P_s = \frac{\left(\sum_{i=1}^{N} \sqrt{P_i \chi_i}\right)^2}{\sum_{i=1}^{N} P_i} \tag{3.33b}$$

where χ_i is the peak-to-average ratio of the ith carrier. All single or multiple (modulated or unmodulated) carriers have a peak-to-average ratio; for a single unmodulated or constant envelope modulated carrier $\chi = 0$ dB, for a nonconstant envelope $\chi = 3$ dB, and for multicarrier complex modulated signals ΔP_s could be greater than 10 dB. Table 3.2 lists four examples of peak-to-average ratios.

Table 3.2 Summary of Various Power Levels and Peak-to-Average Ratios for Four Different Carrier Scenarios

Example	Number of Carriers	Output Power of Each Carrier (W)	χ Ratio	χ dB	Total Average Power (W)	Peak Power (W)	ΔP_s (dB)
1	4 GSM	0.5	1	0	2	8.0	6
2	4 IS136	0.15	2	3	1	11.0	10.4
	1 IS95	0.40	3.162	5			
3	20 IS136	0.1	2	3	2	80.0	16
4	2 WCDMA	15.0	7.943	9	30	476.5	12

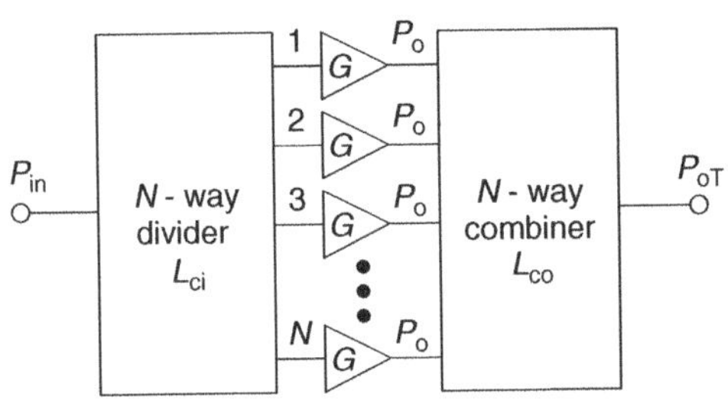

Figure 3.5 An N-way power amplifier combiner configuration.

3.9 COMBINER EFFICIENCY

For an N-way power combiner, shown in Figure 3.5, assuming all HPA units are matched in terms of magnitude and phase, the combining efficiency is defined as

$$\eta_c = \frac{P_{oT}}{N P_o} = L_{co} \tag{3.34}$$

where P_{oT} is the total output power, P_o is the output power for each amplifier, and L_{ci} and L_{co} are the input and output combiner loss. Here the gain, G, of the amplifier is assumed much greater than 10 dB and the isolation between the combining ports is infinite. The amplifiers are well matched to combiner ports. This does not include any mismatch loss between the HPAs and the combiners or the effect of imbalance of phase and gain between amplifiers.

Now consider a case when the gain is less than 10 dB. In this case, the power added efficiency, $\mathrm{PAE_c}$, of the combined HPA is given by

$$PAE_c = \frac{P_{oT} - P_{in}}{P_{DCT}} = \frac{P_{in} L_{ci} G L_{co} - P_{in}}{P_{DCT}} \tag{3.35}$$

If PAE is the power added efficiency and P_{DC} is the DC power for each amplifier, then total DC power P_{DCT} is given by

$$P_{DCT} = N P_{DC}$$

and

$$PAE = P_o\,(1 - 1/G)\,/P_{DC} = \frac{NP_o\,(1 - 1/G)}{P_{DCT}}$$

or

$$PAE = \frac{P_{in}L_{ci}G\,(1 - 1/G)}{P_{DCT}} \tag{3.36}$$

From (3.35) and (3.36), we find

$$PAE_c = PAE\,(L_{ci}GL_{co} - 1)/[L_{ci}(G - 1)] \tag{3.37}$$

When $G \gg 1$,

$$PAE_c = PAE \cdot L_{co} \tag{3.38}$$

EXAMPLE 3.1

Consider a four-way combining scheme using four 10-W HPAs each having 13-dB gain. The divider and combiner have 0.3-dB loss each. If the PAE for each HPA is 40%, calculate the output power and PAE for the combined power amplifier.

SOLUTION Here $L_{ci} = L_{co} = 0.3$ dB or $L_{ci} = L_{co} = 10^{-0.3/10} = 0.9333$ and

$$\text{Gain} = 13 \text{ dB} = 10^{13/10} = 19.95$$

The output power

$$P_{OT} = L_{co}NP_o = 0.9333(4 \times 10)\text{W} = 37.33 \text{ W}$$

or

$$P_{OT} = -0.3 + 6.0206 + 40 \text{ dBm} = 10^{45.7206/10} \text{ mW} = 37{,}330 \text{ mW}$$
$$= 37.33 \text{ W}$$

Using (3.37), we find

$$PAE_c = 40(0.9333 \times 19.95 \times 0.9333 - 1)/[0.9333(19.95 - 1)]\% = 37.04\%$$

Using (3.38), we have

$$PAE_c = 40 \times 0.9333\% = 37.33\%$$

■

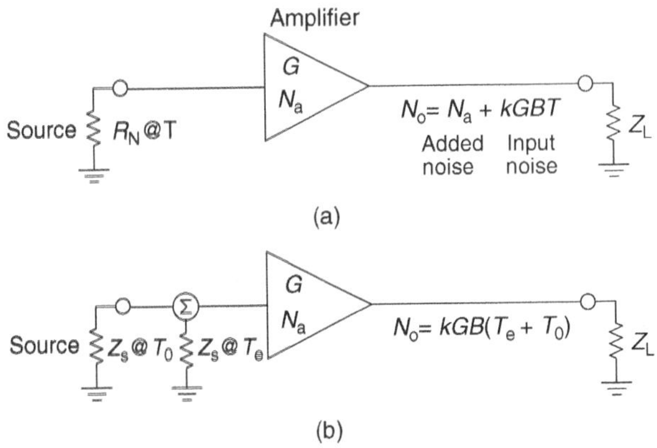

Figure 3.6 Equivalent noise power and noise temperature representation of an amplifier.

3.10 NOISE CHARACTERIZATION

In a stable microwave amplifier, a small output power can be measured, even when there is no input signal. This is the amplifier noise power. The total output noise power consists of amplified input noise entering the amplifier plus the noise generated in the amplifier itself. The model of a noisy two-port microwave amplifier is shown in Figure 3.6. The noise input power can be modeled by a noisy resistor. This noise is caused by random motion of electrons in the resistor due to thermal agitation and therefore is known as *thermal* or *Johnson noise*. The maximum available noise power, N_R, from R_N is

$$N_R = kTB \quad \text{in watts} \tag{3.39}$$

where k is Boltzmann's constant (i.e., $k = 1.38 \times 10^{-23}$ J/K), T is the resistance noise temperature expressed in kelvin units (K), and B is the noise bandwidth in hertz (Hz). At $B = 1$ GHz, $T = 290$ K, $kTB = 4 \times 10^{-9}$ mW $= -84$ dBm. If the bandwidth is decreased to 1 MHz, the noise power is reduced by 30-dB. For 1-Hz bandwidth, $kTB = -174$ dBm. Note that the available noise power is independent of the magnitude of the resistor value but, of course, the amount of noise power actually delivered to a load resistor will decrease as the ratio of the source and load resistor values varies from unity. This is the basic principle of lowering source noise contribution by creating a mismatch at the input of the device in order to design a low-noise amplifier.

Equation (3.39) shows that the thermal noise power depends on the bandwidth but not on a given frequency. Such a distribution of noise is called *white noise*. Obviously no true white noise sources exist because these sources would put out infinite noise power over an infinite bandwidth. Actually the formula in (3.39) breaks down at frequencies well above the millimeter-wave range, where the noise power drops. Over microwave and millimeter-wave frequencies, most dissipative elements in electrical circuits are very well characterized as ideal kTB noise-power sources.

In addition, most resistors exhibit increased noise at low frequencies. This increase is usually referred to as the $1/f$ or flicker noise. Typically, $1/f$ noise does not actually have $1/f$ frequency dependence, but it does increase with decreasing frequency. The frequency at which the $1/f$ power equals kTB is called the *$1/f$ knee frequency*. Flicker noise is not usually important in the design of RF and microwave amplifiers as the knee frequency for transistors is below 100 MHz. But it is an important source of phase noise in microwave oscillators and broadband RF amplifiers used for community antenna television (CATV) applications working below 100 MHz.

In microwave transistors, in addition to Johnson and $1/f$ noise, shot noise is generated by random passage of charges in the modulating channels. The shot noise is relatively constant from DC to 100 MHz and increases with frequency. Thus the measured NF of a transistor between 0.2 and 26 GHz consists of Johnson and shot noise contributions, whereas for applications below 100 MHz, an additional $1/f$ noise contribution must also be considered in the amplifier design.

3.10.1 Noise Figure

The noise figure of any linear two-port network can be defined as

$$\begin{aligned} F &= \frac{\text{signal-to-noise ratio at input}}{\text{signal-to-noise ratio at output}} \\ &= \frac{\text{available noise power at output}}{\text{gain} \times \text{available noise power at input}} = \frac{N_o}{GkTB} \end{aligned} \tag{3.40}$$

where N_o is the available noise power at output, G is the available gain of the network over the bandwidth B, and T is the operating temperature in kelvin units.

If N_a is the noise power added by the amplifier, then

$$F = \frac{GkTB + N_a}{GkTB} = 1 + \frac{N_a}{GkTB} \tag{3.41}$$

An amplifier that contributes no noise to the circuit has $F = 1$.

3.10.2 Noise Temperature

Since Johnson noise is invariant against the resistance and proportional to absolute temperature, it is used to characterize the noise power in units of temperature. Consider the amplifier shown in Figure 3.6b. Typically, the source and load resistors are 50 Ω, but the following definitions apply to arbitrary impedances. The output noise power is given by

$$N_o = Gk(T_0 + T_e)B \tag{3.42}$$

where T_0 is the room temperature (usually $T_0 = 290$ K) and T_e is the equivalent noise temperature of the amplifier with $N_a = 0$. From (3.40) and (3.42) when the source impedance is at T_0,

$$F = 1 + \frac{T_e}{T_0} \tag{3.43}$$

The noise figure is usually expressed in decibels as

$$NF = 10\ \log(F) = 10\ \log\left(1 + \frac{T_e}{T_0}\right) \tag{3.44}$$

3.10.3 Noise Bandwidth

The total noise power from an amplifier is given by

$$NT = \int_{-\infty}^{\infty} T_A(\omega) G_A(\omega)\, d\omega \tag{3.45a}$$

where $T_A(\omega)$ and $G_A(\omega)$ are the noise temperature and power gain of the amplifier, respectively, and ω is the radian frequency. It is often useful to treat the amplifier as having fixed T_A and G_A over a specific noise bandwidth and no gain elsewhere, where the noise bandwidth (NBW) is chosen to make the total noise power correct; that is,

$$NBW \cdot k \cdot T_A \cdot G_A = k \int_{0}^{\infty} T_A(\omega) G_A(\omega)\, d\omega$$

or

$$NBW = \frac{\int_0^\infty T_A(\omega) G_A(\omega)\, d\omega}{T_A G_A} \tag{3.45b}$$

The range of ω is typically limited by other components in the system, or by the gain response of the amplifier.

3.10.4 Optimum Noise Match

In general, any noisy two-port network may be represented by a noise voltage and a noise current source connected at the input of a noiseless two-port network as shown in Figure 3.7. If the circuit has a dominant voltage noise, using a high source impedance will minimize the transmission noise signal, but if the current noise is dominant, connecting a low source impedance will minimize the transmission of the noise signal. When both noise sources are present, a minimum noise figure of the circuit occurs at a specific source admittance or impedance, known as the optimum source admittance. Circles of constant noise figure on the input admittance or impedance plane can be plotted using a Smith chart (described in Appendix F). How the noise figure increases from the minimum value is described by the following relation:

$$F = F_{\min} + \frac{R_n}{G_S} |Y_S - Y_{\text{opt}}|^2 \tag{3.46a}$$

or

$$F = F_{\min} + \frac{R_n}{G_S} [(G_S - G_{\text{opt}})^2 + (B_S - B_{\text{opt}})^2] \tag{3.46b}$$

where

F = noise figure

$Y_S = G_S + jB_S$ = source admittance

$F_{\min}$ = minimum noise figure

$Y_{\text{opt}} = G_{\text{opt}} + jB_{\text{opt}} = 1/Z_{\text{opt}}$

= optimum source admittance that gives minimum noise figure

R_n = equivalent noise resistance

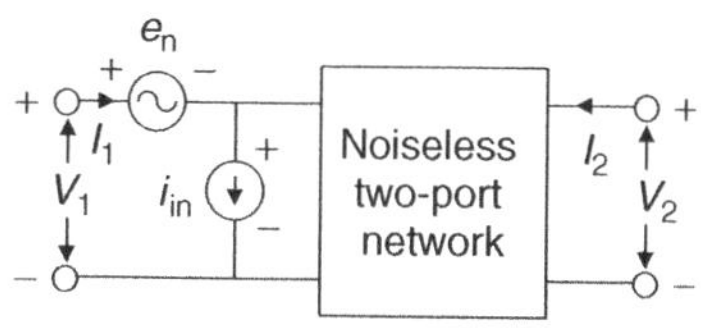

Figure 3.7 Noise equivalent circuit of a two-port network.

The lower the equivalent noise resistance, the less is the sensitivity of the noise figure to an increase in a nonoptimum source. In the above equations, the device's output has been assumed conjugately matched. However, if the device has poor isolation (less than 15 dB) between its input and output terminals, the effect of output match on the noise figure of the circuit must be considered.

We can express Y_S and Y_{opt} in terms of the reflection coefficient Γ_S and Γ_{opt}, and the resultant relation becomes

$$F = F_{min} + \frac{4\bar{R}_n|\Gamma_S - \Gamma_{opt}|^2}{(1 - |\Gamma_S|^2)|1 + \Gamma_{opt}|^2} \tag{3.47}$$

where $\bar{R}_n = R_n/Z_0$ (Z_0 is usually 50 Ω). The quantities F_{min}, R_n, and Γ_{opt} are known as the noise parameters and are given by the manufacturer of the transistor or can be determined experimentally as described in Chapter 22.

3.10.5 Constant Noise Figure and Gain Circles

The low-noise amplifier is designed with input and output matching networks to meet specified noise figure and gain over the bandwidth. The design is facilitated by plotting the constant noise figure and constant gain circles. These are mapped on a Smith chart showing the impedance values that produce a constant gain and noise figure. The constant gain circles are plotted in both Γ_S and Γ_L planes while the constant noise figure circles represent only the Γ_S plane.

Constant Noise Figure Circles

These are mapped on a Smith chart showing the impedance values that produce a constant noise figure. If the parameters F_{min}, R_n, and Γ_{opt} are known, F can be calculated as a function of Γ_S. Equation (3.47) can also be used to construct constant noise factor circles on a Smith chart as given below:

$$\left|\Gamma_S - \frac{\Gamma_{opt}}{1 + N_i}\right|^2 = \frac{N_i^2 + N_i(1 - |\Gamma_{opt}|^2)}{(1 + N_i)^2} \tag{3.48a}$$

which defines a circle with its center at

$$C_F = \frac{\Gamma_{opt}}{1 + N_i} \tag{3.48b}$$

and a radius of

$$R_F = \frac{\sqrt{N_i^2 + N_i(1 - |\Gamma_{opt}|^2)}}{1 + N_i} \tag{3.48c}$$

where

$$N_i = \frac{Z_0}{4R_n}(F - F_{\min})|1 + \Gamma_{\text{opt}}|^2 \tag{3.48d}$$

Constant Gain Circles

These are mapped on a Smith chart showing the impedance values that produce a constant gain. The simplest way to obtain constant gain circles is by considering the unilateral gain equation given by (3.23). The normalized gain factor g_i is defined as

$$g_i = \frac{G_i}{G_{i\max}} = \frac{(1 - |\Gamma_i|^2)}{(|1 - S_{ii}\Gamma_i|^2)}(1 - |S_{ii}|^2) \tag{3.48e}$$

where $i = S$ ($ii = 11$), and $i = L$ ($ii = 22$). The above equations can be rearranged to prove that they represent a family of circles. It is quite often convenient to draw these circles on a Smith chart. The circles have their centers located on the vector drawn from the center of the Smith chart to the point S_{11}^* or S_{22}^*. The distance from the center of the Smith chart to the center of the constant gain circles along the vector S_{11}^* for the source side is given by

$$c_S = \frac{g_S|S_{11}|}{1 - |S_{11}|^2(1 - g_S)}. \tag{3.48f}$$

The radius of the constant gain circles is expressed as

$$r_S = \frac{(1 - |S_{11}|^2)\sqrt{(1 - g_S)}}{1 - |S_{11}|^2(1 - g_S)}. \tag{3.48g}$$

Similar expressions for the constant gain circles for the load side can be derived by replacing the S-parameters of the source side with that of the load side. A typical set of constant gain and noise figure circles in the Γ_S plane are shown in Figure 3.8 for a 300-μm low-noise FET biased at 3 V and 25% I_{dss}.

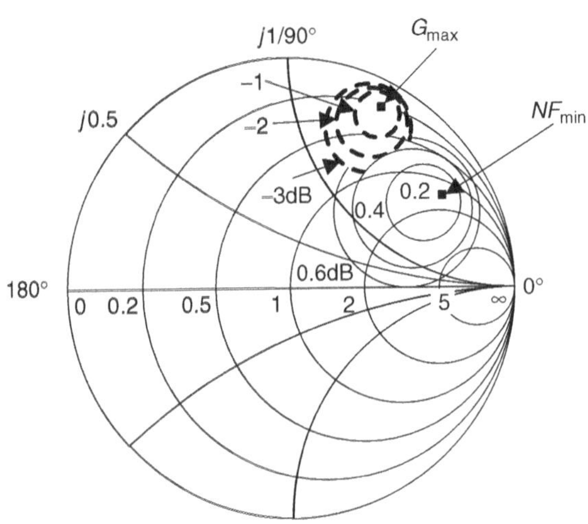

Figure 3.8 Constant gain and noise figure circles in the Γ_S plane for a 300-μm low-noise FET at 3 GHz

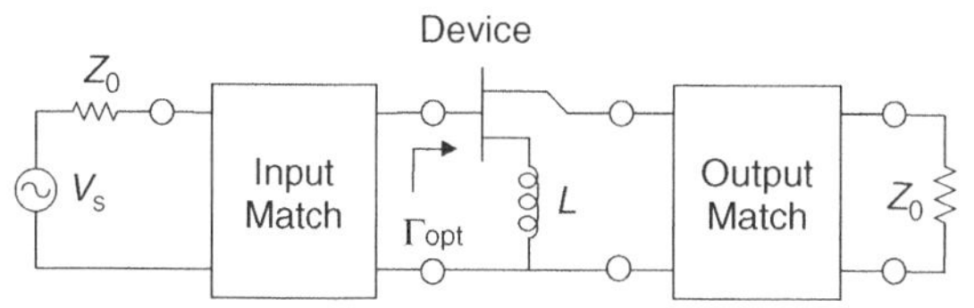

Figure 3.9 Simultaneous input and noise match device configuration.

3.10.6 Simultaneous Input and Noise Match

In general, the input of a low-noise amplifier is matched for minimum noise figure but not for good VSWR. However, simultaneous conjugate input impedance and the optimum noise match impedance can be realized using an inductive series feedback [7] as shown in Figure 3.9. In this case, Γ_{opt} becomes a function of series inductance. The series feedback reduces gain, which helps in achieving a stable operation of the amplifier. Especially at low frequencies, where the devices are potentially unstable, a series inductive feedback provides simultaneously optimum noise figure and good input match, and a stable amplifier operation. Inductive (lossless) series feedback adds no extra noise to the circuit and also makes the circuit more tolerant to device parameter variations, by increasing the device input impedance and decreasing the equivalent noise resistance R_n.

Figure 3.10 shows an amplifier's input and output match conditions and corresponding gain magnitude. The amplifier is assumed to be stable (see Chapter 17). A simultaneous conjugate input and output impedance match condition provides the maximum gain, whereas the other two conditions have in general about 2–3 dB lower gain.

3.11 DYNAMIC RANGE

The range of an input signal that can be detected by a receiver without much distortion is called the dynamic range (DR). The dynamic range of an amplifier is defined as

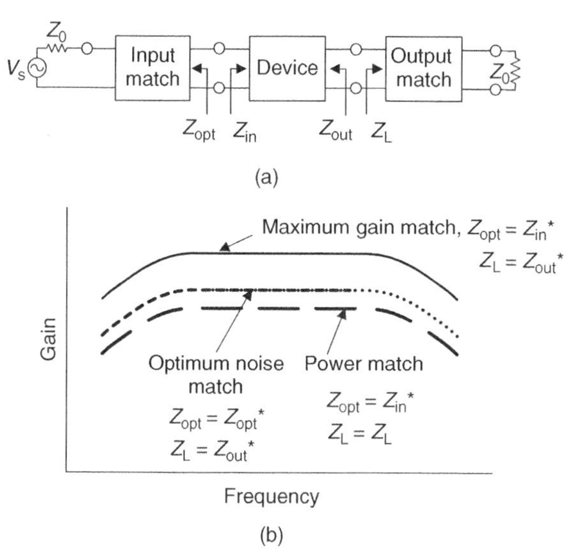

Figure 3.10 (a) Amplifier's typical input and output match conditions and (b) corresponding gain magnitude.

the ratio of the 1-dB gain compressed power output (P_{1dB}) to the amplified minimum detectable signal (P_{in}^{min}).

The output noise of a two-port device with noise figure F can be written from (3.40) as

$$N_o = FGkTB \tag{3.49}$$

If minimum detectable input signal is X (dB) above the noise floor, then

$$P_{in}^{min} = N_o - G + X(\text{dB}) \tag{3.50}$$

$$P_{out}^{min} = P_{in}^{min} + G = N_o + X(\text{dB}) \tag{3.51}$$

Dynamic range is defined as

$$DR = P_{1dB} - P_{out}^{min} \tag{3.52}$$

If a typical value of X is 3 dB, then from (3.49)–(3.52) using $kT + X = -171$ dBm, we find

$$DR = P_{1dB} + 171 - 10\log(\text{B}) - NF - G(\text{dB}) \tag{3.53}$$

where P_{1dB} is in decibels above 1 mW (0 dBm), in other words dBm. All quantities in (3.53) are expressed in decibels.

EXAMPLE 3.2

Determine the dynamic range of a low-noise amplifier with a gain of 30 dB, a noise figure of 2 dB, a 1-dB compression point of 15 dBm, and a noise bandwidth of 1 GHz. Also calculate spurious free dynamic range when IP3 = 25 dBm.

SOLUTION From (3.53),

$$DR = 15 + 171 - 90 - 2 - 30\ dB = 64 \text{ dB}$$

The spurious free dynamic range (DR_f) of an amplifier is defined as the ratio of the fundamental signal power output to the third-order intermodulation product power output, when the third-order intermodulation product is equal to the minimum detectable output signal. The spurious free dynamic range in decibels is given by [5]

$$DR_f = \tfrac{2}{3}[IP3 - P_{out}^{min}] \tag{3.54}$$

From (3.51),

$$\begin{aligned} P_{out}^{min} &= -171 \text{ dBm} + 10\log(B) + NF + G \text{ dBm} \\ &= -171 + 90 + 2 + 30 = -49 \text{ dBm} \end{aligned} \tag{3.55}$$

From (3.54),

$$DR_f = \tfrac{2}{3}[25 + 49] = 49.33 \text{ dB}$$

■

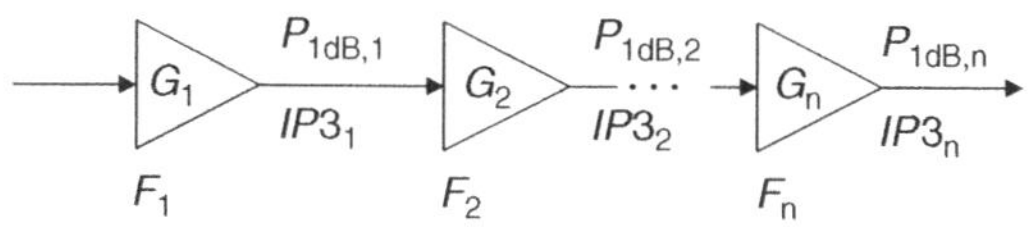

Figure 3.11 Schematic of a multistage cascaded amplifier. Subscript designates the stage number.

3.12 MULTISTAGE AMPLIFIER CHARACTERISTICS

Consider the multistage amplifier configuration shown in Figure 3.11. This schematic shows the gain (G), noise figure (F), P_{1dB}, and third-order intermodulation product ($IP3$) for the various stages. In this section, a multistage amplifier for $IP3$, PAE, and noise figure is characterized.

3.12.1 Multistage IP3

The input IP3 of a multistage amplifier can be calculated by adding powers in parallel at the input. Assuming all intercept points are independent and uncorrelated, the input IP3 (expressed in mW) of a transmitter consisting of n stages is given by

$$IP3_{\text{input}} = \frac{1}{\frac{1}{IP3_1} + \frac{G_1}{IP3_2} + \frac{G_1 G_2}{IP3_3} + \ldots + \frac{G_1 G_2 \ldots G_{n-1}}{IP3_n}} \tag{3.56}$$

where $IP3_n$ and G_n are the input IP3 (expressed in mW) and gain of the nth stage. The output IP3 is the product of input IP3 and total power gain ratio or the sum of input IP3 and total gain, expressed in decibel units. For a three-stage amplifier, the input IP3 expressed in dBm is given by

$$IP3_{\text{input}} = 10 \log \left[\frac{1}{\frac{1}{IP3_1} + \frac{G_1}{IP3_2} + \frac{G_1 G_2}{IP3_3}} \right] \tag{3.57}$$

EXAMPLE 3.3

Consider a three-stage power amplifier, having 10, 9, and 8 dB gain and output IP3 values of 28, 34, and 40 dBm for the first, second, and third stages, respectively. Determine the input and output IP3.

SOLUTION Here $G_1 = 10$, $G_2 = 7.94$, and $G_3 = 6.3$:

$$IP3_1 = 28 - 10 \text{ dBm} = 63.1 \text{ mW} \qquad IP3_2 = 34 - 9 \text{ dBm} = 316.2 \text{ mW} \qquad IP3_3 = 40 - 8 \text{ dBm} = 1584.9 \text{ mW}$$

$$\begin{aligned} IP3_{\text{input}} &= 10 \log \left[\frac{1}{\frac{1}{63.1} + \frac{10}{316.2} + \frac{10 \times 7.94}{1584.9}} \right] \text{dBm} \\ &= 10.1 \text{ dBm} \\ IP3_{\text{output}} &= 10.1 + 27 \text{ dBm} = 37.1 \text{ dBm} \end{aligned}$$

Thus the output IP3 is reduced by 2.9 dB from a single-stage amplifier to a three-stage amplifier due to cascading. In order to minimize this reduction, one needs higher gain per stage or higher IP3 values for the preceding stages. ■

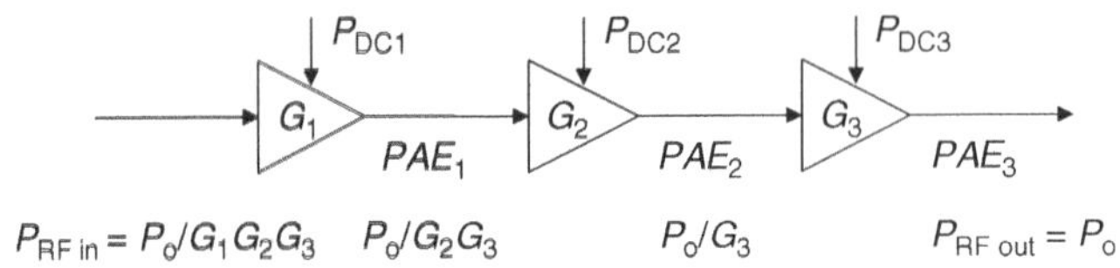

Figure 3.12 Schematic of a three-stage cascaded power amplifier.

3.12.2 Multistage PAE

Power added efficiency of a three-stage power amplifier, shown in Figure 3.12, is given by

$$PAE = \frac{P_o\left(1 - \frac{1}{G_1G_2G_3}\right)}{P_{DC}} \tag{3.58}$$

where $P_{DC} = P_{DC1} + P_{DC2} + P_{DC3}$:

$$P_{DC1} = \frac{P_o\left(1 - \frac{1}{G_1}\right)}{G_2G_3PAE_1} \tag{3.59a}$$

$$P_{DC2} = \frac{P_o\left(1 - \frac{1}{G_2}\right)}{G_3PAE_2} \tag{3.59b}$$

$$P_{DC3} = \frac{P_o\left(1 - \frac{1}{G_3}\right)}{PAE_3} \tag{3.59c}$$

Here PAE_1, PAE_2, and PAE_3 are the PAE efficiencies of the first, second, and third stages, respectively. When $G_1G_2G_3 \gg 1$, from (3.58) and (3.59) we find

$$PAE = \frac{1}{\frac{(G_1-1)}{G_1G_2G_3PAE_1} + \frac{G_2-1}{G_2G_3PAE_2} + \frac{G_3-1}{G_3PAE_3}} \tag{3.60}$$

when G_1, G_2, and $G_3 \gg 1$, we have

$$PAE = \frac{1}{\frac{1}{G_2G_3PAE_1} + \frac{1}{G_3PAE_2} + \frac{1}{PAE_3}} \tag{3.61}$$

3.12.3 Multistage NF

Now consider an amplifier having n stages cascaded in series with gain values G_1, $G_2, \ldots, G_n$ and noise figure values F_1, $F_2, \ldots$, F_n; then the total noise figure of an n-stage amplifier is given by

$$F = F_1 + \frac{F_2 - 1}{G_1} + \frac{F_3 - 1}{G_1G_2} + \ldots + \frac{F_n - 1}{G_1G_2 \ldots G_{n-1}} \tag{3.62}$$

Table 3.3 Noise Figure for a Three-Stage LNA Having Different Gain Values for Each Stage[a]

Stage 1		Stage 2		Stage 3		
G	*NF*	*G*	*NF*	*G*	*NF*	Amplifier *NF*
20	2	20	2	20	2	2.02
13	2	13	2	13	2	2.08
10	2	10	2	10	2	2.17
8	2	8	2	8	2	2.29
6	2	6	2	6	2	2.48
4	2	4	2	4	2	2.81
10	2	6	2	6	2	2.20
6	2	10	2	10	2	2.42

[a]Gain *G* and *NF* in dB.

When $F_1 = F_2 = \ldots = F$ and $G_1 = G_2 = \ldots = G$, as $n \to \infty$, the overall noise figure obtained is called the noise measure, F_M, and is given by

$$F_{\mathrm{M}} = \frac{F - 1/G}{1 - 1/G} \tag{3.63}$$

For most applications, noise measure is a much better figure of merit for an amplifier than noise figure because noise measure accounts for the noise contribution of succeeding amplifier stages. An amplifier with low gain does not guarantee good system noise performance, no matter how low the amplifier noise figure may be. Table 3.3 provides a cascaded analysis of noise figure (NF) for three-stage amplifiers with different gain values. To maintain low NF, the gain of the first stage is more important than other stages.

EXAMPLE 3.4

Consider an example of two amplifiers cascaded in series. If the gain and noise figures are 9 and 2 dB for the first amplifier and 10 and 3 dB for the second amplifier, respectively, calculate the total noise figure of the cascaded amplifier. What happens to the noise figure if the gain of the first amplifier is 15 dB or higher?

SOLUTION The overall noise figure is given by (3.62), that is,

$$F = F_1 + \frac{F_2 - 1}{G_1}$$

Here

$$NF_1 = 2 \text{ dB}, \quad F_1 = 1.58$$
$$NF_2 = 3 \text{ dB}, \quad F_2 = 2$$
$$G_1 = 9 \text{ dB} = 7.94$$
$$G_2 = 10 \text{ dB} = 10$$
$$F = 1.58 + \frac{2 - 1}{7.94} = 1.7$$

Or

$$NF = 10 \log(1.7) = 2.3 \text{ dB}$$

when

$$G_1 = 15 \text{ dB} = 31.6, \quad F = 1.58 + \frac{2-1}{31.6} = 1.6, \quad NF = 2.07 \text{ dB}$$

Thus as the gain of the first amplifier, G_1, gets larger, the relative importance of the noise contribution for the second amplifier becomes less and less. ■

EXAMPLE 3.5

Consider an amplifier preceded by an attenuator when they are cascaded in series. If loss of the attenuator is L, the gain and noise figure of the amplifier are G_A and NF_A, and all are expressed in decibels, what is the overall noise figure of the assembly?

SOLUTION This example is very similar to the previous example, where

$$F_1 = 10^{L/10}, \quad F_2 = 10^{NF_A/10}$$
$$G_1 = 10^{-L/10}, \quad G_2 = 10^{G_A/10}$$

From (3.62), we have

$$F = 10^{L/10} + \frac{10^{NF_A/10} - 1}{10^{-L/10}}$$
$$F = 10^{L/10} \times 10^{NF_A/10}$$
$$NF = 10 \log F = L + NF_A$$

In other words, a matched attenuator at room temperature connected in front of an amplifier increases the noise figure of the amplifier by an amount equal to its loss. ■

3.13 GATE AND DRAIN PUSHING FACTORS

The use of HPAs in many critical applications requires the sensitivity of the HPA's transmission phase and output power to be checked versus gate and drain voltages to simulate the effect of noise jitter on the bias lines. The changes in output power with respect to gate ($\Delta P/\Delta V_g$) and drain ($\Delta P/\Delta V_d$) voltages are known as gate voltage and drain voltage power pushing, respectively, and are expressed in dB/V. Similarly, the changes in transmission phase with respect to gate and drain voltages are known as gate voltage and drain voltage phase pushing, respectively, and are expressed in degrees/V. The changes in transmission phase ϕ and output power P_o with respect to gate and drain voltages are defined as

$$\Delta\phi_g/\Delta V_g = \angle S_{21}|_{V_g} - \angle S_{21}|_{V_g \pm \Delta V_g} \tag{3.64a}$$
$$\Delta\phi_d/\Delta V_d = \angle S_{21}|_{V_g} - \angle S_{21}|_{V_d \pm \Delta V_d} \tag{3.64b}$$

and

$$\Delta P_{og}/\Delta V_g = P_o|_{V_g} - P_o|_{V_g \pm \Delta V_g} \tag{3.65a}$$
$$\Delta P_{od}/\Delta V_d = P_o|_{V_g} - P_o|_{V_d \pm \Delta V_d} \tag{3.65b}$$

where subscripts g and d designate gate and drain, respectively. V_g and V_d are nominal bias voltages and ΔV_g and ΔV_d are the specified changes in voltages. Typical values for ΔV_g and ΔV_d are 0.2 V and 0.5 V, respectively, and expected values for $\Delta\phi_g$ and ΔP_{og} are $5^\circ-10^\circ$ and 0.1 dB. Typical values for $\Delta\phi_d$ and ΔP_{od} are $2-4^\circ$ and 0.5 dB, respectively. Thus the transmission phase is more sensitive to gate voltage whereas the drain voltage affects the output power more.

3.14 AMPLIFIER TEMPERATURE COEFFICIENT

Characteristics such as gain, NF, and power of transistor amplifiers are temperature dependent. The gain and output power of an amplifier decreases with increasing temperature, whereas the noise figure increases with temperature due mainly to reduction in transconductance g_m (described in the next chapter) value with increased junction/channel temperature of the transistor. In other words, the higher the temperature with respect to the reference temperature (room temperature), the lower the gain is and the higher the NF is, and vice versa. For a low-noise amplifier, the change in gain and noise figure with temperature are denoted by ΔG and ΔNF, respectively. For FETs/HEMTs, the typical values for these temperature coefficients are given as

$$\Delta G = 0.01 \text{ dB}/^\circ\text{C/stage} \tag{3.66}$$

$$\Delta NF = 0.01 \text{ dB}/^\circ\text{C} \tag{3.67}$$

For power amplifiers, a typical value for the temperature power coefficient is

$$\Delta P = 0.01 \text{ dB}/^\circ\text{C} \tag{3.68}$$

More accurate values for ΔG, ΔNF, and ΔP may be determined using temperature-dependent device models in the circuits. In comparison to transistors, the temperature dependence of passive components (microstrip, inductors, and capacitors) is usually negligible.

EXAMPLE 3.6

At 25°C, a three-stage low-noise amplifier has gain and NF of 25 dB and 2 dB, respectively. Determine the amplifier parameters at 85°C and −45°C.

SOLUTION Here

$$G = G_{25} \mp \Delta G \times \Delta T \quad \text{and} \quad NF = NF_{25} \pm \Delta NF \times \Delta T \tag{3.69}$$

For gain using (3.66), we have

$$\text{at } 85^\circ\text{C}, \quad G = 25 - \Delta G \times (85 - 25) \times 3 = 25 - 1.8 = 23.2 \text{ dB}$$

$$\text{at } -45^\circ\text{C}, \quad G = 25 + \Delta G \times (45 + 25) \times 3 = 25 + 2.1 = 27.1 \text{ dB}$$

For noise figure using (3.67), we have

$$\text{at } 85^\circ\text{C}, \quad NF = 2.0 + \Delta NF \times (85 - 25) = 2.0 + 0.6 = 2.6 \text{ dB}$$

$$\text{at } -45^\circ\text{C}, \quad NF = 2.0 - \Delta NF \times (45 + 25) = 2.0 - 0.7 = 1.3 \text{ dB}$$

■

3.15 MEAN TIME TO FAILURE

In general, in an amplifier the reliability is dictated by transistors not passive components. Transistors have significantly higher failure rates than passive matching elements because of higher stress levels. Solid state transistors typically have maximum allowed operating channel or junction temperature (150°C; 175°C for GaAs) that cannot be exceeded without affecting its reliability, which is projected in terms of mean time to failure (MTTF). Lowering channel temperature can improve electrical performance and greatly increase the device lifetime. Life testing of transistors demonstrates that for small-signal and power GaAs transistors, the estimated MTTF values at 150°C are greater than 10^7 h and 0^6 h, respectively.

Numerous tests on solid state devices have shown that the device failure mechanisms vary exponentially with temperature [9–13]. MTTF is calculated using a temperature versus lifetime relationship known as an *Arrhenius model* given by

$$MTTF = Ce^{E_a/kT} \tag{3.70}$$

where C is a constant, E_a is the activation energy (expressed in eV), k is Boltzmann's constant (1.38×10^{-23} J/K), and T is the absolute temperature measured in kelvin units. Solid state devices, under normal RF operations, have typical MTTF values ranging from 10 to 100 years. In order to generate meaningful MTTF data for a transistor in a reasonable time, an accelerated temperature test for the device is needed. Thus the MTTF is calculated by using elevated temperature, also known as accelerated stress life testing of transistors or amplifiers. Testing is performed under DC conditions or under RF conditions. The elevated channel temperatures are generally below 325°C and testing is done above 150°C in multiples of 25 or 50°C.

In MESFETs, the failure mechanism that determines the reliability is an ohmic metal migration "wearout" rather than a random failure due to open or short circuits caused by electromigration, lateral shorting of fingers, and vertical shorting of active regions due to metal diffusion. A brief description of electromigration is given in Chapter 15. In transistors, the major failure is the wearout mechanism resulting in open or short circuits caused by electromigration, lateral shorting of fingers, and vertical shorting of active regions due to metal diffusion. In FETs/HEMTs, open circuits in gates give rise to limited control on drain–source current and higher drain–source current values and reduction in gain, output power, and PAE. The ohmic contact diffusion into the GaAs has been evidenced by observing voids in the ohmic contacts close to the gate channel. This causes a gradual reduction in drain current and output power. At higher temperatures, the devices will eventually short circuit due to lowering of the breakdown voltage caused by the metal diffusion.

As a worst-case example of M/A-COM's MSAG FET reliability, Figure 3.13 is an Arrhenius plot of HPAs under CW RF life test at rated bias conditions and output power. This plot includes a very conservative 1.2-eV activation energy life projection. The amplifiers were operated at approximately 1.5–2 dB gain compression. A 1-dB loss of power was viewed as a failure. This plot shows a 10^6-h MTTF under CW conditions at a maximum channel temperature of 170°C. A small-signal MSAG FET Arrhenius plot shows about an order of magnitude longer MTTF than the power plot in Figure 3.13. Recent studies show an activation energy of 1.4–1.8 eV, which would realistically increase the extrapolated life by an order of magnitude over what is shown in Figure 3.13.

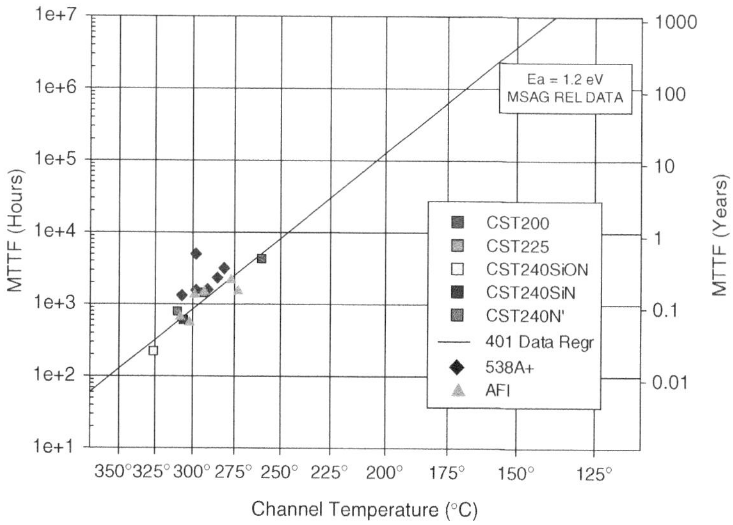

Figure 3.13 Arrhenius plot of MSAG HPAs under RF life test at rated bias conditions and output power.

REFERENCES

1. P. Naraine, Predicting the EVM performance of WLAN power amplifiers with OFDM signals, *Microwave J.*, Vol. 47, pp. 222–226, May 2004.
2. N. Pothecary, *Feedforward Linear Power Amplifiers*, Artech House, Norwood, MA, 1999.
3. T. T. Ha, *Solid-State Microwave Amplifier Design*, John Wiley & Sons, Hoboken, NJ, 1981.
4. H. Fukui (Ed.), *Low-Noise Transistors and Amplifiers*, IEEE Press, New York, 1981.
5. G. Gonzalez, *Microwave Transistor Amplifiers Analysis and Design*, Prentice Hall, Englewood Cliffs, NJ, 1984.
6. I. J. Bahl and P. Bhartia, *Microwave Solid State Circuit Design*, 2nd edition, John Wiley & Sons, Hoboken, NJ, 2003.
7. R. E. Lehmann and D. D. Heston, X-band monolithic series feedback LNA, *IEEE Trans. Microwave Theory Tech.*, Vol. MTT-33, pp. 1560–1566, December 1985.
8. K.C Gupta, *Microwaves*, Wiley Eastern Ltd, New Delhi, India 1979.
9. F. Wilhemsen et al., Temperature vs. reliability in power GaAs FETs and MIC GaAs FET power amplifiers, *Microwave J.*, Vol. 27, pp. 227–242, May 1984.
10. W. Poole and L. Walshak, A five-year quest to prove MW power transistor reliability, *Microwave & RF*, Vol. 23, pp. 71–80, July 1984.
11. W. E. Poole, Perform device—life tests under RF conditions, *Microwave & RF*, Vol. 36, pp. 84–91, October 1997.
12. A. Christou (Ed.), *Reliability of Gallium Arsenide MMICs*, John Wiley & Sons, Hoboken, NJ, 1992.
13. D. Crowe and A. Feinberg (Eds.), *Design for Reliability*, CRC Press, Boca Raton, FL, 2001.

PROBLEMS

3.1 A four-stage power amplifier has 10, 9, 8, and 7 dB gain and output IP3 values of 24, 31, 37, and 42 dBm for the first, second, third, and fourth stages, respectively. Calculate the input and output IP3 of the amplifier.

3.2 Determine the noise figure of the receiver whose block diagram is shown below. Compare its performance with another receiver where the diode mixer is replaced by a dual-gate FET mixer having gain of 7 dB and noise figure of 10 dB.

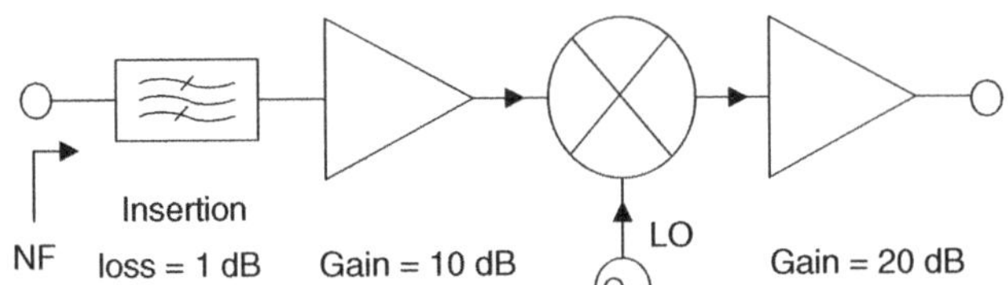

3.3 Derive an expression for the noise figure for n identical amplifiers, with gain G and noise figure F, cascaded in series. If $G = 5$ and $F = 2$, what is the maximum number of amplifiers that can be cascaded to obtain a noise figure less than 2.3?

3.4 S-parameters of a device were measured using a 50-Ω system at 5 GHz and are given as:

$$S_{11} = 0.895\angle -44.9,\ S_{21} = 4.392\angle 141.5,\ S_{12} = 0.057\angle 69.8,\ S_{22} = 0.548\angle -21.5$$

Calculate G_T, G_A, and G_P as defined in Eq (3.22b) and Eq. (3.25).

3.5 A three-stage low-noise amplifier has a noise figure of 3 dB and gain of 30 dB over a 1-GHz bandwidth at room temperature. Calculate the DR and DR_f when the 1-dB compression point for the amplifier is 15 dBm.

3.6 A two-stage power amplifier has a noise figure of 8 dB and gain of 15 dB over a 1-GHz bandwidth at room temperature. Calculate the output noise power in watts and dBm.

3.7 Derive equations for constant gain circles in the Γ_S and Γ_L planes.

3.8 Derive equations for constant noise circles in the Γ_S plane.

Chapter 4

Transistors

The frequency of operation of semiconductor devices has increased remarkably since the invention of the silicon transistor in 1948. This increase has been partly due to use of intrinsically higher performance semiconductor materials, but it has primarily been achieved by reducing critical dimensions in the transistor. Reduced device dimensions reduce electron transit times and also parasitic capacitances increasing operational frequency of the solid state transistor. Advances in the lithographic equipment, mostly driven by the needs of the silicon semiconductor industry, now support routine manufacture of devices with critical dimensions of less than 0.5 μm, operating at 100 GHz and higher. As critical dimensions are reduced, the transistor current and voltage handling capabilities invariably decline with the result that output power decreases. This explains why semiconductors can switch the output power of a generating plant at 60 Hz while achieving 1W of output power at 100 GHz is difficult!

The aim of this chapter is to provide an overview of basic operating principles of several transistors used in the design of amplifiers. Their characteristics such as equivalent circuit models and figures of merit are reviewed.

4.1 TRANSISTOR TYPES

Several solid state devices being used to develop amplifier circuits include silicon bipolar junction transistors (BJTs), metal oxide semiconductor field effect transistors (MOSFETs) and laterally diffused metal oxide semiconductor (LDMOS) transistors, GaAs metal semiconductor field effect transistors (MESFETs), or simply FETs, both GaAs and InP based high electron mobility transistors (HEMTs), both silicon germanium (SiGe) and GaAs based heterojunction bipolar transistors (HBTs), and silicon carbide (SiC) based FETs and GaN HEMTs. These devices operate over a range of power supply voltages from 2 V for small-signal devices up to 48 V for high-power applications. For small-signal applications the devices are usually operated at low drain voltages, 2–3 V; for improved noise performance and for power application these devices are operated at 5–10 V. Very high-power transistors such as BJTs, LDMOS, and SiC based devices are normally operated at much higher supply voltages and these transistors are discussed in Chapter 13. In this chapter, a brief overview of devices, including their basic operations and models, is provided. Extensive transistor model data useful for amplifier designs is given in the next chapter. Each device has several versions that vary in size, gate or emitter structures, and channel formations, designed

Fundamentals of RF and Microwave Transistor Amplifiers. By Inder J. Bahl

for different frequency ranges and amplifier functions. Since it is not possible to give all these variations in one small chapter, only salient features are described. The readers are referred to the references at the end of this chapter [1–53] for greater device details. Several commercial foundries provide GaAs and InP based FET, pHEMT, and HBT and Si based transistor processes for custom designs.

Fabrication of transistors starts with the selection of a wafer type or substrate. Various substrate materials used for active devices are silicon, silicon carbide, sapphire, GaAs, InP, and GaN. Their electrical and physical properties are compared in Table 4.1. Except for Si, all other substrate materials are called compound semiconductors. Silicon dominates the marketplace. GaAs is a distant second with less mature technologies such as InP, SiC, and GaN only now emerging. The semi-insulating property of the substrate material is crucial to providing higher device isolation and lower dielectric loss for monolithic microwave integrated circuits (MMICs). For example, while silicon devices are capable of operating up to about 100 GHz, the relatively low resistivity of bulk silicon precludes the development of high-performance monolithic integration. GaAs semi-insulating substrates provide isolation up to about 100 GHz. InP has been used for millimeter-wave HEMTs up to 280 GHz. Pseudomorphic HEMTs fabricated on InP substrate exhibit much higher performance in terms of gain, noise figure, and power than a GaAs based pHEMT of similar geometry. In this case, the InP substrate supports higher two-dimensional electron gas densities resulting in high current and transconductance values. The high value of transconductance in InP based pHEMTs is responsible for ultralow noise figure, high gain, and high frequency of operation. For high-power and high-temperature applications, wide bandgap materials with relatively high thermal conductivity, such as SiC and GaN, play a significant role. Recent advancements in epitaxial techniques have made it possible to develop active devices on these substrates.

In comparison to Si bipolar transistors, the GaAs FETs and HEMTs have higher frequency of operation, lower noise figure, and higher power and PAE capability at lower operating voltages. The GaAs FET as a single discrete transistor has widely been used in hybrid microwave integrated circuit (MIC) amplifiers for broadband, medium-power, high-power, and high-efficiency applications. This wide utilization of GaAs FETs can be attributed to their high frequency of operation and versatility. However, increasing emphasis is being placed on new devices for better performance and higher frequency operation. HEMT and HBT devices offer potential advantages in

Table 4.1 Comparison of Transistor/Monolithic Integrated Circuit Substrates[a]

Property	Silicon	SiC	GaAs	InP	GaN
Semi-insulating	No	Yes	Yes	Yes	Yes
Resistivity (Ω• cm)	10^3–10^5	$>10^{10}$	10^7–10^9	$\sim 10^7$	$>10^{10}$
Dielectric constant	11.7	9.7	12.9	14	8.9
Electron mobility (cm^2/V•s)	1450	500	8500	4000	800
Saturation electrical velocity (cm/s)	9×10^6	2×10^7	1.3×10^7	1.9×10^7	2.3×10^7
Radiation hardness	Poor	Excellent	Very good	Good	Excellent
Density (g/cm^3)	2.3	3.1	5.3	4.8	6.1
Thermal conductivity (W/cm•°C)	1.45	3.5	0.46	0.68	1.3
Operating temperature (°C)	250	>500	350	300	>500
Energy gap (eV)	1.12	2.86	1.42	1.34	3.39
Breakdown field (kV/cm)	≈300	≥ 2000	400	500	≥ 5000

[a]Pure materials at room temperature.

microwave and millimeter-wave IC applications. HEMTs appear to have a performance edge in ultralow-noise, high-linearity, and high-frequency applications. The pHEMTs have excellent millimeter-wave power performance from Ku- through W-bands and are considered the transistors of choice. HBTs are popular as power devices at RF and low microwave frequencies and are being operated using a single power supply. They offer better linearity and lower phase noise than FETs and HEMTs. Si based bipolar devices also require a single power supply, have low leakage and low $1/f$ noise, and are produced much cheaper on Si. The SiGe HBTs have the low-cost potential of Si BJTs and electrical performance similar to GaAs HBTs except lower breakdown voltage.

Various figure of merit terms at the operating bias conditions are used to evaluate the transistor characteristics including maximum available gain, cutoff frequency (f_T), maximum frequency of oscillation (f_{max}), minimum noise figure (F_{min}), output power density, and power added efficiency (PAE). For amplifier design purposes, the devices are tested for S-parameters, noise parameters, linear and nonlinear models, and source-pull and load-pull data and more details are given in the next chapter.

4.2 SILICON BIPOLAR TRANSISTOR

The silicon (Si) bipolar transistor is a current driven device in which the base current modulates the collector current of the transistor. The three configurations of the bipolar transistor are (a) common base, (b) common emitter, and (c) common collector. For power gain applications, common-emitter and common-base configurations are often employed. In a typical common-emitter configuration, the emitter–base junction is reverse biased and the collector–base junction is forward biased using the same polarity power supply.

The Si bipolar transistor is a pn-junction device and is formed from back-to-back junctions. Since it is a three-terminal device, it can be either pnp or npn. For high-frequency applications, the npn structure, shown in Figure 4.1a, is preferred, since electrons (majority carriers) generally have superior transport characteristics compared to holes. The basic physics responsible for the operation of npn bipolar transistors can be explained through analysis of the common-emitter device shown in Figure 4.1b. Under normal bias, the emitter–base is reverse biased and the collector–base is forward biased. Electrons are injected across the emitter–base junction, travel through the base region, and are extracted at the collector–base junction. Electrons are collected at the collector terminal and holes are collected at the base terminal, resulting in large

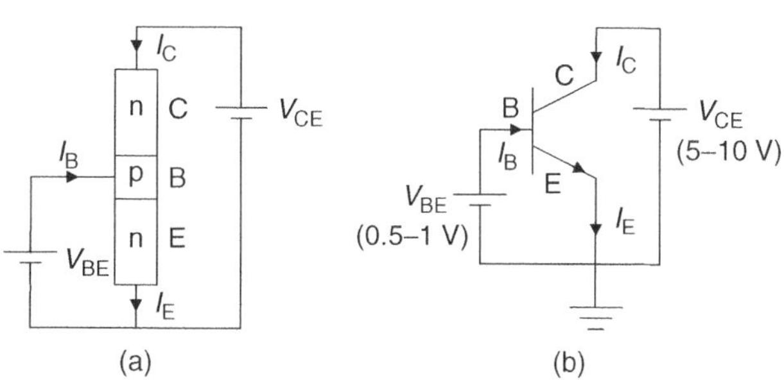

Figure 4.1 Common-emitter npn bipolar transistor configuration showing bias voltages: (a) pn-junction representation and (b) BJT circuit representation.

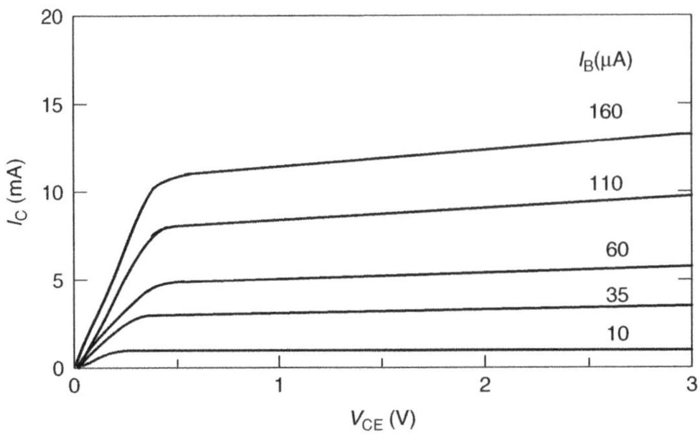

Figure 4.2 $I-V$ representation of a $0.4 \times 100\ \mu m^2$ emitter area bipolar transistor in common-emitter configuration.

electron current between the collector–emitter terminals and small hole current between base–emitter terminals. The ratio of the two terminal currents provides current gain.

Figure 4.2 shows current–voltage ($I-V$) characteristics for an npn $0.4 \times 100\ \mu m^2$ emitter area bipolar transistor in common-emitter configuration. For small-signal operation transistors, typical maximum collector and base current values are 10–15 mA and 20–100 μA, respectively. The value of breakdown voltage between the collector and emitter, BV_{CE}, is in the range of 15–20 V.

The emitter current (I_E) and collector current (I_C) of a bipolar transistor are given by

$$I_E = I_s[\exp(qV_i/kT) - 1] \quad \text{and} \quad \Delta I_C = \alpha \Delta I_E \tag{4.1}$$

The DC current gain β and the transconductance g_m are defined as

$$\beta = \frac{\Delta I_C}{\Delta I_B} \tag{4.2a}$$

$$g_m = \frac{\Delta I_C}{\Delta V_{BE}} = \frac{q}{kT}\alpha I_E = \frac{I_E(\text{in mA})}{26} \tag{4.2b}$$

where I_s = the surface combination depletion current
I_B = base current
q = electron charge = 1.602×10^{-19} C
V_{BE} = base–emitter voltage
V_i = built-in potential
k = Boltzmann constant = 1.38×10^{-23} J/K
T = temperature in kelvin units
α = low-frequency common-base current gain

The physical cross section of a two-cell silicon bipolar transistor is shown in Figure 4.3. When multiple cells are combined in parallel, the most important device dimensions that determine the frequency response are the emitter pitch and the emitter area. The distributed T-equivalent circuit, shown in Figure 4.4, has been found to be

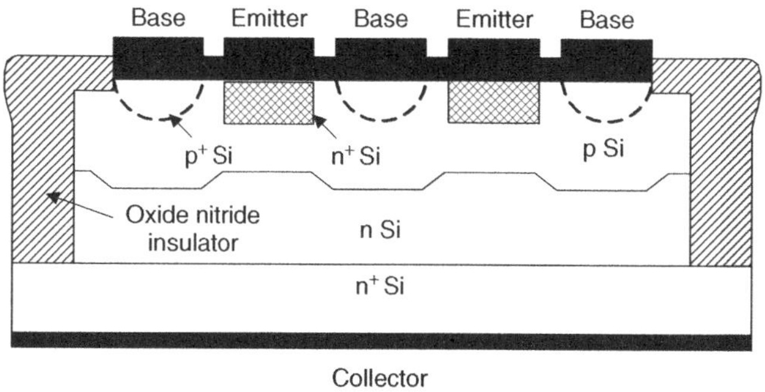

Figure 4.3 Cross-sectional view of a BJT.

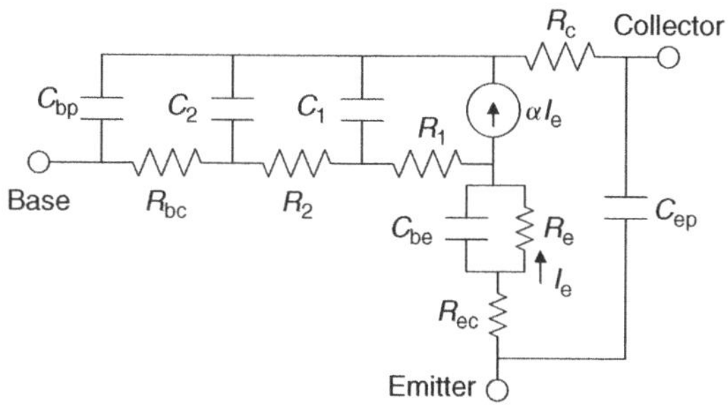

Figure 4.4 Distributed T-equivalent circuit of a BJT.

an effective small-signal model at a fixed bias condition. In this figure R_1, R_2 are the distributed base resistances; C_1, C_2 are the distributed base-to-collector capacitances; C_{bp} is the base bond pad capacitance; C_{ep} is the emitter bond pad capacitance; C_{be} is the base-to-emitter junction capacitance; R_c is the collector resistance; R_{bc} is the base contact resistance; R_e is the emitter resistance; R_{ec} is the emitter contact resistance; I_e is the emitter current; and α is common-base current gain. The bonding inductances at the base, and at the emitter, must be included in RF design for accurate prediction of circuit performance.

4.2.1 Figure of Merit

The figure of merit of a bipolar transistor can be expressed as

$$(f_{\max})^2 = \frac{f_T}{8\pi R_b C_c} \tag{4.3}$$

where $f_{\max}$ is the maximum frequency at which the unilateral gain becomes unity, R_b is the base resistance, C_c is the collector–base capacitance (see Fig. 4.5), and f_T is related to the transit time by the expression

$$f_T = \frac{1}{2\pi \tau_{ec}} \tag{4.4}$$

Here τ_{ec} is the transit time (the delay time from the emitter to collector) given by

$$\tau_{ec} = \tau_e + \tau_b + \tau_d + \tau_c + \tau_{eb} + \tau_{bc} \tag{4.5}$$

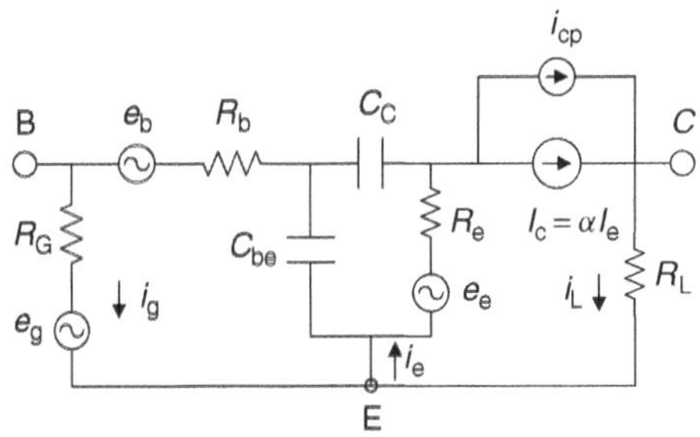

Figure 4.5 High-frequency noise equivalent circuit of a common-emitter BJT.

where τ_e is emitter delay time due to excessive carriers, τ_b is base transit time, τ_c is base–collector capacitance charging time through the collector, τ_{eb} is emitter–base capacitance charging time, and τ_{bc} is base–collector charging time through the emitter. The frequency f_T is also defined as the frequency at which common-emitter current gain reduces to unity.

4.2.2 High-Frequency Noise Performance of Silicon BJT

The high-frequency noise properties of a silicon BJT can be characterized by three noise sources. These dominant noise generators are represented in the noise model of a BJT shown in Figure 4.5. The base thermal noise (e_b) due to the base resistor (R_b), shot noise (e_e) due to the forward-biased emitter–base junction, and the collector partition noise (i_{cp}) are the dominant noise sources. The parameter R_G is the impedance of the signal source e_g, R_e is the emitter resistance, I_e is the forward-biased emitter current, and i_L is the current through the load impedance R_L. The noise generators of the BJT can be expressed as [22]

$$\bar{e}_g^2 = 4kTBR_G \tag{4.6a}$$

$$\bar{e}_b^2 = 4kTBR_b \tag{4.6b}$$

$$\bar{e}_e^2 = 2kTBR_e \tag{4.6c}$$

$$\bar{i}_{cp}^2 = 2kTB\left(\frac{\alpha_0 - |\alpha|^2}{R_e}\right) \tag{4.6d}$$

$$R_e = \frac{kT}{qI_e} \tag{4.7a}$$

$$\alpha = \frac{\alpha_0}{(1 + jf/f_b)} \tag{4.7b}$$

$$f_b = \frac{1}{2\pi\tau_b} \tag{4.7c}$$

where α_0 is the low-frequency common-base current gain, R_G is the source or generator impedance, B is the bandwidth, and f_b is the base cutoff frequency.

As discussed in Section 3.12.3, the device achieves minimum noise figure when it is operated at the optimum input impedance given by

$$Z_{opt} = R_{opt} + jX_{opt} \tag{4.8}$$

Following the analysis of Nielsen [23], expressions for the impedance and minimum noise figure can be written

$$X_{\text{opt}} = \frac{\alpha_0}{|\alpha|^2} \frac{\omega C_{\text{be}} R_{\text{e}}^2}{a} \tag{4.9a}$$

$$R_{\text{opt}} = \sqrt{R_{\text{b}}^2 - X_{\text{opt}}^2 + \frac{\alpha_0}{|\alpha|^2} \frac{R_{\text{e}}(2R_{\text{b}} + R_{\text{e}})}{a}} \tag{4.9b}$$

$$F_{\min} = a \frac{R_{\text{b}} + R_{\text{opt}}}{R_{\text{e}}} + \frac{\alpha_0}{|\alpha|^2} \tag{4.10}$$

where

$$a = \left[\left(1 + \frac{f^2}{f_{\text{b}}^2}\right) \left(1 + \frac{f^2}{f_{\text{e}}^2}\right) - \alpha_0 \right] \frac{1}{\alpha_0} \tag{4.11a}$$

$$f_{\text{e}} = \frac{1}{2\pi \tau_{\text{e}}} \tag{4.11b}$$

The equivalent noise resistance, R_{n}, is given by [24]

$$R_{\text{n}} = \frac{R_{\text{b}}}{\alpha} + \frac{R_{\text{e}}}{2} \left[1 + \left(\frac{R_{\text{b}}}{R_{\text{e}}}\right)^2 \left(\frac{f}{f_{\text{b}}}\right)^2 \right] \tag{4.12}$$

There are two limiting cases of operation for the bipolar transistor. The base limited case when the base time constant τ_{b} is most dominant, and the emitter limited case when the emitter time constant τ_{e} is most dominant. The minimum noise figure of the transistor in a base limited case can be expressed as

$$F_{\min} = \left(1 + \frac{f^2}{f_{\text{b}}^2} - \alpha_0\right) \left(\frac{R_{\text{b}} + R_{\text{opt}}}{\alpha_0 R_{\text{e}}}\right) + \left(1 + \frac{f^2}{f_{\text{b}}^2}\right) \frac{1}{\alpha_0} \tag{4.13a}$$

When a BJT is operated in the emitter limited condition, the minimum noise figure can be expressed as

$$F_{\min} = \left(1 + \frac{f^2}{f_{\text{e}}^2} - \alpha_0\right) \left(\frac{R_{\text{b}} + R_{\text{opt}}}{\alpha_0 R_{\text{e}}}\right) + \frac{1}{\alpha_0} \tag{4.13b}$$

where f_{e} is the emitter cut-off frequency.

The minimum noise figure increases more rapidly in the base limited case than in the emitter limited case. Also note that $F_{\min}$ increases by f^2 above some corner frequency given by f_{e}.

4.2.3 Power Performance

The important electrical performance parameters of a power transistor are power output, power gain, and efficiency. The power is determined by the maximum current and voltage handling capability of the transistor. The maximum current is nothing but full open channel current and is directly proportional to emitter (gate for FET/HEMT) periphery. The maximum voltage is determined by the breakdown voltage of the transistor. Normally, safe operating bias voltage is no more than half of the breakdown voltage. Power gain is obtained from the figure of merit parameters such as f_T and f_{max} as discussed in the next section.

The maximum output power (P_{max}) of a BJT operated as a class-A amplifier (see Chapter 8) is given by

$$P_{max} = \frac{1}{4} V_{CE} I_{max} \tag{4.14}$$

where V_{CE} is the collector–emitter voltage and I_{max} is the maximum allowed device current. A transistor with $I_{max} = 200$ mA and biased at $V_{CE} = 5$ V has a maximum output power of 250 mW(= 24 dBm).

4.3 GaAs MESFET

A MESFET consists of a highly conducting layer of high-quality n-type semiconductor layer placed on a semi-insulating (i.e., high-resistivity) substrate as shown in Figure 4.6. The conducting channel is interfaced with external circuitry through two ohmic contacts (called the source and drain), separated by a distance of 3–4 μm. The gate electrode is constructed by forming a rectifying (Schottky barrier) contact between the two ohmic contacts. The conducting channel's depth is on the order of 0.1–0.3 μm and is optimized so that the depletion region that forms under the Schottky contact (gate) can efficiently control the flow of current in the conducting or active layer. Actual depth depends on the application: thinner for low-noise applications and thicker for power and switching applications. The device therefore behaves as a voltage-controlled current source switch, capable of very high modulation rates.

When a metal and a semiconductor having different Fermi levels are brought into contact, a potential barrier is formed between them to equalize the Fermi levels across the junction. Since the Fermi level for GaAs is greater than that for the Schottky metal, the excess n-type carriers in the GaAs spill over into the metal, leaving behind a bound positive charge that forms the depletion region as shown in Figure 4.7a. A potential barrier (the magnitude at zero bias is defined as the built-in potential and the value is about 0.75 V for MESFETs) exists between the gate contact and the conducting region of the semiconductor. The potential barrier exists over the depletion region, which acts like an insulating region separating the gate contact from the mobile electrons in the conducting channel. When drain and gate bias voltages are applied, the depletion region is modified and affects the device $I-V$ characteristic by constricting the cross-sectional area through which current can flow. When the gate voltage is zero and the drain voltage is increased between the drain and source terminals (source

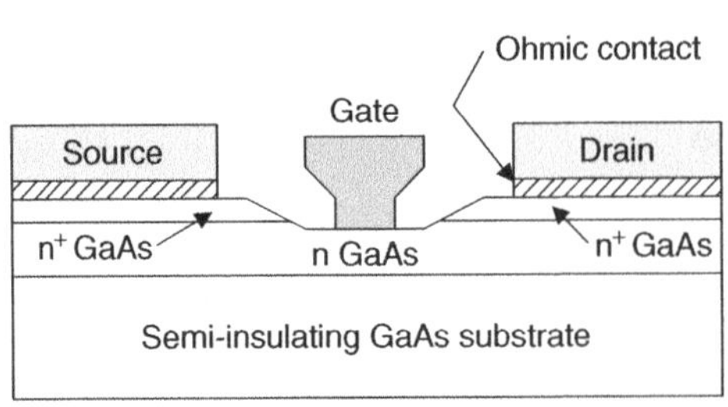

Figure 4.6 Physical cross section of a MESFET.

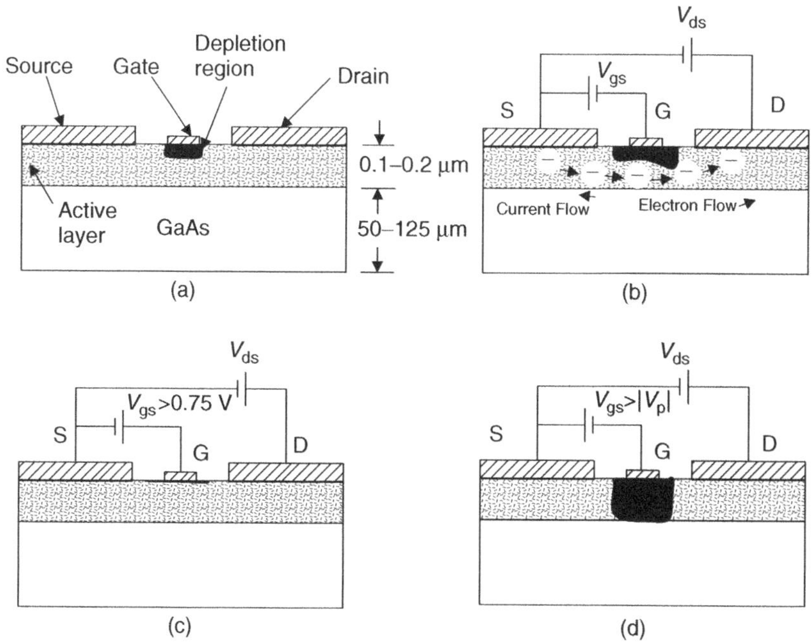

Figure 4.7 Illustration of FET operation under different bias conditions: (a) no bias, (b) normal bias, (c) channel fully open, and (d) channel pinched off.

is normally at zero potential), the electrons between the source and drain accelerate; this causes the electric field under the gate to increase with distance toward the drain because of the potential drop along the channel. This, in turn, results in the saturation field being achieved first at the drain side of the gate. The drain voltage at which the maximum field in the channel achieves the saturation field E_S is defined as V_{dsat}, and the current through the channel is called the saturated current I_{dss}. As V_{ds} is increased beyond V_{dsat}, the depletion region extends toward the drain as shown in Figure 4.7b. Therefore an increasing V_{ds} produces an increase in I_{ds} and a positive slope to the $I_{ds}-V_{ds}$ characteristic is obtained in the saturation region. The channel has a finite positive resistance.

The gate bias between the gate and the source modulates the depletion region: more negative voltage decreases the channel opening and the drain current decreases and more positive voltage increases the channel opening and the drain current increases accordingly. The drain current is thus modulated by the gate voltage. Fully open and fully closed channels are shown in Figure 4.7c and Figure 4.7d, respectively. As an example, measured pulsed $I-V$ characteristics for a GaAs FET are shown in Figure 4.8. The $I-V$ data were obtained using a commercial test system and the pulse width and duty cycle were 200 ns and 1%, respectively. The Q-point drain voltage was 5 V and the gate voltage was applied from −4 V to +0.8 V with a 0.2-V step. The device gate periphery was 0.6 mm. A small change in the gate voltage results in a large change in the drain current, which provides power gain by properly selecting the source and load impedance values. The DC device transconductance is defined as

$$g_m = \left.\frac{\Delta I_D}{\Delta V_G}\right|_{\text{operating } V_{ds}} \tag{4.15}$$

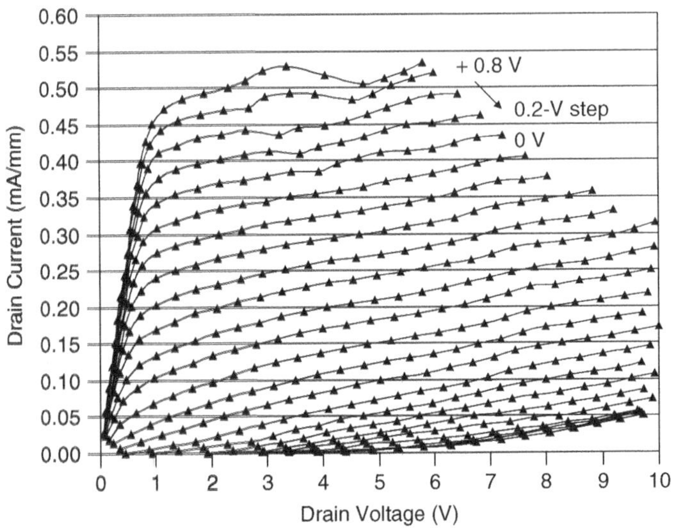

Figure 4.8 Typical measured pulsed (200 ns and 1% duty cycle) $I-V$ data for a MESFET. $V_{ds} = 5$ V and V_{gs} ranged from + 0.8 V to −4 V, with a 0.2-V step.

There are three possible FET configurations: common source (CS), common gate (CG), and common drain (CD), shown in Figure 4.9. Their low-frequency simplified equivalent circuits are also shown. The most commonly used FET configuration is the common-source one and its characteristics are compared with the other two configurations in Table 4.2. The common-source configuration has the highest power gain. In a CG configuration, its input impedance is primarily determined by g_m, is proportional to the transistor's width, and is broadband in nature. In a CD configuration, its output impedance is primarily determined again by g_m. This configuration is commonly used at the output of a multistage amplifier and is also known as a source follower.

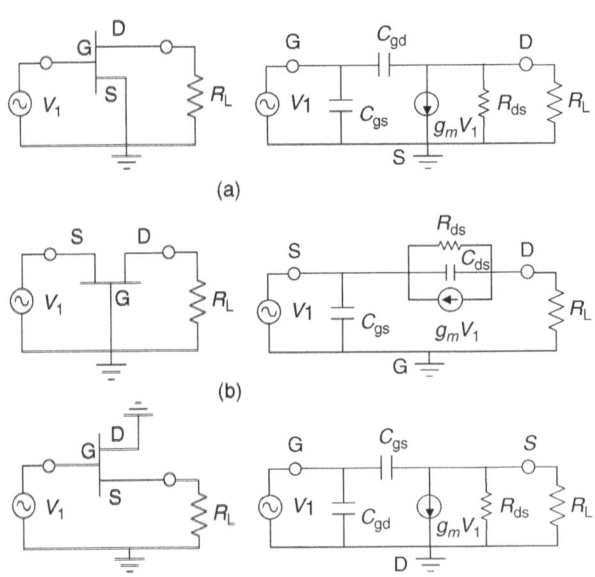

Figure 4.9 FET configurations and their low-frequency equivalent circuits: (a) common source, (b) common gate, and (c) common drain.

Table 4.2 Comparison of Basic FET Configurations[a]

Characteristic	Common Source	Common Gate	Common Drain
Voltage gain, A_v	$-g_m \dfrac{R_L R_{ds}}{R_L + R_{ds}}$, high	$g_m R_L$, higher	$\dfrac{g_m R_L}{1 + g_m R_L} < 1$
C_{in}	$C_{gs} + (1 - A_v)C_{gd}$	$C_{gs} + (1 - A_v)C_{ds}$	$C_{gd} + (1 - A_v)C_{gs}$
Input impedance, Z_{in}	High	$\dfrac{R_{ds} + R_L}{g_m R_{ds} + 1} \cong \dfrac{1}{g_m}$	High
Output impedance, Z_{out}	R_{ds}	$R_{ds} + (g_m R_{ds} + 1)R_L$	$\dfrac{R_{ds}}{g_m R_{ds} + 1} \| \| R_L \cong \dfrac{1}{g_m}$
Phase inversion from input to output	Yes	Yes	No
Noise contribution	Low	High	High
Reverse isolation	Best	Good	Poor

[a] R_L is the load resistance.

4.3.1 Small-Signal Equivalent Circuit

The small-signal equivalent circuit is useful for circuit designs at a lower power level. Figure 4.10 shows the small-signal equivalent circuit and the location of the circuit elements in the FET structure. The various components in the model are defined as follows:

Intrinsic Elements

R_i	Input (channel) resistance
C_{gs}	Gate–source capacitance
C_{gd}	Gate–drain feedback capacitance
R_{ds}	Drain–source resistance
g_m	Transconductance
τ	Phase delay due to carrier transit in channel

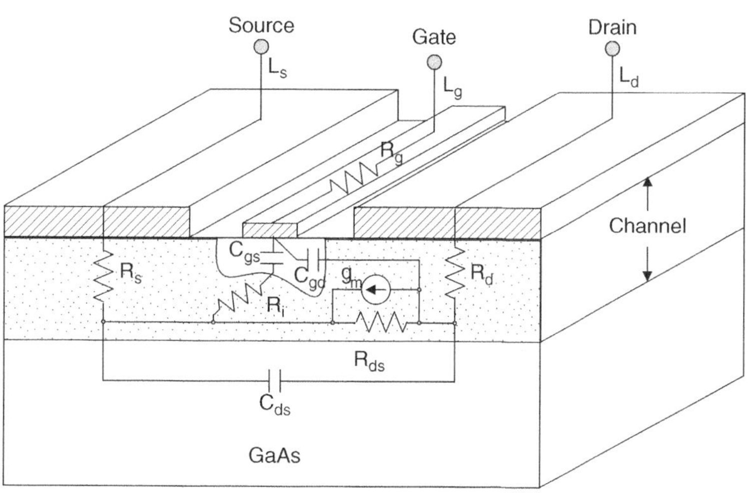

Figure 4.10 Small-signal equivalent circuit of a MESFET and the physical origin of the circuit elements.

Extrinsic Elements

C_{ds}	Drain–source capacitance
R_d	Drain-channel resistance, including contact resistance
R_s	Source-channel resistance, including contact resistance
R_g	Gate metal resistance

Parasitic inductances L_d, L_s, and L_g could be added in series with R_d, R_s, and R_g in the equivalent circuit to account for the effects of device pads and bonding wires in the case of discrete devices. The value of intrinsic elements depends on the channel doping, channel type, material, and dimensions. The large values of the extrinsic resistances will seriously decrease power gain and efficiency and increase the noise figure of the device. Figure 4.11 shows the layout of a MESFET having four fingers and 300-μm total gate periphery. The gate fingers are connected in parallel in an interdigital configuration. Typical values of device parameters for a 300-μm MSAG FET biased at 3 V and 50% I_{dss} are

$$C_{gs}(\text{fF}) = 50 + 1.33W$$
$$C_{ds}(\text{fF}) = 30 + 0.167W$$
$$C_{gd}(\text{fF}) = 8 + 0.05W$$
$$g_m(\text{mS}) = 0.17W$$
$$R_{ds}(\Omega) = 105 \times 10^3/W$$

The parameters are normalized with respect to gate periphery W. The FET has four fingers and W is in micrometers (μm). These relations hold good over 100 μm $< W <$ 800 μm.

4.3.2 Figure of Merit

High-frequency operation of a FET requires high gain. The maximum available gain (MAG) is given by

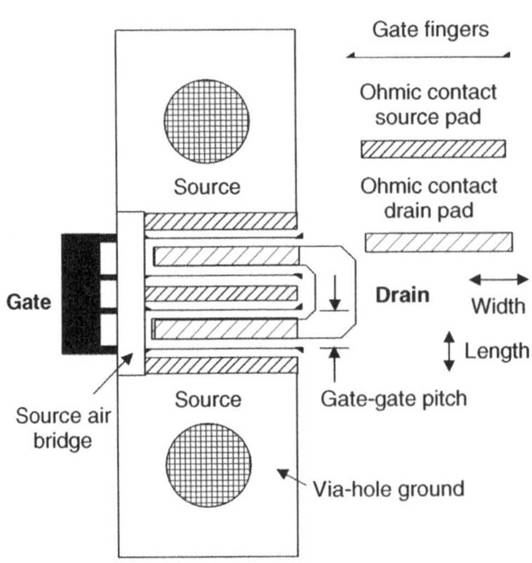

Figure 4.11 Physical layout of a four-finger FET with source via grounds.

$$MAG = \left(\frac{f_T}{f}\right)^2 \frac{1}{4R/R_{ds} + 4\pi f_T C_{gd}(R + R_g + \pi f_T L_s)} \tag{4.16a}$$

where

$$R = R_g + R_i + R_s + \pi f_T L_s \quad \text{and} \quad f_T = \frac{g_m}{2\pi C_{gs}} \tag{4.16b}$$

f_T is known as the cutoff frequency at which the device has unity current gain. This equation shows that MAG rolls off by 6 dB/octave. The frequency at which MAG is unity signifies the maximum frequency of operation and is given by

$$f_{max} = f_T[4R/R_{ds} + 4\pi f_T C_{gd}(R + R_g + \pi f_T L_s)]^{-1/2} \tag{4.16c}$$

This equation illustrates the importance of $R_g + R_i + R_s$, R_{ds}, C_{gd}, and L_s in addition to f_T to extend the operation of FETs to higher frequencies. The above expressions are approximate and must be used with care. The calculated gain might be off by as much as 4 dB; however, the expressions are useful to study the effect of device parameters on gain in a relative sense.

In general, a minimal value of source/emitter inductance is required to maintain high gain in amplifiers. As an example, Figure 4.12 shows the effect of source inductance L_S on maximum available gain at 18 GHz for a 300-μm FET. The device parameters are more sensitive to device current than supply voltage. Figure 4.13 shows the current dependence variations of g_m, C_{gd}, C_{gs}, and R_{ds} of a 300-μm FET, biased at 3 V.

The output power depends on the maximum voltage and current swings for various values of gate length and maximum channel current that the device can provide before breakdown. Shorter gate length implies higher channel doping due to thinner channel thickness because gate length/channel thickness ratio is almost constant. The maximum current swing limit is about 10–20% (depending on the pinch-off voltage) greater than the drain–source saturation current I_{dss} for power devices. This occurs when the gate–source bias is positive and lower than the built-in voltage. The maximum output power (P_{max}) and power added efficiency (PAE) of a FET operated as a class-A

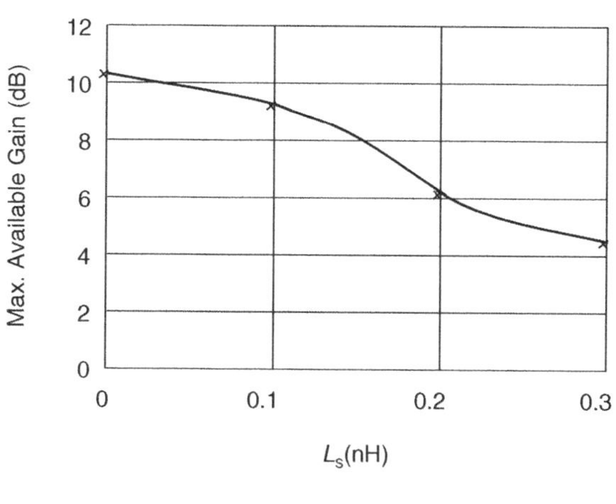

Figure 4.12 Effect of source inductance on maximum available gain at 18 GHz.

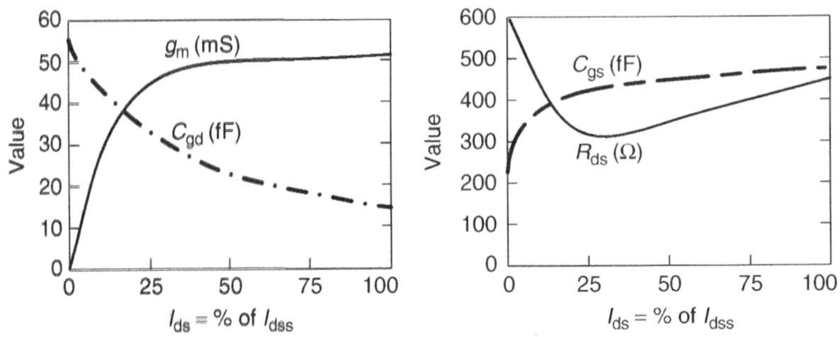

Figure 4.13 The variation of g_m, C_{gd}, C_{gs}, and R_{ds} of a 300-μm MESFET, biased at 3 V, as a function of device current.

amplifier (see Chapter 8) are given by

$$P_{max} = \tfrac{1}{8}(BV_{ds} - V_k)I_{max} \tag{4.17a}$$

$$PAE = \frac{1}{2}\left(1 - \frac{V_k}{V_{ds}}\right)\left(1 - \frac{1}{G}\right) \tag{4.17b}$$

where V_k is the knee voltage, BV_{ds} is the drain–source breakdown voltage, V_{ds} is the drain–source voltage, I_{max} is the full channel open current obtained under forward gate bias, and G is the associated gain of the device. The minimum current swing level is zero since negative current is impossible for positive drain bias with respect to source and gate. The maximum voltage swing between the gate and the drain is limited by avalanche breakdown due to intense local field strength at the drain edge of the gate metal. The power density for FETs operating at 10 V, is on the order of 1 W/mm.

EXAMPLE 4.1

For a power device biased at 10 V and 50% I_{dss}, the device parameters are $R_g = 0.5\ \Omega$, $R_i = 0.6\ \Omega$, $R_s = 1.0\ \Omega$, $R_{ds} = 175\ \Omega$, $g_m = 72$ mS, $C_{gs} = 0.8$ pF, $C_{gd} = 0.035$ pF, and $L_s = 0.02$ nH. Determine: (a) f_T, f_{max}, MAG, P_{max}, and PAE at 10 GHz. For the device $I_{max} = 0.2$ A, $BV_{ds} = 20$ V, and $V_k = 1$ V, and assume $G = MAG$ and the transit time is much smaller than $1/f$. (b) When L_s is increased from 0.02 to 0.05 nH, recalculate MAG and PAE.

SOLUTION

(a)

$$f_T = \frac{g_m}{2\pi C_{gs}} = \frac{72 \times 10^{-3}}{2 \times 3.1416 \times 0.8 \times 10^{-12}}\ \text{Hz} = 14.32\ \text{GHz}$$

$$R = 0.5 + 0.6 + 1 + 3.1416 \times 14.32 \times 0.02 = 3\ \Omega$$

$$f_{max} = 14.32[4 \times 3/175 + 4\pi \times 14.32 \times 0.035 \times 10^{-3}(3 + 0.5 + 0.9)]^{-1/2}$$
$$= 46.16\ \text{GHz}$$

From (4.16) and (4.17), we have

$$MAG = (f_{max}/f)^2 = (46.16/10)^2 = 21.3 = 13.3\ \text{dB}$$

$$P_{max} = \tfrac{1}{8}(20 - 1) \times 0.2 = 0.475\ \text{W} = 26.77\ \text{dBm}$$

$$PAE = \frac{1}{2}\left(1 - \frac{1}{10}\right)\left(1 - \frac{1}{21.3}\right) = 0.43 = 43\%$$

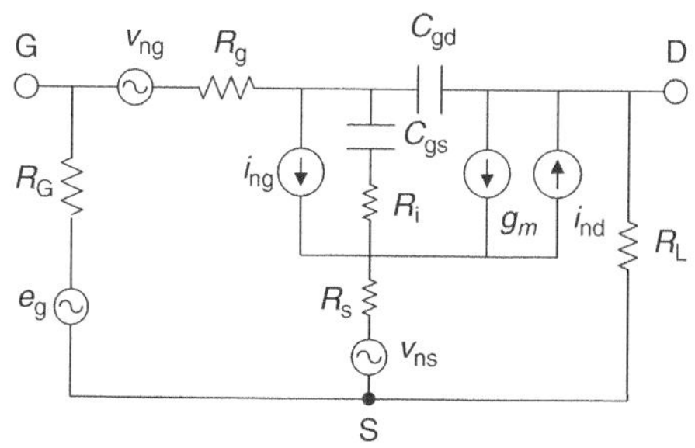

Figure 4.14 High-frequency noise equivalent circuit of a FET.

(b) When $L_s = 0.05$ nH, $MAG = 14.2 = 11.5$ dB and $PAE = 41.8\%$. Thus higher power gain, which requires lower L_s, is important to achieve higher PAE. ■

4.3.3 High-Frequency Noise Properties of MESFETs

Let us now examine the origins of high-frequency noise in FET structures. The noise generators for an FET device as shown in Figure 4.14 can be represented by the following expressions [25, 26]:

$$\bar{i}^2_{\text{nd}} = 4kTBPg_m \tag{4.18}$$

$$\bar{i}^2_{\text{ng}} = 4kTBP\omega^2 C^2_{\text{gs}} \frac{R}{g_m} \tag{4.19}$$

P and R are numerical factors associated with the drain and gate thermal noise generators, and B is the bandwidth.

In addition, there are two more noise voltage generators: one due to the gate resistance R_g and another one due to source resistance R_s. These noise sources can be expressed as

$$\bar{v}^2_{\text{ns}} = 4kTBR_s \tag{4.20}$$

$$\bar{v}^2_{\text{ng}} = 4kTBR_g \tag{4.21}$$

Pucel et al. [27] introduced the concept of carrier diffusion noise in 1975. They were able to predict the noise performance of MESFETs more accurately using this approach. Following the analysis of Pucel and co-workers, it can be shown that the minimum noise figure can be expressed as

$$F_{\min} = 1 + 2\left(\frac{f}{f_{\text{T}}}\right)\sqrt{P}\left(1 - \sqrt{\frac{R}{P}\frac{C_{\text{gs}}}{C_{11}}}\right)\sqrt{g_m(R_g + R_s)} \tag{4.22}$$

where $C_{11} = C_{\text{gs}} + C_{\text{gs1}}$, C_{gs1} is the parasitic capacitance from gate to source, C_{gs} is the intrinsic gate capacitance, and f is the operating frequency.

Fukui [28] measured the MESFET DC and small-signal parameters and derived an empirical relationship for the MESFET noise parameters:

$$F_{\min} = 1 + \left(\frac{f}{f_{\text{T}}}\right)k_1\sqrt{g_m(R_g + R_s)} \tag{4.23}$$

$$R_{\mathrm{n}} = \frac{k_2}{g_m} \tag{4.24a}$$

$$R_{\mathrm{opt}} = k_3\left(\frac{1}{4g_m} + R_{\mathrm{g}} + R_{\mathrm{s}}\right) \tag{4.24b}$$

$$X_{\mathrm{opt}} = \frac{k_4}{f C_{\mathrm{gs}}} \tag{4.24c}$$

where k_1, k_2, k_3, and k_4 are empirical fitting factors and their typical values are 3.0, 0.03, 2.2, and 0.16, respectively. By comparing (4.22) with (4.23), the Fukui factor k_1 can be written

$$k_1 = 2\left(\sqrt{P} - \sqrt{R}\frac{C_{\mathrm{gs}}}{C_{11}}\right) \tag{4.25}$$

Since

$$f_{\mathrm{T}} = \frac{g_m}{2\pi C_{\mathrm{gs}}} \tag{4.26}$$

the Fukui equation becomes

$$F_{\min} = 1 + 2\pi f C_{\mathrm{gs}} k_1 \sqrt{(R_{\mathrm{g}} + R_{\mathrm{s}})/g_m} \tag{4.27a}$$

Fukui's expression is extensively used to predict noise figure performance of FETs.

An expression for k_1 as a function of the bias current I_{ds} is given [29] as follows:

$$k_1 = 3.38\sqrt{\frac{I_{\mathrm{ds}}}{L \cdot g_m}} \tag{4.27b}$$

where I_{ds} is the drain–source current (mA) at which the noise performance is measured, g_m is the transconductance (in mS), and L is the gate length (measured in μm). The above equation along with Eq. (4.27a) calculates $F_{\min}$ with reasonably good accuracy for several FET types including MESFETs and HEMTs [29].

EXAMPLE 4.2

For a low-noise device biased at 3 V and 25% I_{dss}, the device parameters are: gate length $L = 0.3$ μm, $R_{\mathrm{g}} = 0.5\ \Omega$, $R_{\mathrm{s}} = 0.5\ \Omega$, $g_m = 56$ mS, $C_{\mathrm{gs}} = 0.255$ pF, $f_{\mathrm{T}} = 35$ GHz, and $I_{\mathrm{ds}} = 13$ mA. Determine its minimum noise figure in decibels at 10 GHz.

SOLUTION From (4.27b),

$$k_1 = 3.38\sqrt{\frac{13}{0.3 \times 56}} = 2.973$$

At $f = 10$ GHz, from (4.23),

$$F_{\min} = 1 + 2.973 \times \frac{10}{35}[56 \times 10^{-3}(0.5 + 0.5)]^{1/2} = 1.2$$

From (4.27a),

$$F_{\min} = 1 + 2.973 \times 10 \times 2 \times 3.1416 \times 0.255 \times 10^{-3}\sqrt{(0.5+0.5)/56 \times 10^{-3}} = 1.2$$
$$NF_{\min} = 10\ \log 1.2 = 0.8\ \text{dB}$$

■

4.4 HETEROJUNCTION FIELD EFFECT TRANSISTOR

The performance improvement of a heterojunction field effect transistor (HFET) over the GaAs MESFET can be understood by comparing the device material structures (see Figs. 4.6 and 4.15). In HFET devices different materials are grown one on top of the other using epitaxial growth technology like molecular beam epitaxy (MBE) or molecular organic chemical vapor deposition (MOCVD). The HFET is also called a modulated doped FET (MODFET).

There are different variations of HFETs. One of the most commonly used versions is called high electron mobility transistor (HEMT). This device structure, shown in Figure 4.15, has a layer of aluminum gallium arsenide (AlGaAs), which has a larger energy bandgap than GaAs, grown on top of the GaAs layer. The difference in the Fermi energy band between two materials causes band bending at the heterojunction interface (Figure 4.16). This band bending results in a quantum well where a large population of electrons forms a two-dimensional gas very close to the interface of

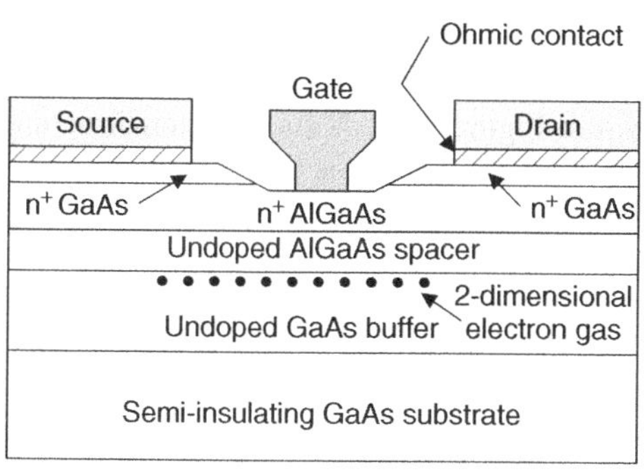

Figure 4.15 Cross-sectional view of an HEMT.

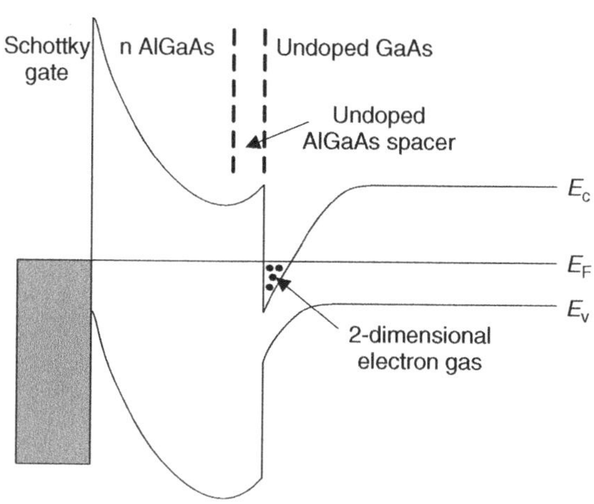

Figure 4.16 HEMT energy band diagram.

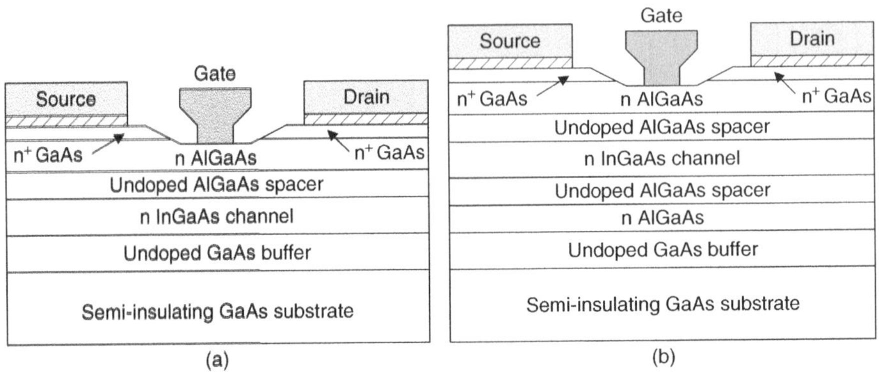

Figure 4.17 Cross-sectional representation of pHEMT structures: (a) single heterostructure and (b) double heterostructure.

the two materials. The AlGaAs layer is doped to provide electrons for the current conduction and the GaAs layer is undoped. The electrons from the donor atoms in the AlGaAs move to the low-energy level on the undoped GaAs. This effectively separates the donors residing in the AlGaAs from the electrons residing in the GaAs layer. Since donors and free electrons are in two different media, chances of collision between the donors and electrons are minimized and the drift velocity of the electrons is increased. These electrons can be modulated very easily by the application of a voltage at the gate terminal. The HEMT devices exhibit higher transconductance and lower noise figure properties and higher frequency operation compared to GaAs MESFETs.

Another variation of the HFET is called pseudomorphic HEMT or pHEMT and its cross-sectional representation is shown in Figure 4.17. A single heterostructure is used for low-power applications and a double heterostructure is used for power applications, respectively. A pHEMT device has superior microwave properties compared to the HEMT. This device is rapidly replacing GaAs HEMT and MESFET devices in many applications. In this structure a thin layer of indium gallium arsenide (InGaAs) is introduced between the undoped GaAs and the doped AlGaAs layers. The InGaAs material has an energy bandgap lower than the AlGaAs and GaAs material. When this layer is sandwiched between the two layers of higher bandgap (InGaAs and GaAs) materials, the lowest energy quantum well states would reside in the InGaAs layer. Therefore the free electrons provided by the donor atoms would move to the energy levels within the thin layer of InGaAs. These electrons are confined to this layer because of the presence of higher energy bandgap materials on both sides. These electrons have much higher drift velocity and can be modulated by the application of a voltage at the gate terminal. The pHEMT device exhibits even higher transconductance and superior RF properties compared to the GaAs HEMT and MESFET.

The small-signal equivalent circuits of the HEMT and pHEMT are very similar to the GaAs MESFET. The circuit designs quite often use the GaAs MESFET equivalent circuit configuration with proper values for the parameters. Detailed discussions of these equivalent circuits are given in the next chapter.

4.4.1 High-Frequency Noise Properties of HEMTs

The noise representation of an HEMT or pHEMT is similar to the MESFET as described previously. Compared to the FET and HEMT, a pHEMT device exhibits

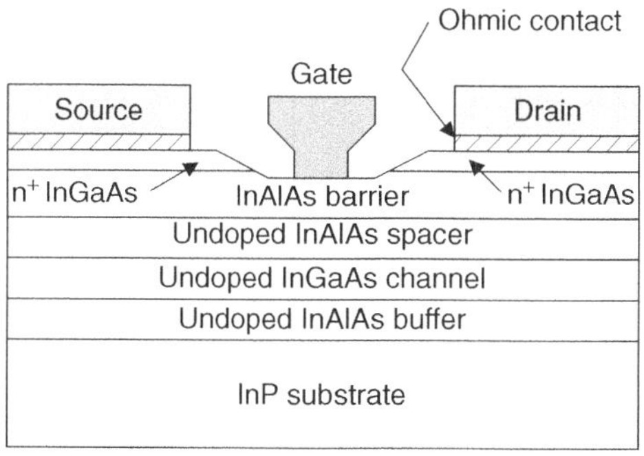

Figure 4.18 Physical cross section of an InP pHEMT.

superior noise figure and power gain performance. A conventional 0.25-μm gate length HEMT device having 2.1-dB noise figure and 7-dB associated gain at 40 GHz has been reported [30]. A pHEMT with the same gate length achieved 1.8-dB noise figure and associated gain of 6.4 dB at 60 GHz [31].

4.4.2 Indium Phosphide pHEMTs

Indium phosphide (InP) technology developed more slowly than GaAs technology due to the difficulty in the growth of the material. InP substrate has a lattice constant that is very close to that of an alloy composed of 50% GaAs and 50% InAs. $Ga_{0.47}In_{0.53}As$ has an energy bandgap of 0.77 eV, $Al_{0.48}In_{0.52}As$ material has a bandgap of 1.48 eV, and InP has a bandgap of 1.358 eV. A cross section of an InP pHEMT structure is shown in Figure 4.18. The high bandgap material AlInAs is used as the donor layer and the low bandgap material is the channel layer. The free electrons supplied by the donor layer reside in the low bandgap channel layer. The InGaAs/InAlAs/InP system offers many advantages over GaAs HEMT structures. The high conduction band discontinuity of the structure allows high two-dimensional electron gas concentration. The high mobility of electrons in InGaAs, coupled with the high density of electrons in the channel, leads to higher conductivity in the active channel. Because of these superior material characteristics, InP based pHEMTs exhibit very high transconductance, lower noise figure, and higher gain than conventional HEMTs fabricated on GaAs material. All the noise sources that contribute to the device noise figure are lower in the GaInAs/AlInAs HEMTs than in GaAs HEMTs. The extremely high conductivity of the two-dimensional (2-D) electron gas lowers the source resistance and associated thermal noise. Due to lower intervalley electron transfer probability, the velocity of electrons in the channel is higher and this leads to higher f_T of the device. Since capacitive coupling of the channel to the gate electrode is lower, coupling of noise sources to the gate is minimized. Figure 4.19 shows the noise figure of HEMTs as a function of frequency [12, 18]. The superior performance of the InP based HEMT at millimeter-wave frequencies is very evident from the figure.

EXAMPLE 4.3

For a low-noise pHEMT biased at 2 V and 10 mA, the device parameters are: gate length $L = 0.25$ μm, $R_g = 0.8\ \Omega$, $R_s = 0.3\ \Omega$, $g_m = 80$ mS, and $C_{gs} = 0.24$ pF. Determine its minimum noise figure in decibels at 20 GHz.

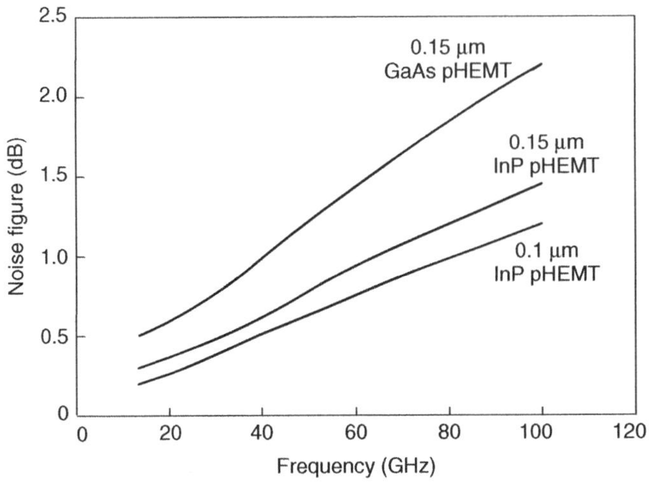

Figure 4.19 Noise figure comparison of GaAs and InP pHEMTs.

SOLUTION From (4.27b),

$$k_1 = 3.38\sqrt{\frac{10}{0.25 \times 80}} = 2.39$$

$$f_T = \frac{g_m}{2\pi C_{gs}} = \frac{80}{2\pi \times 0.24} = 53.05 \text{ GHz}$$

At $f = 20$ GHz, from (4.23),

$$F_{min} = 1 + \frac{20}{53.05} \times 2.39[80 \times 10^{-3}(0.8 + 0.3)]^{1/2} = 1.2673$$

$$NF_{min} = 10 \log(1.2673) = 1.029 \text{ dB}$$

■

GaAs pHEMTs are very popular today and are being used for numerous microwave circuits because of the combination of high performance and low cost. The f_T for 0.15-μm gate GaAs pHEMTs is about 80 GHz. InP based pHEMTs have significantly better gain and NF than that of GaAs based pHEMTs. The f_T values for 0.1-μm gate pHEMTs using GaAs and InP are about 120 GHz and 180 GHz, respectively. The f_{max} values for these transistors are over 220 GHz. Table 4.3 provides a comparison between 0.15-μm GaAs and InP pHEMTs employed in 20-GHz two-stage MMIC amplifiers [32]. Metamorphic HEMT (or MHEMT) combines the affordability of the GaAs process and the superior performance of In pHEMTs. The NF for an MHEMT has been reported as low as 1.24 dB at 120 GHz.

4.5 HETEROJUNCTION BIPOLAR TRANSISTORS

An HBT is an improved version of a BJT. In this structure the emitter has a wider energy bandgap heterostucture and high base doping density resulting in reduced base resistance, lower base–emitter capacitance, higher output resistance, higher f_T and f_{max}, lower $1/f$ noise, higher voltage handling capability, and higher current density.

Table 4.3 Comparison of 0.15-μm GaAs and InP pHEMTs Employed in 20-GHz Two-Stage MMIC Amplifiers with $V_{ds} = 5$ V [32]

Parameter	GaAs	InP
Gain (dB)	18	19
P_{out} (dBm)	26	23.4
PAE (%)	48	55

Thus, in general, the HBT's performance is much better than that of BJTs and FETs and at par with pHEMTs.

Shockley first proposed the concept of the heterojunction bipolar transistor in 1948. The advancement of MBE and MOCVD epitaxial techniques enabled the fabrication of AlGaAs/GaAs heterostructure HBTs. Most of the initial development of HBTs has been with the AlGaAs/GaAs material structure. HBTs with excellent performances have also been demonstrated on InP substrates. InGaP/GaAs is the preferred HBT material system today.

A cross section of an npn HBT is shown in Figure 4.20. When the base–emitter junction is forward biased, carriers are injected from the n-type emitter to the p^+ base region. These injected electrons are swept across the base region by drift and diffusion processes and ultimately are collected by the reverse-biased base–collector junction. Electrons are the minority carriers in the p^+ base region. Due to the short lifetime of electrons, they recombine with the majority carrier holes in the base region, resulting in a current through the base terminal. The transport of electrons across the collector–base space charge region takes place due to the high field electron saturation velocity.

The primary advantage of the HBT is its high collector efficiency and positive supply operation. The forward-biased collector injection efficiency is very high for the HBT since the wider bandgap material (AlGaAs) emitter injects electrons into the GaAs base at a lower energy level. At the same time the holes are prevented from flowing into the emitter by the energy barrier. The base can be doped heavily to reduce the base resistance.

The most important performance parameters in HBT design are the DC current gain, β, and emitter–base doping concentration ratio. These parameters can be

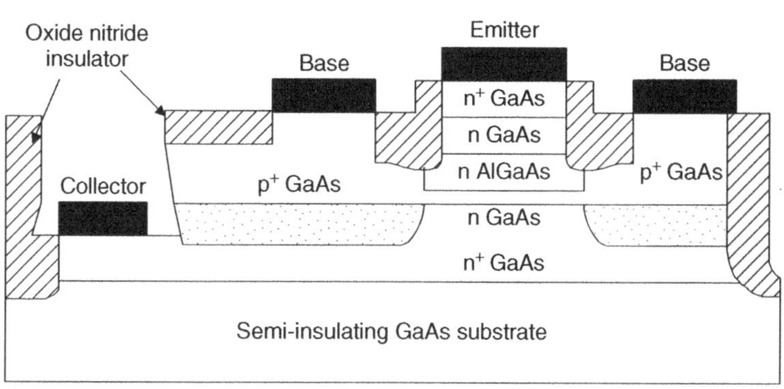

Figure 4.20 Physical cross section of a GaAs npn HBT.

expressed as [33]

$$\beta = \frac{I_c}{I_b} = \frac{(I_n - I_r)}{(I_p + I_r + I_s)} \tag{4.28}$$

$$\frac{I_n}{I_p} = \beta_{max} = \left(\frac{N_e}{P_b}\right)\left(\frac{v_{nb}}{v_{pe}}\right) e^{(\Delta E_g/kT)} \tag{4.29}$$

where I_c = the collector current
I_b = base current
I_p = reverse hole injection current
I_r = bulk recombination current
I_s = surface depletion region recombination current
I_n = injected electron current
β_{max} = maximum value of β in the absence of recombination currents
N_e/P_b = the emitter–base doping concentration ratio
ΔE_g = energy bandgap ratio between the emitter and base material
v_{nb}/v_{pe} = ratio of the electron and hole velocities

The high β_{max} allows a bipolar transistor design with low emitter doping to achieve a low emitter capacitance and high base doping to achieve low base resistance. These two parameters are critical in reducing the RC time constant, which in turn leads to improved speeds. In silicon bipolar transistors, the speed and linearity are relatively low due to relatively low base doping that can be achieved while still maintaining a good value of β. Figure 4.21 shows measured current–voltage (I–V) characteristics for a 2 × 80 μm^2 emitter area InGaP/GaAs HBT in common-emitter configuration.

Compared to the silicon bipolar transistor, the HBT has higher gain–bandwidth product with relaxed geometries due to the higher electron mobility in GaAs/InP and reduced parasitics. Higher common-emitter output impedance resulting from the high base doping permitted by the heterojunction minimizes the base width modulation. This leads to higher linearity and lower harmonic distortion in HBTs. The semi-insulating substrate of GaAs or InP HBTs allows fabrication of miniature MMICs, with better RF performance due to lower dielectric substrate loss compared to silicon based technologies. MESFETs and HEMTs are majority carrier devices with lateral current conduction

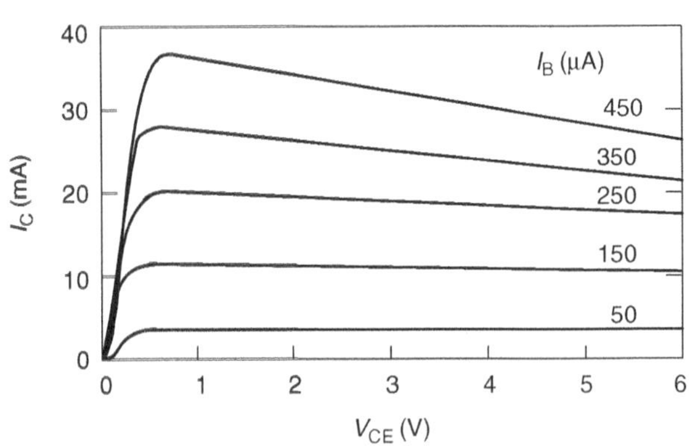

Figure 4.21 I–V representation of a 2 × 80 μm^2 emitter area GaAs HBT in common-emitter configuration.

Table 4.4 Substrate and Layer Materials for Several Versions of HBTs

Substrate	Emitter	Base	Collector
GaAs	AlGaAs	GaAs or AlGaAs	GaAs or AlGaAs
	GaInP	GaAs or AlGaAs	GaAs or GaInP
InP	InP	InGaAs	GaAs or GaInP
	AlInAs	InGaAs	InGaAs or GaInP
Si	Si/poly-Si	SiGe	Si
Sapphire	AlGaN	GaN	GaN

while the HBT is a vertical device that allows electron and hole current conduction. The speed of the HBT device is determined by the transit time through the thin vertical base–collector layers. The maximum speed of the FET is determined by the transit time and is controlled by the gate length defined by the lithographic techniques. HBTs have higher transconductance than FETs due to the exponential output current to input voltage variation. The values of f_T and f_{max} are 35 GHz and 100 GHz for 2-μm HBTs, and 65 GHz and 180 GHz for 1-μm HBTs, respectively.

Other advantages of HBTs are the high output current per unit cell emitter width, high current gain, and lower $1/f$ noise properties. The power handling capability of this device is much higher since the entire emitter area can carry the current because of the lower emitter resistance. Since the current density for such devices is very high due to the lower base resistance, power densities as high as 3–5 W/mm are attainable. The high output power handling capability along with the single power supply operation has made this the most suitable candidate for power amplifiers used in hand-held portable communication devices.

Based on substrate and layer materials, there are several versions of HBTs as summarized in Table 4.4 [45].

4.5.1 High-Frequency Noise Properties of HBTs

At high frequencies the important noise generators are the shot noise at the collector and base of the transistors and the thermal noises due to the emitter and collector series resistance. All other noise sources are negligible and are not considered in the calculation of high-frequency noise figures. In order to analyze the noise behavior, an equivalent circuit consisting of a noiseless transistor with noise generators is considered as shown in Figure 4.22. In this the noise generator at the base due to the thermal noise of the base resistance R_b is represented by e_b and the noise generator due the emitter resistance R_e is represented as e_e. The shot noise current at the base is represented by i_{bs} and the shot noise current at the collector is denoted by i_{cs}. An equivalent composite noise resistance, R_n, and a noise conductance, G_n, can be derived from the circuit model as

$$R_n = \frac{1}{2g_m} + (R_e + R_b) + G_n R_b^2 \tag{4.30a}$$

$$G_n = \frac{g_m}{2}\left(\frac{1}{\beta^2} + \omega^2 \tau_{ec}^2\right) + \eta \frac{g_\pi}{2} \tag{4.30b}$$

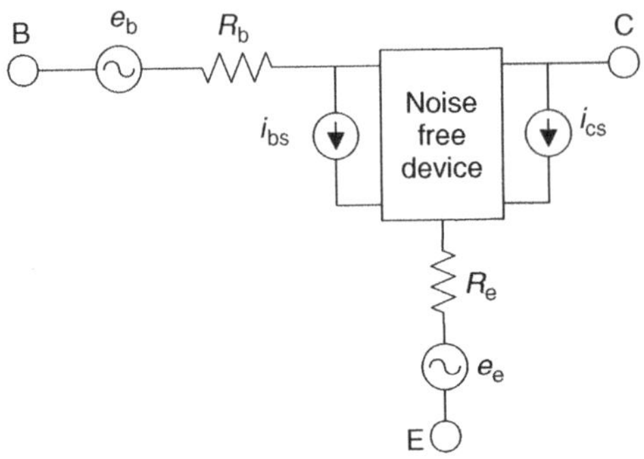

Figure 4.22 A simplified high-frequency noise equivalent circuit of an HBT.

where

$$g_m = \frac{qI_c}{kT} \tag{4.31a}$$

$$g_\pi = \frac{qI_b}{\eta kT} \tag{4.31b}$$

τ_{ec} is the emitter–collector transit time and η is the ideality factor for the gate current (typically between 1 and 2). An expression for the minimum noise figure can be written [34–37]

$$F_{min} = 1 + 2R_n g_{so} + 2r\sqrt{R_n G_n} \tag{4.32}$$

g_{so} is the optimum source conductance and r is the real part of the complex correlation coefficient.

Equation (4.32) can be represented in terms of the network parameters and noise model parameters. At optimum bias point, a simplified expression for the HBT noise figure can be derived:

$$F_{min} = 1 + \sqrt{2}\left(\frac{f}{f_T}\right)\sqrt{g_m(R_e + R_b)} \tag{4.33a}$$

where

$$f_T = \frac{1}{2\pi\tau_{ec}} \tag{4.33b}$$

Comparing this with Eq. (4.22) for the FET/HEMT shows that the noise cancellation occurring between the gate and drain noise sources represented by the factor $\left(\sqrt{P} - \sqrt{R}C_{gs}/C_{11}\right)$ is the reason for the lower noise figure performance of the FET/HEMT.

Table 4.5 provides a comparison for noise figure and power performance between GaAs and InP devices at 10 GHz. The data is typical and power data is taken for a unit cell using a load-pull setup. In this comparison the power dissipations in the transistors are similar.

Several commercial foundries provide GaAs and InP based FET, pHEMT, and HBT processes for custom designs.

Table 4.5 A Comparison for Devices

(a) Power Performance Comparison of Discrete Power Transistors @ 10 GHz

Parameter	GaAs FET	GaAs pHEMT	GaAs HBT	InP HBT
Gate/emitter length (μm)	0.4	0.25	2.0	2.0
Gain (dB)	11	12	11	12
P_{out} (W)	0.5	0.5	0.6	0.5
PAE (%)	65	70	60	70
Voltage (V)	10	8	5	5

(b) Noise Figure Comparison of Discrete Transistors @ 10 GHz

Parameter	GaAs FET	GaAs pHEMT	GaAs HBT	InP HBT
Gate/emitter length (μm)	0.4	0.25	2.0	2.0
Gain (dB)	12.6	16.0	14.7	16.1
NF (dB)	0.8	0.4	1.0	1.5
Voltage (V)	3	2	2	2

4.5.2 SiGe Heterojunction Bipolar Transistors

SiGe technology [16, 38–40] has been evolving over the last decade, driven primarily by the work of a team of scientists from IBM. One of the major advantages of SiGe is that it can easily be integrated into standard Si-bipolar and bipolar-CMOS processes. The performance advantages of SiGe over traditional silicon devices are the very high cutoff frequency and improved power added efficiency at low voltages. An f_T of 130 GHz and f_{max} of 160 GHz have been reported [39, 40]. These numbers are about a factor of 2 better than the best silicon n-type bipolar device but 2–6 times lower than the best GaAs device. The advantage of SiGe technology over the III–V material HBTs is that the standard silicon wafer process line could be used for the fabrication with minimal additional cost. This results in reduced fabrication cost and hence low-cost RFICs with very good performance characteristics.

SiGe heterojunction bipolar transistors are like silicon bipolar transistors with the incorporation of germanium in the base region of the transistor. This reduces the bandgap of the base material. The presence of germanium in the base reduces the base transit time for a given base resistance, thus achieving a device having high f_T and f_{max}. The higher base doping concentration helps to increase the Early voltage and lower the noise figure of the device. These devices have $1/f$ noise characteristics that are much better than GaAs based devices. Corner frequencies down to 100 Hz are possible while GaAs has frequencies of 2 kHz or higher.

One of the process advantages of SiGe is the differential growth of the SiGe layer after a standard oxidation process. SiGe-poly is used for the base contact and for resistors. The emitter has two spacers, one on the inside and the other on the outside, and an additional amorphous layer of silicon. This n^+ silicon amorphous layer is used for collector and emitter contacts. The SiGe process is comparable to the silicon process in masks: the process steps and therefore the cost of fabrication are about the same as that of silicon bipolar technology. Figure 4.23 shows measured current–voltage (I–V) characteristics for a $0.8 \times 10\ \mu m^2$ emitter area SiGe HBT in common-emitter configuration.

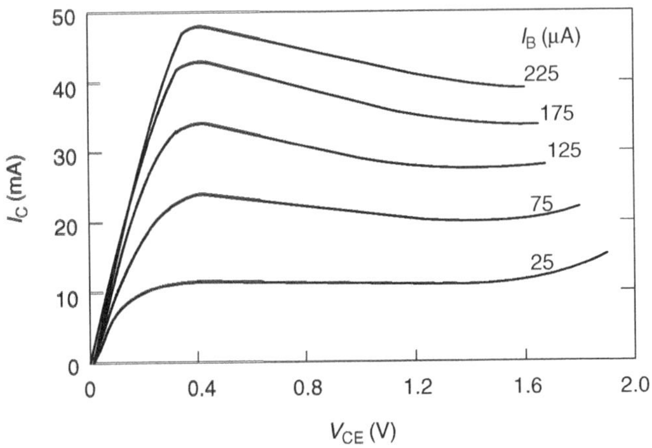

Figure 4.23 $I-V$ representation of a $0.8 \times 10\ \mu m^2$ emitter area SiGe HBT in common-emitter configuration.

SiGe HBTs have the low-cost potential of Si based BJT/CMOS processes and the small-signal performance of GaAs based HBTs. However, they have lower breakdown voltage and much lower output power capability than the GaAs HBTs. At present, SiGe HBTs have been demonstrated with f_T and f_{max} values greater than 150 GHz. The breakdown values are on the order of only 5 V. The low-power performance of SiGe HBTs are comparable to GaAs HBTs. The f_T values for 0.35-μm and 0.18-μm SiGe HBTs are about 62 GHz and 120 GHz, respectively.

Low-noise amplifiers and power amplifiers have been demonstrated using SiGe technology. A minimum noise figure of 0.8 dB with associated gain of 17 dB was achieved at 2.0 GHz when the device was biased at 3.0 V with an operating current of 3.6 mA [38]. Note that the 50-Ω noise figure values are very close to F_{min} and hence matching a SiGe device is easier than the Si BJT or GaAs FETs in this frequency regime. A four-cell SiGe HBT power device, operating at 3.0 V and 900 MHz under class-B bias conditions, achieved 1.1-W power output and 66% power added efficiency [38].

4.6 MOSFET

The metal oxide semiconductor field effect transistor (MOSFET) is widely used in RF and microwave MMICs for commercial and military applications. Both n-type MOSFETs (nMOSFET) and p-type MOSFETs (pMOSFET) are used. They are also referred to as NMOS and PMOS, respectively, and the process supporting these devices is popularly know as complementary MOS or CMOS. Today RF integrated circuits (RFICs) are primary based on complementary MOS transistors. The recent development of submicron-CMOS technology has led to the development of RF, microwave, and millimeter-wave single-function and multifunction ICs. The typical values of f_T for 0.25-, 0.18-, 0.13-, and 0.09-μm gate CMOS transistors are 30, 50, 75, and 110 GHz, respectively. A minimum noise figure of 0.8 dB with associated gain of 13 dB was achieved at 10 GHz for the 0.09-μm gate device.

Figure 4.24 shows a cross-sectional view of an n-type MOSFET. In basic operation, when a positive drain voltage is applied between the drain and source and there

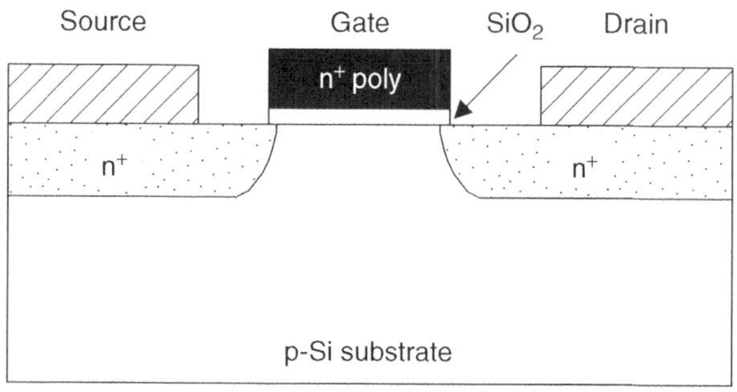

Figure 4.24 Physical cross section of an Si nMOSFET.

is no gate voltage, the pn junction on the drain side is reverse biased. In this case, negligibly small reverse current flows between the drain and source. In other words, I_{dss} is practically zero. Next, when a positive gate–source voltage along with positive drain–source voltage is applied, the positive gate voltage repels holes and attracts electrons and both pn junctions are forward biased. In this case, a conductive path is established between drain and source; a significant current starts flowing that is a function of drain and gate voltages. Thus the amplification operation of this device is similar to the FET/HEMT. Since gate voltage is positive and there is no gate current involved, this device can be operated with a single positive power supply. For a detailed analysis of MOSFET devices, readers are referred to excellent textbooks in the references [18, 41, 42]. A power version of the MOSFET, commonly referred to as the laterally diffused metal oxide semiconductor (LDMOS) field effect transistor, is described in Chapter 13.

Tables 4.6 and 4.7 provide comparisons for noise figure and power performance for various CMOS processes. Up to C-band frequencies, LNAs have achieved noise figures in the 2–3-dB range. Power amplifiers with over 1-W output power and 60% PAE have been demonstrated up to L-band frequencies.

In summary, recent advances in low-cost Si based RF CMOS and SiGe HBT technologies offer great potential for very-high-volume wireless applications. High RF

Table 4.6 Summary of CMOS LNAs

Frequency Range (GHz)	Gain (dB)	NF (dB)	P_{DC} (mW)	Process (nm)
3.1–4.8	12	4.2–4.9	20	250
3.0–7.5	19	3.0–3.8	32	180
2.4–5.4	25	2.2–3.1	4.7	130
3.1–5.9	19.5	2.8–3.8	5.4	90

Table 4.7 Summary of CMOS Power Amplifiers[a]

Frequency (GHz)	Gain (dB)	Output Power (W)	PAE (%)	Process (nm)
0.855	—	1.0	60	350
0.875	30	1.5	62	180
1.7	—	1.3	58	130
2.0	27	1.0	70	65

[a]Nominal operating voltage range is 3–5 V.

Table 4.8 Comparison of GaAs HBT, Si CMOS, and SiGe HBT Technologies for Mobile Wireless Applications

Parameter	GaAs HBT	Si CMOS	SiGe HBT
Thermal conductivity	0.46	1.45	1.45
Dielectric loss/output match loss	Low	High	High
Current density	Medium	Medium	High
Immunity to thermal runaway	No	Yes	Yes
Immunity to load mismatch	Yes	Yes	No
Wafer cost	Medium	Low	Low

performance of CMOS and HBT devices with large-scale integration capabilities make these technologies very attractive for low-cost amplifier solutions. Several commercial foundries provide Si based CMOS and HBTs processes for custom designs. Table 4.8 provides a comparison of GaAs HBT, Si CMOS, and SiGe HBT technologies for mobile wireless applications.

REFERENCES

1. H. A. Watson, *Microwave Semiconductor Devices and Their Circuit Applications*, McGraw-Hill, New York, 1969.
2. H. F. Cooke, Microwave transistors: theory and design, *Proc. IEEE*, Vol. 59, pp. 1163–1181, August 1971.
3. H. Fukui (Ed.), *Low-Noise Transistors and Amplifiers*, IEEE Press, New York, 1981.
4. J. V. Dilorenzo and D. D. Khandelwal (Eds.), *GaAs FET Principles and Technology*, Artech House, Norwood, MA, 1982.
5. R. S. Pengally, *Microwave Field-Effect Transistors—Theory, Design and Applications*, John Wiley & Sons, Hoboken, NJ, 1982.
6. I. J. Bahl and P. Bhartia, *Microwave Solid State Circuit Design*, John Wiley & Sons, Hoboken, NJ, 1988.
7. R. Soares (Ed.), *GaAs MESFET Circuit Design*, Artech House, Norwood, MA, 1989.
8. P. H. Ladbrooke, *MMIC Design: GaAs FETs and HEMTs*, Artech House, Norwood, MA, 1989.
9. R. Goyal (Ed.), *Monolithic Microwave Integrated Circuits: Technology and Design*, Artech House, Norwood, MA, 1989.
10. K. Chang (Ed.), *Handbook of Microwave and Optical Components*, Vol. 2, John Wiley & Sons, Hoboken, NJ, 1990.
11. C. T. Wang (Ed.), *Introduction to Semiconductor Technology: GaAs and Related Compounds*, John Wiley & Sons, Hoboken, NJ, 1990.
12. F. Ali and A. Gupta (Eds.), *HEMTs and HBTs: Devices, Fabrication and Circuits*, Artech House, Norwood, MA, 1991.
13. J. L. B. Walker (Ed.), *High-Power GaAs FET Amplifiers*, Artech House, Norwood, MA, 1993.
14. K. Chang, *Microwave Solid-State Circuits and Applications*, John Wiley & Sons, Hoboken, NJ, 1994.
15. W. Liu, *Handbook of III–V Heterojunction Bipolar Transistors*, John Wiley & Sons, Hoboken, NJ, 1998.
16. J. S. Yuan, *SiGe, GaAs, InP Heterojunction Bipolar Transistors*, John Wiley & Sons, Hoboken, NJ, 1999.
17. K. Chang, I. Bahl, and V. Nair, *RF and Microwave Circuit and Component Design for Wireless Systems*, John Wiley & Sons, Hoboken, NJ, 2002.

18. F. Schwierz and J. J. Liou, *Modern Microwave Transistors*, Wiley-Interscience, Hoboken, NJ, 2003.
19. K. K. Ng, *Complete Guide to Semiconductor Devices*, 2nd edition, Wiley-Interscience, Hoboken, NJ, 2003.
20. R. Singh, D. L. Harame, and M. M. Oprysko, *Silicon Germanium*, John Wiley & Sons, Hoboken, NJ, 2004
21. A. Grebennikov, *RF and Microwave Power Amplifier Design*, McGraw-Hill, New York, 2004.
22. R. J. Hawkins, Limitations of Neilson's and related noise equations applied to microwave bipolar transistors, and new expressions for the frequency and current dependent noise figures, *Solid-State Electron*., Vol. 20, pp. 191–196, 1977.
23. E. G. Nielsen, Behavior of noise figure in junction transistors, *Proc. IRE*, Vol. 45, pp. 233–258, 1957.
24. H. Fukui, The noise performance of microwave transistors, *IEEE Trans. Electron Devices*, Vol. ED-13, pp. 329–341, 1986.
25. A. van der Zeil, Thermal noise in field-effect transistors, *Proc. IRE*, Vol. 50, pp. 1808–1812, 1962.
26. W. Baechtold, Noise behavior of Schottky barrier gate Field-effect transistors at microwave frequencies, *IEEE Trans. Electron Devices*, Vol. ED-18, pp. 97–106, 1971.
27. R. A. Pucel et al., Signal and noise properties of gallium arsenide field-effect transistor, in *Advances in Electronic and Electron Physics*, Vol. 38, Academic Press, New York, 1975, pp. 145–265.
28. H. Fukui, Design of microwave GaAs MESFET's for broad-band low-noise amplifiers, *IEEE Trans. Electron Devices*, Vol. ED-37, pp. 67–78, 1990.
29. D. Delagebeaudeuf et al., A new relationship between the Fukui coefficient and optimal current value for low-noise operation of field-effect transistors, *IEEE Electron Device Lett*., Vol. EDL-6, pp. 444–445, September 1985.
30. P. C. Chao et al., Millimeter-wave low-noise high electron mobility transistors, *IEEE Electron Device Lett*., Vol. EDL-6, pp. 75–77, January 1985.
31. K. H. G. Duh et al., Ultra low noise characteristics of millimeter-wave electron mobility transistors, *IEEE Electron Device Lett*., Vol. EDL-9, pp. 521–523, 1988.
32. A. Oki et al., HEMT and HBT MMIC power amplifiers, in *IEEE MTT-S International Microwave Symposium Workshop*, #WMH, 2006.
33. H. Kroemer, Theory of wide-gap emitter for transistors, *Proc. IRE*, Vol. 45, pp. 1535–1537, 1957.
34. H. Kroemer, Hetero-structure bipolar transistors and integrated circuits, *Proc. IRE*, Vol. 70, pp. 13–25, 1982.
35. A. van der Zeil et al. A theory of Hooge's parameters of solid state devices, *IEEE Trans. Electron Devices*, Vol. ED-32, pp. 667–671, 1985.
36. A. van der Zeil, Noise in junction transistors, *Proc. IRE*, Vol. 46, pp. 1019–1038, 1958.
37. M. B. Das and O. A. Dogha, On the noise properties of bipolar transistors in untuned amplifiers, *Solid-State Electron*., Vol. 19, pp. 827–836, 1976.
38. R. A. Metzger, Is silicon germanium the new "material of the future"?, *Compound Semiconductor Mag*., pp. 21–28, November/December 1995.
39. G. L. Patton et al., 75GHz f_t SiGe base hetero-junction bipolar transistors, *IEEE Electron Devices Lett*., pp. 171–173, 1990.
40. K. Oda et al., 130GHz f_t SiGe HBT technology, *IEDM Tech. Digest*, pp. 791–794, 1997.
41. S. M. Sze, *Physics of Semiconductor Devices*, 2nd edition, Wiley-Interscience, Hoboken, NJ, 1981.
42. R. F. Pierret, *Field Effect Devices*, Addison-Wesley, Reading, MA, 1990.
43. J. M. Golio, *Microwave MESFETs and HEMTs*, Artech House, Norwood, MA, 1991.
44. R. Anholt, *Electrical and Thermal Characterization of MESFETs, HEMTs and HBTs*, Artech House, Norwood, MA, 1995.

45. H. Wang et al., Millimeter-wave integrated circuits, in *Encyclopedia RF and Microwave Engineering*, K. Chang (Ed.), Vol. 4, John Wiley & Sons, Hoboken, NJ, 2005, pp. 3021–3046.
46. P. H. Aaen, J. A. Pla, and J. Wood, *Modeling and Characterization of RF and Microwave Power FETs*, Cambridge University Press, Cambridge, UK, 2007.
47. http://www.tycoelectronics.com.
48. http://www.velocium.com.
49. http://www.winfoundry.com.
50. http://www.triquint.com.
51. http://www.tsmc.com.
52. http://www.ibm.com.
53. http://www.cree.com.

PROBLEMS

4.1 Compare the salient features of GaAs MESFETs, HEMTs, and HBTs. Why does low-resistivity Si have a higher substrate loss than GaAs?

4.2 Consider the MESFET equivalent circuit model given in Example 4.1. Calculate the maximum gain that the device can have at 12 GHz. Also calculate the gain degradation when (a) L_s, C_{gs}, and C_{gd} are increased by 20% and (b) g_m and R_{ds} are decreased by 20%.

4.3 Calculate F_{min} for the MESFET whose equivalent circuit model is given in Example 4.1 and with $I_{ds} = 45$ mA. The gate length is 0.4 μm and the frequency is 10 GHz.

4.4 When the MESFET whose equivalent circuit model parameters are given in Example 4.1 was tested for F_{min} at 100 °C, its value increased by 0.7 dB from the room temperature (25 °C) value of 3 dB. Determine the decrease in g_m from room temperature to 100 °C when other parameter values remain unaffected.

4.5 The I–V data of a transistor is given below. Calculate its R_{ds} and g_m at $V_{ds} = 4$ V.

$V_{ds} = 3\ V:$	$V_{gs} = -2\ V$	$I_{ds} = 5$ mA
	$= -1$ V	$= 100$ mA
	$= 0$ V	$= 200$ mA
$V_{ds} = 5\ V:$	$V_{gs} = -2$ V	$I_{ds} = 7$ mA
	$= -1$ V	$= 110$ mA
	$= 0$ V	$= 205$ mA

4.6 Compare qualitatively the MESFET, pHEMT, and MOSFET devices for medium-power amplifier applications.

Chapter 5

Transistor Models

5.1 TRANSISTOR MODEL TYPES

An electrical model of a transistor represented by an equivalent circuit and/or mathematical relations is used to calculate its behavior under a predetermined set of conditions. These models are used to improve the transistor performance and to design the amplifier circuits. The objective of this chapter is to present a brief overview of transistor modeling and models that are useful in the design of amplifiers.

The application of modern computer-aided design (CAD) tools offers an improved approach to reduce design time for transistor amplifiers (see Chapter 9 for more details). The development of integrated CAD tools and accurate and comprehensive models for transistors was a major activity during the 1980s and 1990s. Both play a key role in the successful development of MICs and MMICs. As the sophistication and accuracy of these tools improve, significant reductions in design cycle time can be realized and "first-pass" design success can be achieved for ICs. Accurate models for BJTs, MOSFETs, FETs, HFETs, HEMTs, and HBTs are an essential part of these tools. Transistors are low-noise, switching, and power types and use both linear and nonlinear (bias and input power-dependent) models. An overview of device models is given in Figure 5.1; equivalent circuit (EC) models are the most popular. More comprehensive information on modeling can be found in books [1–8] and other publications [9–31]. Basically there are three types of models:

1. Physics/electromagnetic theory based models
2. Analytical or hybrid models
3. Measurement based models

These models are briefly described next.

5.1.1 Physics/Electromagnetic Theory Based Models

Development of accurate physics based models for active devices that are derived in terms of doping profile and physical geometry are essential for establishing the link between the process and RF performance and for designing MMICs. These models consist of two parts: intrinsic and extrinsic. The intrinsic part deals with the active channel

Fundamentals of RF and Microwave Transistor Amplifiers. By Inder J. Bahl

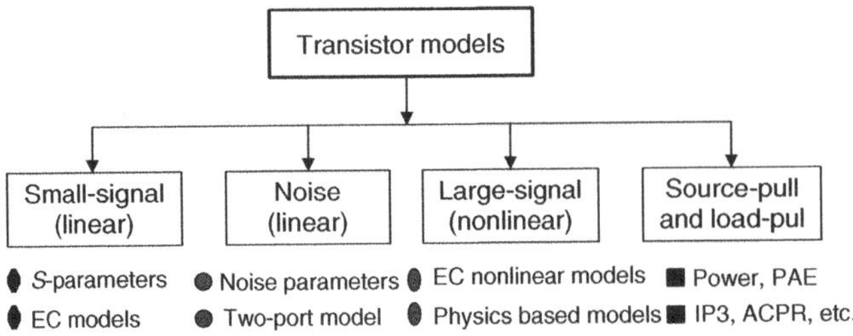

Figure 5.1 An overview of linear and nonlinear transistor models for amplifiers.

of the transistor, whereas the extrinsic part represents device pad/electrode parasitics, which are expressed in resistances, inductances, and capacitances. The intrinsic part of the model, the heart of the device, is obtained by solving device equations using appropriate boundary and bias conditions in the device channel (e.g., under the gate or between the gate, source, and drain electrodes). The semiconductor device equations are derived from the Boltzmann transport equation coupled to the solution of the Poisson equation. These partial differential equations describe carrier transport properties of the device and are solved numerically by using techniques such as finite differences or finite elements. The physical models also include device interface phenomena, quantum effects, effects of temperature and heterostructures, low-noise and high-field phenomena, and electromagnetic interaction effects between electrodes. These models are of a general nature but quite complex. They include accurate parasitic reactance, bias, and temperature and frequency dependence and can be used in time and frequency domains. The physical models are very useful for investigating the physical operation of active devices and predicting device performance as a function of process, material, and geometry. Thus the device model helps in device studies, process control, and circuit yield and optimization. Any adjustment in the device can be achieved using the physics based model without costly fabrication experiments.

Because of lengthy execution times (physics based analysis time increases rapidly with model complexity), the application of physical models is usually limited to device studies. Models are available for FETs, HEMTs, and HBTs. Given sufficient computer resources, these models can become an integral part of microwave CAD tools. Electromagnetic (EM) simulators are used to accurately model the device pad/electrode parasitic reactance, including the coupling capacitance between electrodes.

5.1.2 Analytical or Hybrid Models

Analytical or hybrid models for active devices are based on a simple equivalent circuit (EC) representation. The model parameters are formulated based on simple equations whose values are obtained from the physics of the component, or DC or RF or both measurements. The analytical models result from both physics and measurement based techniques. Nonlinear devices are generally represented by analytical models.

5.1.3 Measurement Based Models

The most commonly used method of developing models for active devices is by measuring their DC characteristics and *S*-parameters. This modeling approach gives quick and accurate results, although they are generally limited to just the devices measured. The component is represented by an equivalent circuit model whose parameter values are extracted by computer correlation to the measured DC and S-parameter data.

The accuracy of the measurement based models depends on the accuracy of the measurement systems, the calibration techniques, and the calibration standards. On-wafer measurements using high-frequency probes provide accurate, quick, nondestructive, and repeatable results up to millimeter-wave frequencies [32]. Various vector network analyzer calibration techniques are being used to determine a two-port error model that de-embeds the device S-parameters. The conventional *short, open, load, and through* (SOLT) calibration technique has proved unsatisfactory because the open and short reference planes cannot be precisely defined. The reference plane uncertainties for a perfect short limit the accuracy of this technique. The *line-reflect-match* (LRM) calibration technique requires a perfect match on each port. The *thru-reflect-line* (TRL) calibration method is based on the transmission line calibration standards, which include nonzero length thru and reflect (open or short) and delay line standards (one or more dictated by the frequency range over which the calibration is performed). The advantage of the TRL calibration lies in simple standards that can be placed on the same substrate as the components, ensuring a common transmission medium. This calibration technique accurately locates the reference planes and minimizes radiative crosstalk effects between the two probes because they are sufficiently far apart during the calibration procedure. On-wafer measurement techniques are described in more detail in Chapter 22.

Measurement based models fall into two groups: linear and nonlinear. In current measurement based modeling, the components are electrically characterized by measuring DC and RF parameters. A lumped-element equivalent circuit model for each component that describes its frequency-dependent electrical characteristics is chosen. The lumped-element model parameter values are extracted by computer optimization to replicate the measured S-parameters. Noise characterization of active devices is obtained by measuring on-wafer S-parameters and noise parameters. The model parameter extraction is generally based on statistical data with average and standard deviation values that will help in centering designs for high yield.

Linear Model

As an example, an equivalent circuit (EC) linear model for a MESFET is shown in Figure 5.2. The parameters C_{gs}, C_{gd}, g_m, τ, and R_{ds} are strongly dependent on device bias conditions. At given bias conditions, this model describes the basic linear operation of an FET, and the model reproduces the small-signal RF terminal characteristics of the device with good accuracy. Analytical equations for model parameters may also be used. The model is widely used to extrapolate S-parameters to frequencies for which experimental data are not available and can be scaled to different sizes for the same device type. The main disadvantages of the EC model are difficulty in scaling to different physical structures, limitation to single bias condition, frequency independence of circuit elements, no time dependence feature, and the inherent limitation to linear circuits. However, by making some incremental changes in the EC model parameter values based on the device's analytical equations, the EC model can be extended, as a first-order approximation, to different bias conditions and sizes. In an FET/HEMT, the most critical EC model parameters are C_{gs}, g_m, and τ. The transit time τ is accurately modeled by fitting the measured phase of S_{21}.

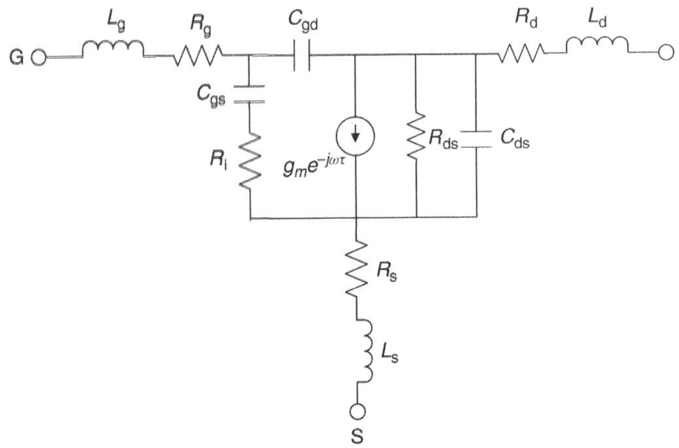

Figure 5.2 An equivalent circuit model of an FET/pHEMT/MOSFET.

The small-signal EC model parameter extraction normally consists of two parts: determination of extrinsic and intrinsic parameter values for transistors. The extrinsic parameters are parasitic elements and are extracted by using measured S-parameters under cold transistor test conditions (no drain/collector bias is applied in this case). These parameters are R_g, R_s, R_d, L_g, L_s, and L_d and are independent of bias conditions. The intrinsic parameter values are extracted from measured S-parameters under operating bias conditions. There are numerous extraction techniques developed for specific transistors and readers are referred to texts dealing with modeling techniques.

Figure 5.3 shows a comparison between the measured small-signal S-parameters and the EC model values for a 300-μm power FET biased at 2.5 V and 50% I_{dss}; R_{gs} is across C_{gs}, not shown in Figure 5.2.

Noise Models

In designing an LNA, one requires an accurate noise model [33–38] for the transistor. A comprehensive description of noise generation in MESFETs and its models are described by Pucel et al. [33] and simplified noise models are given by Fukui [34, 35] and Gupta et al. [36]. The models are accurate only for use in common-source amplifiers and do not provide enough information to design a multiport circuit. A commonly used measurement based noise model for transistors, supported by commercial CAD tools, uses the noise parameters described in the previous chapters. The noise model consists of noise parameters (F_{min}, Γ_{opt}, and R_n) and S-parameters. Determination of the noise model is described in Chapter 22. One of the primary limitations of this model is that it is only valid for the device tested and has no scalability. Since all the noise is lumped at the input of the device, it also does not predict accurately the effect of mismatch at the output in a multistage amplifier or the effect of parallel feedback on the amplifier's noise figure.

A flexible two-port noise model based on a nodal representation of the MESFET has been developed [37]. Four steps are required to obtain the noise model:

1. Measure S-parameters and noise figure across a broad frequency range in a 50-Ω system. The S-parameters and noise figure are typically measured using on-wafer RF probes from 1 to 26 GHz.

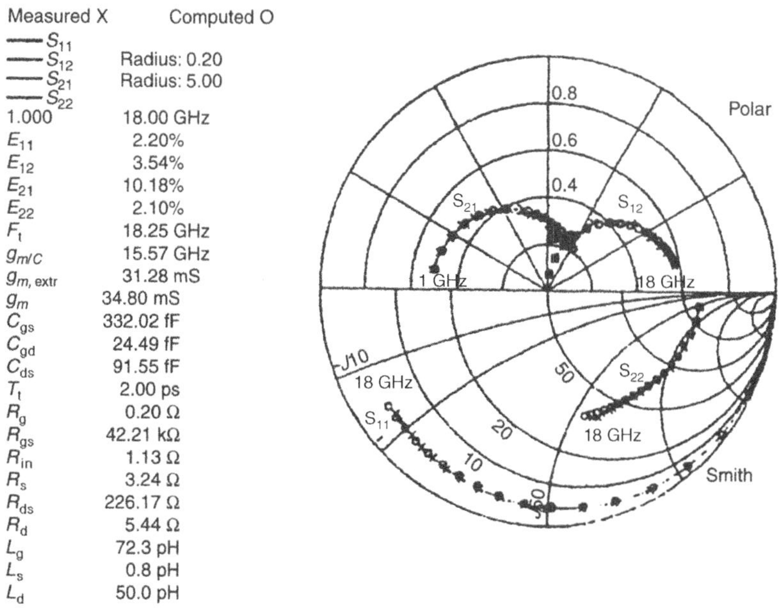

Figure 5.3 Small-signal EC model values for 300-μm power FET biased at 2.5 V and 50% I_{dss}; R_{gs} is across C_{gs}, not shown in Figure 5.2.

2. Fit the parameters of the equivalent circuit FET model shown in Figure 5.4a to the S-parameters. Any commercial or in-house software can be used to do this step.
3. Select the shot-current noise source value, I_{shot}, such that the calculated 50-Ω noise figure for the FET matches the measured data. The equivalent circuit noise model shown in Figure 5.4a is calculated assuming room temperature Johnson noise from each resistor, in conjunction with a constant (frequency-invariant) value of I_{shot} (normalized with respect to $T_{ambient}$ measured in kelvin units), and matches both measured S-parameters and noise parameters of the FET. Here, the excess noise is lumped into I_{shot}. The I_{shot} term can be interpreted as an apparent channel temperature rise affecting the output resistance, R_{ds}. The effective channel temperature is calculated from the I_{shot} term:

$$T_{channel} = T_{ambient}(1 + I_{shot}^2) \tag{5.1}$$

Referring all of the excess noise to the channel is a theoretical swindle. Just as in the case where all noise is referred to the input, the choice of where to assign the excess noise is irrelevant for circuits with no feedback.

4. The transistor two-port noise model may be validated by comparing with the noise figure calculated using noise parameters for the transistor terminated with a 50-Ω source and load. A typical variation of I_{shot} for a low-noise FET as a function of drain–source current is shown in Figure 5.4b. As expected, I_{shot} increases with increasing drain–source current.

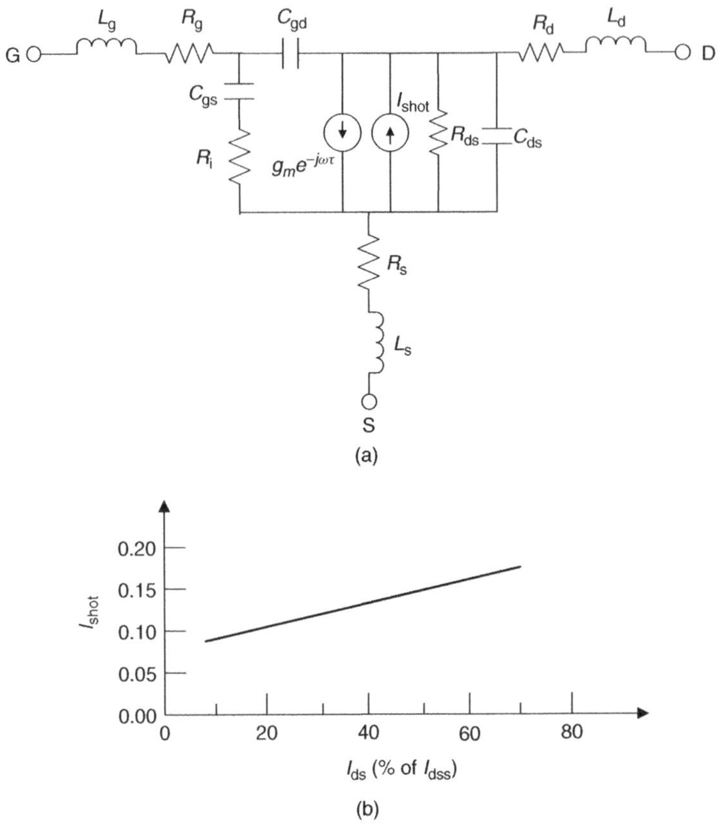

Figure 5.4 (a) An equivalent circuit noise model of an FET/pHEMT. (b) Typical variation of I_{shot} with bias current for a 300-μm low-noise FET.

Nonlinear Model

Commercially supported nonlinear models on CAD tools use time–frequency domain techniques to perform nonlinear simulations for steady state analysis. To perform nonlinear analysis, CAD tools use a powerful harmonic balance (HB) simulation engine or Volterra power series method. The nonlinear models are basically classified [31] into two groups: empirical black box models [23, 24] and EC based models [17–20, 25]. The black box models are based on extensive measured data in the form of look-up tables and are relatively more accurate, but can only be used for the transistors characterized. The EC models use predefined analytical expressions for the model elements and their values are extracted using measured data over large operating conditions. The EC models have greater flexibility in terms of bias and device size scaling but are not as accurate as black box models.

Many nonlinear EC models have been reported in the literature [17–22, 25–28]. All of these models have the same basic configuration as shown in Figure 5.5 plus the drain current generator. The EC model parameters and common-source DC or pulsed I–V curves that are in qualitative agreement with experimental data are obtained. Significant differences occur, however, in the quantitative behavior of the models in comparison with experimental data. The differences in the various models are the expressions used to characterize the drain current generator and gate–source and the

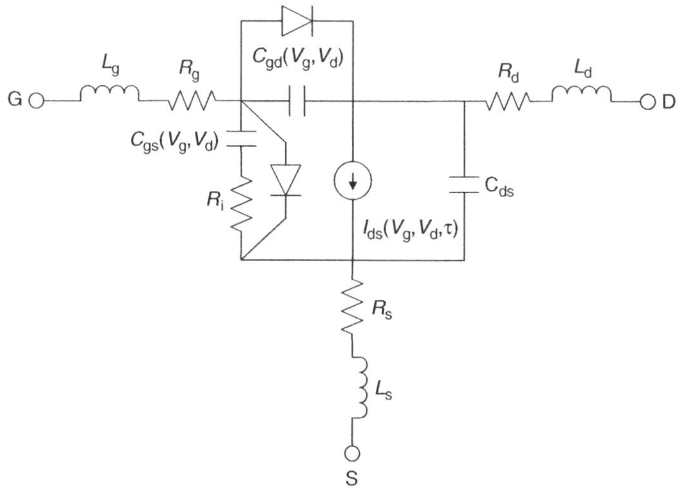

Figure 5.5 A nonlinear or voltage-dependent EC model of an FET/pHEMT.

gate–drain capacitances. Some of the most commonly referenced models are the Curtice, Curtice–Ettenberg, Stratz, Materka–Kacprzak, and TriQuint Own Model (TOM). A comparative study of these models is given in References 4 and 21.

The measurement based models need to include terminal voltage-dependent EC elements such as $I_{ds}(g_m, R_{ds})$, C_{gs}, and C_{gd} as shown in Figure 5.5 to predict accurately the nonlinear behavior of the active device. Commonly used representation of the nonlinear MESFET model is given by

$$I_{ds} = (A_0 + A_1 V_1 + A_2 V_1^2 + A_3 V_1^3) \tanh(\alpha V_{ds}) \tag{5.2}$$

where

$$V_1 = V_{gs}[1 + \beta(V_{dso} - V_{ds})] \tag{5.3}$$

and

$$C_{gs} = C_{gso} \bullet f(V_{gs}, V_{gd}) \tag{5.4}$$

$$C_{gd} = C_{gdo} \bullet g(V_{gs}, V_{gd}) \tag{5.5}$$

Here V_{gs}, V_{gd}, and V_{ds} are the terminal voltages between the gate–source, gate–drain, and drain–source of the device, respectively. The source of the FET is normally grounded. The A_i (where $i = 0, 1, 2, 3$) coefficients and constants α, β, and V_{dso} are evaluated using measured DC or pulsed I–V data. The quantities C_{gso} and C_{gdo} are extracted from the measured S-parameter at the normal operating DC bias conditions, while f and g, which are functions of both V_{gs} and V_{gd}, are determined from measured S-parameters over a large range of DC bias conditions to cover the full range of device operation. Inclusion of the tanh function to describe drain current is the generally accepted technique used to extend operation to the saturation region. Basically there are three steps in the development of nonlinear EC models.

1. Extract coefficients for I_{ds} to match with measured $I-V$ data. Important data appears near the knee of the curves and breaks down near pinch-off.
2. Measure S-parameters, extract small-signal model parameter values, and derive coefficients for gate–source and gate–drain capacitances to describe the model's dependence on gate and drain voltages.
3. Validate the model by comparing measured and simulated data with 50-Ω input and output for P_{1dB} compression point and power levels for other harmonics. The nonlinear models are also validated using load-pull data and some sort of hybrid amplifier measurements. Simulations are generally carried out using harmonic balance analysis.

The main advantage of the EC models is the ease with which they can be integrated into RF circuit simulators. For linear operation (i.e., small-signal), the interface is direct since the entire device and circuit model are simulated in the frequency domain. For nonlinear applications, the device models are formulated in the time domain and are interfaced with the frequency domain linear circuit simulators by means of the harmonic balance method [16, 22]. The RF performance obtained from these simulators can be satisfactory to good for a well-defined circuit, especially for mildly nonlinear applications such as a class-A power amplifiers not operating under hard saturation. The large-signal EC models generally do not scale well with varying operational conditions such as frequency or bias. As the circuit becomes increasingly nonlinear, simulator performance degrades.

The main disadvantage of the EC models is inherent inaccuracy resulting from simplifications in the model formulation, such as neglect of domain capacitance and the interdependencies of the nonlinear elements. In the actual device all of the nonlinear elements are interdependent. For example, in the MESFET it is not possible to change the device transconductance without also changing elements such as the gate–source capacitance. Perhaps the most significant limitation of EC models, however, is the need to experimentally characterize the devices that are to be used. The devices must be designed, fabricated, and characterized before the CAD models can be defined. A simple change in any design parameter (such as gate width or channel impurity concentration) requires a complete recharacterization since scaling techniques are difficult to apply. This limits the designer's flexibility in obtaining optimum performance of integrated circuits where tailoring the device design for special applications would be desirable. Also, it is challenging to develop an accurate, global nonlinear model to predict simultaneously the output power, PAE, and nonlinearities (AM-to-PM, ACPR, IP3, EVM, etc.). Generally, a nonlinear model is fine-tuned for a specific application. To extend the nonlinear model to higher frequencies, the nonlinear behavior of transit time must be accurately modeled.

Extensive work on linear and nonlinear models for transistors has been reported in the literature. This includes model extraction techniques, models, and model validation. In most cases, limited linear and nonlinear transistor models are provided by device vendors and GaAs/SiC MMIC foundry suppliers (TriQuint, WIN, Motorala, Tyco Electronics, GCS, Cree, Nitrox, IBM, etc.) or Si foundary suppliers. Check with the vendors and foundry suppliers for updated models. Next, linear and nonlinear models for FETs, HEMTs, HBTs, and MOSFETs are briefly described and all these models should be considered as examples.

5.2 MESFET MODELS

In this section, linear, noise, and nonlinear models for M/A COM's GaAs MSAG MESFETs [39] are described. Several different transistors and their bias conditions are also included. These are typical of GaAs MESFET technologies to first order.

5.2.1 Linear Models

The fundamental design of an amplifier is based on the linear EC models or small-signal S-parameters obtained over 0.5–40 GHz at the operating bias point. The EC model topology shown in Figure 5.2 is typical of most FET models in commercial simulators. For a low-noise design, measured small-signal S-parameters along with noise parameters are used. For MSAG 0.4-μm FETs, the Q-point V_{ds} and % of I_{dss} values are 3 V and 25%, 5 V and 50%, and 8–10 V and 25–30% for low-noise, linear, and high-PAE amplifier applications, respectively. The Q-point for a power FET is generally selected for a class-AB operation of the device in order to obtain the maximum power output, PAE, and linearity.

The S-parameters and noise parameters of a low-noise 300-μm FET are given in Table 5.1. Here, $\bar{R}_n$ is the normalized R_n with respect to 50 Ω. Typical values of I_{dss}, V_p, breakdown voltage, and f_T for a low-noise FET are 210 mA/mm, −1 V, 8 V, and 34 GHz, respectively.

Table 5.1 S-Parameters and Noise Parameters of a Low-Noise 300-μm FET Biased at 3 V and 25% I_{dss}[a]

(a) S-Parameters

Frequency (GHz)	$\|S_{11}\|$	$\angle S_{11}$	$\|S_{21}\|$	$\angle S_{21}$	$\|S_{12}\|$	$\angle S_{12}$	$\|S_{22}\|$	$\angle S_{22}$
2.0	0.973	−27.87	4.145	159.62	0.045	72.12	0.531	−18.51
4.0	0.930	−52.59	3.762	142.17	0.082	57.51	0.499	−35.45
6.0	0.887	−73.59	3.303	127.24	0.108	45.64	0.461	−49.36
8.0	0.853	−90.65	2.890	114.84	0.125	36.05	0.428	−60.46
10.0	0.820	−103.86	2.527	104.52	0.135	27.88	0.404	−70.04
12.0	0.801	−114.46	2.228	95.85	0.141	22.56	0.390	−77.43
14.0	0.794	−123.76	1.979	87.70	0.145	17.20	0.381	−84.44
16.0	0.776	−131.55	1.779	80.54	0.147	12.11	0.376	−90.71
18.0	0.777	−137.56	1.611	74.24	0.146	8.200	0.377	−94.80

(b) Noise Parameters

Frequency (GHz)	NF_{min} (dB)	Γ_{opt} Magnitude	Γ_{opt} Angle	$\bar{R}_n$
2.0	0.17	0.89	12.10	0.43
4.0	0.33	0.80	25.10	0.46
6.0	0.49	0.75	38.50	0.47
8.0	0.65	0.71	51.90	0.47
10.0	0.81	0.69	65.10	0.46
12.0	0.98	0.68	77.70	0.43
14.0	1.14	0.67	89.50	0.40
16.0	1.30	0.66	100.20	0.36
18.0	1.46	0.63	109.40	0.32

[a]Gate–gate pitch = 20 μm; die is 125 μm thick and has two source ground vias.

Table 5.2 *S*-Parameters and Noise Parameters of a Power 300-μm FET Biased at 10 V and 15% I_{dss}[a]

(a) *S*-Parameters

Frequency (GHz)	$\lvert S_{11}\rvert$	$\angle S_{11}$	$\lvert S_{21}\rvert$	$\angle S_{21}$	$\lvert S_{12}\rvert$	$\angle S_{12}$	$\lvert S_{22}\rvert$	$\angle S_{22}$
2.0	0.951	−25.97	2.379	158.87	0.020	74.72	0.800	−5.50
4.0	0.934	−49.63	2.221	139.89	0.037	62.96	0.783	−12.43
6.0	0.915	−71.06	2.007	123.26	0.049	53.52	0.750	−16.46
8.0	0.864	−88.42	1.776	108.54	0.058	46.97	0.757	−21.65
10.0	0.838	−102.02	1.570	95.47	0.063	39.76	0.745	−26.12
12.0	0.829	−113.24	1.140	84.27	0.067	36.84	0.736	−31.19
14.0	0.814	−122.75	1.272	75.03	0.071	35.13	0.735	−32.69
16.0	0.798	−129.95	1.130	64.93	0.072	30.53	0.755	−40.06
18.0	0.796	−136.39	1.046	55.65	0.074	27.03	0.735	−45.91

(b) Noise Parameters

Frequency (GHz)	NF_{min} (dB)	Γ_{opt} Magnitude	Γ_{opt} Angle	$\bar{R}_n$
2.0	0.96	0.84	14.10	1.27
4.0	1.14	0.74	28.49	1.14
6.0	1.33	0.68	42.79	1.04
8.0	1.51	0.65	56.49	0.96
10.0	1.70	0.65	69.50	0.88
12.0	1.88	0.66	81.40	0.80
14.0	2.06	0.67	91.90	0.71
16.0	2.25	0.66	100.70	0.63
18.0	2.43	0.64	107.40	0.57

[a]Gate–gate pitch = 20 μm; die is 75 μm thick and has two source ground vias.

The *S*-parameters and noise parameters of power 300-μm and 625-μm FETs are given in Tables 5.2 and 5.3, respectively. Typical values of I_{peak}, V_p, breakdown voltage, and f_T for a power FET are 460 mA/mm, −3 V, 20 V, and 21 GHz, respectively. The EC model parameters of power and linear FETs are given in Tables 5.4 and 5.5, respectively. Typical values of I_{dss}, V_p, breakdown voltage, and f_T for a linear FET are 420 mA/mm, −3 V, 10 V, and 20 GHz, respectively. Both tables include four FET sizes and the parameters were extracted at the nominal bias conditions.

Figure 5.6a shows a comparison of measured NF_{min} for a low-noise FET (5N), a power FET (5A), and a linear power FET (5G) as a function of bias current at 10 GHz. The NF_{min} value is constant over a larger bias current range for low-noise devices than power devices because of lower noise contributions of shot noise. Typical I_{dss} values for the 5N, 5A, and 5G FETs are 210, 380, and 420 mA/mm, respectively. Figure 5.6b shows the variation of the transistor's (0.625-mm power FET, Table 5.4) MSG and isolation. Below 13 GHz the device is potentially unstable (see Chapter 17 for stability analysis). Figure 5.6c shows the transistor's input impedance (Z_{in} or S_{11}) and output impedance (Z_D or S_{22}) as a function of frequency.

For small-signal transistors the drain/collector voltage is usually fixed. However, for power transistors the supply voltage may be lowered to perform stability analysis and to reduce the power consumption in the driver amplifier stages. Drain-voltage-dependent EC models for power FETs were obtained and are given in Table 5.6 for 300-μm and 625-μm gate peripheries. The gate–source capacitance C_{gs}, transit time τ, and output resistance R_{ds} increase with drain–source voltage V_{ds},

Table 5.3 S-Parameters and Noise Parameters of a Power 625-μm FET Biased at 10 V and 15% I_{dss}[a]

(a) S-Parameters

Frequency (GHz)	$\|S_{11}\|$	$\angle S_{11}$	$\|S_{21}\|$	$\angle S_{21}$	$\|S_{12}\|$	$\angle S_{12}$	$\|S_{22}\|$	$\angle S_{22}$
2.0	0.922	−51.16	3.898	144.87	0.034	61.75	0.606	−15.61
4.0	0.894	−88.25	3.123	118.38	0.054	43.42	0.561	−28.33
6.0	0.883	−113.03	2.445	99.09	0.062	31.76	0.525	−37.12
8.0	0.854	−129.44	1.972	83.78	0.067	25.99	0.533	−45.41
10.0	0.843	−140.36	1.621	70.96	0.066	20.18	0.534	−52.74
12.0	0.848	−148.73	1.391	59.59	0.067	18.53	0.547	−61.60
14.0	0.846	−155.51	1.211	49.76	0.068	16.99	0.553	−67.84
16.0	0.842	−160.25	1.050	40.24	0.066	14.99	0.600	−75.53
18.0	0.846	−164.34	0.936	30.70	0.064	13.44	0.614	−85.09

(b) Noise Parameters

Frequency (GHz)	NF_{min} (dB)	Γ_{opt} Magnitude	Γ_{opt} Angle	$\bar{R}_n$
2.0	0.86	0.73	30.00	0.58
4.0	1.05	0.60	56.10	0.54
6.0	1.24	0.55	78.20	0.49
8.0	1.43	0.56	96.70	0.45
10.0	1.62	0.61	111.80	0.40
12.0	1.82	0.67	123.90	0.35
14.0	2.01	0.72	133.40	0.29
16.0	2.20	0.74	140.60	0.25
18.0	2.39	0.71	145.80	0.22

[a]Gate–gate pitch = 20 μm; die is 75 μm thick and has two source ground vias.

Table 5.4 EC Model Parameter Values of Power FETs Biased at 9 V and 25% I_{dss} ($V_{gs} = -2$ V)[a]

Model Elements	Units	FET Size (mm) 0.3	0.625	1.0	1.8
C_{gs}	pF	0.36	0.76	1.22	2.2
C_{gd}	pF	0.02	0.04	0.063	0.09
C_{ds}	pF	0.06	0.12	0.2	0.4
g_m	mS	37	76	115	215
τ	ps	6.6	6.6	6.6	6.6
R_g	Ω	1.0	0.5	0.4	0.4
R_i	Ω	1.2	0.6	0.35	0.2
R_s	Ω	2.0	1.0	0.8	0.5
R_d	Ω	2.0	1.0	0.7	0.35
R_{ds}	Ω	355	170	110	61
L_g	nH	0.005	0.005	0.005	0.01
L_s	nH	0.01	0.02	0.02	0.02
L_d	nH	0.005	0.005	0.005	0.01

[a]Gate–gate pitch = 20 μm; die is 75 μm thick and has two source ground vias.

Table 5.5 EC Model Parameter Values of Linear FETs Biased at 5V and 50% I_{dss} ($V_{gs} = -1.2$V)[a]

Model Elements	Units	FET Size (mm)			
		0.15	0.3	0.45	0.6
C_{gs}	pF	0.17	0.35	0.5	0.64
C_{gd}	pF	0.013	0.026	0.039	0.055
C_{ds}	pF	0.04	0.07	0.1	0.135
g_m	mS	27	54	78	99
τ	ps	2	2	2	2
R_g	Ω	1.0	0.5	0.7	0.3
R_i	Ω	4	2	1.3	1.0
R_s	Ω	2.4	1.2	0.8	0.6
R_d	Ω	2.4	1.2	0.8	0.6
R_{ds}	Ω	450	230	150	115
L_g	nH	0.01	0.005	0.007	0.005
L_s	nH	0.005	0.01	0.01	0.02
L_d	nH	0.005	0.005	0.005	0.005

[a]Gate–gate pitch = 20 μm; die is 75 μm thick and has two source ground vias.

whereas the gate–drain capacitance C_{gd} and transconductance g_m decrease with V_{ds}. Also, f_T decreases with V_{ds}.

5.2.2 Nonlinear Models

The nonlinear FET model used to simulate the amplifiers working up to Ku-band [40] is based on a modified Materka and Kacprzak model [17] optimized to predict accurately the output power and PAE. The nonlinear model has I–V equations along with improved capacitance equations compiled into a commercial CAD tool. The model parameters were extracted using extensive S-parameter, pulsed I–V, and load-pull data. The nominal bias Q-point is 10 V and 25% I_{dss}. The nonlinear model was verified for the standard 0.625-mm FET using extensive hybrid measurements at 14.5 GHz. The model equations are given below:

$$C_{gd1} = \text{scale} \times c_{d1} \times [1 + c_{d2} \times \tanh\{c_{d3} \times (V_{gs} - V_{dg0})\}] \tag{5.6}$$

$$C_{gd2} = \text{scale} \times c_{dd1} \times [1 - c_{dd2} \times \tanh\{c_{dd3} \times (V_{gs} - V_{dd0})\}] \tag{5.7}$$

$$C_{gs} = \text{scale} \times c_{g1} \times [1 + c_{g2} \times \tanh\{c_{g3} \times (V_{gs} - V_{gg0})\}] \tag{5.8}$$

$$I_{ds} = \text{scale} \times I_{pk} \times \tanh[a_0 \times (V_1 - V_{th})^{a_1 + a_2 V_1}] \times \tanh(b2 \times V_{ds}) \tag{5.9}$$

$$V_1 = V_{gs}[1 + bb \times (V_{ds0} - V_{ds})] \tag{5.10}$$

$$i_{gd} = \text{scale} \times i_{sd} \times e^{-(V_{gd}+V_{br})/tvs} \tag{5.11a}$$

$$i_{gs} = \text{scale} \times i_{sg} \times e^{-(V_{gs}+V_{bi})/tvd} \tag{5.11b}$$

where $C_{gd} = C_{gd1} + C_{gd2}$ and

$$R_g = r_g/\text{scale} \tag{5.12a}$$

$$R_{in} = r_{in}/\text{scale} \tag{5.12b}$$

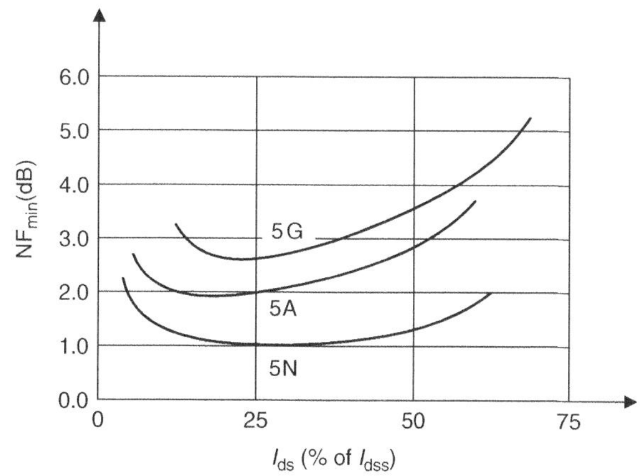

(a)

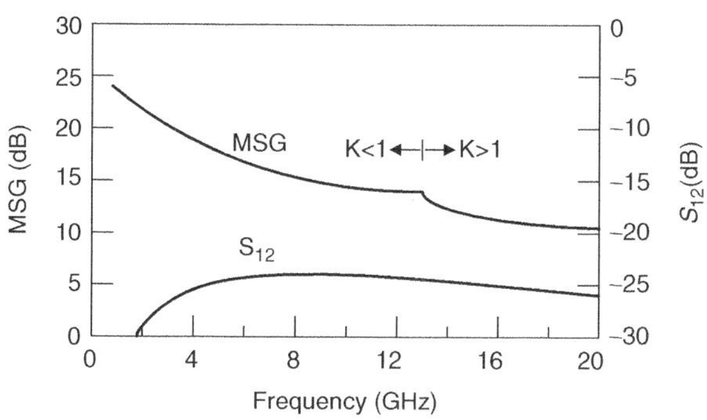

(b)

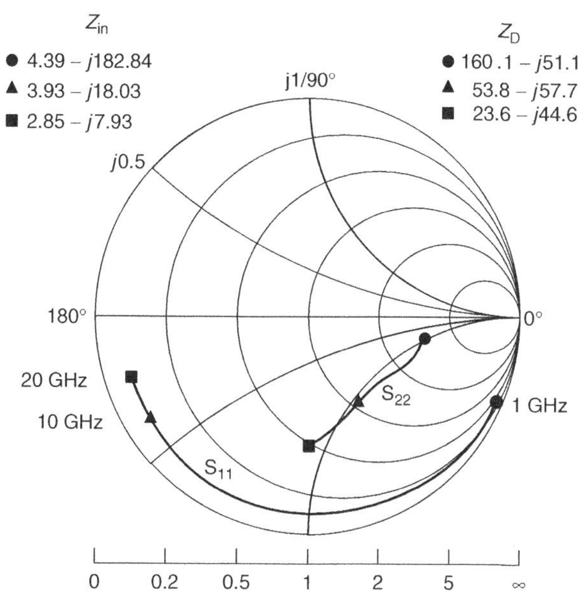

(c)

Figure 5.6 (a) Optimum noise figure as a function of drain current at $V_{ds} = 3$ V and 10 GHz. (b) Variation of stable gain and isolation of a 0.625-mm power FET as a function of frequency at $V_{ds} = 9$ V. (c) Variation of input impedance (Z_{in} or S_{11}) and output impedance (Z_D or S_{22}) of a 0.625-mm power FET as a function of frequency at $V_{ds} = 9$ V.

Table 5.6 EC Model Parameter Values of Power FETs Biased at Fixed Gate Voltage ($V_{gs} = -2$ V) and variable drain voltage

(a) FET Size 300 μm

Model Elements	Units	Drain–source bias voltage (V) 3	5	8	10
C_{gs}	pF	0.37	0.39	0.40	0.42
C_{gd}	pF	0.037	0.024	0.017	0.015
C_{ds}	pF	0.09	0.085	0.08	0.08
g_m	mS	52	45	39	37
τ	ps	2.0	3.0	4.0	5.0
R_g	Ω	1.0	1.0	1.0	1.0
R_i	Ω	1.2	1.2	1.2	1.2
R_s	Ω	2.5	2.5	2.5	2.5
R_d	Ω	2.0	2.0	2.0	2.0
R_{ds}	Ω	180	270	370	390
L_g	nH	0.005	0.005	0.005	0.005
L_s	nH	0.01	0.01	0.01	0.01
L_d	nH	0.005	0.005	0.005	0.005

(b) FET Size 625 μm

Model Elements	Units	Drain–source bias voltage (V) 3	5	8	10
C_{gs}	pF	0.75	0.78	0.80	0.85
C_{gd}	pF	0.076	0.047	0.035	0.032
C_{ds}	pF	0.16	0.155	0.15	0.15
g_m	mS	103	90	78	73
τ	ps	2.0	3.0	4.0	5.0
R_g	Ω	0.5	0.5	0.5	0.5
R_i	Ω	0.6	0.6	0.6	0.6
R_s	Ω	1.2	1.2	1.2	1.2
R_d	Ω	1.0	1.0	1.0	1.0
R_{ds}	Ω	90	132	170	195
L_g	nH	0.005	0.005	0.005	0.005
L_s	nH	0.015	0.015	0.015	0.015
L_d	nH	0.005	0.005	0.005	0.005

[a]Gate–gate pitch = 20 μm; die is 75 μm thick and has two source ground vias.

$$R_s = r_s/\text{scale} \tag{5.12c}$$

$$R_d = r_d/\text{scale} \tag{5.12d}$$

$$C_{ds} = c_{ds} \times \text{scale} \tag{5.12e}$$

$$L_g = l_g/\text{scale} \tag{5.12f}$$

$$L_s = l_s/\text{scale} \tag{5.12g}$$

$$L_d = l_d/\text{scale} \tag{5.12h}$$

In case of scaling, the device parasitic resistor and inductors such as R_g, L_g, L_s, and L_d also depend on the unit gate width and the number of fingers. Table 5.7 provides the nonlinear model parameter values for the 0.625-mm gate periphery power FET.

Table 5.8 summarizes the electrical performance of MSAG FETs at 12 GHz calculated using the nonlinear model. The device gain and PAE are reduced significantly

Table 5.7 Nonlinear Model Parameters of a 625-μm MSAG MESFET[a]

$l_g = 0.0173350$	$c_{ds} = 0.1225e^{-12}$
$l_s = 0.0097140$	$c_{g1} = 0.8e \times 10^{-12}$
$l_d = 0.0000000$	$c_{g2} = 0.32$
$r_s = 0.7781$	$c_{g3} = 0.4$
$r_g = 0.100$	$c_{d1} = 0.0332e \times 10^{-12}$
$r_d = 0.0100$	$c_{d2} = 0.4$
$r_{in} = 2.3821$	$c_{d3} = 3.0$
$i_{pk} = 0.22$	$v_{dg0} = -1.8$
$a_0 = 0.1428$	$v_{dg0} = -1.8$
$a_1 = 1.6231$	$tvd = 0.05$
$a_2 = -0.0486$	$tvs = 0.05$
$b_2 = 1.4543$	$v_{gg0} = -1.4$
$v_{dso} = 10.0$	$v_{po} = -3.4854$
$bb = 0.0385$	$c_{dd1} = 0.0298e^{-12}$
$v_{bi} = 0.0$	$c_{dd2} = 0.4$
$v_{br} = 23.0$	$c_{dd3} = 1.5$
$i_{sg} = 1.0e\text{-}14$	$V_{dd0} = -1.8$
$i_{sd} = 1.0e\text{-}14$	$\tau = 5.4395$
	scale = 1

[a]Units are: Ω for resistors, nH for inductors, F for capacitors, ps for time, and A for current.

Table 5.8 Simulated Data for Several MSAG FETs Using Nonlinear Models at 12 GHz Biased at 10 V

FET Size (mm)	Gate–Gate Pitch (μm)	Number of Fingers	G_{max} (dB)	P_o (dBm)	G_A (dB)	PAE (%)
0.625	30	6	13.8	27.1	9.5	64
0.94	30	10	13.4	28.5	9.3	60
1.5	30	14	12.5	30.6	8.4	58
1.8	24	18	11.5	31.5	8.0	57
2.5	20	24	10.4	32.7	7.5	55

when the device size is increased from 0.625 to 2.5 μm mainly because of increased unit width and higher source inductance value.

5.3 pHEMT MODELS

A pHEMT is the most popular designer's choice for low-noise, power, and millimeter-wave applications; extensive data for linear and nonlinear models have been reported in the literature. In this section, linear, noise, and nonlinear models for these transistors are included as examples.

5.3.1 Linear Models

Linear models for pHEMT devices are available in the literature as well as from device vendors. The *S*-parameters and noise parameters of a 200-μm gate periphery low-noise 0.25-μm gate pHEMT (Fujitsu #FHX13X), biased at 2 V and 10 mA, are given in Table 5.9. Table 5.10 provides typical EC model parameters for two power 0.25-μm gate length and 600-μm gate periphery pHEMTs from two different sources. Linear models for 0.5-μm gate power pHEMT devices suitable for linear operation have been developed. Such devices are very suitable when one needs high PAE under power back-off conditions. Their nominal Q-point is about 10% of I_{dss}. Table 5.11 provides typical EC model parameters for power 300- and 900-μm gate periphery pHEMTs. Linear models for 0.15-μm gate pHEMT devices have also been developed [41] using a unit cell of 160 μm. Both passivated and unpassivated devices were tested. The devices have peak current of 500 mA/mm and a gate–drain breakdown voltage of 9 V. The nominal pinch-off and operating voltages are −0.8 V and 5 V, respectively. The small-signal EC model is the same as shown in Figure 5.2 and the EC parameters are given in Table 5.12 for both passivated and unpassivated devices.

5.3.2 Nonlinear Models

Nonlinear models for 0.15-μm gate pHEMT devices have been developed [41]. The nonlinear model used is Curtice asymmetric and shown in Figure 5.7. The parameter values for passivated and unpassivated devices are summarized in Table 5.13. Both these models were validated using measured data at 60 GHz and good agreement was found between the measured and simulated performance.

Table 5.9 *S*-Parameters and Noise Parameters of a pHEMT

			(a) *S*-Parameters					
Frequency (GHz)	$\|S_{11}\|$	$\angle S_{11}$	$\|S_{21}\|$	$\angle S_{21}$	$\|S_{12}\|$	$\angle S_{12}$	$\|S_{22}\|$	$\angle S_{22}$
0.5	0.999	−4.7	4.894	175.9	0.006	87.7	0.601	−2.3
1.0	0.995	−9.4	4.876	171.9	0.013	85.5	0.599	−4.6
2.0	0.981	−18.6	4.806	163.9	0.025	81.1	0.591	−9.2
3.0	0.958	−27.7	4.696	156.1	0.037	77.0	0.580	−13.5
4.0	0.929	−36.4	4.555	148.6	0.048	73.2	0.565	−17.7
5.0	0.895	−44.9	4.392	141.5	0.057	69.8	0.548	−21.5
6.0	0.860	−53.0	4.215	134.8	0.066	66.8	0.530	−25.0
7.0	0.823	−60.7	4.034	128.4	0.074	64.2	0.512	−28.3
8.0	0.786	−68.1	3.852	122.4	0.080	62.0	0.493	−31.3
9.0	0.751	−75.3	3.675	116.8	0.086	60.2	0.475	−34.0
10.0	0.718	−82.1	3.506	111.5	0.092	58.9	0.458	−36.6
11.0	0.687	−88.7	3.345	106.5	0.096	57.8	0.442	−39.0
12.0	0.659	−95.0	3.194	101.8	0.101	57.1	0.426	−41.3
13.0	0.633	−101.2	3.054	97.3	0.105	56.6	0.412	−43.6
14.0	0.610	−107.2	2.923	93.0	0.108	56.4	0.399	−45.8
15.0	0.590	−113.0	2.801	88.9	0.112	56.4	0.386	−47.9
16.0	0.572	−118.7	2.688	85.0	0.116	56.6	0.375	−50.1
17.0	0.556	−124.2	2.584	81.3	0.120	56.9	0.364	−52.3
18.0	0.543	−129.6	2.487	77.7	0.124	57.3	0.353	−54.6
19.0	0.532	−134.9	2.397	74.2	0.129	57.8	0.344	−56.9
20.0	0.523	−140.0	2.314	70.8	0.133	58.4	0.335	−59.4
21.0	0.516	−145.0	2.236	67.5	0.138	58.9	0.326	−62.0
22.0	0.511	−149.8	2.164	64.4	0.144	59.5	0.318	−64.7
23.0	0.507	−154.6	2.096	61.3	0.150	60.0	0.310	−67.5
24.0	0.505	−159.2	2.033	58.3	0.156	60.5	0.303	−70.5
25.0	0.504	−163.6	1.974	55.3	0.163	60.9	0.296	−73.7
26.0	0.505	−167.9	1.918	52.4	0.170	61.2	0.290	−77.1

Table 5.9 (*Continued*)

(b) Noise Parameters

Frequency (GHz)	F_{min} (dB)	Γ_{opt} Magnitude	Γ_{opt} Angle	$\bar{R}_n$
1.0	0.27	0.96	6	0.69
2.0	0.28	0.92	13	0.65
3.0	0.29	0.88	19	0.61
4.0	0.30	0.84	25	0.54
5.0	0.31	0.80	32	0.47
6.0	0.32	0.77	38	0.41
7.0	0.33	0.74	45	0.36
8.0	0.34	0.71	51	0.31
9.0	0.36	0.68	58	0.27
10.0	0.39	0.66	65	0.23
11.0	0.42	0.64	72	0.20
12.0	0.45	0.61	79	0.17
13.0	0.51	0.60	86	0.14
14.0	0.56	0.58	93	0.12
15.0	0.62	0.57	101	0.11
16.0	0.68	0.56	108	0.09
17.0	0.77	0.55	115	0.08
18.0	0.86	0.54	122	0.07
19.0	0.94	0.53	129	0.07
20.0	1.03	0.52	136	0.07
21.0	1.12	0.51	143	0.07
22.0	1.22	0.50	150	0.07
23.0	1.32	0.48	156	0.07
24.0	1.43	0.46	162	0.07
25.0	1.54	0.44	168	0.08
26.0	1.66	0.41	174	0.08

Table 5.10 EC Model Parameters for two 0.25-μm Gate and 600-μm Gate Periphery Power pHEMTs Biased at 8 V and 25% I_{dss}

Parameter	Units	#1	#2
C_{gs}	pF	1.0	1.07
C_{gd}	pF	0.047	0.07
C_{ds}	pF	0.12	0.14
g_m	mS	178	168
τ	ps	2.0	4.0
R_g	Ω	0.8	0.8
R_i	Ω	1.5	3.2
R_s	Ω	0.5	0.8
R_d	Ω	0.5	0.6
R_{ds}	Ω	200	181
L_g	nH	0.01	0.04
L_s	nH	0.005	0.008
L_d	nH	0.01	0.04

Table 5.11 EC Model Parameters for two 0.5-μm Gate pHEMT Biased at 12 V and 10% I_{dss} for Linear Operation

Parameter	Units	0.3 mm	0.9 mm
C_{gs}	pF	0.5	1.5
C_{gd}	pF	0.032	0.096
C_{ds}	pF	0.07	0.21
g_m	mS	42	126
τ	ps	4.9	4.9
R_g	Ω	0.5	0.5
R_i	Ω	6.0	2.0
R_s	Ω	6.0	2.0
R_d	Ω	0.2	0.1
R_{ds}	Ω	900	300
L_g	nH	0.012	0.012
L_s	nH	0.005	0.01
L_d	nH	0.019	0.019

Table 5.12 EC Model Parameters for a 0.15-μm Gate pHEMT with Cell Size = 160 μm

Parameter	Units	VPA1-1 Device Model Unpassivated	VPA1-2 Device Model Passivated
C_{gs}	pF	0.165996	0.142282
C_{gd}	pF	0.01222	0.017306
C_{ds}	pF	0.046298	0.046449
g_m	S	0.082643	0.074975
τ	ps	1.326755	1.199083
R_{ds}	Ω	255.8411	235.2786
R_i	Ω	2.761873	2.855210
R_g	Ω	1.016306	0.465033
R_d	Ω	0.732158	0.510000
R_s	Ω	0.500142	0.380000
L_g	nH	0.030750	0.023044
L_d	nH	0.030553	0.031925
L_s	nH	0.000399	0.002532

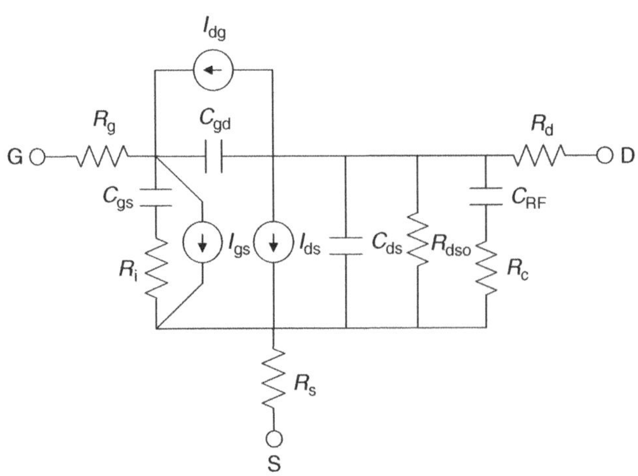

Figure 5.7 Nonlinear model used for a pHEMT.

Table 5.13 Nonlinear Model Parameters for 0.15-μm Gate pHEMT[a]

(a) VPA1-1 GaAs pHEMT Model 2, Unpassivated				
A0 = 0.096068	A1 = 0.058977	A2 = − 0.041950	A3 = −0.024154	
BETA = 0.023429	GAMMA = 2.1651	C_{gs} = 0.151998	C_{gd} = 0.014300	C_{ds} = 0.045489
R_{dso} = 345.95	RIN = 2.761873	R1=140	R2= −74	RF = 48.3
TAU = 1.326755 ps	VBI = 0.76	VBR = 11	VDSO = 5.0	VDSDC = 5.0
R_c = −1050	C_{RF} = 10000	VTO = −1.4		
(b) VPA1-1 GaAs pHEMT Model 2, Passivated				
A0 = 0.087875	A1 = 0.079742	A2 = −0.036464	A3 = −0.36464	
BETA = 0.019506	GAMMA = 2.1651	C_{gs} = 0.142282	C_{gd} = 0.017306	C_{ds} = 0.046449
R_{dso} = 219.92	RIN = 2.85521	R1 = 140	R2 = −74	RF = 48.3
TAU = 1.199083 ps	VBI = 0.76	VBR = 11	VDSO = 5.0	VDSDC = 5.0
R_c = −1050	C_{RF} = 10000	VTO = −1.4		

[a]Cell size = 160 μm. Units are: Ω for resistors, pF for capacitors, and V for voltages.

5.4 HBT MODEL

A small-signal EC model of an HBT is reported by Bayraktaroglu et al. [42] and is shown in Figure 5.8. The EC model parameters for a 60-μm emitter periphery npn device are given in Table 5.14. The emitter unit length and pitch were 15 μm and 4 μm, respectively. The typical bias conditions for this model are $V_{CE} = 4$ V and $I_C = 20$ mA.

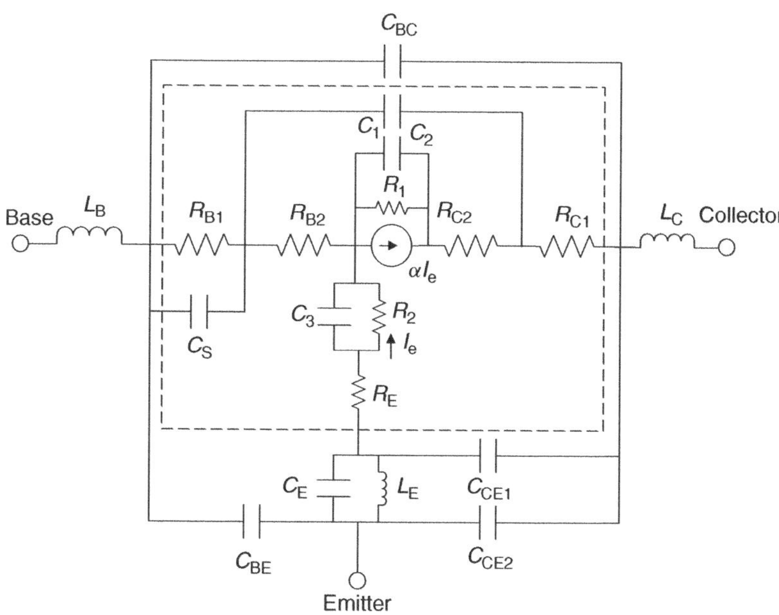

Figure 5.8 Linear model used for an HBT.

Table 5.14 EC Model Parameter Values for a 60-μm Emitter Periphery npn HBT

Parameter	Value	Parameter	Value
f_t	22 GHz	C_s	1.34 pF
f_{max}	40 GHz	R_{C1}	1 Ω
α_0	0.93	R_{C2}	4 Ω
τ	2 ps	R_E	8.5 Ω
f_b	65 GHz	C_{BC}	0.012 pF
C_1	0.06 pF	C_{BE}	0.022 pF
C_2	0.01 pF	C_{CE1}	0.012 pF
C_3	0.4 pF	C_{CE2}	0.06 pF
R_1	1.0×10^6	C_E	0.022 pF
R_2	10 Ω	L_B	0.165 nH
R_{B1}	17 Ω	L_E	0.032 nH
R_{B2}	27.5 Ω	L_C	0.06 nH

Table 5.15 EC Model Parameters for MOSFETs

Parameter	Units	MOSFET1	MOSFET2
C_{gs}	pF	0.055	0.15
C_{gd}	pF	0.015	0.057
C_{ds}	pF	—	0.045
g_m	mS	9.3	29.5
τ	ps	—	—
R_g	Ω	8.4	4.1
R_i	Ω	29.6	1.0
R_s	Ω	3.1	1.1
R_d	Ω	3.1	10.6
R_{ds}	Ω	1000	305
L_g	nH	—	0.0001
L_s	nH	—	0.0174
L_d	nH	—	0.0001

5.5 MOSFET MODELS

The small-signal EC model for the MOSFET is the same as for the MESFET and HEMT, as shown in Figure 5.2. The model parameter values for two devices are summarized in Table 5.15. The MOSFET1 [43] and MOSFET2 [44] have gate lengths of 0.35 μm and 0.25 μm, and gate fingers 1 and 4, respectively. Both devices have a unit gate width of 50 μm. An extensive treatment of MOSFET modeling is given in Reference 45.

Figure 5.9 shows a simplified EC model for a MOSFET having a 0.4-μm gate and 20-μm unit gate width. The effect of device electrode and bond wire inductors is ignored. The model parameter values were extracted for three gate peripheries: 200, 800, and 2400 μm. The devices were biased at $V_{ds} = 2.4$ V and $V_{gs} = 1.2$ V. The large sizes were realized by using more fingers in parallel. Table 5.16 lists the EC model parameter values and shows parameter scaling with gate periphery [46]. The capacitors and g_m are proportional and resistors are inversely proportional to gate periphery.

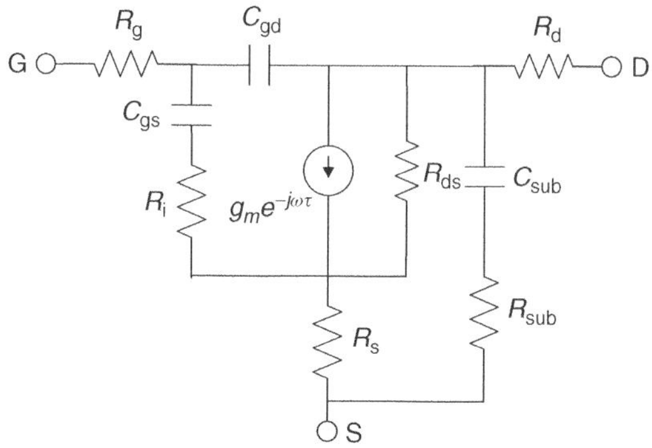

Figure 5.9 A simplified equivalent circuit model of a MOSFET.

Table 5.16 Scalable EC Model Parameters for MOSFETs

Parameter	Units	200 μm	800 μm	2400 μm
C_{gs}	pF	0.246	0.993	2.953
C_{gd}	pF	0.055	0.227	0.703
C_{sub}	pF	0.058	0.226	0.716
g_m	mS	42	170	513
R_g	Ω	12.7	3.3	1.3
R_i	Ω	9.9	2.7	0.8
R_s	Ω	7.8	1.9	0.6
R_d	Ω	6.5	1.8	0.7
R_{ds}	Ω	856	204	63
R_{sub}	Ω	158	82	35

5.6 BJT MODELS

There are several models proposed for Si-based transistors (Si BJTs and CMOS, SiGe HBTs) including Ebers–Moll (EM), Gummel–Poon (GP), vertical bipolar intercompany (VBIC), most exquisite transistor model (MEXTRAM), and high-current model (HICUM) [46, 47]. Among these, MEXTRAM is more advanced and HICUM is developed for both BJTs and HBTs [48]. Many HBT-specific models as well as EM and GP models have been used in the GaAs HBT based circuits. Also, numerous models for Si MOSFETs have been developed. Among these, the Berkeley short-channel insulated gate FET (IGFET) model (BSIM) and the MOS Model developed by Philips Company and their different versions are in use. The most popular model is MOS Model 11, where 11 designates the model version. The methodology used in the model development for Si and GaAs bipolar transistors is that the model is valid for both small-signal and large-signal simulations. Several commercial foundry suppliers provide Si BJT and CMOS, and SiGe HBT models for their processes for custom designs.

The small-signal EC model shown in Figure 4.4 (Chapter 4) for a Agilent/Avantek AT-60500 BJT was determined. The transistor was biased at $V_{CE} = 8$ V and a collector current of 2 mA. The typical EC parameter values are $C_{bp} = 0.055$ pF, $C_1 = 0.01$ pF, $C_2 = 0.039$ pF, $C_{ep} = 0.026$ pF, $C_{be} = 0.75$ pF, $R_{bc} = 4.2\ \Omega$, $R_{ec} = 0.66\ \Omega$, $R_1 = 7.5\ \Omega$, $R_2 = 10.3\ \Omega$, $R_e = 12.9\ \Omega$, $R_c = 5.0\ \Omega$, $\alpha_0 = 0.99$, $\tau_{ec} = 6.9$ ps, and $f_b = 22.7$ GHz.

More device models and their parameter values can be found in Reference [8].

5.7 TRANSISTOR MODEL SCALING

The device model scaling concept has been used for both low-noise and power transistors. The device gate periphery can be increased or decreased by adding or subtracting unit cells or by increasing or decreasing unit gate width. This can be done safely up to 50% scale factor. Large scale factors add resistive losses and large phase imbalance, resulting in reduced gain and power. Adding and subtracting unit cells approach with proper input and output feeds is relatively more accurate.

For example, the EC model parameters for an FET/HEMT can be normalized to 1 mm and scaled to other gate peripheries. If a scale factor is greater than 2, special attention is required to scale extrinsic parameters such as R_g, L_g, L_s, and L_d. These parameters depend on the unit gate width and the number of fingers. When parameters are normalized to 1 mm and the number of fingers remains same, the scaled parameters for other device sizes become capacitors, g_m, inductors, and $R_g \propto$ periphery, and resistors $\propto 1/\text{periphery}$. When the number of fingers is also scaled, the scaled parameters for capacitors and $g_m \propto$ periphery, and inductors, R_g, and resistors $\propto 1/\text{periphery}$. The gate resistance also depends on the number of fingers:

$$R_g^S = R_g\left(\frac{W^S}{W}\right)\left(\frac{n}{n^S}\right)^2 \tag{5.13}$$

where R_g, W, and n are the gate resistance, total periphery, and number of fingers of a modeled FET, and R_g^S, W^S, and n^S are the respective values for the scaled FET.

Consider an example of scaling a 900-μm gate periphery FET, whose EC model is given in Table 5.11, to a 600-μm gate periphery FET. The number of fingers = 8 and only the unit gate width is reduced. As a first-order approximation, the transit time can be considered as independent of gate periphery. The scaled EC model parameter values for the 600-μm gate periphery FET are given in Table 5.17.

An approximate scalable large-signal simplified EC model of a GaAs power FET, biased at 10 V and for class-AB/B operation, is shown in Figure 5.10. Various circuit elements can be approximated for an X-band FET as follows:

$$R_{in} = R_g + R_i + R_S \cong \frac{1.5}{W} \quad (\Omega \cdot \text{mm}) \tag{5.14a}$$

$$C_{gs} \cong 1.2\,W \quad (\text{pF/mm}) \tag{5.14b}$$

$$g_m \cong 55\,W \quad (\text{mS/mm}) \tag{5.14c}$$

$$R_o \cong \frac{110}{W} \quad (\Omega \cdot \text{mm}) \tag{5.14d}$$

Table 5.17 Scaling of EC Model Parameter Values from 900-μm to 600-μm Gate FET Periphery

$C_{gs} = \left(\frac{600}{900}\right) \times 1.5$ pF/mm = 1.0 pF	$R_s = \left(\frac{900}{600}\right) \times 2.0\ \Omega \cdot \text{mm} = 3.0\ \Omega$
$C_{gd} = \left(\frac{600}{900}\right) \times 0.096$ pF/mm = 0.064 pF	$R_d = \left(\frac{900}{600}\right) \times 0.1\ \Omega \cdot \text{mm} = 0.15\ \Omega$
$C_{ds} = \left(\frac{600}{900}\right) \times 0.21$ pF/mm = 0.14 pF	$R_{ds} = \left(\frac{900}{600}\right) \times 300\ \Omega \cdot \text{mm} = 450\ \Omega$
$g_m = \left(\frac{600}{900}\right) \times 126$ mS/mm = 84 mS	$L_g = \left(\frac{600}{900}\right) \times 0.012$ nH/mm = 0.008 nH
$R_g = \left(\frac{600}{900}\right) \times 0.5\ \Omega \cdot \text{mm} = 0.33\ \Omega$	$L_s = \left(\frac{600}{900}\right) \times 0.01$ nH/mm = 0.007 nH
$R_i = \left(\frac{900}{600}\right) \times 2.0\ \Omega \cdot \text{mm} = 3.0\ \Omega$	$L_d = \left(\frac{600}{900}\right) \times 0.019$ nH/mm = 0.013 nH

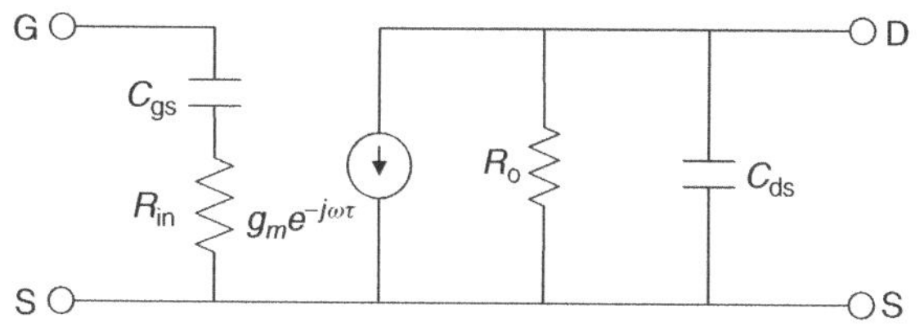

Figure 5.10 A simplified large-signal EC model for a FET.

$$C_{ds} \cong 0.2\ W \quad (\text{pF/mm}) \tag{5.14e}$$

and

$$P_{1dB} \cong 0.7\ W \quad (\text{W/mm}) \tag{5.14f}$$
$$P_{sat} \cong 0.8\ W \quad (\text{W/mm}) \tag{5.14g}$$

where W is the total gate width of the power FET in millimeters. For example, a device delivering about 2 watts of power out at 10 GHz has an input impedance about $0.5 - j4\ \Omega$ and output impedance about $12.59 - j17.4\ \Omega$.

5.8 SOURCE-PULL AND LOAD-PULL DATA

In the design of high-efficiency and linear power amplifiers, one requires an accurate nonlinear device model or accurately measured source-pull and load-pull data [49–55]. The source-pull and load-pull data consist of optimum source and load impedance values for the device under test (DUT) for the specific application as well as numerous other impedance-performance points around the optimum values. This data will enable the design of matching networks for power amplifiers that achieve optimum power and PAE, and good IM3/IM5 and ACPR performance simultaneously by having the right source and load condition for the fundamental as well as for the harmonic frequencies. The load conditions can be obtained theoretically or measured for the transistors being used in the design. These measurements can be made either on-wafer using RF probes or packaged devices as described in Chapter 22. However, there is no simple theoretical way to obtain source-pull data.

5.8.1 Theoretical Load-Pull Data

A simple calculation of load impedance, a ratio of operating voltage and device current, for power amplifiers has been discussed in Chapter 8. Figure 5.11 shows a representation of device load for output power. Here R_{opt} is the optimum load based on the maximum current and voltage swings allowed for a device. If Z_D is the impedance looking into the device, then the load impedance is given by

$$Z_L = Z_D^* \tag{5.15}$$

where the asterisk denotes a complex conjugate value. The device impedance can be written

$$Z_D = \frac{R'_{opt} X_C}{R'_{opt} + X_C} + R_d + j\omega L_d \tag{5.16a}$$

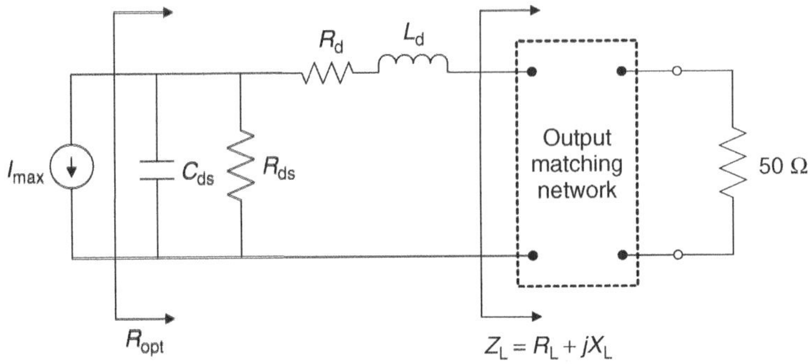

Figure 5.11 Optimum load representation of a device.

where

$$R'_{opt} = \frac{R_{opt}R_{ds}}{R_{opt} + R_{ds}} \quad \text{and} \quad X_C = -j/\omega C_{ds} \tag{5.16b}$$

Due to the Miller effect, a more accurate value of X_C can be used in (5.16a) by replacing C_{ds} in (5.16b) by C'_{ds} given by

$$C'_{ds} = C_{ds} + C_{gd}(1 + 1/G) \tag{5.17}$$

where C_{gd} is the gate–drain capacitance and G is the device gain.

EXAMPLE 5.1

Consider a 1-mm gate periphery power FET whose EC model parameters at $V_{ds} = 9$ V are given in Table 5.4. When the maximum current I_{max} for this device is 0.27 A, determine its load impedance at 2 GHz. The power gain of this device at 2 GHz is 13 dB.

SOLUTION Here, $C_{gd} = 0.063$ pF, $C_{ds} = 0.2$ pF, $R_{ds} = 110\ \Omega$, $R_d = 0.7\ \Omega$, and $L_d = 0.005$ nH.

$$\begin{aligned}
\text{Gain} &= 13\text{dB} = 20 \\
C'_{ds} &= 0.2 + 0.063(1 + 1/20) = 0.266 \text{ pF} \\
X'_C &= \frac{-j1000}{2\pi \times 2 \times 0.266} = -j299.16\Omega \\
\omega L_d &= 2\pi \times 2 \times 0.005 = 0.063\ \Omega
\end{aligned}$$

Using (5.15) and (5.16a), we have

$$\begin{aligned}
Z_L &= Z_D^* = 41.425 + j5.59\ \Omega \\
Y_L &= 0.0237 - j0.0032 \text{ S}
\end{aligned}$$

When a load is represented by a parallel combination of R_L and C_L (see Fig. 5.17),

$$\begin{aligned}
R_L &= 1/0.0237 = 42.2\ \Omega \\
C_L &= -0.0032/\omega = -0.255 \text{ pF}
\end{aligned}$$

■

This method of obtaining load impedance was extended for assessing load-pull data. The load-pull method works well at low frequencies where the load impedance has negligible reactive component.

Next, we describe how to obtain load-pull contours. Consider the real part of an optimum load as R_{LO}, and the high and low values of load for $-X$ dB lower power are R_{LH} and R_{LL}, respectively. By converting the X dB into a power ratio ($= 10^{X/10}$), the R_{LH} and R_{LL} can be expressed as

$$R_{LH} = R_{LO} 10^{X/10} \tag{5.18a}$$

$$R_{LL} = R_{LO}/10^{X/10} \tag{5.18b}$$

On a Smith chart, follow the constant resistance contour from R_{LL} and the constant conductance contour from R_{LH} and locate the intersection. The X-dB contour is defined by the periphery of the intersecting lines. The aforementioned method does not include the effect of the reactive portion of the load. In this analysis it has been assumed that the load capacitance C_L is absorbed into the output matching network. A more accurate procedure is described by Cripps [56].

EXAMPLE 5.2

Plot Z_L obtained in the previous example on the Smith chart. Also plot −1-dB and −2-dB load contours.

SOLUTION Here,

$$Z_L = 41.425 + j5.59\ \Omega \quad \text{and} \quad \overline{Z}_L = Z_L/50 = 0.8285 + j0.1118$$

$$Y_L = 0.0237 - j0.0032\ \ \text{S} \quad \text{and} \quad \overline{Y}_L = Y_L/0.02 = 1.185 - j0.16$$

Since the real part of Z_L is much greater than the imaginary part, as a first-order approximation, the imaginary part may be ignored. In this case,

$$\begin{aligned} \overline{R}_{LH} &= 0.8285 \times 10^{0.1} &&= 1.043 \ \ (-1\ \text{dB}) \\ \overline{R}_{LL} &= 0.8285 \times 10^{-0.1} &&= 0.658 \ \ (-1\ \text{dB}) \\ \overline{R}_{LH} &= 0.8285 \times 10^{0.2} &&= 1.313 \ \ (-2\ \text{dB}) \\ \overline{R}_{LL} &= 0.8285 \times 10^{-0.2} &&= 0.5227\ (-2\ \text{dB}) \end{aligned}$$

Figure 5.12 shows Z_L and −1-dB and −2-dB power contours. ■

Many systems require improved performance of IM3/IM5 intermodulation distortion and ACPR performance in addition to output power and PAE. Several source-pull and load-pull techniques can now characterize devices for ACPR and IM3/IM5 along with power and PAE. Thus one can determine optimum source and load conditions for active devices based on a specific application: output power, PAE, IP3, or ACPR.

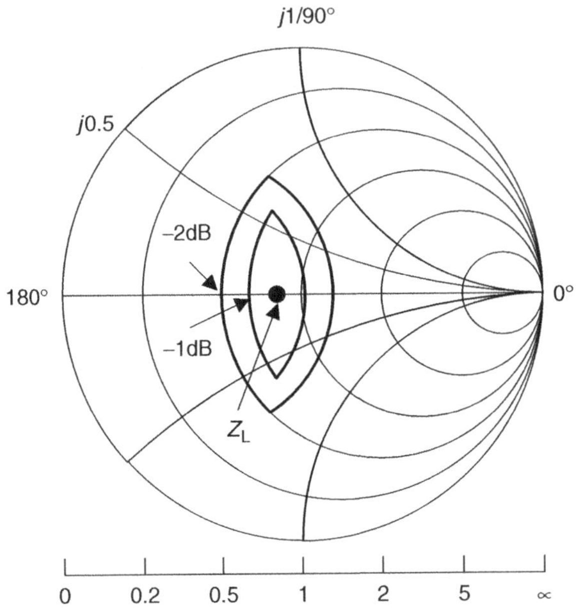

Figure 5.12 Plots of −1-dB and −2-dB power contours using the Cripps method.

5.8.2 Measured Power and PAE Source Pull and Load Pull

The source-pull and load-pull measurements start with accurate test setup calibration to establish exact reference impedances and power levels presented to the DUT at the operating frequency. Different sizes of high PAE FETs were source-pulled and load-pulled for maximum power and PAE at 3.8, 8, and 12.5 GHz. An average normalized value for source and load impedances at 10 V were obtained for optimum power and PAE as given below:

$$R_S = 6\ \Omega \cdot \text{mm} \quad \text{and} \quad C_S = -1.4\ \text{pF/mm}\ \text{(series)}$$
$$R_L = 59\ \Omega \cdot \text{mm} \quad \text{and} \quad C_L = -0.3\ \text{pF/mm}\ \text{(shunt)}$$

Figure 5.13 shows load-pull contours (1 dB/step for power and 5%/step for PAE) for a 3.488-mm FET at 3.8 GHz and $P_{in} = 22$ dBm. The P_{out} and PAE versus input power plot is shown in Figure 5.14 and their values at optimum PAE point were 34 dBm and 62%, respectively. Generally, the value of output R_L is higher for an optimum PAE match than for an optimum power match.

Source-pull and load-pull data were measured for a 1.7 × 30μm^2 HBT for maximum power and PAE at 20 GHz. Figure 5.15 shows power and PAE contours for source impedance (0.1 dB/step for power and 0.2%/step for PAE) and load impedance (0.5 dB/step for power and 1%/step for PAE). At $P_{in} = 10$ dBm, $V_{CE} = 5$ V, and mA, the measured values for output power and PAE were 21.8 dBm and 63.2%, respectively. The optimum source and load impedance values are [57]

$$R_S = 3.391\ \Omega \quad \text{and} \quad C_S = -0.034\ \text{pF}\ \text{(series)}$$
$$R_L = 152.6\ \Omega \quad \text{and} \quad C_L = -0.1225\ \text{pF}\ \text{(shunt)}$$

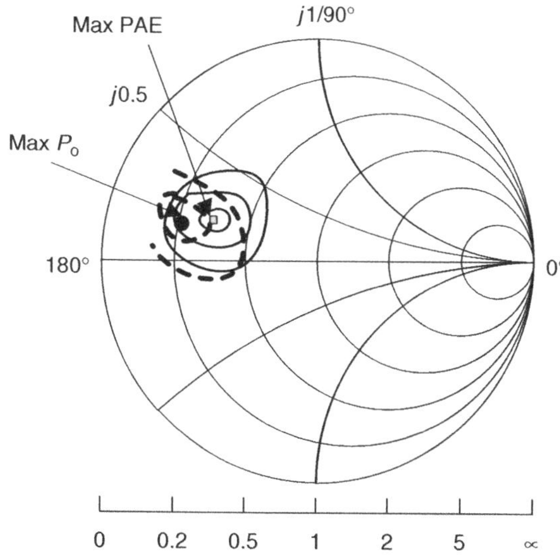

Figure 5.13 Load-pull data for a 3.488-mm FET biased at 10 V and 20% I_{dss}. $f = 3.8$ GHz; 0.5 dB/step for power and 5%/step for PAE.

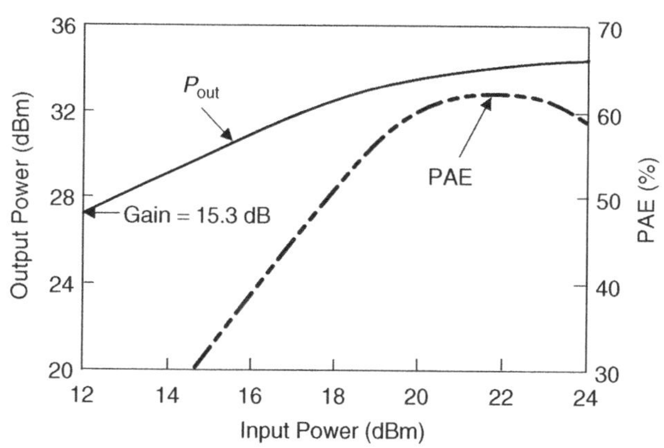

Figure 5.14 Output power and PAE versus input power at optimum PAE tune point at 10 V, 20% I_{dss}, and $f = 3.8$ GHz.

Load Impedance Examples

Examples of measured load impedance values for various devices are listed below:

Device Type	Device Size (mm)	Bias V, mA	Frequency (GHz)	R_L (Ω)	C_L (pF)
MESFET	3.488	10, 200	3.8	21	−1.0
MESFET	0.625	10, 45	10	85	−0.19
MESFET	0.625	8, 50	20	72	−0.20
pHEMT	0.6	5,10	30	28.3	−0.20
HBT	0.03	5, 8.3	20	152.6	−0.123

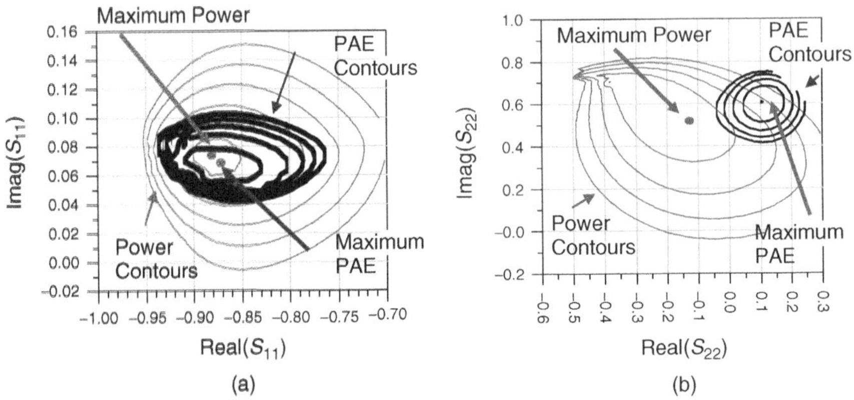

Figure 5.15 HBT measured data: (a) source pull and load pull. (Courtesy of A. K. Sharma, Northrop Grumman.)

5.8.3 Measured IP3 Source and Load Impedance

For linear MESFETs whose EC model parameters are given in Table 5.5, source-pull and load-pull data for optimum IP3 were measured. Several device sizes and frequencies were used. The bias Q-point was $V_{ds} = 5$ V and $I_{ds} = 200$ mA/mm. Figure 5.16 shows the source-pull and load-pull data for a 0.6-mm FET at 3.8 GHz. The source impedance was represented by a series combination of resistor (R_S) and capacitor (C_S) and the load impedance was equivalent to a parallel combination of resistor (R_L) and capacitor (C_L), as shown in Figure 5.17b.

For the linear MESFET, normalized (with respect to 1-mm device size) optimum IP3 source impedance, load impedance, and IP3 values are

$$R_S = 12\Omega \cdot \text{mm} \quad \text{and} \quad C_S = -1.5\text{pF/mm (series)} \tag{5.19}$$

$$R_L = 23\Omega \cdot \text{mm} \quad \text{and} \quad C_L = -0.33\text{pF/mm (shunt)} \tag{5.20}$$

$$IP3 = 41\text{dBm/mm} = 12.59\text{W/mm} \tag{5.21}$$

The IP3 model is reasonably accurate up to about IP3 of 43 dBm and 15 GHz.

5.8.4 Source and Load Impedance Scaling

The measured source impedance may be scaled up and down with device periphery, and load impedance may be scaled up and down with both periphery and drain bias voltage. Next, the scaling of source and load impedance is described.

Source Impedance Scaling

Consider the example of a high IP3 device whose source impedance is represented by a series combination of a resistor (R_S) and a capacitor (C_S) of values given by (5.19). If this device has to be scaled to a larger size of periphery W_g (in mm), the source impedance for the scaled version may be calculated as follows:

$$R_S = 12/W_g\,\Omega \quad \text{and} \quad C_S = -1.55 \times W_g\,\text{pF}$$

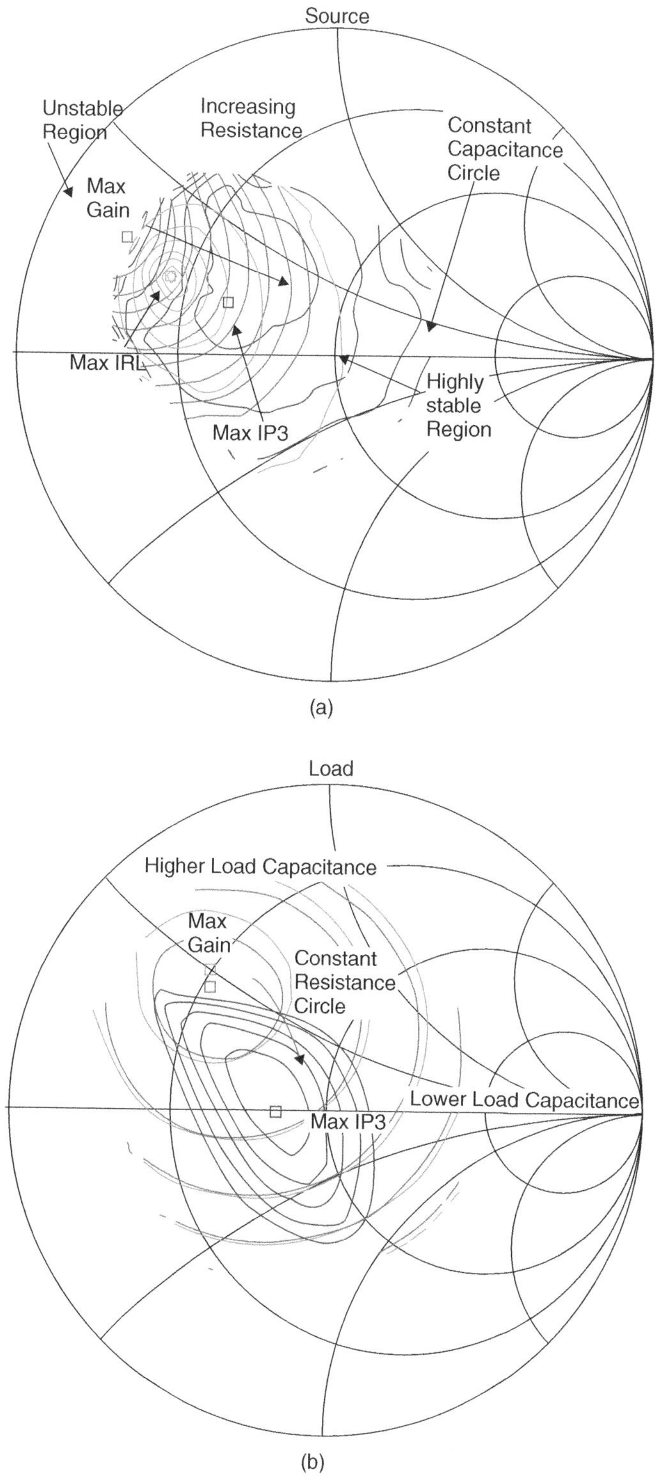

Figure 5.16 (a) Source-pull data and (b) load-pull data for a 600-mm FET biased at 5 V and 50% I_{dss}. $f = 3.8$ GHz.

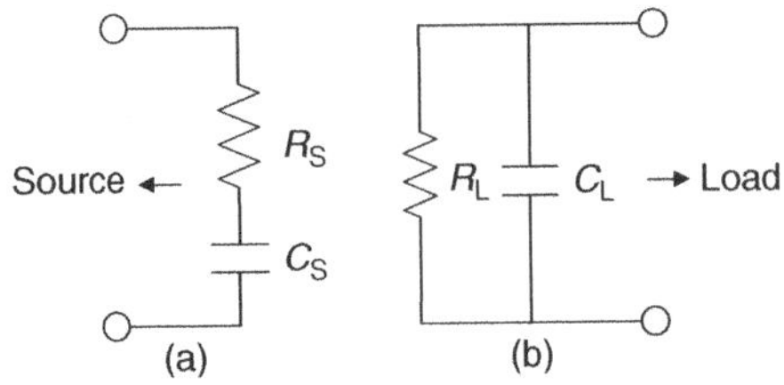

Figure 5.17 Optimum IP3 impedance representations of a device: (a) source and (b) load.

Load Impedance Scaling

As an example, if the measured load impedance for a 625-μm FET biased at 10 V is represented as equivalent to a parallel combination of a resistor (R_L) and a capacitor (C_L) of values 90 Ω and—0.19 pF, respectively, the load for other FET peripheries may be calculated using the following scaling relationships:

$$R_L = \frac{90 \times 0.625 \times V_{ds}}{W_g \times 10} \quad \Omega \tag{5.22}$$

$$C_L = -\frac{0.19 \times W_g}{0.625} \quad \text{pF} \tag{5.23}$$

where W_g is the total FET periphery (in mm) and V_{ds} is the operating drain voltage (in volts). As a first-order approximation, the reactive part of the load is assumed to be independent of the drain voltage. The negative sign in (5.20) represents inductive reactance. For example, for a 1-mm FET operating at 8 V, the values of R_L and C_L are

$$R_L = 45\ \Omega \quad \text{and} \quad C_L = -0.304\text{pF}$$

This scaling method for obtaining the load impedance works reasonably well over 6–10 V operation, gate peripheries less than 3 mm, and up to 20 GHz when the FET's output feed is designed or scaled accordingly.

Next, consider a high IP3 device as described in the previous section. Suppose we have to design a single-stage linear amplifier using a 5-V supply with an output IP3 = 37 dBm (5.0 W). In this case using (5.21), the transistor periphery is given by

$$W_g = 5.0/12.59 \text{ mm} = 0.4 \text{ mm}$$

The load is calculated using (5.20) as

$$R_L = 23/0.4\ \Omega = 57.5\ \Omega \quad \text{and} \quad C_L = -0.33 \times 0.4 \text{ pF} = -0.132 \text{ pF}$$

5.9 TEMPERATURE-DEPENDENT MODELS

The temperature dependence in amplifiers can be calculated by using measured S-parameters at different temperatures or temperature-dependent models. Many device models supported by foundries also have built-in temperature dependence, especially for bipolar transistors. The primary parameter in the transistor EC model which dependends on temperature is g_m. The amplifier gain varies linearly with temperature;

therefore as a first-order approximation g_m can be considered a linear function of temperature, that is,

$$g_{mT} = g_{m25}[(25 - T)a + 1] \tag{5.24}$$

where g_{mT} and g_{m25} are the g_m values at temperatures T and 25 °C, respectively, and a is the temperature coefficient. For a low-noise device, described in Section 5.2.1, the value of a is 0.0011. The effect of temperature is calculated by using g_m as a variable in the EC model. As discussed in Chapters 3 and 16, the temperature also affects the transistor's built-in voltage and correspondingly the drain current.

In this chapter, transistor models including ECs and source-pull and load-pull data pertaining to amplifier design have been described. Scaling of model parameters from one size to another size has also been discussed. The data presented in this chapter will be used in design examples and problems throughout the book.

REFERENCES

1. H. Fukui (Ed.), *Low-Noise Transistors and Amplifiers*, IEEE Press, New York, 1981.
2. P. H. Ladbrooke, *MMIC Design: GaAs FETs and HEMTs*, Artech House, Norwood, MA, 1989.
3. J. M. Golio, *Microwave MESFETs and HEMTs*, Artech House, Norwood, MA, 1991.
4. J. L. B. Walker (Ed.), *High-Power GaAs FET Amplifiers*, Artech House, Norwood, MA, 1993, Chapter 3.
5. R. Anholt, *Electrical and Thermal Characterization of MESFETs, HEMTs and HBTs*, Artech House, Norwood, MA, 1995.
6. P. J. C. Rodrigues, *Computer-Aided Analysis of Nonlinear Microwave Circuits*, Artech House, Norwood, MA, 1998.
7. A. Grebennikov, *RF and Microwave Power Amplifier Design*, McGraw-Hill, New York, 2004.
8. G. D. Vendelin et al., *Microwave Circuit Design Using Linear and Nonlinear Techniques*, John Wiley & Sons, Hoboken, NJ, 2005.
9. Advanced Design System, *ADS Users' Manual*, Agilent EEsof EDA, Palo Alto, CA.
10. *Microwave Office Users' Manual*, Applied Wave Research Inc., El Segundo, CA.
11. *Ansoft Designer Users' Manual*, Ansoft Corp., Pittsburgh, PA.
12. *APLAC 7.91 Users' Manual*, APLAC Solutions Corp., Espoo, Finland.
13. *LINMIC 6.2 USERS' MANUAL*, AC Microwave GmbH, Aachen, Germany.
14. *MIMICAD Users' Manual*, Optotek Ltd., Ottawa, Ontario, Canada.
15. *WaveMaker*, Barnard Microsystems Ltd., London, United Kingdom.
16. V. Rizzoli, A. Lipparini, and E. Marazzi, A General-purpose program for nonlinear microwave circuit design, *IEEE Trans. Microwave Theory Tech.*, Vol. 31, pp. 762–770, September 1983.
17. A. Materka and T. Kacprzak, Computer calculation of large-signal GaAs FET amplifier characteristics, *IEEE Trans. Microwave Theory Tech.*, Vol. MTT-33, pp. 129–135, February 1985.
18. W. R. Curtice and M. Ettenberg, A nonlinear GaAs FET model for use in the design of output circuits for power amplifiers, *IEEE Trans. Microwave Theory Tech.*, Vol. MTT-33, pp. 1383–1394, December 1985.
19. D. Estreich, Nonlinear modeling for MMICs, *IEEE Microwave Millimeter-Wave Monolithic Circuits Symp. Dig.*, pp. 93–96, 1987.
20. W. R. Curtice, GaAs MESFET modeling and nonlinear CAD, *IEEE Trans. Microwave Theory Tech.*, Vol. 36, pp. 220–230, February 1988.

21. R. J. Trew, MESFET models for microwave CAD applications, *Int. J. Microwave Millimeter-Wave Computer-Aided Eng.*, Vol. 1, pp. 143–158, April 1991.
22. M. B. Steer, C.-R. Chang, and G. W. Rhyne, Computer-aided analysis of nonlinear microwave circuits using frequency-domain nonlinear analysis techniques: the state of the art, *Int. J. Microwave Millimeter-Wave Computer-Aided Eng.*, Vol. 1, pp. 181–200, April 1991.
23. D. E. Root, S. Fan, and J. Meyer, Technology independent non quasi-static FET models by direct construction from automatically characterized device data, in 21st *European Microwave Conference Proceedings*, pp. 927–932, 1991.
24. F. Filicori, G. Vannini, and V. A. Monaco, A nonlinear integral model of electron devices for HB circuit analysis, *IEEE Trans. Microwave Theory Tech.*, Vol. 40, pp. 1456–1465, July 1992.
25. I. Angelov, H. Zirath, and N. Rosman, A new empirical nonlinear model for HEMT and MESFET devices, *IEEE Trans. Microwave Theory Tech.*, Vol. 40, pp. 2258–2266, December 1992.
26. G. Crupi et al., On the small signal modeling of advanced microwave FETs: a comparative study, *Int. J. RF and Microwave Computer-Aided Eng.*, Vol. 18, pp. 417–425, September 2008.
27. Special Issue on Computer-Aided Design of Nonlinear Microwave Circuits, *Int. J. Microwave Millimeter-Wave Computer-Aided Eng.*, Vol. 6, 1996.
28. Special Issue on Optimization-Oriented Microwave Computer-Aided Design, *Int. J. Microwave Millimeter-Wave Computer-Aided Eng.*, Vol. 7, 1997.
29. F. Bonani et al., Physics-based large-signal sensitivity analysis of microwave circuits using technological parametric sensitivity from multidimensional semiconductor device model, *IEEE Trans. Microwave Theory Tech.*, Vol. 45, pp. 846–854, 1997.
30. I. J. Bahl, Monolithic microwave integrated circuits, in *Encyclopedia of RF and Microwave Engineering*, K. Chang (Ed.), Vol. 4, John Wiley & Sons, Hoboken, NJ, 2005, pp. 3213–3238.
31. P. Colantonio, F. Giannini, and E. Limiti, Microwave power amplifiers, in *Encyclopedia RF and Microwave Engineering*, K. Chang (Ed.), Vol. 4, John Wiley & Sons, Hoboken, NJ, 2005, pp. 2871–2905.
32. I. Bahl, G. Lewes, and J. Jorgenson, Automatic testing of MMIC wafers, *Int. J. Microwave Millimeter-Wave Computer-Aided Eng.*, Vol. 1, pp. 77–89, January 1991.
33. R. A. Pucel et al., Signal and noise properties of gallium arsenide field-effect transistor, in *Advances in Electronic and Electron Physics*, Vol. 38, Academic Press, New York, 1975, pp. 145–265.
34. H. Fukui, Design of microwave GaAs MESFET's for broad-band low-noise amplifiers, *IEEE Trans. Microwave Theory Tech.*, Vol. MTT-27, pp. 643–650, July 1979. Also see addendum, Vol. MTT-29, p. 1119, October 1981.
35. H. Fukui, The noise performance of microwave transistors, *IEEE Trans. Electron Devices*, Vol. ED-33, pp. 329–341, 1986.
36. M. S. Gupta et al., Microwave noise characterization of GaAs MESFETs: evaluation by on-wafer low-frequency output noise current, *IEEE Trans. Microwave Theory Tech.*, Vol. 35, pp. 1208–1218, December 1987.
37. D. Willems, I. Bahl, and E. Griffin, Accurate design of multi-port low noise MMICs up to 20GHz, *SPIE*, Vol. 1475, pp. 55–61, 1991.
38. J. Gao et al., A new method for pHEMT noise-parameter determination based on 50-Ω noise measurement system, *IEEE Trans. Microwave Theory Tech.*, Vol. 51, pp. 2079–2089, October 2003.
39. A. E. Geissberger, I. J. Bahl, E. L. Griffin, and R. A. Sadler, A new refractory self-aligned gate technology for GaAs microwave power FETs and MMICs, *IEEE Trans. Electron Devices*, Vol. 35, pp. 615–622, May 1988.
40. I. Bahl, Ku-band MMIC power amplifiers developed using MSAG MESFET technology, *Microwave J.*, Vol. 49, pp. 56–82, February 2006.

41. A. K. Sharma et al., A V-band high-efficiency pseudomorphic HEMT monolithic power amplifier, *IEEE Trans. Microwave Theory Tech.*, Vol. 42, pp. 2603–2609, December 1994.
42. B. Bayraktaroglu, N. Camilleri, and S. A. Lambert, Microwave performances of npn and pnp AlGaAs/GaAs heterojunction bipolar transistors, *IEEE Trans. Microwave Theory Tech.*, Vol. 36, pp. 1869–1873, December 1988.
43. A. E. Schmitz et al., A deep-submicrometer microwave/digital CMOS/SOS technology, *IEEE Electron Device Lett.*, Vol. 12, pp. 16–17, January 1991.
44. M. H. Hanes et al., MICROX^{TM}—an all-silicon technology for microwave integrated circuits, *IEEE Electron Device Lett.*, Vol. 14, pp. 219–221, 1993.
45. R. S. Gupta, M. Gupta, and M. Saxena, MOSFET Modeling, in *Encyclopedia RF and Microwave Engineering*, K. Chang (Ed.), Vol. 4, John Wiley & Sons, Hoboken, NJ, 2005, pp. 3278–3317.
46. A. Raghavan, N. Srirattana, and J. Laskar, *Modeling and Design Techniques for RF Power Amplifiers*, John Wiley & Sons, Hoboken, NJ, 2008.
47. P. H. Aaen, J. A. Pla, and J. Wood, *Modeling and Characterization of RF and Microwave Power FETs*, Cambridge University Press, Cambridge, UK, 2007.
48. J. McMacken et al., HBT modeling, *IEEE Microwave Mag.*, Vol. 9, pp. 48–72, April 2008.
49. R. S. Tucker and P. D. Bradley, Computer-aided error correction of large-signal load-pull measurements, *IEEE Trans. Microwave Theory Tech.*, Vol. MTT-32, pp. 296–300, March 1984.
50. B. Hughes, A. Ferrero, and A. Cognata, Accurate on-wafer power and harmonic measurements of MM-wave amplifiers and devices, *IEEE MTT-S Int. Microwave Symp. Dig.*, pp. 1019–1022, 1992.
51. M. Demmler, B. Hughes, and A. Cognata, A 0.5–50GHz on-wafer, intermodulation, load-pull and power measurement system, *IEEE MTT-S Int. Microwave Symp. Dig.*, pp. 1041–1044, 1995.
52. F. M. Ghannouchi, G. Zhao, and F. Beauregard, Simultaneous load-pull of intermodulation and output power under two-tone excitation of accurate SSPA's design, *IEEE Trans. Microwave Theory Tech.*, Vol. 42, pp. 929–934, June 1994.
53. P. Berni et al., An experimental study of the effects of harmonic loading on microwave MESFET oscillators and amplifiers, *IEEE Trans. Microwave Theory Tech.*, Vol. 42, pp. 963–950, June 1994.
54. M. Nebus et al., Improvement of the active load-pull technique for the optimization of high power communication SSPAs, *Int. J. Microwave Millimeter-Wave Computer-Aided Eng.*, Vol. 5, pp. 169–160, July 1995.
55. K. Chang, I. Bahl, and V. Nair, *RF and Microwave Circuit and Component Design for Wireless Systems*, John Wiley & Sons, Hoboken, NJ, 2002.
56. S. C. Cripps, *RF Power Amplifiers for Wireless Communications*, Artech House, Norwood, MA, 1999.
57. M. V. Aust, A. K. Sharma, A. L. Gutierrez-Aitken, M. Demmler, B. Hughes, and A. Cognata, A 0.5–50GHz on-wafer, intermodulation, load-pull and power measurement system, *IEEE MTT-S Int. Microwave Symp. Dig.*, pp. 1041–1044, 1995.

PROBLEMS

5.1 Describe the pros and cons of an EC model technique.

5.2 Extract EC model (Fig. 5.2) parameter values for a 300-μm low-noise transistor whose S-parameters are given in Table 5.1a.

5.3 The EC model parameters of a 600-μm gate periphery device are given in Table 5.5. Determine the EC parameters for a 1-mm gate device. Also compare gain (in dB) and K factor for these two devices at 5, 10, 15, and 20 GHz. For K-factor calculations see Chapter 17. Assume the same number of fingers in both FETs.

5.4 As an example, a load for a 600-μm gate periphery FET/HEMT, biased at 10 V, is represented as a parallel combination of a resistor (R_L) and a capacitor (C_L) of values 85 Ω and −0.20 pF, respectively. Determine the load impedance (Z_L) and load admittance (Y_L) at 10 GHz. Also calculate Z_L and Y_L at 5 GHz and a drain bias of 8 V.

5.5 The EC model parameters of a 600-μm gate periphery device are given in Table 5.5 and its IP3 model is described in Section 5.8.3. Determine the device size for output IP3 = 43 dBm and its EC parameters. Also calculate Z_S and Z_L at 5 GHz.

5.6 Discuss the pros and cons of device model scaling in the design of amplifiers.

5.7 Discuss the pros and cons of the EC model versus S-parameters in the design of amplifiers.

5.8 When the EC model shown in Figure 5.2 is connected between a 50-Ω source and a load impedance and the effects of resistors, capacitors, and inductors are ignored, show that

$$|S_{21}| = 100 g_m$$

Calculate S_{21} (in dB) using 100 g_m and for the EC model parameters for the different FET sizes given in Table 5.5 at 2 GHz.

Chapter 6

Matching Network Components

In addition to transistors, the other important building blocks for an amplifier are impedance matching components. Their primary function is to transform 50 Ω to the required impedance at the input and output interface of the transistor over the desired frequency range. This chapter deals with such components and their models.

6.1 IMPEDANCE MATCHING ELEMENTS

An impedance matching network is generally comprised of distributed elements, shown in Figure 6.1a, or lumped elements, shown in Figure 6.1b, or their combinations [1–6]. Sections of microstrip lines or coplanar waveguide (CPW) constitute the basic distributed circuit element building blocks of microwave integrated circuits. Most of other passive circuits such as hybrids, couplers, baluns, and power combiners used in amplifiers are commonly based on the microstrip transmission line. These components are briefly described in Chapter 19. To realize compact amplifier circuits, lumped-element matching networks or lumped distributed circuit elements are utilized to transform device impedance to 50 Ω. In addition to these components, metamaterials and microelectromechanic system (MEMS) may also be used in the design of amplifiers.

The development of accurate and comprehensive models for passive circuit elements and their integration with CAD tools was a major activity during the 1980s and 1990s. Both models and CAD tools play a key role in the successful development of RF and microwave circuits including amplifiers. An overview of linear models (independent of bias conditions and input power) for these circuit elements is given in Figure 6.2. The electromagnetic (EM) simulators for modeling of passive components have become the CAD tools of choice and are briefly described in Chapter 15. Readers should consult References 4 and 6 for comprehensive design models for matching passive components.

6.2 TRANSMISSION LINE MATCHING ELEMENTS

For a transmission structure to be suitable as a matching circuit element in MICs and MMICs, one of the principal requirements is that the structure should be *planar* in configuration. A planar geometry implies that the characteristics of the element can be determined from the dimensions in a single plane. Microstrip and coplanar lines are the basic building blocks of matching networks. The microstrip is the most

Fundamentals of RF and Microwave Transistor Amplifiers. By Inder J. Bahl

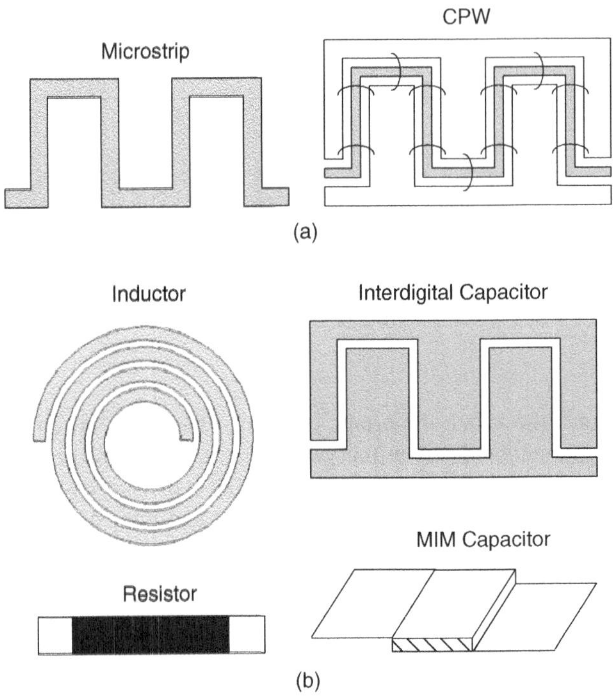

Figure 6.1 (a) Microstrip and coplanar waveguide distributed matching elements. (b) Inductor, capacitors, and resistor lumped matching elements.

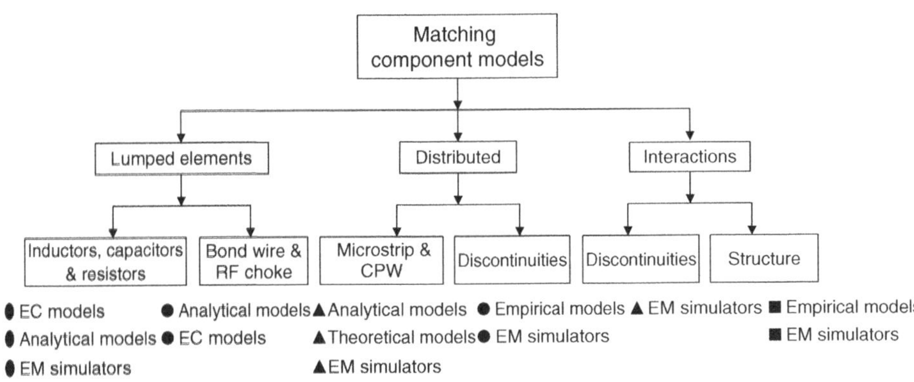

Figure 6.2 Overview of amplifier matching circuit elements.

important transmission line used in the design of components, impedance matching, and component connections.

6.2.1 Microstrip

The microstrip line is the most commonly used transmission medium in RF and microwave circuits, due to its quasi-TEM nature and excellent layout flexibility. When

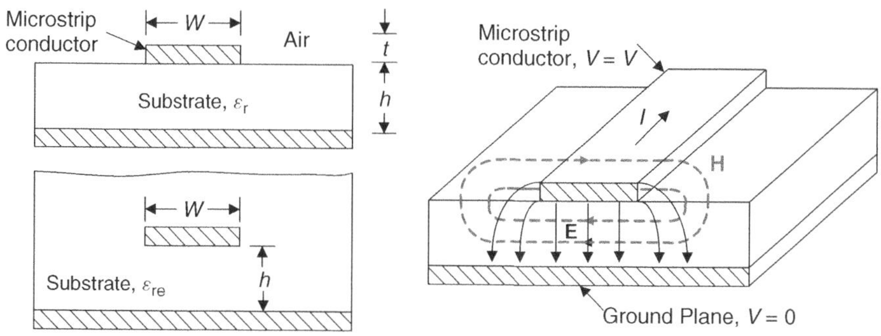

Figure 6.3 Microstrip line configuration showing effective dielectric constant concept. Electric and magnetic field distributions are also shown.

the size of the microstrip section is reduced to dimensions much smaller than the wavelength, it can be used as a lumped element. Thus the basic design of many lumped elements is based on a microstrip structure. A cross-sectional view of a microstrip line with physical parameters is shown in Figure 6.3. The concept of effective dielectric constant (ε_{re}) and the electric and magnetic field distributions in the microstrip are also shown in this figure. The design parameters are conductor width W and thickness t, substrate dielectric constant ε_r, and height h. The important parameters for designing these transmission lines are the characteristic impedance (Z_0), effective dielectric constant, attenuation constant (α), and discontinuity reactance. A complete set of design information for the microstrip is presented in References [4–6]. Next, closed form expressions are described for the characteristic impedance, effective dielectric constant, and attenuation constant.

Characteristic Impedance and Effective Dielectric Constant

Closed form expressions for Z_0 and ε_{re} when conductor thickness $t = 0$ are given below:

$$Z_0 = \frac{\eta}{2\pi\sqrt{\varepsilon_{re}}} \ln\left(\frac{8h}{W} + 0.25\frac{W}{h}\right) \qquad (W/h \le 1) \qquad (6.1a)$$

$$Z_0 = \frac{\eta}{\sqrt{\varepsilon_{re}}}\left\{\frac{W}{h} + 1.393 + 0.667 \ln\left(\frac{W}{h} + 1.444\right)\right\}^{-1} \qquad (W/h \ge 1) \qquad (6.1b)$$

where $\eta = 120\,\pi$ ohms and

$$\varepsilon_{re} = \frac{\varepsilon_r + 1}{2} + \frac{\varepsilon_r - 1}{2} F(W/h) \qquad (6.2)$$

$$F(W/h) = \begin{cases} (1 + 12h/W)^{-1/2} + 0.041(1 - W/h)^2 & (W/h \le 1) \\ (1 + 12h/W)^{-1/2} & (W/h \ge 1) \end{cases} \qquad (6.3)$$

The microstrip line capacitance, C (F/unit length), and inductance, L (H/unit length), per unit length are given by

$$C = \sqrt{\varepsilon_{re}}/(Z_0 c) \qquad (6.4a)$$

$$L = Z_0\sqrt{\varepsilon_{re}}/c \qquad (6.4b)$$

respectively where c is the speed of light. A low-impedance (wide) line is capacitive and a high-impedance (narrow) line is inductive.

EXAMPLE 6.1

Two microstrip sections of widths 25 and 500 μm are 1 mm long. Calculate their capacitance and inductance on 250-μm thick alumina substrate ($\varepsilon_r = 9.9$).

SOLUTION For $W/h = 25/250 = 0.1$, using (6.2) and (6.3), we find

$$F(W/h) = 0.1322 \quad \text{and} \quad \varepsilon_{re} = 6.04 \text{ and } \sqrt{\varepsilon_{re}} = 2.4573$$

From (6.1a), $Z_0 = 107\ \Omega$.

From (6.4a), using $c = 3 \times 10^{11}$ mm/s, the capacitance C (F/mm) is calculated and multiplied by 1 mm to get total capacitance C_t, that is,

$$C_t = C \times 1 \text{ mm} = 2.4573/(107 \times 3 \times 10^{11}) \text{ F} = 0.077 \text{ pF}$$

From (6.4b), the inductance L (H/mm) is calculated and multiplied by 1 mm to get total inductance L_t, that is,

$$L_t = L \times 1 \text{ mm} = 107 \times 2.4573/(3 \times 10^{11}) \text{ H} = 0.876 \text{ nH}$$

Similarly, for $W/h = 500/250 = 2$,

$$\text{F}(W/h) = 0.378, \quad \varepsilon_{re} = 7.13, \quad \sqrt{\varepsilon_{re}} = 2.6706, \quad \text{and} \quad Z_0 = 33.468\ \Omega$$

$$C_t = C \times 1 \text{ mm} = 2.6706/(33.468 \times 3 \times 10^{11}) \text{ F} = 0.266 \text{ pF}$$

$$L_t = L \times 1 \text{ mm} = 33.468 \times 2.6706/(3 \times 10^{11}) \text{ H} = 0.2979 \text{ nH}$$

Thus a wide line has about 3 times more capacitance and 3 times less inductance compared to a narrow line. ■

The maximum frequency of operation of a microstrip transmission line is limited as a result of several factors, including excitation of spurious modes, higher losses, pronounced discontinuity reactance effects, low Q caused by radiation from discontinuities, effect of dispersion on pulse distortion, tight fabrication tolerances, handling fragility, and, of course, technological processes. The frequency at which significant coupling occurs between the dominant quasi-TEM mode and the lowest order surface wave spurious mode is given by [4]

$$f_T = \frac{150}{\pi h}\sqrt{\frac{2}{\varepsilon_r - 1}} \tan^{-1} \varepsilon_r \tag{6.5a}$$

where f_T is in gigahertz, h is in millimeters, and the inverse tangent is expressed in radians.

EXAMPLE 6.2

Calculate the maximum frequency of operation for microstrip circuits designed on a 25-mil thick alumina substrate ($\varepsilon_r = 9.9$).

SOLUTION The maximum frequency of operation for microstrip circuits is defined when it no longer supports quasi-TEM mode and is given by (6.5a):

$$f_T = \frac{150}{3.1416 \times 0.635}\sqrt{\frac{2}{9.9-1}}\tan^{-1} 9.9$$
$$f_T = 75.19 \times 0.474 \times 1.47 = 52.4 \text{ GHz}$$

The excitation of higher order modes in a microstrip can be avoided by operating it below the cutoff frequency of the first higher order mode, which is given approximately by

$$f_c = \frac{300}{\sqrt{\varepsilon_r}(2W + 0.8h)} \tag{6.5b}$$

where f_c is in gigahertz, and W and h are in millimeters. This limitation is mostly applicable for low-impedance lines that have wide microstrip conductors. ■

Several different versions of the microstrip to improve the amplifier's performance have been reported. These include a multilayer microstrip to lower output matching network loss and defected ground structure (DGS) to improve PAE by harmonic termination. The multilayer microstrip will be discussed in Chapter 10; DGS is not popular due to higher loss and complex assembly.

Microstrip Losses

Attenuation in a microstrip structure is caused by two loss components: conductor loss and dielectric loss. The conductor loss occurs due to finite resistance of the metal strip and ground plane, and the dielectric loss is a result of the loss tangent of the substrate. Closed form expressions for the conductor (α_c) and dielectric (α_d) attenuation constants, expressed in dB per unit length, are given below:

$$\alpha_c = 0.072\frac{\sqrt{f}}{WZ_0} \tag{6.6a}$$

$$\alpha_d = 27.3\frac{\varepsilon_r}{\varepsilon_r - 1}\frac{\varepsilon_{re} - 1}{\sqrt{\varepsilon_{re}}}\frac{\tan\delta}{\lambda_0} \tag{6.6b}$$

where λ_0 is the free-space wavelength, f is the frequency of operation (in GHz), $\tan\delta$ is the loss tangent of the dielectric substrate, and W is the strip width. Equation (6.6a) is an approximation; a more accurate expression for α_c is given in Reference 4.

The dielectric loss is normally very small compared to the conductor loss for dielectric substrates such as alumina, AlN, and BeO. The dielectric loss in silicon substrates (used for monolithic MICs), however, is usually of the same order, or even larger than the conductor loss. This is because of the lower resistivity available in silicon wafers. However, higher resistivity can be obtained in GaAs semi-insulator substrate, and therefore the dielectric loss is lower for this material. At a given frequency the total loss can be obtained by adding the two values, that is,

$$\alpha_T = \alpha_c + \alpha_d \tag{6.6c}$$

Table 6.1 summarizes microstrip characteristics at 10 GHz for several line widths and substrate materials. The values of conductor thickness (t), substrate thickness (h), and loss tangent ($\tan\delta$) were 5 μm, 381 μm, and 0.0005, respectively.

Microstrip Discontinuities

In microwave circuits, discontinuities between distributed elements, between lumped elements, and between distributed and lumped circuit elements always exist [4]. These discontinuities occur at the junctions due to the perturbation of electric and magnetic field distributions. When two passive components (e.g., microstrip and wire, microstrip and MIM capacitor, microstrip and CPW, or series and shunt microstrip sections) are connected together, a junction discontinuity occurs due to different field configurations between them even though they support the same TEM or quasi-TEM mode. If the magnetic field dissimilarity is more pronounced, the junction

Table 6.1 Microstrip Data

(a) $\varepsilon_r = 2.2$

W (μm)	W/h	Z_0	ε_{re}	α (dB/cm)	Line Capacitance (pF/100 μm)	Line Inductance (nH/100 μm)
50	0.13	183.9	1.66	0.122	0.002	0.079
75	0.20	166.7	1.68	0.099	0.003	0.072
100	0.26	154.2	1.69	0.086	0.003	0.067
125	0.33	144.3	1.70	0.077	0.003	0.063
150	0.39	136.2	1.71	0.070	0.003	0.059
200	0.52	123.3	1.73	0.062	0.004	0.054
250	0.66	113.4	1.74	0.056	0.004	0.050
300	0.79	105.3	1.75	0.052	0.004	0.046
400	1.05	92.6	1.78	0.047	0.005	0.041
500	1.31	83.0	1.80	0.045	0.005	0.037
750	1.97	66.6	1.84	0.041	0.007	0.030
1000	2.62	55.9	1.87	0.039	0.008	0.025
1500	3.94	42.6	1.92	0.037	0.011	0.020
2000	5.25	34.5	1.96	0.036	0.014	0.016
3000	7.87	25.2	2.00	0.035	0.019	0.012
4000	10.50	19.8	2.04	0.035	0.024	0.009

(b) $\varepsilon_r = 4.3$

W (μm)	W/h	Z_0	ε_{re}	α (dB/cm)	Line Capacitance (pF/100 μm)	Line Inductance (nH/100 μm)
25	0.07	164.6	2.75	0.214	0.003	0.091
50	0.13	142.2	2.82	0.159	0.004	0.080
75	0.20	128.5	2.86	0.130	0.004	0.072
100	0.26	118.6	2.89	0.112	0.005	0.067
125	0.33	110.9	2.92	0.100	0.005	0.063
150	0.39	104.5	2.95	0.092	0.005	0.060
200	0.52	94.4	2.99	0.081	0.006	0.054
250	0.66	86.6	3.03	0.074	0.007	0.050
300	0.79	80.3	3.07	0.069	0.007	0.047
400	1.05	70.4	3.13	0.064	0.008	0.042
500	1.31	63.0	3.19	0.061	0.009	0.038
750	1.97	50.2	3.31	0.057	0.012	0.030
1000	2.62	42.0	3.41	0.054	0.015	0.026
1500	3.94	31.8	3.55	0.052	0.020	0.020
2000	5.25	25.7	3.65	0.051	0.025	0.016

Table 6.1 *(Continued)*

(c) $\varepsilon_r = 6.7$

W (μm)	W/h	Z_0	ε_{re}	α (dB/cm)	Line Capacitance (pF/100 μm)	Line Inductance (nH/100 μm)
10	0.03	159.4	3.87	0.377	0.004	0.105
25	0.07	136.8	4.02	0.258	0.005	0.091
50	0.13	118.0	4.14	0.192	0.006	0.080
75	0.20	106.5	4.21	0.157	0.006	0.073
100	0.26	98.2	4.27	0.136	0.007	0.068
125	0.33	91.7	4.33	0.122	0.008	0.064
150	0.39	86.4	4.37	0.112	0.008	0.060
200	0.52	78.0	4.45	0.099	0.009	0.055
250	0.66	71.5	4.52	0.091	0.010	0.051
300	0.79	66.2	4.58	0.085	0.011	0.047
400	1.05	58.0	4.70	0.079	0.012	0.042
500	1.31	51.8	4.81	0.076	0.014	0.038
750	1.97	41.3	5.02	0.071	0.018	0.031
1000	2.62	34.4	5.19	0.068	0.022	0.026
1500	3.94	26.0	5.45	0.065	0.030	0.020
2000	5.25	21.0	5.63	0.064	0.038	0.017

(d) $\varepsilon_r = 9.9$

W (μm)	W/h	Z_0	ε_{re}	α (dB/cm)	Line Capacitance (pF/100 μm)	Line Inductance (nH/100 μm)
10	0.03	134.6	5.50	0.447	0.006	0.105
25	0.07	115.3	5.73	0.307	0.007	0.092
50	0.13	99.3	5.92	0.229	0.008	0.080
75	0.20	89.5	6.04	0.187	0.009	0.073
100	0.26	82.5	6.14	0.162	0.010	0.068
125	0.33	77.1	6.22	0.146	0.011	0.064
150	0.39	72.6	6.29	0.134	0.012	0.061
200	0.52	65.5	6.42	0.118	0.013	0.055
250	0.66	60.0	6.53	0.108	0.014	0.051
300	0.79	55.6	6.63	0.101	0.015	0.048
400	1.05	48.7	6.83	0.096	0.018	0.042
500	1.31	43.5	7.00	0.093	0.020	0.038
750	1.97	34.6	7.35	0.087	0.026	0.031
1000	2.62	28.8	7.62	0.084	0.032	0.027
1500	3.94	21.7	8.03	0.080	0.043	0.021
2000	5.25	17.5	8.32	0.079	0.055	0.017

discontinuity is an inductive type, while in the case of electric field dissimilarity, the junction discontinuity is a capacitive type.

Typical discontinuities, as shown in Figure 6.4, are (a) open circuits, (b) gap between conductors, (c) step change in dimensions (introduced for a change in impedance level), (d) bends (right-angled and others), (e) T-junctions, and (f) cross-junction. In most cases, discontinuities are basically undesirable circuit reactances. Some of the effects of discontinuities on circuit performances are:

- Frequency shift in narrowband circuits
- Degradation in input and output voltage standing-wave ratios (VSWRs)

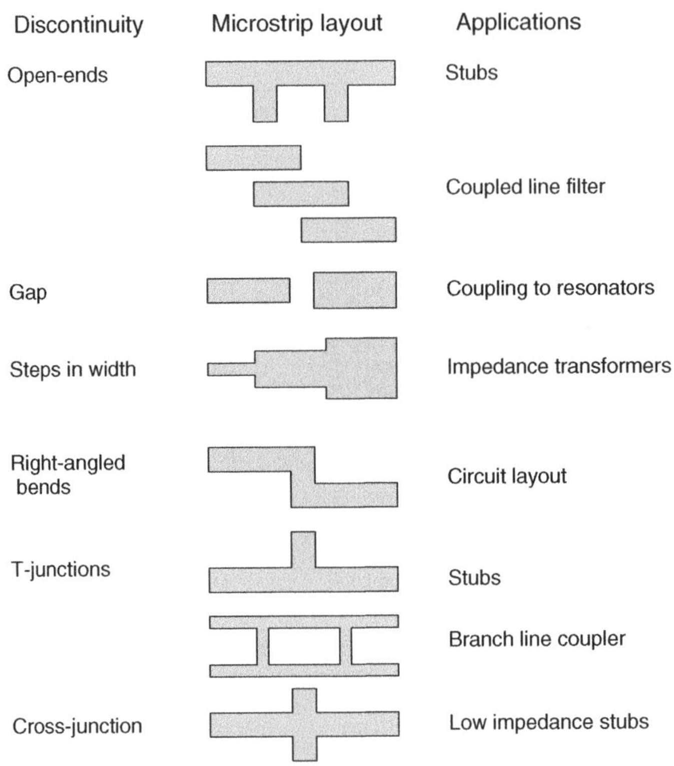

Figure 6.4 Microstrip discontinuities.

- Higher ripple in gain flatness of broadband ICs
- Interfacing problem in multistage amplifiers
- Lower circuit yield due to degradation in circuit performance
- Surface wave and radiation couplings that may cause oscillations in high-gain amplifiers

The effect of discontinuities becomes more critical at higher frequencies. The discontinuities should be either taken into account or compensated for at the final stage of design. A complete understanding of the design of microwave circuits requires characterization of the discontinuities present in these circuits. Figure 6.5a shows commonly used discontinuity compensation techniques. In general, compensated discontinuities improve the circuit performance as well as the bandwidth. Usually the chamfered discontinuity technique is used, in which the discontinuity reactances are minimized by removing appropriate portions of the microstrip conductor near the discontinuity location. The effect of discontinuities may also be reduced by using high-impedance lines in a circuit, as shown in Figure 6.5b. High-impedance (70–90 Ω narrow microstrip lines have smaller discontinuity effects and result in compact matching networks as compared to wide microstrip (impedance 20–40 Ω) lines.

As shown in Figure 6.6, interaction effects take place between the closely spaced discontinuities due to interaction of fields from one discontinuity to another. The interaction between two discontinuities affects the isolated discontinuity reactance and must be accurately determined using EM simulators. The minimum distance between two

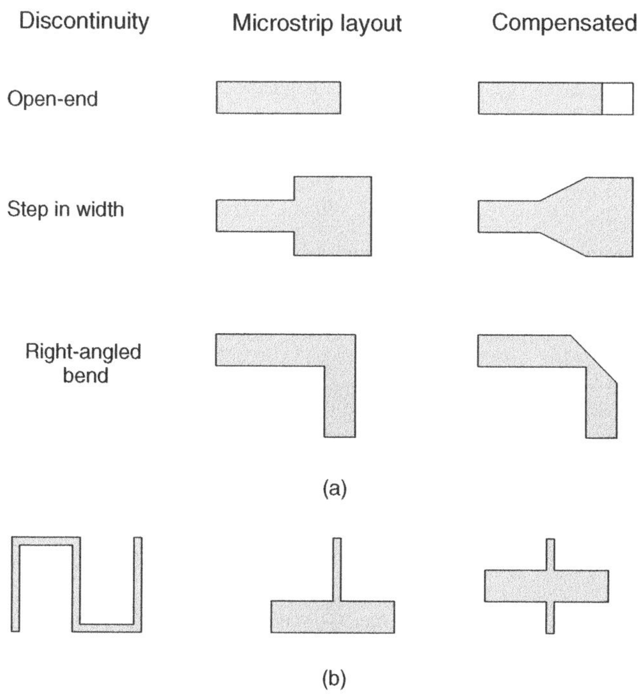

Figure 6.5 (a) Compensated discontinuities and (b) low-discontinuity reactance effect structures using high-impedance line junctions.

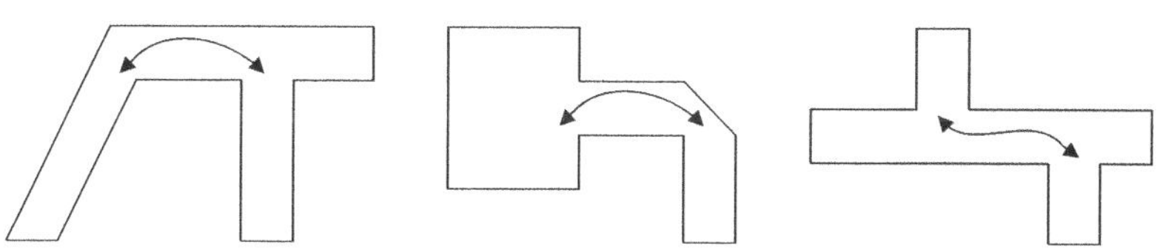

Figure 6.6 Examples of interactions between discontinuities. Arrows point at junction discontinuities.

discontinuities with the potential for coupling is $2W$ or $2h$, whichever is greater. For T-junction and cross-junction discontinuities on thick substrates ($\geq$100 μm for GaAs and $\geq$250 μm for alumina), EM simulators are commonly used to accurately include their effects in the circuit design.

Coupling Effect

Generally, in MMICs, matching networks are packed very closely together and have layout and junction discontinuity reactances. Coupling occurs between the matching elements due to close proximity and can be accounted for by using EM simulators. For long lines on the order of a quarter wavelength, spacing between the lines must be $3W$ (width of line) or $3h$ (height of substrate) whichever is greater, in order to reduce coupling effects. For short lines on the order of 1/10 wavelength, $1W$ or $1h$ spacing between the lines is adequate for most applications.

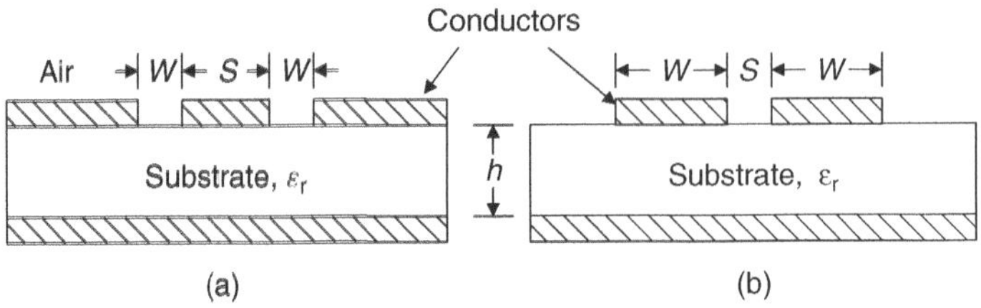

Figure 6.7 (a) Coplanar waveguide (CPW) and (b) coplanar strips (CPSs).

Table 6.2 Expressions for Coplanar Line Characteristic Impedance and Effective Dielectric Constant

Structure	Characteristic Impedance (Ω)	Effective Dielectric Constant
Coplanar waveguide	$Z_0 = \dfrac{30\pi}{\sqrt{\varepsilon_{re}}}\dfrac{K(k')}{K(k)}$	$\varepsilon_{re} = 1 + \dfrac{\varepsilon_r - 1}{2}\dfrac{K(k')K(k_1)}{K(k)K(k_1')}$
Coplanar strips	$Z_0 = \dfrac{120\pi}{\sqrt{\varepsilon_{re}}}\dfrac{K(k)}{K(k')}$	$\varepsilon_{re} = 1 + \dfrac{\varepsilon_r - 1}{2}\dfrac{K(k')K(k_1)}{K(k)K(k_1')}$
	$k = \dfrac{a}{b}, \quad a = \dfrac{S}{2}, \quad b = \dfrac{S}{2} + W$	$k_1 = \dfrac{\sinh(\pi a/2h)}{\sinh(\pi b/2h)}$

6.2.2 Coplanar Lines

Coplanar waveguide (CPW) and coplanar strips (CPSs) also used in the MIC and MMIC designs. Cross-sectional views of these lines are shown in Figure 6.7; they support quasi-TEM modes. Expressions for Z_0 and ε_{re} of CPWs and CPSs are given in Table 6.2, where $k' = \sqrt{1 - k^2}$ and

$$\frac{K(k)}{K'(k)} = \frac{1}{\pi}\ln\left\{2\frac{1+\sqrt{k}}{1-\sqrt{k}}\right\} \qquad \text{for } 0.707 \le k \le 1$$
$$= \frac{\pi}{\ln\left[2\frac{1+\sqrt{k'}}{1-\sqrt{k'}}\right]} \qquad \text{for } 0 \le k \le 0.707 \tag{6.7}$$

The CPW and CPS configurations are not as popular as the microstrip. However, the CPW is used exclusively in flip-chip MMICs and at millimeter-wave frequencies.

6.3 LUMPED ELEMENTS

A lumped element in microwave circuits is defined as a passive component whose size across any dimension is much smaller than the operating wavelength so that there is no appreciable phase shift between the input and output terminals. Generally, keeping the maximum dimension less than λ/20 is a good approximation. Lumped elements for use at RF and microwave frequencies are designed on the basis of this consideration, and three basic lumped-element building blocks are capacitors, inductors, and resistors. Figure 6.8 shows basic microstrip line inductors and a capacitor. Lumped inductor transformers and baluns are also commonly used in many circuits.

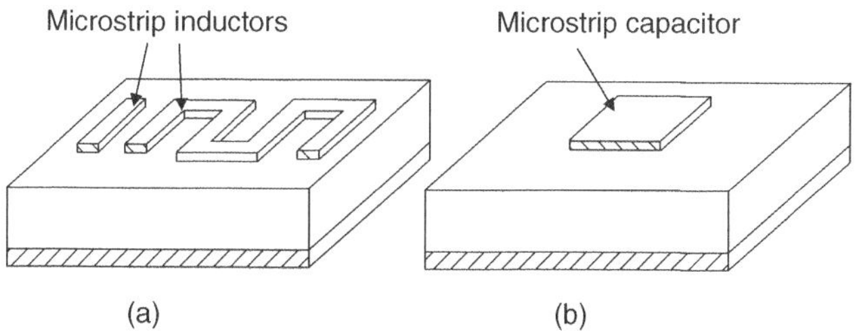

Figure 6.8 Microstrip (a) inductors and (b) capacitor.

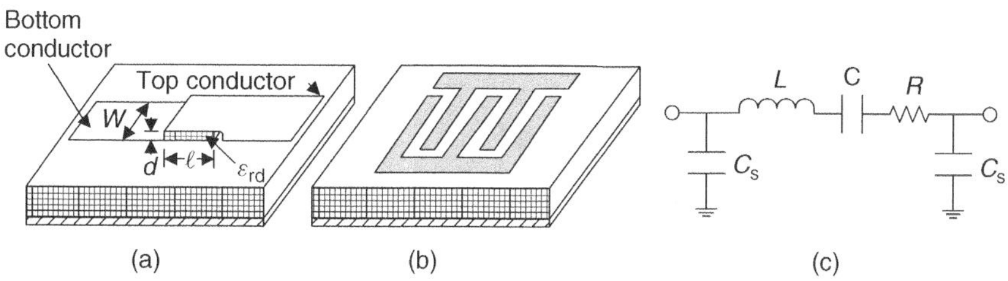

Figure 6.9 (a) MIM capacitor, (b) interdigital capacitor, and (c) equivalent circuit.

At radiofrequencies, lumped discrete spiral inductors, MIM capacitors, and thin-film resistors are commonly used in matching networks. Also, the use of a discrete spiral inductor as a bias choke depends on the frequency of operation and the current handling capacity. Lumped-element circuits will have lower Q than distributed circuits, but they offer the advantage of smaller size, lower cost, and larger bandwidth characteristics. These are especially suitable for monolithic MICs and for broadband hybrid MICs, where real-estate requirements are of prime importance. Large impedance transformations can easily be accomplished using the lumped-element approach. Therefore high-power devices that have very low impedance values can easily be matched to 50 Ω with impedance transformers realized using lumped elements. At low frequencies (below C-band) amplifiers designed using lumped inductors and capacitors have an order of magnitude smaller size compared to amplifiers designed using the microstrip or CPW.

6.3.1 Capacitors

RF and microwave capacitors are classified into three categories: microstrip (Fig. 6.8b), metal–insulator–metal (MIM), and interdigital, as shown in Figure 6.9. A small length of an open circuited microstrip section can be used as a lumped capacitor with a low capacitance value (<0.2 pF) per unit area due to thick substrates. MIM capacitors are fabricated using a multilevel process and provide the largest capacitance value (0.1–50 pF for monolithic on GaAs) per unit area because of a very thin dielectric layer sandwiched between two electrodes. The interdigital geometry has applications where one needs moderate capacitance (0.1–0.5 pF) and high Q values. Both microstrip and interdigital configurations are fabricated using conventional MIC techniques. A detailed treatment of these components can be found in Reference [6]. MIM capacitors are generally of discrete type and are described next.

The MIM capacitor structure might have two or more conductors. The capacitance, C (in farads), of a capacitor structure consisting of two conductors as shown in Figure 6.9a is expressed as

$$C = \varepsilon_0 \varepsilon_{rd} \frac{A}{d} = \varepsilon_0 \varepsilon_{rd} \frac{W \times \ell}{d} \tag{6.8}$$

where W and ℓ are the width and length of one of the plates, ε_{rd} is the dielectric constant of the capacitor dielectric film, and ε_0 is the free-space permittivity. The above equation does not include the effect of fringing field. Equation (6.8) can be expressed in commonly used units as follows:

$$C = 0.2249 \varepsilon_{rd} \frac{W \times \ell}{d} \quad \text{(pF)}, \quad W, \ell, \text{and } d \text{ in inches} \tag{6.9a}$$

$$C = 8.85 \times 10^{-3} \varepsilon_{rd} \frac{W \times \ell}{d} \quad \text{(pF)}, \quad W, \ell, \text{ and } d \text{ in millimeters} \tag{6.9b}$$

$$C = 8.85 \times 10^{-6} \varepsilon_{rd} \frac{W \times \ell}{d} \quad \text{(pF)}, \quad W, \ell, \text{ and } d \text{ in micrometers} \tag{6.9c}$$

A unit of capacitance is the farad (F) and depending on its value, it can be expressed in microfarad (μF or 10^{-6} F), nanofarad (nF or 10^{-9} F), picofarad (pF or 10^{-12} F), or sometimes very small values as femtofarad (fF or 10^{-15} F).

6.3.2 Inductors

Planar inductors may be classified into two-dimensional and three-dimensional structures. Two-dimensional inductors include a section of high-impedance line, loop, and coil. The coil may have a circular, rectangular, hexagonal, or octagonal shape. Figures 6.10 and 6.11 show some of these configurations, and their lumped-element equivalent circuit models are shown in Figure 6.12. The three-dimensional inductors consist of multiple layers of two-dimensional coils. The input impedance of a small section ($\gamma\ell \ll 1$) short-circuited ($Z_L = 0$) line is expressed as

$$Z_{in} = Z_0 \tanh(\gamma\ell) \cong Z_0 \gamma\ell = R + j\omega L \tag{6.10}$$

Thus a small short-circuited transmission line behaves as an inductor in series with a resistor R as shown in Figure 6.12. The resistance R represents the conductor loss, which is negligible for short sections of copper or gold conductors.

Approximate expressions for inductance L, resistance R, and parasitic capacitances for microstrip section, circular loop, and circular coil are given [5, 6] below.

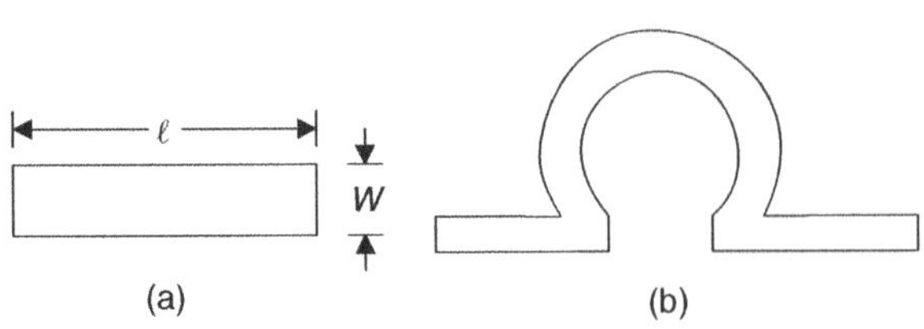

Figure 6.10 Microstrip inductors: (a) strip and (b) loop.

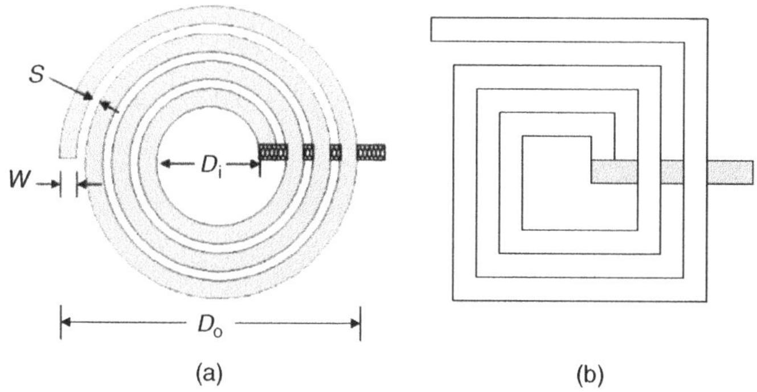

Figure 6.11 Coil inductors: (a) circular and (b) rectangular.

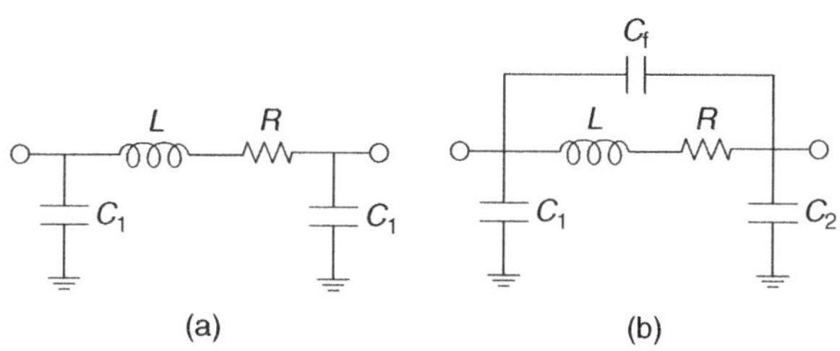

Figure 6.12 Equivalent circuit of inductors: (a) microstrip and (b) loop and coil.

Microstrip Section

$$L(\text{nH}) = 2 \times 10^{-4}\ell\left[\ln\left(\frac{\ell}{W+t}\right) + 1.193 + \frac{W+t}{3\ell}\right]\cdot K_g \tag{6.11}$$

$$R(\Omega) = \frac{KR_s\ell}{2(W+t)} \tag{6.12a}$$

$$C_1(\text{pF}) = 16.67 \times 10^{-4}\ell\sqrt{\varepsilon_{re}}/Z_0 \tag{6.12b}$$

where all dimensions are in μm and

$$K_g = 0.57 - 0.145\ln\frac{W}{h}, \quad \frac{W}{h} > 0.05 \tag{6.13}$$

$$K = 1.4 + 0.217\ln\left(\frac{W}{5t}\right), \quad 5 < \frac{W}{t} < 100 \quad \text{(for a ribbon)} \tag{6.14}$$

$$K = 1 + 0.333\left(1 + \frac{S}{W}\right)^{-1.7} \quad \text{(for a spiral)} \tag{6.15}$$

The term K_g accounts for the presence of a ground plane and decreases as the ground plane is brought nearer. Another term K is a correction factor that takes into account the crowding of the current at the corners of the conductor. W, t, h, ℓ, and R_s are the line width, line thickness, substrate thickness, length of the section, and sheet resistance per square of the conductor. The microstrip parameters Z_0 and ε_{re} are calculated as discussed in Section 6.2.1.

Circular Loop

$$L(\text{nH}) = 1.257 \times 10^{-3} a \left[\ln\left(\frac{a}{W+t}\right) + 0.078 \right] \cdot K_g \qquad (6.16a)$$

$$R(\Omega) = \frac{K R_s}{W+t} \pi a \qquad (6.16b)$$

$$C_1(\text{pF}) = 33.33 \times 10^{-4} \pi a \sqrt{\varepsilon_{re}} / Z_0 \qquad (6.16c)$$

where a is the mean radius of the loop.

Circular Spiral

$$L(\text{nH}) = 0.03937 \frac{a^2 n^2}{8a + 11c} \cdot K_g \qquad (6.17)$$

$$a = \frac{D_o + D_i}{4}, \qquad c = \frac{D_o - D_i}{2} \qquad (6.18)$$

$$R(\Omega) = \frac{K \pi a n R_s}{W} \qquad (6.19a)$$

$$C_1 = C_2(\text{pF}) = 3.5 \times 10^{-5} D_o + 0.06 \qquad (6.19b)$$

where n is the number of turns, D_i the inductor's inside diameter, D_o the inductor's outside diameter, and S the spacing between the turns. All dimensions are in μm. The effect of the ground plane on the inductance value can be reduced by keeping $S < W$ and $S \ll h$.

EXAMPLE 6.3

A three-turn coil inductor is printed on a 250-μm thick alumina substrate ($\varepsilon_r = 9.9$). The line width W, spacing between the turns S, and inside coil diameter dimensions are 25 μm, 25 μm, and 200 μm, respectively. If the coil conductor is gold and 5 μm thick, calculate its inductance, resonance frequency, and Q at 4 GHz.

SOLUTION Here

$$R_s = 1/(\sigma t) = 1/(4.1 \times 10^7 \times 5 \times 10^{-6}) = 0.00488\ \Omega/\text{square}$$
$$W/h = 0.1$$
$$D_o = D_i + 2nW + (2n-1)S = 200 + 150 + 125 = 475\ \mu\text{m}$$
$$a = (475 + 200)/4 = 168.75\ \mu\text{m}$$
$$c = (475 - 200)/2 = 137.5\ \mu\text{m}$$

From (6.13) and (6.15), we have

$$K_g = 0.57 - 0.145\ \ln(0.1) = 0.904$$
$$K = 1 + 0.333(1+1)^{-1.7} = 1.1$$

Table 6.3 Coefficients for General Inductance Expression [7, 8]

Inductor Geometry	c_1	c_2	c_3	c_4
Square	1.27	2.07	0.18	0.13
Hexagonal	1.09	2.23	0.00	0.17
Octagonal	1.07	2.29	0.00	0.19
Circle	1.00	2.46	0.00	0.20

From (6.17), we find

$$L(\text{nH}) = 0.03937 \times 168.75^2 \times 9 \times 0.904/(8 \times 168.75 + 11 \times 137.5) = 3.19$$
$$C_t = C_1 + C_2 = 2 \times [3.5 \times 10^{-5} \times 168.75 + 0.06] = 0.122 \text{ pF}$$
$$R = 1.1 \times \pi \times 168.75 \times 3 \times 0.00488/25 = 0.3415\ \Omega$$
$$Q = \omega L/R = 2\pi \times 4 \times 10^9 \times 3.19 \times 10^{-9}/0.3415 = 235$$
$$f_0 = \frac{1}{2\pi\sqrt{LC_t}} = \frac{1}{2\pi\sqrt{3.19 \times 10^{-9} \times 0.122 \times 10^{-12}}} \text{Hz} = 8.06 \text{ GHz}$$

■

A more general expression for inductance of arbitrary shape has been reported in the literature [7, 8] and reproduced as follows:

$$L = \frac{\mu_0 n^2 D_{\text{av}} c_1}{2}[\ln(c_2/\chi) + c_3\chi + c_4\chi^2] \tag{6.20}$$

where coefficients c_i for various geometries are given in Table 6.3, χ is the fill ratio, and D_{av} is the average diameter of the inductor, and their expressions are given below:

$$\chi = \frac{D_o - D_i}{D_o + D_i} \tag{6.21a}$$
$$D_{\text{av}} = \frac{1}{2}(D_o + D_i) \tag{6.21b}$$

PCB Inductors

Multilayer PCB technology is quite suitable for realizing high-value inductances suitable for up to 1–2-GHz applications. These inductors (50–200 nH) can carry currents up to 3–5 A and can have Q values in the range of 100 for applications up to 100–200 MHz.

6.3.3 Resistors

As discussed earlier, the short-circuited transmission line behaves as an inductor when the conductors have low resistance. However, when a section of good conductor is replaced with a thin lossy conductor such as NiCr (typical sheet resistance $\cong$ 50 Ω/square compared to 0.005 Ω/square for gold) or tungsten nitride resistive metal WN (typical sheet resistance $\cong$ 10 Ω/square), the resistive part becomes dominant and the microstrip section behaves as a resistor with negligible parasitic inductive and capacitive reactance. Figure 6.13a shows the microstrip resistor configuration.

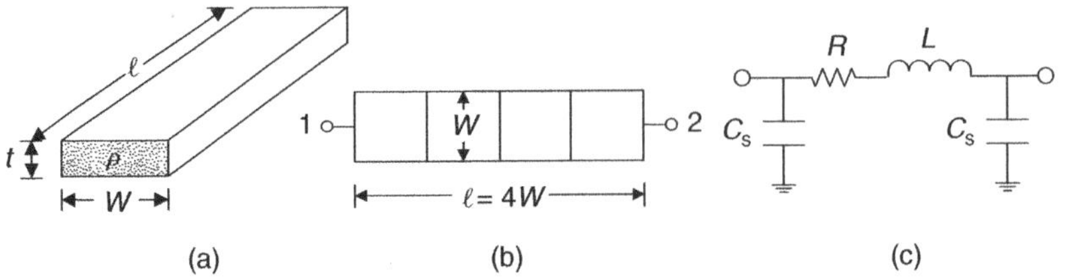

Figure 6.13 (a) Geometry of a planar resistor, (b) resistance calculation representation of a resistor from sheet resistance, and (c) equivalent circuit.

The resistance R value of a planar resistor shown in Figure 6.13a depends on the resistor material properties and its dimensions and is given by

$$R = \rho\frac{\ell}{A} = \rho\frac{\ell}{Wt} = \frac{\ell}{\sigma Wt} \tag{6.22a}$$

where ρ is the bulk resistivity of the material expressed in Ω·m, σ is the bulk conductivity expressed in S/m, ℓ is the length of the resistor along the direction of current flow, W is the width, t is the thickness, and A is the cross-sectional area. The resistance can also be calculated from the sheet resistance R_S (Ω/□) of the resistive film (for given thickness t) using the following relationship:

$$R = R_S\frac{\ell}{W}, \quad \text{where} \quad R_S = \frac{\rho}{t} = \frac{1}{\sigma t} \tag{6.22b}$$

For given material R_S, the resistance can easily be calculated from the number of squares of width in total length. For example, a line of width 50 μm and 1000 μm long has 2.5 times less resistance than a line of width 20 μm and 1000 μm long. Thus the key in increasing the resistance is to keep the number of metallization squares as large as possible in a given length. Figure 6.13b shows four squares between terminals 1 and 2 and if the R_S value is 10 Ω/□, the total resistance is 40 Ω.

EXAMPLE 6.4

A thin-film resistor is printed on an alumina substrate. The bulk resistivity of the film is 5×10^{-6} Ω·m. If the resistor width W is 20 μm and thickness t is 0.1 μm, calculate its length for a resistance value of 250 Ω.

SOLUTION Here $R_S = \rho/t = 5 \times 10^{-6}/(0.1 \times 10^{-6}) = 50$ Ω/□. From (6.21b), we find

$$\ell = R\frac{W}{R_S} = \frac{250 \times 20}{50} = 100\ \mu\text{m}$$

■

Table 6.4 compares the microstrip with lumped elements. The microstrip has low loss and is more suitable for high-performance amplifier applications, whereas lumped elements are compact in size and are more suitable for high impedance transformation ratio and broadband amplifier applications.

Table 6.4 Comparison of Distributed and Lumped-Element Matching Elements

	Microstrip	Lumped Elements
Major advantages	1. Well characterized and has flexibility in design 2. Low loss and high performance 3. Has better harmonic tuning capability for high-efficiency applications	1. Capable of transforming high impedance ratios 2. Minimum interactions between elements due to small size 3. Compact in size
Major disadvantages	1. Strong interaction effects between elements and discontinuities in packed design 2. Large in size even with interactions included	1. Low Q, limited DC power handling capability 2. Substandard performance due to low Q at the output of a power amplifier in terms of P_o and PAE

Coupling Effects Between Lumped Elements

Low-cost RF and microwave amplifiers mandate a higher level of integration and more circuit functions in a smaller size. In other words, one needs to integrate RF/microwave circuits in a compact size to reduce the area and cost. When lumped elements are placed in close proximity to each other, a fraction of the power present on the main component is coupled to the secondary element. The power coupled is a function of the physical dimensions of the structure, the mode of propagation, the frequency of operation, and the distance between the components. In these structures, there is a continuous coupling between the electromagnetic fields of the two components.

The coupling effects between various lumped elements such inductor–inductor, inductor–capacitor, and inductor–viahole have been studied [9]. It was found that the effect of coupling between inductors is less than 1% for inductors having reactance of about 50 Ω and separated by 20 μm on a 75-μm thick GaAs substrate. Thus a distance of 20 μm or greater than $h/4$ (for high-K substrates) between lumped elements is quite safe for hybrid or monolithic circuits at radiofrequencies.

6.4 BOND WIRE INDUCTORS

Most RF and microwave circuits and subsystems use bond wires to interconnect components such as lumped elements, planar transmission lines, transistors, and ICs. These bond wires have 0.5–2-mil diameters and their lengths are electrically short compared to the operating wavelength. Bond wires are accurately characterized using simple formulas in terms of their inductances and series resistances. As a first-order approximation, the parasitic capacitance associated with bond wires can be neglected.

6.4.1 Single Wire

In hybrid MICs, bond wire connections are used to connect active and passive circuit components, and in MMICs bond wire connections are used to connect the MMIC chip to other circuitry. The free-space inductance L (in nH) of a wire of diameter d and length ℓ (in microns) is given by

$$L = 2 \times 10^{-4}\ell \left[\ln\left\{ \frac{2\ell}{d} + \sqrt{1 + \left(\frac{2\ell}{d}\right)^2} \right\} + \frac{d}{2\ell} - \sqrt{1 + \left(\frac{d}{2\ell}\right)^2} + C \right] \qquad (6.23)$$

where the frequency-dependent correction factor C is a function of bond wire diameter and its material's skin depth δ, expressed as

$$C = 0.25 \tanh(4\delta/d) \qquad (6.24a)$$

$$\delta = \frac{1}{\sqrt{\pi \sigma f \mu_0}} \qquad (6.24b)$$

where σ is the conductivity of the wire material. For gold wires, $\delta = 2.486 f^{-0.5}$, where the frequency f is expressed in GHz. When δ/d is small, $C = \delta/d$.

For $\ell \gg d$, (6.23) reduces to

$$L = 2 \times 10^{-4}\ell \left(\ln\frac{4\ell}{d} + 0.5\frac{d}{\ell} - 1 + C \right) \qquad (6.25a)$$

The wire resistance R (in ohms) is given by

$$R = \frac{R_s \ell}{\pi d} \qquad (6.25b)$$

where R_s is the sheet resistance in ohm per square ($\Omega/\square$). Taking into account the effect of skin depth, (6.25b) may be written

$$R = \frac{4\ell}{\pi \sigma d^2} \left[0.25\frac{d}{\delta} + 0.2654 \right] \qquad (6.25c)$$

6.4.2 Ground Plane Effect

The effect of the ground plane on the inductance value of a wire can also be calculated. If the wire is at a distance h above the ground plane, as shown in Figure 6.14a, it sees its image at $2h$ from it. The wire and its image result in a mutual inductance L_{mg}. Since the image wire carries a current opposite to the current flow in the bond wire, the effective inductance of the bond wire becomes

$$L_e = L - L_{mg} \qquad (6.26a)$$

$$L_{mg} = 2 \times 10^{-4}\ell \left[\ln\left\{ \frac{\ell}{2h} + \sqrt{1 + \left(\frac{\ell}{2h}\right)^2} \right\} + \frac{2h}{\ell} - \sqrt{1 + \left(\frac{2h}{\ell}\right)^2} + C \right] \qquad (6.26b)$$

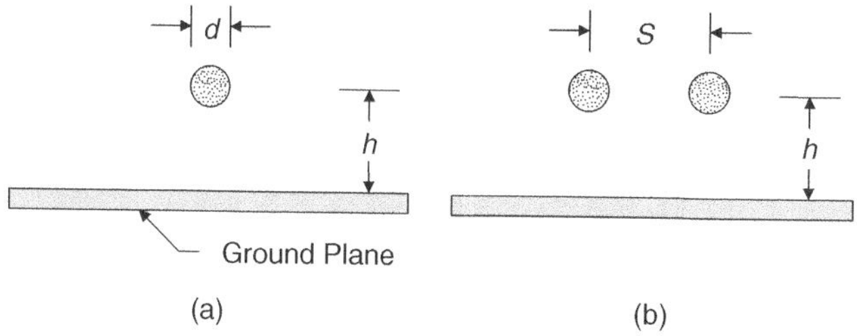

Figure 6.14 Wires above ground plane: (a) single and (b) twin.

From (6.23) and (6.26), we find

$$L_e = 2 \times 10^{-4}\ell \left[\ln\frac{4h}{d} + \ln\left(\frac{\ell + \sqrt{\ell^2 + d^2/4}}{\ell + \sqrt{\ell^2 + 4h^2}}\right) + \sqrt{1 + \frac{4h^2}{\ell^2}} - \sqrt{1 + \frac{d^2}{4\ell^2}} - 2\frac{h}{\ell} + \frac{d}{2\ell} \right] \tag{6.27}$$

Here L_e is in nH, and ℓ, h, and d are in μm.

6.4.3 Multiple Wires

In many applications, multiple wires are required to carry higher currents, reduce wire inductance, or improve wire reliability. For example, in the case of two wires placed parallel to each other at a distance S between their centers (Fig. 6.14b), above the ground plane, the total inductance of the pair becomes

$$L_{ep} = (L_e + L_m)/2 \tag{6.28}$$

Here both wires carry current in the same direction, therefore the mutual inductance L_m and self-inductance add together. In this case L_m is given by

$$L_m = 2 \times 10^{-4}\ell \left[\ln\left\{ \frac{\ell}{S} + \sqrt{1 + \left(\frac{\ell}{S}\right)^2} \right\} - \sqrt{1 + \left(\frac{S}{\ell}\right)^2} + \frac{S}{\ell} \right] \tag{6.29}$$

Equations (6.26b) and (6.29) are identical when $2h = S$ and $C = 0$. The same procedure can be carried out to calculate the mutual inductance for multiple wires.

Table 6.5 lists inductance of 1-mil diameter gold wires. So far we have treated uniformly placed horizontal wires above the ground plane. However, in practice, the wires are curved, nonhorizontal, and not parallel to each other. In such situations, an average value of S and h can be used. Also wires have shunt capacitance, which must be taken into account.

The inductance of a bond wire connection is generally reduced by connecting multiple wires in parallel. However, as shown in Table 6.5, the inductance of multiple wires in parallel depends on the separation between them. For a very large distance between two wires, the net inductance of two wires is half that of a single wire. When the distance between two wires is 4–6 times the diameter of the wires, the net

Table 6.5 Wire Inductance of 1-mil Diameter Gold Wires

Length (mil)	Number of Wires	Loop Height (mil)	Spacing Between Wires (mil)	Inductance Value (nH)
19	1	7	—	0.28
34	1	—	—	0.49
45	1	—	—	0.67
34	2	—	2	0.39
34	2	—	6	0.31
34	2	—	15	0.27
45	2	—	2	0.54
45	2	—	6	0.44
45	2	—	15	0.37
45	3	—	2	0.46
45	3	—	6	0.34
45	3	—	15	0.26
57	2	20	—	0.93
57	3	20	—	0.73
57	9	20	—	0.43
57	13	20	—	0.38
93	2	20	—	1.22
93	8	20	—	0.60
93	12	20	—	0.40
93	14	20	—	0.42

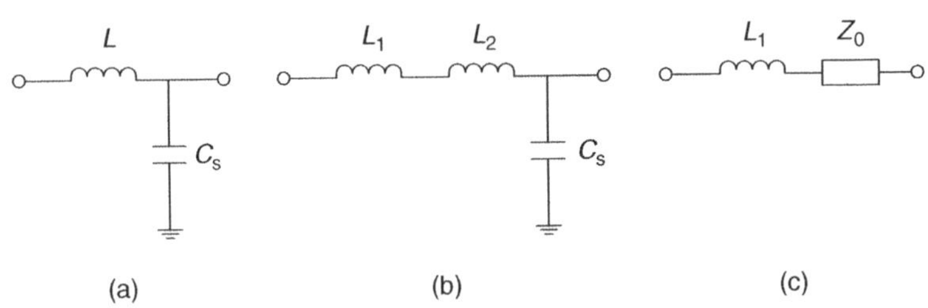

Figure 6.15 Simplified bond wire models.

inductance is only $1/\sqrt{2}$ times that of a single wire, and for n wires it is approximately $1/\sqrt{n}$ times that of a single wire.

When a wire bond inductor is modeled from the measured S-parameter data, it might result in a lower value than the actual value if one is not careful. This can be explained by using Figure 6.15. A simple model of a short wire bond is shown in Figure 6.15a. The series inductance may be split into two parts as shown in Figure 6.15b. A part of the series inductance L_2 with shunt capacitance C_s is equivalent to a 50-Ω line, that is,

$$Z_0 = \sqrt{L_2/C_s} = 50\ \Omega \tag{6.30}$$

This is shown in Figure 6.15c. Thus, during de-embedding, a part of the inductance is absorbed in the de-embedding impedance, which lowers the series inductance value. In order to obtain an accurate model, one must carefully compare both the magnitude and phase of the modeled response with the measured S-parameter data. Also, by measuring the SRF, one can de-embed the shunt capacitance C_s. The SRF is given by

$$f_{\mathrm{res}} = \frac{1}{2\pi\sqrt{LC_s}} \tag{6.31}$$

For example, two 30-mil long wires have $L \cong 0.4$ nH, $C_s = 0.06$ pF, and $SRF = 32.49$ GHz.

6.4.4 Maximum Current Handling of Wire

When a large current is passed through a wire, there is a maximum current value that the wire can withstand due to its finite resistance. At this maximum value, known as the *fusing current*, the wire will melt or burn out due to metallurgical fatigue. The factors affecting the fusing mechanism in a wire are its melting point, resistivity, thermal conductivity, and temperature coefficient of resistance. The fusing current is given by

$$I_{\mathrm{f}} = Kd^{1.5} \tag{6.32}$$

where K depends on the wire material and the surrounding environment and d is the diameter of the wire. Thicker wires have larger current carrying capability than thinner wires. When the wire diameter d is expressed in mm, I_{f} is expressed in A; the K values for gold, copper, and aluminum wires are 183, 80, and 59.2, respectively. For 1-mil diameter wires, I_{f} values for gold, copper, and aluminum wires are 0.74, 0.32, and 0.24 A, respectively. A safer maximum value for the current in wires used in assemblies is about half of the fusing current value. For example, to apply 1-A current one requires three 1-mil diameter wires of gold. When a wire is placed on a thermally conductive material such as Si or GaAs, its fusing current value is higher than the value given above. Longer wires will take a longer time to fuse than shorter wires because of the larger area for heat conduction.

6.5 BROADBAND INDUCTORS

A broadband amplifier working from 10 MHz to 3 GHz requires an RF choke with a high value of inductance (6–10 μH) that is resonance free across the desired frequency range. However, with typical inductors, due to helical construction, high inductance values in the 50-nH to 10-μH range also have very low parallel resonance frequencies (PRFs), which limit their usable bandwidth as chokes. To perform RF choking from 10 MHz to 3 GHz, conical inductors [10], available from Piconics, Inc.[11], can provide both high inductance values for RF to millimeter-wave choking and simultaneous large resonance-free bandwidth. An example of the conical SMT inductor is shown in Figure 6.16. Conical inductors have demonstrated resonance-free bandwidths from 10 MHz to 40 GHz and range in value from 0.5 to 8 μH with current handling capabilities up to 1 A. Table 6.6 lists typical commercially available conical inductors.

The large value of inductance is realized by using a high-permeability ferrite material inside the conical coil. The resonance frequency is increased by reducing the interturn capacitance by using the conical geometry. A typical lumped-element equivalent circuit of a conical inductor is shown in Figure 6.12b. The useful bandwidth of an inductor is determined by its primary inductance and the associated parallel and

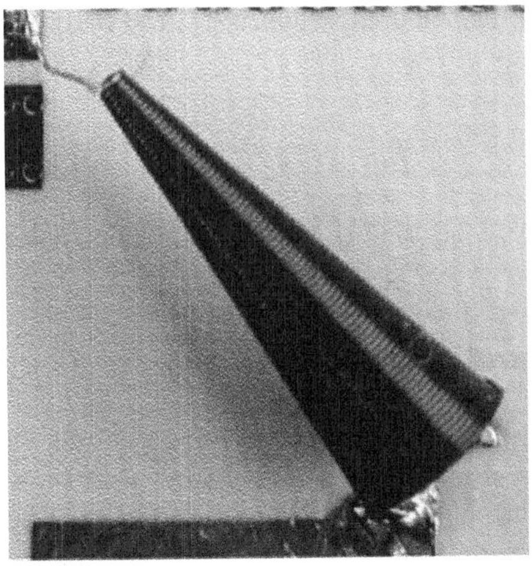

Figure 6.16 Conical inductor from Piconics, Inc. This inductor is 400 mil long and can handle 800 mA of current. The resonance frequency is greater than 32 GHz.

Table 6.6 Standard Piconics Wideband Conical Inductor Selection

Inductor Model	Frequency (GHz)	L (μH)	I_{max} (mA)
CC21T36K240G5	0.01–36	0.531	1000
CC45TH47K240G5	0.01–40	0.84	100
CC50T44K240G5	0.01–20	4.7	250
CC75T36K240G5	0.01–32	6.5	800
CC75T38K240G5	0.01–37	6.93	430
CC82T44K240G5	0.01–36	5.87	230
CC110T47K240G5	0.01–40	8	100

shunt parasitic capacitances. The first limitation is the net feedback capacitance, C_f, that exists across the winding inductance, shown in Figure 6.12b. The second limitation is due to the shunt capacitances, C_1 and C_2, to ground associated with the inductor windings and attach pads. The net parallel winding capacitance, C_f, across an inductor will resonate with the primary inductance and determine the first parallel resonance frequency (PRF). At operating frequencies higher than the PRF, an inductor's dominant reactance is capacitive,

$$PRF = \frac{1}{2\pi\sqrt{C_f L}} \quad \text{and} \quad Q = \frac{\omega L}{R} \tag{6.33}$$

and could no longer be used as a tuning inductor or a choke. At higher frequencies than the PRF, an inductor can have other resonance modes, which are not represented with the simple model in Figure 6.12b. Standard surface mount technology (SMT) inductors have a wide range of inductance values, current handling capability, Q (dissipative loss), and mounting requirements. One of the main trade-offs required when using SMT inductors is between the choice of inductance and usable bandwidth. A simplified de-embedded 6-GHz model for the CC75T36K240G5 conical inductor with leads soldered to an FR-4 substrate was determined [10]. The model values are $L = 6.6$ μH, $R = 0.8$ Ω, $C_f = 0.00007$ pF, $C_1 = 0.1$ pF, and $C_2 = 0.1$ pF. The measured values for Q at 10 MHz and PRF were 518 and 7.4 GHz, respectively. The FR-4 board degrades the inductor's PRF.

The conical shape of the inductor creates its ultrabroad bandwidth feature and an embedded ferrite material increases its total inductance. However, the use of a conical inductor at microwave and millimeter-wave frequencies requires careful assembly. The high-impedance, high-frequency end of the inductor is the end with the smallest coil diameter. In order not to introduce any packaging related parasitic reactance and therefore resonances in the inductor performance, the inductor attach pads must have very low shunt capacitance and the inductor leads must be carefully attached to the pads. The inductor must also be placed sufficiently above its supporting substrate so that no extra interwinding feedback or shunt capacitance is created that could lower the PRF. The conical shape combined with careful assembly to low-capacitance mounting pads allows usable bandwidth of these inductors from 10 MHz to 40 GHz. The conical inductor overcomes the limited bandwidth seen in standard SMT inductors by virtue of its conical design and careful assembly.

In this chapter, basic matching elements and their models useful for the design of amplifier matching networks have been described. The design of matching network topologies is treated in the next chapter.

REFERENCES

1. M. Caulton and H. Sobol, Microwave integrated circuit technology—a survey, *IEEE J. Solid-State Circuits*, Vol. SC-5, pp. 292–303, December 1970.
2. H. Sobol, Applications of integrated circuit technology to microwave frequencies, *Proc. IEEE*, Vol. 59, pp. 1200–1211, August 1971.
3. H. Sobol, Technology and design of hybrid integrated circuits, *Solid State Technol.*, Vol. 13, pp. 49–57, February 1970.
4. K. C. Gupta, R. Garg, I. Bahl, and P. Bhartia, *Microstrip Lines and Slotlines*, 2nd edition, Artech House, Norwood, MA, 1996.
5. I. Bahl and P. Bhartia, *Microwave Solid State Circuit Design*, 2nd edition, John Wiley & Sons, Hoboken, NJ, 2003, Chapter 2.
6. I. Bahl, *Lumped Elements for RF and Microwave Circuits*, Artech House, Norwood, MA, 2003.
7. S. S. Mohan et al., Simple accurate expressions for planar spiral inductances, *IEEE J. Solid-State Circuits*, Vol. 34, pp. 1419–1424, October 1999.
8. R. L. Bunch, D. I. Sanderson, and S. Raman, Quality factor and inductance in differential IC implementations, *IEEE Microwave Mag.*, Vol. 3, pp. 82–92, June 2002.
9. I. J. Bahl, Coupling effects between lumped elements, *IEEE Microwave Mag.*, Vol. 2, pp. 73–79, September 2001.
10. T. A. Winslow, Conical inductors for broadband applications, *IEEE Microwave Mag.*, Vol. 6, pp. 68–72, March 2005.
11. Piconics, Inc., 26 Cummings Rd., Tyngsboro, MA 01879.

PROBLEMS

6.1 A microstrip of characteristic impedance 20 Ω is used at a frequency of 10 GHz with a load that is a microwave transistor's output impedance with conductance 0.05 S in parallel with a 1-pF capacitor. The line is one-eighth wavelengths long at the design frequency. Find the reflection coefficient at the load and the input admittance. If the dielectric constant of the substrate is 2.5, what is the physical length of the line?

6.2 Calculate various parameters such as Z_0, β, and α for a microstrip with $\varepsilon_r = 10$, $W = 1$ mm, and $h = 0.6$ mm at 10 GHz. Here, gold conductors are 6 μm thick. Also, calculate Q for a $\lambda/2$ resonator.

6.3 A microstrip line of characteristic impedance Z_{01} is connected to another microstrip of higher characteristic impedance Z_{02}. If the step discontinuity reactance is represented by a series inductance (jX), determine the S-matrix for the junction.

6.4 A microstrip line having a characteristic impedance of 50 Ω has a width $W = 1$ mm. The dielectric constant of the substrate is $\varepsilon_r = 10$. What is the limiting frequency at which radiation starts? Is it necessary to correct for dispersion when operating at that frequency?

6.5 Since modes on microstrip lines are only quasi-TEM, derive from the basic theory of a lossless line that the inductance L and capacitance C per unit length of a microstrip line are given by

$$L = \frac{Z_0}{v} = \frac{Z_0\sqrt{\varepsilon_{re}}}{c}$$

$$C = \frac{1}{Z_0 v} = \frac{\sqrt{\varepsilon_{re}}}{Z_0 c}$$

where Z_0 = characteristic impedance of the microstrip line
v = wave velocity in the microstrip line
$c = 3 \times 10^8 \text{m}/s$, the velocity of light in vacuum
ε_{re} = effective dielectric constant

Also, determine L, C, and the time delay per unit meter for the microstrip described in Problem 6.4.

6.6 Transistors are biased through 1-mil diameter gold wires. If the supply current is 1 A, determine the number of wires requires. If wires are connected in parallel, 30-mil long, separated by 5 mil, and are located 25-mil above the ground plane, determine the inductance of the wires.

6.7 In an amplifier working over 4–6 GHz, the transistor bias is applied through a 1-mil bond wire, which acts as an RF choke (see Chapter 18). When the system impedance is 50 Ω, Determine the length of the wire so that its loading effect measured in the form of return loss is greater than 20 dB.

6.8 Calculate physical length at 2 GHz of a quarter-wavelength long 500-μm wide microstrip on FR-4 having $\varepsilon_r = 4$ and $h = 250$ μm. Also, determine its total capacitance and inductance. When its substrate thickness is doubled, what effect does it has on length, impedance, capacitance, and inductance?

6.9 Repeat the calculations done for Problem 6.8 for an alumina substrate, $\varepsilon_r = 9.9$, and show which line has maximum capacitance and maximum inductance.

6.10 A power device in a chip form needs to be connected to a bias line using gold wires. The maximum rated current for the device is 5 A. Determine how many 1-mil diameter gold wires are needed for safe operation. If wires are 40 mil long, calculate their total inductance.

Chapter 7

Impedance Matching Techniques

Impedance transformation or matching is an essential part of an amplifier design. Impedance matching is important to achieve optimum noise figure, gain, power, and PAE performance over the required bandwidth by efficiently transferring the signal from a generator to a device, from a device to a load, or between devices. There are a variety of matching networks [1–11] that can be used for this purpose. The important factors to be considered in the selection of matching networks are (a) bandwidth, (b) frequency response, (c) complexity, and (d) ease of implementation. The implementation also depends on the required fabrication technology, such as a printed circuit board, hybrid, internally matched, or monolithic. In this chapter, impedance matching techniques for narrowband and wideband applications are discussed. Mostly, the discussion is generic and an appropriate reference to a particular technology will be made wherever applicable.

7.1 ONE-PORT AND TWO-PORT NETWORKS

The usual impedance transformation problem is considered to be one-port in nature where a complex or a real load Z_L is to be matched to a real source impedance Z_S as shown in Figure 7.1a. Bode [12] and Fano's results [13], which are discussed in detail in Section 7.3, show that an ideal impedance matching network for a complex load would be a bandpass filter that cuts off sharply at the band edges. Alternatively, a lowpass or a highpass filter design can be used to match load impedance. Lowpass filter structures can provide a relatively broadband impedance match for microwave loads that can be approximated by an inductance and resistance in series, or by a capacitance and conductance in parallel. Similarly, highpass networks may be used for loads that can be approximated by a capacitance and resistance in series or by an inductance and conductance in parallel. But both lowpass and highpass networks are suitable for limited applications compared with bandpass networks because of some inherent disadvantages.

A more generalized matching network synthesis is based on the network theory as described in Chapter 2. Here the transistor is represented by a two-port (input and output) Y, Z, or scattering coefficient (S) matrix. The increased complexity of matching network design based on an active two-port is necessary in several cases because of the coupling of the input and output circuits through the device. For example, in

Fundamentals of RF and Microwave Transistor Amplifiers. By Inder J. Bahl

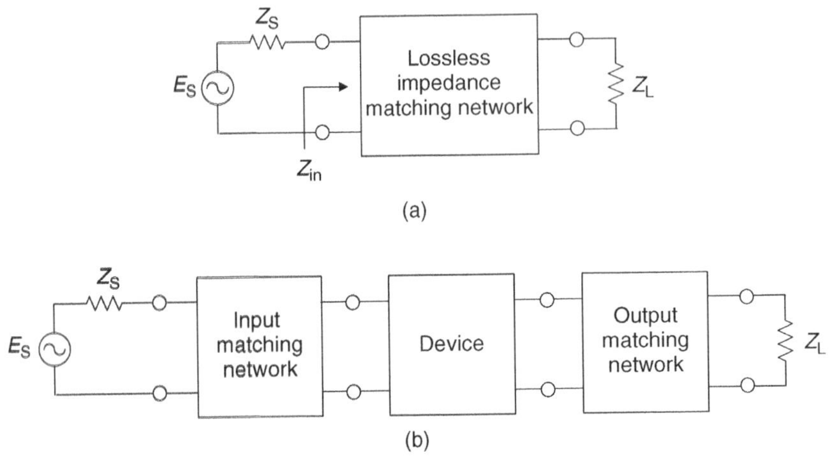

Figure 7.1 (a) One-port and (b) two-port impedance matching networks.

an amplifier design the input matching network is dependent on the output matching network (Fig. 7.1b), and vice versa due to limited device isolation. The transistor parameters, however, may be measured under actual operating conditions for a meaningful matching network design. Active devices or microwave networks, which require impedance match, can be completely characterized in terms of parameters measured at the network terminals without regard to the contents of the network. Although a particular device may have any number of ports, the two-port representation is applicable only between the input and the output ports, while all the other ports are terminated in known impedances. For example, a two-port representation of a three-terminal FET may characterize the device behavior between the gate as input and drain as the output port, while the source is assumed to be RF and DC grounded.

In this chapter, narrowband and wideband matching techniques using a one-port network representation will be discussed. For the sake of simplicity, mostly lossless matching elements are considered. The design of lossless matching networks consisting of lumped, distributed, and combined lumped and distributed elements will be presented in the following sections. Since the transformers are reciprocal in nature, reverse transformation is also valid.

7.2 NARROWBAND MATCHING TECHNIQUES

In this section, narrowband matching techniques are discussed with examples. The bandwidth is defined as the frequency range over which the tolerable reflection coefficient (specified as a design goal, usually with a value of 0.33) is attained. Therefore the matching is accomplished perfectly only at the desired frequency range, which is nominally around 10–20%, but less than 40%.

7.2.1 Lumped-Element Matching Techniques

Matching networks for RF and microwave circuits are generally designed to provide a specified electrical performance over the required bandwidth. To realize compact circuits, lumped-element matching networks are utilized to transform the transistor impedance to 50 Ω. At RF, lumped elements are commonly used in matching networks. The treatment of these components is already provided in the previous chapter. Large impedance transformation ratios can easily be accomplished using the lumped-element approach. Therefore high-power transistors that have very low impedance values can easily be matched using lumped elements. This section describes various inductor–capacitor (LC) configurations suitable for designing matching networks.

Impedance can be represented by a parallel combination of a resistance and a reactance or an equivalent series combination of a resistance and reactance as shown in Figure 7.2. Thus one can convert a "parallel network" to an equivalent "series network" using the following relations:

$$R_S = R_P X_P^2/(R_P^2 + X_P^2) \tag{7.1a}$$

$$X_S = X_P R_P^2/(R_P^2 + X_P^2) \tag{7.1b}$$

Similarly, a "series network" can be converted into a "parallel network" using the following equations:

$$R_P = (R_S^2 + X_S^2)/R_S \tag{7.2a}$$

$$X_P = (R_S^2 + X_S^2)/X_S \tag{7.2b}$$

L-Network

Inductors and capacitors connected in an L-section configuration are widely used as impedance matching circuit elements. As shown in Figure 7.3, there are eight possible arrangements of inductors and capacitors which can be utilized. The range of impedance transformation depends on the value of the original inductors or capacitors in the L-configuration. Rearranging Eq. (7.1a) or (7.2a), we have

$$\frac{R_P}{R_S} = 1 + Q^2 \tag{7.3a}$$

where

$$Q = X_S/R_S \quad \text{or} \quad R_P/X_P \quad \text{or} \quad [R_P/R_S - 1]^{1/2} \tag{7.3b}$$

In Figure 7.2, the circuits are equivalent and for a given Z, $R_P > R_S$. Thus when a reactance X_S is added in series with a resistor R_S, and converted to an equivalent parallel combination, the resistance increases by a $1 + Q^2$ factor. Conversely, when a reactance X_P is added in parallel with a resistor R_P, and converted to an equivalent series combination, the resistance decreases. These properties of network conversion

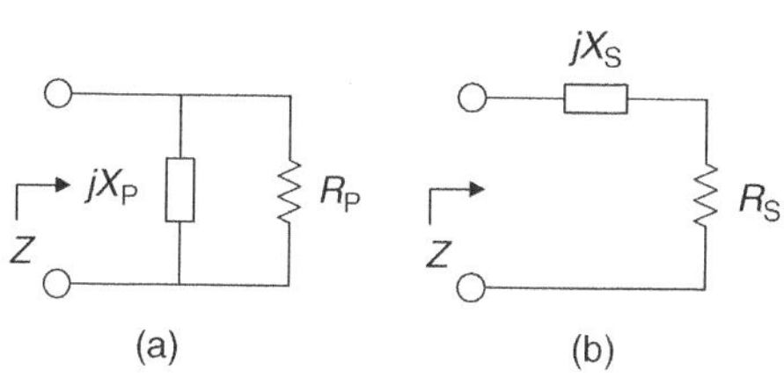

Figure 7.2 One-port impedance network configurations: (a) shunt and (b) series.

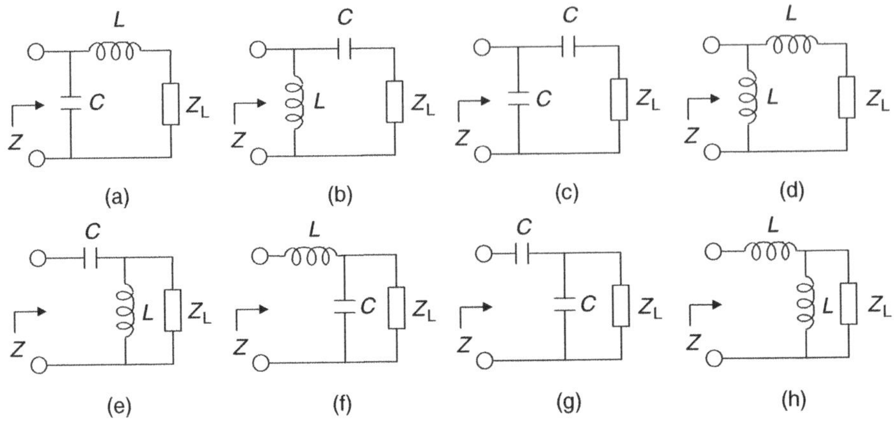

Figure 7.3 L-network impedance matching topologies.

are used to design L-section matching networks for step-up or step-down impedance transformation.

This is illustrated by considering an example of transforming a real source impedance R'_S to real load impedance R'_L. Here $R'_L < R'_S$. In this case, a matching network shown in Figure 7.4 is selected. By equating a parallel circuit (Fig. 7.2a) to a parallel combination of R'_S and C'_S ($R_P = R'_S$ and $X_P = 1/\omega C'_S$) and a series circuit (Fig. 7.2b) to a series combination of L'_S and R'_L ($R_S = R'_L$ and $X_S = \omega L'_S$), from (7.3), we have

$$Q = \omega L'_S/R'_L \text{ or } R'_S\omega C'_S \tag{7.4a}$$

$$C'_S = \frac{Q}{\omega R'_S} \tag{7.4b}$$

$$L'_S = QR'_L/\omega \tag{7.4c}$$

$$Q = [R'_S/R'_L - 1]^{1/2} \tag{7.4d}$$

EXAMPLE 7.1

Design a matching network to transform a 50-Ω source impedance to a 10-Ω load impedance at 10 GHz.

SOLUTION Referring to Figure 7.4, we have

$$R'_S = 50\ \Omega,\ R'_L = 10\ \Omega,\ Q = [50/10 - 1]^{1/2} = 2,\ \omega = 2\pi \times 10 \times 10^9 = 6.283 \times 10^{10} \text{ rad/s}$$

From (7.4b), we find

$$C'_S = \frac{2}{6.283 \times 10^{10} \times 50} F = 0.637 \text{ pF}$$

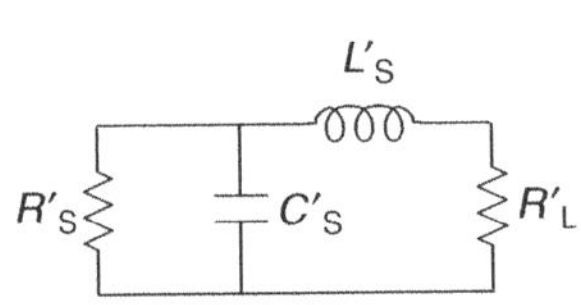

Figure 7.4 L-section impedance matching topology.

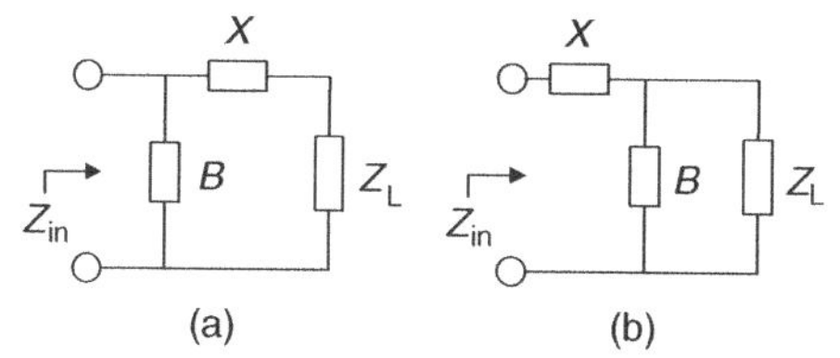

Figure 7.5 Basic lumped-element based impedance matching topologies.

From (7.4c), we have

$$L'_S = 2 \times 10/(6.283 \times 10^{10})H = 0.318 \text{ nH} \qquad \blacksquare$$

Two basic topologies of L-section matching networks are shown in Figure 7.5. These networks consist of purely reactive elements. The impedances of the series reactance and shunt susceptance elements are represented by jX and jB, respectively. The matching areas for the above configurations are shown in Figure 7.6. The areas in gray show the locations of loads that cannot be matched with the respective network [11]. Consider a design problem where the complex load impedance $Z_L (= R_L + jX_L)$ is required to be transformed to a real impedance $Z_{in} (= Z_0)$.

When $R_L < Z_{in}$, the configuration in Figure 7.5a is used and

$$Z_{in} = Z_0 = \left[jB + \frac{1}{R_L + j(X + X_L)} \right]^{-1} \tag{7.5}$$

By equating real and imaginary parts, we find

$$X = \pm\sqrt{R_L(Z_0 - R_L)} - X_L \tag{7.6a}$$

$$B = \pm\frac{\sqrt{(Z_0 - R_L)/R_L}}{Z_0} \tag{7.6b}$$

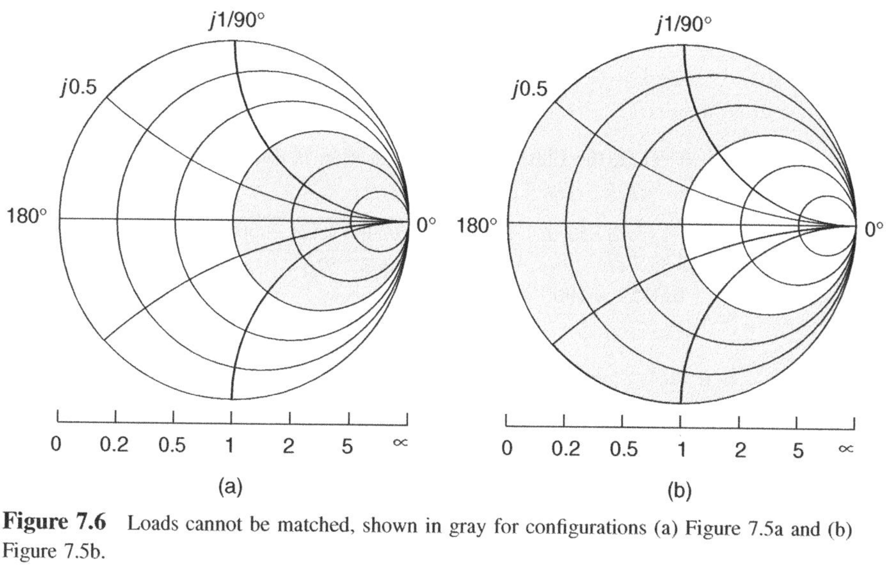

Figure 7.6 Loads cannot be matched, shown in gray for configurations (a) Figure 7.5a and (b) Figure 7.5b.

The positive and negative signs imply that there are two possible solutions. Positive reactance corresponds to an inductor and negative reactance implies a capacitor. Similarly, a positive susceptance corresponds to a capacitor and a negative susceptance implies an inductor.

If $R_L > Z_{in}$, the configuration in Figure 7.5b is used and

$$Z_{in} = Z_0 = jX + \frac{1}{jB + (R_L + jX_L)^{-1}} \tag{7.7}$$

Again, by equating real and imaginary parts,

$$B = \frac{X_L \pm \sqrt{R_L/Z_0}\sqrt{R_L^2 + X_L^2 - Z_0 R_L}}{R_L^2 + X_L^2} \tag{7.8}$$

$$X = \frac{1}{B} + \frac{X_L Z_0}{R_L} - \frac{Z_0}{B R_L} \tag{7.9}$$

Again, there are two possible solutions. Once the values of X and B are determined, one could proceed to calculate the values of the lumped elements needed at the design frequency. The above design equations are valid over a narrow frequency range (10–20%) around the center frequency.

EXAMPLE 7.2

Consider a device whose load impedance (Z_L) is represented by a parallel combination of resistor and capacitor and their values for a 625-μm device, operating at 10 V, are 90 Ω and −0.19 pF, respectively. Synthesize an LC network to match the load to $Z_{in} = Z_0 = 50\ \Omega$ at 5 GHz.

SOLUTION Using (7.1), we have

$$Z_L = R_L + jX_L = 90\ \Omega \| -0.19\ \text{pF} = 69.84 + j37.52\ \Omega$$

Since $R_L > Z_{in} = 50\ \Omega$, the Figure 7.5b configuration is used. From (7.8) and (7.9), we have

$$B = \omega C = \frac{37.52 \pm 1.1819\sqrt{69.84^2 + 37.52^2 - 50 \times 69.84}}{69.84^2 + 37.52^2} = 0.0159\ \text{S}$$

$$C = \frac{0.0159 \times 1000}{2\pi \times 5}\ \text{pF} = 0.506\ \text{pF}$$

$$X = \omega L = \frac{1}{0.0159} + \frac{37.52 \times 50}{69.84} - \frac{50}{0.0159 \times 69.84} = 44.72\ \Omega$$

$$L = \frac{44.72}{2\pi \times 5}\ \text{nH} = 1.42\ \text{nH}$$

The calculated fractional bandwidth for return loss >20 dB or $\rho_m < 0.1$ is 34%. ■

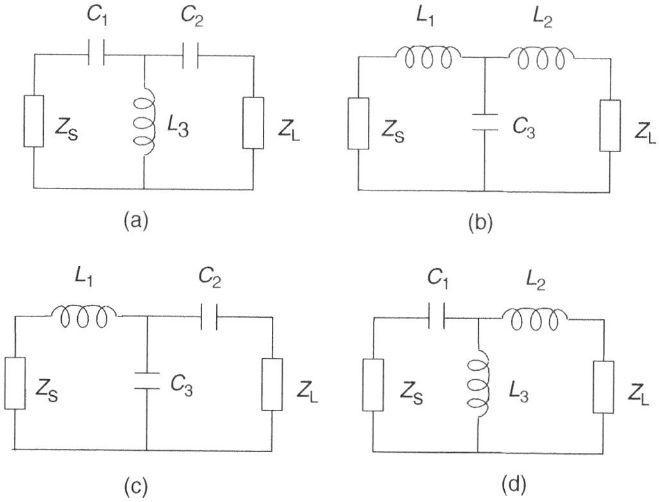

Figure 7.7 Impedance matching T-network configurations.

Tee and Pi Networks

In order to get a larger bandwidth and realize large impedance ratios, more elements are required in the matching networks. The "tee" and "pi" arrangements of lumped elements are commonly used. Such arrangements are simply considered back-to-back L-section networks. Figure 7.7 shows the T-type matching network. The addition of one more element to the simple L-section matching circuit gives the designer much greater control over the bandwidth and also permits the use of more practical circuit elements. For the T-network, circuit configurations and circuit element values, shown in Figure 7.7, are given below.

For Figure 7.7a,

$$C_1 = (\omega_0 Z_S \sqrt{N-1})^{-1} \tag{7.10a}$$

$$C_2 = (\omega_0 Z_L \sqrt{N/M-1})^{-1} \tag{7.10b}$$

$$L_3 = N Z_S / [\omega_0 (\sqrt{N-1} + \sqrt{N/M-1})] \tag{7.10c}$$

For Figure 7.7b,

$$L_1 = (Z_S \sqrt{N-1})/\omega_0 \tag{7.11a}$$

$$L_2 = (Z_L \sqrt{N/M-1})/\omega_0 \tag{7.11b}$$

$$C_3 = [(\sqrt{N-1} + \sqrt{N/M-1})]/(\omega_0 N Z_S) \tag{7.11c}$$

For Figure 7.7c,

$$L_1 = (Z_S \sqrt{N-1})/\omega_0 \tag{7.12a}$$

$$C_2 = (\omega_0 Z_L \sqrt{N/M-1})^{-1} \tag{7.12b}$$

$$C_3 = [(\sqrt{N-1} - \sqrt{N/M-1})]/(\omega_0 N Z_S) \tag{7.12c}$$

For Figure 7.7d,

$$C_1 = (\omega_0 Z_S \sqrt{N-1})^{-1} \tag{7.13a}$$

$$L_2 = (Z_L \sqrt{N/M - 1})/\omega_0 \tag{7.13b}$$

$$L_3 = N Z_S / [\omega_0 (\sqrt{N-1} - \sqrt{N/M - 1})] \tag{7.13c}$$

Here,

$$M = \frac{Z_L}{Z_S} > 1 \quad \text{and} \quad N > M \tag{7.14}$$

M is the impedance transformation ratio and N is a variable. By properly selecting N, a compromise is obtained in terms of bandwidth and realizable circuit element values.

When $N/M \cong 1$, the networks provide the largest bandwidth, responses of the networks in Figure 7.7a and 7.7d are identical, and the networks provide improved bandwidth at higher frequencies. On other hand, responses of networks in Figure 7.7b and 7.7c are identical, and they provide improved bandwidth at lower frequencies. These trends remain the same for higher values of N/M. However, for $N/M \cong 1$, the value of C_2 becomes very large and L_2 becomes very small.

EXAMPLE 7.3

Determine network element values of Figure 7.7 when $Z_s = 10\ \Omega$, $Z_L = 50\ \Omega$, $N/M = 1.05$, and the design center frequency is 10 GHz. Compare reflection responses of these matching networks at the load port over 8 to 12 GHz.

SOLUTION In this case,

$$M = Z_L/Z_S = 5$$

The calculated element values using (7.10)–(7.14) are

(a) $C_1 = 0.722$ pF, $C_2 = 1.424$ pF, $L_3 = 0.366$ nH
(b) $L_1 = 0.328$ nH, $L_2 = 0.178$ nH, $C_3 = 0.693$ pF
(c) $L_1 = 0.328$ nH, $C_2 = 1.424$ pF, $C_3 = 0.557$ pF
(d) $C_1 = 0.772$ pF, $L_2 = 0.178$ nH, $L_3 = 0.455$ nH

A plot of reflection coefficient magnitude as a function of frequency is shown in Figure 7.8. ■

7.2.2 Transmission Line Matching Techniques

In this section the commonly used impedance transforming properties of transmission lines for designing narrowband as well as broadband matching networks are reviewed. A standard transmission line, such as a microstrip, can be used as a series transmission line, as an open-circuited or a short-circuited stub, and as a quarter-wavelength transformer section for impedance matching.

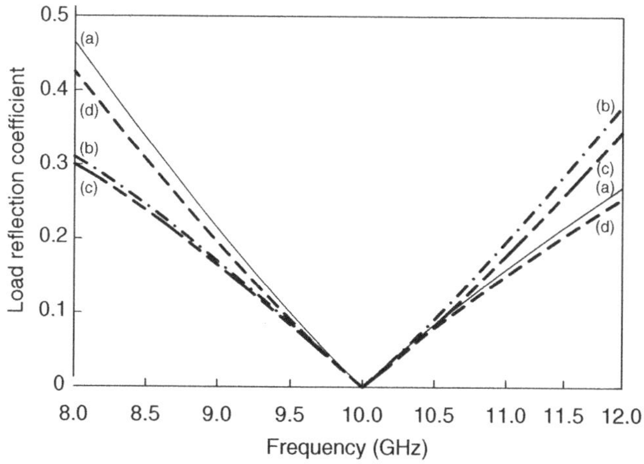

Figure 7.8 Magnitude of load reflection coefficient versus frequency for four cases described in Example 7.3.

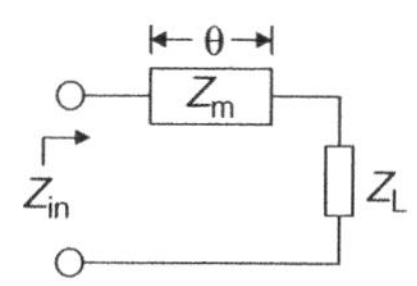

Figure 7.9 A quarter-wave transmission line section impedance transformer.

Quarter-Wavelength Transformer

The most important class of impedance matching network is the quarter-wave impedance transformer, which is used to match a real impedance to another real impedance. The input impedance of a lossless network shown in Figure 7.9 is given by

$$Z_{in} = Z_m \frac{Z_L + jZ_m \tan\theta}{Z_m + jZ_L \tan\theta} = Z_0 \frac{1+\Gamma}{1-\Gamma} \tag{7.15}$$

where Z_m is the characteristic impedance of the line, $\theta = \beta l$, and Γ is the reflection coefficient.

When $Z_{in} = R_S = Z_0$ at $\theta = \pi/2$,

$$Z_0 = \frac{Z_m^2}{Z_L} \tag{7.16}$$

If Z_m is chosen equal to $\sqrt{Z_0 Z_L}$, the load impedance Z_L is matched to Z_0 at the center frequency, where the transformer length corresponds to $(\lambda/4 + n\lambda/2)$. For TEM transmission lines, $\theta = \pi f/(2f_0)$, where f_0 is the frequency for which $\theta = \pi/2$. In this case, the fractional bandwidth can be expressed as

$$\frac{\Delta f}{f_0} = 2\left(1 - \frac{f}{f_0}\right) = 2\left(1 - \frac{2}{\pi}\theta_m\right) \tag{7.17a}$$

where

$$\theta_m = \cos^{-1}\left(\frac{2\Gamma_m}{\sqrt{1-\Gamma_m^2}}\frac{\sqrt{Z_0 Z_L}}{Z_L - Z_0}\right) \tag{7.17b}$$

and Γ_m is the maximum value of reflection coefficient in the required passband. The fractional bandwidth, FBW or simply BW, is defined as

$$BW = \frac{\Delta f}{f_0} \tag{7.18a}$$

$$\Delta f = f_2 - f_1, \qquad f_0 = \sqrt{f_1 f_2} \tag{7.18b}$$

where f_1 and f_2 are the lower and upper edges of the frequency band.

EXAMPLE 7.4

A load impedance of 100 Ω is to be matched to 50 Ω using a $\lambda/4$ microstrip. Determine its characteristic impedance, physical length, and bandwidths for maximum reflection coefficients of 0.1 and 0.2. The frequency is 10 GHz and the effective dielectric constant of microstrip $\varepsilon_{re} = 6.25$.

SOLUTION Using (7.16), we have

$$Z_m = \sqrt{100 \times 50} = 70.7\ \Omega$$

$$\ell = \frac{\lambda}{4} = \frac{\lambda_0}{4\sqrt{\varepsilon_{re}}} = \frac{3 \times 10^{10}}{4 \times 10 \times 10^9 \times \sqrt{6.25}}\ \text{cm} = 3\ \text{mm}$$

From (7.17), the bandwidth for $\Gamma_m = 0.1$ is given by

$$BW = 2 - \frac{4}{\pi}\cos^{-1}\left[\frac{2 \times 0.1}{\sqrt{1-0.1^2}}\frac{70.7}{50}\right] = 0.37 \text{ or } 37\%$$

For $\Gamma_m = 0.2$,

$$BW = 2 - \frac{4}{\pi}\cos^{-1}\left[\frac{2 \times 0.2}{\sqrt{1-0.2^2}}\frac{70.7}{50}\right] = 0.78 \text{ or } 78\%$$

Thus for double the reflection coefficient, the BW is approximately doubled. ■

Series Single-Section Transformer

Simplest of all matching networks is a series transmission line of electrical length θ and characteristic impedance Z_m used to match a complex load $Z_L (= R_L + jX_L)$ to a real resistance $Z_S = R_S$ as shown in Figure 7.10a. In this case (7.15) becomes

$$R_S = Z_{in} = Z_m \frac{R_L + j(X_L + Z_m \tan\theta)}{Z_m - X_L \tan\theta + jR_L \tan\theta} \tag{7.19}$$

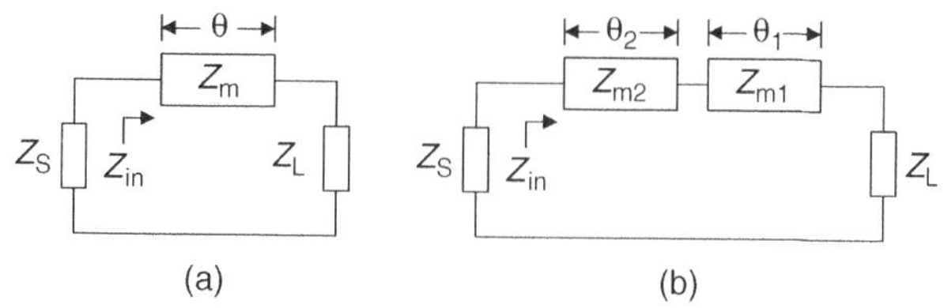

Figure 7.10 Series transmission line section impedance transformers: (a) single section and (b) two sections.

Equating real and imaginary parts, we have

$$R_S(Z_m - X_L \tan\theta) = Z_m R_L \tag{7.20a}$$

$$R_S R_L \tan\theta = Z_m(X_L + Z_m \tan\theta) \tag{7.20b}$$

or

$$Z_m = \frac{\sqrt{R_S R_L - (R_L^2 + X_L^2)}}{1 - R_L/R_S} \tag{7.21a}$$

and

$$\tan\theta = \frac{\sqrt{(1 - R_L/R_S)[R_S R_L - (R_L^2 + X_L^2)]}}{X_L} \tag{7.21b}$$

It is a narrowband matching technique with limited use, since only those impedances can be matched that result in a real value of Z_m in (7.21a).

In a practical design, the characteristic impedance of the transformer is limited by the type of transmission line used. For example, the impedance values lie between 20 and 100 Ω when microstrip lines or strip lines are used. Using (7.21), one can show that with these restrictions on the impedance values tuning can be done over approximately 22% of the Smith chart [7]. This range can be enhanced by using two transmission lines in series, as shown in Figure 7.10b. One section brings the impedance into the matchable area and the other one matches it to the source impedance. The two-line section transformer with characteristic impedances limited within the 20- to 100-Ω range can match loads over 75% of the Smith chart.

When $\theta = \lambda/8 = 45^\circ$,

$$Z_m = |Z_L| = [R_L^2 + X_L^2]^{1/2} \tag{7.22a}$$

$$R_S = \frac{Z_m R_L}{Z_m - X_L} \tag{7.22b}$$

Thus a $\lambda/8$ transmission section transforms complex impedance to real impedance given by (7.22b).

Quarter-Wave and $\lambda/8$ Transformer

The complex load impedance (Z_L) can also be converted into a real impedance (R_S) by using two transmission line sections as shown in Figure 7.11. Here the matching section (45°) transforms the complex impedance into a real impedance and the $\lambda/4$ section transforms the real intermediate impedance into a real impedance R_S.

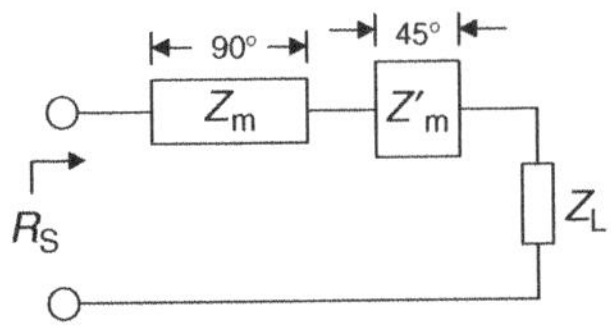

Figure 7.11 Two series transmission line section impedance transformer.

The initial section (45°) of this network has characteristic impedance equal to the magnitude of the terminating impedance,

$$Z'_m = \sqrt{R_L^2 + X_L^2} \tag{7.23a}$$

$$R'_S = \frac{Z'_m R_L}{Z'_m - X_L} \tag{7.23b}$$

and the final section (90°) has the characteristic impedance

$$Z_m = \sqrt{\frac{R_S R_L Z'_m}{Z'_m - X_L}} \tag{7.23c}$$

EXAMPLE 7.5

Consider a device whose load impedance (Z_L) is represented by a parallel combination of a resistor and a capacitor and their values for a 1.25-mm device, operating at 8 V, are 36 Ω and −0.38 pF, respectively. The load impedance is to be matched to 50 Ω using a series combination of $\lambda/8$ and $\lambda/4$ microstrip sections.

Determine its characteristic impedance, physical length, and bandwidths for maximum reflection coefficient of 0.1 and 15-mil alumina substrate. The frequencies are 5 and 10 GHz.

SOLUTION The load impedance values are

$$\begin{aligned} Z_L &= 30.39 + j13.06\ \Omega \quad \text{at 5 GHz} \\ &= 20.7 + j17.8\ \Omega \quad \text{at 10 GHz} \end{aligned}$$

At 5 GHz, for a $\lambda/8$ section, using (7.23a), we find

$$Z'_m = \sqrt{30.39^2 + 13.06^2} = 33.08\ \Omega$$

$$R'_S = \frac{33.08 \times 30.39}{33.08 - 13.06} = 50.2\ \Omega$$

Since impedance looking into the $\lambda/8$ section is very close to 50 Ω, there is no need for a $\lambda/4$ matching section in this case. Using a CAD, the calculated values of bandwidth, line width, and length are 67%, 30.3 mils, and 108.3 mils, respectively.

At 10 GHz, using (7.23a), we find

$$Z'_m = \sqrt{20.7^2 + 17.8^2} = 27.3\ \Omega$$

$$R'_S = \frac{27.3 \times 20.7}{27.3 - 17.8} = 59.49\ \Omega$$

For a $\lambda/4$ section, using (7.23c), we find

$$Z_m = \sqrt{50 \times 59.49} = 54.54\ \Omega$$

Using a CAD, the calculated value of bandwidth is 34% and calculated line width and length are 41 mils and 53 mils for the $\lambda/8$ section, and 11.8 mils and 115 mils for the $\lambda/4$ section, respectively. Thus bandwidth of the transformer at twice the operating frequency became approximately half because of the reactance part of the load. ■

Complex to Complex Impedance Transformer

Complex to complex impedance transformation is needed for an amplifier's interstage and can be achieved using the techniques already discussed in this section. Consider Figure 7.9. By replacing Z_{in} in (7.15) with the complex conjugate of the source impedance $Z_S = R_S + jX_S$, the transformer design equations are obtained. These are

$$Z_m = \left(\frac{R_S|Z_L|^2 - R_L|Z_S|^2}{R_L - R_S}\right)^{1/2} \tag{7.24a}$$

$$\tan\theta = \frac{Z_m(R_L - R_S)}{R_L X_S - R_S X_L} \tag{7.24b}$$

where $Z_S^2 = R_S^2 + X_S^2$ and $Z_L^2 = R_L^2 + X_L^2$. A three-section complex to complex impedance transformer similar to Figure 7.11 is shown in Figure 7.12.

Stub Matching

Alternatively, a shunt single stub or multiple stubs are used to match complex impedances between the generator and the input of an active device or output of the device and the load. Double-stub matching is needed in cases where it is impractical to place a single stub physically in the ideal location. Location and length of open-circuited or short-circuited stubs is easily calculated using the Smith chart [11].

Single-Stub Matching

Since the stub matching problems involve parallel connections on the transmission lines, it is easier to design the circuit using admittance values rather than impedances. In a single-stub matching network, the total admittance of the terminated line and the stub is matched with the admittance of the source for maximum power transfer. For example, in Figure 7.13, the stub is located at AA in such a way that

$$Y_{AA} = Y_0 = Y_S + Y_D \tag{7.25}$$

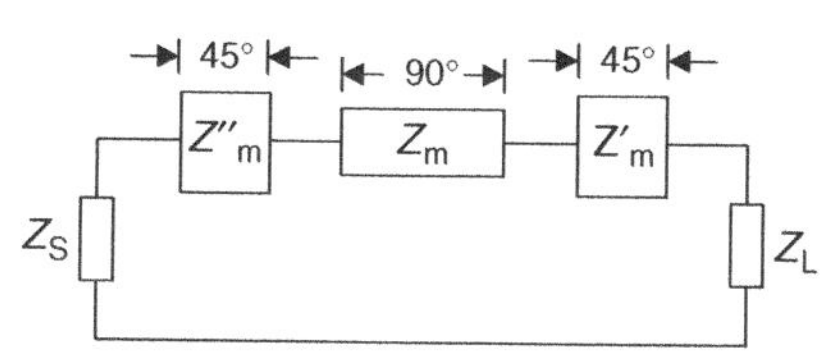

Figure 7.12 Three series transmission line section impedance transformer.

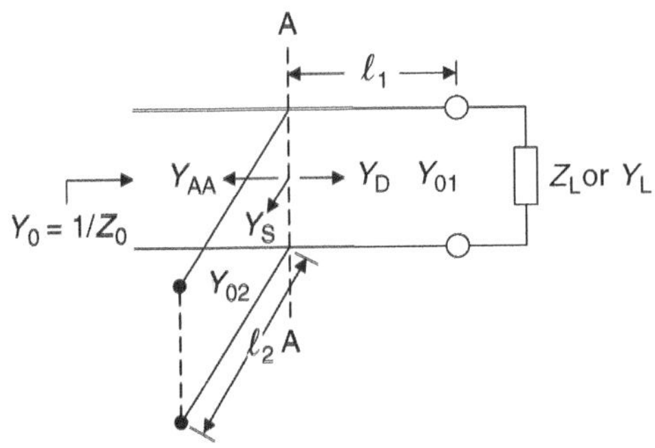

Figure 7.13 Single-stub impedance matching problem.

where Y_S is the admittance of the stub (short-circuited or open-circuited) of length ℓ_2, and Y_D is the admittance of the load transformed at AA location. The expressions for Y_S and Y_D are given as

$$Y_S = -jY_{02}\cot\theta_2 \quad \text{(short-circuited)} \tag{7.26a}$$

$$= jY_{02}\tan\theta_2 \quad \text{(open-circuited)} \tag{7.26b}$$

$$Y_D = Y_{01}\frac{Y_L + jY_{01}\tan\theta_1}{Y_{01} + jY_L\tan\theta_1} = G_D + jB_D \tag{7.27}$$

where $\theta_i = \beta\ell_i = (2\pi/\lambda)\ell_i$ ($i = 1, 2$) and λ is the guide wavelength. Under matched conditions,

$$Y_0 = G_D \tag{7.28a}$$

and

$$Y_S + B_D = 0 \tag{7.28b}$$

These equations are solved to determine ℓ_1, ℓ_2 and Y_{01}, Y_{02}. Generally, the lengths are kept less than $\lambda/4$. When $Y_{01} = Y_0$, expressions for ℓ_1 and ℓ_2 (short-circuited stub) are given by

$$\ell_1 = \frac{1}{\beta}\tan^{-1}\left[\frac{X_L \pm [R_L\{(Z_0 - R_L)^2 + X_L^2\}/Z_0]^{1/2}}{R_L - Z_0}\right] \tag{7.29}$$

$$\ell_2 = \frac{1}{\beta}\cot^{-1}\left[\frac{Z_{02}[R_L^2 tt - (Z_0 - X_L tt)(X_L + Z_0 tt)]}{Z_0[R_L^2 + (X_L + Z_0 tt)^2]}\right] \tag{7.30}$$

where $tt = \tan\theta_1$.

Double-Stub Matching

A double-stub matching network consists of two stubs (short-circuited sections preferred, because they are easier to obtain than a good open circuit) connected in parallel with a fixed length between them (Fig. 7.14). Usually the length of the transmission line between the stubs is 1/8, 3/8, or 5/8 of a wavelength. Lengths and impedances

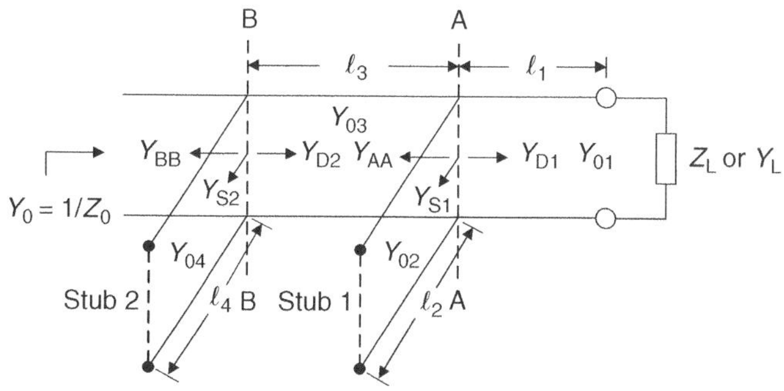

Figure 7.14 Double-stub impedance matching problem.

of both stubs and the location of the stub closest to the load can be adjusted to get a perfect match. In Figure 7.14,

$$Y_{BB} = Y_{S2} + Y_{D2} = Y_0 \tag{7.31}$$

where Y_{BB} is total admittance at the left of the second stub, Y_{S2} is the admittance of the second stub, Y_{D2} is the admittance $Y_{AA}(= Y_{S1} + Y_{D1})$ at location AA transformed to location BB by the line section between the two stubs, Y_{S1} is the admittance of stub one, and Y_{D1} is the load admittance transformed to the location of the first stub. There is no definite procedure for solving a double-stub matching problem. One can go through an iterative design, starting with an arbitrary location for the first stub (closest to the load).

Figure 7.15 shows a configuration for a quarter-wavelength double-stub transformer. The electrical lengths of the stubs are

$$\theta_1 = \tan^{-1}[-b_L \mp \sqrt{g_L(1 - g_L)}] \tag{7.32a}$$

$$\theta_2 = \mp \tan^{-1}[(1 - g_L)/g_L] \tag{7.32b}$$

where $R_S = Z_m$ and $g_L + jb_L = Z_m/Z_L$.

EXAMPLE 7.6

A load impedance, $Z_L = 25 + j25\ \Omega$ is to be transformed to an impedance of 25 Ω. Design a matching network consisting of an open-circuited stub having characteristic impedance of 50 Ω and a quarter-wave transformer to accomplish the transformation. If the load is to be transformed to a system impedance of 50 Ω, does one need the quarter-wave transformer? Use the Smith chart, described in Appendix F, to determine the matching networks.

SOLUTION The normalized (with respect to 50 Ω) load impedance is $Z_L' = (25 + j25)/50 = 0.5 + j0.5$ and is point A on the Smith chart shown in Figure 7.16a. Draw the constant-reflection coefficient circle that passes through point A and point B. Point B is at a distance of $\lambda/2$ from A, representing a normalized load admittance $Y_L' = 1 - j1$. A normalized susceptance of $+1$ is needed across the load to cancel the load normalized susceptance of -1. The susceptance of $+1(= 1/50)$ is to be obtained by means of an open-circuited stub. The length (ℓ_1) of the open-circuited stub needed is 0.125 λ. After connecting this stub across the load, the effective normalized load admittance will be 1 or the unnormalized load impedance will be 50 Ω. Thus, in this case, one stub is sufficient to match the load impedance to 50 Ω. The effective load impedance of 50 Ω can be transformed to 25 Ω by means of a quarter-wave impedance transformer having a $Z_{0T} = \sqrt{50 \times 25} = 35.35$ Ω and $\ell_2 = 0.25\lambda$ Figure 7.16b shows the matching network. ■

7.3 WIDEBAND MATCHING TECHNIQUES

In the previous section narrowband impedance matching techniques were presented. These techniques provide a perfect impedance match (zero reflection coefficient) at a single frequency and a tolerable reflection coefficient over the designed frequency range. Traditionally the bandwidth can be increased using multiple sections of impedance transformers as discussed in Section 7.2. In the multiple-section transformers, the impedances are gradually transformed. In this section commonly used wideband impedance matching techniques will be described.

7.3.1 Gain–Bandwidth Limitations

In designing wideband matching networks it is important to know the greatest possible bandwidth that can be obtained. This can be explained with the help of the Bode–Fano criterion [12, 13], which gives for certain types of load impedances a theoretical limit on the minimum reflection coefficient magnitude that can be obtained with an arbitrary matching network. The Bode–Fano criterion thus gives the optimum solution that can be ideally achieved, even though such a solution may only be approximated in practice. Such an optimum solution is important because it gives the designer the upper limit of performance and provides a benchmark for comparisons of practical designs.

In general, the input and output impedances of active devices are approximated by lumped elements. For example, input and output impedances of transistors are approximated by four general types of equivalent circuits shown in Figure 7.17. Only circuits in Figure 7.17a and 7.17c are fundamentally different, because circuits in Figure 7.17b and 7.17d are duals of the circuits in Figure 7.17a and 7.17c, respectively.

For the simple case shown in Figure 7.17, Bode [12] was the first to derive the fundamental limitation of the matching network. The gain–bandwidth limitation known as the Bode–Fano criterion is given as

$$\int_0^{\infty} \ln\left|\frac{1}{\Gamma}\right| d\omega \leq \frac{\pi}{RC}$$

where Γ is the input reflection coefficient of the matching network. If Γ_m is the minimum reflection coefficient and is constant over the desired band (ω_a to ω_b) and unity everywhere else, as shown in Figure 7.18, then the best reflection coefficient that

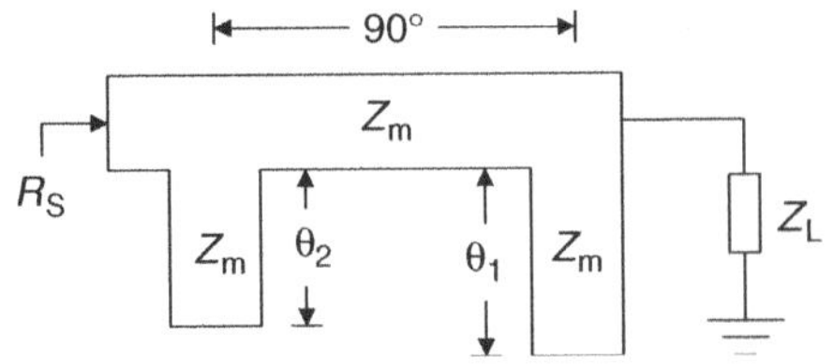

Figure 7.15 A simplified double-stub impedance matching.

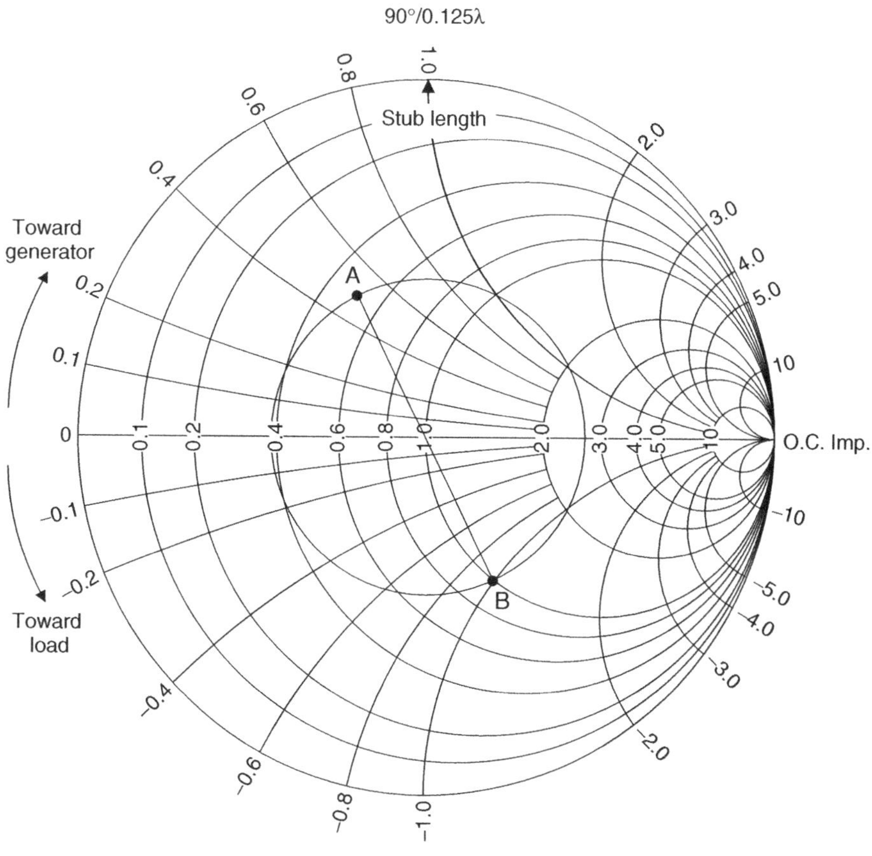

(a)

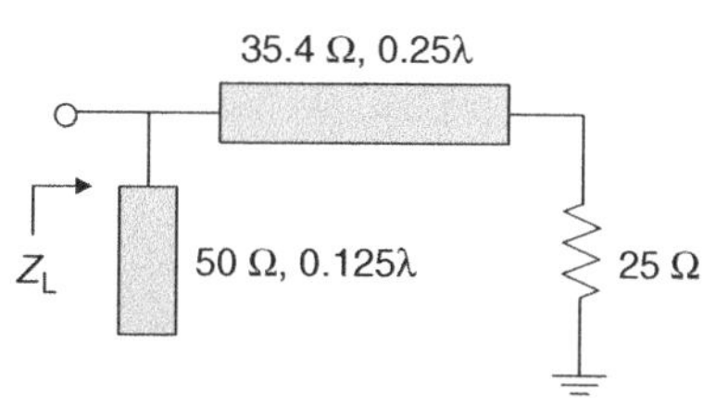

(b)

Figure 7.16 (a) Smith chart solution for load match. (b) Schematic of the matching network.

can be achieved is given by

$$\int_{w_a}^{w_b} \ln\left|\frac{1}{\Gamma}\right| d\omega = \frac{\pi}{RC} = (\omega_b - \omega_a)\ \ln\left|\frac{1}{\Gamma_m}\right| \tag{7.33a}$$

or

$$\Gamma_m = e^{-\pi Q_C/Q_L} \tag{7.33b}$$

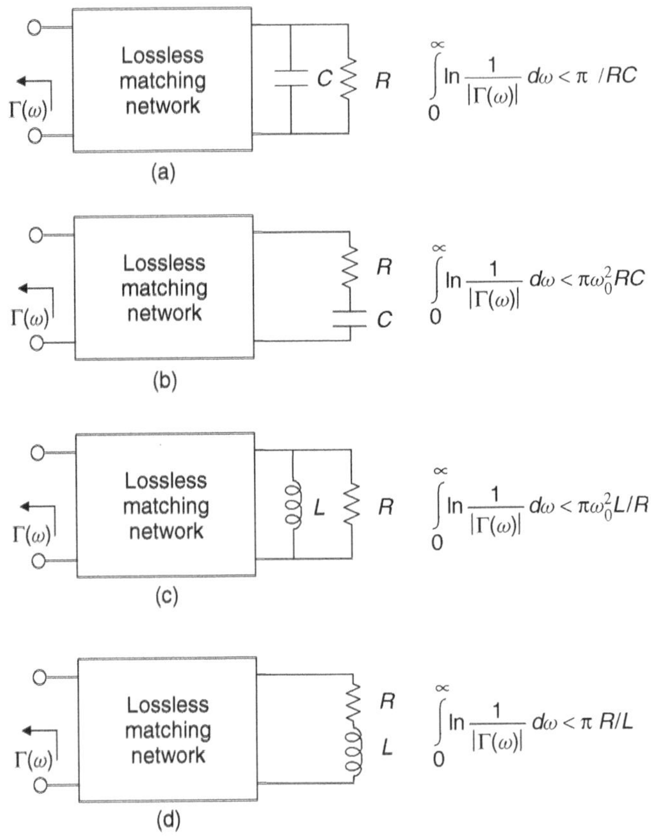

Figure 7.17 Bode–Fano limits for *RC* and *RL* loads: (a) parallel *RC*, (b) series *RC*, (c) parallel *RL*, and (d) series *RL*.

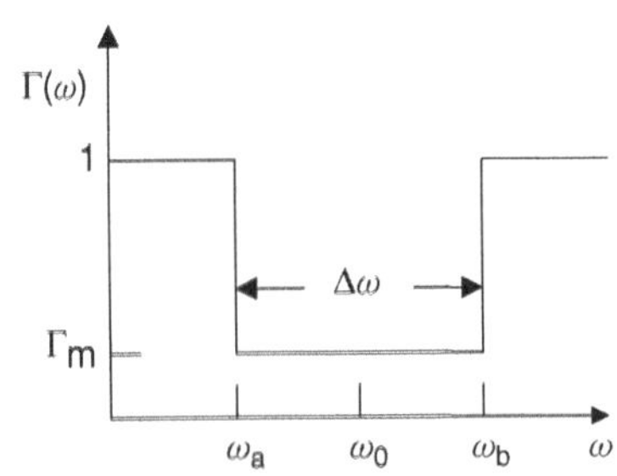

Figure 7.18 Optimum reflection coefficient response of a transformer.

where Q_C and Q_L are the circuit quality factor and load quality factor given by

$$Q_C = \frac{\omega_0}{\Delta\omega}, \quad \Delta\omega = \omega_b - \omega_a, \quad BW = 1/Q_C \tag{7.34a}$$

$$Q_L = \omega_0 RC \tag{7.34b}$$

$$\omega_0 = \sqrt{\omega_a \omega_b} \tag{7.34c}$$

and

$$Q_L BW = \frac{\pi}{\ln(1/\Gamma_m)} \tag{7.34d}$$

Similarly, Eq. (7.34d) is valid for other loads, when Q_L is defined appropriately.

The following conclusions can be drawn from Eq. (7.33b):

1. For a given load, the ratio of bandwidth and reflection coefficient is fixed. The bandwidth can only be increased at the expense of the reflection coefficient.
2. The passband reflection coefficient (Γ_m) cannot be zero unless the bandwidth ($\Delta\omega$) is zero. Thus a perfect match (zero reflection coefficient) can be accomplished only at a finite number of frequencies.
3. When the value Q_L is decreased, a better bandwidth is realized for a given return loss and vice versa. In this case, a relatively lower reactance is absorbed in the matching network.
4. From (7.33) and (7.34), $\Delta\omega$ or $1/\Gamma_m$ decreases with increasing value of R or C. Thus higher Q_L loads are more difficult to match than are lower Q_L loads. Larger Q_C/Q_L ratios result in lower Γ_m values.
5. When $C = 0$, Q_L and Γ_m are zero. Multiple-section transformers with infinite numbers have infinite theoretical bandwidth. However, in practice, having more than 4–5 sections becomes impractical.
6. When $R = \infty$, Q_L and Γ_m become infinite. Thus a reactive load cannot be matched with a lossless network.

Bode–Fano analysis basically treats the impedance transformation from a complex impedance to a real impedance that might not be to 50 Ω. It is assumed that the real load impedance can be matched to 50-Ω generator impedance by using an additional matching network, as shown in Figure 7.19a. A simplified relationship for the load Q–bandwidth product ($Q_L BW$) in terms of the reflection coefficient Γ, using a lossless n-pole tuned matching network, is given by Lopez [14–16]:

$$Q_L BW = \frac{1}{b_n \sinh\left[\frac{1}{a_n}\ln\left(\frac{1}{\Gamma}\right)\right] + \frac{1-b_n}{a_n}\ln\left(\frac{1}{\Gamma}\right)} \tag{7.35a}$$

where the coefficients a_n and b_n for various values of n are given by Lopez [15] and are reproduced in Table 7.1. The above relationship for different n values becomes:
Single-tuned matching

$$Q_L BW = \frac{2\Gamma}{1-\Gamma^2} \qquad n = 1 \tag{7.35b}$$

Double-tuned matching

$$Q_L BW = \frac{2\sqrt{\Gamma}}{1-\Gamma} \qquad n = 2 \tag{7.35c}$$

Infinite-tuned matching

$$Q_L BW = \frac{\pi}{\ln(1/\Gamma)} \qquad n = \infty \tag{7.35d}$$

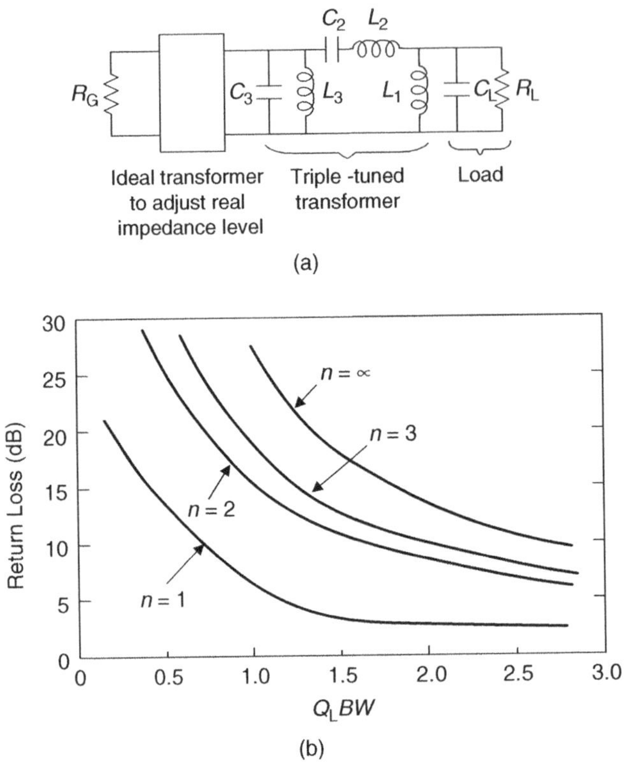

Figure 7.19 (a) Load impedance matching networks using triple-tuned topology. (b) $Q_L BW$ product versus return loss for different tuned matching networks.

Table 7.1 Coefficients a_n and b_n for Various Values of n

n	a_n	b_n
1	1	1
2	2	1
3	2.413	0.678
4	2.628	0.474
5	2.755	0.347
6	2.838	0.264
7	2.896	0.209
8	2.937	0.160
∞	π	0

Equations (7.34d) and (7.35d) are the same. Table 7.2 gives $Q_L BW$ product values for $n = 1$, 2, 3, and ∞, and two values of VSWR. For a practical value of VSWR, the $Q_L BW$ product is finite; even an infinite number of matching elements may be used. For VSWR $= 1.2$ and $n = \infty$, the $Q_L BW$ product is 1.31. If the value of Q_L is 1, the fractional bandwidth is 1.31, which is slightly over one octave. Figure 7.19b shows $Q_L BW$ product versus return loss for different tuned matching networks.

The Bode–Fano criterion provides an absolute limit on bandwidth for a given device Q and achievable return loss. There are several assumptions used, including

Table 7.2 $Q_L BW$ Product Values for Two Values of VSWR

Matching	VSWR = 1.2	VSWR = 2
Single tuned, $n = 1$	0.183	0.750
Double tuned, $n = 2$	0.663	1.732
Triple tuned, $n = 3$	0.901	2.146
Infinite tuned, $n = \infty$	1.310	2.860

that the matching networks are ideal, and one needs another transformer from real impedance to 50 Ω. Here load impedance is transformed using reactance absorbing filter sections. Thus, in practice, the achievable bandwidth is lower than the predictable bandwidth.

EXAMPLE 7.7

Consider a 1-mm gate periphery transistor whose load is equivalent to a parallel combination of a resistor (R_L) and a capacitor (C_L) of values 56 Ω and −0.32 pF, respectively. The device is to be matched to 50 Ω. If the size of this transistor is increased by four, that is, the load is 14 Ω and −1.28 pF, determine which device has a better bandwidth at 10 GHz.

SOLUTION For device 1, from (7.34b), we have

$$\begin{aligned} Q_L &= 2\pi f R_L |C_L| = 2 \times 3.1416 \times 10 \times 10^9 \times 56 \times 0.32 \times 10^{-12} = 1.1259 \\ Q_L BW &= Q_L \Delta f / f_0 = 0.1259 \Delta f \end{aligned}$$

For device 2, we have

$$\begin{aligned} Q_L &= 2\pi f R_L |C_L| = 2 \times 3.1416 \times 10 \times 10^9 \times 14 \times 1.28 \times 10^{-12} = 1.1259 \\ Q_L BW &= Q_L \Delta f / f_0 = 0.1259 \Delta f \end{aligned}$$

Thus both devices have the same $Q_L BW$ product. According to Bode–Fano, for a specified return loss, the bandwidths (Δf) are the same for both transistors. Using a multituned matching circuit, the reactive absorbing bandwidths are the same. However, a transformer from 56 Ω to 50 Ω is much simpler than a transformer from 14 Ω to 50 Ω. Therefore, in actual practice, a larger size transistor has smaller bandwidth than a smaller size transistor. ■

7.3.2 Lumped-Element Wideband Matching Techniques

The wider bandwidth networks are synthesized using conventional filter network theory or by cascading L-networks. Analysis of the latter technique is presented next.

Consider an n-section L impedance matching network as shown in Figure 7.20. Here $R_S > R_L$ and $R_1, R_2, \ldots, R_{n-1}$ are known as virtual resistors with their values being between the R_S and R_L values. These resistors are not connected, but are used to extend the analysis described in Section 7.2.1 to n-section topology. In this case, the impedance transformation is performed step-by-step from R_S to R_1 to $R_2 \ldots$ to

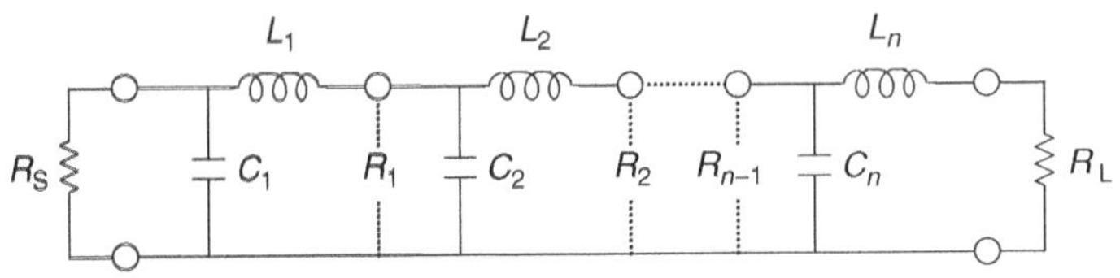

Figure 7.20 An n-stage impedance matching using L-section type configuration.

finally R_L. Optimum bandwidth is obtained if the ratios of each of the two succeeding resistances are equal:

$$\frac{R_S}{R_1} = \frac{R_1}{R_2} \cdots = \frac{R_{n-1}}{R_L} \tag{7.36}$$

$$R_1 = (R_S R_2)^{1/2} \tag{7.37a}$$

$$R_2 = (R_1 R_3)^{1/2}$$

$$\vdots$$

$$R_{n-1} = (R_{n-2} R_L)^{1/2} \tag{7.37b}$$

For example, for a three-section transformer,

$$R_1 = R_S^{2/3}\, R_L^{1/3} \tag{7.38a}$$

$$R_2 = R_S^{1/3}\, R_L^{2/3} \tag{7.38b}$$

The C and L values are obtained using the following expressions:

$$C_1 = \frac{Q_1}{\omega R_S}, \quad C_2 = \frac{Q_2}{\omega R_1}, \quad C_3 = \frac{Q_3}{\omega R_2} \cdots \tag{7.39}$$

$$L_1 = Q_1 R_1/\omega, \quad L_2 = Q_2 R_2/\omega, \quad \cdots, \quad L_n = Q_n R_L/\omega \tag{7.40}$$

where

$$Q_1 = [R_S/R_1 - 1]^{1/2}, \quad Q_2 = [R_1/R_2 - 1]^{1/2}, \quad \cdots, \quad Q_n = [R_{n-1}/R_L - 1]^{1/2} \tag{7.41}$$

EXAMPLE 7.8

The input impedance of a device to be matched to 50 Ω is represented by a series combination of 2.5 Ω and 1.0 pF. Synthesize matching networks and determine matching element values at 10 GHz.

SOLUTION In this case, the 50 Ω is transformed to a series combination of 2.5 Ω and −1.0 pF. Here,

$$Z_{in}^* = R_{in} + jX_{in} = 2.5 - j\frac{1000}{-1 \times 2\pi \times 10} = 2.5 + j15.9\ \Omega$$

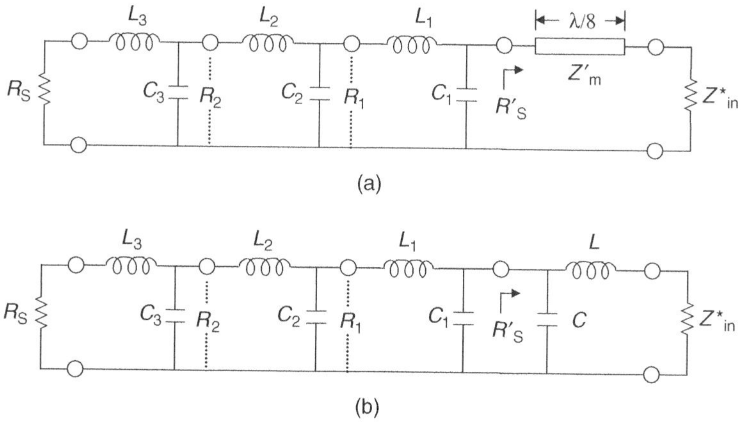

Figure 7.21 Four-section impedance transformer.

Two approaches, as shown in Figure 7.21, are selected for the impedance transformer.

For Figure 7.21a,

$$Z'_{\text{m}} = \sqrt{R_{\text{in}}^2 + X_{\text{in}}^2} = \sqrt{2.5^2 + 15.9^2} = 16.1\ \Omega$$

$$R'_{\text{S}} = \frac{Z'_{\text{m}} R_{\text{in}}}{Z'_{\text{m}} - X_{\text{in}}} = \frac{16.1 \times 2.5}{16.1 - 15.9} = 201.3\ \Omega$$

Next, the 201.3-Ω impedance is matched to $R_{\text{S}} = 50\ \Omega$ by a three-section transformer. Using (7.38), we have

$$R_1 = 201.3^{2/3}\ 50^{1/3} = 126.5\ \Omega$$

$$R_2 = 201.3^{1/3}\ 50^{2/3} = 79.5\ \Omega$$

$$Q_1 = \left[\frac{201.3}{126.5} - 1\right]^{1/2} = 0.769 = Q_2 = Q_3$$

$$\omega = 2\pi f = 62.83 \times 10^9 \quad \text{rad/s}$$

Next, we calculate the C and L values. In this case,

$$C_1 = \frac{Q_1}{\omega R'_{\text{S}}} = \frac{0.769 \times 10^3}{62.83 \times 201.3}\ \text{pF} = 0.061\ \text{pF}$$

$$C_2 = \frac{Q_2}{\omega R_1} = 0.0968\ \text{pF} \quad \text{and} \quad C_3 = \frac{Q_3}{\omega R_2} = 0.154\ \text{pF}$$

$$L_1 = Q_1 R_1/\omega = 1.548\ \text{nH}, \quad L_2 = Q_2 R_2/\omega = 0.973\ \text{nH}, \quad \text{and} \quad L_3 = Q_3 R_{\text{S}}/\omega = 0.612\ \text{nH}$$

Using a CAD, the calculated value of bandwidth is 26%. For the $\lambda/8$ microstrip section on 15-mil alumina substrate, the calculated line width and length are 85 mils and 50.8 mils, respectively.

Next, Figure 7.21b is considered. First the L-section component values are determined. Using (7.6), we find

$$X = \omega L = \pm\sqrt{R_{\text{in}}(R'_{\text{S}} - R_{\text{in}})} - X_{\text{in}}$$

$$B = \omega C = \pm\frac{\sqrt{(R'_{\text{S}} - R_{\text{in}})/R_{\text{in}}}}{R'_{\text{S}}}$$

Assuming $R'_S = 150\ \Omega$ and using positive signs, we find

$$\omega L = \sqrt{2.5(150 - 2.5)} - 15.9 = 3.3029\ \Omega, \quad L = 0.0526\ \text{nH}$$

$$\omega C = \frac{\sqrt{(150 - 2.5)/2.5}}{150} = 0.0512\ \text{S}, \quad C = 0.815\ \text{pF}$$

Following the above procedure, the calculated values for the three-section transformer to match 150 Ω to 50 Ω are as follows:

$$R_1 = 104.06\ \Omega, \quad R_2 = 72.1\ \Omega$$
$$Q_1 = Q_2 = Q_3 = 0.664$$
$$C_1 = 0.0705\ \text{pF}, \quad C_2 = 0.102\ \text{pF}, \quad \text{and} \quad C_3 = 0.1467\ \text{pF}$$
$$L_1 = 1.1\ \text{nH}, \quad L_2 = 0.762\ \text{nH}, \quad \text{and} \quad L_3 = 0.5287\ \text{nH}$$

Using a CAD, the calculated value of bandwidth is 33%, which is about 27% larger than the previous case. ■

7.3.3 Transmission Line Wideband Matching Networks

Quarter-wavelength multisection impedance transformers are commonly used to match real impedances over wide bandwidths. The reflection coefficient at the input of the matching network is given by a sum of the series of multiple partial reflections arising at each of the impedance discontinuities. For small reflections, it can be approximated to be

$$\Gamma = |\Gamma_1| + |\Gamma_2|\, e^{-j2\theta} + |\Gamma_3|\, e^{-j4\theta} \cdots |\Gamma_{N+1}|\, e^{-2jN\theta} \tag{7.42}$$

where Γ_i is the first-order reflection at the ith discontinuity in an N-section transformer as shown in Figure 7.22a. For a transformer designed with symmetrical steps ($|\Gamma_i| = |\Gamma_{N+2-i}|$), the expression in (7.42) is simplified to be

$$\Gamma = 2e^{-jN\theta}[\rho_1 \cos N\theta + \rho_2 \cos{(N-2)\theta} + \cdots + A] \tag{7.43}$$

where

$$\rho_i = |\Gamma_i| = \left|\frac{Z_{i+1} - Z_i}{Z_{i+1} + Z_i}\right| \quad \text{or} \quad \frac{Z_{i+1}}{Z_i} = \frac{1 + \rho_i}{1 - \rho_i} \tag{7.44}$$

and

$$A = \begin{Bmatrix} \rho_{(N+1)/2} \cos\theta, & N \text{ odd} \\ \frac{1}{2}\rho_{(N/2)+1}, & N \text{ even} \end{Bmatrix} \tag{7.45}$$

A minimum passband VSWR for a given input–output impedance ratio and bandwidth is possible when the reflection coefficients, ρ_i in (7.44), are chosen to have ratios corresponding to the like terms in a Nth-order Chebyshev or binomial polynomial as discussed in the literature [1–10] or simply following the procedure as described next.

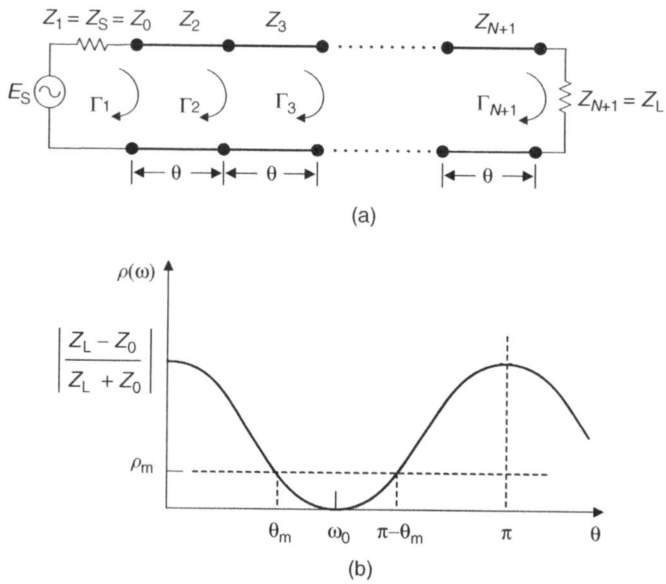

Figure 7.22 N-section quarter-wave impedance transformer. (b) Reflection coefficient response of a binomial transformer.

Binomial Transformer

In a binomial transformer, a maximally flat passband characteristic for the transformer is obtained. The maximally flat passband characteristic of a binomial transformer is shown in Figure 7.22b. The maximum tolerable reflection coefficient in the passband is given as $\rho_{\rm m}$. The transformer sections will be a quarter-wavelength long at the band center frequency. The parameter $\theta_{\rm m}$ (Fig. 7.22b) is given by

$$\theta_{\rm m} = \cos^{-1}\left|\frac{2\rho_{\rm m}}{\ln(Z_{\rm L}/Z_0)}\right|^{1/N} \tag{7.46}$$

In the case of transmission line sections, $\theta = \pi f/2f_0$, and therefore the fractional bandwidth is given by

$$\frac{\Delta f}{f_0} = \frac{2(f_0 - f_{\rm m})}{f_0} = 2 - \frac{4}{\pi}\cos^{-1}\left|\frac{2\rho_{\rm m}}{\ln(Z_{\rm L}/Z_0)}\right|^{1/N} \tag{7.47}$$

since $\theta_{\rm m} = \pi f_{\rm m}/2f_0$. Note that the solution to (7.46) is chosen such that $\theta_{\rm m} < \pi/2$. The maximum out-of-band reflection coefficient occurs for $\theta = 0, \pi$ and is given by

$$\rho_{\max} = \left|\frac{Z_{\rm L} - Z_0}{Z_{\rm L} + Z_0}\right| \tag{7.48}$$

Figure 7.23 shows a four-section quarter-wave impedance transformer. Table 7.3 lists the impedance data for two, three, and four sections [1, 8]. The table lists data for $Z_{\rm L}/Z_0 > 1$. For $Z_{\rm L}/Z_0 < 1$, the results for $Z_0/Z_{\rm L}$ should be used with Z_1 beginning from the load end.

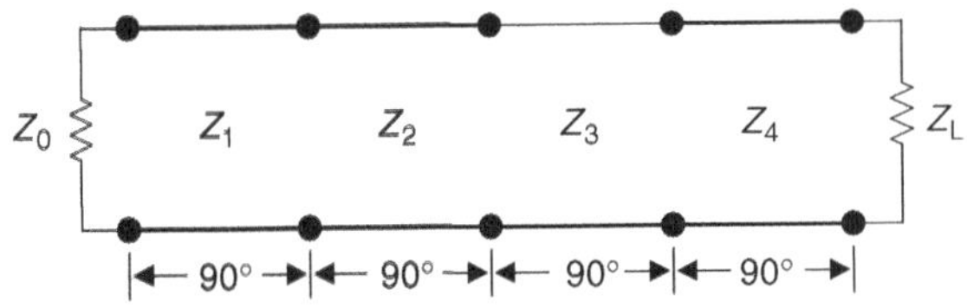

Figure 7.23 A four-section quarter-wave impedance transformer.

Table 7.3 Binomial Transformer Design Data

	$N = 2$		$N = 3$			$N = 4$			
Z_L/Z_0	Z_1/Z_0	Z_2/Z_0	Z_1/Z_0	Z_2/Z_0	Z_3/Z_0	Z_1/Z_0	Z_2/Z_0	Z_3/Z_0	Z_4/Z_0
1.0	1.0000	1.0000	1.0000	1.0000	1.0000	1.0000	1.0000	1.0000	1.0000
1.5	1.1067	1.3554	1.0520	1.2247	1.4259	1.0257	1.1351	1.3215	1.4624
2.0	1.1892	1.6818	1.0907	1.4142	1.8337	1.0444	1.2421	1.6102	1.9150
3.0	1.3161	2.2795	1.1479	1.7321	2.6135	1.0718	1.4105	2.1269	2.7990
4.0	1.4142	2.8285	1.1707	2.0000	3.3594	1.0919	1.5442	2.5903	3.6633
6.0	1.5651	3.8336	1.2544	2.4495	4.7832	1.1215	1.7553	3.4182	5.3500
8.0	1.6818	4.7568	1.3022	2.8284	6.1434	1.1436	1.9232	4.1597	6.9955
10.0	1.7783	5.6233	1.3409	3.1623	7.4577	1.1613	2.0651	4.8424	8.6110

For a two-section binomial impedance transformer.

$$Z_1 = Z_L^{1/4} Z_0^{3/4} \tag{7.49a}$$

$$Z_2 = Z_L^{3/4} Z_0^{1/4} \tag{7.49b}$$

EXAMPLE 7.9

Design a two-section binomial transformer to match a load $Z_L = 10\ \Omega$ to $Z_0 = 50\ \Omega$. Determine the maximum fractional bandwidth for $\rho_m = 0.2$ and frequency range when the center frequency is 10 GHz. Also, determine the physical dimensions on a 15-mil thick alumina substrate ($\varepsilon_r = 9.9$).

SOLUTION In this case,

$$\frac{\Delta f}{f_0} = 2 - \frac{4}{\pi}\cos^{-1}\left|\frac{2\rho_m}{\ln(Z_L/Z_0)}\right|^{1/N} = 2 - \frac{4}{\pi}\cos^{-1}\left|\frac{2x0.2}{\ln(10/50)}\right|^{1/2} = 0.6645$$

Thus a fractional bandwidth for this case is 0.6645 or 66.35%.

Also, from (7.18), we have

$$f_2 - f_1 = 0.6645 f_0$$
$$f_2 f_1 = f_0^2$$

Solving for f_1 and f_2, we find

$$f_1 = 7.215 \text{ GHz} \quad \text{and} \quad f_2 = 13.86 \text{ GHz}$$

From (7.49), we have

$$Z_1 = Z_L^{1/4} Z_0^{3/4} = 33.44\ \Omega$$
$$Z_2 = Z_L^{3/4} Z_0^{1/4} = 14.95\ \Omega$$

We select a microstrip as the transmission medium. The physical dimensions are

$$33.44\ \Omega, \quad W_1 = 0.76\ \text{mm}, \quad \varepsilon_{re1} = 7.35, \quad \ell_1 = 2.77\ \text{mm}$$
$$14.95\ \Omega, \quad W_2 = 2.3\ \text{mm}, \quad \varepsilon_{re2} = 8.5, \quad \ell_2 = 2.57\ \text{mm}$$

Low-impedance lines have higher ε_{re} values and thus $\lambda/4$ lines are slightly shorter than high-impedance lines. ■

Chebyshev Transformer

In a Chebyshev transformer, ρ is allowed to vary between 0 and ρ_m in a periodic manner in the passband. Thus the transformer has an equal-ripple characteristic, as shown in Figure 7.24. A transformer of this type provides a considerable improvement in bandwidth over the binomial transformer.

The Chebyshev quarter-wave transformer design data for three sections are given in Table 7.4 [8]. More extensive tables can be found elsewhere [1].

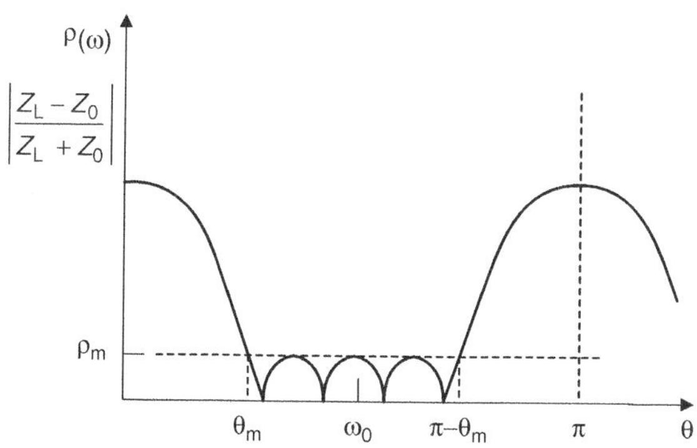

Figure 7.24 Reflection coefficient response of a Chebyshev transformer.

Table 7.4 Chebyshev Transformer Design Data[a]

	$\Delta f/f_0 = 0.2$		$\Delta f/f_0 = 0.4$		$\Delta f/f_0 = 0.6$	
Z_L/Z_0	Z_1/Z_0	k^2	Z_1/Z_0	k^2	Z_1/Z_0	k^2
2	1.09247	1.19×10^{-7}	1.09908	7.98×10^{-6}	1.1083	9.57×10^{-5}
4	1.19474	5.35×10^{-7}	1.20746	3.55×10^{-5}	1.23087	4.31×10^{-4}
10	1.349	1.92×10^{-7}	1.37482	1.28×10^{-4}	1.4232	1.55×10^{-3}
20	1.48359	4.29×10^{-7}	1.52371	2.85×10^{-4}	1.60023	3.45×10^{-3}
100	1.87411	2.33×10^{-6}	1.975	1.55×10^{-3}	2.17928	1.87×10^{-2}

[a]Note that $Z_2 = \sqrt{Z_L Z_0}$; $Z_3 = Z_L Z_0 / Z_1$.

Using Table 7.4, for a given Z_L/Z_0, k^2, and bandwidth, the impedances of the sections can be determined. The passband tolerance k^2 and maximum tolerable passband ripple ρ_m are related by

$$\rho_m = \left(\frac{k^2}{1+k^2}\right)^{1/2} \tag{7.50}$$

Tapered Transmission Lines

An alternative approach to realizing a broadband transformer is to use a tapered line section with characteristic impedance that is variable with longitudinal distance, as shown in Figure 7.25. Some of the tapers that have been used for impedance matching are linear, exponential, and Chebyshev designs [8, 17, 18]. The tapered length is normally greater than $\lambda/2$. These represent a practical and effective solution to impedance matching.

The synthesis of a tapered line transformer is done at the lowest frequency of interest because the derivative of the reflection coefficient decreases rapidly with frequency. The length of the exponential taper is given as [18]

$$\ell = \frac{\lambda}{\rho_m}\,\frac{\ln(\bar{Z}_L)}{4\pi} \tag{7.51}$$

where ρ_m is the maximum reflection coefficient, $\bar{Z}_L(= Z_L/Z_0)$ is the normalized (with respect to source impedance) value of the load impedance to be matched, and λ is the guide wavelength at the lowest frequency of interest. The normalized impedance profile as a function of length parameter z is given as

$$\bar{Z}(z) = e^{(z/\ell)\ln \bar{Z}_L} \tag{7.52}$$

The design of linear and Chebyshev tapered line transformers is described in References 8 and 18.

EXAMPLE 7.10

Design an exponential TEM line taper for matching a 50-Ω source to a 100-Ω load when the maximum reflection coefficient is $\rho_m = 0.1$ and the frequency of interest is 4 GHz.

SOLUTION From (7.51) one gets

$$\ell = \frac{\lambda}{0.1}\cdot\frac{\ln(2)}{4\pi} = 0.55\lambda$$

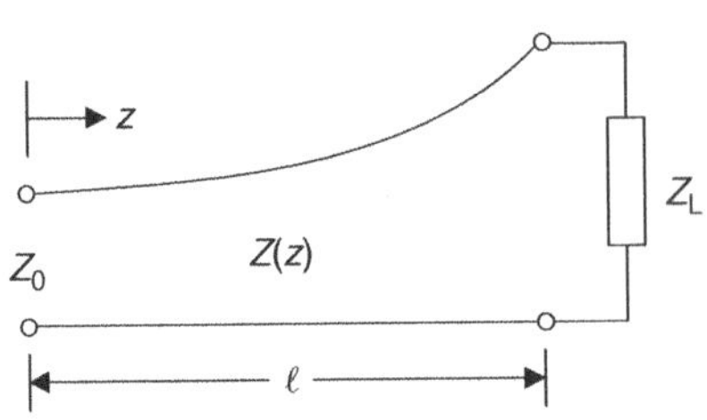

Figure 7.25 A tapered line transformer.

and $\bar{Z}(z) = e^{1.26z/\lambda}$. When z/ℓ takes on the values 0, $\frac{1}{8}$, $\frac{1}{4}$, $\frac{1}{2}$, $\frac{3}{4}$, and 1, the corresponding characteristic impedance values are 50, 54.5, 59.5, 70.7, 84.1, and 100 Ω.

Assuming an effective dielectric constant of 6.25, we have

$$\lambda = \frac{30}{4} \times \frac{1}{\sqrt{6.25}} = 3 \text{ cm}$$
$$\ell = 1.65 \text{ cm}$$

■

7.3.4 Balun-Type Wideband Matching Techniques

Transmission line transformers (TLTs) using straight or coiled sections of coupled transmission lines are frequently used to realize multi-octave impedance transformation at RF and lower microwave frequencies [19]. This type of a transformer can be designed employing any multilayer fabrication technology such as a printed circuit board, LTCC, HTCC, and monolithic Si or GaAs ICs. Figure 7.26 shows a 4:1 impedance matching transformer ($Z_S = 4Z_L$), where Z_S and Z_L are the source and load impedance, respectively. Here transmission lines A and B are not electromagnetically coupled and their length (L) is typically $\lambda/4$. In this configuration, each line has the characteristic impedance Z_0. If the lines are coupled, such as the side-coupled microstrip lines, shown in Figure 7.27 (top view), the bandwidth increases and the transformer length decreases with stronger coupling between the conductors. In asymmetric broadside-coupled microstrip lines, shown in Figure 7.28 (side view), the coupling coefficient is much stronger. Therefore this configuration results in better bandwidth and smaller transformer size. In this case, the length (L) is typically $\lambda/8$. When ports 1 and 2 are switched, the transformer is designated as a 1:4 TLT.

The design of an asymmetric broadside-coupled TLT cannot be performed accurately using conventional circuit simulators, because the substrate and the conductor are of multilayer type. However, accurate solutions can be obtained by using an electromagnetic simulator. A 50-Ω to 12.5-Ω (4:1) asymmetric broadside-coupled TLT was designed [19] using "em^{tm}" by Sonnet software. The substrate parameters for the impedance transformer are given in Table 7.5. The conductors have a width W and a length L. The characteristics of this transformer were compared with three other transformers: a single section quarter-wave microstrip, uncoupled TLT, and side-coupled TLT. For these transformers, Figures 7.29 and 7.30 show the reflection and transmission coefficients versus frequency, respectively, and Table 7.6 summarizes the bandwidth performance. Here, port 1 is terminated in 50 Ω and port 2 is terminated in 12.5 Ω.

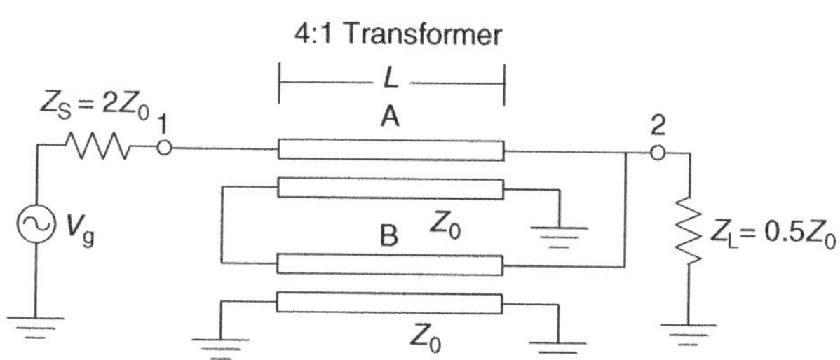

Figure 7.26 Impedance matching transformer configurations using two uncoupled transmission lines A and B.

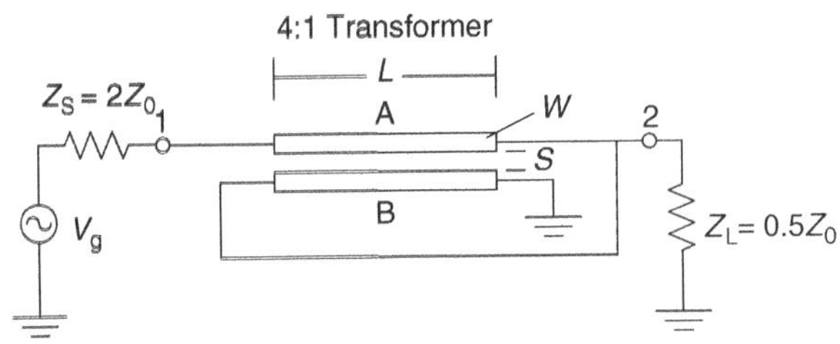

Figure 7.27 Impedance transformer configurations using side-coupled microstrip lines. Ground plane not shown.

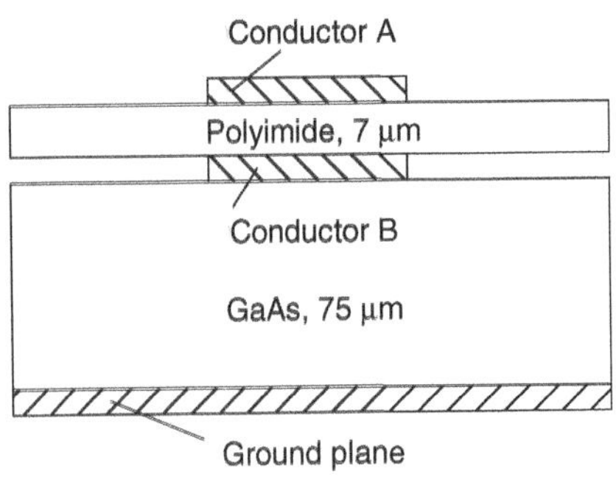

Figure 7.28 Asymmetric broadside-coupled microstrip lines.

Table 7.5 Substrate Parameters for Broadband Impedance Transformers

GaAs substrate, $\varepsilon_r = 12.9$
Substrate thickness, $h = 75$ μm
Polyimide, $\varepsilon_{rd} = 3.2$
Polyimide thickness, $d = 7$ μm
Gold conductors thickness, $t = 4.5$ μm

Among these four transformers, the broadside-coupled TLT has the largest bandwidth and shortest line length. Table 7.6 compares bandwidths for three cases of return loss: 10, 15, and 20 dB. The bandwidth of a transformer is defined in terms of return loss (RL), that is the frequency range over which the RL is equal to or greater than a specified value. The fractional bandwidth, FBW, is defined in (7.18a). Although this example of a 50-Ω to 12.5 − Ω (both real impedances) transformer demonstrates the unique features, such as largest bandwidth and shortest line length of an asymmetrical broadside-coupled microstrip TLT, this transformer can also be used to transform a complex impedance to real impedance or vice versa. It may also be used to transform one complex impedance to another complex impedance.

The size of these transformers is further reduced by folding the lines in a coil or loop shape because the line width is much narrower than the length. By cascading two sections of 4:1 transformers in series, one can match 50 Ω to 3.1 Ω over larger bandwidths. The selection of suitable structure parameters for a transformer is very important to provide the minimum loss and the required impedance transformation over the desired frequency range. Some of these parameters are described next.

The effect of polyimide thickness, d, on the bandwidth of a TLT was studied by varying its value and keeping other parameters constant. The microstrip width and length are 30 μm and 1400 μm, respectively. Figure 7.31 shows the maximum FBW

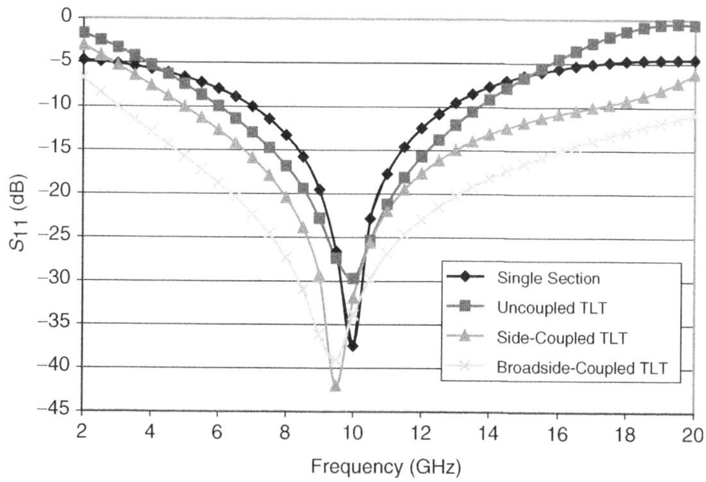

Figure 7.29 Reflection coefficient response of four types of 4:1 transformers. $Z_S = 50\ \Omega$.

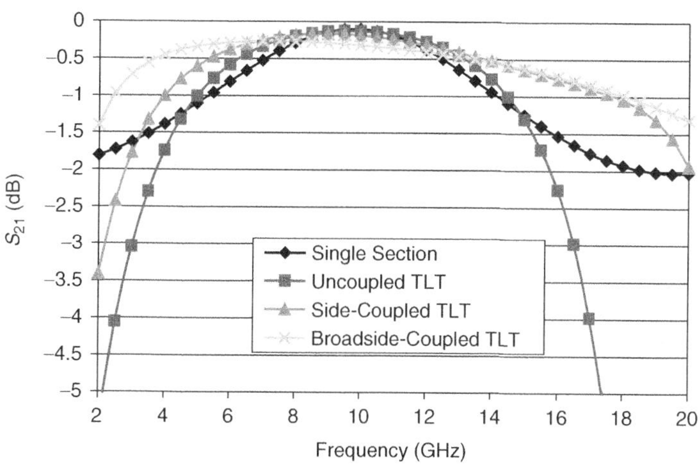

Figure 7.30 Transmission coefficient response of four types of 4:1 transformers. $Z_S = 50\ \Omega$.

and corresponding source impedance of a 4:1 transformer as a function of polyimide thickness between the two broadside-coupled conductors. It may be noted that the tighter coupling between the conductors results in larger bandwidth and lower input impedance.

Next, we consider the effect of microstrip width on the optimum source impedance to be matched and the corresponding bandwidth. The substrate parameters are given in Table 7.5. The line length in this case is 1400 μm. Table 7.7 gives the fractional bandwidth for 4:1 transformers having different microstrip widths. Here Z_S (Fig. 7.26) is the source impedance and W is microstrip width. As the microstrip width decreases, or the characteristic impedance increases, the bandwidth decreases. In this example, to match 50 Ω to 12.5 Ω, a line width of about 15 μm or a characteristic impedance of 74 Ω is required and the resultant fractional bandwidth is about 130%.

Table 7.6 Bandwidth Comparison of Several 4:1 Impedance Transformers: 50 Ω to 12.5 Ω

Configuration	Performance		
	Return Loss (dB)	f_0 (GHz)	FBW (%)
Single section microstrip $W = 190$ μm, $L = 2500$ μm	20 15 10	9.86 9.77 9.47	16.1 31.3 61.3
Uncoupled microstrip lines $W = 200$ μm, $L = 2500$ μm	20 15 10	9.81 9.61 9.07	26.4 47.9 83.8
Coupled microstrip lines $W = 130$ μm, $S = 20$ μm, $L = 2300$ μm	20 15 10	9.50 9.35 9.25	36.8 67.0 130.9
Broadside-coupled microstrip lines $W = 20$ μm, $d = 7$ μm, $L = 1400$ μm	20 15 10	9.13 8.74 8.01	72.6 130.0 227.5

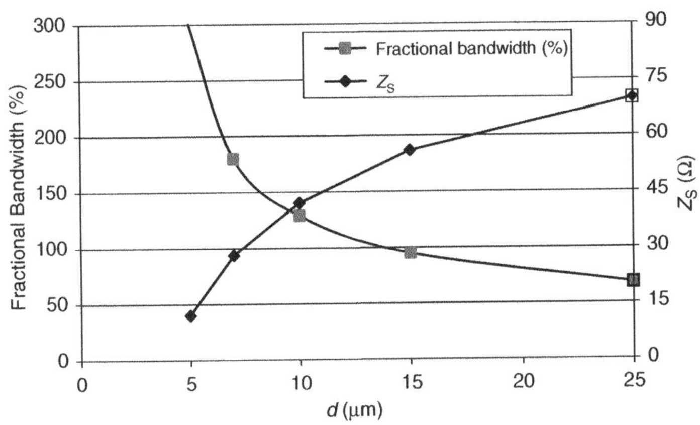

Figure 7.31 Maximum fractional bandwidth and source impedance of a 4:1 TLT versus polyimide thickness.

The fractional bandwidth as a function of source impedance of a 4:1 transformer was calculated for three microstrip widths: 20, 40, and 60 μm. This is shown in Figure 7.32. For each microstrip width there is a maximum FBW and it decreases for other impedance value.

The advantages of the TLT technique are its compact size, low loss, and wider bandwidth capability. When this transformer is used as part of an active circuit, a DC block capacitor is required between the transformer and ground connection. The design of a high-power TLT is described in Reference 19.

7.3.5 Bridged-T Matching Network

A lossy matching network using bridged-T topology is capable of providing greater than an octave bandwidth with flat gain response. As an example, a commonly used

Table 7.7 Maximum Bandwidth as a Function of Line Width for Several TLTs Where Z_S is the Source Impedance and the Load Impedance $Z_L = Z_S/4$[a]

Line Width, W (μm)	Z_S (Ω)	Frequency Range (GHz)	Center Frequency, f_0 (GHz)	Fractional Bandwidth, FBW (%)
10	60	7–21	12.1	115.5
20	40	5–20	10.0	150.0
40	23	3.5–20	8.37	197.2
60	16	3.0–20	7.75	219.5
80	12	2.5–20	7.07	247.5
100	10.5	2.1–20	6.48	276.2
120	8.0	2.0–20	6.32	284.6

[a]RL = 15 dB.

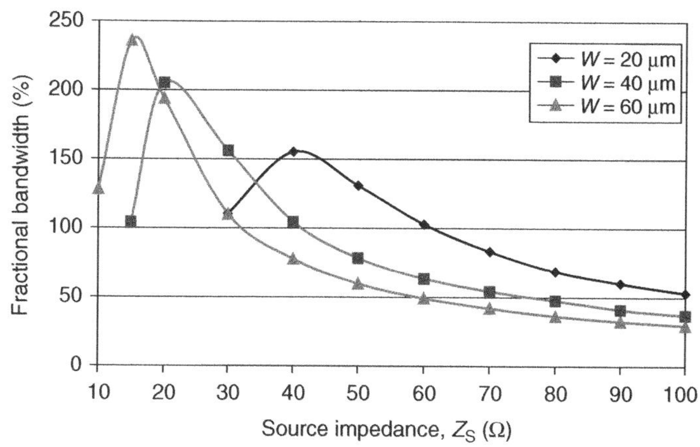

Figure 7.32 Fractional bandwidth of a 4:1 TLT as a function of source impedance.

lossy matching network with bridged-T [20, 21] for broadband matching is shown in Figure 7.33. Here the transistor's input is matched to R_G, which is usually 50 Ω. Since the network has very low Q, it can be used for ultra-broadband applications. Using network analysis, the matching element values can be obtained as follows:

$$L_1 = (1 - RR)R_G^2 C_{gs}/2 \tag{7.53a}$$

$$L_2 = (1 + RR)R_G^2 C_{gs}/2 \tag{7.53b}$$

$$C_1 = (1 - RR^2)C_{gs}/4 \tag{7.53c}$$

where

$$R_1 = R_G$$

$$RR = R_{in}/R_G < 1$$

A brief description of the basic narrowband and broadband matching techniques has been included in this chapter. These techniques will be applied to the design of amplifiers as described in several succeeding chapters. The aforementioned discussion

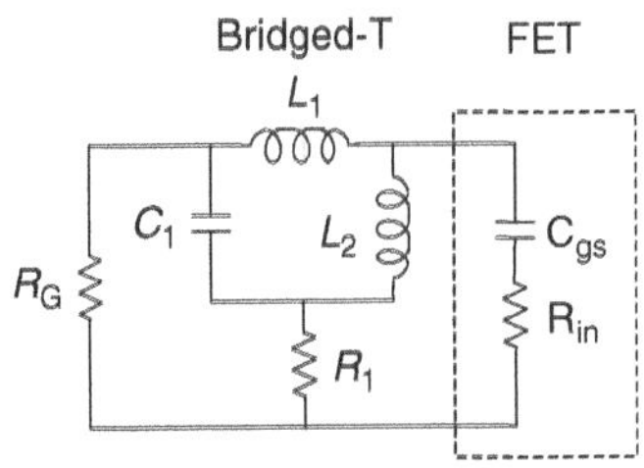

Figure 7.33 Configuration of a lossy input matching network using bridged-T topology.

on the matching techniques is limited to analysis only. No synthesis methods were discussed as they are integral parts of most commercial CAD tools.

REFERENCES

1. G. L. Matthaei, L. Young, and E. M. T. Jones, *Microwave Filters, Impedance-Matching Networks and Coupling Structures*, McGraw-Hill, New York, 1964.
2. J. K. Skwirzynski, *Design Theory and Data for Electrical Filters*, Van Nostrand, London, 1965.
3. H. Howe, Jr., *Stripline Circuit Design*, Artech House, Norwood, MA, 1974.
4. J. D. Rhodes, *Theory of Electrical Filters*, John Wiley & Sons, Hoboken, NJ, 1976.
5. J. A. G., Malherbe, *Microwave Transmission Line Filters*, Artech House, Norwood, MA, 1979.
6. I. J. Bahl and P. Bhartia, *Microwave Solid State Circuit Design*, John Wiley & Sons, Hoboken, NJ, 1988, Chapter 4.
7. R. E. Collin, *Foundations for Microwave Engineering*, John Wiley & Sons, Hoboken, NJ, 2001, Chapter 5.
8. I. J. Bahl and P. Bhartia, *Microwave Solid State Circuit Design*, 2nd edition, John Wiley & Sons, Hoboken, NJ, 2003, Chapter 4.
9. K. Chang (Ed.), *Handbook of RF/Microwave Components and Engineering*, John Wiley & Sons, Hoboken, NJ, 2003, Chapter 3.
10. D. M. Pozar, *Microwave Engineering*, John Wiley & Sons, Hoboken, NJ, 1998, Chapter 3.
11. K. Wu, D. Deslandes, and Y. Cassivi, Impedance transformers and matching networks, in *Encyclopedia RF and Microwave Engineering*, K. Chang (Ed.), Vol. 3, John Wiley & Sons, Hoboken, NJ, 2005, pp. 2079–2092.
12. H. W. Bode, *Network Analysis and Feedback Amplifier Design*, Van Nostrand, New York, 1945.
13. R. M. Fano, Theoretical limitations on the broad-band matching of arbitrary impedances, *J. Franklin Inst.*, Vol. 249, pp. 57–83, January 1950 and pp. 139–154, February 1950.
14. A. R. Lopez, Review of narrowband impedance-matching limitations, *IEEE Antennas Propag. Mag.*, Vol. 46, pp. 88–90, August 2004.
15. A. R. Lopez, Rebuttal to Fano limits on matching bandwidth, *IEEE Antennas Propag. Mag.*, Vol. 47, pp. 128–129, October 2005.
16. A. R. Lopez, Wheeler and Fano impedance matching, *IEEE Antennas Propag. Mag.*, Vol. 49, pp. 116–119, August 2007.
17. W. C. Grunau and C. R. Mason, A simplified solution for tapered transformer lines, *Microwaves*, Vol. 20, pp. 82–83, October 1981.
18. A. M. Khilla, Optimum continuous microstrip tapers are amenable to computer-aided design, *Microwave J.*, Vol. 26, pp. 221–224, May 1983.
19. I. J. Bahl, Broadband and compact impedance transformers for microwave circuits, *IEEE Microwave Mag.*, Vol. 7, pp. 56–62, August 2006.

20. R. Goyal (Ed.), *High-Frequency Analog Integrated Circuit Design*, John Wiley & Sons, Hoboken, NJ, 1995, p. 179.
21. P. K. Ikalainen, An *RLC* matching network and application in 1–20GHz monolithic amplifier, *IEEE MTT-S Int. Microwave Symp. Dig*., pp. 1115–1118, 1989.

PROBLEMS

7.1 A $(70 + j50)$-Ω impedance of a solid state device has to be matched to a 50-Ω system. Calculate the characteristic impedance and length of a single line matching section at 9 GHz. What is the bandwidth if the maximum acceptable reflection coefficient is 0.1?

7.2 What will be the impedance and length of the matching section if the impedance in Problem 7.1 has to be matched to a complex load of $45 + j45$ Ω ?

7.3 Calculate the length of a short-circuited stub to match a $(70 + j50)$-Ω load to a 50-Ω system at 9 GHz. Assume a characteristic impedance of 50 Ω for all transmission lines.

7.4 Design a three-section quarter-wave transformer to match a 50-Ω source to a 125-Ω load at 9 GHz with a reflection coefficient less than 0.1. What is the bandwidth of this transformer?

7.5 If the load in Problem 7.4 has to be matched to the source (50 Ω) using an exponential tapered line, design the tapered section for maximum reflection coefficient $\rho_m = 0.1$.

7.6 The input impedance of a device is represented by a shunt *RC* network where $R = 10$ Ω and $C = 5$ pF. The device is to be matched to a 50-Ω source at 9 GHz over a 20% bandwidth using a third-order network. Select an appropriate network, and then calculate element values and the gain-bandwidth limitation of the network.

7.7 Recalculate the element values in Problem 7.6 for the case that the matching network has to have an insertion loss slope of 3 dB/octave.

7.8 A parallel *RC* load is to be matched to 50 Ω over 3 to 6 GHz. When $R = 100$ Ω and $C = 0.5$ pF, what is the minimum reflection coefficient that can be achieved by using ideal matching elements?

7.9 Consider the case using a series L and a shunt C to match Z_L to Z_0, where $Z_L < Z_0$. Use a Smith chart approach to show the matching steps.

Chapter 8

Amplifier Classes and Analyses

In an amplifier design it is important to select a bias condition known as a DC quiescent point, or simply Q-point, for the transistor, and the source impedance at the input and the load impedance at the output of the device. These three requirements characterize the amplifier operation for the application and also determine the noise figure, gain, bandwidth, output power, PAE, and its linearity performance. The characteristics of various amplifier classes using an ideal transistor's transfer characteristics are discussed in this chapter. Since amplifier gain is not included, the analysis presented in this chapter is limited to drain or collector efficiency.

8.1 CLASSES OF AMPLIFIERS

Amplifier circuits are generally identified for various amplifier types with a class designation depending on the relation between the input and output signals. At RF/microwave frequencies, amplifiers are defined to operate in class A, B, AB, C, D, E, and F [1–33]. These classes are implemented by selecting bias conditions of the active device and by designing an input and output match, to meet system requirements in terms of noise figure, output power, efficiency (or DC power consumption), linearity conditions (modulation schemes), frequency range, size, weight, and cost. Small-signal amplifiers are implemented using class-A operation, while other classes are mainly for power amplifiers with the exception of a low-noise amplifier, which is biased for class AB. In any class, there is a trade-off between the efficiency and linearity; higher efficiency means poorer linearity and vice versa.

Class-A Amplifier

In class A, the active device is conducting at all times for a full cycle (360°) of the input sinusoidal signal. In this case the quiescent point (Fig. 8.1) is selected approximately at the center of the device current and the linear output power is maximized by properly selecting the load match. The class-A amplifier is designed essentially in the same manner as a small-signal linear amplifier and has the same characteristics in that the output signal is an amplified replica of the input signal.

In a small-signal amplifier, the transistor is conjugately matched at the input and output for maximum gain and good input and output VSWR. A class-A power amplifier differs from a small-signal amplifier in the power level at which the amplifier operates. The class-A amplifier can be operated to produce high linear output power or it can

Fundamentals of RF and Microwave Transistor Amplifiers. By Inder J. Bahl

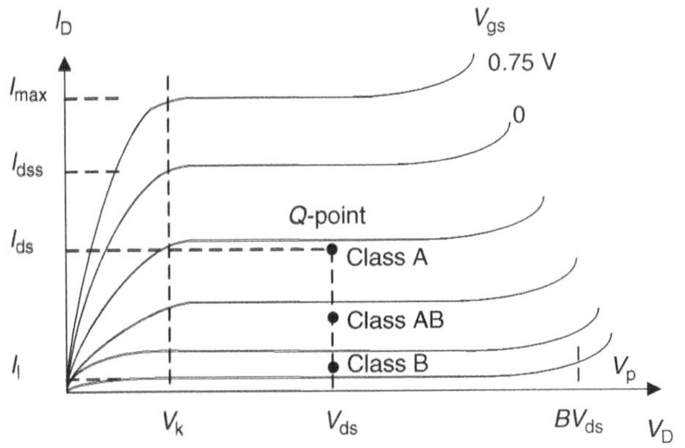

Figure 8.1 Biasing an FET for various classes of operation. I_{max}, I_{dss}, I_{ds}, I_l, V_{gs}, V_{ds}, V_D, BV_{ds}, V_p, and V_k are the peak current, drain–source saturated current, drain–source bias current, leakage current, gate–source voltage, drain–source voltage, drain supply voltage, drain–source breakdown voltage, pinch-off voltage, and knee voltage, respectively.

be designed to operate in saturation by properly selecting the output match for the device. Generally, more RF output power is obtained when the amplifier is operated at saturation. If the amplifier is not driven far into saturation (e.g., 1 dB into gain compression), harmonics and distortion will be low. The maximum drain efficiency of a class-A amplifier, assuming there is no saturation, is limited to 50% and provides the highest signal linearity as long as the active device operates in the linear portion of its transfer function of the device. In this class, the input and output waveforms of the signals are the same.

Class-B Amplifier

In this class the Q-point is set at the cutoff of the device current (at pinch-off, i.e., negligible current without RF signal, Fig. 8.1) and the active device conducts only for one half-cycle, that is, 180° of the input sinusoidal signal. In this case the load match is designed to obtain the best gain, maximum output power, and maximum efficiency. Single-ended class-B amplifiers have poor linearity performance. At RF, class-B amplifiers are realized in push–pull configurations (uses two single-ended stages) in which two devices share the total output load. During the positive cycle, the transistor pushes the current into the load, but pulls the current from the load during the negative cycle. In this case one amplifier stage operates only during the positive cycle of the sine wave while the other stage operates only during the negative cycle of the sine wave. This scheme provides higher output power and linear operation under ideal conditions. However, the linearity characteristics are not as good as for class A due to the nonlinear transition between the two stages during a full RF cycle. Since each device is turned on for only one half-cycle, this class has higher efficiency than class A. The maximum drain efficiency of class B is 78.5%.

Class-AB Amplifier

In class AB, the device bias conditions are chosen somewhat in between the class-A and class-B operations. In this case the Q-point current is 5–30% of the total device current depending on the frequency of operation, noise figure, efficiency, and linearity requirements. This class is commonly used in single-ended power amplifiers and has greater efficiency than class A and better linearity than class B. Cellular and point-to-point radio power amplifiers are operated under class-AB conditions to achieve the highest efficiency with improved linearity.

Class-C Amplifier

In a class-C amplifier the Q-point is set well below the cutoff point of the device current. In this case the device conducts over 25–45% of the RF cycle, and only some portion of the positive half of the sine wave is actually amplified at the output, which is similar to an impulse function. These amplifiers are very nonlinear because the output waveform is quite different from the input signal. Since the device is turned on for less than half a cycle, the efficiency of this class is higher than class A, B, or AB. Class-C amplifiers are often used at RF and lower microwave frequencies, where transistors have high gain values and device linearity is not required in the modulation schemes such as CW, pulse, FM, AM, and PM. High-voltage power BJTs are normally operated in class C.

Class-D or Class-E Amplifier

In this class the transistor works as a switch; with very low resistance in the "ON" state and very high impedance in the "OFF" state with respect to the load impedance value. In this situation the voltage and current waveforms at the output show a 180° phase difference. If the ON resistance is negligible, there is no power dissipation in the device, and if the OFF impedance is very large, there is no current flow through the device. In an ideal switching amplifier one can achieve 100% drain or collector efficiency. The difference between class D and class E is that a class-E amplifier has a high Q tuned circuit at the output of the device to provide designed reactive load at the fundamental frequency for zero power dissipation and open termination at second and third harmonics; in class D the even harmonics are short circuited and the odd harmonics are open circuited.

In practice, the active devices have finite ON resistance and also there is significant transition time from the ON state to the OFF state and vice versa, which leads to reduced efficiency in class-E amplifiers. This effect is more pronounced at higher frequencies.

Class-F Amplifier

In the class-F amplifier configuration, the reduced power dissipation is achieved by employing impedance matching technique (such as resonant circuits) to terminate harmonic frequencies in desired loads. In a class-F amplifier, the efficiency is improved by using a multiple-resonator output matching network to control the harmonic power levels to shape the drain voltage and/or drain current waveforms. In an ideal design, at the device output all even harmonics are short circuited to reproduce a half sine-wave current waveform while the odd harmonics are open circuited to shape the output voltage to a square wave. Again, the active device is basically operated as a switch and the theoretical drain or collector efficiency approaches 100%. Efficiency degradation results from the fact that in practice only a finite number of harmonics can be terminated. Class-F circuits, in particular, are currently of interest since they appear to be feasible at microwave frequencies. A class-F amplifier is designed in essentially the same manner as a class-AB/B amplifier, except that the output circuit is designed to

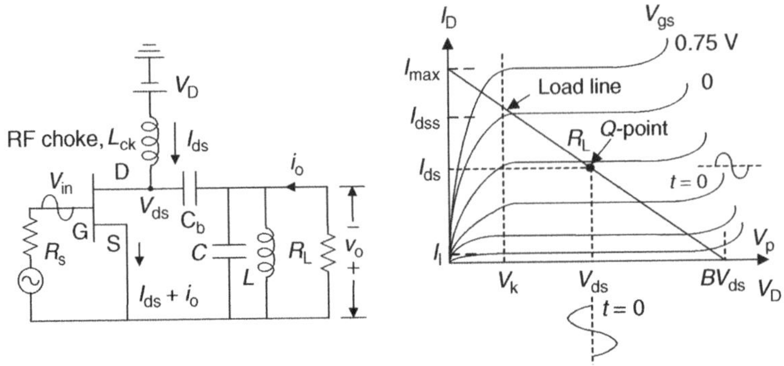

Figure 8.2 Circuit topology for a class-A amplifier and its load line representation.

present a short circuit to the second harmonic and an open circuit to the third harmonic. In practice, often only the short at the second harmonic is employed due to the difficulty in designing suitable matching circuits over extremely wide bandwidths.

As discussed earlier, RF power amplifiers can be designed to operate under a variety of efficiency and linearity conditions with the various amplifier types identified with a class designation. In practice, all amplifier classes operate at reduced efficiencies due to inherent device parasitic losses and nonideal operating conditions, loss in the matching circuit due to bandwidth and output return loss trade-off, and the inability to realize simultaneously optimum load and harmonic terminations. The interest in other modes of operation results from a desire to obtain greater efficiency and/or RF output power. All of the classes beyond class A are designed to operate with efficiency greater than the 50% available from a class-A amplifier. Realization of most of the high-efficiency modes of operation is very difficult at microwave frequencies. The reactive output impedance characteristics of microwave transistors result in difficulties in achieving the switching mode of operation. Generally, high-efficiency operation at microwave frequencies is currently limited to the saturating class-A, class-AB, and class-B modes. Class C is possible using bipolar transistors but not FETs due to higher leakage current. Class F shows considerable promise for obtaining high-efficiency operation.

Next, we discuss the basic operation of various classes of transistor amplifiers. In these simple analyses, the transistor as well as the matching networks have been assumed ideal.

8.2 ANALYSIS OF CLASS-A AMPLIFIERS

A schematic of a device in class-A operation is shown in Figure 8.2. The load presented at the output of the device is R_L. Here, a parallel LC resonator is used which allows the fundamental frequency signal only to pass. The RF choke and DC block capacitor C_b are selected for RF open and RF short, respectively. The RF output voltage and current may be expressed as [1, 2]

$$v_o = V_o \sin\theta \tag{8.1a}$$

$$i_o = I_o \sin\theta \tag{8.1b}$$

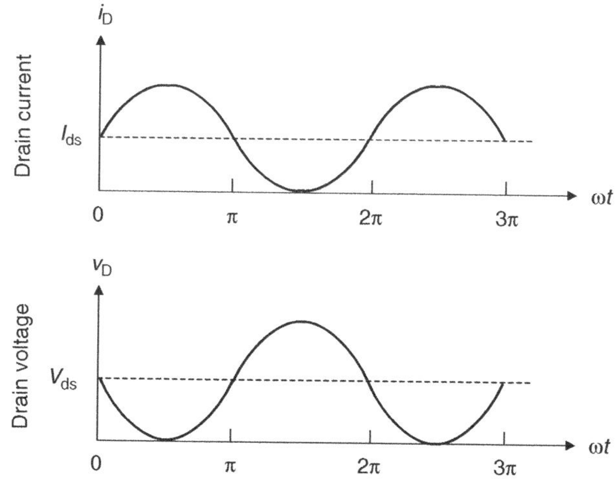

Figure 8.3 Voltage and current waveforms of an ideal class-A amplifier.

Referring to Figure 8.3, the equations for the circuit drain voltage v_D and drain current i_D waveforms may be expressed as

$$v_D(\theta) = V_{ds} - V_o \sin\theta \tag{8.2a}$$

$$i_D(\theta) = I_{ds} + I_o \sin\theta \tag{8.2b}$$

where $\theta = 2\pi f_0 t$ and f_0 is the fundamental frequency (input signal frequency) of the desired output. The coefficients V_o, I_o are for the fundamental frequency and represent peak values. I_{ds} is purely a DC term. The opposite sign for the fundamental frequency coefficients signifies that the voltage and current are opposite in phase; that is, the voltage at the device terminal is positive when the current comes out of the device. They are related by the following relationship:

$$V_o = I_o R_L \tag{8.3}$$

where R_L is the load resistance. The DC voltage and current are blocked by the blocking capacitor C_b, and the RF output power is given by

$$P_o = v_o i_o = V_o I_o \sin^2\theta$$

The average output power becomes

$$P_o = \frac{1}{2\pi}\int_0^{2\pi} V_o I_o \sin^2\theta \; d\theta = \tfrac{1}{2} V_o I_o$$

The maximum fundamental frequency RF output power is given by

$$P_{om} = \frac{V_{om} I_{om}}{2} = \frac{V_{om}^2}{2R_L} \quad \text{where} \quad V_{om} = I_{om} R_L \tag{8.4}$$

and I_{om} and V_{om} are the maximum drain–source current and voltage, respectively. Their values are $I_{max}/2$ and V_{ds}, respectively. I_{max} is also known as I_p, the peak current in the device when the channel is open. The bias point is selected such that I_{ds} is about $I_{max}/2$ and V_{ds} is about $BV_{ds}/2$. Thus Eqs. (8.3) and (8.4) become

$$R_L = 2V_{ds}/I_{max} \tag{8.5a}$$

$$P_{om} = \frac{V_{ds}I_{max}}{4} \tag{8.5b}$$

and the DC power P_{DC} in the device is given by

$$P_{DC} = V_{ds}I_{ds} = \frac{V_{ds}I_{max}}{2} \tag{8.5c}$$

The drain efficiency, η_D, is the ratio of the RF power delivered to the load at the fundamental to the DC power in the device; that is,

$$\eta_D = \frac{P_{om}}{P_{DC}} = 0.5 \tag{8.6}$$

Thus the maximum theoretical drain efficiency of a class-A amplifier is 50%. This is under ideal conditions, that is, no power dissipation in the device due to its parasitics (e.g., ignoring knee voltage V_k), no matching network loss, and a resistive load. This is a linear analysis only. The above equations do not apply to saturated power amplifiers.

EXAMPLE 8.1

A transistor having $I_{max} = 2$ A and breakdown voltage of 20 V is biased at $I_{max}/2$ and $V_{ds} = 10$ V and is operated as a class-A amplifier at 1 GHz. The amplifier has very high gain and output power is 5 W. Determine the load value required and the net power dissipation in the amplifier.

SOLUTION Here $P_{DC} = 10 \times 1 = 10$ W and $P_{om} = 5$ W. Also, $V_{om} = V_{ds}$. From (8.4), we have

$$R_L = \frac{V_{om}^2}{2P_{om}} = \frac{10 \times 10}{2 \times 5} = 10\,\Omega \quad \text{or} \quad R_L = \frac{2V_{ds}}{I_{max}} = \frac{2 \times 10}{2} = 10\,\Omega$$

Net power dissipation $= P_{DC} - P_{om} = 10 - 5 = 5$ W. ■

8.3 ANALYSIS OF CLASS-B AMPLIFIERS

A schematic of a device in class-B operation is shown in Figure 8.4. In class-B amplifiers, the transistors are biased such that they only conduct over half of the RF cycle. This configuration can be analyzed using a single transistor known as a single-ended amplifier or two transistors commonly referred to as a push–pull amplifier. In class B, the even harmonics are short circuited. Both single-ended and push–pull configurations are discussed next.

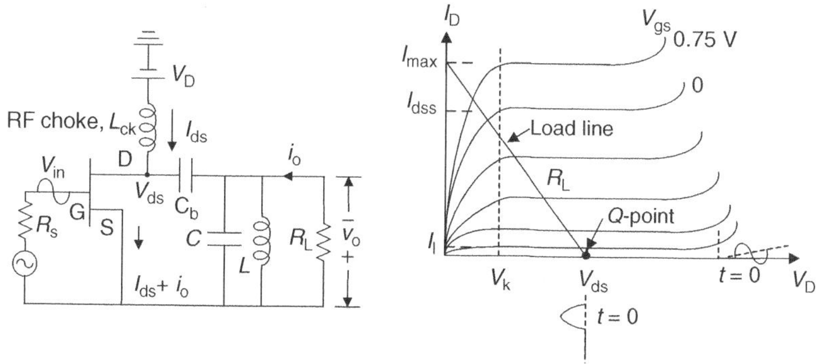

Figure 8.4 Circuit topology for a class-B amplifier and its load line representation.

8.3.1 Single-Ended Class-B Amplifier

Figure 8.5 shows the transistor's voltage and current waveforms. In this case the drain voltage and current waveforms may be expressed as

$$v_D(\theta) = V_{ds} - V_o \sin\theta \tag{8.7}$$

$$\begin{aligned} i_D(\theta) &= I_o \sin\theta, \quad & 0 \le \theta \le \pi \\ &= 0, & \pi \le \theta \le 2\pi \end{aligned} \tag{8.8}$$

The half sine-wave current can be expanded in series as

$$i_D(\theta) = I_o\left[\frac{1}{\pi} + \frac{1}{2}\sin\theta - \frac{2}{\pi}\sum_{n=2,4,6,\cdots}\frac{1}{n^2-1}\cos n\theta\right] \tag{8.9}$$

The first term is the DC current and the current equation contains no odd harmonics. Since the maximum value of I_o is I_{max}, the DC power dissipated in the device

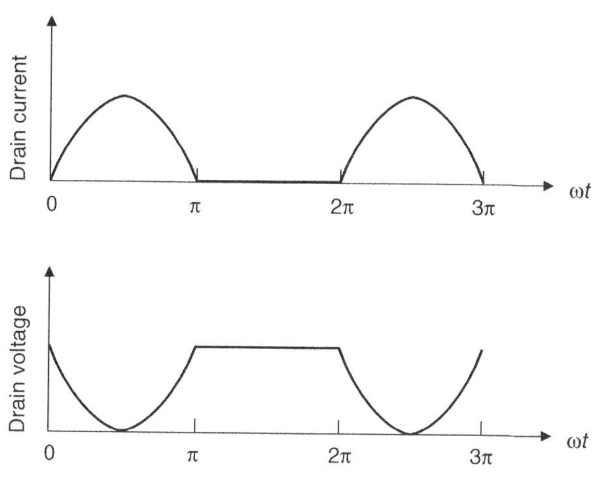

Figure 8.5 Voltage and current waveforms of an ideal class-B amplifier.

is given by

$$P_{\text{DC}} = V_{\text{ds}} I_{\max}/\pi \tag{8.10}$$

Assuming all harmonic components in Eq. (8.9) are filtered out, or the even harmonics are short circuited, the output signal is a pure half sine wave. The average output power may be written

$$P_{\text{o}} = \frac{1}{2\pi}\int_0^{\pi} V_{\text{o}} I_{\text{o}} \sin^2\theta \, d\theta = \tfrac{1}{4} V_{\text{o}} I_{\text{o}} \tag{8.11}$$

In this case the maximum values of V_{o} and I_{o} are V_{ds} and $I_{\max}$. Thus the load and the maximum RF output power become

$$\begin{aligned} R_{\text{L}} &= v_{\text{o}}/i_{\text{o}} = V_{\text{o}}/I_{\text{o}} = V_{\text{ds}}/I_{\max} \\ P_{\text{om}} &= \tfrac{1}{4} V_{\text{ds}} I_{\max} \end{aligned} \tag{8.12}$$

From (8.10) and (8.12), the drain efficiency is given by

$$\eta_{\text{D}} = \frac{P_{\text{om}}}{P_{\text{DC}}} = \frac{\pi}{4} = 0.785 \tag{8.13}$$

Thus the drain efficiency of class-B amplifiers is 78.5%, considerably higher than class-A amplifiers.

8.3.2 Push–Pull Class-B Amplifier

In a single-ended class-B amplifier, only the positive half sine wave is amplified, giving rise to distorted output. When two single-ended class-B amplifiers are combined using 180° baluns at the input and output, both positive and negative half-waves are amplified and combined at the output, maintaining the full sine wave with no distortion. For each transistor the I–V data and R_{L} are the same as shown in Figure 8.4.

A push–pull class-B configuration is shown in Figure 8.6. The input balun splits the input RF signal into two halves having the same magnitude but a 180° phase difference. Transistor 1 conducts for the first half-cycle and transistor 2 conducts for the second half-cycle. The output balun combines the output signals. In this case the drain currents are

$$i_{\text{D1}}(\theta) = I_{\text{o}}\left[\frac{1}{\pi} + \frac{1}{2}\sin\theta - \frac{2}{\pi}\sum_{n=2,4,\cdots}\frac{1}{n^2-1}\cos n\theta\right] \tag{8.14a}$$

$$i_{\text{D2}}(\theta) = I_{\text{o}}\left[\frac{1}{\pi} + \frac{1}{2}\sin(\theta+\pi) - \frac{2}{\pi}\sum_{n=2,4,\cdots}\frac{1}{n^2-1}\cos n(\theta+\pi)\right] \tag{8.14b}$$

Since the output balun adds another 180° phase difference, the total drain current is given by

$$i_{\text{D}}(\theta) = i_{\text{D1}}(\theta) - i_{\text{D2}}(\theta) = I_{\text{o}}\sin\theta \tag{8.15}$$

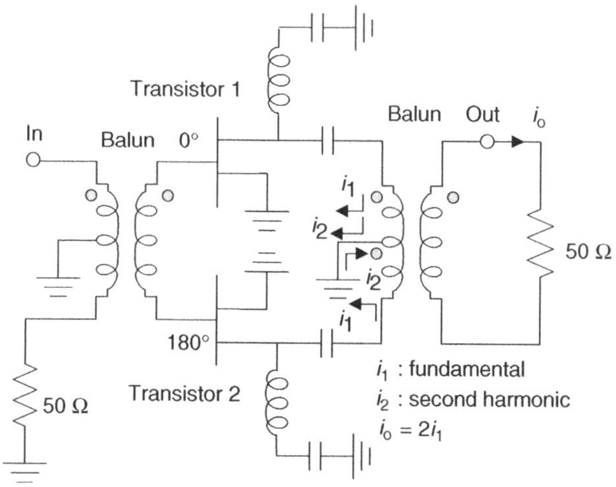

Figure 8.6 Configuration of a push–pull amplifier.

This shows that in a push–pull configuration all harmonics are cancelled out and a pure sine wave similar to a class-A amplifier is produced at the output of the balun. In a MMIC topology, where both single-ended amplifiers are balanced, the push–pull configuration has linearity performance similar to a class-A amplifier. The maximum RF output power is

$$P_{\text{om}} = \frac{V_{\text{om}} I_{\text{om}}}{2} = \frac{V_{\text{ds}} I_{\text{max}}}{4} \tag{8.16}$$

The average DC current is

$$I_{\text{DC}} = \frac{1}{2\pi} \int_0^{2\pi} I_{\text{o}} \sin\theta \, d\theta = \frac{1}{\pi} I_{\text{o}} \tag{8.17}$$

The maximum DC power is

$$P_{\text{DC}} = \frac{V_{\text{ds}} I_{\text{max}}}{\pi} \tag{8.18}$$

The drain efficiency, from (8.16) and (8.18), is

$$\eta_{\text{D}} = \frac{P_{\text{om}}}{P_{\text{DC}}} = \frac{4}{\pi} \tag{8.19}$$

which is the same as for the single-ended configuration. However, it does not include the loss in the balun. High loss in baluns above radio frequencies (above 1 GHz) precludes the realization of push–pull amplifiers. When the balun loss is about 1 dB, the output power is reduced by about 20.5% and the drain efficiency is reduced from 78.5% to 62.4%.

However, push–pull amplifiers have several distinct features including higher device impedances, high gain, low second harmonic levels, and broadband linear characteristics. In push–pull amplifiers, the device's input and output impedance values are four times those of equivalent power device impedances in a single-ended configuration. Thus for a push–pull amplifier, the realizable power level is much higher than a single-ended version. The effect of any inductance common to source connections for the two identical devices is cancelled because of currents flowing in opposite directions. Higher device impedances and the elimination of common-source inductances result in higher gain in push–pull amplifiers. In push–pull amplifiers, the second harmonics are cancelled out in the center tap of the baluns. This provides much lower second harmonic levels at the output of the balun and linear operation of the push–pull amplifier over larger bandwidths.

Since printed circuit baluns have much higher loss than the coaxial baluns, almost all low microwave frequency (less than 5 GHz), high-power (greater than 200 W) push–pull amplifiers are being realized using coaxial baluns. A detailed description of such amplifiers is given in Chapter 10. A major difference between push–pull and balanced amplifier configurations is that a push–pull topology is more suitable for low distortion power combining while the balanced configuration provides power combining with excellent input and output match even for single-ended amplifiers that might have a poor match. On the other hand, the push–pull topology does not alter the single-ended amplifier's match.

8.3.3 Overdriven Class-B Amplifier

Snider [32] reported an analysis of an idealized overdriven class-B amplifier. The analysis was carried out using Fourier series representation of the current and voltage waveforms, shown in Figure 8.7, as a function of angular parameter θ_1 and overdrive factor k. For the sake of comparison, the maximum current and the voltage swings for the class-B and overdriven amplifiers are assumed the same. For $k = 1/\sin\theta_1$, Fourier expansion of voltage and current are given below [2, 32].

For the fundamental frequency and odd harmonics ($n = 3, 5, \ldots$), expressions for the voltage components are

$$v_1 = \frac{2V_{ds}}{\pi}\left(\frac{\theta_1}{\sin\theta_1} + \cos\theta_1\right) \tag{8.20a}$$

$$v_n = \frac{2V_{ds}}{\pi}\left(\frac{\sin(\theta_1 - n\theta_1)}{(1-n)\sin\theta_1} - \frac{\sin(\theta_1 + n\theta_1)}{(1+n)\sin\theta_1} + \frac{2\cos n\theta_1}{n}\right) \tag{8.20b}$$

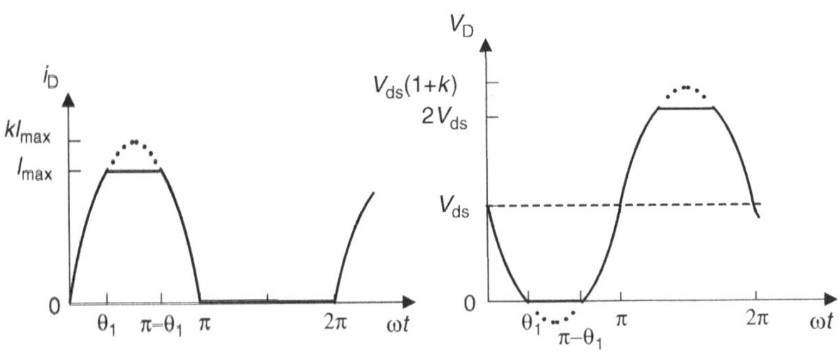

Figure 8.7 Overdriven class-B drain current and voltage waveforms.

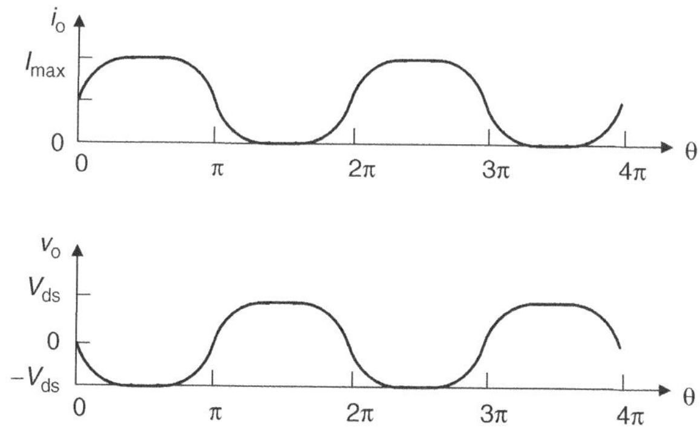

Figure 8.8 Output current and voltage waveforms for an overdriven amplifier.

Expressions for the fundamental frequency and odd harmonic ($n = 3, 5, \ldots$) current components are

$$i_1 = \frac{I_{max}}{\pi}\left(\frac{\theta_1}{\sin\theta_1} + \cos\theta_1\right) \tag{8.21a}$$

$$i_n = \frac{I_{max}}{\pi}\left(\frac{\sin(\theta_1 - n\theta_1)}{(1-n)\sin\theta_1} - \frac{\sin(\theta_1 + n\theta_1)}{(1+n)\sin\theta_1} + \frac{2\cos n\theta_1}{n}\right) \tag{8.21b}$$

The output power at the fundamental frequency is given by

$$P_{o1} = \frac{v_1 i_1}{2} = \frac{V_{ds} I_{max}}{\pi^2}\left[\frac{\theta_1}{\sin\theta_1} + \cos\theta_1\right]^2 \tag{8.22}$$

The DC power in the device is also a function of the angular parameter θ_1 and is given by

$$P_{DC} = \frac{V_{ds} I_{max}}{\pi}\left[\frac{\pi}{2} - \theta_1 + \tan\frac{\theta_1}{2}\right] \tag{8.23}$$

Figure 8.8 shows output current and voltage waveforms as a function of θ for an overdriven amplifier. Since the waveforms are not square waves, the drain efficiency for an overdriven amplifier is less than 100%. The drain efficiency is given by

$$\eta_D = \frac{P_{o1}}{P_{DC}} = \frac{1}{\pi}\left[\frac{\theta_1}{\sin\theta_1} + \cos\theta_1\right]^2\left[\frac{\pi}{2} - \theta_1 + \tan\frac{\theta_1}{2}\right]^{-1} \tag{8.24}$$

$$\begin{aligned}
\eta_D &= \frac{8}{\pi^2} = 0.81 \quad \text{for } \theta_1 = 0 \text{ (class-A overdriven)}\\
&= 0.886 \quad \text{for } \theta_1 = 32.4^\circ \text{ (class-B overdriven)}\\
&= 0.867 \quad \text{for } \theta_1 = 50.4^\circ\\
&= \frac{\pi}{4} = 0.785 \quad \text{for } \theta_1 = 90^\circ \quad \text{(class-B operation)}
\end{aligned}$$

The analysis of overdriven class-A amplifiers is similar to overdriven class-B amplifiers. Every amplifier starts to compress and eventually saturates with input power. If the devices are designed and biased for class-A operation, the amplifier's output current and voltage waveforms start clipping symmetrically under higher input power drive levels.

8.4 ANALYSIS OF CLASS-C AMPLIFIERS

In class-C amplifiers the devices are biased below cutoff or pinch-off and only conduct over less than half of the RF cycle. In this case, as shown in Figure 8.9, the current waveform may be considered a biased sine wave. The drain current is given by [1]

$$i_D(\theta) = \begin{cases} 0, & 0 \le \theta < \pi/2 - \phi \\ I_Q + I_o \sin\theta, & \pi/2 - \phi \le \theta \le \pi/2 + \phi \\ 0, & \pi/2 + \phi \le \theta < 2\pi \end{cases} \tag{8.25}$$

where 2ϕ is the conduction angle. At $\theta = \pi/2 - \phi$, $i_D(\theta) = 0$; that is, $I_Q = -I_o \cos\phi$ and (8.25) becomes

$$i_D(\theta) = I_o[\sin\theta - \cos\phi] \tag{8.26}$$

The DC current is the average value given by

$$I_{DC} = \frac{1}{2\pi}\int_{\pi/2-\phi}^{\pi/2+\phi} I_o[\sin\theta - \cos\phi]\, d\theta = \frac{I_o}{\pi}[\sin\phi - \cos\phi] \tag{8.27}$$

Assuming that the class-C amplifier has the same schematic as shown in Figure 8.4, the output tank circuit allows only the fundamental frequency and the output power is

$$P_o = \frac{1}{2\pi}\int_{\pi/2-\phi}^{\pi/2+\phi} V_o I_o[\sin\theta(\sin\theta - \cos\phi)]\, d\theta = \frac{V_o I_o}{2\pi}[2\phi - \sin 2\phi] \tag{8.28}$$

When $V_o = V_{ds}$, the maximum power is given by

$$P_{om} = \frac{V_{ds} I_o}{4\pi}[2\phi - \sin 2\phi] \tag{8.29}$$

The drain efficiency is

$$\eta_D = \frac{P_{om}}{I_{DC} V_{ds}} = \frac{2\phi - \sin 2\phi}{4[\sin\phi - \phi\cos\phi]} \tag{8.30}$$

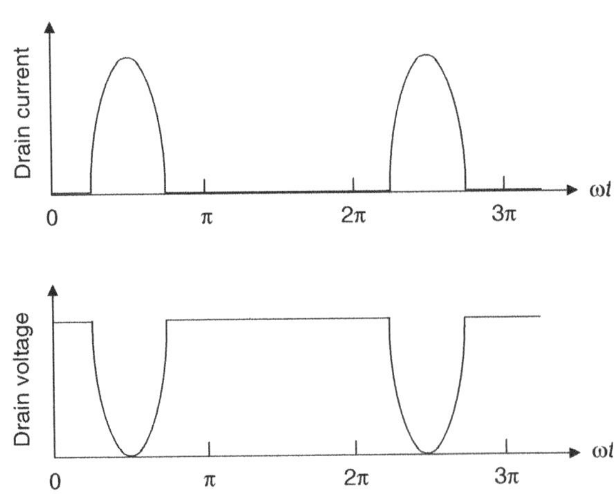

Figure 8.9 Voltage and current waveforms of an ideal class-C amplifier.

When $\phi \cong \pi/2$,

$$\eta_D \cong \frac{\pi}{4}, \quad \text{class B} \tag{8.31}$$

When $\phi \cong \pi/4$,

$$\eta_D \cong 0.939 \tag{8.32}$$

When ϕ approaches zero, η_D is almost 100%. However, the device power gain and output power also approach zero. This makes the class-C amplifier impractical at microwave frequencies because of low gain resulting in low PAE.

8.5 ANALYSIS OF CLASS-E AMPLIFIERS

Class-E amplifiers have been studied extensively [1–19]. It was reported [6] that at low radio frequencies class-E amplifiers have higher efficiency and better linearity than class-B, class-C, and class-F amplifiers. Until now, the class-E amplifiers were limited to the VHF band; however, recent interest in wireless applications has demonstrated [11] that FETs can be used as class-E devices at much higher radio frequencies including the low end of microwave frequencies. A brief discussion on the design of such amplifiers is given next.

Figure 8.10 shows a basic configuration of a class-E amplifier in which the device is represented by an ideal switch S. The device's output capacitance C_d is in parallel with the switch. At the output, a series L_0–C_0 tuned circuit, a reactive component jX, and a load R_L are all connected in series. By suitably selecting the bias conditions and input drive power level, the device is approximated as a switch, which turns on and off periodically at the RF input frequency. The series tuned circuit is resonant at the input frequency and thus passes the fundamental frequency signal to load R_L. The resonator helps in maintaining the output signal to be sinusoidal. If the switch is closed during the first half-cycle ($v_S = 0$) and open during the second half-cycle this results in maximum output power. When the switch is open, the current I_{ds} splits between the capacitor path i_C and the load path i_o. The time-varying current i_C charges the capacitor C_d and produces the voltage across the switch. In the next instant, when the switch is closed, the charge on the capacitor leaks to the ground, giving rise to power loss. This can be minimized by designing the output matching networks such that the voltage across the switch reaches zero at its turn-on time.

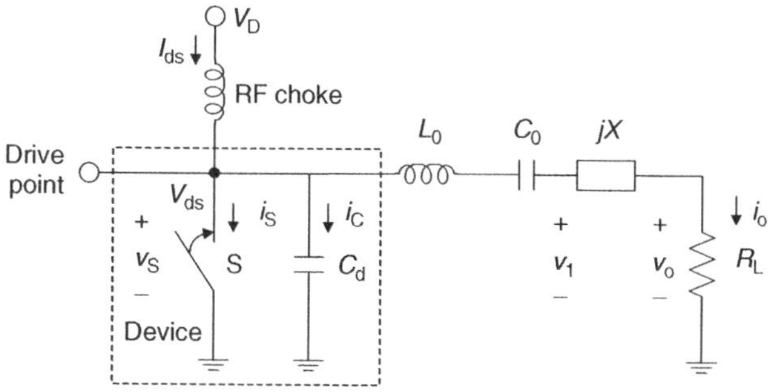

Figure 8.10 Configuration of an ideal class-E amplifier.

Analyses of several different configurations of a class-E amplifier have been reported in the literature [2]. Next, we describe a basic operation of a class-E amplifier. For $0 \leq \omega t \leq \pi$, the switch is closed, the current through the capacitor $i_C = 0$, and the current through the switch may be written

$$i_S = I_{ds} - I_o \sin(\omega t + \phi) \tag{8.33}$$

where $i_o = I_o \sin(\omega t)$. At $t = 0$, $i_S = 0$, and (8.33) becomes

$$I_{ds} = I_o \sin\phi \tag{8.34a}$$

$$i_S = I_o[\sin\phi - \sin(\omega t + \phi)] \tag{8.34b}$$

For $\pi \leq \omega t \leq 2\pi$, the switch through the capacitor is given by

$$i_C = I_{ds} - I_o \sin(\omega t + \phi) = I_o[\sin\phi - \sin(\omega t + \phi)] \tag{8.35}$$

and the voltage across the switch can be written

$$v_S = \frac{1}{\omega C_d}\int_{\pi}^{\omega t} i_C \, d\omega t = \frac{I_o}{\omega C_d}[\cos(\omega t + \phi) + \cos\phi + (\omega t - \pi)\sin\phi] \tag{8.36}$$

At $\omega t = 2\pi$, $v_C = 0$ and from (8.36),

$$\phi = \tan^{-1}\left[-\frac{2}{\pi}\right] = -32.482^\circ$$

or

$$\sin\phi = \frac{-2}{\sqrt{\pi^2 + 4}} \quad \text{or} \quad \cos\phi = \frac{\pi}{\sqrt{\pi^2 + 4}} \tag{8.37}$$

Using (8.34a) and (8.37) in (8.36), the voltage across the switch takes the form

$$v_S = \frac{I_{ds}}{\omega C_d} F(\omega t) \tag{8.38a}$$

where

$$F(\omega t) = \omega t - \frac{3\pi}{3} - \frac{\pi}{2}\cos\omega t - \sin\omega t \tag{8.38b}$$

The supply voltage $V_D = V_{ds}$ when the RF choke is considered to be lossless. In terms of supply current and capacitance C_d, it is expressed as

$$V_{ds} = \frac{1}{2\pi}\int_0^{2\pi} v_S \, d\omega t = \frac{I_{ds}}{\pi \omega C_d} \tag{8.39}$$

The peak switch current and voltage are given by

$$i_{sp} = 2.862 I_{ds} \tag{8.40}$$

and

$$v_{sp} = 3.562 V_{ds} \tag{8.41}$$

The reactive component jX adjusts the phase between the output v_o and switch v_S voltage waveforms as shown in Figure 8.11. In this case, there is no overlap between the output voltage and current waveforms resulting in no power dissipation in the device, and the drain efficiency of an ideal class-E amplifier is 100%. Therefore DC power and RF power (P_o) are equal; that is,

$$P_{DC} = V_{ds} I_{ds} = \frac{I_o^2}{2} R_L = P_o \tag{8.42}$$

From (8.34a), (8.37), and (8.42), we have

$$I_{ds} = \frac{8}{\pi^2 + 4} \frac{V_{ds}}{R_L} = 0.5768 \frac{V_{ds}}{R_L} \tag{8.43}$$

and using (8.39) in (8.43),

$$R_L = \frac{8}{\pi^2 + 4} \frac{1}{\pi \omega C_d} = \frac{0.1836}{\omega C_d} \tag{8.44}$$

EXAMPLE 8.2

For a device having $C_d = 1.0$ pF, $I_{max} = 2$ A, and $V_{ds} = 10$ V, operated as a class-E amplifier at 1 GHz, determine the load value, maximum output current and voltage, and output power.

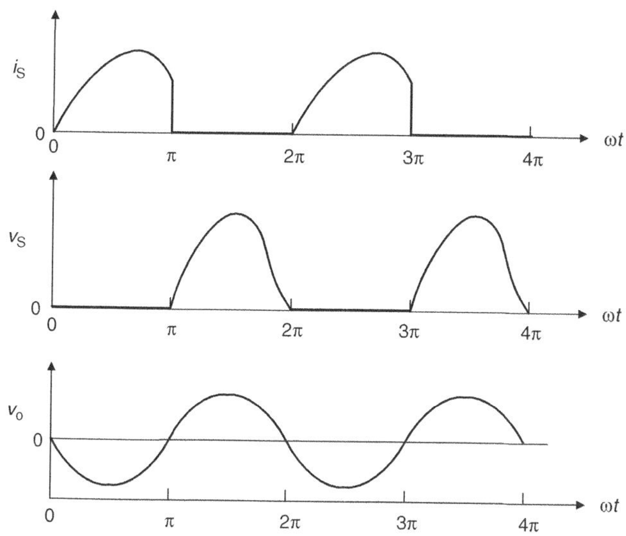

Figure 8.11 Voltage and current waveforms of an ideal class-E amplifier.

SOLUTION From (8.44),

$$R_L = \frac{0.1836}{2 \times \pi \times 1 \times 1 \times 10^{-3}} = 29.22\ \Omega$$

Maximum output current is obtained from (8.34a), (8.37), and (8.39):

$$i_o = |I_o| = \frac{\sqrt{\pi^2+4}}{2}\pi\omega C_d V_{ds} = 367.564\,\text{mA}$$

Maximum output voltage is

$$v_o = |V_o| = i_o R_L = \frac{4}{\sqrt{\pi^2+4}} V_{ds} = 1.0741 V_{ds} = 10.741\ \text{V}$$

$$P_o = P_{DC} = \frac{I_o V_o}{2} = 1.974\,\text{W}$$

■

At radio frequencies, the finite on-resistance of the switching device, the finite time delay between the turn-on and turn-off states, and the output matching loss result in reduced efficiency. As the frequency of operation increases, the efficiency of the amplifier further decreases. The design steps for the class-E amplifier are as given as follows, below.

1. The switching transistors are selected with small on-resistance and output capacitance. The on-resistance affects the efficiency while the output capacitance affects the maximum frequency of class-E operation. The drain efficiency, η_D, and maximum frequency, f_{max}, are given by [10]

$$\eta_D = \frac{1 + (\pi/2 + \omega_0 C_d R_s)^2}{\left(1 + \pi^2/4\right)(1 + \pi\omega_0 C_d R_s)^2} \tag{8.45a}$$

$$f_{max} = \frac{I_{max}}{56.5\ C_d V_{ds}} \tag{8.45b}$$

   where R_s is the on-resistance of the device, V_{ds} is the supply voltage, I_{max} is the open channel current of the device, and ω_o is the angular operating frequency. For example, an FET with $I_{max} = 1$ A, $C_d = 0.6$ pF, and $V_{ds} = 3$ V can be operated up to 9.8 GHz if its f_T is greater than f_{max}. For $R_s \cong 0.4\Omega$, $\eta_D = 92.6\%$ at 9.8 GHz and $\eta_D = 98.2\%$ at 1.9 GHz. However, at $V_{ds} = 10$ V, this device's operation is limited to below C-band.

2. The class-E design sets a relation between the load impedance and the C_d capacitance given by [9]

$$Z_L = (0.183 + j0.211)/(\omega C_d) = R_L + jX_L \tag{8.46}$$

   This is a modified version of the load impedance given by (8.44) and includes the reactive portion of the load.

3. The output power P_o is given by

$$P_o = 0.5768\frac{V_{ds}^2}{R_L} = 1.7337\ I_{ds}^2 R_L \tag{8.47}$$

   where I_{ds} is the drain–source current.

In class-E amplifier operation the output match is designed in such a way that there is no voltage and current overlap; that is, the drain capacitance C_d is fully discharged before the next cycle begins. During the charging of capacitor C_d, the peak voltage can be as high as 3–4 times of supply voltage. This mandates the devices with very high breakdown voltage or multiple low breakdown transistors connected in series (stacking configuration) as discussed in Chapter 13.

8.6 ANALYSIS OF CLASS-F AMPLIFIERS

In a class-F amplifier, the efficiency is improved by using a multiple-resonator output matching network to control the harmonic power levels of the drain voltage and/or drain current waveforms. In an ideal design, at the device output all even harmonics are short circuited to realize sinusoidal current waveform while the odd harmonics are open circuited to obtain a square voltage waveform. However, the efficiency can be improved by controlling the second or third or both harmonics [20–30]. Due to device parasitic reactance and other unknown reactances of the matching networks at higher frequencies, accurate knowledge of harmonic termination under nonlinear operation of the device is difficult to determine. Various harmonic peaking techniques have been described by Raab [23] to improve the efficiency of power amplifiers. As an example, third-harmonic peaking is discussed in this section.

Referring to Figure 8.12, the equations for the drain voltage and current waveforms may be expressed as

$$v_D(\theta) = V_{ds} + V_{om}\sin\theta + V_{3m}\sin 3\theta + V_{5m}\sin 5\theta + \cdots \tag{8.48}$$

$$i_D(\theta) = I_{ds} - I_{om}\sin\theta - I_{2m}\sin 2\theta - I_{4m}\sin 4\theta - \cdots \tag{8.49}$$

where $\theta = 2\pi f_0 t$ and f_0 is the fundamental frequency (input signal frequency) of the desired output. The coefficients V_{om}, I_{om} are for the fundamental frequency, I_{2m} is for the second harmonic, V_{3m} is for the third harmonic, and so on. The opposite sign for the fundamental frequency coefficients signifies that the voltage and current are opposite in phase. At low drive levels, the harmonic levels are small, and the drain voltage waveform remains sinusoidal and the output voltage is given by

$$V_{om} = I_{om}R_L \tag{8.50}$$

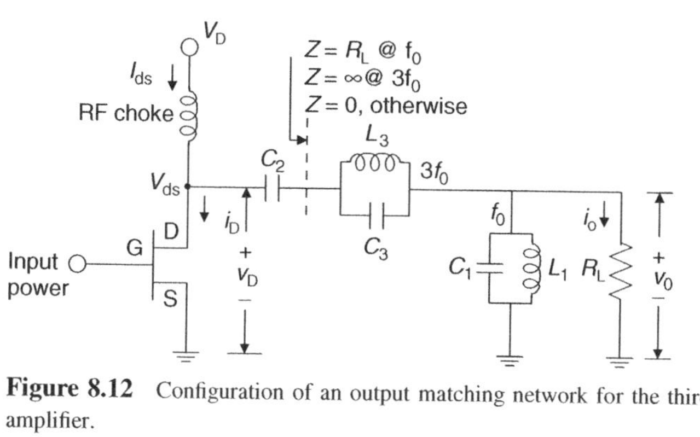

Figure 8.12 Configuration of an output matching network for the third-harmonic peaking class-F amplifier.

where R_L is the load resistance. The fundamental frequency output power is given by

$$P_o = \frac{V_{om} I_{om}}{2} = \frac{V_{om}^2}{2R_L} \tag{8.51}$$

As the input power level increases, the solid state devices start saturating and at some point output RF voltage $v_o = V_{om} = V_{ds}$. Consider all even harmonics are short circuited and only the third harmonic is open circuited. For a maximally flat drain voltage waveform as shown in Figure 8.13, $V_{3m} = V_{om}/8$. Under this condition the output voltage becomes

$$v_o = V_{om} + V_{om}/8 = V_{ds} + V_{ds}/8 = \tfrac{9}{8} V_{ds} \tag{8.52}$$

By using (8.51), the peak output power can be expressed as

$$P_{om} = \frac{v_o^2}{2R_L} = \frac{81}{128} \frac{V_{ds}^2}{R_L} \tag{8.53}$$

The device DC current is as in a class B: that is, $I_{ds} = I_{max}/\pi = 2I_{om}/\pi$. Therefore DC power becomes

$$P_{DC} = V_{ds}\, I_{ds} = \frac{2}{\pi} V_{ds}\, I_{om} = \frac{2}{\pi} V_{ds} \left(\frac{v_o}{R_L}\right) = \frac{9}{4\pi} V_{ds}^2 / R_L \tag{8.54}$$

and the drain efficiency is

$$\eta_D = \frac{P_{om}}{P_{DC}} = \frac{9\pi}{32} = 0.884 \tag{8.55}$$

Thus by third-harmonic peaking one can get a drain efficiency of about 88% (i.e., 10% points higher than class B), and output power about 27% greater than class B.

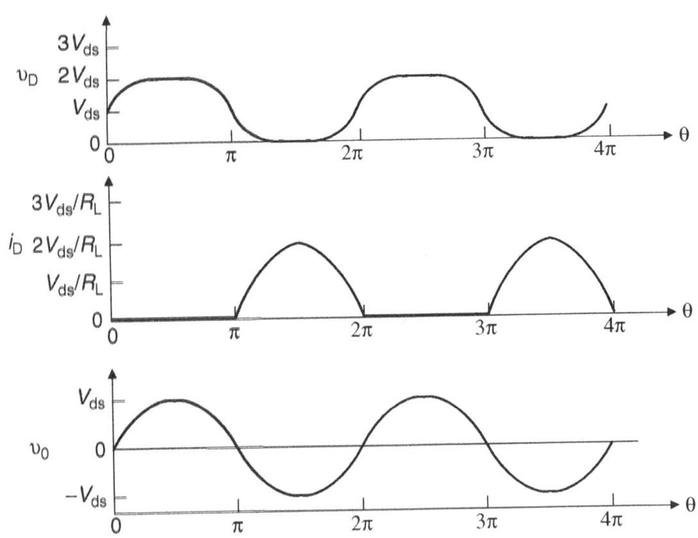

Figure 8.13 Voltage and current waveforms for the third-harmonic peaking class-F power amplifier.

Table 8.1 Percentage of Drain Efficiency for Various Combinations of Harmonic Terminations for Class-F Power Amplifiers[a]

Harmonic	1	1, 3	1, 3, 5	1, 3, 5, ..., ∞
1	50.0, class A	56.3	58.6	63.7
1, 2	66.7	75.0, HRA[b]	78.1	84.9
1, 2, 4	71.1	80.0	83.3	90.5
1, 2, 4, ..., ∞	78.5, class B	88.4, 3 HP[c]	92.0, 3 + 5 HP	100.0, class D

[a]The even harmonics are short circuited and the odd harmonics are open circuited.
[b]HRA = harmonic reaction amplifier.
[c]HP = harmonic peaking.

Table 8.1 provides the percentage of drain efficiency for various combinations of even-and odd-harmonic terminations for class-F power amplifiers [23].

EXAMPLE 8.3

A transistor having $I_{\text{max}} = 2$ A and breakdown voltage of 25 V is biased for class B at $V_{\text{ds}} = 10$ V and operated as a class-F amplifier (only third harmonic terminated) at 1 GHz. When the amplifier has a very high gain value, determine the load value and the output power (in dBm) for the amplifier.

SOLUTION Here $V_{\text{o}} = \frac{9}{8} V_{\text{ds}}$ and $I_{\text{o}} = I_{\text{max}}/2$.

$$R_{\text{L}} = V_{\text{o}}/I_{\text{o}} = \tfrac{9}{8}(2V_{\text{ds}}/I_{\text{max}}) = 1.125 \times 2V_{\text{ds}}/I_{\text{max}} = 11.25\,\Omega$$

The load value is about 1.125 times the class-A load value. The output power is given by

$$P_{\text{om}} = \frac{81}{128}\frac{V_{\text{ds}}^2}{R_{\text{L}}} = \frac{81 \times 100}{128 \times 11.25} = 5.625\text{ W} = 37.5\,\text{dBm}$$ ■

The design steps for a class-F amplifier are as follows:

1. The input is matched for maximum gain.
2. The load impedance is the same as obtained for class-B amplifier design. R_{L} is the real part.
3. The parallel L_1, C_1 network (shunt) is resonant at f_0 and the parallel L_3, C_3 network (series) is resonant at the third harmonic $3f_0$. Capacitor C_2 is a blocking capacitor and can also be used as a matching element. At f_0 the third-harmonic resonant circuit represents a small inductance, which can be a matching element along with C_2, to provide desirable load at the output of the device. The values of these elements are calculated using the following relationships:

$$C_1 = \frac{\alpha}{2\pi f_0 R_{\text{L}}(1-\alpha^2)}, \quad \alpha = 1 - 0.5BW/f_0 \tag{8.56a}$$

$$L_1 = 1/(\omega_0^2\, C_1) \tag{8.56b}$$

$$L_3 = 1.9753\, L_1 R_{\text{L}}^2/[9R_{\text{L}}^2 + 4\omega_0^2 L_1^2] \tag{8.56c}$$

$$C_3 = 1/(9\omega_0^2 L_3) \tag{8.56d}$$

$$C_2 = 8C_3 \tag{8.56e}$$

where $\omega_0 = 2\pi f_0$ and BW is the fractional design bandwidth.

4. In a multistage amplifier, harmonic termination can also be selected. Several interstage loading networks have been described by Trask [25].

During the past two decades there has been significant progress in harmonic tuned monolithic power amplifiers to achieve extraordinarily high PAE values. A C-band class-F single-stage MMIC amplifier with 70% PAE, 8-dB gain, and 1.7-W power output has been demonstrated [20]. In this amplifier the harmonic tuning was used at the output. Another C-band single-stage class-F MMIC amplifier with 60% PAE, 9-dB gain, and 13-W power output was developed [31]. In the later design, harmonic tuning was applied to both the input and output of the FET. The original design was performed using a nonlinear model and fine-tuned using the Taguchi empirical approach described in Chapter 9.

8.7 COMPARISON OF VARIOUS AMPLIFIER CLASSES

Sowlati et al. [6] compared the characteristics of class B, C, F, and E operating at 800 MHz and delivering 0.25 W of output power. In the comparison two supply voltages, 2.5 V and 5.0 V, were considered. The matching networks were assumed lossless and the transistors have a square-law characteristic with a knee voltage of 0.5 V. The topologies studied are shown in Figure 8.14: part (a) for both class B and class C and part (b) for class F. The configuration and design for class E are the same as discussed previously. The operating conditions for other classes are summarized below:

1. The parallel tuned L_0–C_0 circuit (Fig. 8.14a) is resonant at fundamental input signal frequency and short circuits all the higher harmonic signals.

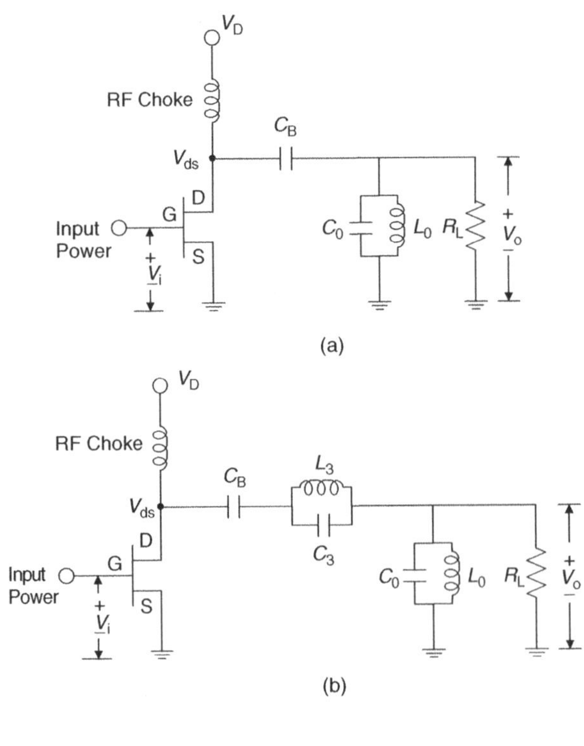

Figure 8.14 Amplifier circuit topologies: (a) class B or class C and (b) class F.

2. The conduction angles of the transistors for class B and class C are 180° and 120°, respectively.
3. The parallel tuned L_0–C_0 circuit (Fig. 8.14b) is resonant at the fundamental input signal frequency while the parallel tuned L_3–C_3 circuit is resonant at the third harmonic.
4. The level of the third-harmonic voltage at the drain of the transistor is set to 1/9 of its first-harmonic voltage level.

The amplifier's important parameters such as the optimum load R_L, the maximum transistor current I_{max}, the output power P_o, the dissipated power P_{DC}, and the drain efficiency η_D are summarized in Table 8.2a for the supply voltage of 5 V and in Table 8.2b for the supply voltage of 2.5 V. The important observations are the following:

1. At low voltage, R_L for class E is much higher than for other classes. The higher the R_L value, the lower is the loss in the matching network.
2. The I_{max} value for class E is the lowest while I_{max} is highest for class C. The lower I_{max} value means lower $I_{max}^2 R$ loss.
3. At low voltage, the reduction in drain efficiency (η_D) is more pronounced in class F and minimum in class E.

Table 8.3 shows a comparison of RF performance measured using LDMOS transistors at 0.5 GHz [33]. It may be noted that both class-E and class–F amplifiers have similar performance. The PAE and power level capabilities for class-E amplifiers are adversely affected at higher voltages and frequencies.

A summary of load impedance and drain efficiency for various classes of amplifiers is given in Table 8.4. Here $R_L = V_{max}/I_{max}$ and n is the harmonic number. V_{max} and I_{max} are the maximum or peak voltage and current, respectively. Figure 8.15 depicts input signal and output current and voltage waveforms for several amplifier classes. In the above comparisons, an ideal device and circuit have been assumed.

Table 8.2 Comparison of Important Parameters of Various Amplifier Classes

	(a) $V_{ds} = 5$ V			
	Class			
Parameter	B	C	F	E
R_L (Ω)	40	40	50	59
I_{max} (mA)	265	320	240	154
P_o (mW)	250	250	250	250
P_{DC} (mW)	330	307	290	274
η_D (%)	75	80	85	91
	(b) $V_{ds} = 2.5$ V			
	Class			
Parameter	B	C	F	E
R_L (Ω)	8	8	10	21
I_{max} (mA)	590	780	525	294
P_o (mW)	250	250	250	250
P_{DC} (mW)	360	326	320	290
η_D (%)	68	74	76	86

Table 8.3 Summary of Measured Performance of Class E and F LDMOS Hybrid Amplifiers

Amplifier	Gain (dB)	P_o (W)	η_D (%)
Class F	14	25	78
Class E	15	22	67

Table 8.4 Summary of RF and Microwave Amplifier Modes of Operation

Class of Operation	Bias	Load Impedance Fundamental Frequency	Even Harmonics	Odd Harmonics	Drain Efficiency (%)
Class A—linear	A	R_L	—[a]	—[a]	50
Class A—overdriven		R_L	—[a]	R_L	81
Class B—linear(pushpull)	B	R_L	Short	—[a]	78.5
Class B—overdriven		R_L	—[a]	R_L	88.0
Class B—single ended		0.73 R_L	—[a]	—[a]	58.0
Class C—linear	C	R_L	—[a]	—[a]	93.9
Class E—switched mode	B	0.58 R_L	—[a]	—[a]	100
Class F—saturated	B	1.27 R_L	Short	Open	100
Class F—saturated		1.1 R_L	Short	Open ($n = 3$)	88.4

[a]No particular harmonics are terminated.

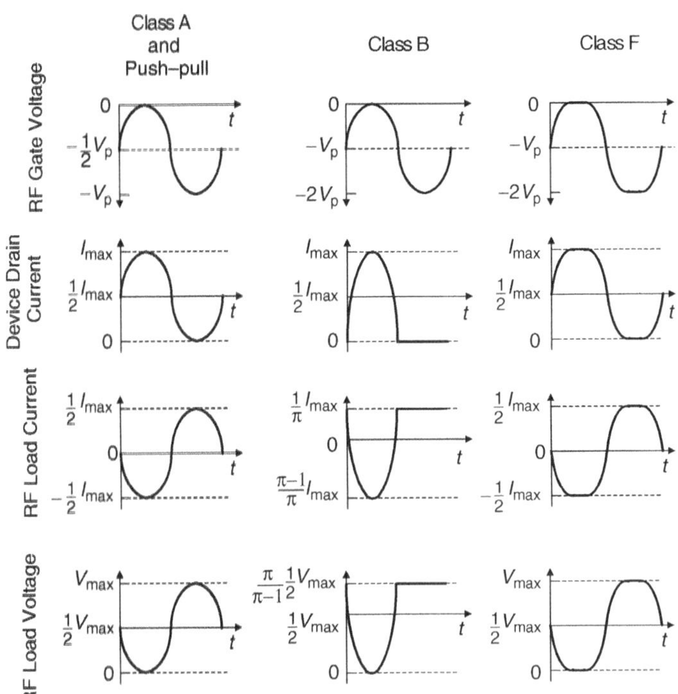

Figure 8.15 Voltage and current relationships for various classes of operation for ideal transistors. V_p, V_{max}, and I_{max} are the pinch-off voltage, maximum or peak voltage, and maximum or peak current, respectively.

In this chapter, simple analyses for various classes of amplifiers have been provided. Such analyses are valid at UHF and VHF; however, at microwave frequencies they become approximate due to transistor parasitic reactances and must be used with care. For example, the voltage and current waveforms are not exactly square or sine wave as described previously. They get distorted. At UHF and VHF, device and matching network parasitic reactances are negligible and transistors have very high gain (20 dB or higher). At these frequencies, the transistors can be operated at class B and still have high gain (15 dB or higher). By terminating even harmonics, and third and fifth, a drain efficiency on the order of 90% may be obtained. A similar efficiency value is possible by using class-E operation, however, at lower output power levels. At RF and microwave frequencies, all devices have significant parasitic reactance, power dissipation, and transient or switching time that is a significant fraction of the RF cycle time period. They cannot be ignored in the analyses. However, at higher frequencies and for a given device, unless it is specifically designed for a particular application, the device behavior is similar under 2–4-dB compression. Due to same circuit limitations, amplifiers of different classes may behave very similarly. Practical aspects of several of these amplifier classes are treated in the following chapters.

REFERENCES

1. H. Krauss, C. Bostian, and F. Raab, *Solid-State Radio Engineering*, John Wiley & Sons, Hoboken, NJ, 1980.
2. A. Grebennikov, *RF and Microwave Power Amplifier Design*, McGraw-Hill, New York, 2004.
3. N. O. Sokal and A. D. Sokal, Class E—a new class of high-efficiency tuned single-ended switching power amplifiers, *IEEE J. Solid-State Circuits*, Vol. SC-10, pp. 168–176, June 1975.
4. F. H. Raab, Idealized operation of the class E tuned power amplifier, *IEEE Trans. Circuits Systems*, Vol. CS-24, pp. 725–735, December 1977.
5. T. B. Mader and Z. B. Popovic, The transmission-line high-efficiency class-E amplifier, *IEEE Microwave Guided Wave Lett.*, Vol. 5, pp. 290–292, September 1995.
6. T. Sowlati et al., Low voltage, high efficiency GaAs class E power amplifiers for wireless applications, *IEEE J. Solid State Circuits*, Vol. 30, pp. 1074–1079, October 1995.
7. J. F. Davis and D. B. Rutledge, A low-cost class-E power amplifier with sine-wave drive, *IEEE MTT-S Int. Microwave Symp. Dig.*, pp. 1113–1116, 1998.
8. N. O. Sokal, Class E high-efficiency power amplifiers, from HF to microwave, *IEEE MTT-S Int. Microwave Symp. Dig.*, pp 1109–1112, 1998.
9. F. J. Ortega-Gonzalez et al., High-efficiency load-pull harmonic controlled class-E power amplifier, *IEEE Microwave Guided Wave Lett.*, Vol. 8, pp. 348–350, October 1998.
10. T. B. Mader et al., Switched-mode high efficiency microwave power amplifiers in a free-space power–combiner array, *IEEE Trans. Microwave Theory Tech.*, Vol. 46, pp. 1391–1397, October 1998.
11. M. Markovic, A. Kain, and Z. Popovic, Nonlinear modeling of class-E microwave power amplifiers, *Int. J. RF Microwave Computer-Aided Design*, Vol. 9, pp. 93–103, March 1999.
12. V. S. Rao Gudimetla and A. Z. Kain, Design and validation of the load networks for broadband class E amplifiers using nonlinear device models, *IEEE MTT-S Int. Microwave Symp. Dig.*, pp. 823–826, 1999.
13. A. Mediano and P. Molina, Frequency limitation of a high-efficiency class E tuned RF power amplifier due to a shunt capacitance, *IEEE MTT-S Int. Microwave Symp. Dig.*, pp. 363–366, 1999.

14. D. K. Choi and S. I. Long, A physically based analytic model of FET class-E power amplifiers—designing for maximum PAE, *IEEE Trans. Microwave Theory Tech.,* Vol. 47, pp. 1712–1720, September 1999.
15. W. H. Cantrell, Tuning analysis for the high-Q class-E power amplifier, *IEEE Trans. Microwave Theory Tech.*, Vol. 486, pp. 2397–2402, December 2000.
16. C. Yoo and Q. Huang, A common-gate switched 0.9-W class-E power amplifier with 41% PAE in 0.25-μm CMOS, *IEEE J. Solid State Circuits*, Vol. 36, pp. 823–830, May 2001.
17. A. Grebennikov, Switched-mode RF and microwave parallel-circuit class E power amplifiers, *Int. J. RF Microwave Computer-Aided Design,* Vol. 14, pp. 21–35, January 2004.
18. D. P. Kimber and P. Gardner, Power series analysis of the class E power amplifier, in *European Microwave Conference Proceedings*, pp. 1461–1464, 2004.
19. Y. Qin et al., Design of low-cost broadband class-E power amplifier using low voltage supply, *Microwave Opt. Tech. Lett.*, Vol. 44, pp. 103–106, January 2005.
20. I. J. Bahl et al., Class-B power MMIC amplifiers with 70 percent power added efficiency, *IEEE Trans. Microwave Theory Tech.*, Vol. 37, pp. 1315–1320, September 1989.
21. C. Duvanaud, S. Dietsche, G. Pataut, and J. Obrgon, High-efficient class F GaAs FET amplifiers operating with very low bias voltages for use in mobile telephones at 1.75 GHz, *IEEE Microwave Guided Wave Lett.*, Vol. 3, pp. 268–270, August 1993.
22. F. H. Raab, Introduction to class-F power amplifiers, *R.F. Design*, Vol. 19, pp. 79–84, May 1996.
23. F. H. Raab, Class-F power amplifiers with maximally flat waveforms, *IEEE Trans. Microwave Theory Tech.*, Vol. 45, pp. 2007–2012, November 1997.
24. P. Colantonio et al., On the class-F power amplifier design, *Int. J. RF Microwave Computer-Aided Design,* Vol. 9, pp. 129–149, March 1999.
25. C. Trask, Class-F amplifiers loading network: a unified design approach, *IEEE MTT-S Int. Microwave Symp. Dig.,* pp. 351–354, 1999.
26. A. Grebennikov, Circuit design technique for high efficiency class F amplifiers, *IEEE MTT-S Int. Microwave Symp. Dig.*, pp. 771–774, 2000.
27. P. Colantonio et al., Class G approach for low-voltage, high-efficiency PA design, *Int. J. RF Microwave Computer-Aided Design*, Vol. 10, pp. 366–378, November 2000.
28. F. H. Raab, Maximum efficiency and output of class F power amplifiers, *IEEE Trans. Microwave Theory Tech.*, Vol. 47, pp. 1162–1166, June 2001.
29. F. H. Raab, Class-E, class-C, and class-F power amplifiers based upon a finite number of harmonics, *IEEE Trans. Microwave Theory Tech.*, Vol. 49, pp. 1462–1468, August 2001.
30. S. Gao, High efficiency class F RF/microwave power amplifiers, *IEEE Microwave Mag.*, Vol. 7, pp. 40–48, February 2006.
31. W. L. Pribble and E. L. Griffin, An ion-implanted 13 watt C-band MMIC with 60% peak power added efficiency, *IEEE Microwave Millimeter Wave Monolithic Circuits Symp. Dig.*, pp. 25–28, 1996.
32. D. M. Snider, A theoretical analysis and experimental confirmation of the optimally loaded and overdriven RF power amplifiers, *IEEE Electron Devices*, Vol. 14, pp. 851–857, December 1967.
33. A. Grebennikov and N. O. Sokal, *Switchmode RF Power Amplifiers*, Elsevier, Burlington, MA, 2007.

PROBLEMS

8.1 Design a class-A power amplifier that delivers 5 W to a 50-Ω system. The supply voltage is 10 V and the device is rated for a maximum current of 2 A. Determine the output matching element values at 5 GHz (consult Chapter 7 for matching network design). Also determine the inductance of the RF choke, which is $20R_L$, in order to minimize its effect on the circuit. R_L is the real part of the load impedance.

8.2 Design a class-E power amplifier that delivers 2 W to a 50-Ω system. The device's output capacitance is 1 pF. Determine the maximum voltage and current ratings for the device. Calculate the output matching element values at 2 GHz (consult Chapter 7 for matching network design).

8.3 Design a class-F (shown in Fig. 8.12) power amplifier that delivers 5 W to a 50-Ω system. The supply voltage is 10 V and the device is rated for a maximum current of 1 A. Determine the output matching element values at 5 GHz (consult Chapter 7 for matching network design). Also determine the net power dissipated in the device.

8.4 Design a push–pull power amplifier at 5V that delivers 2 W to a 50-Ω system. Determine the maximum voltage and current ratings for the device. Calculate the output matching element values and the balun's/transformer's parameters at 2 GHz (consult Chapter 7 for matching network design). Baluns are assumed lossless.

8.5 Design an output network of a class-A amplifier that delivers 2 W to a 50-Ω load at 1 GHz. The power supply voltage is 15 V. Also determine device's output power ratings.

8.6 Design an output matching network of a single-ended class-B amplifier that delivers 20 W at 1 GHz. The power supply voltage is 28 V. The output match uses a $\lambda/4$ transmission section to transform the 50-Ω system impedance to the required load at the device output. Also determine the device's output power ratings.

8.7 Discuss the pros and cons of various classes of amplifier operation in terms of linearity, efficiency, and power. Describe the methods to validate the class of operation of power amplifiers.

8.8 Discuss the basic limitations of class-F amplifiers at microwave frequencies.

Chapter 9

Amplifier Design Methods

The design of amplifiers for a particular application and frequency range is complex in the sense that they have to meet the physical, electrical, thermal, and cost requirements. The amplifier performance requirements in terms of frequency band, gain, noise figure, power output, PAE, linearity, input and output VSWR, stability, ruggedness to mismatch, and so on are determined by the transistor type and sizes, the circuit design topology, matching networks, the number of gain stages, the aspect ratio for the devices between the stages, design methodology, biasing schemes, thermal design, fabrication technology, and packaging. More often it involves trade-offs in terms of size, electrical performance, reliability, and cost. In this chapter we discuss, in general, various aspects of amplifier design methodology with examples, and specific designs are described in detail in several succeeding chapters.

9.1 AMPLIFIER DESIGN

The design of an amplifier may be for low-power or high-power application, but basic design steps are similar. The low power amplifier may be a low-noise or a buffer or a variable gain amplifier and is generally designed using linear design methods. However, power amplifiers are based on nonlinear design methods and may be for high-power or high-efficiency and/or high-linearity applications. These designs are not as trivial as low-power ones and require accurate CAD tools to describe nonlinearities. Thus the focus of this chapter is toward power amplifiers although the methods are also applicable for low-power designs. The design methodology includes the selection of circuit topology, number of stages, syntheses of matching networks, biasing networks and realization of amplifiers' circuitry, simulation and optimization of amplifier performance, and power combining for higher power level if required. The amplifier design could be narrowband, broadband, single, or multistage. The amplifier design may be grouped into six steps:

1. Transistor type and fabrication technology
2. Transistor sizes
3. Design method
4. Selection of circuit topology and components

Fundamentals of RF and Microwave Transistor Amplifiers. By Inder J. Bahl

5. Circuit analysis and optimization
6. Stability and thermal analyses

9.1.1 Transistor Type and Fabrication Technology

First and foremost, one must select the proper transistor type to meet amplifier design requirements (available power supply, noise figure, power output, frequency range, etc.). A device type may be selected from CMOS, BJT, FET, HBT, and HEMT. The next step is to determine the fabrication technology, which again depends on the application, development cycle, cost target, volume, and so on.

Several RF/microwave manufacturing technologies are being used to reduce component counts, size, and cost. These are printed circuit board (PCB), thin- and thick-film hybrid, low- and high-temperature cofired ceramic, and monolithic integrated circuit. The printed circuit board approach is used at RF and limited to low-power levels. Thin-film hybrid technology is commonly used for discrete devices, single-stage internally matched amplifiers, and high-power combining methods. Generally, for low-volume applications, discrete devices are used while high-volume and low-cost targets are met with monolithic IC solutions. The monolithic approach is also a preferred choice at millimeter-wave frequencies. Plastic packages offer a low-cost solution for power amplifiers (less than 5 W) operating up to Ku-band frequencies, while for high-power applications at L- through Ku-band frequencies, ceramic packages are commonly used. At millimeter-wave frequencies, the chip on carrier approach is often utilized. Chapter 14 exclusively deals with hybrid amplifier design. Monolithic technology is being widely used from RF through millimeter-wave frequencies and is discussed in more detail in Chapter 15.

9.1.2 Transistor Size Selection

For low-noise applications, the selection of transistor size is not as critical as for power applications. For low-noise applications larger transistor size is used at lower frequencies. For low-power and low-noise applications, the device size is selected based on frequency range and other requirements such as P_{1dB} and TOI (or IP3). Larger transistor size is easier to match at RF and lower microwave frequencies but has a higher power consumption, P_{1dB}, and TOI. Smaller transistor size is preferred at millimeter-wave frequencies with its lower power consumption. At low microwave frequencies, low-noise transistor size ranges from about 0.6 to 1.0 mm, at X-band frequencies it is about 0.2–0.3 mm, and above 20 GHz the preferred size is 0.05–0.1 mm.

The selection of power transistor size depends on the power levels and the bandwidth to meet design requirements. For narrowband applications about 20–30% margin should be included from the device P_{out} to amplifier P_{out}. The amplifier circuit should be designed to function within its safest operating conditions: current $\leq I_{max}$ and voltage $\leq$ breakdown voltage. Recommended bias conditions for transistors are generally provided by vendors. Assuming that the output matching circuit loss can be kept to within 0.5 dB and another 0.5 dB due to mismatch between the desired and designed load impedance at the output of the device, a transistor must provide 10-W output power to realize an 8-W power amplifier. In a multistage amplifier, designing a proper device aspect ratio (ratio of output to input stage transistor sizes) plays a very important role in realizing a high PAE or a linear amplifier. In a two-stage amplifier design, the transistor aspect ratio depends on the amplifier's compressed gain at the operating frequency, the PAE or linearity requirements, and the bandwidth. In narrow-band applications, a margin of 2–3-dB gain loss due to dissipative loss and mismatch between the input and output transistors is considered in the amplifier design, whereas in broadband application this loss is 3–4 dB. For example, FETs having 10 dB of compressed gain per stage at C-band frequencies require about 4:1 output to input stage FET size ratio for high-PAE applications. However, for high-linearity applications this ratio may be reduced to 3:1, and even 2:1 depending on the requirements.

9.1.3 Design Method

A power amplifier can be designed based on the load-line method, low-loss match (LLM) method using load-pull data, and small-signal *S*-parameters obtained at the operating bias point or employing a nonlinear device model. In the load-line method, analytical equations are used to determine the circuit parameters and output power and PAE are calculated for only the fundamental frequency. In the LLM method, optimum load impedances at the output of each device are realized using load-pull data and *S*-parameters. The output power and PAE are calculated empirically. Both these methods do not provide other nonlinear performance such as TOI. In the nonlinear design method, a nonlinear device model is used to calculate the power amplifier performance including power compression of each stage, and the output power and PAE as a function of input power. This method is very comprehensive and also allows one to calculate harmonic levels and their reactive harmonic terminations for high PAE, and other nonlinearity performance as per design requirements.

9.1.4 Circuit Topology

The circuit topology and matching networks are generally designed to provide the required bandwidth with suitable guard bands. In order to reduce size, biasing circuitry is generally incorporated as a part of the matching network. The matching networks usually include the package lead frame and bond wire parasitic reactance. For high-power applications, cluster type topology is generally used. Microstrip sections are used as matching elements in high-performance amplifiers. To realize compact circuits, lumped-element matching networks or lumped-distributed circuit elements are utilized to transform device impedance to 50 Ω Also, low-loss circuit elements at the output of the amplifier are desirable since the efficiency and power are reduced by a loss in the output matching network. At millimeter-wave frequencies, the microstrip is usually used.

Another advantage of using lumped elements in RF and microwave circuits lies in the fact that several design techniques used in circuits at lower radio frequencies, which are not practical at microwave frequencies using microstrip, can now be successfully applied up to X-band frequencies. The circuit configurations include true differential, push–pull, and feedback amplifiers, high-voltage and phase-splitting amplifiers, direct-coupled amplifiers, bridged T-coil amplifiers, and series and shunt gain peaked broadband amplifiers.

In broadband applications, lumped elements play a significant role in achieving the required circuit performance. In order to tune out the transistor capacitance, one needs an inductance with minimum possible parasitic capacitance. At microwave frequencies, one generally uses high-impedance lines that are inductive in nature (see Chapter 6). However, these lines have associated shunt capacitance, which reduces the gain–bandwidth product of the circuit. For example, a 12-μm wide microstrip conductor on a 75-μm thick GaAs substrate requires a 1160-μm long line to realize an inductance value of 0.8 nH. The associated shunt capacitance value is about 0.13 pF. On the other hand, a lumped-element inductor with 12-μm trace width and 2.5 turns will result in 0.8-nH inductance and only 0.04 pF of shunt capacitance. Therefore using lumped inductors with much lower parasitic capacitance will result in wider bandwidth circuits.

RF chokes using lumped inductors have a distinct advantage in terms of size and bandwidth in comparison to the $\lambda/4$ line transformers commonly used in microwave circuits to bring bias to active or passive solid state devices. For example, a compact inductor having 5-nH value and series resonant frequency above 20 GHz can be used as an RF choke from 5 to 20 GHz, whereas one needs two to three sections of $\lambda/4$ transformers to realize the same bandwidth (see Chapter 18 for more details). In summary, lumped elements in comparison to conventional distributed elements have smaller size and lower cost, large impedance transformation ratio capability, smaller interaction effects between circuit elements, lower associated complementary reactance, and wider bandwidth capability.

Lowpass matching networks provide good rejection for high spurious frequency and harmonic frequencies but have a tendency toward high gain (and hence instability) at very low frequencies. Thus in multistage amplifiers, a combination of bandpass at the input stage and interstages, and lowpass at the output stage would result in the required frequency response.

9.1.5 Circuit Analysis and Optimization

The electrical device parameters available in vendor catalogs or foundry libraries include S-parameters or equivalent circuit models at recommended bias conditions, nonlinear models, optimum input and output impedances for maximum output power or PAE, and associated gain, at nominal frequencies. Data for the best linearity design is hardly available. Small-signal S-parameters/EC models and nonlinear models are required to calculate the stability and electrical performance of an amplifier. Source-pull and load-pull data are essential to accurately characterize power devices for optimum amplifier design. Power amplifiers are commonly designed using the load-line method [1–3].

Accurate linear and nonlinear models are required to design multistage amplifiers and to predict an amplifier's performance in terms of P_{1dB}, PAE, gain, second- and third-harmonic levels, and third-order intercept point, and to arrive at the best compromise between them. Commercial CAD tools are available for circuit analysis and optimization. Since it is very difficult to realize the required load impedances over wider bandwidths and to optimize a circuit using nonlinear models, a combination of load-line optimization and nonlinear simulation is used. The above design process is repeated so that an optimum solution for the simultaneous match for load impedances at the drain of each FET and the best gain, power, and PAE are achieved. However, a simple single-stage power amplifier design may be optimized using a nonlinear model with a loose convergence tolerance setting in the harmonic balance simulator. Final circuit optimization includes EM simulated data, stability analysis, sensitivity analysis, and packaging effects.

9.1.6 Stability and Thermal Analyses

The design must be conditionally stable and also odd-mode, loop, parametric, and low-frequency oscillation conditions (see Chapter 17 for details) must be prevented in power amplifiers. Any amplifier with power gain can be made to oscillate by applying an external positive feedback, for example, biasing circuitry in a multistage high-gain amplifier or a high-gain MMIC chip in a plastic or ceramic package with poor isolation. At RF and microwave frequencies, unavoidable parasitic reactances are often sufficient to cause oscillations if care is not taken in the design and fabrication of the amplifier. The stability analysis of power amplifiers is discussed in Chapter 17.

Power transistors, depending on the technology, have a maximum allowed operating channel temperature (the range is 150–200° C) that cannot be exceeded in order to maintain its RF performance as well as reliability. However, lowering the channel temperature can improve electrical performance and greatly increase the device lifetime. Thus in a power amplifier product design, it is very important to effectively take out the excessive heat generated in the device. This can be achieved by using small-size devices, spreading heat over a larger area, minimizing high-resistance interfaces between the device and heat sink, and using high-conductivity heat sinks. For high-power, high-efficiency, and high power-density devices, diamond heat spreaders are desirable. The thermal design of transistors and amplifiers is discussed in Chapter 16.

9.2 AMPLIFIER DESIGN TECHNIQUES

In a single-stage amplifier, the input matching network is designed for good VSWR, stability, gain flatness, and noise figure or linearity (IP3), while the output matching network is designed for maximum gain, output power, and PAE or IP3. The input matching network for an LNA is designed to be low loss to achieve the minimum possible noise figure. The output matching network for a power amplifier that is designed to deliver maximum power should also be low loss to achieve the maximum possible efficiency. Matching circuits should limit the gain outside the desired frequencies.

A multistage power amplifier is designed in three parts: design of input, interstages, and output matching networks. In an LNA, the input match is designed first, then the interstage, and finally the output match. In the HPA design, the sequence is reversed; that is, the output match is designed first and the input match is designed last. The design of input and output matching networks is similar to the single-stage amplifier design. The interstage matching networks usually provide the gain shaping required to compensate for the 6-dB/octave gain roll-off of each device and also provide enough output power to the succeeding stage for maximum output power from the next stage. The design of the interstages is very critical at high frequencies and for broadband amplifiers. A multistage amplifier has more flexibility in designing input match and gain flatness than a single-stage amplifier. At higher frequencies, where the gain is minimum over the band, the matching must provide maximum gain as well as power to drive the gate of the next stage for maximum output power. The output and interstage matching networks are designed to provide the desired load impedances for maximum output power. Small-signal simulators do a good job in terms of input match and gain but are not adequate to simulate power performance. Nonlinear simulators and/or extensive source-pull and load-pull data are required to design high-efficiency and wider bandwidth power amplifiers.

There are several methods that can be used to accurately determine a device's input and output impedance at large-signal conditions, an essential requirement for successful power amplifier circuit design. The classical source-pull and load-pull techniques, small-signal and large-signal S-parameters, equivalent circuit linear and nonlinear models, and physics-based models are currently being used to determine device input and output impedances. The nonlinear equivalent circuit and physics-based models incorporated in nonlinear CAD tools can be used to design power amplifiers and accurately simulate their performance. For narrowband applications, source-pull and load-pull data are sufficient to design amplifiers. However, for octave or higher bandwidths and very-high-efficiency applications, nonlinear device models are essential. Commonly used methods in the design of power amplifiers are briefly discussed next.

9.2.1 Load-Line Method

Traditionally, a power amplifier can be designed based on the load-line method [1–4]. In a power amplifier design the voltage and current waveforms limit the circuit performance. Thus the amplifier's characteristics can be calculated based on voltage and current limits at the output of the device. Considering a common-source MESFET amplifier, as shown in Figure 9.1, the basic voltage and current relationships may be written

$$v_{\mathrm{d}} = V_{\mathrm{ds}} - i_{\mathrm{o}} R_{\mathrm{L}} \tag{9.1}$$

$$i_{\mathrm{d}} = I_{\mathrm{ds}} + i_{\mathrm{o}} \tag{9.2}$$

where V_{ds} and I_{ds} are the supply DC voltage and current, v_{d} is the total voltage at the drain of the device, i_{o} is the RF current passing through load R_{L}, and i_{d} is the total current passing through the device. The gate and drain bias inductors (L_{G}, L_{D}) are assumed to be ideal chokes ($\omega L_{\mathrm{G}} \gg R_{\mathrm{S}}$, $\omega L_{\mathrm{D}} \gg R_{\mathrm{L}}$) and the drain capacitor C_{b} is assumed to be an ideal DC blocking capacitor ($1/\omega C_{\mathrm{b}} \ll R_{\mathrm{L}}$).

Assume the device has an ideal gate voltage controlled current source, shown in Figure 9.2a, having minimum current value of zero and maximum current value of I_{max}. Consider an input RF voltage at the gate of the device sufficiently large so that the device drain current is approximately a square wave having minimum and maximum values of zero and I_{max}. This situation is more suitable for high-PAE or high-power

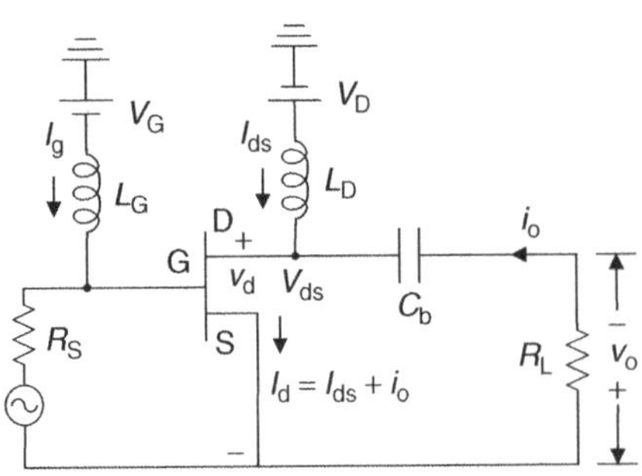

Figure 9.1 Schematic for common-source FET amplifier.

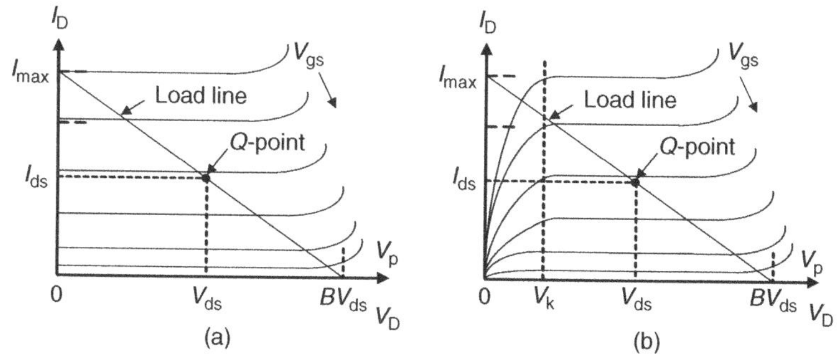

Figure 9.2 $I-V$ curves: (a) ideal and (b) showing knee voltage.

output. In this case the supply current is $I_{max}/2$ and from (9.1) and (9.2), the drain voltage becomes

$$v_d = V_{ds} + \left(\frac{I_{max}}{2} - i_d\right) R_L \tag{9.3}$$

Since the current is square wave at the operating frequency, the drain voltage v_d becomes square wave but 180° out of phase; that is, v_d is minimum when i_d is maximum and equal to I_{max}. The current flows through the device only when the drain voltage $\geq$ 0; that is,

$$(v_d)_{min} = V_{ds} + \left(\frac{I_{max}}{2} - I_{max}\right) R_L \geq 0 \tag{9.4a}$$

or

$$R_L \leq \frac{2V_{ds}}{I_{max}} \tag{9.4b}$$

Equation (9.4b) represents the load impedance and $2V_{ds}$ is the safe voltage one can apply to the device ($2V_{ds} < BV_{ds}$). When R_L is less than $2V_{ds}/I_{max}$, the drain current's range is zero to I_{max} and the drain voltage varies over $V_{ds} \pm (I_{max}R_L)/2$. This condition is called *current clipping*. In this case, the output signal can be increased by selecting devices with higher I_{max}, while increasing V_{ds} would not increase the output power. On the other hand, if R_L is greater than $2V_{ds}/I_{max}$, the drain current's range is from zero to $2V_{ds}/R_L$, and the drain voltage varies from zero to $2V_{ds}$; this condition is known as *voltage clipping*. A condition with $R_L = 2V_{ds}/I_{max}$ results in both current and voltage clipping.

The DC power, P_{DC}, is given by

$$P_{DC} = V_{ds}\frac{I_{max}}{2} \tag{9.5}$$

Since we assumed a square-wave current and the drain voltage is 180° out of phase with the drain current, an ideal device does not dissipate any power. The total RF output power, P_{oT}, delivered to the load, R_L, is given by

$$P_{oT} = \tfrac{1}{4} I_{max}^2 R_L \tag{9.6}$$

The output square-wave current contains all odd harmonics. Calculating the Fourier integral, the fundamental frequency RF output power, P_o, delivered to load, R_L, is given by

$$P_o = \frac{2}{\pi^2} I_{max}^2 R_L \tag{9.7}$$

From (9.5) and (9.7), the drain efficiency η_D is given by

$$\eta_D = \frac{P_o}{P_{DC}} = \frac{4}{\pi^2}\left(\frac{I_{max}}{V_{ds}}\right) R_L \tag{9.8}$$

If we select a load $R_L = 2V_{ds}/I_{max}$, we have

$$\eta_D = \frac{8}{\pi^2} \quad or \quad 81\% \tag{9.9}$$

The drain efficiency of 81% is the same as for an overdriven class-A amplifier, described in the previous chapter. In this analysis, the harmonics are assumed to be terminated in 50 Ω.

As shown in Figure 9.2b, the FET exhibits a knee voltage below which the device works as a resistor. This can be approximately implemented in the above equations as follows:

$$R_L \leq \frac{2(V_{ds} - V_k)}{I_{max}} \tag{9.10}$$

$$P_o = \frac{4}{\pi^2} I_{max}(V_{ds} - V_k) \tag{9.11}$$

$$\eta_D = \frac{8}{\pi^2}\left(\frac{V_{ds} - V_k}{V_{ds}}\right) \tag{9.12}$$

Thus the DC $I-V$ data can approximately predict the amplifier performance. At $V_{ds} =$ 10 V, the M/A-COM's MSAG 625-μm gate periphery FET has an I_{max} value of 180 mA and $V_k = 1.2$ V. For such a device the calculated value of $P_o = 0.64$ W and $\eta_D =$ 71.3%, whereas corresponding measured values at 14 GHz are 0.56 W and 70%, respectively. Despite drastic assumptions, the agreement between the simple load-line method and measurements in general agree, provided device parasitic reactance at the operating frequency does not have significant effect on the device calculations. The above described simple theory also does not take harmonic contributions into account. This method can easily be implemented in a linear simulator to design power amplifiers [4].

9.2.2 Low Loss Match Design Technique

The low-loss match (LLM) design technique has been applied successfully to multistage driver and high-power amplifier designs [5–9]. In this scheme both the resistive or dissipative loss (DL) and mismatch loss (ML) for each stage are calculated and adjusted as required in the design. Generally, DL and ML for the output match are kept at a minimum and ML for the interestage(s) is minimized. The controlling factors for DL and ML for each interstage include stability criteria and electrical performance. The interstage matching network provides enough output power to the output stage FET over a large drain voltage range. The dissipative loss is for the individual passive

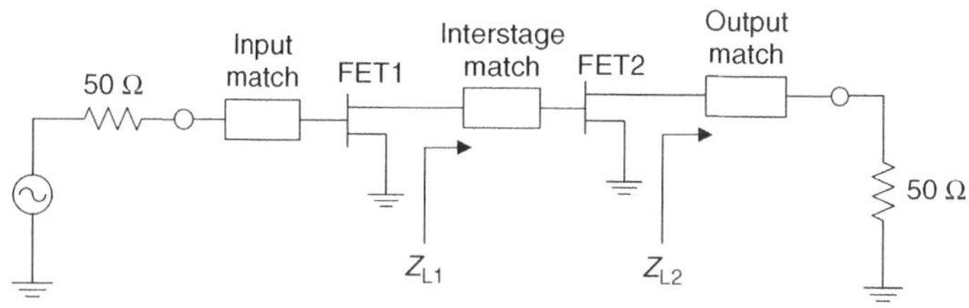

Figure 9.3 Schematic of a two-stage amplifier showing optimum load impedances.

stage, (i.e., interstage, output, etc.) and the mismatch loss is the difference between the required device's optimum load impedance and the transformed 50-Ω output impedance at the drain terminal of the FET. Consider the two-stage design shown in Figure 9.3. In this case, the total loss (TL) values for interstage and output matching networks are given by

$$TL_1 = D_{\text{int}} + M_{L1} \tag{9.13}$$

$$TL_2 = D_{\text{out}} + M_{L2} \tag{9.14}$$

where the D's and M's are the dissipative and mismatch losses, respectively, and are positive quantities expressed in dB. The calculation of these losses is discussed next.

Dissipative Loss

The dissipative loss for each passive stage is calculated by connecting complex conjugate impedance at ports connected to FETs. For the input and output matching networks, Figure 9.4a, the dissipative losses are given by

$$D_{\text{in}} = -10\log\frac{|S_{21i}|^2}{1-|S_{22i}|^2} \tag{9.15a}$$

$$D_{\text{out}} = -10\log\frac{|S_{21o}|^2}{1-|S_{11o}|^2} \tag{9.15b}$$

where the S's are the S-parameters for the respective networks. For the interstage, Figure 9.4b, the dissipative loss is calculated using

$$D_{\text{int}} = \frac{|S_{21}|^2(1-|\Gamma_S|^2)(1-|\Gamma_L|^2)}{|(1-S_{11}\Gamma_S)(1-S_{22}\Gamma_L)-S_{12}S_{21}\Gamma_S\Gamma_L|^2} \tag{9.16}$$

where the S's are the S-parameters of the interstage, and Γ_S and Γ_L are the simultaneous conjugate match input and output reflection coefficients of the interstage, respectively. For a passive network this is calculated using the following relationships:

$$\Gamma_S = \frac{B_1 - \sqrt{B_1^2 - 4|C_1|^2}}{2C_1} \tag{9.17a}$$

$$\Gamma_L = \frac{B_2 - \sqrt{B_2^2 - 4|C_2|^2}}{2C_2} \tag{9.17b}$$

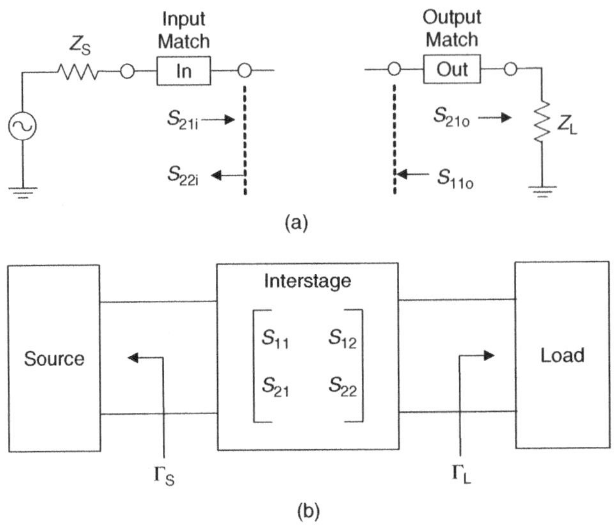

Figure 9.4 Dissipative loss calculation representations: (a) input and output matching networks and (b) simultaneous conjugate match representation of an interstage.

$$B_1 = 1 + |S_{11}|^2 - |S_{22}|^2 - |\Delta|^2 \tag{9.17c}$$

$$B_2 = 1 + |S_{22}|^2 - |S_{11}|^2 - |\Delta|^2 \tag{9.17d}$$

$$C_1 = S_{11} - \Delta S_{22}^* \tag{9.17e}$$

$$C_2 = S_{22} - \Delta S_{11}^* \tag{9.17f}$$

$$\Delta = S_{11}S_{22} - S_{12}S_{21} \tag{9.17g}$$

Mismatch Loss

Consider the one-port network shown in Figure 9.5, in which the source impedance is Z_S and the load impedance is Z_L. In this case, the reflection coefficient Γ and the mismatch loss M_L are given by

$$\Gamma = \frac{Z_L - Z_S}{Z_L + Z_S}, \quad M_L = 10 \log \frac{1}{1 - |\Gamma|^2} \tag{9.18}$$

Now consider Figure 9.3. If Z_{L1} and Z_{L2} are the required optimum load impedance values for the first- and second-stage FETs, and Z'_{L1} and Z'_{L2} are the measured or calculated impedance values at the drains of first- and second-stage FETs looking toward the load, the mismatch losses are given by

$$M_{Li} = 10 \log \frac{1}{1 - |\Gamma_i|^2}, \quad i = 1, 2 \tag{9.19}$$

where

$$\Gamma_i = \frac{Z'_{Li} - Z^*_{Li}}{Z'_{Li} + Z^*_{Li}} \tag{9.20}$$

and Z^*_{Li} is the complex conjugate of Z_{Li}. When $Z^*_{Li} \cong Z'_{Li}$, $M_{Li} \cong 0$ dB.

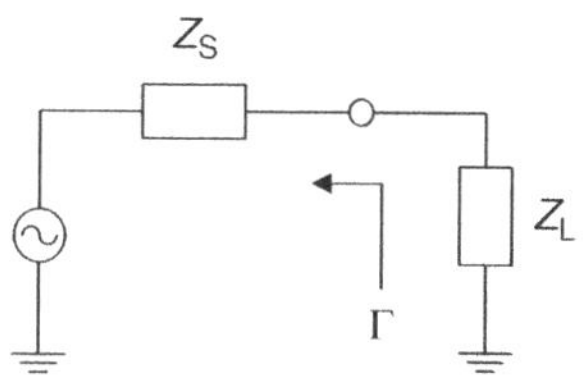

Figure 9.5 One-port network representation for mismatch calculation.

The above method is based on the assumption that the device input impedance depends strongly on the load connected at the drain terminal rather than its large-signal parameters. For FETs and HEMTs, this assumption is fairly accurate and is the cornerstone of the proposed method. However, for accurate calculation of mismatch loss, large-signal *S*-parameters must be used. A narrowband power amplifier design based on source-pull and load-pull data is still more accurate; however, the LLM technique is more comprehensive and useful for broadband power amplifier designs. This technique uses device *S*-parameters or an EC model and load-pull data.

9.2.3 Nonlinear Design Method

The application of a nonlinear device model with modern computer-aided design tools [10] offers an improved approach to reducing the design time for power amplifiers. As the sophistication and accuracy of these tools improve, significant reductions in design cycle time can be realized and "first-pass" design success can be achieved for ICs. The development of integrated CAD tools and accurate nonlinear models for active devices plays a key role in the successful development of MMIC amplifiers for narrowband, broadband, linear, and high-PAE applications. Accurate models for FETs, HFETs, HEMTs, and HBTs are an essential part of these tools. As discussed in Chapter 5, both linear and nonlinear models for transistors are based on an equivalent circuit (EC) representation and are generally provided by the device manufacturers/vendors.

Circuit optimization is an important step in CAD. In power amplifiers it is done during circuit topology selection and final design. It converts an initial design into an optimized final design meeting the given specifications. Optimization procedures involve iterative modifications of the initial design, followed by circuit analysis and comparison with the specified performance.

Since in nonlinear analysis, calculations are done in both the frequency and time domains, the execution times are very long and for most circuits optimization is not possible using nonlinear analysis. Initially the PA design is completed by using the load-line method and small-signal optimization. In the beginning, no discontinuity effects are included and ideal lumped elements are used to speed up the selection of the amplifier topology. After the amplifier topology is finalized, the circuit is optimized including all parasitic reactance. At this moment, all other analyses such as stability, sensitivity, and EM are also performed. Finally, the PA design is tweaked using nonlinear analysis to optimize the specified performance, provided the nonlinear models are considered accurate. However, a simple circuit with a few variables affecting the nonlinear performance may be optimized using a nonlinear optimization methodology. Nonlinear models are normally used to calculate input power dependent RF performance. They are seldom used to optimize multistage power amplifiers because of the enormous memory and time requirements and model nonconvergence.

In order to speed up the nonlinear simulation times, a single-device nonlinear model for each stage is used.

9.2.4 Taguchi Experimental Method

There are several statistical techniques, known as *design of experiments* (DoE), available to realize an optimum design in terms of its performance, which are more tolerant to process variations. These techniques are generally applied to optimize the circuit performance of an existing design and are not used for new design synthesis. The Taguchi technique [11–14] is one of many DoE concepts. The Taguchi technique utilizes a standard orthogonal array approach to optimize the circuit's electrical performance and make the design robust against manufacturing variation. This technique maximizes the amount of extractable information for a minimum number of experiments. In this technique, the variables that can affect the LNA or HPA performance, such as line length, capacitor, inductor, and device size, are called *factors* or *variables* and their assigned values are known as *levels*. The orthogonal array is designated by L_n, where n is the number of experiments. Table 9.1 compares the numbers of experiments required for full factorial and orthogonal array for several examples of variables. The Taguchi L_8 and L_{18} have a manageable number of experiments and are commonly used. The next two examples using L_4 and L_{18} illustrate the Taguchi experiment design methodology.

L_4 Example

Consider an example of L_4 Taguchi experiments applied to a single-stage power amplifier. The output match has three variables to improve the PAE and fabrication yield. For each variable, two levels were used. The variables can be the lengths of transmission lines, or the values of inductors and capacitors, or their combinations. Instead of a full factorial of eight experiments, the Taguchi analysis allows four experiments. Table 9.2 provides the summary of four experiments including variables and their levels, and measured PAE values. Variables 1 and 3 were chosen to improve PAE and variable 2 was selected to improve yield. Table 9.3 shows the orthogonal array experiment analysis. In the analysis, an average PAE for variable 1 and level 1 was

Table 9.1 Comparison of Examples of Full Factorial Experiments Versus Orthogonal Array Experiments

Variable States	Number of Levels	Full Factorial Experiments	Orthogonal Array Experiments	Orthogonal Array
3	All 2 levels	8	4	L_4
7	All 2 levels	32	8	L_8
5	Four 2 levels One 4 levels	64	8	L_8
4	All 3 levels	81	9	L_9
15	All 2 levels	2,048	16	L_{16}
8	One 2 levels Seven 3 levels	4,374	18	L_{18}
13	All 3 levels	1,594,323	27	L_{27}

Table 9.2 Measured PAE Data for Orthogonal Array Experiments

Experiment	VAR 1	VAR 2	VAR 3	Measured PAE
1	1	1	1	30%
2	1	2	2	25%
3	2	1	2	30%
4	2	2	1	35%

Table 9.3 Orthogonal Array Experiment Analysis

	VAR 1	VAR 2	VAR 3
Average PAE—Level 1	27.5%	30%	32.5%
Average PAE—Level 2	32.5%	30%	27.5%
Std Dev PAE—Level 1	3.5	0	3.5
Std Dev PAE—Level 2	3.5	7	3.5

obtained by adding experiments 1 and 2, and for variable 2 and level 1 was obtained by adding experiments 1 and 3, and so on. Similarly, standard deviation values were obtained. This analysis predicts that a selection of variable 1 with level 2 and variable 3 with level 1 improves PAE by 5%, whereas the variable 2 with level 1 has negligible effect on the PAE. Figure 9.6 shows the L_4 orthogonal array PAE response plots. Therefore the solution for variables 1, 2, and 3 are levels 2, 1, and 1, respectively.

L_{18} Example

The first step in the Taguchi design process is creation of the nominal baseline amplifier circuit. Here a narrowband single-stage MMIC HPA for Taguchi experimental study was considered [14]. After establishing the baseline design for the power amplifier, an extensive circuit sensitivity analysis is performed in terms of device size, matching capacitors, and transmission line lengths. We selected 7 variables (known as L_{18}) and designed 18 different circuits, which were different from each other in a very structured way. One variable was left blank. In this case, the variables included within the orthogonal array were FET sizes, two line lengths, and four MIM matching capacitors. The goal was to empirically select the key design variables to both optimize the HPA performance and center the design for best manufacturing yield.

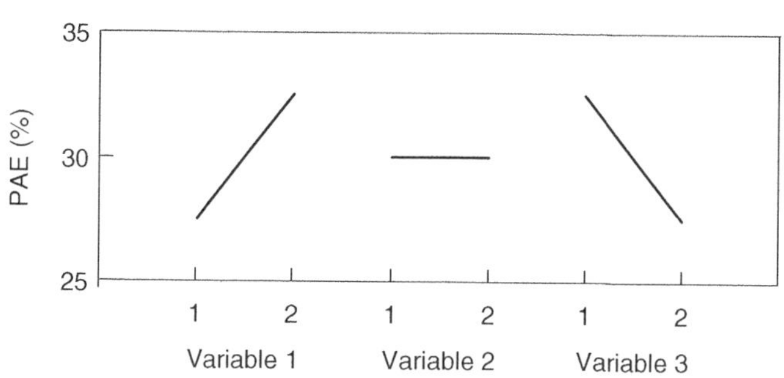

Figure 9.6 L_4 orthogonal array PAE response plots.

The improved performance may be in terms of bandwidth, output power, PAE by harmonic tuning, or higher linearity. The transmission line and capacitor values and FET sizes were selected to produce about a 10–20% variation in amplifier performance. The circuit elements chosen for variation within the orthogonal experiments are shown Figure 9.7 and their designation as variable/factor and their range is given in Table 9.4. The 18 orthogonal experiments are listed in Table 9.5. A column from the orthogonal array has been left blank (variable G) in order to approximate the variable interactions [12, 14].

All 18 different power amplifier circuits were packaged and tested. The measured data were analyzed by applying known noise levels in the experiment and optimizing a signal-to-noise ratio [14]. Here PAE was the parameter for optimization for a given output power over the desired frequency range. Figure 9.8 shows the L_{18} orthogonal array PAE response plots. These plots may be obtained at a single frequency or averaged over the desired bandwidth. The plots show that variable A (FET size),

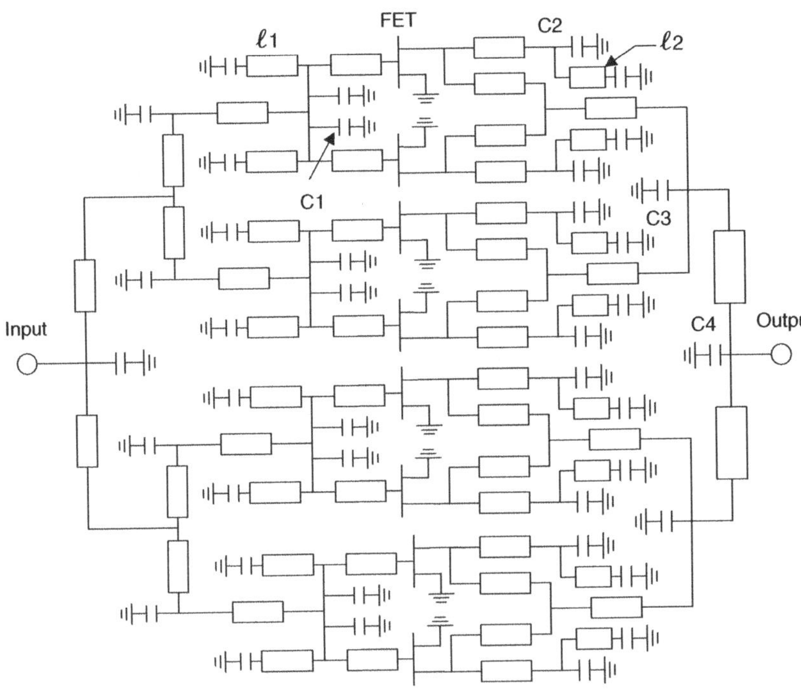

Figure 9.7 Schematic of the HPA showing seven Taguchi variables.

Table 9.4 Orthogonal Array Circuit Variables

A. FET Size	2.5 mm	1.8 mm	N/A
B. C1	2.3 pF	2.0 pF	2.6 pF
C. ℓ1	Nominal	+35 μm	−35 μm
D. C2	1.1907 pF	1.47 pF	0.9408 pF
E. C3	1.1907 pF	0.9919 pF	1.3669 pF
F. C4	1.2409 pF	1.0092 pF	1.5123 pF
G. ℓ2	Nominal	+125 μm	−125 μm
H. Blank			

Table 9.5 Orthogonal Array

Variable/Experiment	A	B	C	D	E	F	G	H
1	1	1	1	1	1	1	1	1
2	1	1	2	2	2	2	2	2
3	1	1	3	3	3	3	3	3
4	1	2	1	1	2	2	3	3
5	1	2	2	2	3	3	1	1
6	1	2	3	3	1	1	2	2
7	1	3	1	2	1	3	2	3
8	1	3	2	3	2	1	3	1
9	1	3	3	1	3	2	1	2
10	2	1	1	3	3	2	2	1
11	2	1	2	1	1	3	3	2
12	2	1	3	2	2	1	1	3
13	2	2	1	2	3	1	3	2
14	2	2	2	3	1	2	1	3
15	2	2	3	1	2	3	2	1
16	2	3	1	3	2	3	1	2
17	2	3	2	1	3	1	2	3
18	2	3	3	2	1	2	3	1

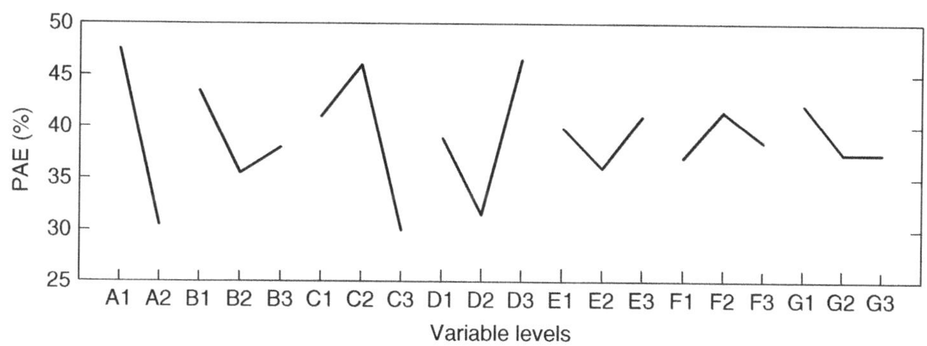

Figure 9.8 L_{18} orthogonal array PAE response plots.

variable C (input stub length, l1), and variable D (output capacitor, C3) are the dominant variables. After analyzing the data, optimum variable levels are selected to design the final circuit known as the Taguchi optimum design.

In this example the Taguchi optimum design was fabricated and tested. The measured data showed an improvement of 5–10% in PAE. At C-band, the measured peak PAE was about 60% with a power output of 14 W.

This technique has been applied successfully to improve both device and MMIC performance [14–17]. However, this approach only works with an established design and requires engineering insight into factor determination and level setting, and involves laying out, fabricating, and testing all experimental designs. Also, extra care is needed in multistage amplifier designs to take into account the interaction effects between the stages.

9.3 MATCHING NETWORKS

There are several matching network schemes available for amplifier designs. These are reactive, reactive/resistive, series feedback, shunt feedback, and distributed (traveling wave). These techniques have advantages and disadvantages in terms of frequency bandwidth. The reactive and reactive/resistive configurations are used for narrowband applications, whereas shunt feedback and distributed topologies are commonly used in broadband amplifiers and are discussed in Chapter 11 in greater detail. The reactive impedance matching techniques were described in Chapter 7.

It is very desirable to lower the dissipative loss in the power amplifier's output matching network (using lumped inductors and microstrip lines) in order to improve the output power and PAE performance. The dissipative loss in the microstrip matching networks can be improved by using a modified microstrip structure (discussed in Chapter 10). This structure is compatible with the standard MMIC fabrication process. The strip conductor is fabricated on a thin polyimide dielectric layer, which is placed on top of the GaAs substrate in order to reduce the dissipative loss by half [18].

9.3.1 Reactive/Resistive

The reactive/resistive matching technique is invariably used in power amplifier design. Figure 9.9 shows a simple form of a reactive/resistive matching scheme. When the resistance value is zero, it becomes reactive, which was described in Chapter 7. The purpose of resistance is twofold: stabilize the circuit and provide necessary gain compensation at lower frequencies where the device gain value is higher. Gain compensation (also known as gain slope equalization) techniques are discussed in detail in Chapter 11 and stabilization techniques are described in Chapter 17.

Single Stage

The design of a narrowband single-stage power amplifier basically consists of input and output matching networks, as shown in Figure 9.9. The input match provides optimum noise figure, good VSWR, any required gain flatness, input harmonic termination, and circuit stabilization. The output match is designed for maximum gain, power, PAE, and linearity requirements. The T's are matching transmission line sections or inductors.

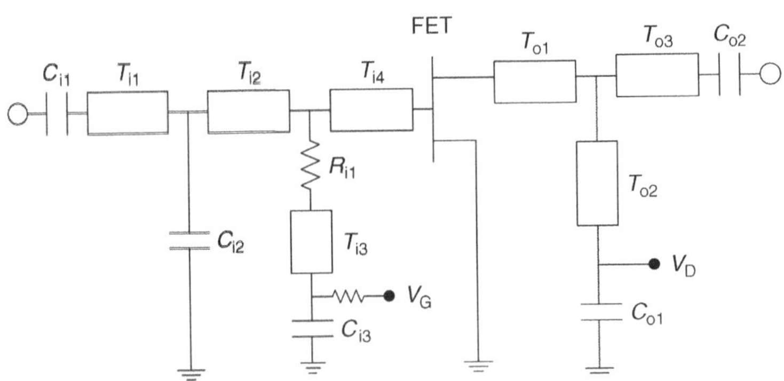

Figure 9.9 Schematic of a single-stage amplifier using reactive/resistive matching networks.

The resistor R_{i1} provides the gain flatness and amplifier stabilization. The capacitors C_{i1} and C_{o2} are for DC blocking at the input and output, respectively. The capacitors C_{i3} and C_{o1} are for RF bypass at the DC gate and drain bias terminals, respectively. The selection of DC blocking and RF bypass capacitors is discussed in Chapter 18. The DC block/RF bypass capacitors can also be used as a part of matching networks. To facilitate the application of bias voltages, shunt lines or inductors are used and usually they are integral parts of matching networks.

Figure 9.10 shows an input matching network that provides a good input match to the 50-Ω source impedance over a wide frequency range. In Figure 9.10, by a proper choice of L_1, C_1, L_2, and L_5, the input impedance of the device is matched to 50 Ω and gain compensation is realized by properly choosing the values of C_2, L_3, L_4, and R_1 (see Section 7.3.5). A typical value of R_1 is 50 Ω; however, its value can be adjusted to obtain an optimum solution in terms of input match and gain flatness. The gate bias is applied through a resistor connected between R_1 and C_{bp}. Capacitor C_{bp} is for RF bypass. Figure 9.11 shows another input matching network that provides the required gain compensation over a wide frequency range. However, the circuit in Figure 9.10 is capable of realizing good input match to 50 Ω over a very wide frequency range. The gate bias is applied through a resistor connected between R_2 and C_{bp}. The device biasing is discussed in detail in Chapter 18.

Figure 9.12 shows an output matching network. No resistors are normally used in the matching sections. Any resistor in the DC path reduces the voltage applied to the

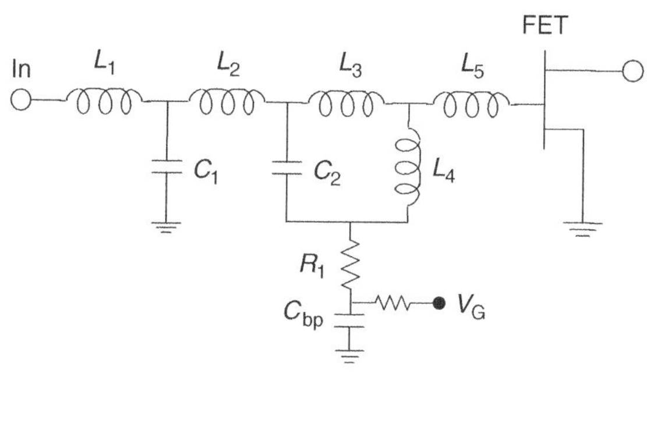

Figure 9.10 Schematic of input matching network of an amplifier for a broad bandwidth.

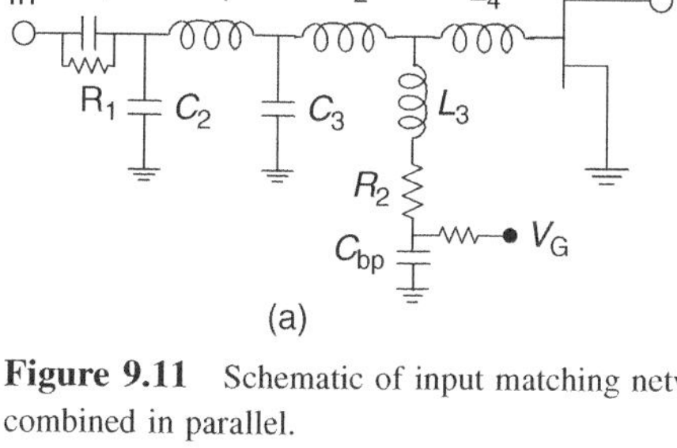

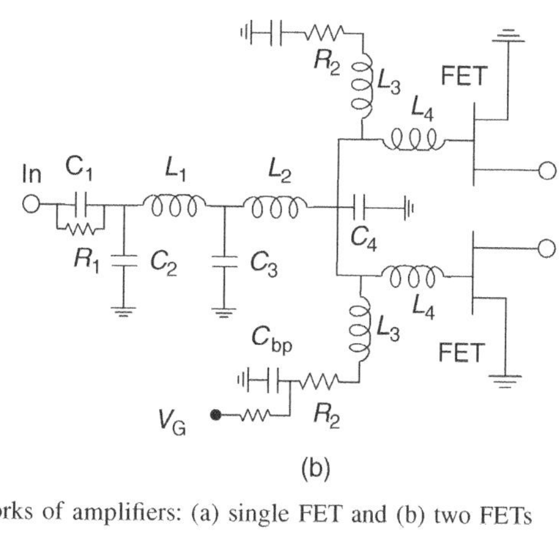

Figure 9.11 Schematic of input matching networks of amplifiers: (a) single FET and (b) two FETs combined in parallel.

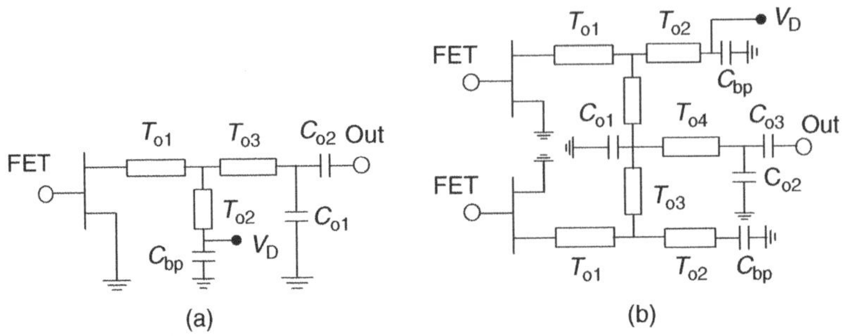

Figure 9.12 Schematic of amplifier examples of output matching networks: (a) single FET and (b) two FETs combined in parallel for higher power.

device. Any resistive loss in the output match also reduces the output power as well as the PAE. The output match transforms the output system impedance (normally 50 Ω) or antenna impedance to the required load impedance at the device's output terminal. The drain bias is applied between T_{o2} and C_{bp}. At low microwave frequencies, the microstrip section T_{o2} is replaced by an inductor or an RF bias choke, which can be a part of the matching network.

Two Stages

In a multistage amplifier there is more flexibility in obtaining flat gain and good input match than in a single-stage design. The input and output matching networks are similar to the single stage. Figure 9.13 shows commonly used lowpass interstage matching networks. The inductor L_{i2} can be an RF bias choke or a part of the matching network. The interstage transforms the input stage device's output impedance (or required load impedance for power, PAE, or linearity) to the output stage device's input impedance. The FET1 drain bias is applied between L_{i2} and C_{bp} and the FET2 gate bias is applied through a resistor connected between R_{i1} and C_{bp}.

9.3.2 Cluster Matching Technique

In MICs and MMICs, requirements for high power and high PAE over greater than 20% bandwidth preclude the conventional single-step/stage reactive matching scheme used in internally matched amplifiers discussed in Chapter 14. This is because it is very hard to match a high-power amplifier's low impedance value to 50 Ω over a wide bandwidth. In order to achieve high power and wide bandwidth, a cluster matching technique is routinely used. In this scheme, as shown in Figure 9.14, four transistors are partially matched and finally combined using another set of matching networks. In a cluster/tree matching scheme, the device impedance is transformed to an intermediate impedance level and then two cells are combined. This continues until all parallel devices are matched to the desired impedance, for example, output match to 50 Ω. Thus two devices need two stages, four devices need three stages, and eight devices use four stages. In this case the size of individual devices is small and they have much higher impedance levels than the aggregate devices. Therefore, in this topology, it is easier to match devices over a wide bandwidth with better performance. In this scheme the impedance levels are much higher and the effective thermal resistance is lower in the individual devices than in the aggregate device.

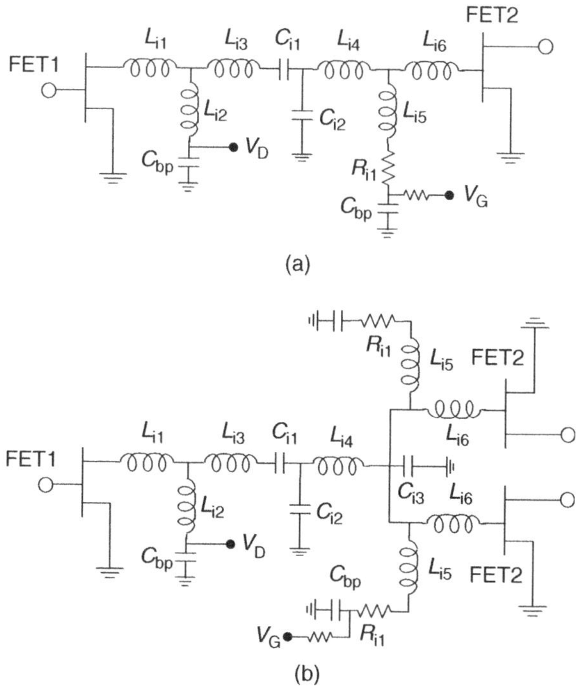

Figure 9.13 Schematic of interstage matching networks of amplifiers: (a) single output stage FET and (b) two output stage FETs combined in parallel for higher power.

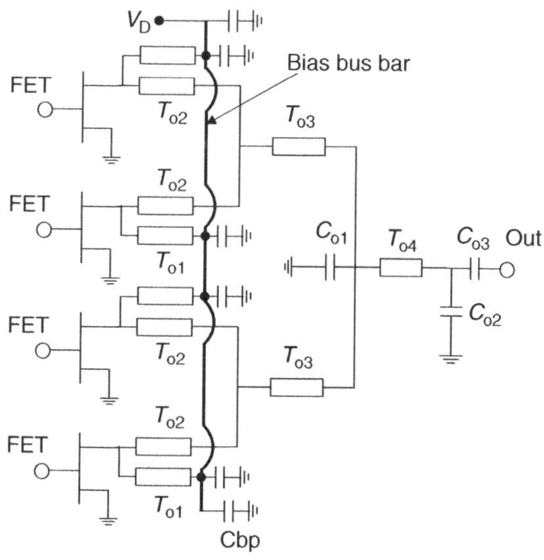

Figure 9.14 Schematic of a four-FET cluster-type output matching network of an amplifier.

The second advantage of this technique lies in the fact that all devices are fed in more uniform phase during combining for the final amplifier. Symmetry in the physical layout of the amplifier design results in efficient power combining.

The third advantage is its lower thermal resistance due to the distributions of heat dissipating devices over a larger area. Some of the recommendations for cluster matching networks are as follows:

1. Divide the total device size into an even number of identical smaller devices to achieve higher impedance levels that can be matched easily.
2. Distribute all power devices evenly in order to achieve lower thermal resistance. Larger separation between the devices results in lower thermal resistance.
3. Apply supply voltage from both sides to ensure symmetrical operation.

The major disadvantages of the cluster matching technique are the large circuit size, complex matching networks, possibility of odd- or loop-mode oscillations, and higher even-harmonic levels. The FET drain bias is applied between T_{o1} and C_{bp} and as shown in Figure 9.14 a low-resistance bus bar may be used.

9.4 AMPLIFIER DESIGN EXAMPLES

Next, we describe several amplifier design examples. These are the low-noise amplifier, maximum gain amplifier, power amplifier, and multistage driver amplifier. High PAE and broadband amplifiers are treated in Chapters 10 and 11, respectively. The first three examples do not include the effect of bond wires, junction discontinuity reactance, and DC blocking/RF bypass capacitor's parasitic reactance. In order to avoid the adverse effect of these on the circuit performance, they must be included in the final design.

9.4.1 Low-Noise Amplifier Design

The low-noise amplifier (LNA) at the receiver front end sets the system noise figure or sensitivity. Thus the LNA must have a low noise figure over its operational band. Both narrowband and broadband LNAs are required depending on the system application. For most radar applications the LNA falls into the narrowband category—bandwidth less than an octave—with 20–30 dB of flat gain and noise figure (NF) on the order of 1.5 dB. If the MMIC LNA's NF is 1.5 dB, the transmit/receive (T/R) module's NF will be about 3 dB at room temperature; these levels are achievable up to X-band. The LNA noise figure is further reduced by using a hybrid approach or employing superior NF transistors such as InP pHEMTs.

Desirable characteristics of an LNA are low noise figure and high gain, high dynamic range (in order to obtain a spurious-free signal), high third-order intercept point, low input VSWR, wide bandwidth, compact size, and low power dissipation. MESFET or HEMT technologies cannot meet all the requirements simultaneously. For a given technology and frequency range, a circuit topology is selected to meet the most important system requirements. Among the possible GaAs transistor based LNA configurations, the common-source amplifier configuration is commonly used. In a multistage amplifier the input matching is designed for minimum noise figure, the interstage matching is designed for flat gain, and the output matching is designed for maximum gain and power output. A series-source feedback single-ended configuration provides good input match and minimum noise figure simultaneously over narrow bands, while a balanced topology is generally used for good input match and minimum noise figure over an octave bandwidth. The distributed amplifier configuration gives best noise figure over multi-octave bandwidth.

In multistage amplifiers the FETs or HEMTs have gate peripheries (or gate width that is proportional to transistor size) increasing progressively, in order to achieve higher dynamic range and higher third-order intercept. However, hybrid LNAs have the same device size in all stages. The input stage is usually biased at low current level for best noise figure and succeeding stages are biased for higher gain and higher current/power output. In multistage LNAs, usually all FETs/HEMTs except in the first stage are biased approximately at half I_{dss}. Shorter gate length (e.g., below 0.25 μm) devices provide lower noise figure and larger bandwidths as compared to longer gate length (e.g., 0.25 μm) devices, but the former are generally more expensive. HEMT LNAs have the lowest noise figure from microwave to millimeter-wave frequencies as compared with FETs, and FETs, in turn, are better than BJTs and HBTs.

The design of a single-stage narrowband low-noise amplifier can be carried out step-by-step as follows:

1. Select a suitable transistor with a noise figure (0.2–0.5 dB) lower and a gain (0.5–1.0 dB) higher than the design value.
2. Calculate its stability factor K.
3. If $K > 1$, select suitable input and output matching networks that include biasing circuitry and complete their design.
4. If $K < 1$, plot the regions of instability on the reflection planes and select matching networks that avoid the unstable regions.
5. Calculate the amplifier performance using analytical methods or CAD tools. Check stability of the amplifier in the band as well as outside the band.
6. Work out the realization of the amplifier.

EXAMPLE 9.1

Design a low-noise single-stage amplifier to be used as a predriver for a power amplifier, using a microstrip on a 0.25-mm thick alumina substrate ($\varepsilon_r = 9.9$) with the following specifications:

Frequency	10 GHz
Bandwidth	5%
Gain (min)	7 dB
Noise figure (max)	2.0 dB
Output VSWR (max)	1.2:1

SOLUTION The FET listed in Table 5.1 satisfies the requirements. In this case $K < 1$ and the amplifier designed is conditionally stable. A simple amplifier configuration consisting of two matching elements at the input and two at the output, as shown in Figure 9.15a, can be selected. At 10 GHz, the input matching circuit must transform 50 Ω to $Z_{opt} = 30 + j63$ Ω, and the output matching circuit must transform 50 Ω to $Z_{DM}^* = 99 + j110$ Ω. These matching circuit elements can be determined using either a Smith chart or the analytical methods described in Chapter 7 or the following equations:

$$Z_{opt} = 30 + j63 = Z_2 \frac{Z + jZ_2 \tan \beta\ell_2}{Z_2 + jZ \tan \beta\ell_2}$$

where

$$Z = 50 || jZ_1 \tan \beta\ell_1$$

and

$$Z_{DM}^* = 99 + j110 = Z_{01} \frac{Z' + jZ_{01} \tan \beta\ell_{01}}{Z_{01} + jZ' \tan \beta\ell_{01}}$$

where

$$Z' = 50 || jZ_{02} \tan \beta\ell_{02}$$

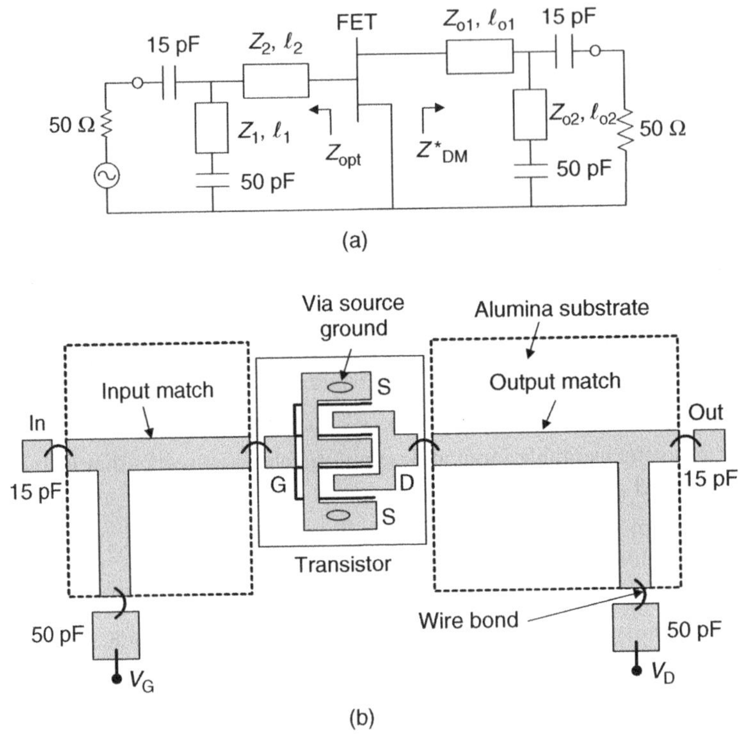

Figure 9.15 (a) Schematic and (b) layout of 10-GHz low-noise amplifier.

Here the effect of DC blocking/RF bypass capacitors is not taken into account. By separating the real and imaginary parts, the transmission line parameters are calculated. Here 50 Ω as the characteristic impedance of microstrip lines has been used. Physical dimensions for matching elements on alumina substrate ($\varepsilon_r = 9.9$, $h = 0.25$ mm) may be calculated using expressions given in Chapter 6. The line width is 0.24 mm. The values of ℓ_1, ℓ_2, ℓ_{o1}, and ℓ_{o2} are 0.98, 1.0, 1.6, and 1.0 mm, respectively. In order to neglect the effect of DC blocking and bypass capacitors, their values are chosen so that their reactance values are less than 1 Ω at the lowest operating frequency. Figure 9.15b shows the layout of the LNA. The calculated performance of this amplifier is shown in Figure 9.16. Since the input is matched for optimum NF, in most cases the return loss is less than 10 dB.

Figure 9.16 shows the simulated gain, noise figure, and output return loss as a function of frequency. ■

9.4.2 Maximum Gain Amplifier Design

The design procedure for a maximum gain amplifier is the same as described for the low-noise one, except that all the stages are matched for maximum gain. The following example illustrates the design of a maximum gain amplifier.

EXAMPLE 9.2

Design a maximum gain amplifier at 18 GHz with a gain of 10 dB and VSWR better than 1.5.

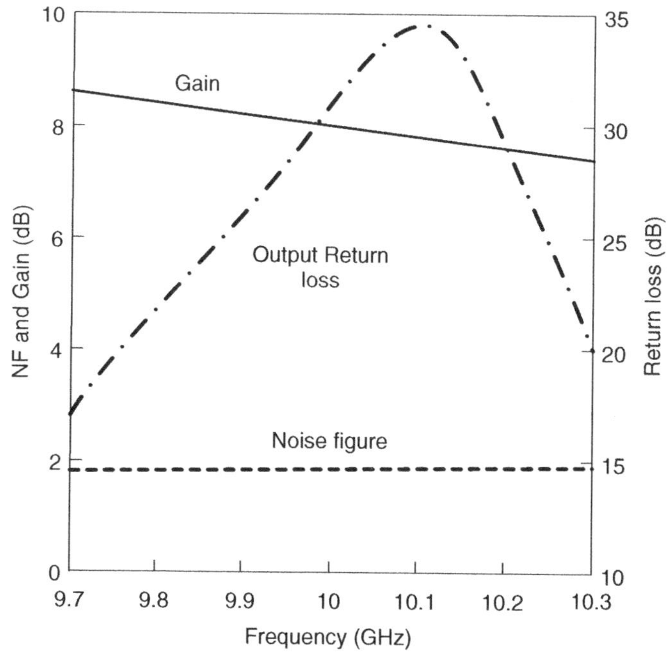

Figure 9.16 Simulated gain, noise figure, and output return loss versus frequency.

SOLUTION The 0.6-mm gate periphery FET listed in Table 5.5 satisfies the requirements, and $K = 1.178$. Equations (17.6) and (17.7) can be used to compute the simultaneous conjugate matched source and load impedances for maximum gain as follows:

$$MAG = MG = 11.0\text{dB}$$
$$\Gamma_{\text{SM}} = 0.958\underline{/159^\circ}, \quad \Gamma_{\text{LM}} = 0.856\underline{/113^\circ}$$
$$Z_{\text{SM}} = 1.109 + j9.252\ \Omega, \quad Z_{\text{LM}} = 5.563 + j32.815\ \Omega$$
$$Y_{\text{SM}} = 0.0128 - j0.1066\ \text{S}, \quad Y_{\text{LM}} = 0.0050 - j0.0296\ \text{S}$$

A simple amplifier configuration consisting of two matching elements at the input and two at the output, as shown in Figure 9.17a, can be selected. At 18 GHz, the input matching circuit must transform 50 Ω to $1.109 + j9.252\ \Omega$, and the output matching circuit must transform 50 Ω to $5.563 + j32.815\ \Omega$. The matching circuit elements may be determined using a Smith chart or the technique described in Chapter 7 or the following equations:

$$Z_{\text{SM}} = 1.109 + j9.252 = 50 || \frac{1}{j\omega C_1} + j\omega L_1 \tag{9.21}$$

$$Y_{\text{LM}} = 0.005 - j0.0296 = \frac{1}{j\omega L_{\text{o1}}} + \frac{1}{j\omega L_{\text{o2}} + 50} \tag{9.22}$$

where the effect of DC blocking/RF bypass capacitors is not taken into account and $\omega = 2\pi f = 113.097$. By separating the real and imaginary parts, the calculated values of C_1, L_1, L_{o1}, and L_{o2} are 1.2365 pF, 0.1438 nH, 0.422 nH, and 0.7637 nH, respectively. Figure 9.17b shows the simulated gain, input return loss, and output return loss as a function of frequency. At 18 GHz the calculated amplifier's gain and VSWR are 11 dB and 1.1, respectively. Since the FET selected is a linear power device, when simultaneously matched it is expected to have good output TOI. The amplifier is conditionally stable below 17 GHz. ■

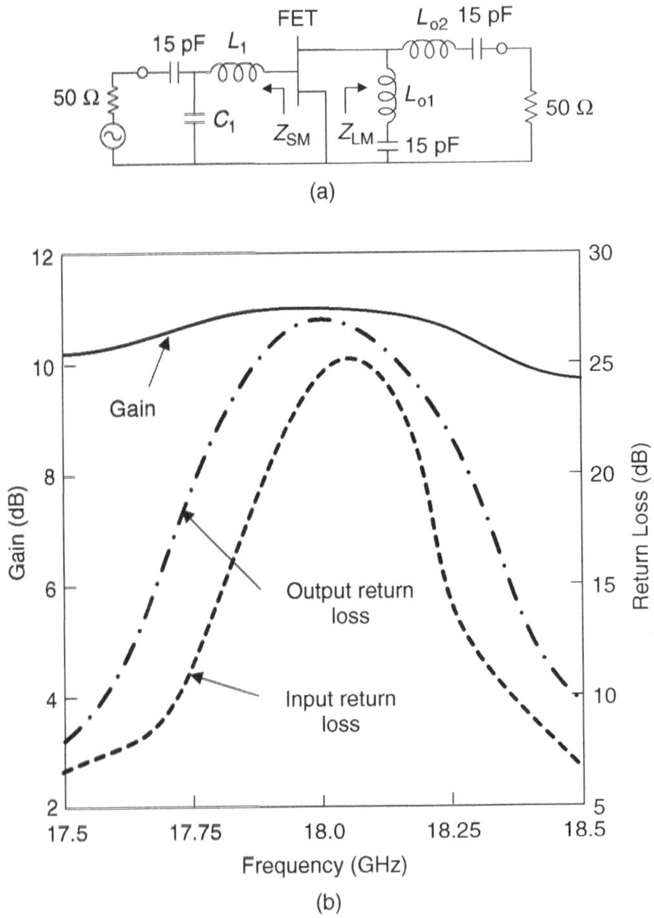

Figure 9.17 (a) Schematic of a 18-GHz maximum gain amplifier. (b) Simulated gain and input and output return loss versus frequency of 18-GHz amplifier.

9.4.3 Power Amplifier Design

Requirements for power amplifiers vary drastically from one application to another. Basic requirements for such amplifiers are high gain, higher linearity, high power added efficiency (PAE), high reliability, small size, and low cost. Power amplifier requirements also include high pulsed-power output with phase and amplitude stability, and linear transmission phase with input power up to the 1–3-dB power compression point. Usually communication applications require linear operation, while for radar applications high PAE is of prime importance. Personal communication systems working in the 800-MHz to 2.5-GHz range use different digital modulation and access schemes. They require high efficiency and linear power amplifiers for hand-held devices as well as for the base station applications.

The design of narrowband and wideband amplifiers using power transistors requires several device and circuit considerations.

1. Select a suitable power device that meets the design objectives (power output and frequency range). Si bipolar transistors, for example, can deliver more RF

power than GaAs FETs in the L- and S-bands. About 20–30% margin should be included from the device P_{out} to amplifier P_{out}.

2. Power transistors with higher breakdown voltage are desirable. Use transistors that are close to industry standards. Transistors on thin substrates with via holes have low series inductance and better heat dissipation.
3. Operate the amplifier circuit within its safest operating bias. Do not exceed the maximum supply voltage and current ratings specified by vendors. As an example, for an FET/HEMT, use $V_{ds} \leq \frac{1}{2} BV_{ds}$.
4. Junction-to-case thermal resistance should be as low as possible for better performance and reliability.
5. Load-pull measurements and nonlinear model are essential to accurately characterize power devices for optimum amplifier design.
6. Internally matched transistors help in reducing the effect of package parasitic reactance. They offer higher output power and efficiency.
7. Design the input matching network for maximum power transfer, while designing the output matching network for maximum power out. Matching circuits should offer minimum gain outside the desired frequencies.
8. Use lumped elements or lumped-distributed circuit elements for matching low impedance to 50 Ω in order to realize a compact circuit. Also, use low-loss circuit elements at the output since the efficiency is reduced more by a given amount of loss in the output than in the input.
9. Use low-loss and 85–90% efficient power-combining techniques such as traveling-wave and Lange couplers for high-power modules.
10. For broadband amplifiers use low-Q matching networks for the input and interstage. The low-Q lumped elements are more suitable than the high-Q distributed elements because they are capable of providing wider bandwidths and higher stabilization.

Typically power amplifiers are operated in three basic modes: class A, class B, and class C. These classes, as discussed in Chapter 8, are implemented by suitably biasing the active devices. For power amplifiers, the input signal level is high, and consequently the output current is either in the cutoff or saturation region during a portion of the input signal cycle. In the basic modes of operation, the harmonics at the device output are resistively terminated in the load. There are two other commonly used classes of power amplifiers known as class E and class F. In class E, the active device works as a switch and has a high-Q tuned circuit at the output of the device to provide a designed reactive load at the fundamental frequency. In the class-F amplifier configuration, the reduced power dissipation is achieved by employing some sort of impedance matching technique (such as resonant circuits) to terminate harmonic frequencies in desired loads (short circuits for even harmonics and open circuits for odd harmonics). Both class E and F are usually operated at class B bias conditions and the theoretical efficiency approaches 100%.

As discussed in Chapter 8, RF power amplifiers can be designed to operate under a variety of power added efficiency and linearity conditions with the various amplifier types identified with a class designation. In practice, all amplifier classes operate at reduced efficiencies due to inherent parasitic losses and nonideal operating conditions. Thus selection of the power amplifier technology is very important to meet system requirements in terms of output power, DC power consumption, and linearity conditions.

In order to transfer maximum power, low-loss matching networks must be used to match the source impedance (usually 50 Ω) to the input of the device and to match the desired load impedance of the device to the system impedance (usually 50 Ω). The FET's/HEMT's input and output impedances decrease as the power level of the device goes up (as the gate width becomes larger). For high output power, a large gate periphery is required, decreasing the input and output impedances to low levels. However, with low impedance it is difficult to maintain a good match over broad bandwidths. In such cases the power amplifier is usually split into cells that can be conveniently prematched and then combined to produce the final output power. The input impedance is the most difficult to match. Most of the time, a combination of lumped and distributed elements is used in realizing impedance matching networks.

Three parameters that are very useful for comparing different power amplifier configurations and classes are the power output, power added efficiency, and linearity. Class A is usually chosen for maximum linear power output, whereas class B is used for best output power and PAE. The circuit topologies are similar to those of small-signal amplifiers. The design of a narrowband power amplifier is described step-by-step in the following.

EXAMPLE 9.3

Design a power amplifier at 5.5 GHz with a 1.7-W power output, 25% power added efficiency, and 6-dB gain at 1-dB compression point. The power FET to be used has the following characteristics measured at 5.5 GHz and $P_{in} = 26$ dBm: $V_{ds} = 9$ V, $I_{max} = 1000$ mA, Gain = 7 dB, $P_{out} = 2$ W, PAE = 35%, and

$$S_{11} = 0.89\angle{-157^\circ}, \quad S_{21} = 1.5\angle 84^\circ, \quad S_{12} = 0.049\angle 54^\circ, \quad S_{22} = 0.31\angle{-145^\circ}$$
$$\Gamma_S = 0.73\angle 150^\circ, \quad \Gamma_L = 0.32\angle 160^\circ$$

SOLUTION From Eq. (17.3) in Chapter 17, $K = 1.267$. Therefore the transistor is unconditionally stable at 5.5 GHz. Substituting Γ_S and Γ_L into (17.9) yields $G = 8.1$ dB. Thus the device has enough gain to design a 6-dB gain power amplifier when circuit losses are incorporated.

From $\Gamma_S = 0.73\underline{/150^\circ}$, $Z_S = 8.35 + j13.05$ Ω and $Y_S = 0.0348 - j.00544$ S.

A short-circuited stub has an input admittance of

$$Y = \frac{-jY_0}{\tan\,\beta\ell} = -0.0544$$

when $Y_0 = 0.02$ and $\beta\ell = 20.2^\circ$. Thus the characteristic impedance of the stub is 50 Ω. To match the parallel conductance of 0.0348 S to the source admittance of 0.02 S, a quarter-wave transformer of characteristic impedance $Z_0\sqrt{1/(0.02\times0.0348)} = 37.9$ Ω is used. The complete input matching network is shown in Figure 9.18.

For the output matching network, $\Gamma_L = 0.32\underline{/160^\circ}$, $Z_L = 26.34 + j6.42$, and $Y_L = 0.0358 - j0.00874$.

A short-circuited stub has an input admittance of $Y = -jY_0/\tan\beta\ell = -j0.00874$ when $Y_0 = 0.02$ and $\beta\ell = 66.5^\circ$. In this case the characteristic impedance of the stub is also 50 Ω. To match the parallel conductance of 0.0358 S to the load admittance of 0.02 S, a quarter-wave transformer of characteristic impedance $Z_0 = \sqrt{1/(0.02\times0.0358} = 37.4$ Ω is used. The complete output matching network is shown in Figure 9.18. ■

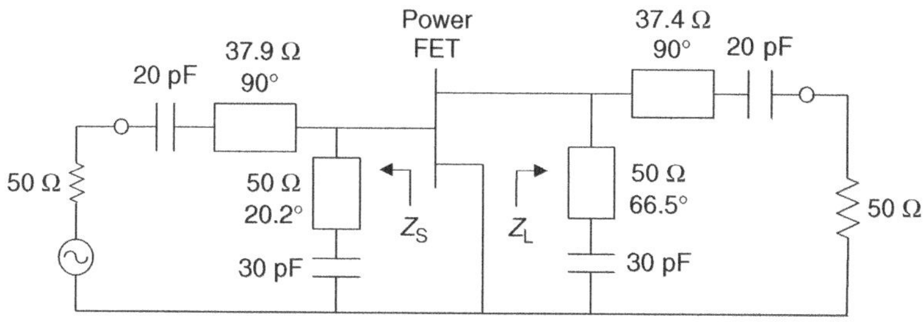

Figure 9.18 Schematic of a 1.7-W power amplifier. Electrical line lengths are at 5.5 GHz.

EXAMPLE 9.4

A transistor requires optimum source and load impedance values of $25 + j25\ \Omega$ and $100 + j50\ \Omega$, respectively. Design the input match using series transmission lines and the output match using a single short-circuited stub having characteristic impedance of 100 Ω. Use a Smith chart, described in Appendix F, to determine the matching networks.

SOLUTION For the ***input***, the normalized (with respect to 50 Ω) source impedance is $Z'_S = (25 + j25)/50 = 0.5 + j0.5$; this is point A on the Smith chart shown in Figure 9.19 Draw the constant-reflection coefficient circle that passes through point A. This circle intersects the zero

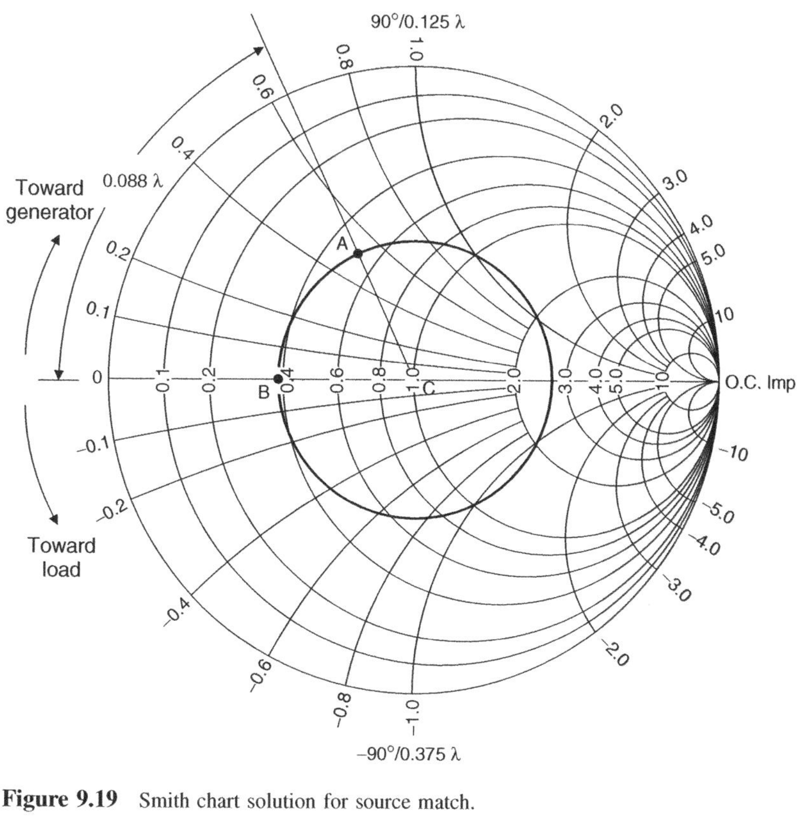

Figure 9.19 Smith chart solution for source match.

reactance line (the horizontal line passing through the center of the Smith chart) at point B. The shortest length of transmission line is obtained by moving from A to B toward load. The transformed source impedance (normalized) at point B is 0.38. The length (ℓ) of the transmission line of characteristic impedance 50 Ω needed to accomplish this can be determined by moving along the constant-reflection coefficient circle from point A to point B. In this case, $\ell = 0.088\lambda$. Then a quarter-wave transformer is used to transform point B impedance, which is $0.38 \times 50 =$ 19 Ω to 50 Ω (at point C, center of the Smith chart). The characteristic impedance of the quarter-wave transformer is given by

$$Z_{0T} = \sqrt{19 \times 50} = 30.8\ \Omega$$

For the ***output***, the normalized (with respect to 50 Ω) load impedance is $Z'_L = (100 + j50)/50 = 2 + j1$; this is point A on the Smith chart shown in Figure 9.20. Draw the constant-reflection coefficient circle that passes through point A; point B, which is $\lambda/2$ from A, represents normalized load admittance $Y'_L = 0.4 - j0.2$. Moving toward the generator on the constant-reflection coefficient circle, which intersects the constant-resistance circle at point C, the normalized admittance is

$$Y'_{LL} = 1 + j1.0$$
$$Y_{LL} = \frac{1}{50} + j\frac{1}{50}\text{mho}$$

The length (ℓ_1) of line needed to move from point B to C is 0.2 λ. Therefore a susceptance of $-j(1/50)$ mho is needed from the short-circuited stub to cancel $+j(1/50)$ mho. When the

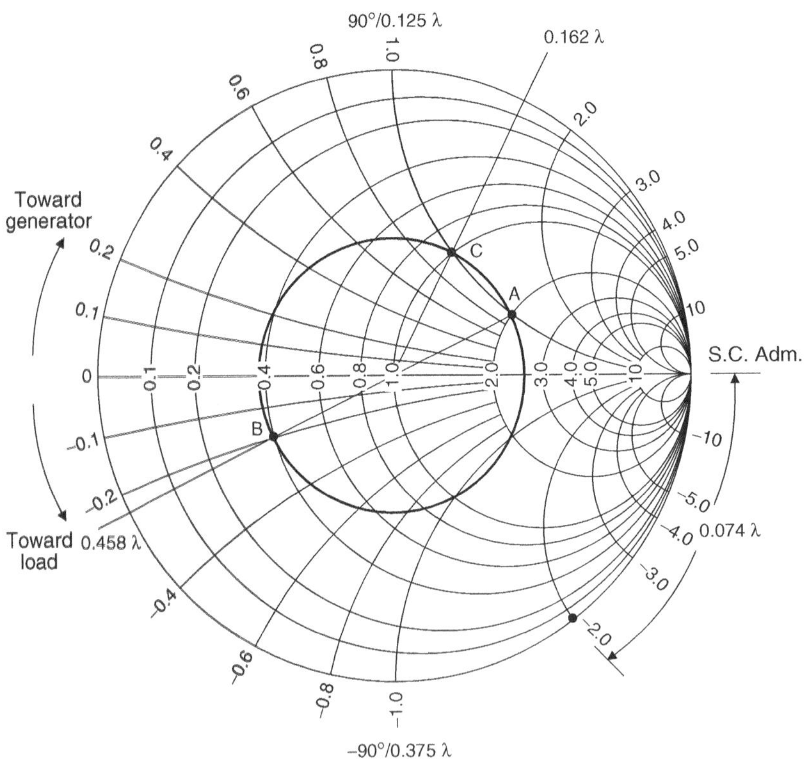

Figure 9.20 Smith chart solution for load match.

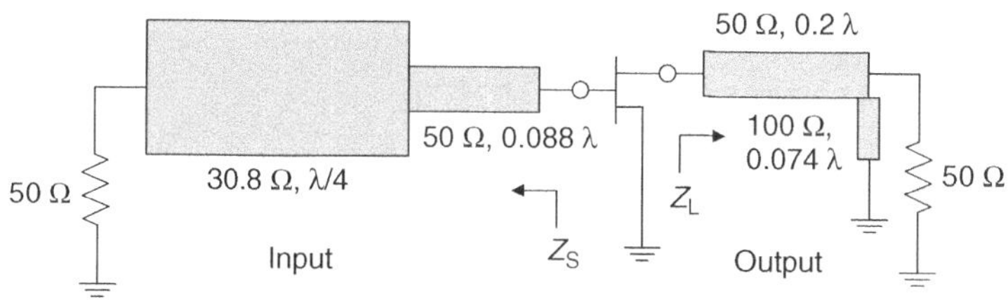

Figure 9.21 Schematic of input and output matching networks.

characteristic impedance of the stub is 100 Ω, $Y'_{\text{stub}} = -j(1/50)/(1/100) = -j2.0$ (normalized with reference to the characteristic admittance (= 1/100 mho) of the stub). The length (ℓ_2) of short-circuited stub needed to produce $Y'_{\text{stub}} = -j2.0$ *is* 0.074λ. Figure 9.21 shows the design values of input and output matching networks. ■

9.4.4 Multistage Driver Amplifier Design

Next, an example of a two-stage variable drain voltage MMIC power amplifier is described. The amplifier is designed from 4.5 to 8.5 GHz, using a nominal 8-V supply voltage and the LLM technique described in Section 9.2.2. The gain, output power, and PAE targets are 15 dB, 2 W, and 30%, respectively. The amplifier is matched to 50 Ω at the input and the output is matched for maximum output power and PAE. The load value was selected at 10 V to optimize the PAE and gain over the 6–10-V operation of the MMIC amplifier. The measured load for the device at $V_{ds} = 10$ V and class-AB operation is equivalent to a parallel combination of resistor and capacitor of values 56.25 Ω·mm and -0.304 pF/mm, respectively. The 625-μm gate periphery FET's typical power gain, output power, and PAE values at 8 V and 10 GHz are 10 dB, 0.4 W (0.64 W/mm), and 63%, respectively. The EC model (Fig. 5.2, Chapter 5) for the 625-μm FET (20-μm gate–gate pitch and six fingers) biased at 8 V and 30% I_{dss} is as follows:

$$R_g = 0.5\ \Omega,\quad R_i = 0.6\ \Omega,\quad R_s = 1.0\ \Omega,\quad R_d = 1.0\ \Omega,\quad R_{ds} = 150\ \Omega$$
$$C_{gs} = 1.0\ \text{pF},\quad C_{gd} = 0.045\ \text{pF},\quad C_{ds} = 0.16\ \text{pF}$$
$$g_m = 100\ \text{mS},\quad \tau = 4.5\ \text{ps}$$
$$L_g = 0.005\ \text{nH},\quad L_s = 0.02\ \text{nH},\quad L_d = 0.005\ \text{nH}$$

In this broadband design a power density of 0.4 W/mm output power for the MMIC is used, to determine the total FET periphery needed in the output stage of this power amplifier. This is a scaled down power density from 0.64 W/mm for the transistor, to account for the output match loss of about 1 dB and power combining loss of about 1 dB. For the 2-W HPA this translates to eight 625-μm FETs. This design uses two 625-μm gate periphery FETs at the input driving eight 625-μm FETs at the output resulting in a 4:1 output to input FET aspect ratio. The optimum load impedances Z_{L1} and Z_{L2} at the drain of the first- and second-stage FETs, respectively, to realize maximum output power and PAE, are shown in Figure 9.22. The values of Z_{L1} and Z_{L2} are determined from $R_L = 56.25$ Ω·mm and $C_L = -0.304$ pF/mm. The circuit was optimized using a CAD and the schematic is shown in Figure 9.23.

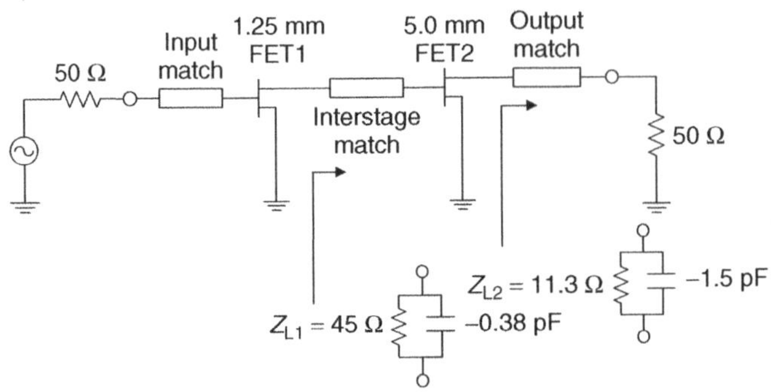

Figure 9.22 Two-stage power amplifier configuration depicting the load required at the drain of each FET stage.

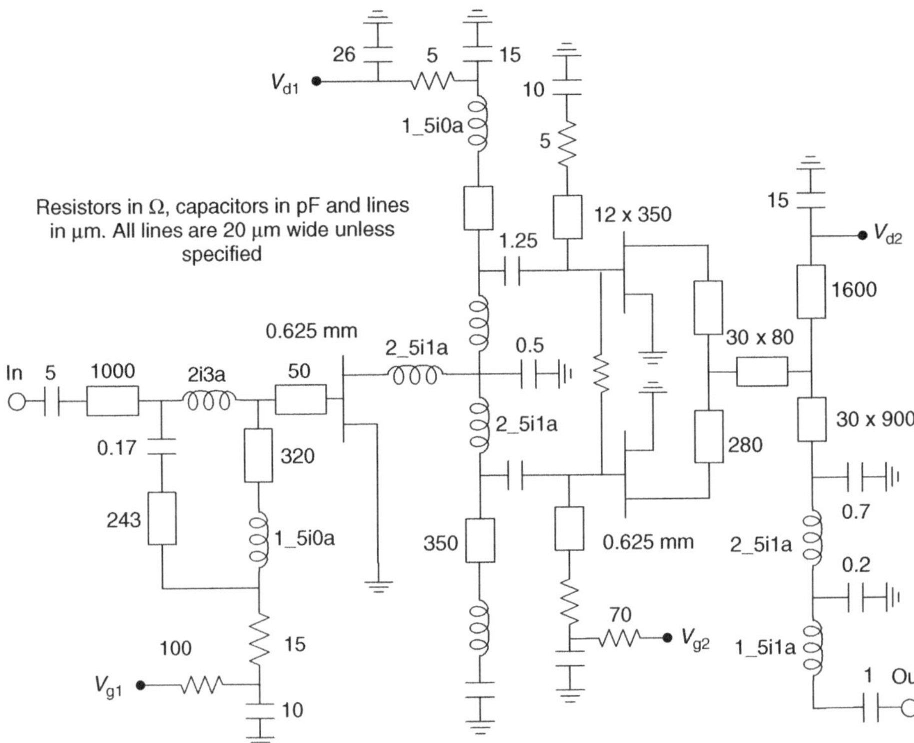

Figure 9.23 Half schematic of the two-stage 2-W power amplifier. Two of them are combined in parallel.

Basically, the amplifier design is based on lumped inductors and capacitors. The microstrip sections are used to connect the components. The input stage, which has a limited gain compensation network, was designed for good input match as well as for maximum power transfer at the high-frequency end. The input matching consists of a bridged T-coil network [19] and is designed to match to 50 Ω. The interstage matching network is designed to provide flat gain response by adjusting the loss so that there is enough output power to succeeding stage FETs for achieving maximum output power and PAE. The interstage matching is comprised of *RLC* lumped-element based topology using thin-film resistors, single-layer inductors, and MIM capacitors. Figure 9.24 shows the simulated dissipative loss and the total loss for the interstage matching network. The dissipative loss is adjusted by sizing the resistors to obtain the unconditional stable operation of each stage. The mismatch loss over the 4.5–8.5-GHz range is much lower than the dissipative loss. The output matching elements were selected to provide an optimum load match with minimum possible insertion loss, since the efficiency is reduced to a greater extent by a given amount of loss due to decreased power out, gain, and available DC power at the FET drain pads. In the output the drain bias lines use 9-μm thick conductors to carry the current with minimum voltage drop [20–22]. Figure 9.25 shows the dissipative loss and total loss for the output matching network. The dissipative loss is about 0.5 dB over 4–7 GHz and within 1 dB over

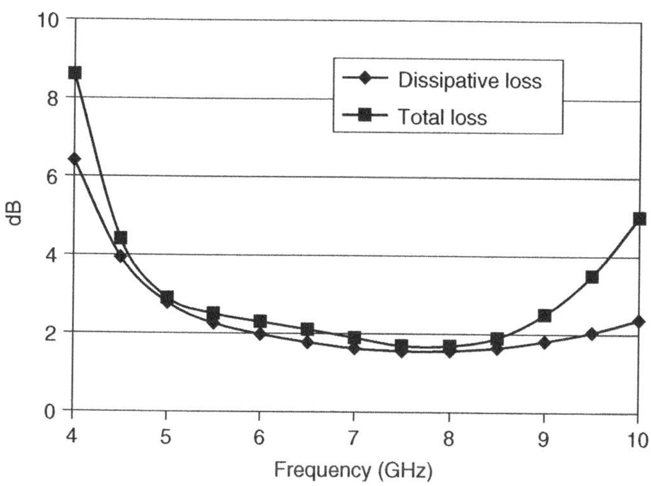

Figure 9.24 Interstage matching network's dissipative loss and total loss of the two-stage 2-W MMIC power amplifier.

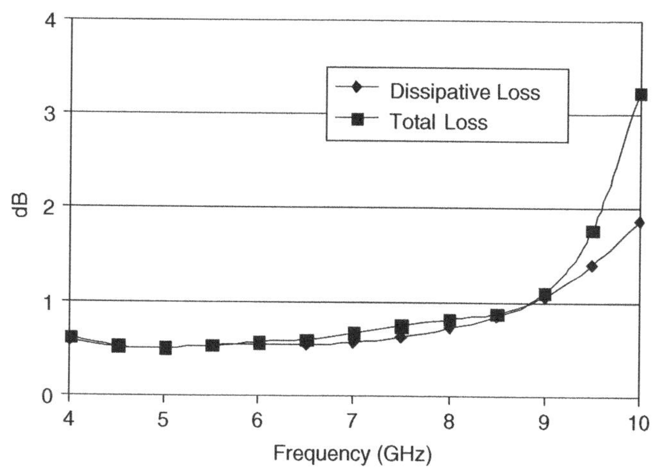

Figure 9.25 Output matching network's dissipative loss and total loss of the two-stage 2-W MMIC power amplifier.

4–9 GHz. In this case the mismatch loss is negligible over 4–9 GHz. The simulated gain, input return loss, and output return loss are better than 17 dB, 13 dB, and 7 dB, respectively, over the 5–9-GHz frequency range.

The power amplifier was fabricated using M/A-COM's multifunction self-aligned gate (MSAG) MESFET IC process [5–23]. High-Q and high-current-carrying capacity inductors [20–22] were used in the design. These inductors use two levels of 4.5-μm thick gold conductors. The substrate thickness used is 75 μm. Figure 9.26 shows a photograph of the two-stage 2-W C/X-band power amplifier.

The MMIC amplifier chips were tested using gold-plated CuW carriers for RF characterization. The ICs were prescreened using on-wafer pulsed power measurements. The MMIC chips were die attached using gold–tin (AuSn) preforms at 300°C on a pedestal in order to keep bond wire lengths to a minimum. They were characterized for power output and PAE as a function of frequency and input power at several drain power supply voltages.

Figure 9.27 shows typical small-signal gain and input VSWR performance. The small-signal gain was about 18–20 dB and VSWR was better than 1.6:1. The CW measured P_{out} and PAE at $V_{ds} = 8$ V and $P_{in} = 18$ dBm are shown in Figure 9.28 as a function of frequency. The amplifier has greater than 33-dBm output power, 15-dB gain, and better than 31% PAE over the 5–8.5-GHz frequency range meeting the design target values. Over 5.5–7.0 GHz, the PAE was greater than 40%. A plot of output power and PAE as a function of drain voltage at 7.5 GHz is shown in Figure 9.29. This plot shows that these chips exhibit more than 30% PAE over a large range of drain bias voltages (5–10 V) and the output power is approximately linear with drain bias.

Figure 9.30 depicts the P_{1dB} values as a function of frequencies at $V_{ds} = 4$, 6, 8, and 10V. At 8 V, P_{1dB} was better than 30 dBm from 4.5 to 9.5 GHz. No oscillations were observed when the chips were biased from 4 to 10 V.

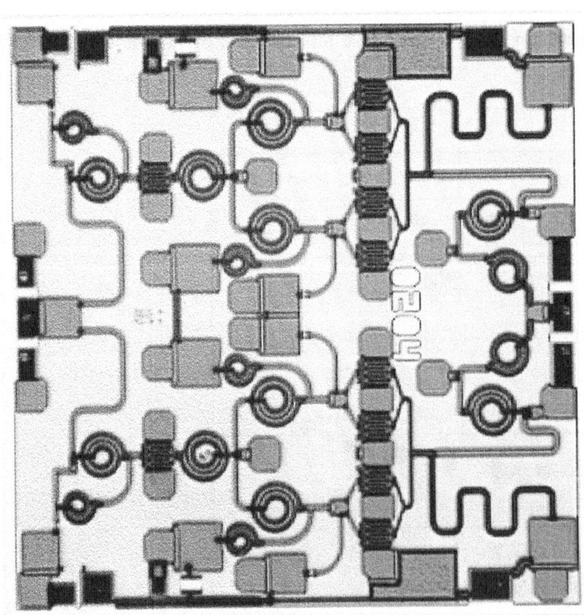

Figure 9.26 Photograph of a two-stage 2-W C/X-band power amplifier. Chip size is 3 × 3 mm^2.

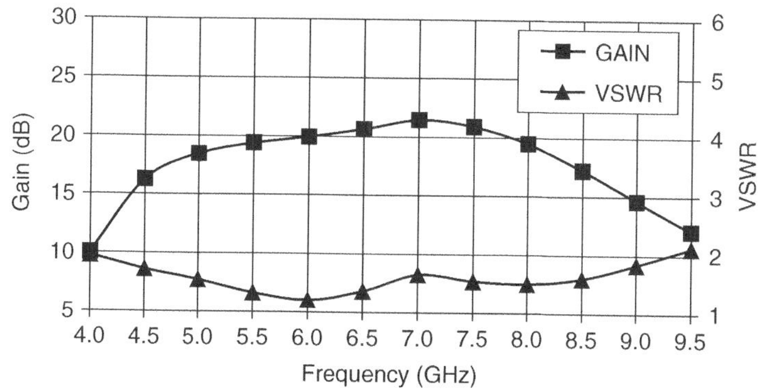

Figure 9.27 Small-signal gain and input VSWR versus frequency at $V_{ds} = 8$ V.

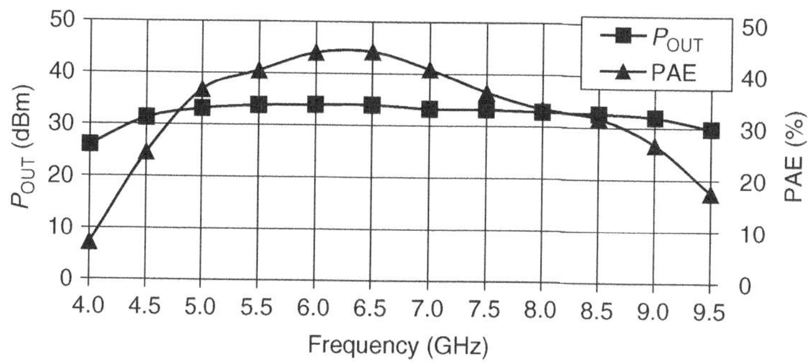

Figure 9.28 Output power and power added efficiency versus frequency at $V_{ds} = 8$ V and $P_{in} = 18$ dBm.

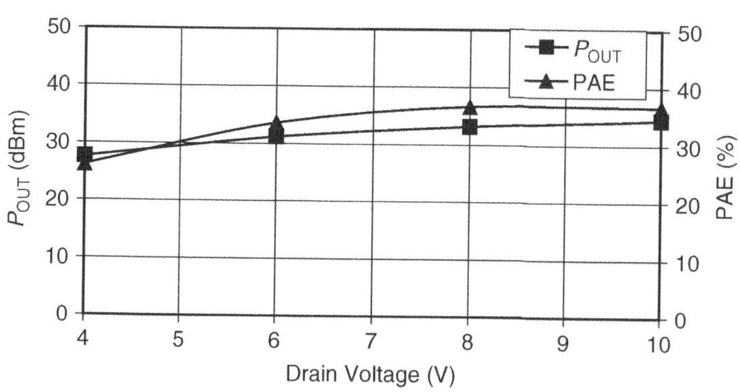

Figure 9.29 Saturated output power and PAE versus drain voltage at $f_0 = 7.5$ GHz.

Designing an amplifier for VSWR, stability, noise figure, gain flatness, power, and PAE or IP3 many times has conflicting requirements. A balanced topology, described in Chapter 11, facilitates the design of an amplifier circuit meeting simultaneously the aforementioned requirements in a compromised sense.

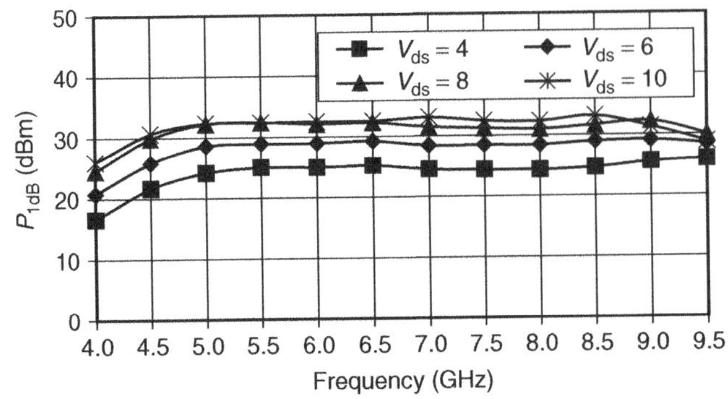

Figure 9.30 The 1-dB gain compression point versus drain voltage.

9.4.5 GaAs HBT Power Amplifer

This example is for a narrowband, three-stage, GaAs HBT amplifier designed for use in cordless telephones operating in the 5.7–5.9-GHz range. The goals for the design were greater than 26 dB gain, 400 mW of saturated power, 35% power added efficiency, on-chip transmit enable control, and operation from an unregulated supply varying from 3 to 5 V. Because of the requirement for the transmit enable control voltage, a bipolar design with an active bias circuit was selected. Design trades-offs between GaAs HBT and SiGe HBT were considered. Because of the supply voltage requirement of 3–5 V, a low-cost commercial GaAs HBT process was selected. The breakdown of the SiGe processes did not support reliable operation at 5 V. Also, the through wafer vias of the GaAs HBT allowed for more gain per stage.

The GaAs HBT power amplifier design consisted of three stages. The cell design was based on a 2-μm emitter with two fingers each 40 μm wide. The output cell size was selected to deliver 600 mW at 3.3-V nominal voltage. The device sizes for the first, second, and third stages were 160, 480, and 1280 μm^2, respectively. The on-chip output match loss was estimated at 1 dB. Another 0.5–1-dB loss is attributed to the wire bonds, package parasitic, board, and connectors (reference plane at the connector). The output matching network consisting of a lowpass circuit with second-harmonic trap was designed based on load-pull data, which was taken for a 1600-μm^2 device at 3.3 V. Then the load impedance was scaled for each device size as discussed in Chapter 5. The equivalent load impedance normalized with respect to emitter width was $R_L = 10.66\ \Omega\cdot$mm and $C_L = -1.79$ pF/mm.

Another important design consideration was the use of the ballast resistor in the unit cell. The IC schematic for a ballasted 160-μm^2 cell is shown in Figure 9.31. To protect the unit cell from thermal runaway, the ballast resistor must be used in the base and/or emitter. In this design, every standard cell (160μm^2) was DC biased through a 300-Ω base ballast resistor. Too high a resistor value reduces the gain of the PA, while too low a resistor will not produce a stabilizing effect for thermal stability. Most foundry suppliers have recommended values for the required ballast resistance. Typical $I-V$ data for the standard cell HBT without the ballast resistor is shown in Figure 9.32.

Another key feature of the PA design was the use of a low-current, active bias circuit. The active bias circuit in this case is essentially an emitter follower circuit, as shown in lower portion of the amplifier IC schematic in Figure 9.33. The IC layout

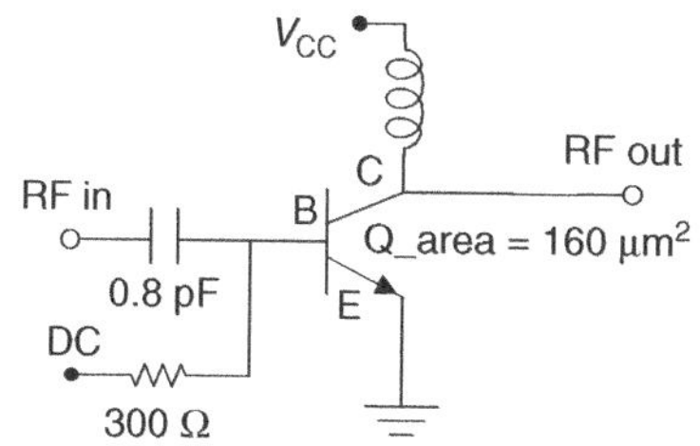

Figure 9.31 A 160-μm² unit cell showing the connection for a base ballast resistor.

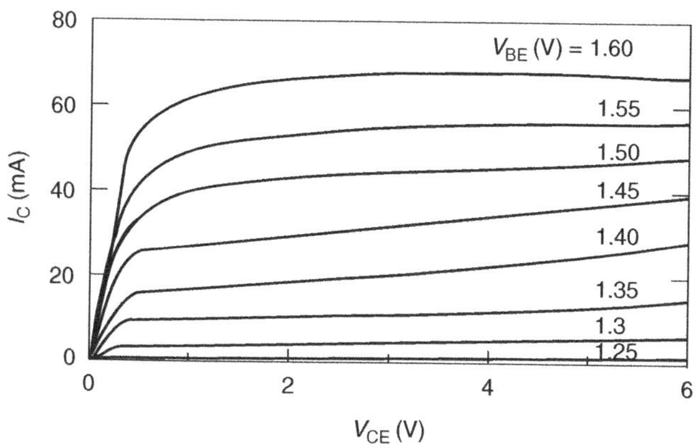

Figure 9.32 Standard HBT 160-μm² cell's $I-V$ curve without the base ballast resistor.

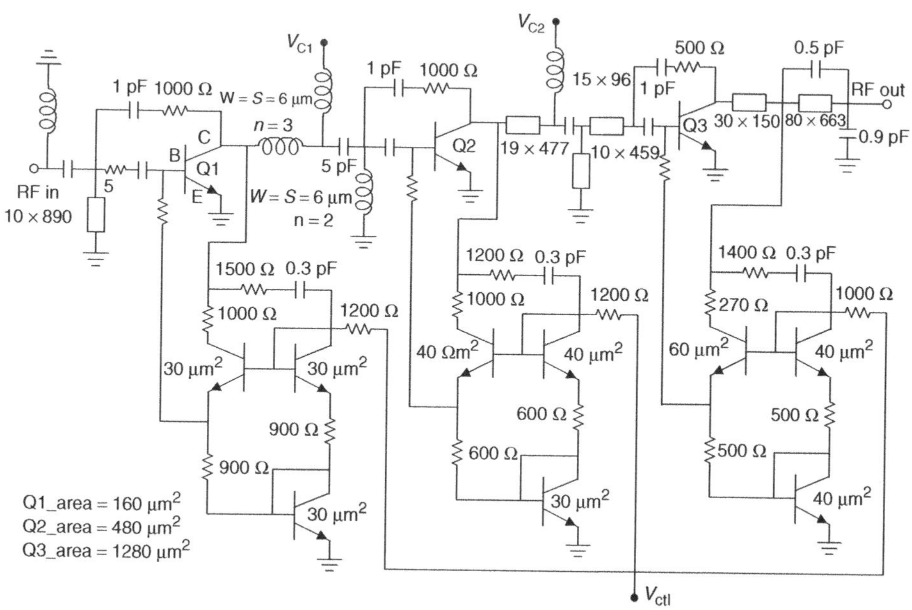

Figure 9.33 Schematic of three-stage HBT PA with all on-chip matching and active bias circuits. V_{C3} is applied through the off-chip ckoke. Transmission line dimensions (width × length) are in μm.

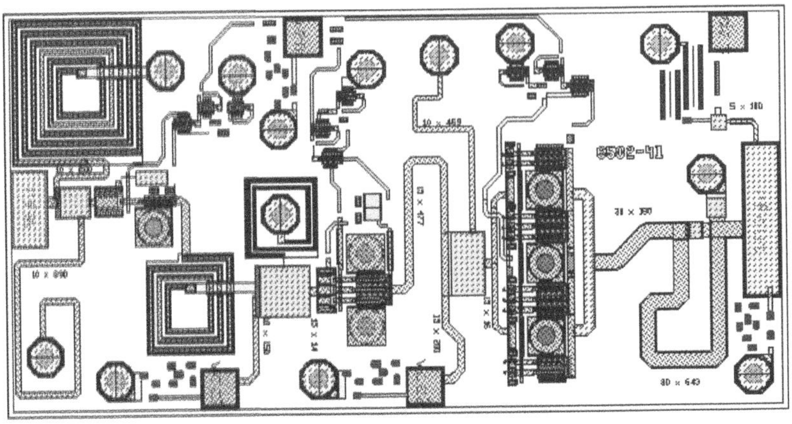

Figure 9.34 GaAs HBT IC layout.

of the HBT PA is shown in Figure 9.34. The GaAs HBT process includes two global metals for wiring, which make it convenient for spiral inductors. Each unit cell has its own through wafer ground vias. The IC was assembled in a leadless 4-mm PQFN 16-lead package and then tested on a printed circuit board (PCB). The PCB was a three-dielectric-layer board. The top layer to second layer dielectric thickness was 10 mils. The microstrip lines on the board have a 50-Ω characteristic impedance value. Under the paddle of the package, the PCB has through board vias to the second layer, which is the RF ground. The package paddle, lead frame, and via inductance were taken into account in the design of the PA. Because the paddle and board inductance to the PCB ground (second layer) can affect the stability and performance of the amplifier, good through board via connections are recommended at this frequency. Also, the wire bond inductance and pad parasitics were included in the RF design. The wirebond assembly is shown in Figure 9.35. The lead frame electrical model used in the design is described in Chapter 21.

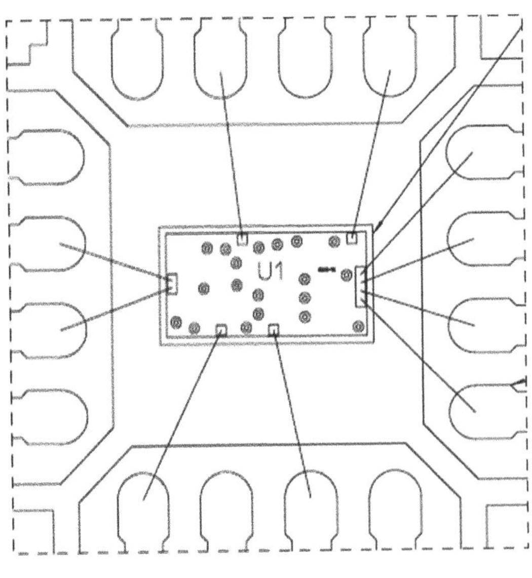

Figure 9.35 Assembly drawing of the GaAs HBT IC die in a leadless package.

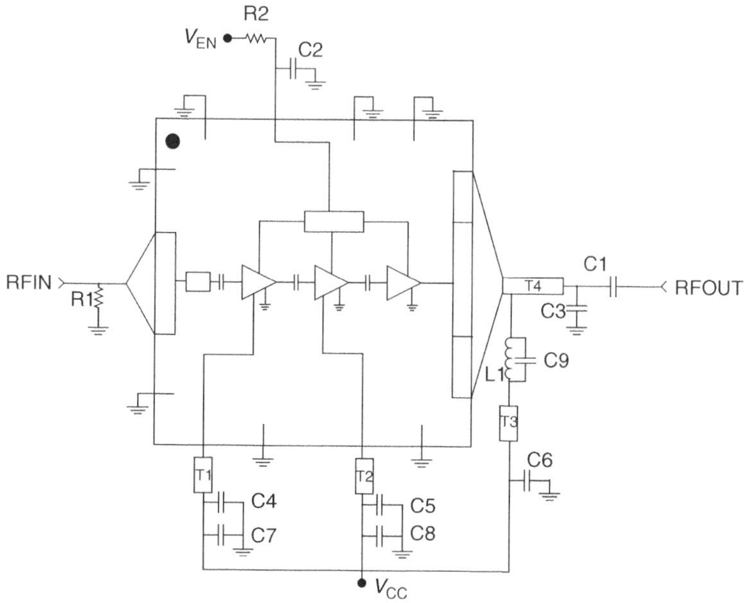

Figure 9.36 Printed circuit board schematic.

The printed circuit board schematic and details of the off-chip matching are shown in Figure 9.36. On the output, four wire bonds to four different pads were used to reduce the output matching loss from the IC to the PCB. Two key components are C3 and C9 (0.2 pF), which help in improving the $2f_0$ performance. C1 = 100 pF is a DC block and is not critical. Also, the five capacitors on the supply are for decoupling and not critical. L1 = 3.9 nH is the DC choke for the third stage and R1 = 250 Ω helps with the input return loss and is not critical. The enable resistance, R2 (600 Ω), can be varied somewhat (10%) to vary gain and current while the values of R2 and C2 (4.7 nF) determine the rise and fall time (<10 μs). Nine via holes were used for the paddle ground.

The typical measured output power, gain, and PAE, as a function of input power at 5.8 GHz, of a packaged device are shown in Figure 9.37. The amplifier met the key

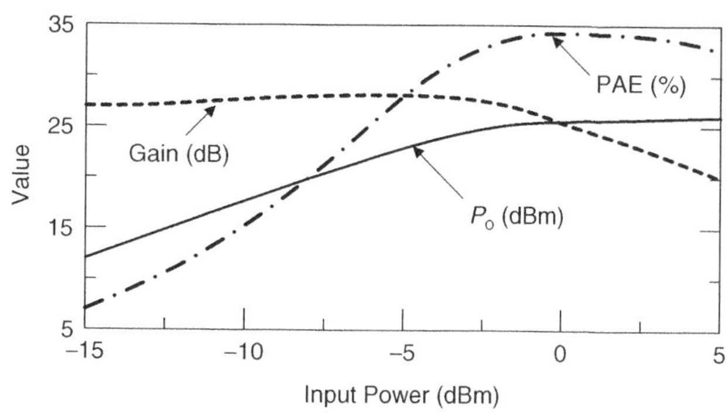

Figure 9.37 Typical measured output power, PAE, and gain versus input power of a packaged device. Supply voltage = 3.3 V.

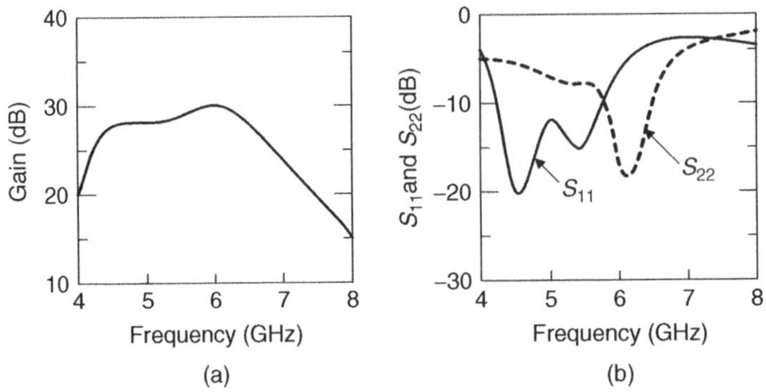

Figure 9.38 Typical measured performance versus frequency of a packaged device. Supply voltage = 3.3 V. (a) gain and (b) S_{11} and S_{22}.

of the design targets. The small-signal gain was approximately 27 dB. The P_{1dB} was over 25 dBm with a corresponding PAE of 34.5%. The small-signal gain and S_{11} and S_{22} versus frequency are shown in Figure 9.38. The input return loss was greater than 10 dB. Reducing the output matching network loss on the IC and in the package and board would further improve the output power up to 0.5 dB and efficiency by 5–10%.

9.5 SILICON BASED AMPLIFIER DESIGN

Low-cost solutions for wireless communications require cheaper and more efficient front-end components including power amplifiers integrated with baseband control circuitry using Si CMOS technology. Thus Si based RF components including low-noise, buffer, and power amplifiers seem a natural choice for integrated circuits, compared to multichip based modules using Si and GaAs devices, to reduce overall cost. Si MOSFETs combined with SiGe HBTs have great potential for building Si based systems. Both devices have demonstrated acceptable performance for medium- and low-power applications, respectively. They have inferior performance in comparison to GaAs HBTs. However, smart design concepts are being implemented in Si transistor based amplifiers to bring their performance at par with GaAs devices. However, for power levels greater than 1 W, GaAs transistors are still used exclusively.

The handset amplifier design is generally based on the best compromise of various transistor and fabrication technologies, available as described in Chapters 4, 14, and 15. They usually employ a combination of PCB including microstrip and CPW, surface-mounted passives, and transistor/MMICs. The output matching network is realized by using off-chip high-Q components. Next, a brief description of design methods for low-noise and power amplifiers is given.

9.5.1 Si IC LNA

Si bipolar transistor based low-noise amplifiers for narrowband and ultra-wideband applications have been developed recently [24–27]. Two basic circuit configurations—feedback and cascode—for a low-noise amplifier are shown in Figure 9.39. Both topologies use source follower at the output for good match and are capable of providing low noise figure over wide bandwidths. The cascode configuration provides high

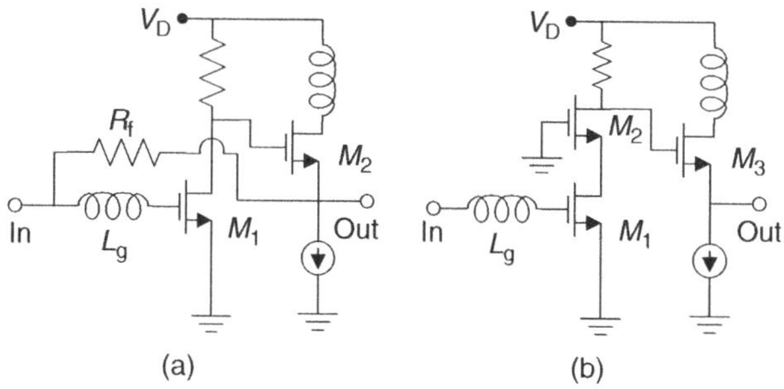

Figure 9.39 Simplified schematics of a low-noise amplifier: (a) feedback with source follower and (b) cascode with source follower.

isolation between the two stages. Since the CMOS transistor's noise match is very close to power match, the circuit's input is matched for good VSWR. The inductor L_g placed in series with the gate also helps in getting higher gain at higher frequencies. At low microwave frequencies, noise figures better than 2 dB over narrowband and less than 4 dB over ultra-wideband have been demonstrated.

9.5.2 Si IC Power Amplifiers

The main function of the power amplifier is to produce the required amount of linear power (high P_{1dB} compression point) with the highest possible PAE in order to increase the battery life of the system. The power amplifier in the RF block is the largest user of battery power and is generally turned on only when it is transmitting. In the efficient and linear power amplifier design, several different schemes have been proposed [28–33]. These include switching Q-point, stage bypassing, and adjustable transistor size. In the switching Q-point technique, the improvement in efficiency is more in the class-A than class-AB/B amplifiers. The stage bypassing scheme has better efficiency but has reduced gain and increased circuit complexity. More discussions on this technique are included in Chapter 20. In the adjustable transistor size amplifier topology, the output stage consists of two different sized transistors connected in parallel. The transistors are switched between the low-power and high-power states. There are several factors to be considered carefully in its implementation. These factors are phase compensation between the two paths, the effect of load (the load designed for the large transistor must be considered in the load design for the smaller device and vice versa), and stabilization. The disadvantage of this scheme is that it requires a complex output matching network.

The adjustable transistor size amplifier design consists of two to three stages and only the last stage has two different sized transistors. In a two-stage design, the input and output stage small transistor sizes are usually the same. The output stage large transistor size is about four times larger than the small transistor size. The input and output matching networks are realized off-chip using lumped and distributed matching elements. Next, several medium-power design examples for Si based transistor PAs are described.

SiGe HBT Cordless PA

This design example is for a low-voltage, high-efficiency, saturated power amplifier for cordless telephone applications. The design requirements for this amplifier are 2.4-V supply voltage, low cost, and plastic packaging. The target specifications are 26-dBm saturated power, 26-dB power gain (30-dB small signal), PAE 40%, operating band 1880–1930 MHz, and microamp shutdown current with an on-chip transmit enable control using 2.4 V. The competing requirements for low-cost amplification at 1.9 GHz and the use of 2.4 V for both voltage supply and bias enable made the choice of semiconductor technology a challenge. Both GaAs and SiGe processes and MESFET and HBT types were considered. Table 9.6 shows a comparison of the three technologies considered. Cordless phone systems require single supply amplifiers with very low leakage, typically 10 μA in the off state. GaAs MESFETs are depletion mode devices and are self-biased for single supply operation and therefore require an external drain switch for shutoff, which adds to the cost. For this reason, bipolar technologies were selected. When comparing bipolar technologies in SiGe and GaAs, SiGe has an advantage when the control voltage or operating supply voltage is in the 2–3-V range and power is relatively low. This is due to the fact that most bipolar amplifier designs use an active bias circuit for DC bias so as not to consume too much current from the transceiver control ports (design goal of < 1 mA). The smaller the control current, the larger the series resistance that can be placed in the control line for RF isolation between the transceiver and power amplifier. CMOS PAs have an advantage for this design requirement. For bipolar designs, there are at least two base–emitter voltage drops from the control voltage to ground including the bias network. The SiGe and GaAs HBT base–emitter forward voltage drops are 0.7 V and 1.3 V, respectively. Therefore GaAs HBT power amplifiers typically require a minimum control voltage of 2.8 V. The control voltage requirement for two-cell cordless systems is 1.8–2.4 V. The GaAs HBT has the advantage of higher breakdown voltage, through wafer vias, and higher substrate resistance. Smart design techniques in SiGe can offset these differences, at least for this application. SiGe also offers the lowest production wafer pricing, which is an important consideration. Based on these factors, the SiGe HBT technology was selected for the design of this circuit.

The design is based on a commercial SiGe process with 1.6-μm emitter geometries and 25-GHz f_T. The process includes passive elements, inductors, capacitors, and three types of resistors. For electrostatic discharge protection, a series of protective diodes is available. The process offers three metal layers with dielectric isolation between the layers. The substrates are 1000 Ω·cm conductive and use p^+ channel isolation to improve device isolation. As mentioned, silicon processes do not include through wafer vias like GaAs, and so all grounding is implemented by multiple down bonds from the top side ground pads to the package paddle. Another design technique would have been to use a differential design so that the RF ground is virtual and not sensitive to down bonds. This requires either an external or on-chip balun. The balun adds to the cost and, more importantly, adds an additional loss to the output matching network.

Table 9.6 Comparison of GaAs MESFET, GaAs HBT, and SiGe HBT IC Technologies

Technology	V_{CC}	Shutdown	Off I_C	PAE	Sub resistivity	Wafer Cost
GaAs MESFET	2–5 V	External	mA	Excellent	High	Medium
GaAs HBT	3–5 V	Internal	μA	Good	High	Medium to low
SiGe HBT	2–4 V	Internal	μA	Good	Low	Low

For the gain requirement of over 30 dB, three amplifier stages were chosen (approximately 10–12 dB per stage of stabilized gain). A simplified schematic of the amplifier is shown in Figure 9.40 with off-chip components. The last stage was sized to deliver approximately 26 dBm of saturated power at 2.4 V and consisted of 1248 μm^2 of emitter area. First-, second-, and third-stage transistors have 2, 7, and 26 cells, respectively, and each cell emitter area is $1.6 \times 30\ \mu m^2$. Each stage was biased for class-AB operation. Under RF operation, the cells were kept at current densities less than 20 kA/cm^2. Each stage has *RC* feedback to improve the overall amplifier stability. The PGNDs were paddle grounds with the wire bonds from the top of the IC connected to the paddle of the package with multiple wires.

As shown in Figure 9.40, some of the RF matching components were off-chip in order to save IC cost. For example, the interstage match for stages 1 and 2 were partially determined by the wire bond and board inductance in conjunction with the series capacitors between transistors. A critical part of the amplifier design was the bias circuit. The bias circuit supplies the DC bias to the base of each RF cell. Each stage has its own independent bias circuit. The bias circuit had good isolation from the RF cell, which enhances the stability performance of the PA. It was also small in size and had good temperature compensation. The turn-on and turn-off performance of the amplifier was set by the off-chip resistance and capacitance. For the cordless application, they were chosen to give a rise and fall time of 3 μs.

The IC layout was designed for minimal die size and high-volume manufacturability. The compact IC layout, shown in Figure 9.41, was less than 1.0 mm^2 in size. The amplifier was assembled in a leadless plastic package, $3 \times 3 \times 0.9\ mm^3$, PQFN, 12L. The chip wiring drawing is shown in Figure 9.42. The circuit board developed

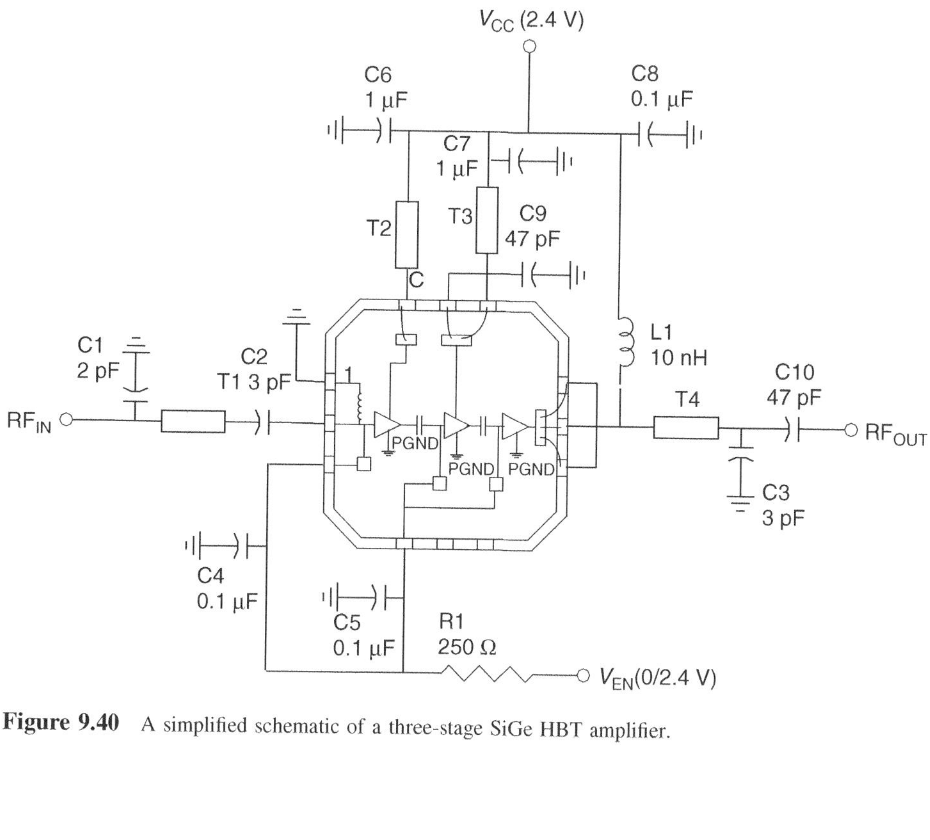

Figure 9.40 A simplified schematic of a three-stage SiGe HBT amplifier.

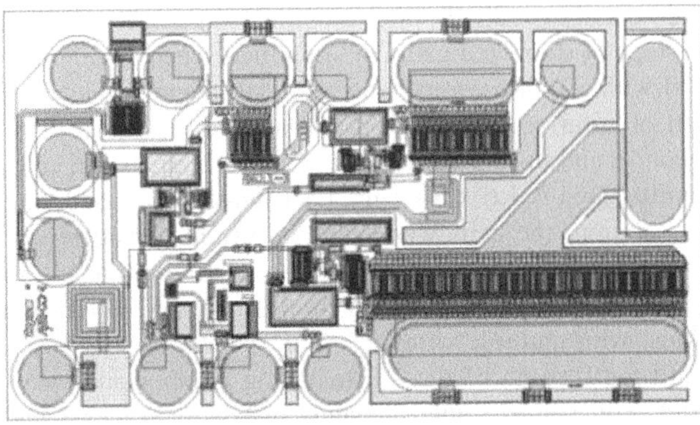

Figure 9.41 SiGe integrated circuit layout showing compact 0.925-mm^2 die area.

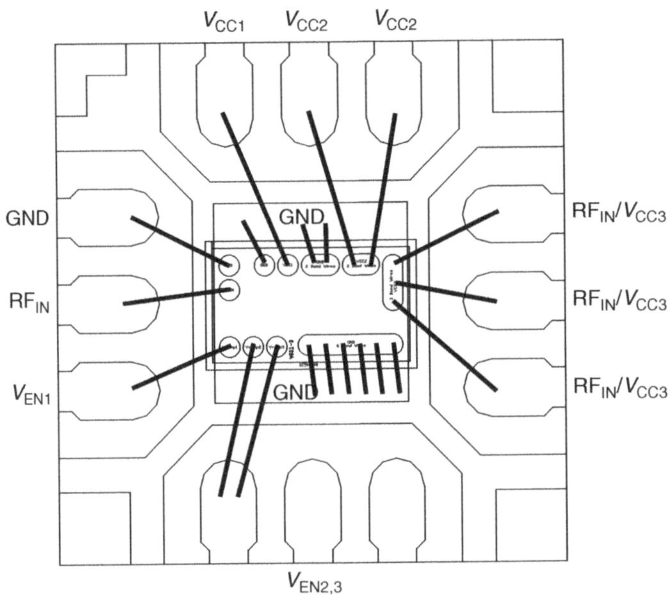

Figure 9.42 Assembly drawing of the three-stage SiGe HBT power amplifier.

for this circuit is shown in Figure 9.43. The typical measured output power, gain, and PAE versus frequency are shown in Figure 9.44. The output power, gain, and PAE values were better than 26 dBm, 30 dB, and 45%, respectively, and met the design requirements for the cordless phone application.

1.9-Gigahertz Si CMOS Power Amplifier

A 1.9-GHz Si CMOS power amplifier, based on the device size-adjustable concept, has been developed [33]. The schematic of an adjustable transistor size power amplifier using Si CMOS is shown in Figure 9.45. The input stage and output stage small transistor size is 400 μm, and the output stage large transistor size is 1600 μm. The measured gains in low-power and high-power modes are 16 and

Figure 9.43 Sample board for the SiGe power amplifier. Standard FR4 material with solder-on RF edge connectors. All off-chip components are commercially available 0402 surface mount.

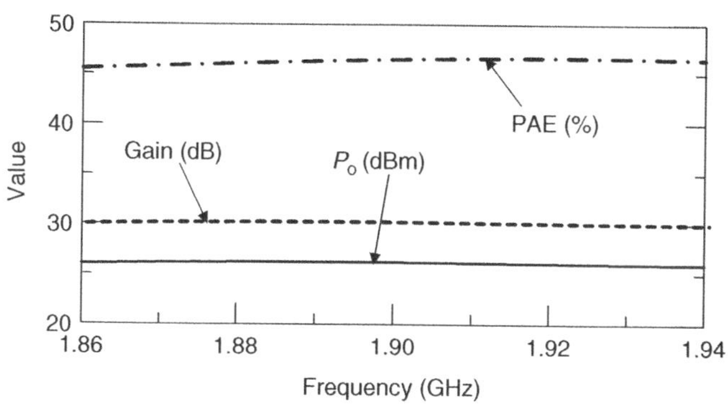

Figure 9.44 Measured performance for the three-stage SiGe HBT amplifier, at $V_{CC} = 2.4$ V and $V_{EN} = 2.4$ V.

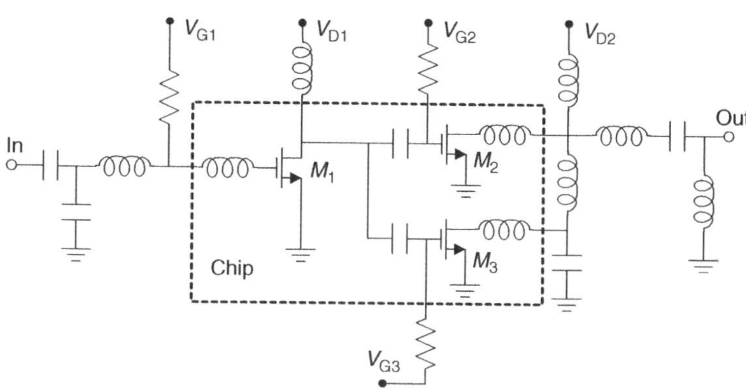

Figure 9.45 Simplified schematic of the 1.9-GHz Si CMOS amplifier with two-step power control.

21 dB, respectively. The corresponding power levels at P_{1dB} are 18 and 23 dBm, respectively. The PAE in the low-power mode is 15%, which is about two times better than the PAE for a class-B amplifier under power back-off condition. The IM3 values are below −30 dBc and −25 dBc for the low- and high-power cases, respectively.

1.9-Gigahertz SiGe HBT Power Amplifier

The schematic of an adjustable transistor size power amplifier using SiGe HBT, for CDMA/PCS applications, is shown in Figure 9.46. The input stage and output stage small transistor emitter area is 480 μm^2, and the output stage large transistor emitter area is 1920 μm^2. The total chip size is 1 mm^2, housed in a 4 × 4 mm 16-pin leadless plastic package. The measured gain G, P_{1dB}, and PAE values for the high-power mode are 21 dB, 27 dBm, and 27%, respectively. When the amplifier is switched to low-power mode, the values of G, P_{1dB}, and PAE are 19 dB, 18 dBm, and 15%, respectively. Both the high-power and low-power modes met ACPR1 and ACPR2 specifications of −44 dBc and −53 dBc, respectively [33].

2.4-Gigahertz SiGe HBT Power Amplifier

For digital wireless communications, the 2.4-GHz ISM band has several applications including DECT, WLL, Bluetooth, and Home RF. Both Si and GaAs based power amplifiers have been developed. The schematic of a three-stage SiGe HBT based power amplifier is shown in Figure 9.47. The amplifier was designed with off-chip output matching network and harmonic suppression filter. The off-chip circuitry was implemented in LTCC [33]. To achieve linearity and efficiency performance, advanced design techniques were employed in the amplifier circuit to overcome the Si transistor's inferior performance in order to extend the talk time for the handset. The amplifier has 27.5-dBm output power and 47% PAE.

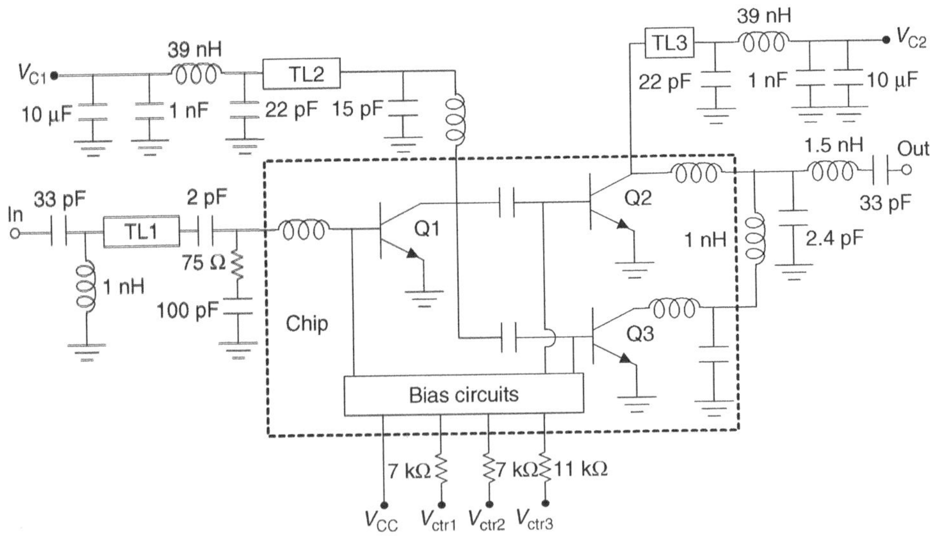

Figure 9.46 Simplified schematic of the 1.9-GHz SiGe HBT amplifier with two-step power control. Emitter areas of Q1 = Q3 and Q2 are 480 μm^2 and 1920 μm^2, respectively.

Future mobile wireless applications are expected to operate over wide-band/multiband frequencies. For example, for data communication WLAN is operating at 2.4 GHz and WiMAX at 3.3–3.9 and 5.1–5.9 GHz. These applications pose a serious challenge to designers to develop power amplifiers with accepted output power and PAE over multibands for seamless operation. Several techniques for dual-band and triple-band PAs have been reported. However, new design concepts in the matching networks to design ultra-wideband or band switchable/reconfigurable output stage matching networks are required. A key element in these techniques is to keep the output match loss within 1 dB.

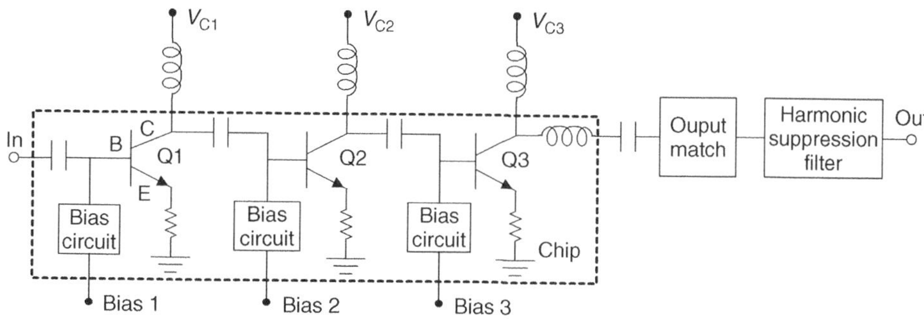

Figure 9.46 Simplified schematic of the 2.4-GHz SiGe HBT amplifier.

Readers are advised that the amplifier's RF parameters are defined at the input and output as reference planes, DC bias conditions are at the suggested DC terminal points, and thermal interface (in the case of power amplifiers) is at the back of the chip or shim or package. Other factors, such as various support circuits including couplers, filters, circulators, antennas, bias lines, fixtures, and connectors, depending on the amplifier type, will affect its performance. For HPAs, any resistive loss or mismatch loss at the output and the thermal setting can reduce the output power and PAE significantly. Under such conditions, the amplifier performance must be reevaluated and sufficient margins in the product design must be considered.

REFERENCES

1. S. C. Cripps, *RF Power Amplifiers for Wireless Communications*, Artech House, Norwood, MA, 1999.
2. E. L. Griffin, Application of loadline simulation to microwave high power Amplifiers, *IEEE Microwave Mag.*, Vol. 1, pp. 58–66, June 2000.
3. I. Bahl and P. Bhartia, *Microwave Solid State Circuit Design*, 2nd edition, John Wiley & Sons, Hoboken, NJ, 2003.
4. KELDIST, M/A-COM's Internal Software.
5. I. J. Bahl, Low loss matching (LLM) design technique for power amplifiers, *IEEE Microwave Mag.*, Vol. 5, pp. 66–71, December 2004.
6. I. Bahl, MESFET process yields MMIC Ka-band PAs, *Microwaves and RF*, Vol. 44, pp. 96–112, May 2005.

7. I. Bahl, Ku-band MMIC power amplifiers developed using MSAG MESFET technology, *Microwave J.*, Vol. 49, pp. 56–82, February 2006.
8. I. J. Bahl, 0.7–2.7GHz 12-watt power amplifier MMIC developed using MLP technology, *IEEE Trans. Microwave Theory Tech.*, Vol. 55, pp. 222–229, February 2007.
9. I. Bahl, 1.6W K-band MMIC power amplifiers developed using low loss matching technique for point-to-point radio applications, *Microwave Opt. Technol. Lett.*, Vol. 49, pp. 1521–1525, July 2007.
10. G. D. Vendelin et al., *Microwave Circuit Design Using Linear and Nonlinear Techniques*, Wiley-Interscience, Hoboken, NJ, 2005.
11. G. Tauguchi and M. S. Phadke, Quality engineering through design optimization, in *Conference Record, GLOBECOM 84 Meeting*, IEEE Communications Society, Atlanta, GA, November 1984, pp. 1106–1113.
12. M. S. Phadke, *Quality Engineering Using Robust Design*, Prentice Hall, Englewood Cliffs, NJ, 1989.
13. E. L. Griffin, Introduction to Taguchi technique applied to MMIC design, in *IEEE MTT-S International Microwave Symposium Workshop on Statistical-Based MMIC and Module Design Techniques*, June 21, 1996.
14. W. L. Pribble and E. L. Griffin, An ion-implanted 13 watt C-band MMIC with 60% peak power added efficiency, *IEEE Microwave MillimeterWave Monolithic Circuits Symp. Dig.*, pp. 25–28, 1996.
15. D. C. Miller, R. A. Sadler, and I. J. Bahl, Low-noise GaAs MESFET optimization using an orthogonal array approach, *Int. J. Microwave Millimeter-Wave Computer-Aided Eng.*, Vol. 4, pp. 396–410, October 1994.
16. J. Carroll and K. Chang, Statistical computer-aided design for microwave circuits, *IEEE Trans. Microwave Theory Tech.*, Vol. 44, pp. 24–32, January 1996.
17. G. N. Henderson and D. Wu, GaAs HBT profile optimization using the Taguchi method, *IEEE MTT-S Int. Microwave Symp. Dig.*, pp. 1463–1468, 1997.
18. I. J. Bahl et al., Low loss multilayer microstrip line for monolithic microwave integrated circuits applications, *Int. J. RF Microwave Computer-Aided Eng.*, Vol. 8, pp. 441–454, November 1998.
19. R. Goyal (Ed.), *High-Frequency Analog Integrated Circuit Design*, John Wiley & Sons, Hoboken, NJ, 1995, p. 179.
20. I. Bahl, *Lumped Elements for RF and Microwave Circuits*, Artech House, Norwood, MA, 2003.
21. I. J. Bahl, High current capacity multilayer inductors for RF and microwave circuits, *Int. J. RF Microwave Computer-Aided Eng.*, Vol. 10, pp. 139–146, March 2000.
22. I. J. Bahl, High-Q and low-loss matching network elements for RF and microwave circuits, *IEEE Microwave Mag.*, Vol. 1, pp. 64–73, September 2000.
23. I. J. Bahl et al., Multifunction SAG process for high-yield, low cost GaAs microwave integrated circuits, *IEEE Trans. Microwave Theory Tech.*, Vol. 38, pp. 1175–1182, September 1990.
24. T. H. Lee, *The Design of CMOS Radio-Frequency Integrated Circuits*, 2nd edition, Cambridge Press, Cambridge, UK, 2004.
25. M. T. Reiha and J. R. Long, A 1.2-V reactive feedback 3.1–10.6-GHz low-noise amplifier in 0.13 μm CMOS, *IEEE J. Solid State Circuits*, Vol. 42, pp. 1023–1033, May 2007.
26. A. Bevilacqua and A. M. Niknejad, An ultrawideband CMOS low-noise amplifier for 3.1–10.6-GHz wireless receivers, *IEEE J. Solid State Circuits*, Vol. 39, pp. 2259–2268, December 2004.
27. T. Chang et al., *ESD-protected wideband CMOS LNAs using modified resistive feedback techniques with chip-on-board packaging*, *IEEE Trans. Microwave Theory Tech.*, Vol. 56 pp. 1817–1826, August 2008.

28. M. Ranjan et al., Microwave power amplifiers with digitally-controlled power supply voltage for high efficiency and high linearity., *IEEE MTT-S Int. Microwave Symp. Dig.*, pp. 493–496, 2000.
29. J. Staudinger, Applying switched gain stage concepts to improve efficiency and linearity for mobile CDMA power amplification, *Microwave J.*, Vol. 43, pp. 152–162, September 2000.
30. J. H. Kim et al., An InGaP/GaAs HBT MMIC smart power amplifier for W-CDMA mobile handsets, *IEEE J. Solid State Circuits*, Vol. 38, No. 6, pp. 905–910, June 2003.
31. H.-M. Park et al., Demonstration of on-chip appended power amplifier for improved efficiency at low power region, *IEEE MTT-S Int. Microwave Symp. Dig.*, pp. 691–694, 2003.
32. J. H. Kim et al., A power efficient W-CDMA smart power amplifier with emitter area adjusted for output power levels, *IEEE MTT-S Int. Microwave Symp. Dig.*, pp. 1165–1168, 2004.
33. A. Raghavan, N. Srirattana, and J. Laskar, *Modeling and Design Techniques for RF Power Amplifiers*, John Wiley & Sons, Hoboken, NJ, 2008.

PROBLEMS

9.1 Design a low-noise single-stage amplifier at 12 GHz with 8-dB gain and 2.5-dB NF using a microstrip on a 0.25-mm thick alumina substrate ($\varepsilon_r = 9.9$). The output VSWR desired is 1.2:1. The transistor has the following parameters at 12 GHz:

$$S_{11} = 0.77\angle{-122^\circ},\quad S_{21} = 1.36\angle 98^\circ,\quad S_{12} = 0.07\angle 40^\circ,\quad S_{22} = 0.67\angle{-32^\circ}$$
$$NF_{\min} = 2.11\ \text{dB},\quad G_A = 8.7\ \text{dB},\quad \Gamma_{opt} = 0.55\angle 118^\circ,\quad R_n = 13.1\ \Omega$$

9.2 Design a maximum gain single-stage amplifier at 14 GHz with 8-dB gain and VSWR better than 1.5. The transistor has the following parameters at 14 GHz:

$$S_{11} = 0.75\angle{-132^\circ},\quad S_{21} = 1.20\angle 92^\circ,\quad S_{12} = 0.07\angle 40^\circ,\quad S_{22} = 0.66\angle{-37^\circ}$$

9.3 Design a high gain amplifier at 15 GHz using Table 5.5 EC model parameters for a 600-mm FET. The amplifier is unconditionally stable. If required, use circuit stabilization as described in Chapter 17. Use ideal LC elements to calculate gain, S_{11} and S_{22}.

9.4 The S-parameters for a low-noise pHEMT are given in Table 5.9, Chapter 5. Design a conditionally stable low-noise amplifier at 11 GHz using an open-circuited single stub at the input and output. Design goals are NF less than 1 dB, gain greater than 10 dB, and output return loss better than 10 dB. The substrate is 15-mil alumina.

9.5 Design a 5-W MMIC power amplifier at 3.5 GHz using the FET model described in Section (5.7). The substrate is (a) 3-mil GaAs and (b) 15-mil alumina. Show that the (b) case has higher gain than the (a) case.

9.6 Design a linear amplifier with the following specifications:

Frequency	5.8 GHz
Output IP3	40 dBm
Gain, minimum	10 dB

Use the FET model given in Section 5.8 and the EC model parameters in Table 5.5. Use ideal LC elements to calculate gain and return loss.

9.7 Design a power amplifier with the following specifications:

Frequency	30 GHz,
Output P_o	27 dBm
Gain, minimum	7 dB
VSWR	2:1

Use the pHEMT model given in Table 5.12 for the passivated device and 4-mil GaAs substrate.

9.8 Design a 27-dBm output power amplifier using a CMOS transistor whose EC model is given in Table 5.16. The frequency is 2.4 GHz and the power density is 23 dBm/mm. The substrate is 10-mil FR-4.

Chapter 10

High-Efficiency Amplifier Techniques

Power amplifiers are a key component for digital cellular phones with ever increasing linearity and efficiency requirements. These amplifiers consume much of the total battery DC power and occupy significant area. Therefore the design of high-efficiency amplifiers is important for reducing DC power consumption. High-efficiency operation of power amplifiers is becoming one of the most important factors for reducing prime power and cooling requirements for advanced microwave and millimeter-wave radar, communication, and electronic warfare systems.

Over the past 30 years several different techniques, summarized in Figure 10.1, have been investigated to realize maximum efficiency in RF/microwave power amplifiers. These are overdriven, class AB/B, class E, class F, rectangularly driven class B (rB), harmonic reaction amplifier (HRA), harmonic control amplifier (HCA), and harmonic injection technique (HIT). The overdriven and harmonic control amplifiers are operated in a class-A bias to realize high gain, whereas the rest are operated in class-AB/B bias to realize high efficiency. The basic theory of operation of various classes, including class AB/B, class E, class F, and overdriven class A, has been treated in Chapter 8. In this chapter, several design examples of high-efficiency power amplifiers [1–69] and their practical aspects are described.

10.1 HIGH-EFFICIENCY DESIGN

The instantaneous dissipated power in a device is obtained by multiplying voltage and current at a given instant. Referring to the voltage and current waveforms (Fig. 8.11, Chapter 8), the dissipated power in the device is lowest at the minimum current or voltage time period range and thus maximum DC to RF conversion efficiency occurs. The purpose of several aforementioned classes is to maintain the minimum dissipated power time period length as large as possible. There are basically three approaches to obtain minimum dissipation power:

1. Use a device as a switch, class E and overdriven.
2. Use a device with reduced conduction angle, class B and class C.
3. Shape the voltage and current waveforms as square waveforms, class F and class F^{-1}.

Fundamentals of RF and Microwave Transistor Amplifiers. By Inder J. Bahl

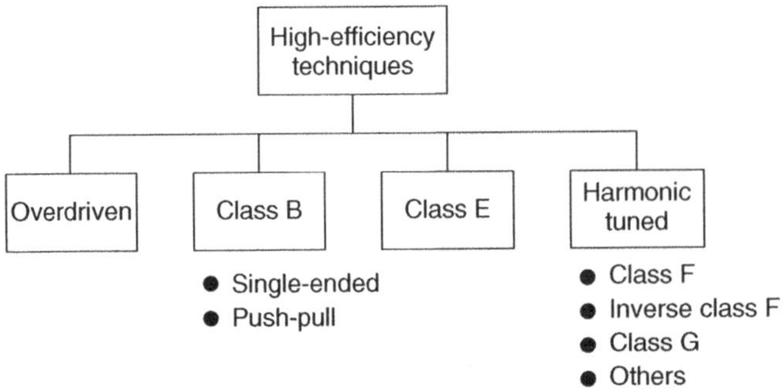

Figure 10.1 High-efficiency techniques for HPAs.

The fundamental source of compression in a power amplifier is current and voltage waveform clipping. When an amplifier is driven into compression, nonlinearities in the input capacitor C_{gs} and output resistance R_{ds} of the transistor also occur. Waveform clipping and other device nonlinearities result in harmonic generation at the input and output of the amplifier. Converting the fundamental frequency signal into harmonic signals degrades the output power and PAE. As the voltage and current waveforms become distorted in "overdriven" operation, RF power is generated at harmonics of the fundamental input frequency. This power can be reflected back into the device to "shape" the waveforms to reduce the dissipated power and increase the fundamental output power. If these harmonic signals are reactively terminated properly, that is, superimposed on fundamental voltage and current waveforms with desired phase, at the input and output, the PAE of the amplifier is enhanced by shaping the sinusoidal wave input signal into square-wave-like signals. As discussed in Chapter 8, the most desirable termination conditions are second-harmonic short circuited and third-harmonic open circuited at the internal port of the device. Because of the device pad's parasitic reactance and Miller effect (device's input impedance depends on its output impedance and vice versa due to feedback between the input and output) at the input and output, the short-circuited and open-circuited conditions are modified. Such changes are obtained accurately by load-pull measurements performed at harmonic frequencies independent of fundamental frequency.

There are numerous matching networks that may be used to provide the desired harmonic termination condition; however, extra care must be exercised such that these termination conditions are achieved without adding extra circuit loss at the fundamental frequency and complexity. Such terminations are sometimes independent of fundamental frequency match. At the output, a lowpass matching topology is preferred to a bandpass network because of lower loss. In general, the harmonic termination network required has very high Q and thus has very narrow bandwidth. As a result, harmonic terminated amplifiers are very sensitive to the harmonic termination and have narrow bandwidth (5–15%). Since high-efficiency power amplifiers are operated near peak efficiency level, they are somewhat insensitive to the input RF drive level.

Most high-PAE work has been from UHF through the Ku-band. Table 10.1 provides an overview of some high-PAE power amplifier examples. The output power and PAE values are typical. The products include PCB, hybrid, and monolithic based fabrication technologies. Hybrid technologies usually result in highest PAE. The commonly used devices are LDMOS, HBT, MESFET, and pHEMT. Figure 10.2a shows typical variations of gain, output power, and PAE versus input power for a transistor

Table 10.1 Some Typical Narrowband High-Efficiency Power Amplifier Performance Parameters

Frequency (GHz)	Number of Stages	Gain (dB)	Power (W)	PAE (%)	Device	Technology	Reference
0.8	1	12	12	82	MESFET	PCB	40
2.1–2.2	2	21	50	50	pHEMT	GaAs monolithic	62
2.9	1	10	31	49.5	HBT	Hybrid	63
3.8	1	12	70	51	HFET	Hybrid	53
3.85	1	14	27	70.6	HFET	Hybrid	64
4.5–5.4	1	10	14	55	MESFET	GaAs monolithic	6
8–10	3	24	12	40	MESFET	GaAs monolithic	8
8–10	3	24	20	35	MESFET	GaAs monolithic	65
12–15	3	18	8	25	MESFET	GaAs monolithic	12
13.5–15	3	22	8	22	pHEMT	GaAs monolithic	66
29–31	3	20	4	25	pHEMT	GaAs monolithic	67
42–46	2	17	2.8	24	pHEMT	GaAs monolithic	68
95	2	15	0.43	19	pHEMT	InP monolithic	69

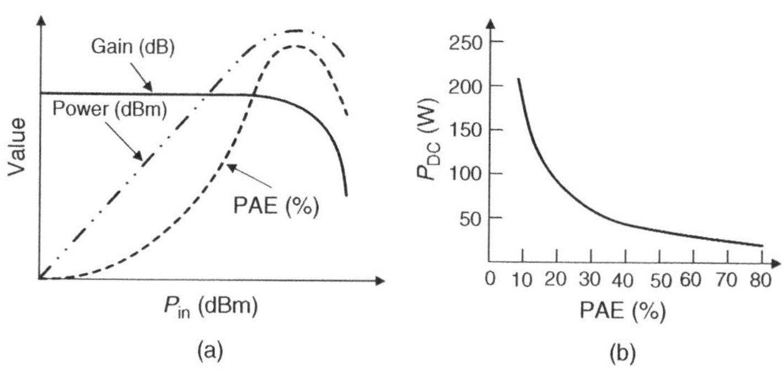

Figure 10.2 Variations of power amplifier performance: (a) gain, output power, and PAE versus input power and (b) P_{DC} versus PAE for a 20-W output power and 10-dB gain.

amplifier. At higher gain compression levels, both power and PAE are reduced. P_{DC} versus PAE for a 20-W output power and 10-dB gain is shown in Figure 10.2b. The DC power requirement increases from 36 to 180 W for PAE ranging from 50% to 10%. Thus PAE is the most critical requirement for HPAs.

10.1.1 Overdriven Amplifier Design

When transistor amplifiers are operated in the 2–3-dB gain compressed mode to improve PAE and no particular harmonics are terminated, such amplifiers are known as *overdriven*. In such amplifiers extra care is required to terminate the harmonics. When the devices are operated at class B or class AB, one-fourth to one-half of the input power is lost due to limited device conduction. This reduces the gain of the amplifier from 3 to 6 dB in comparison to class A. At high microwave frequency, the associated power gain is only in the 7–9-dB range. Therefore any gain reduction in class-AB/B amplifier significantly affects its PAE. In such cases, for overdriven power amplifiers, a class A is preferred to class AB/B. At 13-dB or higher power gain, the PAE becomes insensitive to gain value.

In class-A power amplifiers, when the output power is increased beyond a fixed level, the voltage and current waveforms are limited and become clipped or distorted. As waveforms start distorting, the gain begins to decrease with increasing input RF drive. This gain compression phenomenon causes the PAE to peak and then to decrease as the amplifier is overdriven beyond its operating limits, where there is no increase in output power with increasing input power. In class-A overdriven HPAs, the waveforms are clipped symmetrically due to the limited range of gate voltage between a little over built-in voltage and pinch-off voltage. At these two extreme gate voltage points, the devices generate maximum second-harmonic levels, thus lowering output power in the fundamental frequency.

In the overdriven amplifier the harmonics are ignored. In the case of terminating the first two harmonics into 50 Ω, the HPA design must be 3:1 broadband. However, this increase in bandwidth must not add significantly to the matching network loss at the fundamental frequency and the circuit complexity in general. In most cases, it will be impossible to design high-efficiency broadband matching networks as discussed in Chapter 11. In a simple overdriven HPA design, an optimum load match (same as for class A and class AB/B) at the device output is provided by the output matching network and the device input has a conjugate match.

EXAMPLE 10.1

A 10-W MMIC class-A amplifier was designed [2] at 5.5 GHz using four 4-mm FETs. Two 5-W single-ended amplifiers were reactively combined in parallel on a single chip. Each of the single-ended amplifiers used two 4-mm FETs for a net 8 mm of FET gate periphery. The following EC model parameters for the 4-mm FETs were used in the design:

$$R_g = 0.6\ \Omega,\quad R_i = 0.6\ \Omega,\quad R_s = 0.4\ \Omega,\quad R_d = 0.4\ \Omega,\quad R_{ds} = 20\ \Omega$$
$$C_{gs} = 5.0\ \text{pF},\quad C_{gd} = 0.1\ \text{pF},\quad C_{ds} = 1.07\ \text{pF}$$
$$g_m = 300\ \text{mS},\quad \tau = 3\ \text{ps}$$
$$L_g = 0.05\ \text{nH},\quad L_s = 0.02\ \text{nH},\quad L_d = 0.05\ \text{nH}$$

The devices were operated at a V_{ds} of 8.5 V and I_{ds} of $40\% I_{dss}$. The I_{dss} and the breakdown voltage values for a 4-mm FET are 1.5 A and 20 V, respectively. The load impedance (Z_L) for the FET is equivalent to a parallel combination of resistor (R_L) and capacitor (C_L). The real part of the load impedance R_L used for 4-mm FET is 13 Ω. The imaginary part of the load impedance (C_L) is $-C_{ds}$.

SOLUTION For each 5-W amplifier, the input impedance of the device was matched to 100 Ω while the output matching network transformed the output 100 Ω impedance to 6.5-Ω load impedance required for maximum power output condition at the output of the 8-mm device. When two of these amplifiers are connected in parallel, the resultant amplifier's input and output impedance becomes 50 Ω. As shown in Figure 10.3a, both distributed and lumped elements were used in the matching networks. The MIM capacitance density is 300 pF/mm^2 and the substrate thickness is 75 μm. The elements in the input and output matching networks were selected for unconditional stabilization and minimum possible loss, respectively. Finally, two of these single-ended designs were combined on chip to complete the 10-W power amplifier.

A photograph of the overdriven 10-W power amplifier MMIC is shown in Figure 10.3b. The 10-W amplifier chips were assembled on gold plated Cu-W carriers for RF characterization. Typical measured output power and PAE as a function of input power at 5.5 GHz are shown in Figure 10.4. At the 2-dB gain compression point, the output power and PAE are 10 W and

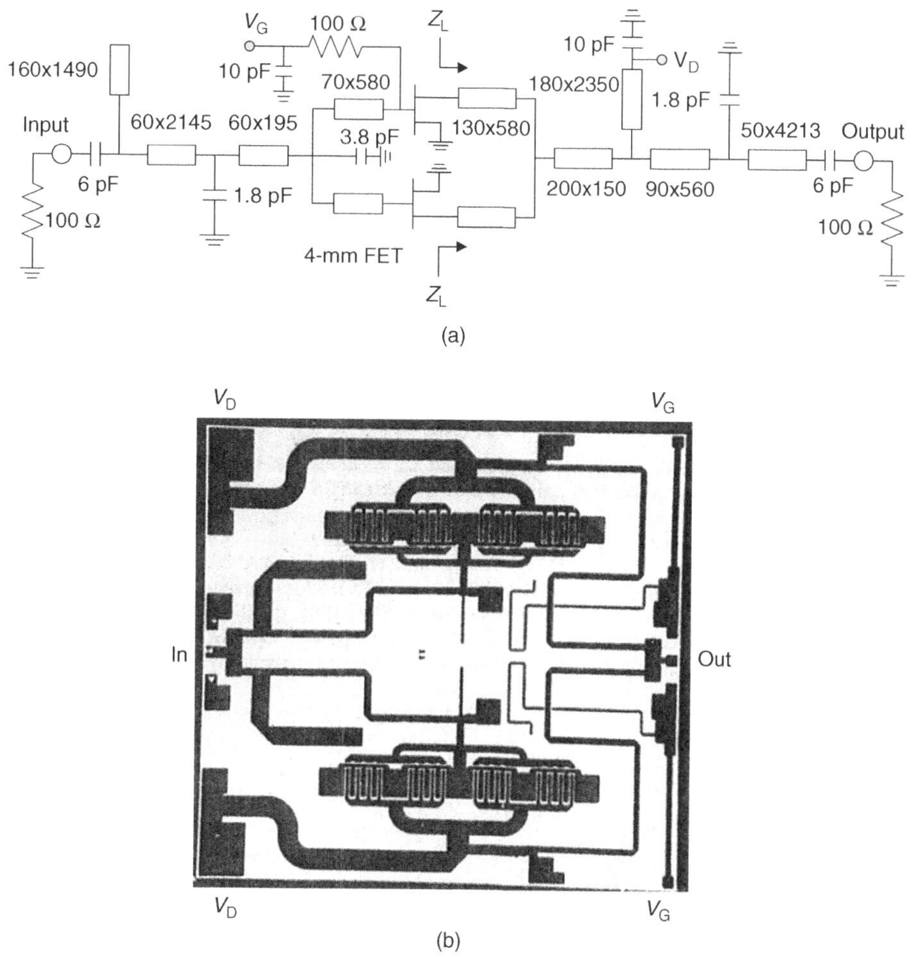

Figure 10.3 (a) Schematic of a 5-W overdriven power amplifier. Two of these amplifiers were combined to obtain 10-W HPA. Microstrip dimensions (width, length) are in micrometers (μm). (b) Photograph of the overdriven 10-W power amplifier MMIC. Chip size is 4.14×4.48 mm^2.

38%, respectively. Other measured performance values include noise figure less than 5.2 dB and AM-to-PM conversion at the P_{1dB} level about 1°/dB. In general, for class-B type HPAs, the AM-to-PM conversion values at the P_{1dB} level are much higher than 1°/dB. ■

10.1.2 Class-B Amplifier Design

The design of a class-B amplifier is similar to an overdriven HPA design except that the devices are biased for lower conduction angle values, that is, 180°–270° range. The optimum load value is close to class A. At RF and lower microwave frequencies, a push–pull configuration using low-loss broadband 180° baluns is frequently employed to reactively terminate even harmonics to reproduce the full sine waveform at the output. In principle, no intentional harmonic terminations are done. In this case, small-signal gain is lower than an overdriven HPA. In class-B amplifiers, gain expansion is observed frequently. Since the gain for a class-B amplifier is 6 dB

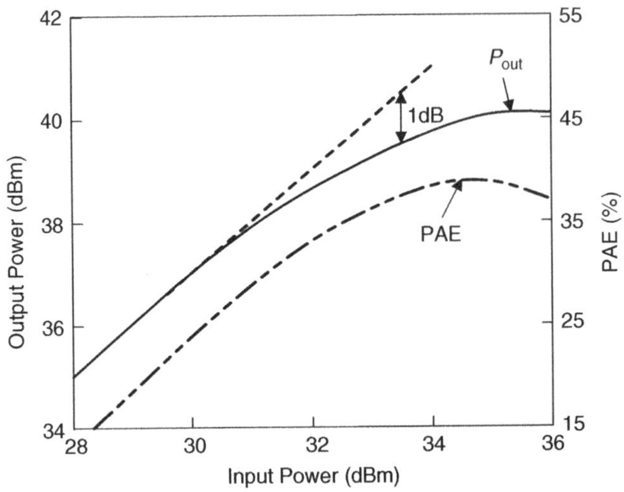

Figure 10.4 Output power and PAE of the overdriven 10-W power amplifier MMIC.

lower than for a class-A amplifier, at microwave frequencies, class AB is normally used. Design-wise both these amplifiers are similar. The design of class-B amplifiers has been reported extensively in the literature [3–15].

EXAMPLE 10.2

This is a design example of a personal communication network (PCN) power amplifier. Typical specifications for this type of amplifier are similar to those shown in Table 10.2.

SOLUTION The amplifier consists of two stages. Next, the amplifier design is described step by step. ■

FET Selection The design process begins with the selection of an appropriate FET type. The MESFET process used in this example is capable of sufficient power and

Table 10.2 PCN Power Amplifier Specifications Using Dual Power Supply

Frequency range	1.7–2.2	GHz
Gain	20	dB
Output power	34.5	dBm
PAE	45	%
Input return loss	2:1	VSWR
Load mismatch	($V_D = 5.5$ V, VSWR = 10:1, P_{IN} = Max)	No degradation in power output
Harmonics ($P_{OUT} = 34.5$ dBm)	−35	dBc
Stability	($P_{OUT} = -46$ to $+34.5$ dBm, P_{IN} controlled, $V_D = 2$–5.5 V, Load VSWR = 8:1)	All nonharmonically related outputs more than 70 dB below desired signal
Noise power	−64	dBm
Drain voltage, V_{ds}	3.5	Volts

efficiency to meet the above specifications. The ratio of the second- and first-stage FET peripheries is selected to be 4.2:1. This ratio is low enough to ensure that there is sufficient power from the first stage to saturate the final stage of the amplifier. A 21-mm MESFET is sufficient as a final stage to yield 34.5 dBm (134 mW/mm) at maximum operating efficiency. Load-pull results on smaller FET cells indicate that the MESFETs are capable of nearly 260-mW/mm power density with 13-dB gain (under compression) and 68% PAE at $V_{ds} = 5$ V. It is expected and is typical that as a smaller FET is scaled by 5–20 times, the power density drops significantly. The sensitivity of a large FET cell to source inductance, input matching, loss in the output match, and the phase of each gate finger cause a cumulative loss in gain and power density compared to a smaller FET cell.

The power transistor used in this design example is a self-aligned gate MESFET, which has a typical pinch-off voltage of −2.9 V. It has a peak current capability of 440 mA/mm and a 0.9-V knee voltage. Figure 10.5 shows the measured S-parameters, and S_{21} and calculated maximum available gain are shown in Figure 10.6. At 1.9 GHz the FET is capable of a stable small-signal gain of 22 dB. Figure 10.7 shows the measured load-pull contours for a 1-mm FET. At 3.5 V the MESFET is capable of a maximum power density of approximately 186 mW/mm as measured using an ATN load-pull system. The load impedance of a 1-mm FET at 3.5 V tuned for maximum power is $0.411\angle 161.5^\circ$. The maximum efficiency tuning impedance is $0.191\angle 47.1^\circ$. The input impedance required for maximum efficiency or power is $0.809\angle 77^\circ$.

Design Considerations The basic design here is a two-stage amplifier matched partially on and off chip. The off-chip match was chosen to reduce the GaAs chip size and to save cost. It also allows using off-chip components such as bias chokes and DC block and RF bypass capacitors. The off-chip matching technique is practical up to 3 GHz. The load of the 21-mm MESFET is a lowpass network supplied by the series inductance from the bond wires, package pins, and PC board output trace followed by a high-Q shunt capacitor. The design uses a 16-pin plastic thin-shrink small-outline package (TSSOP). Figure 10.9 shows the lumped-element equivalent model of the 40-mil wirebond and lead frame combination.

The load impedance from first- to second-harmonic frequencies is shown in Figure 10.10. The load impedance of the first-stage FET is shown in Figure 10.11 as a function of input power.

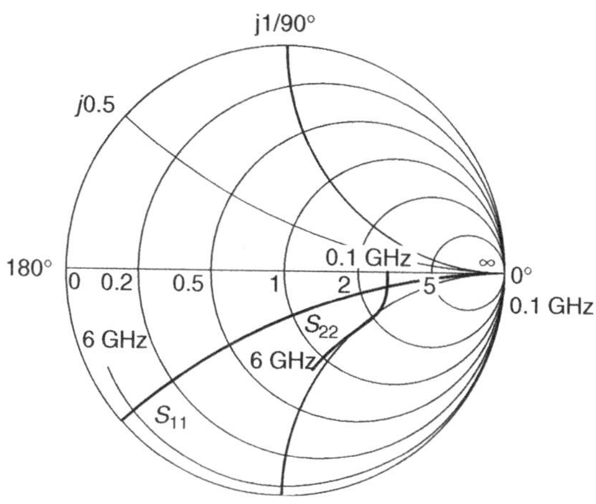

Figure 10.5 Simulated S-parameters of 1-mm power FET.

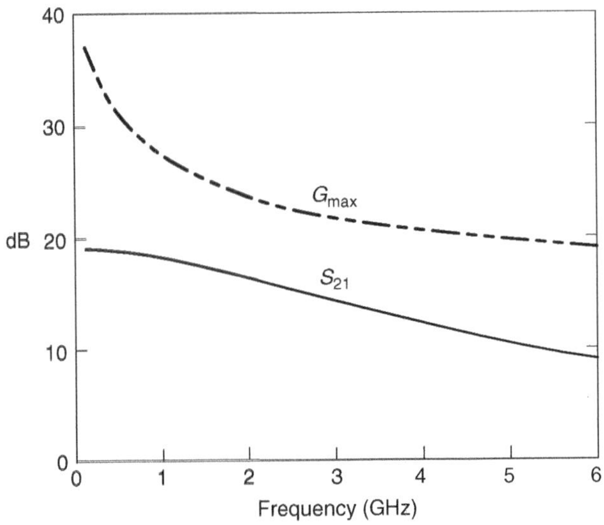

Figure 10.6 Simulated S_{21} and G_{max} of 1-mm power FET.

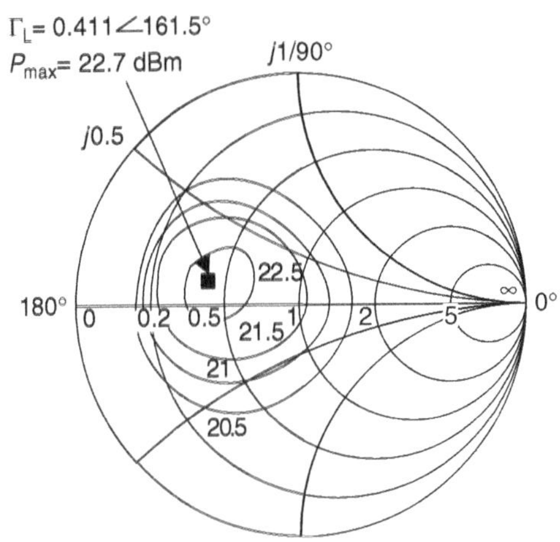

Figure 10.7 Measured power load-pull contours using 1-mm FET model.

The PA has an off-chip input match and a partial off-chip interstage matching network. This allows for minimum chip size and maximum tuning flexibility. This works for amplifiers operating below C-band frequencies. The input match can be constructed by a sufficient amount of line length and a shunt input capacitor. The interstage matching network shown below is generated by a series blocking capacitor and a shunt inductance off the first-stage drain. The shunt inductance is developed by multiple wirebonds in series with multiple lead frame pins terminated with a high-Q RF bypass capacitor. By selecting the appropriate number of wirebonds and pins, the amount of shunt inductance can be controlled to optimize the PA tune for specific requirements. The input and output tuning elements can also be optimized to maximum PA performance. Figure 10.12 shows the schematic diagram of the two-stage power amplifier and Figure 10.13 shows this amplifier with off-chip circuit elements. Gate–drain feedback is used to stabilize the first-stage FET. The MMIC chip layout is shown in Figure 10.14.

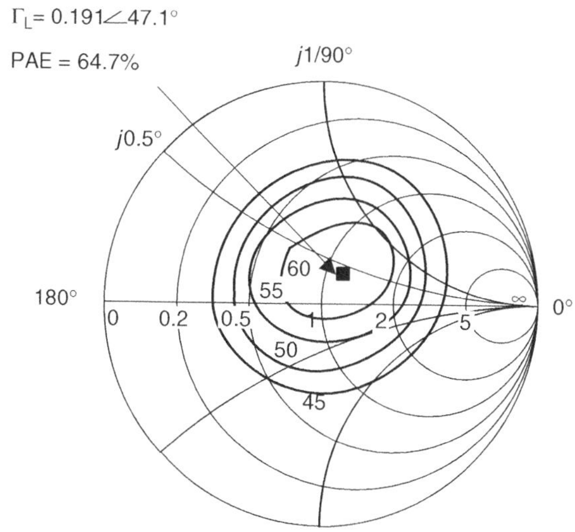

Figure 10.8 Measured PAE load-pull contours using 1-mm FET model.

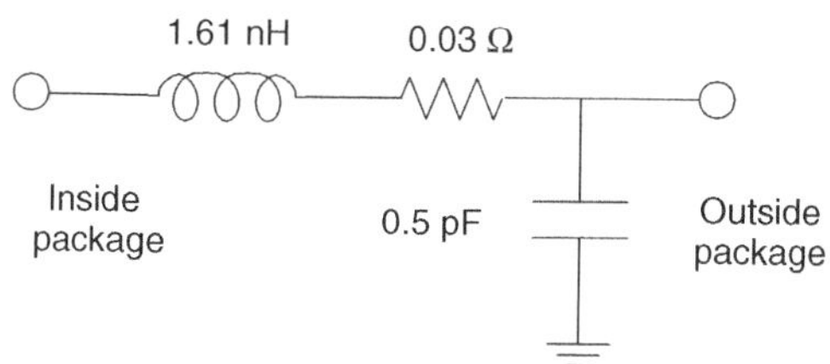

Figure 10.9 EC model for TSSOP wirebond/pin combination.

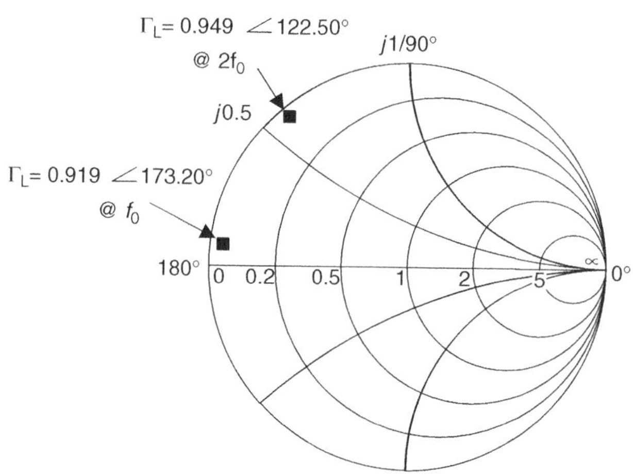

Figure 10.10 Simulated load impedance of the 21-mm FET.

Packaging Selection The MESFET amplifier chip is epoxied into a full downset TSSOP 16-pin plastic package. A high thermal conductivity epoxy is used to bind the chip to the lead frame paddle. The thermal conductivity of the epoxy is 80 W/m·°C. The chip is wire bonded to the lead frame with 1-mil diameter gold wire. The lead frame is soldered to an FR-4 prototyping board. Each lead frame pin sits on an 18-mil wide

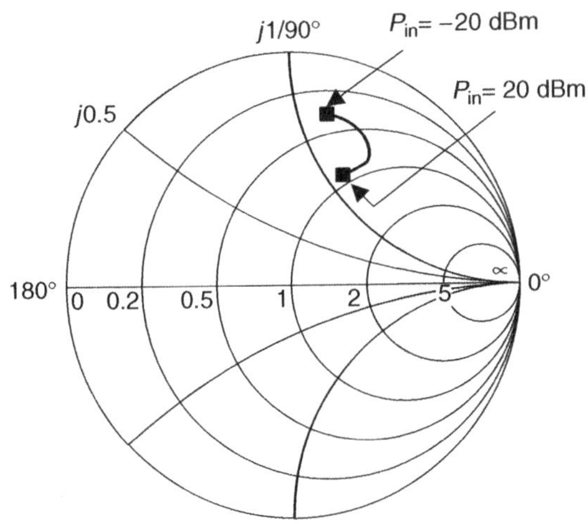

Figure 10.11 Simulated first-stage MESFET load as a function of input power.

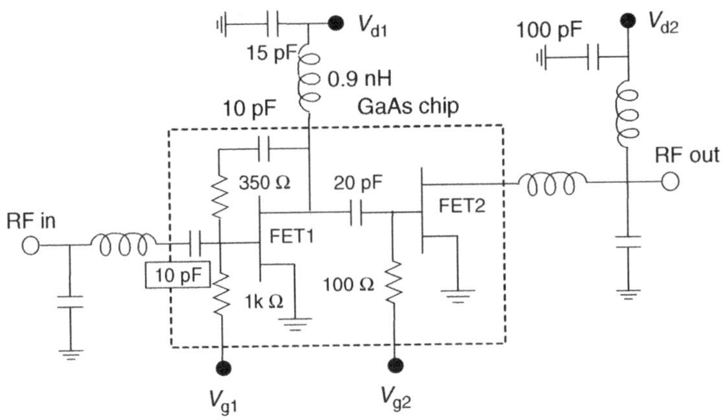

Figure 10.12 Functional schematic for the PCN two-stage PA.

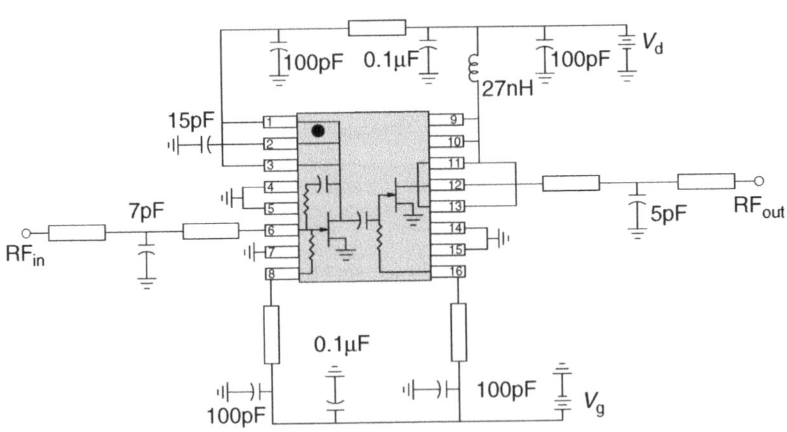

Figure 10.13 Schematic of PCN PA with off-chip elements shown.

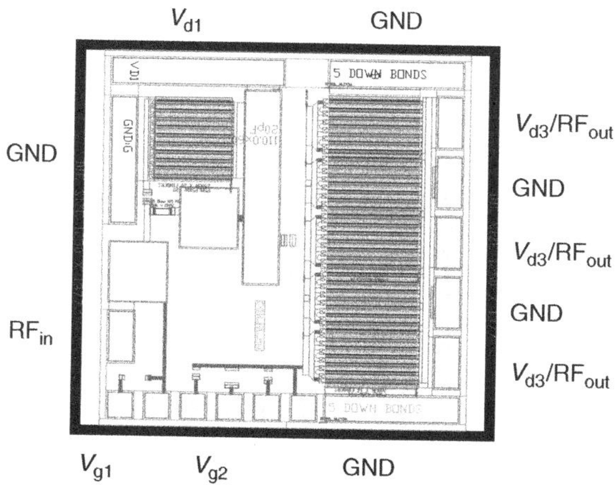

Figure 10.14 Physical layout of PCN PA.

by 50-mil pad from which microstrip or coplanar transmission lines are connected. Figure 10.15a shows the wiring of the MMIC chip into the plastic package. The open-faced packaged PA is shown in Figure 10.15b.

Board Properties The RF prototyping board is made from multilayer FR-4. The top dielectric layer is 10-mil thick. The top metal layer is made from 1-oz Cu (1.4 mils thick). The ground plane 10 mils below is made from 2-oz Cu (2.8 mils). The 10-mil thickness between the RF layer and the ground layer sets the width of a 50-Ω microstrip line to 17.5 mils wide. The total board thickness is set to 62 mils thick to be compatible with standard RF connectors. The 62-mil thick FR-4 is very rigid for withstanding benchtop tuning.

The permitivitty of FR-4 is approximately $\varepsilon_r = 4.3$ with a metal thickness ~ 1.4 mils. The loss tangent of microstrip lines on 10-mil thick FR-4 is $\tan \delta = 0.012$. Typical drill holes are specified to 14 mils in diameter. The actual prototyping board is shown in Figure 10.16. The board is solder plated for ease of assembly. The exposed backside paddle of the lead frame is solder pasted to the prototype board. The FR-4 board has a land area filled with 14-mil diameter vias to help conduct heat away from the lead frame. This helps keep the chip cooler within the package.

Simulated RF drain current, power, and PAE performance of the amplifier are shown in Figures 10.17, 10.18, and 10.19, respectively. The nature of the package/board provides enough simulated harmonic tuning that 68% PAE is achievable.

10.1.3 Class-E Amplifier Design

It was reported [16] that at low radiofrequencies class-E amplifiers have higher efficiency and better linearity than class-B, class-C, and class-F amplifiers. Until now, the class-E amplifiers were limited to the VHF band; however, recent interest in wireless applications has demonstrated that FETs/pHEMTs can be used as class-E devices at much higher radiofrequencies including the low end of microwave frequencies. The design of class-E amplifiers has been treated in numerous publications [16–35]. The theoretical analysis of class-E amplifier operation has been described in Chapter 8. A brief discussion on the design of such amplifiers is given next.

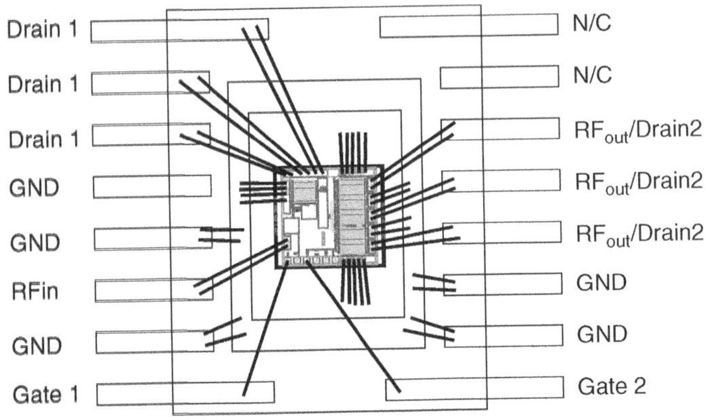

(a)

(b)

Figure 10.15 (a) Two-stage PCN PA bonding diagram. (b) Packaged PCN power amplifier in open cavity full downset TSSOP package.

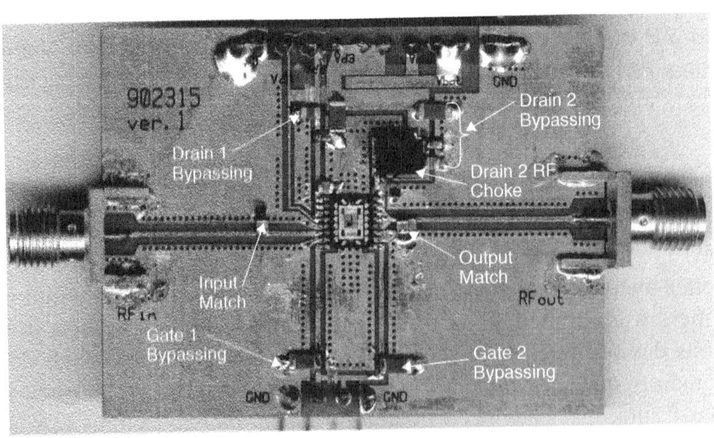

Figure 10.16 Example prototype board for the PCN power amplifier.

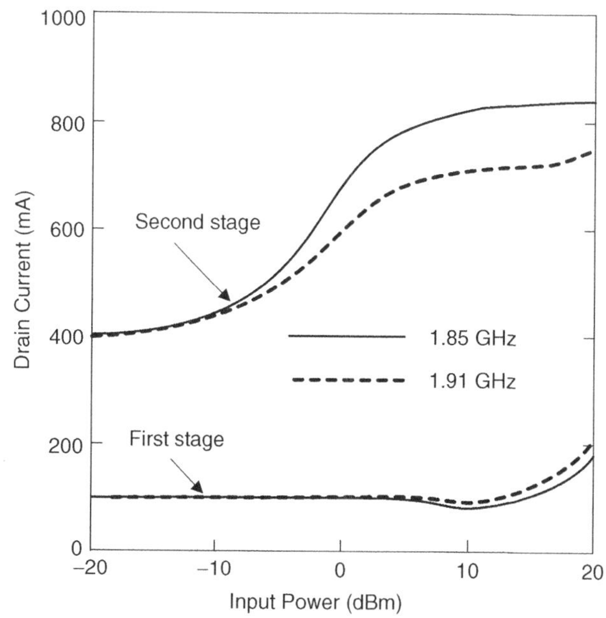

Figure 10.17 Simulated drain current of the PCN PA.

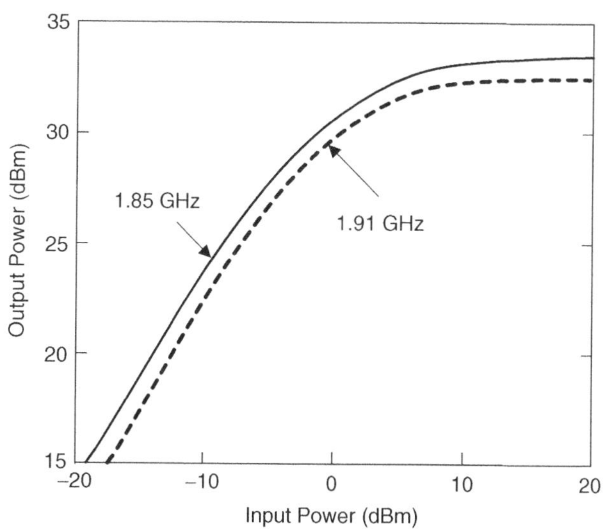

Figure 10.18 Simulated output power of the PCN PA.

Figure 10.20 shows a basic configuration of a class-E amplifier in which the device is represented by an ideal switch S. The device's output capacitance C_{ds} (this is the same as C_d used in Chapter 8) is in parallel with the switch. At the output, a series $L_0 - C_0$ tuned circuit, a reactive component jX, and a load R_L are all connected in series. By suitably selecting the bias conditions and input drive power level, the device is approximated as a switch, which turns on and off periodically at the RF input frequency. The series tuned circuit is resonant at the input frequency and thus passes

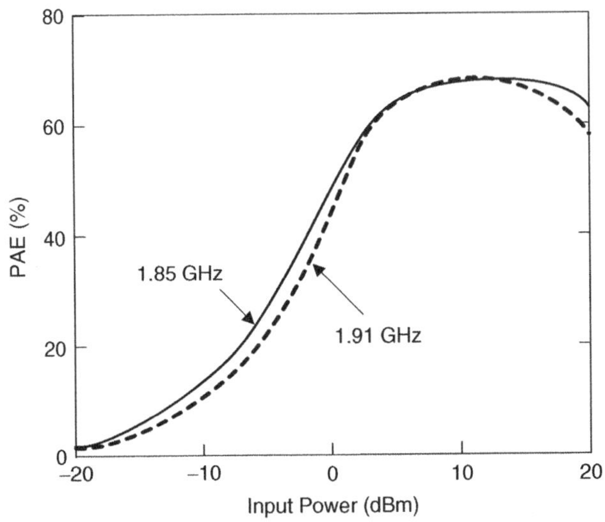

Figure 10.19 Simulated PAE of the PCN PA.

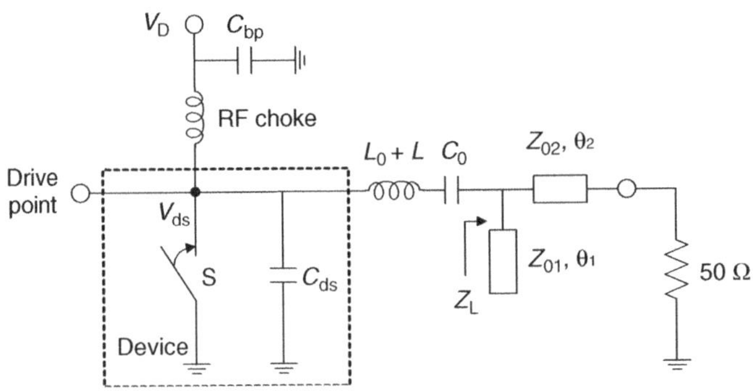

Figure 10.20 Schematic of a classical class-E amplifier impedance matching network.

the fundamental frequency signal to the load R_L. The reactive component jX adjusts the phase between the output voltage and switch voltage waveforms to get maximum efficiency. The drain efficiency of an ideal class-E amplifier is close to 100%.

The design equations for the class-E amplifiers are given in Chapter 8. The device's on-resistance R_s, output capacitance, maximum current, and operating voltage affect the amplifier's performance and the maximum frequency of class-E operation. Expressions for the maximum frequency f_{max}, load impedance Z_L, output power P_o, series inductor L, and drain efficiency η_D are given by [22, 23]

$$f_{max} = \frac{I_{max}}{56.5\, C_{ds} V_{ds}} \tag{10.1}$$

$$Z_L = (0.183 + j0.211)/(\omega_0 C_{ds}) = R_L + jX_L \tag{10.2}$$

$$P_o = 0.577 \frac{V_{ds}^2}{R_L} = 1.7337\, I_{ds}^2 R_L \tag{10.3}$$

$$L = 1.1525\ R_L/\omega_0 \tag{10.4}$$

$$\eta_D = \frac{1 + \left(\frac{\pi}{2} + \omega_0 C_{ds} R_s\right)^2}{\left(1 + \frac{\pi^2}{4}\right)(1 + \pi\ \omega_0 C_{ds} R_s)^2} \tag{10.5}$$

where V_{ds} is the supply voltage, I_{ds} is the device current, I_{max} is the open channel current of the device, and ω_0 is the angular operating frequency. The harmonics are assumed to be open circuited.

EXAMPLE 10.3

Figure 10.20 shows a schematic of a classical class-E amplifier with impedance matching network. The purpose of the impedance matching network is to transform 50 Ω to an optimum load impedance Z_L given by (10.2). The inductance L_0 with C_0 provides a series resonance at the fundamental frequency f_0. A single-stage class-E amplifier was designed which demonstrates very high PAE. The single-ended design consists of a single stage using a 1.0-mm FET. The FET was biased close to class B at 8 V and 15% I_{dss}. For this device, the C_{ds} and I_{max} values are 0.12 pF and 0.32 A, respectively. The f_T and f_{max} values are 16 GHz and 5.9 GHz, respectively. Note that f_T is for transistor and f_{max} is for class-E operation. The operating frequency f_0 was set to 5 GHz. The design was simulated using the nonlinear model given in Table 5.7 (Chapter 5), and the substrate was 75-μm thick GaAs. The design parameters are

$$Z_L = (0.183 + j0.211)/\omega_0 C_{ds} = 48.54 + j55.97\ \Omega$$
$$L = 1.1525 \times 48.54/31.4 = 1.7816\ \text{nH}$$

Total inductance $(L + L_0)$ is 3 nH. In this case,

$$L_0 = 1.2184\ \text{nH} \quad \text{and} \quad C_0 = 1/[(2\pi f_0)^2 L_0] = 0.8324\ \text{pF}$$
$$W_1 = 50\ \mu\text{m} \quad \ell_1 = 4000\ \mu\text{m}$$
$$W_2 = 14\ \mu\text{m} \quad \ell_2 = 4600\ \mu\text{m}$$

Using these design parameters the calculated output power and efficiency values were too low. So the value of L was tuned to obtain the optimum values of output power and efficiency. The tuned value of L is 0.7316 nH. In this design, the 1-mm FET was matched conjugately at the input. The simulated values for power gain, output power, drain efficiency, and PAE are 8.7 dB, 24.7 dBm, 77.3%, and 66.9%, respectively. The output power value is about 4 dB lower than the predicted value of 28.7 dBm. It has been reported that the low-frequency load given in (10.2) overestimates power at higher frequencies [22].

Next, a measured load impedance value of 56.25 Ω in parallel with $-C_{ds}$, for maximum PAE, was selected for the aforementioned transistor. The design parameters in this case are

$$Z_L = 53.83 + j11.42\ \Omega$$
$$L = 1.1525 \times 53.83/31.4 = 1.9758\ \text{nH}$$

In this case,

$$L_0 = 1.2184\ \text{nH} \quad \text{and} \quad C_0 = 0.8324\ \text{pF}$$
$$W_1 = 35\ \mu\text{m}, \quad \ell_1 = 5150\ \mu\text{m}$$
$$W_2 = 35\ \mu\text{m}, \quad \ell_2 = 2300\ \mu\text{m}$$

Using these design parameters the calculated output power and efficiency values were still too low. So the value of L was tuned to obtain the optimum values of output power and efficiency. The tuned value of L is 1.1816 nH. The simulated values for power gain, output power, drain efficiency, and PAE are 8.5 dB, 26.5 dBm, 79.2%, and 68%, respectively. In this case, the output power value is only about 2 dB lower than the predicted value of 28.7 dBm. ■

Table 10.3 provides a comparison between the measured and predicted performance of three class-E power amplifiers [23]. The device used was a 5-mm gate periphery FET biased at 6-V power supply. For this device, the C_{ds} and I_{max} values are 2.6 pF and 1.2 A, respectively. The load was calculated using (10.2).

There is reasonably good agreement between the measured and predicted performance except the output power at 2 GHz. This is due to the fact that the transistor's f_{max} value of 1.4 GHz is lower than the operating frequency.

The efficiency of the class-E amplifier was further enhanced by controlling the second and third harmonics [22]. This includes, as shown in Figure 10.21, quarter-wave open-circuited stubs at second- and third-harmonic frequencies. Here, the values of $\theta_1, \theta_2, \theta_3$, and θ_4 are selected such that the matching network at f_0 provides the desired load impedance at the switch location 1. At $2f_0$, θ_4 provides a short at location 3, which is transformed as a short at location 1 when $\theta_1 + \theta_3 = \lambda/2$. At $3f_0$, θ_2 provides a short at location 2, which is transformed as an open at location 1 when $\theta_1 = \lambda/4$.

10.1.4 Class-F Amplifier Design

At microwave frequencies, switched-mode class-E transistor amplifiers do not have good output power and efficiency due to the finite switching speed and the fact that voltage and current waveforms overlap. The output power decreases quickly with frequency as compared with other classes of amplifiers. Class-F amplifiers are frequently

Table 10.3 A Comparison of Measured and Predicted Performance of Three Class-E Amplifiers

Frequency (GHz)		0.5	1.0	2.0
Measured gain (dB)		15.3	14.7	9.1
P_o (W)	Predicted	0.77	1.35	2.07
	Measured	0.55	0.94	0.53
η_d (%)	Predicted	85	73	56
Measured		83	75	62
Measured PAE (%)		80	73	54

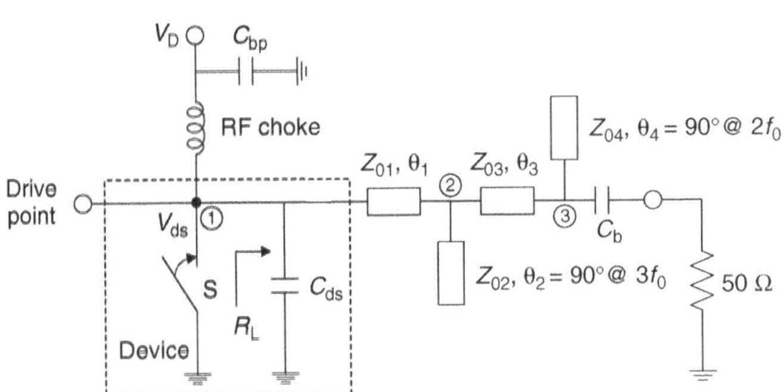

Figure 10.21 Harmonic tuned transmission line based class-E amplifier.

used to obtain high PAE at microwave frequencies. The theoretical analysis of class-F amplifier operation has been described in Chapter 8. A brief discussion on the design of such amplifiers is given next.

Class-F amplifiers dealing with harmonic tuning or termination have been studied extensively in the literature [36–56]. However, it is only possible to terminate two to three higher-order harmonics to obtain high efficiency at high frequencies mostly below 18 GHz. In addition to impedance matching networks, the harmonic tuned resonators for several harmonics result in additional loss in the output match. Since the contribution of harmonics higher than third order is not significant to warrant additional complexity and loss, in most practical cases only second and third harmonics are considered in high-efficiency amplifier design.

Harmonic Termination

In principle, in power amplifiers with suitable harmonic tuning at the input and output, one can increase both output power by 0.5–1.0 dB and PAE by 10–15%. The PAE can be improved by controlling the second or third or both harmonics. Due to device parasitic reactance and other parasitic reactance (e.g., discontinuities) of the matching networks at higher frequencies, accurate knowledge of harmonic termination under nonlinear operation of the device is required to determine the correct phase termination for harmonic signals. In this situation, a simple approach is used in which the load seen by the nonlinear output current source of the transistor is considered to be resistive; that is, its reactive part must be tuned out at all harmonics. In this case, the load is connected at the internal port of the transistor as shown in Figure 10.22. There are three conditions to be met to obtain maximum power and PAE:

1. Conjugate match at the input port A for maximum gain
2. Required load at the fundamental frequency transformed at the internal port for maximum power and PAE
3. Required load at harmonic frequencies transformed at the internal port for maximum PAE

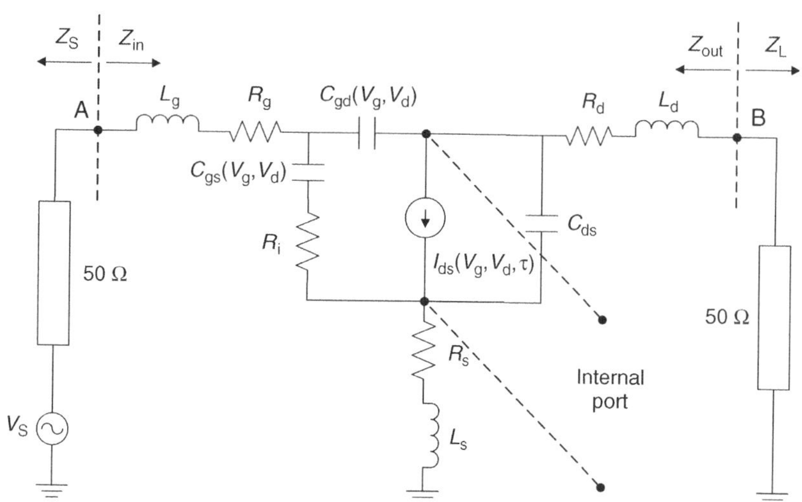

Figure 10.22 A nonlinear transistor representation for load.

Also, all three conditions have to be met over the operating frequency range. At the input port, short-circuit terminations for all even harmonics result in a square-wave voltage at the transistor's gate, which also provides higher PAE.

The output of the device with matching network representation is shown in Figure 10.23. Here Z_L is the transformed value of R_{opt} to be seen at the internal node. For given R_{opt}, C_{ds}, R_{ds}, R_d, and L_d, the value of Z_L may also be calculated. The function of the matching network is to transform 50 Ω to required optimum load Z_L. Output of an amplifier using a simple matching network without and with bias circuitry is shown in Figure 10.24.

There are several implementations of class-F amplifiers as shown in Figure 10.25. Both input and output harmonic terminations are required to enhance the amplifier's efficiency. There are numerous matching networks that may be used to provide the

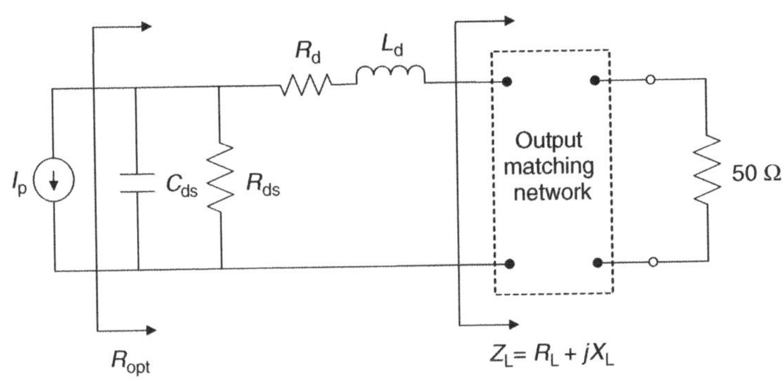

Figure 10.23 The device output with matching network representation.

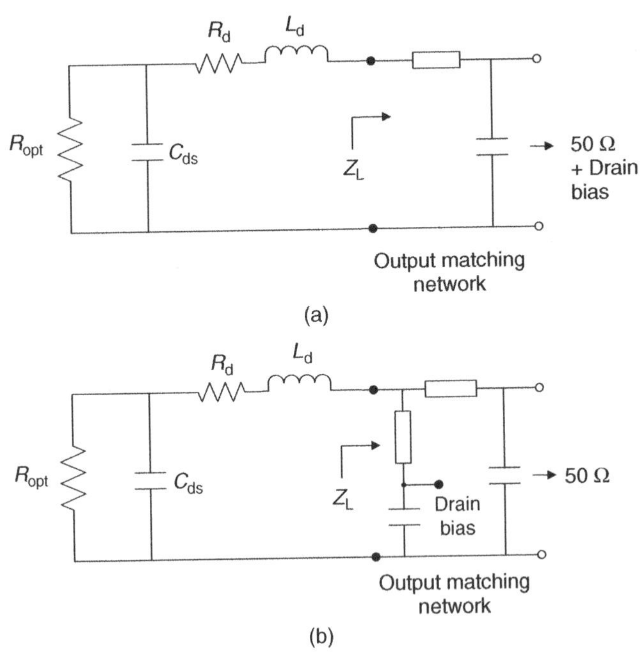

Figure 10.24 The output of the device with matching network representations: (a) without bias network and (b) with biasing network.

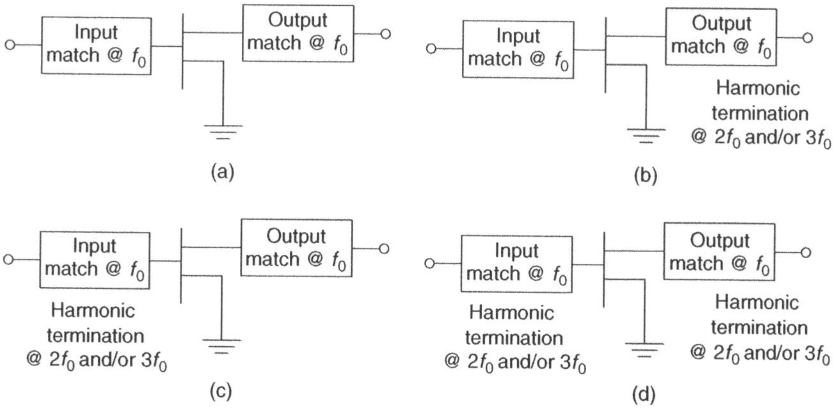

Figure 10.25 Harmonic tuned topologies: (a) no harmonic terminations, (b) second- and/or third-harmonic terminations at output, (c) second- and/or third-harmonic terminations at input, and (d) second- and or third-harmonic terminations at input and output.

desired termination conditions; however, extra care must be exercised such that these termination conditions are achieved without adding extra circuit loss and complexity. At C-band, one can improve PAE by 3–5% using second-harmonic termination at the input and 7–10% by using second- and third-harmonic terminations at the output. Altogether, 10–15% improvement in PAE is possible.

Figure 10.26 shows several possible output matching networks with harmonic terminations. In a class-F amplifier, the PAE is improved by using a multiple-resonator output matching network to control the harmonic power levels to shape the drain voltage and/or drain current waveforms. In an ideal design, at the device internal port, all even harmonics are short circuited to reproduce a half sine wave while the odd harmonics are open circuited to shape the output voltage to a square wave. Since there is no overlap between the output voltage and current, 100% drain efficiency is possible in an ideal case.

In Figure 10.26a, the output match consists of a series $\lambda/4$ @ f_0 line and a parallel inductor–capacitor resonator connected in shunt configuration. The value of C_0 is selected such that it provides a resonance with L_0 at the fundamental frequency f_0 and very low impedance at its harmonics. Above C-band, $C_0 = 5$ pF or higher meets this requirement. At f_0 the resonator provides an open circuit and the $\lambda/4$ line transforms 50 Ω to the required load impedance Z_L. At harmonic frequencies the resonator provides a short circuit and the $\lambda/4$ line transforms it to a short circuit at even harmonics and an open circuit at odd harmonics at the drain location. In Figure 10.26b, the output match consists of a short-circuited $\lambda/4$ @ f_0 stub and a series $\lambda/4$ @ f_0 line. The short-circuited line provides a short circuit at even harmonics only at the drain location. Figure 10.26c shows another scheme, which provides a required impedance match at f_0, a short (series resonance) at $2f_0$, and an open (parallel resonance) at $3f_0$. This scheme is good for the first two harmonics and may be sufficient for most practical cases. A single stub of any length cannot be used to provide a required termination at both even and odd harmonics. However, a tight coupled line approach, shown in Figure 10.26d, with a proper value of C may be used to provide a required impedance match at f_0, a short (series resonance) at $2f_0$, and an open (parallel resonance) at $3f_0$.

Figure 10.27 shows broadband (up to one octave) harmonic termination circuits: Figure 10.27a using a 180° coupler at the input and output and Figure 10.27b using a

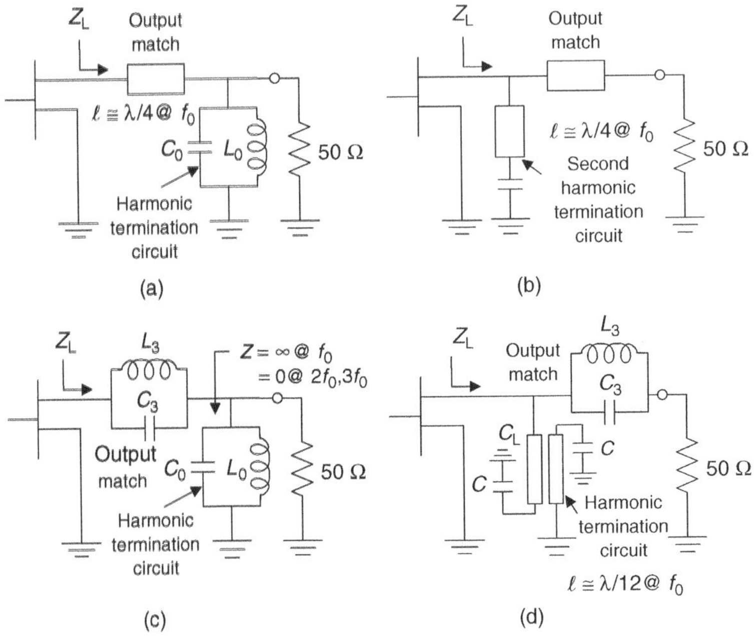

Figure 10.26 Output harmonic tuned configurations. Harmonic termination circuits: (a) second and third harmonics, (b) second harmonic, (c) second and third harmonics, and (d) second and third harmonics.

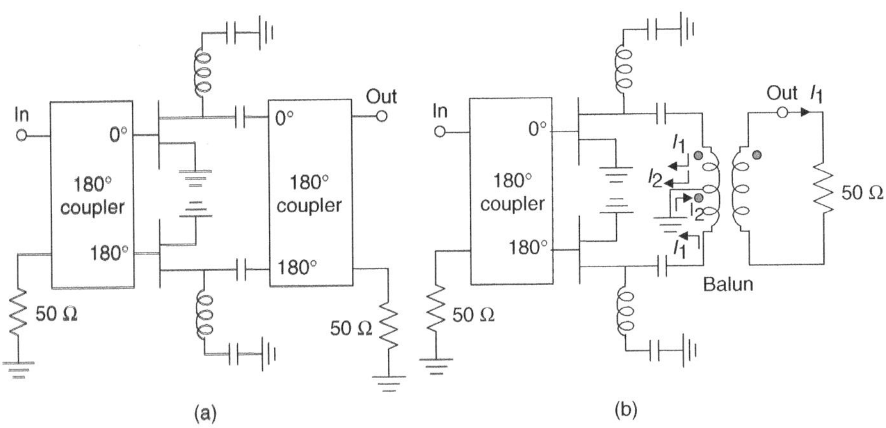

Figure 10.27 Broadband harmonic termination circuits: (a) using 180° coupler at input and output (b) using 180° coupler at input and balun at output.

180° coupler at the input and a balun/transformer at the output. The center-tap transformer at the output provides a short for the second- and higher-order even harmonic currents depending on the transformer design.

The design steps for class-F amplifiers are as follows:

1. The input is conjugate matched for maximum gain.
2. The conduction angle of the transistors for class-B amplifiers is between 180° and 270°.

3. The load impedance at the fundamental frequency is the same as for a class-B amplifier.
4. The parallel L_0, C_0 network (shunt) is resonant at f_0 and the parallel L_3, C_3 network (series) is resonant at the third harmonic $3f_0$. At f_0 the third-harmonic resonant circuit represents a small inductance that can be a matching element.
5. The level of the third-harmonic voltage at the drain of the transistor is set to 1/9 of its first-harmonic voltage level. See Section 8.6 for more details.
6. In a multistage amplifier, harmonic termination can also be selected. Several interstage loading networks have been described by Trask [43].

EXAMPLE 10.4

At RF and low microwave frequencies, for class-F amplifiers, the maximum achievable drain/collector efficiency is about 88.4%. Figure 10.28 shows a possible transmission line based implementation of such an amplifier. At the fundamental frequency the various electrical lengths are given by [45]

$$\theta_1 = \frac{\pi}{2}, \quad \theta_2 = \frac{1}{3}\tan^{-1}\left(\frac{1}{3Z_{02}\omega_0 C_{ds}}\right), \quad \theta_3 = \frac{\pi}{6} \tag{10.6}$$

where $\omega_0 = 2\pi f_0$. The first line short circuits the even harmonics and the third line provides the third-harmonic peaking. The characteristic impedances of these lines along with the fourth line transforms the 50 Ω to the required optimum load at the device output. The parameters of these lines are generally adjusted to achieve maximum PAE.

A single-stage class-F amplifier (Fig. 10.28) was designed at 5 GHz demonstrating very high PAE. The single-ended design consists of a 1.0-mm FET and the output matching network with second- and third-harmonic reactive terminations. The FET was biased close to class B at 8 V and 15% I_{dss}. The load impedance used was a parallel combination of 56 Ω and -0.3 pF. The design was simulated using the nonlinear model given in Table 5.7 and the substrate was 75-μm thick GaAs. The design parameters are

$$W_1 = 20\ \mu\text{m}, \quad \ell_1 = 4500\ \mu\text{m}$$
$$W_2 = 20\ \mu\text{m}, \quad \ell_2 = 3600\ \mu\text{m}$$
$$W_3 = 35\ \mu\text{m}, \quad \ell_3 = 1300\ \mu\text{m}, \quad \ell_4 = 0$$

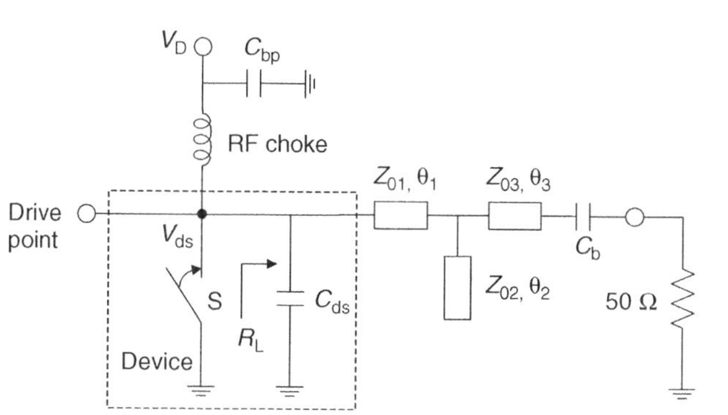

Figure 10.28 Harmonic tuned transmission line based class-F amplifier.

In this design, the 1-mm FET was matched conjugately at the input. The simulated values for power gain, saturated output power, and PAE are 10.1 dB, 27.1 dBm, and 65.3%, respectively.

EXAMPLE 10.5

Next, a class-F design example demonstrating very high PAE is described. The single-ended design consists of a single stage using 2.5-mm FET [3] having reactive terminations of higher-order harmonics. The FET was biased close to class B. The lumped-element model parameters (Fig. 5.2) for the 2.5-mm FET biased at 12 V and 5% I_{dss} are

$$R_g = 2\ \Omega,\ \ R_i = 0.4\ \Omega,\ \ R_s = 0.6\ \Omega,\ \ R_d = 0.3\ \Omega,\ \ R_{ds} = 70\ \Omega$$
$$C_{gs} = 2.5\text{ pF},\ \ C_{gd} = 0.1\text{ pF},\ \ C_{ds} = 0.2\text{ pF}$$
$$g_m = 150\text{ mS},\ \ \tau = 5\text{ ps}$$
$$L_g = 0.05\text{ nH},\ \ L_s = 0.08\text{ nH},\ \ L_d = 0.07\text{ nH}$$

An RF bypass capacitor terminated by two short-circuited stubs ($\lambda/4$ and $\lambda/6$ at 5.5 GHz) connected at the drain location reactively short the second and third harmonics to increase the overall efficiency of the circuit. The $\lambda/4$ stub does not affect the fundamental frequency performance. However, at twice the fundamental frequency the line becomes $\lambda/2$ long, providing low impedance to the second harmonic. The short-circuited $\lambda/6$ long stub is inductive at the fundamental frequency and is part of the matching network. At three times the fundamental frequency, this line becomes $\lambda/2$ long and reactively terminates the third-harmonic component. In this design, the 2.5-mm FET was matched to 50-Ω input and output.

The amplifier was designed as an MMIC on 75-μm thick GaAs substrate with via hole source grounds. The MIM capacitance density is 300 pF/mm^2. A schematic of the 1.5-W amplifier with design parameters is shown in Figure 10.29. Typical measured performance, at 5.5 GHz and 12 V power supply, for this amplifier is shown in Figure 10.30. The measured values of output power, PAE, gain, and second- and third-harmonic power levels are 1.7 W, 70%, 8 dB, −26 dBc, and −28 dBc, respectively.

In most cases, it is difficult to design load impedances that provide exactly the right termination conditions at input and output for all required harmonic frequencies. At high frequencies, for example, at Ku-band and above, in practice the improvement in PAE due to harmonic terminations is only a few percent. This improvement diminishes

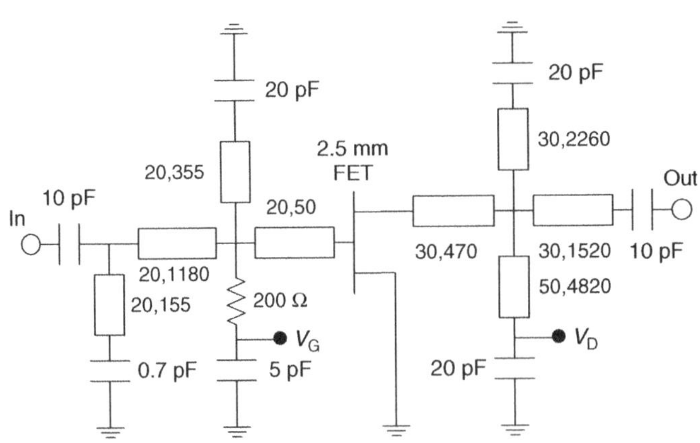

Figure 10.29 Schematic of 1.5-W class-F MMIC amplifier. All line dimensions (width, length) are in micrometers (μm).

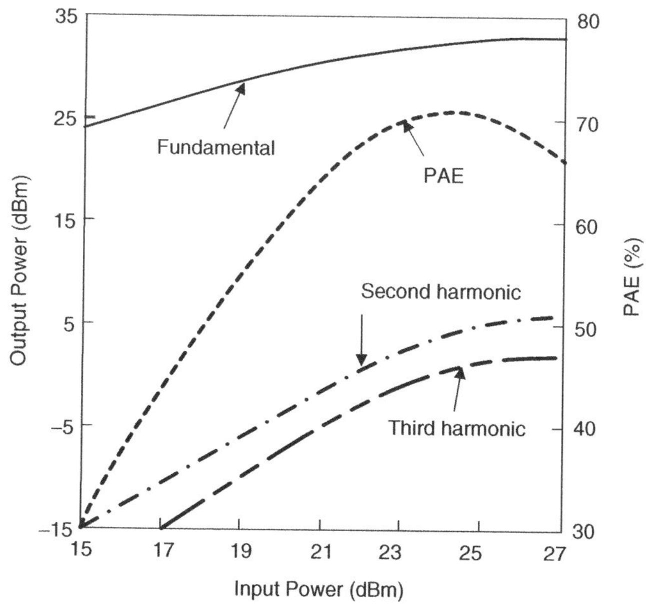

Figure 10.30 Output power for fundamental, second-, and third-harmonic frequencies, and PAE versus input power of a 1.5-W class-F MMIC amplifier.

due to loss and filter actions in the matching networks at second- and third-harmonic frequencies. Since the drain–source parasitic capacitive reactance is significant at high microwave frequencies, it is very difficult to tune out across the bandwidth and also simultaneously at fundamental, second-, and third-harmonic frequencies. However, at even-harmonic frequencies the device capacitive reactance provides some sort of short, which automatically results in higher PAE.

In the design of high-PAE power amplifiers, there are several factors that affect the PAE. At low frequencies, harmonic termination is very critical, while at high frequencies, low-loss match and accurate load are very important. In order to perform harmonic termination, one needs to generate the harmonic levels required to shape the drain voltage or current. This is achieved by operating devices close to pinch-off. Since at low frequencies the device gain is quite high and the device is operated for class AB (close to B), one still gets small-signal power amplifier gain larger than 12–14 dB. Also, to achieve high PAE one has to compress devices 3–4 dB. Such conditions generate enough harmonic levels for wave shaping and result in high PAE. However, at high frequencies the device is operated for class AB (more likely close to class A) in order to get high gain to realize high PAE. At high frequencies the amplifier gain is in the 8–10-dB range and the devices are compressed only 1–2 dB. Such conditions do not generate enough harmonic levels for wave shaping and to improve PAE. This is probably the reason why, for most transistors above Ku-band frequencies, the contribution in PAE due to harmonic termination is very small as compared to amplifiers operating at C-band and below.

Inverse Class F

An inverse class F, or class F^{-1}, is a dual version of class F. Here, the role of harmonic terminations is switched; that is, odd harmonics are open circuited and even harmonics

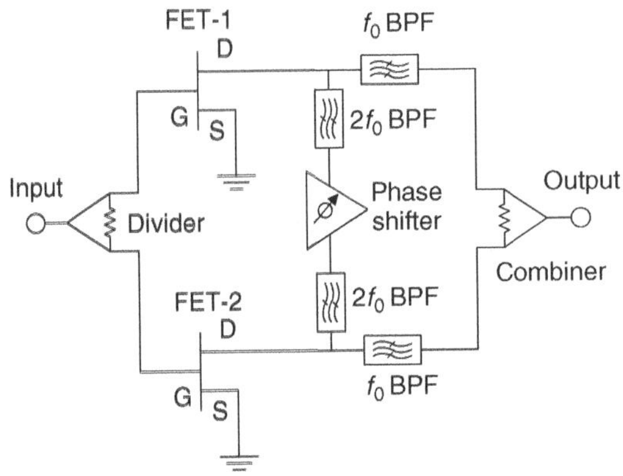

Figure 10.31 Schematic of the harmonic reaction amplifier.

are short circuited. In an ideal design, at the device output, all even harmonics are open circuited to shape the output current to a square wave, while the odd harmonics are short circuited to reproduce the output voltage as a half sine wave. Again, in an ideal class F^{-1}, 100% drain efficiency is possible. It has been reported that class F^{-1} has better efficiency than class F [55]. However, with device parasitic reactance and matching network loss and discontinuity effects, it may be difficult to demonstrate this finding at microwave frequencies.

There are several other techniques that have been employed to improve the PAE of power amplifiers. These are discussed briefly next.

10.2 HARMONIC REACTION AMPLIFIER

A basic circuit configuration for a harmonic reaction amplifier (HRA) is shown in Figure 10.31 [57, 58]. It consists of two single-ended amplifiers combined using reactive or Wilkinson power dividers/90° hybrids. The outputs of two FETs are connected by a transmission line path that has two second-harmonic bandpass filters. The output of each FET is matched to the desirable load impedance at the fundamental frequency and the transmission line path is matched at the second-harmonic frequency. When these devices are biased for class-AB/B operation, under RF drive conditions a large second-harmonic signal is generated at each FET's output and only flows into the second-harmonic path. Thus in this configuration, second-harmonic signals are mutually injected into other FETs without reflection. When both single-ended amplifiers have balanced characteristics, by adjusting the transmission line path phase, a voltage null or short circuit at each drain at the second-harmonic frequency is realized to achieve high efficiency. A PAE of 70% for a 2-GHz 5-W MESFET HRA has been demonstrated.

10.3 HARMONIC INJECTION TECHNIQUE

The PAE of an amplifier can also be improved by using a harmonic injection technique [59] as shown in Figure 10.32a. In this method, a sample of the input signal is taken, multiplied, amplified, and reinjected in proper phase into the drain side of the amplifier. The injected harmonic signals allow one to create multiharmonic standing waves to shape the RF output voltage at fundamental frequency to achieve maximum possible PAE.

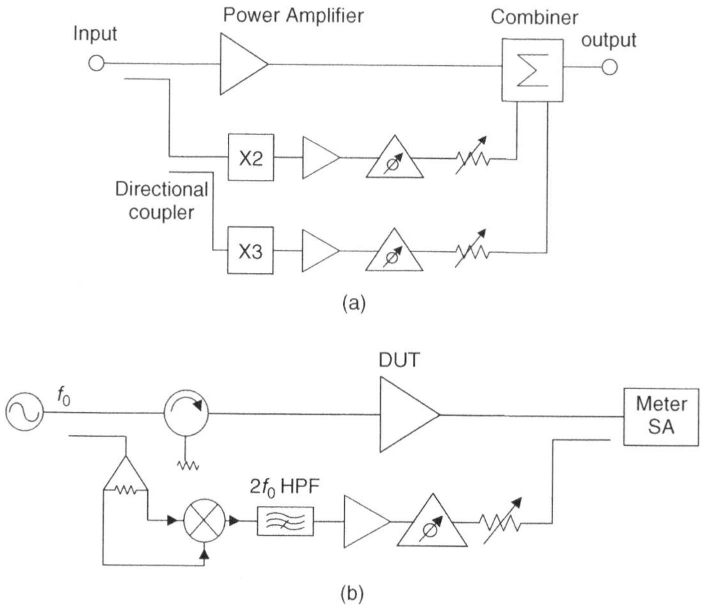

Figure 10.32 (a) Schematic of the harmonic injection amplifier in which individual phase and amplitude controlled paths are provided for the fundamental, the second, and the third harmonics prior to summation at the output. (b) Test setup used for second-harmonic injection HPA.

Figure 10.32b shows an experimental setup used at 1.5 GHz to demonstrate the validity of the harmonic injection technique. The device under test (DUT) is a hybrid amplifier using a 2.5-mm FET designed for maximum gain, output power, and PAE at 1.5 GHz and biased at 10 V. In this test system, a sample of the input signal is taken, its frequency is doubled, and it is reinjected into the drain side of the amplifier. The phase and amplitude of this harmonic are adjustable; that is, one can simulate any impedance level of the second-harmonic which is presented to the drain of the FET. The results of the measurement were dramatic [59] and given in Table 10.4. The second-harmonic levels are referenced to the output power when the second harmonic is zero. Utilizing a −12-dBc harmonic level, one could vary the PAE from 59% to 84%, confirming that proper termination of harmonics in high-efficiency amplifiers is extremely important. Inherently, this technique is narrowband and needs additional components for its realization.

10.4 HARMONIC CONTROL AMPLIFIER

This technique takes the advantageous features of class A and class B to realize high PAE with low distortion. The operation of the harmonic control amplifier (HCA) is described by Ingruber et al. [60, 61] and salient features are given here.

Figure 10.33 shows a schematic of the HCA that uses two stages. The first stage is biased for class B while the second stage is biased for class A. The output of the first stage is designed for diplexer action; that is, it has two output signal ports—f_0 and $2f_0$. This separates the fundamental and second-harmonic signals. The tuning network consists of attenuators and phase shifters for adjusting the correct magnitude and phase for f_0 and $2f_0$ signals, which are fed to the input of the second-stage amplifier through

Table 10.4 A Summary of Second-Harmonic Injection Experiments at 10 V

Second Harmonic Level		Q-Point I_{ds}	P_{out}	PAE	I_{ds}
(-dBc)	(W)	(% of I_{dss})	(W)	(%)	(mA)
0	0	5	0.851	62	131
9	0.107	5	1.148	81	137
12	0.054	5	1.023	75	133
15	0.027	5	0.977	72	132
0	0	10	0.955	59	158
9	0.120	10	1.259	84	147
12	0.060	10	1.148	77	147
15	0.030	10	1.096	72	151
0	0	25	0.977	55	174
9	0.123	25	1.259	62	202
12	0.062	25	1.230	74	166
15	0.031	25	1.122	67	163

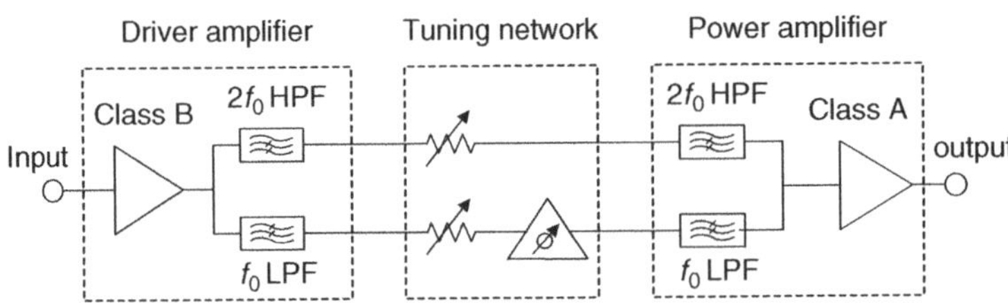

Figure 10.33 Schematic of the two-stage harmonic control amplifier.

another diplexer. This shapes the second-stage gate voltage to a square wave. This leads to class-F operation of the HCA with lower distortion. In two-tone measurements at $P_{1\,dB}$, the amplifier's single-carrier PAE was 64% and IM3 was −29 dBc. The saturated output power and PAE were 27.9 dBm and 71%, respectively.

10.5 HIGH-PAE DESIGN CONSIDERATIONS

In high-PAE power amplifier designs, there are several design features that must be considered carefully. These are harmonic terminations, high-PAE load match, and low-loss output matching networks. Several harmonic termination techniques were discussed in the preceding sections. These techniques work well over narrow bandwidth designs. However, if proper harmonic impedance termination is not provided, they might affect adversely the PAE, especially in broadband amplifiers. In such cases, the harmonics must be terminated in 50 Ω or the devices are operated close to class A, so that the magnitude of generated harmonics is very low. The selection of low-loss matching networks that provide the required high-PAE load match at the device interface is equally important for both narrowband and broadband amplifiers.

10.5.1 Harmonic Tuning Bench

Several techniques for harmonic tuning have been described previously. Next, we describe a harmonic tuning bench, which could be a test setup or a CAD with accurate

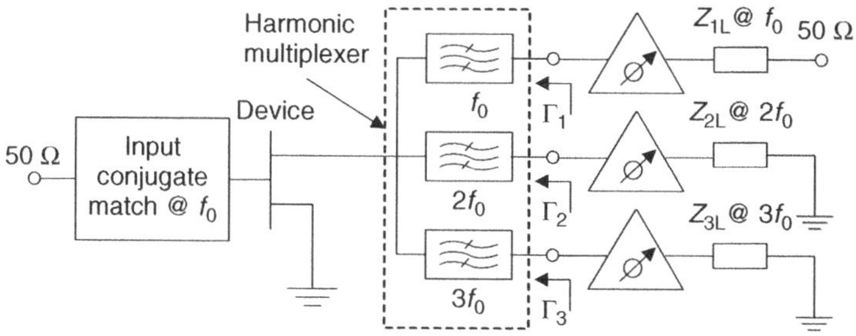

Figure 10.34 Schematic of a harmonic tuned power amplifier using a four-port multiplexer.

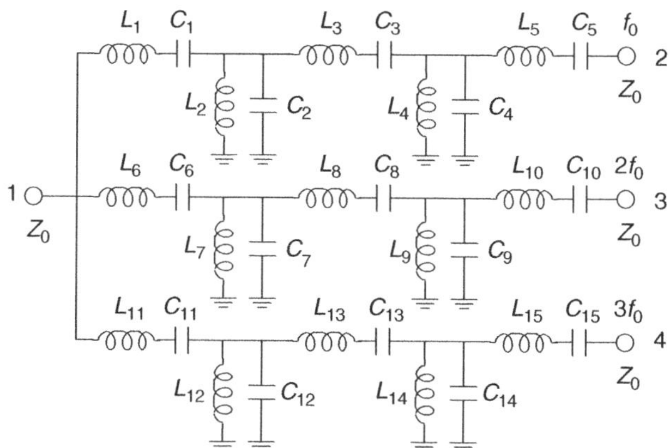

Figure 10.35 Schematic of a four-port multiplexer using lumped elements. All ports are matched to $Z_0 = 50\ \Omega$.

nonlinear model predicting the harmonic generations and their effect on the PAE. The purpose is to check independently the effect on the power amplifier's efficiency and the capability of the individual transistor for harmonic tuning. This can be done by using a four-port amplifier configuration, as shown in Figure 10.34. The four-port multiplexer may be implemented using ideal lumped-element filters or using ideal four-port S-parameters. Figure 10.35 shows a schematic of a four-port multiplexer that may be designed using lumped elements. All ports are matched to $Z_0 = 50\ \Omega$.

The four-port multiplexer is designed using ideal lumped elements and the element values at a fundamental frequency of 7 GHz are

$$L_1 = 0.001623,\ L_2 = 0.686835,\ L_3 = 19.74598,\ L_4 = 12.90148,\ L_5 = 1.373211$$
$$C_1 = 0.545355,\ C_2 = 0.533216\ C_3 = 0.026556,\ C_4 = 0.248627,\ C_5 = 9.817777$$
$$L_6 = 0.521705,\ L_7 = 0.262430,\ L_8 = 1.321280,\ L_9 = 0.077168,\ L_{10} = 0.444648$$
$$C_6 = 9.857342,\ C_7 = 0.715008,\ C_8 = 9.724642,\ C_9 = 1.854019,\ C_{10} = 9.736774$$
$$L_{11} = 0.589948,\ L_{12} = 0.032772,\ L_{13} = 3.721135,\ L_{14} = 20.04767,\ L_{15} = 2.976402$$
$$C_{11} = 0.414891,\ C_{12} = 1.859923,\ C_{13} = 0.296352,\ C_{14} = 0.037720,\ C_{15} = 9.638715$$

The units for inductors and capacitors are nH and pF, respectively. The simulated insertion loss of the multiplexer is shown in Figure 10.36. The input was matched at

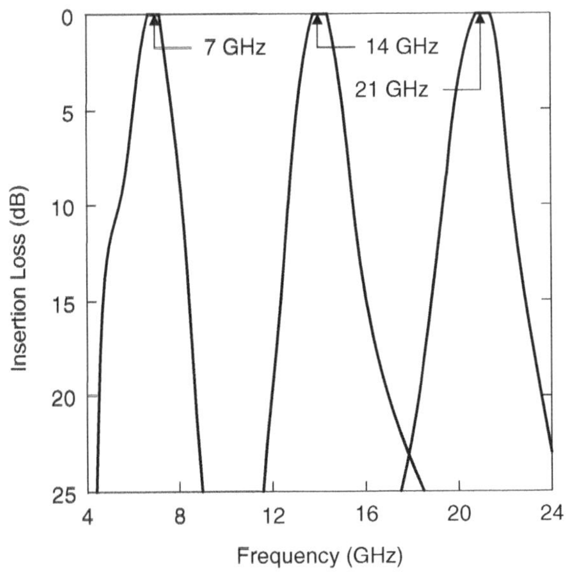

Figure 10.36 Simulated insertion loss of the four-port multiplexer.

all three frequencies, that is, at 7, 14, and 21 GHz, and the return loss was greater than 30 dB.

The above-mentioned test bench was used to study the effect of harmonic termination on an amplifier performance. At the fundamental frequency the output of the device was matched to optimum load impedance for power and PAE and the input was matched conjugately. The simulated performance of a 1.25-mm FET at 7 GHz biased at 8 V and 25% I_{dss} is

$$G_A = 10 \text{ dB}$$
$$P_o = 28.2 \text{ dBm}$$
$$\text{PAE} = 62\% \ @ \ \Gamma_2 = \Gamma_3 = 0$$
$$\text{PAE} = 70\% \ @ \ \Gamma_2 = 1 \angle -100^\circ \text{ and } \Gamma_3 = 1 \angle 140^\circ$$

Thus there is about 8% improvement in PAE due to second- and third-harmonic terminations. This device was measured at 9 GHz and measured gain and output power were 9.4 dB and 28.0 dBm, respectively. Measured and simulated PAE under various harmonic termination conditions at 9 GHz are compared in Table 10.5.

Table 10.5 Comparison Between Harmonic Tuned Simulated and Measured PAE

Simulated			Measured		
Γ_2	Γ_3	PAE (%)	Γ_2	Γ_3	PAE (%)
0	0	61	0.1	0.035	58
$1.0\angle -60^\circ$	0	66	$0.95\angle -80^\circ$	0.035	62.5
0	$1\angle 160^\circ$	63			
$1\angle -60^\circ$	$1\angle 160^\circ$	67.5			

Instead of the four-port multiplexer shown in Figure 10.35, one can use ideal four-port S-parameters for the multiplexer. The S-parameters at f_0, $2f_0$, and $3f_0$ are

$$[S]_{f_0} = \begin{bmatrix} 0 & 1 & 0 & 0 \\ 1 & 0 & 0 & 0 \\ 0 & 0 & 1 & 0 \\ 0 & 0 & 0 & 1 \end{bmatrix} \tag{10.7a}$$

$$[S]_{2f_0} = \begin{bmatrix} 0 & 0 & 1 & 0 \\ 0 & 1 & 0 & 0 \\ 1 & 0 & 0 & 0 \\ 0 & 0 & 0 & 1 \end{bmatrix} \tag{10.7b}$$

$$[S]_{3f_0} = \begin{bmatrix} 0 & 0 & 0 & 1 \\ 0 & 1 & 0 & 0 \\ 0 & 0 & 1 & 0 \\ 1 & 0 & 0 & 0 \end{bmatrix} \tag{10.7c}$$

In order to get some bandwidth one can repeat the S-parameters at $f_0 \pm \Delta f_0$, $2(f_0 \pm \Delta f_0)$, and $3(f_0 \pm \Delta f_0)$ corresponding to $f_0, 2f_0$, and $3f_0$, respectively. The use of S-parameters is more generic than using an actual multiplexer circuit.

10.5.2 Matching Network Loss Calculation

In a power amplifier design, the loss of the output matching network has a profound effect on the output power and PAE. When the device's impedance is low, that is, much less than Z_S or Z_L, the current densities become high, thus leading to high conductor loss in the matching networks. In other words, the very low impedance of the transistor output in a very high power amplifier results in high output match loss, which lowers the efficiency. Thus special care must be exercised to maintain a low conductor loss by using high-Q matching elements. One must select matching network elements as well as the circuit topology to realize a desired low-loss load impedance match. The output match loss is preferred to be within 0.2–0.3 dB but no greater than 0.5 dB for narrowband and less than 1 dB for ultra-wideband applications. In this section we will study the effect of circuit Q and impedance transformation ratio on the loss of the matching network [70].

Distributed Matching Networks

As discussed in Chapter 7, consider a quarter-wave impedance transformer with a characteristic impedance of Z_0. If the impedance transformation ratio is n and $n > 1$, then

$$n = Z_L/Z_S \tag{10.8a}$$

$$Z_0 = \sqrt{Z_L Z_S} = Z_S/\sqrt{n} \tag{10.8b}$$

Although the impedances are matched, there exist voltage standing waves on the quarter-wave line. If S_i is the input VSWR and S_o is the VSWR at the output, they can be expressed as

$$S_i = \frac{Z_0}{Z_S} = \sqrt{n} \tag{10.9a}$$

$$S_o = \frac{Z_L}{Z_0} = \sqrt{n} \tag{10.9b}$$

Or

$$S_i = S_o = S = \sqrt{n} \tag{10.9c}$$

Depending on the n value, the existence of standing waves increases the attenuation constant of the line. If α is the attenuation constant of the matched line, that is, the $\lambda/4$ line is terminated in Z_0 at the input and output, and α_s is the attenuation constant of the unmatched line (not terminated in its characteristic impedance), we have

$$\alpha_s = \alpha\frac{S^2+1}{2S} = \alpha\frac{n+1}{2\sqrt{n}} \tag{10.10}$$

When $n = 4$, $\alpha_s = 1.25\alpha$ and when $n = 16$, $\alpha_s = 1.875\alpha$. Thus the attenuation in the $\lambda/4$ line is increased by about 50%.

The power dissipated in the line of length ℓ when 1 watt of power is incident is given by

$$P_{diss} = 1 - e^{-2\alpha_s\ell} = 1 - e^{-\alpha_s\lambda/2} \tag{10.11}$$

When $\alpha_s\lambda/2 \ll 1$,

$$P_{diss} = \alpha_s\lambda/2 = \alpha\lambda\frac{n+1}{4\sqrt{n}} \tag{10.12}$$

Power dissipated (in dB) is given by

$$P_{diss} = 10\log\left[1 - \alpha\lambda\frac{n+1}{4\sqrt{n}}\right]^{-1} \tag{10.13}$$

where α is expressed in neper/cm and λ is expressed in cm. One neper is equal to 8.686 dB. Equation(10.12) can also be expressed in terms of quality factor Q of the line, which is given by [71]

$$Q = \frac{\pi}{\lambda\alpha} \quad \text{or} \quad \lambda\alpha = \frac{\pi}{Q} \tag{10.14}$$

From (10.12) and (10.13), we find

$$P_{diss} = 10\log\left[1 - \frac{\pi}{Q}\frac{n+1}{4\sqrt{n}}\right]^{-1} \tag{10.15}$$

For a $\lambda/4$ microstrip section having unloaded Q of 100 and $n = 10$, $P_{diss} = 0.12$ dB as compared to a loss of 0.068 dB for the $\lambda/4$ line terminated in its characteristic impedance. When $Q = 100$ and $n = 50$ and when $Q = 25$ and $n = 50$, P_{diss} becomes 0.253 dB and 1.11 dB, respectively.

Lumped-Element Matching Networks

Similar to (10.15) an approximate expression for the dissipated loss for a lowpass L-section lumped-element matching network (as discussed in Chapter 7) is reported in Reference 70 and is given as

$$P_{\text{diss}} = 10\log\left[1 - \left(\frac{1}{Q_L} + \frac{1}{Q_C}\right)\frac{n+1}{\sqrt{n}}\right]^{-1} \tag{10.16}$$

Where Q_L and Q_C are the quality factors of the series inductor and shunt capacitor, respectively. When $Q_L = Q_C = Q$,

$$P_{\text{diss}} = 10\log\left[1 - \frac{2}{Q}\frac{n+1}{\sqrt{n}}\right]^{-1} \tag{10.17}$$

For $Q_L = 100$, $Q_C = 30$, and $n = 10$, $P_{\text{diss}} = 0.71$ dB. A lower value of Q results in higher loss.

Comparison of (10.15) and (10.17) suggests that, for similar Q values, a distributed network has lower loss than a lumped-element network. Therefore for HPAs, a distributed type output matching network is generally used. Next, we describe how to reduce the loss in matching networks.

10.5.3 Matching Network Loss Reduction

Another area that has a lot of potential for PAE performance improvement is loss reduction in the passive matching structures. As discussed earlier, loss in the matching networks, especially the output matching network, reduces amplifier efficiency severely. Although it might be possible to use lumped elements to design a matching network, distributed elements have less loss at high frequencies. Loss in such distributed elements is due to the same mechanisms as loss in plain microstrip transmission lines: due to ohmic loss in the conductors, and loss in the dielectric, in this case, GaAs for MMICs and alumina for hybrid MICs.

Conductor Loss Reduction

Table 10.6 shows bulk properties of gold, silver, and copper. At high frequencies, current flows only on the surface of conductors, so making the plated conductor lines thicker will not reduce the RF loss. However, changing them from gold to copper/silver should decrease their loss. As shown in the table, the surface resistivity is about 20% lower for copper/silver than for gold. The dissipation loss in the microstrip matching networks at radiofrequencies can also be reduced using relatively thick conductors ($t > 5\delta$). Thick conductors also improve their current and power handling capabilities.

Table 10.6 Properties of Gold, Copper, and Silver

Element	Conductivity, σ (S/m)	Permeability, μ_0 (H/m)	Skin Depth, δ, at 14 GHz (μm)	Surface Resistivity, $1/(\sigma\delta)$, at 14 GHz (ohm/□)
Gold	4.1×10^7	$4\pi \times 10^{-7}$	0.66	0.037
Copper	5.8×10^7	$4\pi \times 10^{-7}$	0.56	0.031
Silver	6.17×10^7	$4\pi \times 10^{-7}$	0.54	0.030

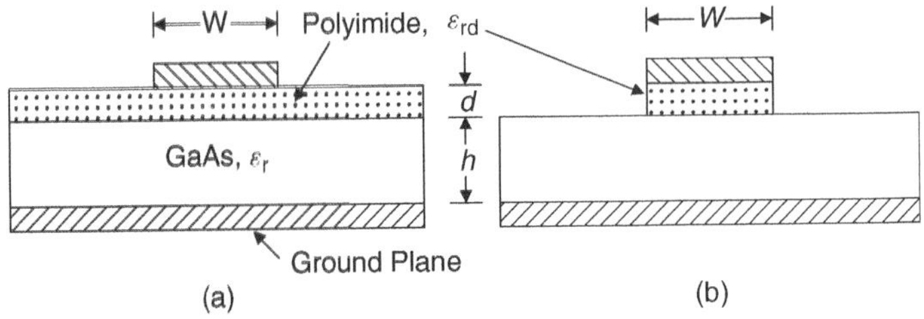

Figure 10.37 Low-loss microstrip configurations.

Substrate Thickness Increase

Another way to decrease the loss of microstrip lines is to increase the substrate thickness of MICs and MMICs. In GaAs MMICs, the microstrip loss decreases by about 35% when a 125-μm thick substrate is used instead of 75 μm. The problem with this approach is its impact on FET thermal resistance. Without any other modifications, the thermal resistance of a 625-μm FET increases from 138 °C/W to 167 °C/W, according to the Cooke model (Chapter 16). This means that to keep a maximum channel temperature of 150 °C, the base plate temperature must be reduced from a maximum of 80 °C to a maximum of 65.3 °C. However, for small-signal applications the thicker substrates result in lower noise figure.

Microstrip Structure Modification

In order to improve dissipative loss in the microstrip matching networks, a ridged microstrip structure [72] as shown in Figure 10.37 may be used. In this structure, the strip conductor is fabricated on a thin polyimide dielectric layer, which is placed on top of the GaAs substrate. In this case the electric flux lines are more in the air and the structure resembles a suspended microstrip line, which has much lower dissipative loss than the conventional microstrip. Microstrip line loss has been reported to be reduced by a factor of 2 [72] by placing a microstrip conductor on a thin, low-dielectric-constant material layer (14-μm thick polyimide) between the conductor and the GaAs substrate. Such lines have 25–30% lower line capacitance in comparison to their counterpart on GaAs. This is another desirable feature required to tune out the transistor capacitance over larger bandwidths.

This modified structure also helps in fabrication of the feed structure of the offset via FETs [11]. In HPA designs, low-loss microstrip lines on 10-μm thick polyimide have been used to lower the loss, especially in the output matching network. Since the lines on polyimide have poor thermal conductivity, a thermal design for such lines must also be performed [73, 74].

REFERENCES

1. D. M. Snider, A theoretical analysis and experimental confirmation of the optimally loaded and overdriven RF power amplifiers, *IEEE Electron Devices*, Vol. 14, pp. 851–857, December 1967.
2. I. J. Bahl et al., C-band 10W MMIC class-A amplifier manufactured using the refractory SAG process, *IEEE Trans. Microwave Theory Tech.*, Vol. 37, pp. 2154–2158, December 1989.

3. I. J. Bahl et al., Class-B power MMIC amplifiers with 70 percent power-added efficiency, *IEEE Trans. Microwave Theory Tech.*, Vol. 37, pp. 1315–1320, September 1989.
4. T. Yokoyama et al., High-efficiency low adjacent channel leakage GaAs power MMIC for 1.9GHz digital cordless phones, *IEEE Trans. Microwave Theory Tech.*, Vol. 42, pp. 2623–2628, December 1994.
5. K. Tateoka et al., A GaAs MCM power amplifier of 3.6V operation with high efficiency of 49% for 0.9GHz digital cellular phone system, *IEEE MTT-S Int. Microwave Symp. Dig.*, pp. 569–572, 1994.
6. W. L. Pribble and E. L. Griffin, An ion-implanted 13 watt C-band MMIC with 60% peak power added efficiency, *IEEE Microwave Millimeter Wave Monolithic Circuits Symp. Dig.*, pp. 25–28, 1996.
7. C. F. Campbell, A fully integrated Ku-band Doherty amplifier MMIC, *IEEE Microwave Guided Wave Lett.*, Vol. 9, pp. 114–116, March 1999.
8. E. L. Griffin, X-band GaAs MMIC size reduction and integration, *IEEE MTT-S Int. Microwave Symp. Dig.*, pp. 709–712, 2001.
9. I. Bahl, Design of a generic 2.5W, 60 percent bandwidth, C-band MMIC amplifier, *Microwave J.*, Vol. 45, pp. 54–70, August 2002.
10. T. Shimura et al., A high power density, 6W MMIC for Ku/K-band applications, *IEEE MTT-S Int. Microwave Symp. Dig.*, pp. 851–854, 2003.
11. I. Bahl, Ku-band MMIC power amplifiers developed using MSAG MESFET technology, *Microwave J.*, Vol. 49, pp. 56–82, February 2006.
12. M/A-COM, *RF and Microwave Product Solutions*, 1011 Pawtucket Blvd., Lowell, MA.
13. TriQuint Semiconductor, 13510 North Central Expressway, Dallas, TX.
14. Mimix Broadband Inc. 10795 Rockley Road, Houston, TX.
15. Hittite Microwave Corporation, 20 Alpha Road, Chelmsford, MA.
16. N. O. Sokal and A. D. Sokal, Class E—a new class of high-efficiency tuned single-ended switching power amplifiers, *IEEE J. Solid-State Circuits*, Vol. SC-10, pp. 168–176, June 1975.
17. F. H. Raab, Idealized operation of the class E tuned power amplifier, *IEEE Trans. Circuits Syst.*, Vol. CS-24, pp. 725–735, December 1977.
18. T. B. Mader and Z. B. Popovic, The transmission-line high-efficiency class-E amplifier, *IEEE Microwave Guided Wave Lett.*, Vol. 5, pp. 290–292, September 1995.
19. T. Sowlati et al., Low voltage, high efficiency GaAs class E power amplifiers for wireless applications, *IEEE J. Solid State Circuits*, Vol. 30, pp. 1074–1079, October 1995.
20. J. F. Davis and D. B. Rutledge, A low-cost class-E power amplifier with sine-wave drive, *IEEE MTT-S Int. Microwave Symp. Dig.*, pp. 1113–1116, 1998.
21. N. O. Sokal, Class E high-efficiency power amplifiers, from HF to microwave, *IEEE MTT-S Int. Microwave Symp. Dig.*, pp 1109–1112, 1998.
22. F. J. Ortega-Gonzalez et al., High-efficiency load-pull harmonic controlled class-E power amplifier, *IEEE Microwave Guided Wave Lett.*, Vol. 8, pp. 348–350, October 1998.
23. T. B. Mader et al., Switched-mode high efficiency microwave power amplifiers in a free-space power-combiner array, *IEEE Trans. Microwave Theory Tech.*, Vol. 46, pp. 1391–1397, October 1998.
24. M. Markovic, A. Kain, and Z. Popovic, Nonlinear modeling of class-E microwave power amplifiers, *Int. J. RF Microwave Computer-Aided Design*, Vol. 9, pp. 93–103, March 1999.
25. V. S. Rao Gudimetla and A. Z. Kain, Design and validation of the load networks for broadband class E amplifiers using nonlinear device models, *IEEE MTT-S Int. Microwave Symp. Dig.*, pp. 823–826, 1999.
26. A. Mediano and P. Molina, Frequency limitation of a high-efficiency class E tuned RF power amplifier due to a shunt capacitance, *IEEE MTT-S Int. Microwave Symp. Dig.*, pp. 363–366, 1999.
27. D. K. Choi and S. I. Long, A physically based analytic model of FET class-E power amplifiers—designing for maximum PAE, *IEEE Trans. Microwave Theory Tech.*, Vol. 47, pp. 1712–1720, September 1999.

28. W. H. Cantrell, Tuning analysis for the high-Q class-E power amplifier, *IEEE Trans. Microwave Theory Tech.*, Vol. 486, pp. 2397–2402, December 2000.
29. C. Yoo and Q. Huang, A common-gate switched 0.9-W class-E power amplifier with 41% PAE in 0.25-μm CMOS, *IEEE J. Solid State Circuits*, Vol. 36, pp. 823–830, May 2001.
30. F. H. Raab, Class-E, class-C, and class-F power amplifiers based upon a finite number of harmonics, *IEEE Trans. Microwave Theory Tech.*, Vol. 49, pp. 1462–1468, August 2001.
31. T. K. Quach et al., Ultrahigh efficiency power amplifier for space radar applications, *IEEE J. Solid State Circuits*, Vol. 37, pp. 1126–1134, September 2002.
32. A. Grebennikov, Switched-mode RF and microwave parallel-circuit class E power amplifiers, *Int. J. RF Microwave Computer-Aided Design*, Vol. 14, pp. 21–35, January 2004.
33. A. Grebennikov, *RF and Microwave Power Amplifier Design*, McGraw-Hill, New York, 2004.
34. D. P. Kimber and P. Gardner, Power series analysis of the class E power amplifier, in *European Microwave Conference Proceedings*, pp. 1461–1464, 2004.
35. Y. Qin et al., Design of low-cost broadband class-E power amplifier using low voltage supply, *Microwave Opt. Tech. Lett.*, Vol. 44, pp. 103–106, January 2005.
36. T. Nojima, S. Nishiki, and K. Chiba, High-efficiency quasimicrowave GaAs FET power amplifier, *Electron. Lett.*, Vol. 23, pp. 512–513, May 1987.
37. M. Easton, R. Basset, D. S. Day, C. Hua, C. S. Chang, and J. Wie, A 3.5 watt high-efficiency GaAs FET amplifier for digital telephone communications, *IEEE MTT-S Int. Microwave Symp. Dig.*, pp. 1183–1184, 1992.
38. C. Duvanaud, S. Dietsche, G. Pataut, and J. Obrgon, High-efficient class F GaAs FET amplifiers operating with very low bias voltages for use in mobile telephones at 1.75GHz, *IEEE Microwave Guided Wave Lett.*, Vol. 3, pp. 268–270, August 1993.
39. S. Toyoda, High-efficiency single and push-pull power amplifiers, *IEEE MTT-S Int. Microwave Symp. Dig.*, pp. 277–280, 1993.
40. S. Toyoda, High efficiency amplifiers, *IEEE MTT-S Int. Microwave Symp. Dig.*, pp. 253–256, 1994.
41. F. H. Raab, Introduction to class-F power amplifiers, *R.F. Design*, Vol. 19, pp. 79–84, May 1996.
42. F. H. Raab, Class-F power amplifiers with maximally flat waveforms, *IEEE Trans. Microwave Theory Tech.*, Vol. 45, pp. 2007–2012, November 1997.
43. C. Trask, Class-F amplifiers loading network: a unified design approach, *IEEE MTT-S Int. Microwave Symp. Dig.*, pp. 351–354, 1999.
44. P. Colantonio et al., On the class-F power amplifier design, *Int. J. RF Microwave Computer-Aided Design*, Vol. 9, pp. 129–149, March 1999.
45. A. Grebennikov, Circuit design technique for high efficiency class F amplifiers, *IEEE MTT-S Int. Microwave Symp. Dig*. pp. 771–774, 2000.
46. P. Colantonio et al., Class G approach for low-voltage, high-efficiency PA design, *Int. J. RF Microwave Computer-Aided Design*, Vol. 10, pp. 366–378, November 2000.
47. P. Colantonio et al., Multiharmonic manipulation for highly efficient microwave power amplifier, *Int. J. RF Microwave Computer-Aided Design*, Vol. 1, pp. 366–384, November 2001.
48. F. H. Raab, Maximum efficiency and output of class F power amplifiers, *IEEE Trans. Microwave Theory Tech.*, Vol. 47, pp. 1162–1166, June 2001.
49. P. Colantonio et al., Theoretical facet and experimental results of harmonic tuned PAs, *Int. J. RF Microwave Computer-Aided Design*, Vol. 13, pp. 459–472, November 2003.
50. H. Jager et al., Broadband high-efficiency monolithic InGaP/GaAs HBT power amplifiers for wireless applications, *Int. J. RF Microwave Computer-Aided Design*, Vol. 13, pp. 496–510, November 2003.
51. M. Spirito et al., Power amplifier PAE and ruggedness optimization by second-harmonic control, *IEEE J. Solid-State Circuits*, Vol. 38, pp. 1575–1583, September 2003.

52. S. Gao et al., High-efficiency power amplifier design including input harmonic termination, *IEEE Microwave Wireless Components Lett.*, Vol. 16, pp. 81–83, February 2005.
53. A. Wakejima et al., C-band GaAs FET power amplifiers with 70-W output power and 50% PAE for satellite communication use, *IEEE J. Solid-State Circuits*, Vol. 40, pp. 2054–2060, October 2005.
54. S. Gao, High efficiency class F RF/microwave power amplifiers, *IEEE Microwave Mag.*, Vol. 7, pp. 40–48, February 2006.
55. Y. Y. Woo et al., Efficiency comparison between highly efficient class-F and inverse class-F amplifiers, *IEEE Microwave Mag.*, Vol. 8, pp. 100–110, June 2007.
56. C. Roff, J. Benedikt, and P. J. Tasker, Desgn approach for realization of very high efficiency power amplifiers, *IEEE MTT-S Int. Microwave Symp. Dig.*, pp. 143–146, 2007.
57. S. Nishiki and T. Nojima, Harmonic reaction amplifier—a novel high-efficiency and high-power microwave amplifier, *IEEE MTT-S Int. Microwave Symp. Dig.*, pp. 963–966, 1987.
58. T. Nojima and S. Nishiki, High-efficiency microwave harmonic reaction amplifier, *IEEE MTT-S Int. Microwave Symp. Dig.*, pp. 1007–1010, 1988.
59. D. A. Willems, E. L. Griffin, I. J. Bahl, and M. D. Pollman, High efficiency harmonic injection power amplifier, US Patent #5,172,072, December 15, 1992.
60. B. Ingruber et al., High efficiency harmonic control amplifier, *IEEE Trans. Microwave Theory Tech.*, Vol. 46, pp. 857–862, June 1998.
61. B. Ingruber et al., Rectangularly driven class-A harmonic-control amplifier, *IEEE Trans. Microwave Theory Tech.*, Vol. 46, pp. 1667–1672, November 1998.
62. M. Akkul et al., 50 Watt MMIC power amplifier design for 2GHz applications, *IEEE MTT-S Int. Microwave Symp. Dig.*, pp. 1355–1358, 2004.
63. S. Piotrowics et al., High power and high efficiency 30W compact S-band HBT power chips with gold or diamond heat spreaders, *IEEE MTT-S Int. Microwave Symp. Dig.*, pp. 1527–1530, 2004.
64. H. Minamide et al., 70% High efficiency C-band 27W hetero-structure FET for space application, *IEEE MTT-S Int. Microwave Symp. Dig.*, pp. 621–623, 2002.
65. D. Conway, M. Fowler, and J. Redus, New process enables wideband high-power GHz amplifiers to deliver up to 20W, *Defense Electronics*, pp. 8–11, February 2006.
66. Q. Zhang and S. A. Brown, Fully monolithic 8W Ku-band high power amplifier, *IEEE MTT-S Int. Microwave Symp. Dig.*, pp. 1161–1165, 2004.
67. K. Kong et al., Ka-band MMIC power amplifier (4W at 30GHz) with record compact size, *IEEE Compound Semiconductor IC Symp. Dig.*, pp. 232–235, 2005.
68. M. V. Aust et al., A 2.8-W *Q*-band high-efficiency power amplifier, *IEEE J. Solid State Circuits*, Vol. 41, pp. 2241–2247, October 2006.
69. Y. C. Chen et al., A 95-GHz InP HEMT MMIC amplifier with 427-mW power output, *IEEE Microwave Guided Wave Lett.*, Vol. 8, pp. 399–401, November 1998.
70. J. V. Dilorenzo and D. D. Khandelwal (Eds.), *GaAs FET Principles and Technology*, Artech House, Norwood, MA, 1982.
71. K. C. Gupta, R. Garg, I. Bahl, and P. Bhartia, *Microstrip Lines and Slotlines*, 2nd edition, Artech House, Norwood, MA, 1996.
72. I. J. Bahl et al., Low loss multilayer microstrip line for monolithic microwave integrated circuits applications, *Int. J. RF Microwave Computer-Aided Eng.*, Vol. 8, pp. 441–454, November 1998.
73. KELDIST, M/A-COM's Internal Software.
74. I. J. Bahl, Average power handling capability of multilayer microstrip lines, *Int. J. RF Microwave Computer-Aided Eng.*, Vol. 11, pp. 385–395, November 2001.

PROBLEMS

10.1 Describe the pros and cons of various harmonic tuning matching networks shown in Figure 10.26.

10.2 Determine the ratio of the drain current to the Q-point current for a class-B FET power amplifier biased at 10% of I_{max}. The amplifier is driven for maximum output power.

10.3 A single-stage class-E amplifier is designed using two pHEMT devices connected in parallel. The EC model for the pHEMT is given in Table 5.10, #1.The device set is biased at 8 V, 100 mA and has a peak current of 600 mA. The input is matched for maximum gain and the output match is shown in Figure 10.20. Calculate the small-signal gain and maximum possible output power and PAE at 3.5 GHz. The amplifier is conditionally stable. Use 15-mil thick alumina substrate.

10.4 Repeat Problem 10.3 when the Figure 10.21 output match is used.

10.5 A single-stage class-F power amplifier is designed at 2.4 GHz using a 0.9-mm pHEMT whose EC model is given in Table 5.11. The device is biased at 12 V and 10% I_{dss}. The output stage design uses Figure 10.26a topology. The second harmonic is short circuited and the third harmonic is open circuited at the internal port of the device. The equivalent load impedance at the fundamental frequency is $R_L = 50\ \Omega$ and $C_L = -0.28$ pF. Determine output match parameters using 15-mil alumina substrate.

10.6 In Problem 10.5, calculate the gain, maximum output power, and PAE when the input is matched for maximum gain. The peak current (I_{max}) is 400 mA and is 30% higher than I_{dss}.The output power for the device is 1 W.

10.7 Repeat Problem 10.5 for the output stage design (a) in Figure 10.26c and (b) in Figure 10.26d.

10.8 The equivalent load impedance of a MESFET at 5 GHz is $R_L = 56\ \Omega$-mm and $C_L = -0.32$ pF/mm. The output matching networks were designed for three amplifiers using different transistor sizes: 1 mm, 10 mm, and 50 mm. Calculate the loss for each output matching network on a 15-mil alumina substrate.

Chapter 11

Broadband Amplifier Techniques

Many systems require broadband amplifiers. The purpose of this chapter is to describe basic types of broadband amplifier configurations and provide a critical assessment of each type in terms of noise, power, and frequency range performance.

Over the past two decades tremendous progress has been made in the design of transistor low-noise and power amplifiers. The amplifiers discussed up to this point have narrow bandwidth, less than 50% bandwidth. Several applications such as electronic warfare and broadband communications require multi-octave amplifiers. The design of high-power amplifiers with bandwidth greater than 50% poses a significant challenge because of low device impedances and thermal limitations due to low PAE. Although hybrid MIC technology, described in Chapter 14, can be used to develop broadband power amplifiers, power MMIC amplifiers, in general, offer smaller size and light weight, higher gain, wider bandwidth, higher reliability, lower cost, and much better unit-to-unit amplitude and phase tracking capability when manufactured in large volume. Monolithic technology, treated in Chapter 15, is particularly beneficial to broadband amplifiers due to the elimination of the parasitic effects of bond wires and discrete components used in hybrid MICs. Progress in LNAs is covered in Section 15.5, Chapter 15, and advances in broadband MMIC power amplifiers [1–10] is summarized in Table 11.1.

11.1 TRANSISTOR BANDWIDTH LIMITATIONS

In this section we briefly discuss the transistor bandwidth limiting factors encountered in the design of LNAs and HPAs.

11.1.1 Transistor Gain Roll-off

The transistor's gain decreases with frequency as discussed in Chapter 5. The gain slope is usually 4–6 dB per octave. However, in most amplifiers one requires flat gain

Fundamentals of RF and Microwave Transistor Amplifiers. By Inder J. Bahl

Table 11.1 Summary of Broadband MMIC Power Amplifiers[a]

Frequency Range (GHz)	Number of Stages	Gain (dB)	P_o (W)	PAE (%)	Device Technology	Year
2–8	1	9	1.4	18	GaAs HBT	2000
2.5–5.5	2	17	2	30	GaAs MESFET	2003
4.5–9	2	17	2	25	GaAs MESFET	2003
4.7–10	1	7	5	8	GaN on Sapphire	2000
6–18	3	22	2.3	20	GaAs pHEMT	2002
0.7–2.7	2	20	12	22	GaAs MESFET	2006
1.35–2.8	2	23	12	28	GaAs MESFET	2006
2.0–6.0	2	15	10	26	GaAs MESFET	2006
2.0–8.0	2	13.5	8	16	GaAs MESFET	2006

[a] Performance listed is minimum over the frequency band. MMICs with greater than 1 W were selected for this comparison.

response over the designed bandwidth. Thus in power amplifiers, one has to design the matching networks that provide the desired load at the output of each device as well as the required insertion loss slope for gain compensation to realize flat gain and maximum output power and PAE. In low-noise amplifiers, the input match is designed for optimum noise figure, interstage is matched for flat gain, and the output is designed for good match. In other words, the matching networks should have higher loss at lower frequencies with a loss slope matching the device's gain slope. Since, in power amplifiers, the output matching network is designed usually for output power matching, the input stage and interstages take care of all required gain compensation with minimum loss in the total gain at the highest operating frequency in the band. For example, a two-stage power amplifier working over an octave band requires 8–12-dB gain compensation. Usually half of such gain compensation is realized in the input and the other half in the interstage. At low frequencies the input stage may be designed with negative feedback for constant gain. The matching elements are selected so that the gain loss at the high end of the band is made as low as possible. Thus the LLM technique described in Chapter 9 is a very suitable method in designing broadband HPAs. The insertion loss required for each stage to compensate for the device gain slope and to obtain a flat amplifier gain is adjusted by maintaining the power and PAE at optimum values.

11.1.2 Variable Device Input and Output Impedance

In addition to gain slope dependence on frequency, the transistor's input and output impedances also change with frequency and decrease with increasing frequency as described in Chapter 5. The impedance variations have also to be matched over the designed frequency range. As discussed in Chapter 7, a lower Q matching network has wider bandwidth. Thus the low-Q gain compensation circuits in the input and interstage matching networks have high loss and make it possible to design such networks for wide bandwidths. In a multistage power amplifier, however, this requires larger device sizes in the preceding stages that lower the PAE. On the other hand, the output matching network requires high-Q components for high power and PAE and poses a serious challenge to design an output match to transform the 50 Ω to required load impedance over multi-octave bandwidth.

11.1.3 Power–Bandwidth Product

The additional loss due to gain compensation in the interstage matching network in a multistage amplifier and the loss in the output matching network over the wide bandwidth reduce the device aspect ratio and output power, resulting in lower PAE. Thus high PAE and high output power levels possible over a narrow frequency band are not realizable for broadband HPAs and result in lower power–bandwidth product. For a given device size, as compared to a narrowband design, a multistage multi-octave bandwidth HPA design may have reduced output power and PAE by a factor of 2. Similarly, a multi-octave bandwith LNA has about 50% higher noise figure in comparison to a narrowband LNA. For example, the NF of a narrowband LNA is 1.0 dB at 10 GHz and 2.7 dB for a 2–18 GHz LNA. Thus nonconventional matching techniques are needed to effectively design multi-octave band LNAs and high-power and high-PAE amplifiers with improved performance.

11.2 BROADBAND AMPLIFIER TECHNIQUES

There are various circuit topologies, as shown in Figure 11.1, used to realize broadband amplifiers. These techniques include traditional reactive/resistive matching, shunt or parallel resistive feedback, balanced configuration, and the traveling-wave approach. The reactively matched amplifier uses purely reactive matching networks at the input and output of the transistor; either lumped inductors and capacitors or transmission lines can be used. The technique gives moderate bandwidth and good noise and power performance. However, because of the transistor's inherent instability and gain roll-off, wideband design is difficult. The reactive/resistive technique uses resistors within its matching networks to lower their Q and to achieve flat gain over a broad bandwidth. The most typical topology is to employ resistors in series with shunt lines or inductors. At low frequencies the shunt lines have little electrical length, and the resistors load the transistor and lower its gain. At high frequencies the resistors have little effect on

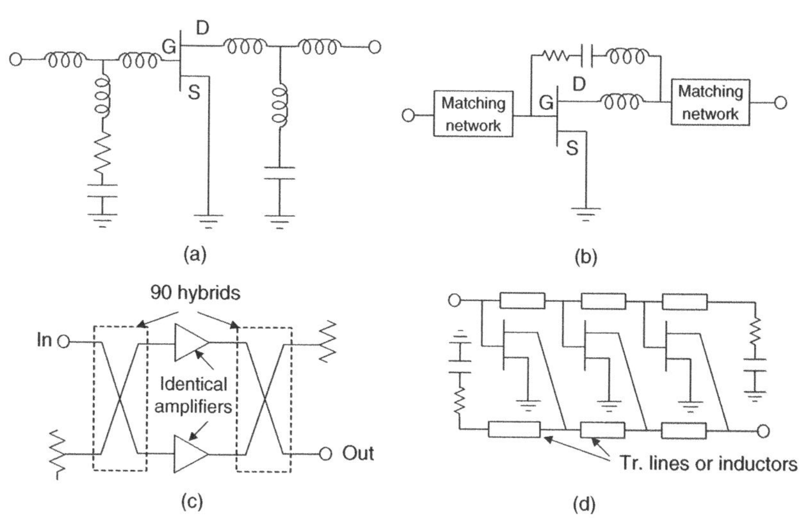

Figure 11.1 Broadband amplifier configurations: (a) reactive/resitive, (b) feedback, (c) balanced, and (d) distributed or traveling wave.

the transistor because of the inductive effect of the shunt lines. Hence the matching networks can introduce a positive gain slope to compensate the transistor's gain roll-off. The reactive/resistive topologies are suitable for bandwidths less than one octave and require more matching elements to implement ultra-wide bandwidths. They also have poor gain flatness and VSWR performance.

The feedback amplifier uses a resistance (on the order of hundreds of ohms) from the gate to the drain. This has the effect of stabilizing the device and can make the input and output impedances much closer to the desired 50 Ω. The resistive feedback configuration results in higher noise figure and lower power density per device due to an additional loss in the feedback resistor. However, this scheme is good for gain flatness, VSWR, and multi-octave bandwidth performance. The balanced configuration has good gain flatness and VSWR performance up to about a two-octave bandwidth. The amplifier noise figure is slightly higher and PAE is slightly lower than a single-ended version due to the additional 0.3–0.8 dB-loss in the coupler. The traveling-wave or distributed power amplifier approach has excellent gain–bandwidth characteristics with flat gain and low VSWR, and has the capability of multi-octave bandwidth—however, with limited power output (2–4 W up to 18 GHz) and poor PAE (on the order of 10%). The distributed approach provides the best gain–bandwidth product while the reactive/resistive matching method is more suitable for high power and high PAE. These techniques are discussed in this chapter.

11.2.1 Reactive/Resistive Topology

The reactive/resistive matching techniques for broadband applications are similar to those described in Section 9.3.1 except they require more matching elements for wider bandwidths. Power and buffer amplifiers require gain compensation networks (GCNs) in the input and interstage. However, for low-noise amplifier applications one might use GCNs in the interstage and output matching networks. Figure 11.2a shows a schematic for a two-stage amplifier. In most applications both the matching network (MN) and GCN are blended together. Figure 11.2b shows some commonly used gain compensation circuits. Component values chosen are simply to demonstrate broadband operation over 1–8 GHz.

Gain performance of the compensation circuits shown in Figure 11.2b are illustrated in Figure 11.3. The values of circuit elements are selected so that the attenuation is about 6–10 dB over a three-octave bandwidth. Some of the attenuation is due to reflected power. However, at the high end GCNs are reasonably well matched. Networks (i) and (iv) have similar responses. Networks (ii) and (iii) have higher attenuation values but their slopes are different. The behavior of network (v), not shown in Figure 11.3, is similar to (iv). Network (vi) has a maximum attenuation at the series resonance of the inductor–capacitor combination. The behavior of such networks may vary slightly depending on the frequency and the type of *LRC* component values and their parasitic reactances. The shape and the attenuation as per design required can be adjusted by a proper selection of the component values. First-hand analysis helps in selecting MN and GCN topologies; however, in most cases one has to depend on CAD tools to finalize the circuit component values.

EXAMPLE 11.1

This example provides some salient features of a broadband 12-W MMIC HPA. The design was done using the MSAG MESFET based MMIC process [9]. The HPA design was selected

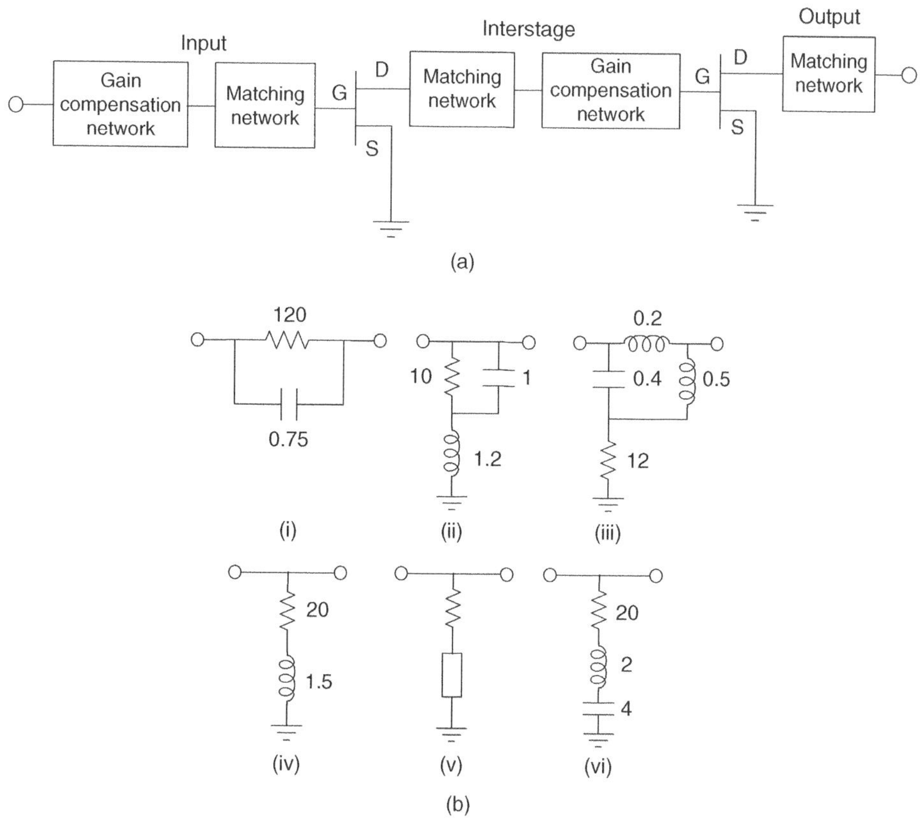

Figure 11.2 (a) Schematic of a two-stage amplifier with gain compensation networks. (b) Commonly used gain compensation circuits. Resistors are in ohms (Ω), capacitors in picofarads (pF), and inductors in nanohenries (nH).

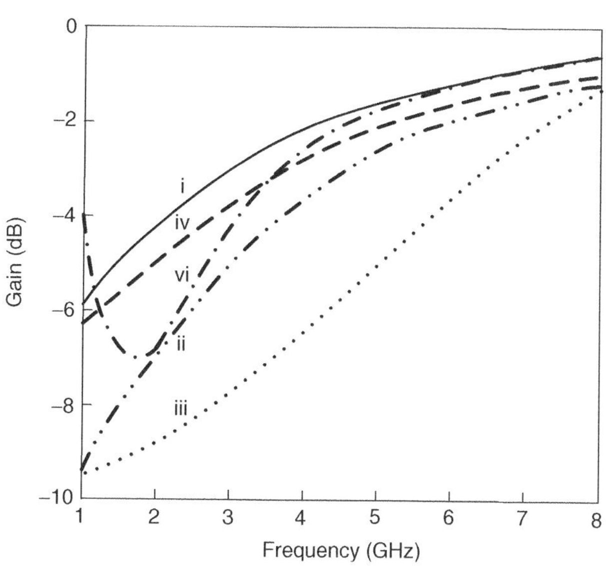

Figure 11.3 Gain response of compensation circuits shown in Figure 11.2b.

as a two-stage fully monolithic class AB operated at a nominal power supply voltage of 10 V. The two-stage amplifier is designed to cover the frequency range from 0.7 to 2.7 GHz and to have good input match. The MSAG MESFET technology was used to develop broadband power amplifiers because of its high across-wafer uniformity, excellent linearity, and low-cost capabilities [9].

SOLUTION The design of the MMIC power amplifier starts with the selection of unit FET sizes based on the gain, PAE/linearity, and output power requirements. The two-stage HPA design consists of four 2.0-mm FETs at the input, driving sixteen 2.0-mm FETs at the output. The 2-mm FET cell has a gate–gate pitch of 54 μm, 6 fingers, and unit gate width of 333.3 μm. The calculated thermal resistance of the FET is 35.8 °C/W. The measured load (Z_L) for this FET is equivalent to a parallel combination of resistor (R_L) and capacitor (C_L) of values 28 Ω and −0.61 pF, respectively. The design was performed using the LLM technique and load-line method described in Chapter 9 and small-signal model for the 2-mm FET obtained at the operating bias point. In this method, the load-line technique is used to optimize the circuit parameters. For example, in a two-stage HPA, the optimum load impedances $Z_{L1}$and Z_{L2} at the drain of the first- and second-stage FETs, respectively, to realize maximum output power and PAE, are shown in Figure 11.4a. The value of Z_{L1} and Z_{L2} used in the design are (7 Ω, −2.44 pF) and (1.75 Ω, −9.76 pF), respectively. In the low-loss matching (LLM) design technique, both the resistive or dissipative loss (DL) and mismatch loss (ML) for each stage are calculated and controlled as required in the design. Generally, DL and ML for the output match are kept

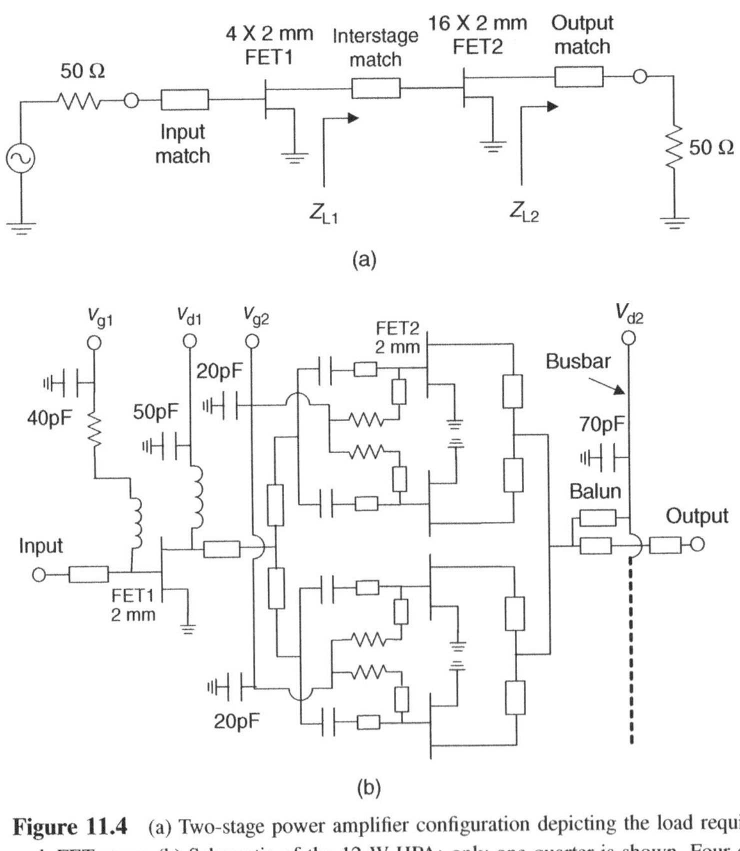

Figure 11.4 (a) Two-stage power amplifier configuration depicting the load required at the drain of each FET stage. (b) Schematic of the 12-W HPA; only one-quarter is shown. Four of them are combined in parallel.

at a minimum and ML for the interestage is minimized. The controlling factors for DL and ML for the interstage include stability criteria and electrical performance. This also helps in optimizing the FET aspect ratio. The dissipative loss is for the individual passive stage, that is, input, interstage, and output. The mismatch loss is the difference between the required device's optimum load impedance and the transformed 50-Ω output impedance at the drain terminal of the FET. The above method is based on the assumption that the device is input impedance depends strongly on the load connected at the drain terminal rather than its large-signal parameters.

The broadband amplifier design is carried out by considering the power gain roll-off of the FET with frequency (usually 6 dB/octave), the gain–bandwidth limitations of the input and output of the FET, and the overall amplifier stability versus frequency. In the output, a 16-way reactive cluster combining matching topology was used. Both lumped elements and distributed circuit elements were used for impedance matching networks. Figure 11.4b shows a simplified schematic of the 12-W HPA (only one-quarter of the circuit is shown). Four quarters are combined in parallel.

The input stage, which has a gain compensation network, shown in Figure 11.4b, was designed for good input match as well as for maximum power transfer at the high-frequency end. Figure 11.5a shows the simulated dissipative loss and the total loss (the sum of dissipative and mismatch losses) for the input matching network. The higher dissipative loss at lower frequencies is for device gain slope compensation. It may be noted that the mismatch loss, the difference between the total loss and dissipative loss, over the 0.75–3-GHz range is less than 0.5 dB. This confirms that the amplifier is matched to 50 Ω. The interstage matching network was designed to provide flat gain response and enough output power to succeeding stage FETs for achieving maximum output power and PAE. The interstage matching is comprised of *RLC* lumped—element based topology. Figure 11.5b shows the simulated dissipative loss and the total loss for the interstage matching network. The dissipative loss is monotonically decreasing with frequency and is adjusted to obtain the unconditional stable operation of each stage. It may be noted that the mismatch loss over the 0.75–3-GHz range is lower than 3 dB. However, for high output power and PAE, a mismatch loss much lower than 3 dB is desired.

The output matching elements were selected to provide an optimum load match with minimum possible insertion loss, since the efficiency is reduced to a greater extent by a given amount of loss due to decreased power out, gain, and available DC power at the FET drain pads. The output match uses thick lines, asymmetric broadside coupled line transformer (described in Section 7.5), and busbar for biasing drain of second-stage FETs. Figure 11.5c shows the dissipative loss and total loss for the output matching network. The dissipative loss is about 1.2 dB over the 0.7–2.5 GHz. In this case the mismatch loss is less than 0.3 dB over most of the band.

Biasing low microwave frequency HPA MMICs generally need off-chip RF chokes. In this design the use of an asymmetric broadside-coupled line transformer in the output matching network facilitates both biasing drain as well as realizing broadband impedance matching. Both the gate bias and drain bias were applied using corporate feed configurations as discussed in Chapter 18. Each FET's gate is individually biased through a gate resistor. The drain bias is applied to each drain using low-resistance busbar topology employing a top conductor of the MIM capacitors. Both the gate bias resistor and MIM capacitor values provide robust stability. Furthermore, exercising special care in maintaining the symmetry in the amplifier's layout and properly selecting isolation resistors prevent odd-mode oscillations. Each stage as well as the complete amplifier were designed to be unconditionally stable. Figure 11.6 shows a photograph of the 12-W broadband HPA.

Average measured output power and PAE of the two-stage broadband power amplifier are shown in Figure 11.7. At 22 dBm of input power, the power gain was greater than 19 dB over the 0.7–2.7-GHz frequency range. The output power was greater than 12 W and the PAE was better than 22%. Over 0.8–2.0 GHz, the output power and PAE were better than 14 W and 27%, respectively. The second-harmonic levels were measured below −20 dBc, demonstrating state-of-the-art second-harmonic performance for a multi-octave bandwidth HPA.

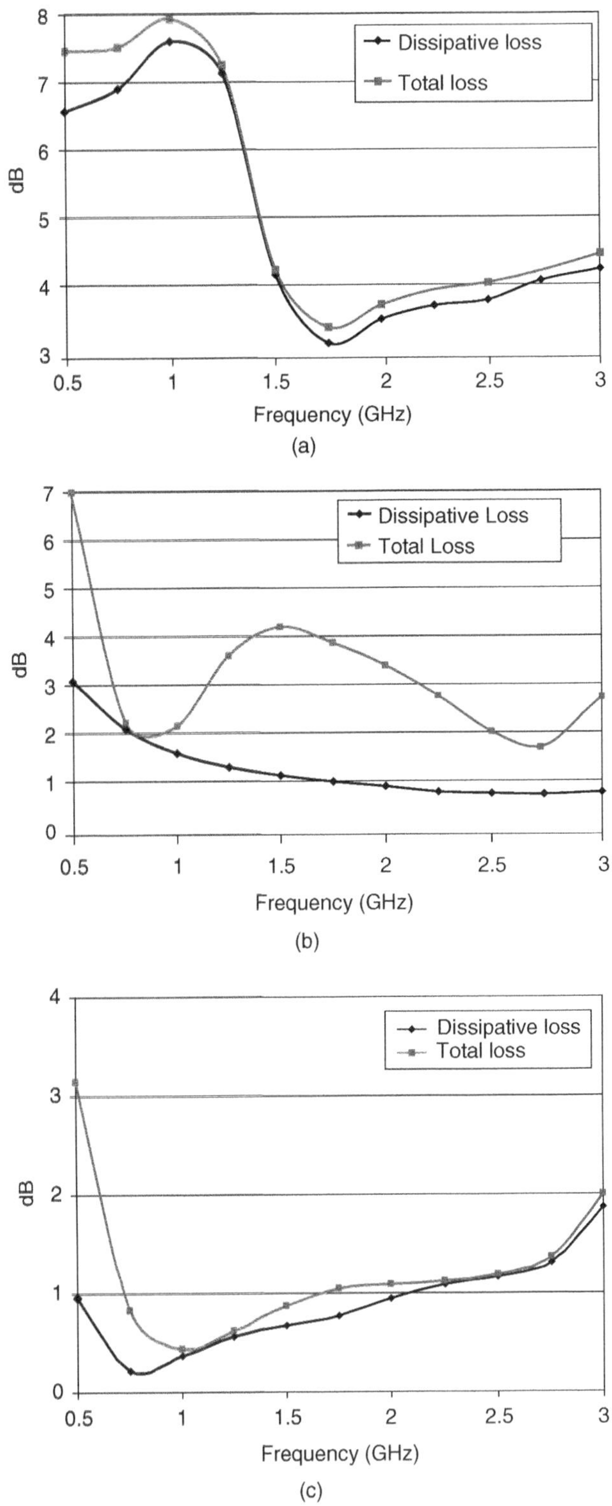

Figure 11.5 (a) Dissipative loss and total loss of the two-stage 12-W MMIC power amplifier: (a) for input matching network, (b) for interstage matching network, and (c) for output matching network. ■

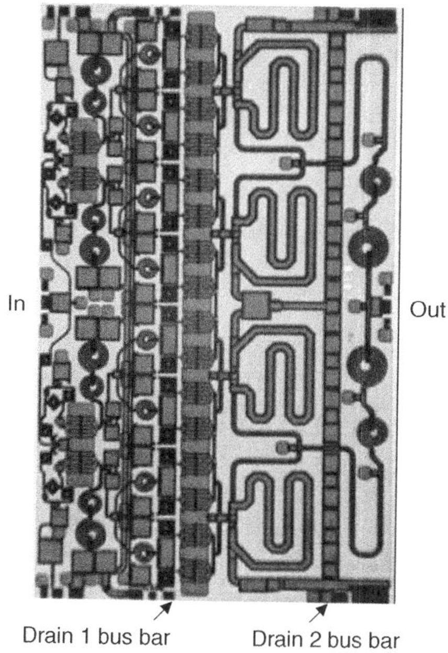

Figure 11.6 Photograph of the UHF/L/S-band 12-W power amplifier. Chip size is 5.0 mm × 8.0 mm^2.

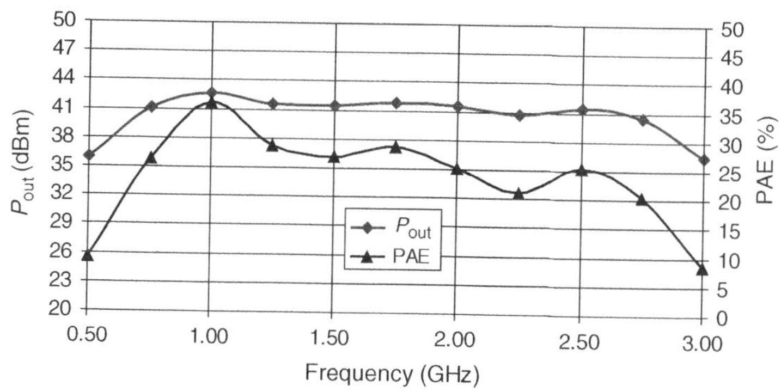

Figure 11.7 Output power and power added efficiency at $V_{ds} = 10$ V, $P_{in} = 22$ dBm, and 25% I_{dss}.

For broadband and high-efficiency applications, one can use class-A/AB bias and class-E load. However, class E has limited bandwidth and limited power above S-band. A suitable solution in this case is to combine the desirable features of these two classes by employing variable load, that is, use the class-E load at lower frequencies and class-B load at higher frequencies.

11.2.2 Feedback Amplifiers

A generic feedback configuration is shown in Figure 11.8, consisting of an *LRC* feedback network. The feedback loop elements control the gain and bandwidth. The feedback resistor R_{fb} controls the gain and L_f extends the frequency range to higher frequencies. The inductors L_1 through L_4 set the bandwidth performance of the amplifier. Primary function of the C_f capacitor is to provide a DC block between the supply

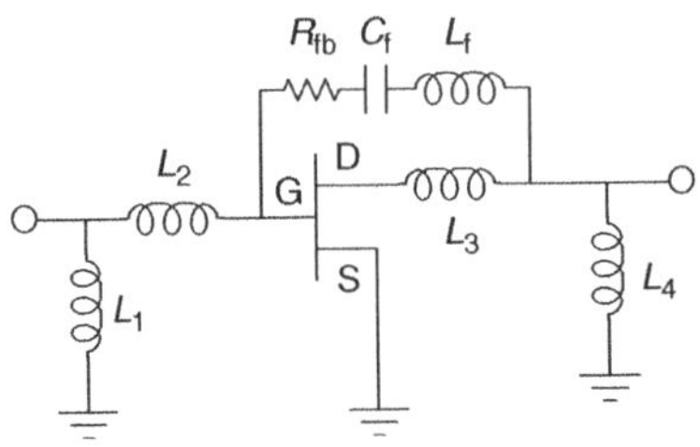

Figure 11.8 Feedback amplifier configuration.

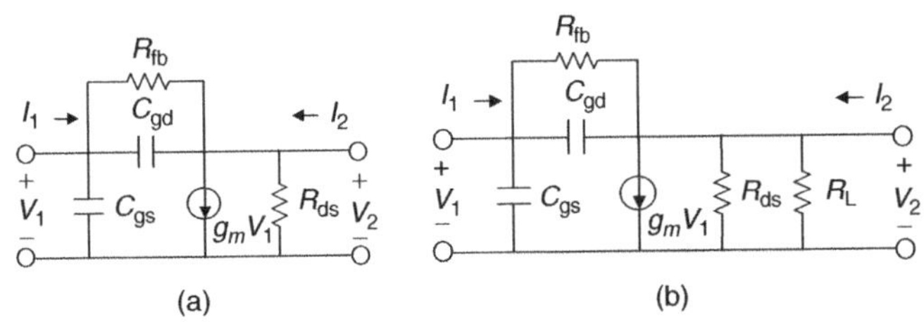

Figure 11.9 (a) A "pi" equivalent circuit representation of (a) a transistor with feedback and (b) a transistor with feedback and load.

voltages applied at the gate and drain terminals. These amplifiers can be designed with optimum performance by using MMIC technology rather than the hybrid MIC approach. A feedback configuration increases bandwidth, improves linearity or reduces distortion, and makes the design less sensitive to process variations.

Figure 11.9a shows a basic "pi" equivalent circuit of a transistor with feedback resistor R_{fb}. Here the effect of device parasitic reactance is neglected. Using Eqs. (2.13)–(2.15), Chapter 2, the admittance matrix is written [11, 12]

$$[Y] = \begin{bmatrix} j\omega C_{gs} + j\omega C_{gd} + \dfrac{1}{R_{fb}} & -j\omega C_{gd} - \dfrac{1}{R_{fb}} \\ g_m - j\omega C_{gd} - \dfrac{1}{R_{fb}} & \dfrac{1}{R_{ds}} + j\omega C_{gd} + \dfrac{1}{R_{fb}} \end{bmatrix} \tag{11.1}$$

Using the equations in Table 2.3, the Y matrix can be converted into the more useful scattering parameter matrix $[S]$. As a first-order approximation, one may neglect capacitor effects; $C_{gs} = 0$ and $C_{gd} = 0$, and (11.1) becomes

$$[Y] = \begin{bmatrix} \dfrac{1}{R_{fb}} & -\dfrac{1}{R_{fb}} \\ g_m - \dfrac{1}{R_{fb}} & \dfrac{1}{R_{ds}} + \dfrac{1}{R_{fb}} \end{bmatrix} \tag{11.2}$$

In this case the scattering parameters may be written

$$S_{11} = \frac{1}{D}\left[\frac{R_{fb}}{Z_0}\left(1 + \frac{1}{R_{ds}} Z_0\right) - \left(g_m + \frac{1}{R_{ds}}\right) Z_0\right] \tag{11.3a}$$

$$S_{12} = \frac{2}{D} \tag{11.3b}$$

$$S_{21} = \frac{-2}{D}[g_m R_{fb} - 1] \tag{11.3c}$$

$$S_{22} = \frac{1}{D}\left[\frac{R_{\text{fb}}}{Z_0}\left(1 - \frac{1}{R_{\text{ds}}}Z_0\right) - \left(g_m + \frac{1}{R_{\text{ds}}}\right)Z_0\right] \tag{11.3d}$$

where

$$D = 2 + \left(g_m + \frac{1}{R_{\text{ds}}}\right)Z_0 + \frac{R_{\text{fb}}}{Z_0}\left(1 + \frac{1}{R_{\text{ds}}}Z_0\right)$$

and Z_0 is the system impedance. These equations are further simplified [13] when perfect-match conditions are applied, that is,

$$S_{11} = S_{22} = 0 \tag{11.4}$$

In this case, from (11.3a)

$$R_{\text{fb}} = g_m Z_0^2 \tag{11.5a}$$

$$S_{12} = \frac{1}{g_m Z_0 + 1} \tag{11.5b}$$

$$S_{21} = -(g_m Z_0 - 1) \tag{11.5c}$$

and the power gain, G, becomes

$$G = |S_{21}|^2 = (g_m Z_0 - 1)^2 \tag{11.6}$$

If the device is not matched perfectly but has an input and output VSWR of $S:1$, then the value of the feedback resistor and the power gain are given by [12]

$$R_{\text{fb}} = SZ_0(1 + g_m Z_0) - Z_0 \tag{11.7}$$

and

$$G = [2(1 - Sg_m Z_0)/(1 + S)]^2 \tag{11.8}$$

Next, let us calculate the loss due to the feedback resistor. Consider an optimum load R_{L} is connected at the output of the device, as shown in Figure 11.9b. Here,

$$V_2 = V_1 + R_{\text{fb}} I_1 \tag{11.9}$$

or

$$V_2 = -(g_m V_1 - I_1)\frac{R_{\text{L}} R_{\text{ds}}}{R_{\text{L}} + R_{\text{ds}}} \tag{11.10}$$

When $R_{\text{ds}} \gg R_{\text{L}}$,

$$V_2 = -(g_m V_1 - I_1)R_{\text{L}} \tag{11.11}$$

From (11.9) and (11.11), we find

$$V_2 = -V_1 g_m R_{\text{L}}\frac{1 - 1/g_m R_{\text{fb}}}{1 + R_{\text{L}}/R_{\text{fb}}} \tag{11.12}$$

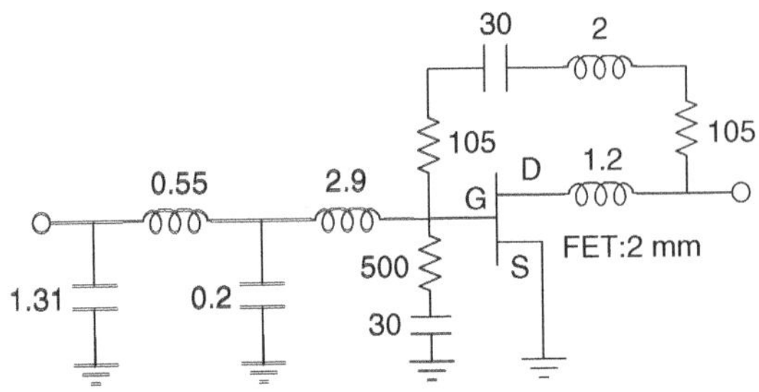

Figure 11.10 Schematic of the 1-W power amplifier. The units for capacitors are picofarads (pF), for inductors nanohenries, (nH), and for resistors ohms (Ω).

With no feedback, the output power is given by

$$P_o = V_2^2/R_L = (g_m V_1)^2 R_L \tag{11.13}$$

With feedback,

$$P_{ofb} = V_2^2/R_L = (g_m V_1)^2 R_L \left(\frac{1 - 1/g_m R_{fb}}{1 + R_L/R_{fb}}\right)^2 \tag{11.14}$$

The ratio P_{ofb}/P_o provides the power loss due to feedback and is given by

$$\frac{P_{ofb}}{P_o} = \left(\frac{1 - 1/g_m R_{fb}}{1 + R_L/R_{fb}}\right)^2 \tag{11.15}$$

For $g_m R_{fb} \gg 100$ and $R_{fb} \gg R_L$, we find

$$\frac{P_{ofb}}{P_o} \cong 1 - \frac{2R_L}{R_{fb}} \tag{11.16}$$

Consider a device with $R_L = 100\ \Omega$ and $R_{fb} = 1000\ \Omega$. The power loss due to feedback is $P_{ofb}/P_o \cong 1 - 200/1000 = 0.8$ or about 1 dB.

EXAMPLE 11.2

A 1-W device has a g_m of 100 mS and a load impedance of 40 Ω. When a feedback 400 Ω resistor is connected across the device determine its output power.

SOLUTION Using (11.15), we find

$$\frac{P_{ofb}}{P_o} = \left[\frac{1 - 1/(0.1 \times 400)}{1 + 40/400}\right]^2 = 0.7856$$

$$P_{ofb} = 1 \times 0.7856\ \text{W} = 0.7856\text{W}$$

■

EXAMPLE 11.3

Design a 1-W amplifier working over 30 MHz to 3 GHz. This example provides some salient features of a broadband HPA. The design was done using an MESFET based MMIC process.

SOLUTION A feedback topology selected for this example is shown in Figure 11.10. The input match is of lowpass type and a shunt resistor is used for circuit stabilization. The output

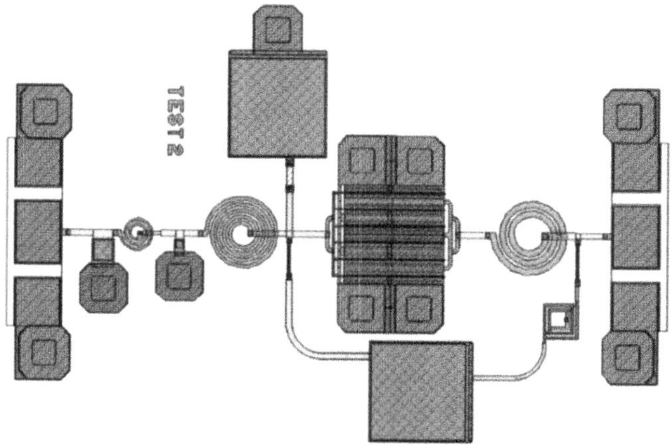

Figure 11.11 Physical layout of the ultra-wideband 1-W power amplifier.

inductor is for gain peaking at the high-frequency end. This design has no output matching network. The size of the transistor used is 2.0 mm, the output power and PAE at 3 GHz are about 1.6 W and 60%, respectively. Figure 11.11 shows the physical layout of the amplifier. The circuits were "on-wafer" tested using external bias tees. Circuits can also be tested on board or on carriers using conical chokes as described in Chapter 6.

Measured output power and PAE using on-wafer pulsed power measurements (pulse width $= 10\ \mu s$ and 10% duty cycle) are shown in Figure 11.12. The CW measured power and PAE were also performed over 0.5–3 GHz and were about 1 dB lower and 5% points higher, respectively, in comparison to the on-wafer measured value. The difference in pulsed and CW PAE values is attributed to accurate CW current measurement. ■

11.2.3 Balanced Amplifiers

In balanced configurations a single-ended amplifier is designed using the reactive/resistive or feedback topologies as described in the previous sections and a pair of

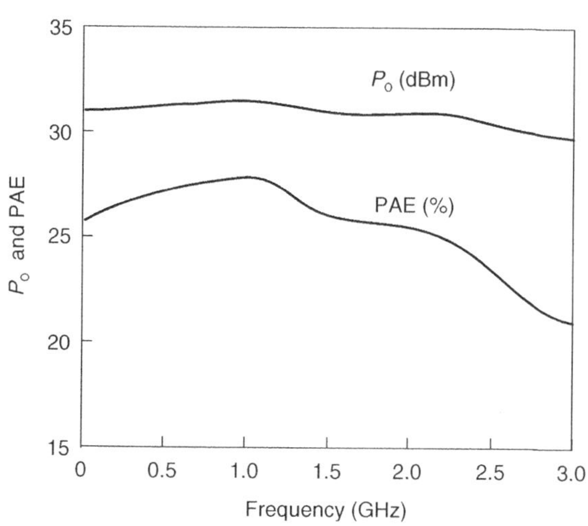

Figure 11.12 Measured pulsed output power and PAE versus frequency of an ultra-wideband 1-W amplifier.

such identical amplifiers are combined employing two 3-dB quadrature couplers [14]. The quadrature couplers used are usually broadband Lange couplers. The single-ended amplifier is mismatched for flat gain, low noise figure (in the case of LNA), high output power and PAE, and good stability. The reflections from the single-ended amplifiers are terminated in 50 Ω, which usually guarantees stability. If one stage fails, the overall gain drops about 6 dB, which may provide useful fault tolerance for some applications. The major advantages of the balanced topology vis-à-vis reactive/resistive topology alone are good input and output VSWR, good gain flatness, better stability, and insensitivity to mismatch in the subsystem. Balanced amplifiers have much better stability than their single-ended versions because of built-in resistor terminations of the hybrids and wideband isolation from the other circuitry.

This topology is very suitable for monolithic or hybrid implementations, where two matched MMIC HPA chips are combined. Since the implementation of Lange couplers on thin GaAs substrate has higher loss and reduced bandwidth compared to hybrid versions, MMIC balanced power amplifiers are limited to power levels less than 10 W. The Lange couplers use $\lambda/4$ long lines; the balanced power amplifiers will unfavorably result in larger size at lower frequencies. MMIC balanced power amplifiers are usually designed above S-band frequencies and are commonly realized above 20 GHz.

Consider a balanced configuration as shown in Figure 11.13. As a first-order approximation the couplers are assumed ideal (i.e., no loss in the couplers); the isolated port has infinite isolation and the split signals have 90° phase difference. The S-parameters of a single-ended amplifier and balanced amplifier are given as follows:

$$S_V = \begin{bmatrix} S_{11V} & S_{12V} \\ S_{21V} & S_{22V} \end{bmatrix}, \quad S_T = \begin{bmatrix} S_{11T} & S_{12T} \\ S_{21T} & S_{22T} \end{bmatrix} \tag{11.17}$$

where V and T represent amplifier A or B and balanced amplifier, respectively. The balanced power amplifier (BPA) is between ports 1 and 7.

The input signal splits between ports 2 and 3 and the split signals have 90° phase difference. The ports on the same end have the same phase whereas those diagonally opposite have 90° phase difference. At port 1, the BPA's reflected signal is

$$S_{11T} = \frac{1}{2}[S_{11A} - S_{11B}] \tag{11.18}$$

If $S_{11A} = S_{11B}$, $S_{11T} = 0$; that is, BPA is matched at port 1. At port 7,

$$S_{21T} = \frac{1}{2}j[S_{21A} + S_{21B}] \tag{11.19a}$$

$$S_{22T} = \frac{1}{2}[S_{22B} - S_{22A}] \tag{11.19b}$$

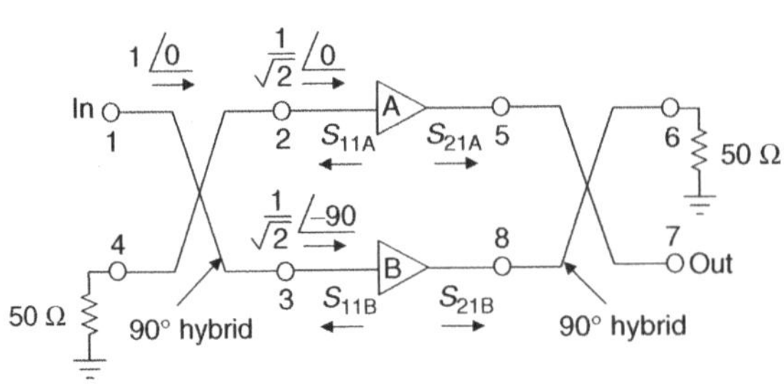

Figure 11.13 Configuration of a balanced amplifier using 90° Lange couplers.

Again, if $S_{22A} = S_{22B}$, $S_{22T} = 0$; that is, BPA is matched at port 7, which is an output port. The output signal is an average of two forward going signals. At port 4, the combined reflected signal is given by

$$\tfrac{1}{2}j[S_{11A} + S_{11B}] \tag{11.20}$$

and is absorbed by the resistor. Similarly, at port 6 the combined reflected signal is absorbed. If both amplifiers have the same gain but $\Delta\phi$ phase difference between them, then S_{21T} becomes

$$S_{21T} = \tfrac{1}{2}|S_{21}|\,|1 + e^{j\Delta\phi}| = \tfrac{1}{2}|S_{21}|\,[(1 + \cos\Delta\phi)^2 + \sin^2\Delta\phi]^{1/2} \tag{11.21}$$

When $\Delta\phi = 0$, $S_{21T} = S_{21}$ and for $\Delta\phi = 22.5^\circ$, $S_{21T} = 0.9808S_{21}$.

A balanced configuration can also be realized using a scheme as shown in Figure 11.14. In this design, one uses a Wilkinson divider/combiner with 90° phase offset 50-Ω line instead of a Lange coupler. In this case, the reflected signals from two single-ended amplifiers have 180° phase difference across the isolation resistor $R = 100\ \Omega$. Thus the out-of-phase reflected signals are absorbed in the isolation resistor. Similarly, the reflected output signals are absorbed by the output isolation resistor. The through signals in the two paths have equal phase and add at the output. This topology minimizes the reflected signals at both the input and output terminals and provides good VSWR, but has larger size due to quarter-wave lines and Wilkinson divider/combiner. The quarter-wave sections also result in narrower bandwidth than using Lange couplers, that is, 40–50% versus octave bandwidth. In the balanced configuration the effect of mismatch on the output power and PAE, due to bond wires and package lead frame, is minimum.

EXAMPLE 11.4

Design a three-stage balanced 2-W HPA working over 12–16 GHz. The design can use two 1.5-W MMIC amplifiers with external Lange couplers or the 2-W amplifier can be implemented fully in monolithic topology as reported in Reference [15].

SOLUTION The 2-W HPA design was based on MSAG FET technology using the 75-μm GaAs MMIC process. Since Lange couplers have higher loss than the Wilkinson divider on a 75-μm GaAs substrate, a balanced configuration, shown in Figure 11.14, was used. The three-stage single-ended design was comprised of a 0.625-μm FET at the input driving two

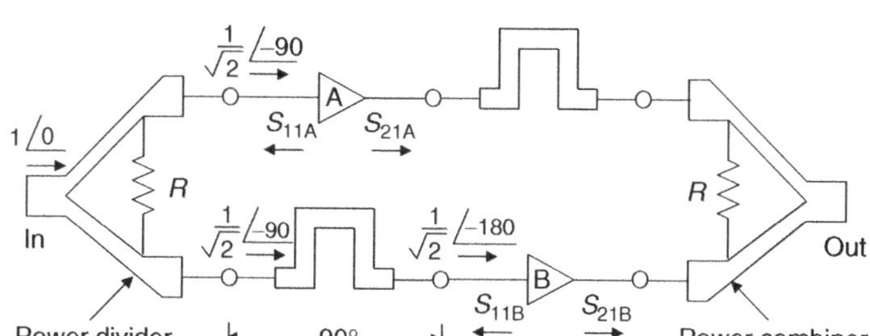

Figure 11.14 An alternate balanced amplifier configuration using Wilkinson dividers/combiners and 90° offset lines.

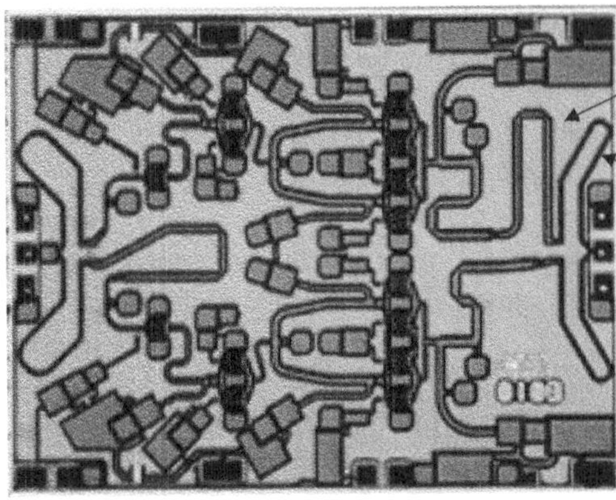

Figure 11.15 Photograph of the Ku-band balanced 2 W amplifier. Chip size is 4.4 mm × 3.4 mm^2.

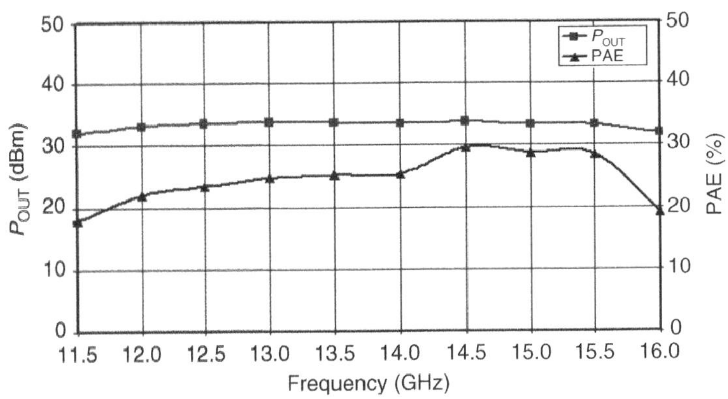

Figure 11.16 Output power and power added efficiency versus frequency at $V_{ds} = 8$ V and $P_{in} = 18$ dBm.

0.625-μm FETs and driving four 0.625-μm FETs at the output. The FET 2:1 aspect ratio, in this case, is required to obtain high power and PAE over a larger bandwidth. This design also uses a single drain pad and single gate pad supply operation. Figure 11.15 shows the photograph of the balanced 2-W amplifier. This design requires bias supply from both sides.

Figure 11.16 shows typical CW measured P_{out} and PAE for the balanced amplifier MMIC at $V_{ds} = 8$ V and $P_{in} = 18$ dBm. The amplifier has greater than 33-dBm output power and better than 22% PAE over the 12–15.5-GHz frequency range. When the supply voltage is increased to 10 V, the output power is better than 34.5 dBm. The input VSWR is better than 1.5:1 over 11.5–15 GHz [15]. ■

11.2.4 Distributed Amplifiers

Distributed amplifiers have been treated extensively in the literature [1, 16–20]. An n-section distributed amplifier (DA) topology is shown in Figure 11.17. A simplified equivalent circuit representation of a DA is shown in Figure 11.18. The gate line inductor and drain line inductor form artificial transmission lines with the C_{gs} and C_{ds} of

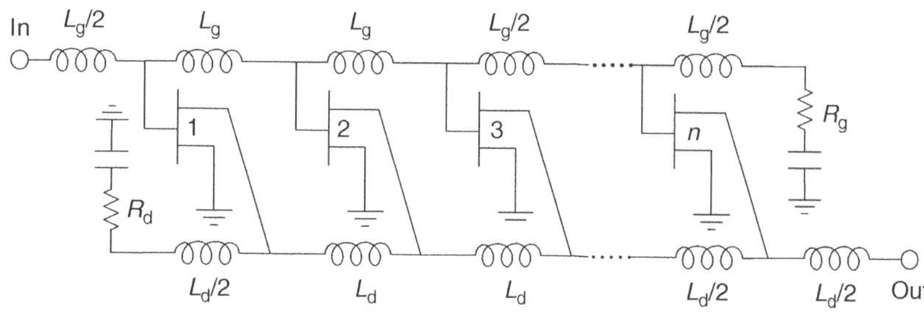

Figure 11.17 A simplified distributed amplifier topology using n sections.

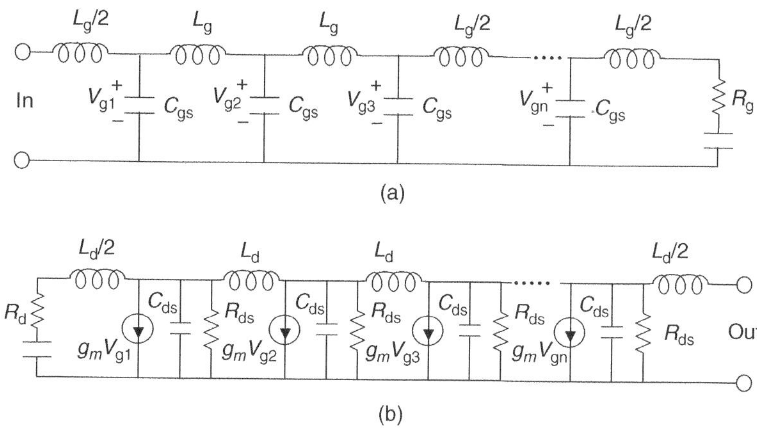

Figure 11.18 Equivalent circuits for artificial lines: (a) gate and (b) drain.

the devices, respectively. Thus the device capacitances are absorbed into the transmission lines. These inductors are usually replaced with high-impedance microstrip lines for low-noise and power amplifier applications. The inductors L_g and device input capacitances of the gain cells, and inductors L_d and device output capacitances form artificial transmission lines. Such lines are designated as artificial equivalent gate and drain lines of characteristic impedances Z_{0g} and Z_{0d}, respectively, and are matched when they are terminated into their respective characteristic impedance value. When a signal travels down the gate line, each transistor is excited and the signal is amplified by its transconductance and added along the drain line. The leftover signal on the gate line and reflected signal on the drain lines are absorbed by the termination resistors R_g and R_d, respectively. Since these lines are lowpass structures and have very high cutoff frequencies, the traveling-wave topology results in very wide bandwidth performance. For most applications, the number of cells or sections (n) are 4 to 6.

The input line has an impedance given by

$$Z_{0g} = (L_g/C_{gs})^{1/2} \tag{11.22a}$$

The drain line impedance is given by

$$Z_{0d} = (L_d/C_{ds})^{1/2} \tag{11.22b}$$

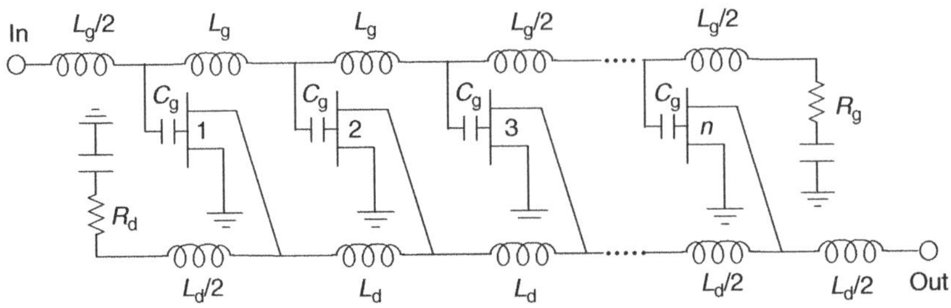

Figure 11.19 A modified distributed amplifier topology using n sections.

Since $C_{gs} > C_{ds}$, the bandwidth of a DA is limited due to gate capacitance. The highest frequency of operation is known as the cutoff frequency of the equivalent gate line and is given by

$$f_c = \frac{1}{\pi Z_{0g} C_{gs}} \tag{11.23}$$

If a capacitance C_g is inserted in series with the gate, as shown in Figure 11.19, the effective gate capacitance is reduced and is given by [17]

$$C'_{gs} = \frac{q}{1+q} C_{gs} \tag{11.24}$$

$$q = C_g / C_{gs}$$

Here it is assumed that the series capacitance is the same for all FETs. In this case the FET is considered modified having the transconductance and cutoff frequency given by

$$g'_m \cong \frac{q}{1+q} g_m \tag{11.25a}$$

$$f'_c = \frac{1+q}{q} f_c \tag{11.25b}$$

Thus using a series capacitor in series with a device increases the bandwidth and decreases the gain of the DA. This technique is commonly used in power DAs to improve power–bandwidth capability.

An approximate expression for the small-signal gain of a distributed amplifier (Fig. 11.17) is given by [1]

$$G \cong \frac{g_m^2 n^2 Z_{0g}^2}{4} [1 - \alpha_g \ell_g n/2]^2 \tag{11.26}$$

where n = number of FETs

g_m = transconductance per FET

Z_{0g} = characteristic impedance of the gate line

α_g = attenuation constant of the gate line

ℓ_g = length of gate line per unit cell

In this expression, constant characteristic impedance and constant unit line length have been assumed. Also, drain line losses are neglected. Equation (11.26) shows that for higher gain one needs higher Z_{0g} and lower α_g.

An approximate expression for the calculation of an optimum value for the number of cells, n_{opt}, is given by

$$n_{opt} = \frac{\ln[\alpha_g \ell_g / \alpha_d \ell_d]}{\alpha_g \ell_g - \alpha_d \ell_d} \tag{11.27a}$$

where

$$\alpha_g \ell_g = \frac{R_g Z_0 (2\pi f C_{gs})^2}{2} \tag{11.27b}$$

$$\alpha_d \ell_d = \frac{Z_0}{2R_{ds}} \tag{11.27c}$$

Here, R_g, C_{gs}, and R_{ds} are the FET/HEMT EC model parameters, f is the maximum operating frequency, and $Z_0 = 50\ \Omega$.

At the circuit level, improvements in performance can be achieved using very-low-loss and high-impedance lines in the gate and drain equivalent transmission lines of a distributed amplifier. These lines are realized using thick conductors, thick substrates, and ridged microstrip as discussed in the previous chapter in Section 10.5.3. In GaAs MMICs, these lines are realized on a 10-μm thick polyimide layer atop the GaAs substrate and fabricated by employing a multilevel plating (MLP) process [20].

As an example of the performance benefit that MLP provides for wideband MMICs, Figure 11.20 shows the predicted gain of a pair of 2–20-GHz distributed amplifiers, one designed using a standard process and the other implemented in MLP [20]. As is easily noted, ~1.5-dB additional gain is provided by the MLP design at 20 GHz. The return loss was better than 10 dB. Thus MLP can be used to either increase gain for a given bandwidth or provide additional bandwidth for a given amount of gain.

Next, we describe several examples of distributed amplifiers developed using MSAG FET with the MLP process. In these MMICs, the gate and drain microstrip conductors are on a 10-μm thick polyimide layer atop the GaAs substrate. The circuits were designed using small-signal models for FET devices and matched to 50 Ω input and output.

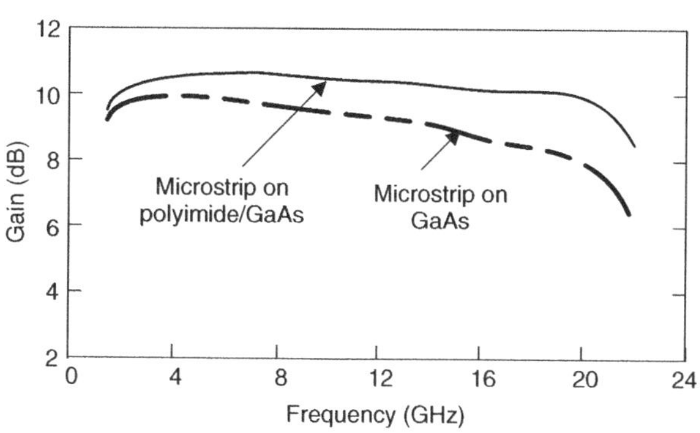

Figure 11.20 Comparison of predicted performance of a 2–20-GHz distributed amplifier using a standard process and MLP processing.

EXAMPLE 11.5

The rapid increase of data traffic requires large-capacity optical communication systems. Such systems demand data rates as high as 40 Gb/s. The most speed-limited component in such systems is a preamplifier in the receiver, which requires a low noise figure, flat gain response, and ultralarge bandwidth (DC to 40 GHz). Various amplifier topologies including traveling-wave and transistor technologies including MESFET, pHEMT, and HBT are being pursued to develop high-speed preamplifiers in hybrid or monolithic form.

A six-cell monolithic distributed amplifier was designed using M/A-COM's MSAG process [20] for 20-Gb/s optical communication applications. The schematic of the amplifier is shown in Figure 11.21 and it requires both positive and negative power supplies. The 1000-pF, 150-nF, and 0.01-μF capacitors were connected externally to extend the frequency range down to 500 kHz. Figure 11.22 shows the measured gain and input and output return losses. The amplifier typically has 10-dB gain and a maximum VSWR of 2:1 in the 500-kHz to 20-GHz band. The gain flatness was within ±1 dB. The measured noise figure (NF) was less than 4.5 dB over the 0.1–20-GHz frequency range. The NF of such amplifiers is improved by about 1 dB using pHEMT devices.

EXAMPLE 11.6

Figure 11.23 shows the physical layout of a self-biased 2–20-GHz distributed LNA. It incorporates six 150-μm gate periphery low-noise FETs. The circuit was designed for low noise figure and high gain using a low-noise FET model. Single supply operation is provided through the use of "on-chip" self-biasing networks. The value of the resistor between the source of the device

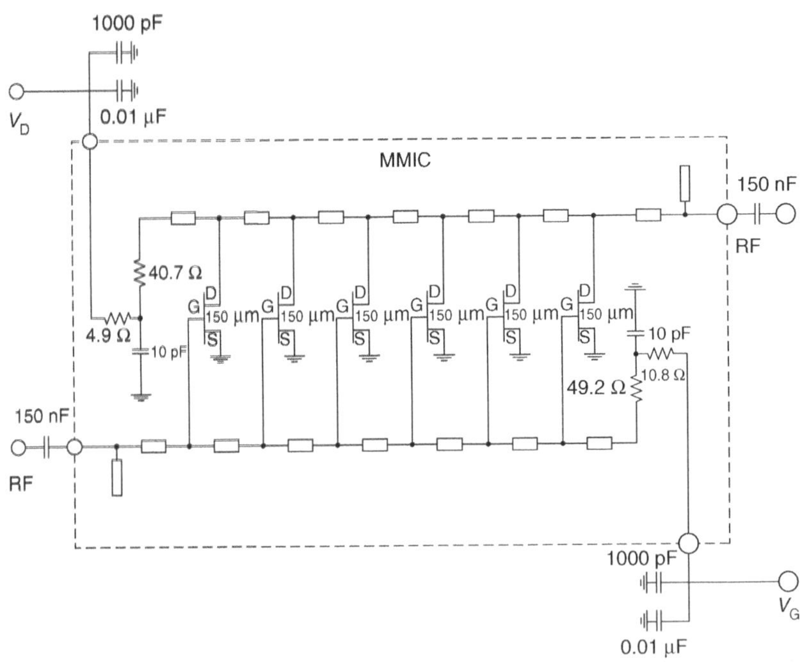

Figure 11.21 Schematic of a 0–20-GHz distributed LNA.

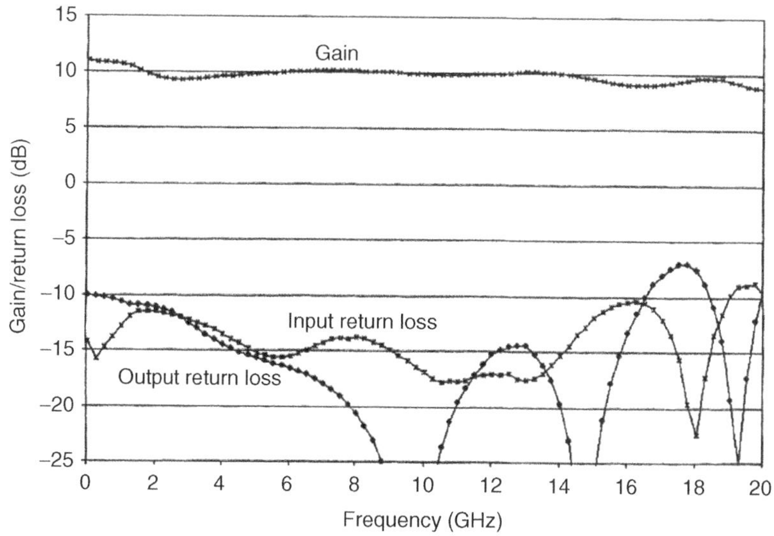

Figure 11.22 Measured gain, and input and output return losses of 0–20-GHz distributed LNA.

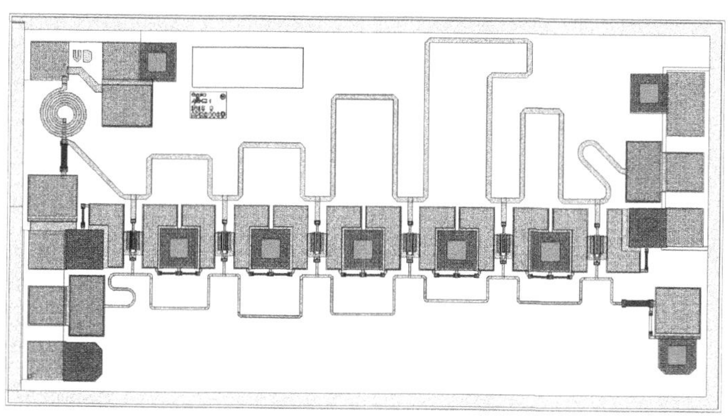

Figure 11.23 Layout of a 2–20-GHz self-biased distributed LNA. Chip size is 3.0 × 1.6 mm^2.

and ground is selected to bias the device at 25% I_{dss} to provide a trade-off between minimum noise figure and high P_{1dB}. The drain supply voltage is 5 V.

Figure 11.24 shows the measured LNA gain and noise figure. Current under drive increases from 75 mA at the Q-point to about 110 mA at the P_{1dB} compression point. The measured return loss was better than 10 dB over the 2–20-GHz frequency range. Measured P_{1dB} compression point values were 20 dBm at lower frequencies and 17 dBm at higher frequencies.

Single-Stage DA Example

This is a high-power version of the low-noise design, which was discussed previously. It uses the same self-bias approach. Figure 11.25 shows the photograph of the distributed driver amplifier designed for the 2–18-GHz frequency band. This topology employs five 300-μm periphery low-noise FETs. The device design parameters are given in

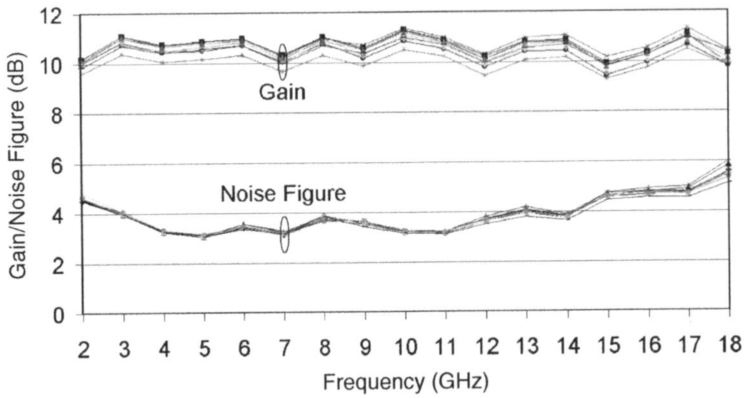

Figure 11.24 Measured gain and noise figure of self-biased 2–20-GHz distributed LNA.

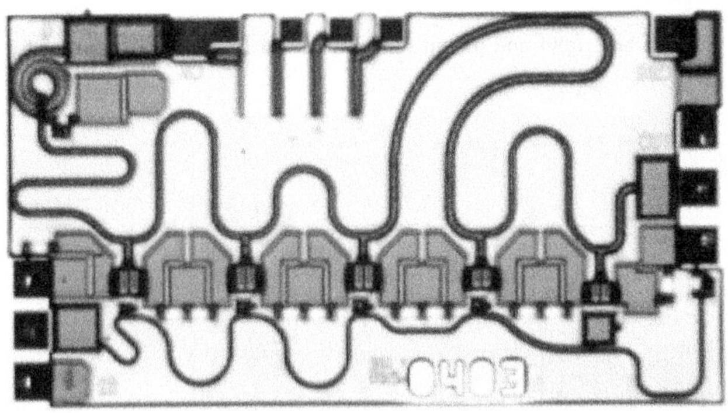

Figure 11.25 Photograph of a 2–18-GHz self-biased distributed driver amplifier. Chip size is 3.0 × 1.7 mm^2.

Table 5.1, Chapter 5. The circuit was designed for high P_{1dB} and high gain while providing good match to 50 Ω at the input and output. Single supply operation is provided through the use of "on-chip" self-biasing networks. The nominal drain supply voltage is 5 V; however, on-chip voltage drop resistors allow variable supply voltage from 5 to 8 V.

Figures 11.26 and 11.27 show the measured gain and VSWR, and saturated output power at different drain biases, respectively.

The DAs provide excellent noise and gain performance over multi-octave bandwidths. However, their power and PAE performance is limited when they are designed for maximum gain and good input and output match in the 50 Ω system impedance. This is due to the fact that each device does not see the correct load impedance required for the maximum power and PAE. This may be explained by using an example of a simplified equivalent circuit for a four-cell DA showing current combining as depicted in Figure 11.28. Let us assume that all four devices have the same size and require a real load value of R_L. The current combining takes place from left to right [21]. Based on an equivalent transmission line theory the characteristic impedance of the line looking from node 1 toward the right should be R_L or $\sqrt{L_1/C_1} \cong R_L$. At node 2,

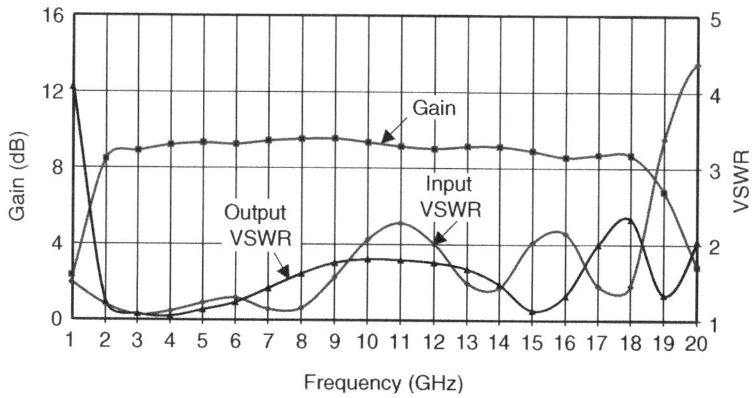

Figure 11.26 Small-signal gain and input VSWR and output VSWR versus frequency at $V_D = 5$ V.

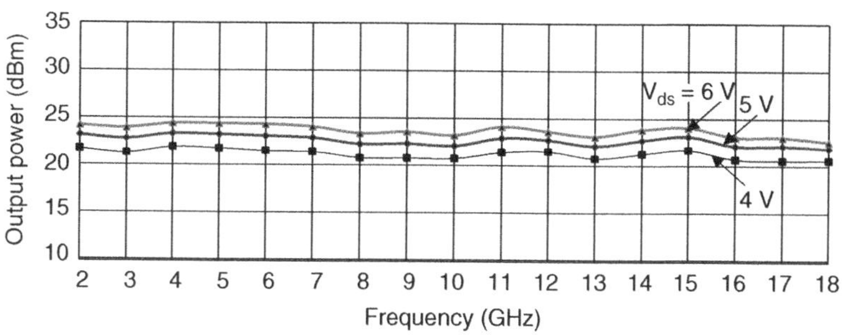

Figure 11.27 Saturated output power versus frequency and drain voltage.

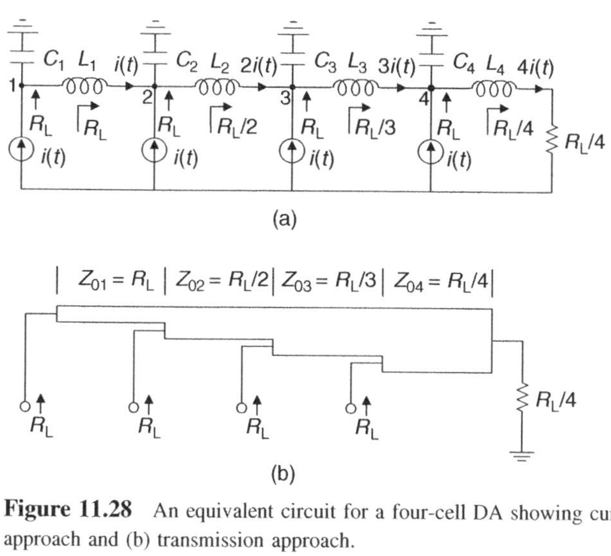

Figure 11.28 An equivalent circuit for a four-cell DA showing current combining: (a) lumped-element approach and (b) transmission approach.

when both currents enter in phase, the total current is now $2i(t)$. In this case, the characteristic impedance of the line looking from node 2 toward the right should be $R_L/2$ or $\sqrt{L_2/C_2} \cong R_L/2$. Using similar analysis, at node 4 looking toward the right one must have $R_L/4$ or $\sqrt{L_4/C_4} \cong R_L/4$ and the required system impedance is $R_L/4$. Thus for high power and high PAE, for DAs one needs tapered drain lines with decreasing characteristic impedance. The values of such impedances may also be varied by using tapered transistor sizes—the largest size at the input and the smallest size at the output. If we have to maintain $R_L/4 = 50\ \Omega$, there is a limit on the power output (due to small transistor sizes) that one can achieve using the DA approach.

Thus the power output of DAs is limited because of lower gate cutoff frequency due to larger transistor size and lower system impedance requirements. The low system impedance may be addressed as shown in Figure 11.29, where the high-power distributed power amplifier is matched to a 3–12-Ω source and load impedance, and an ultra-wideband balun (described in Chapter 7) is used to transform the 3–12-Ω impedance to 50 Ω. The value of Z_T is estimated to be between 3 and 12 Ω.

Single-Stage Power DA Example

This is another DA, designed for higher power level. The 1-W single-stage power amplifier topology was selected as shown in Figure 11.29. In this configuration, a DA is designed for source and load impedance values less than 50 Ω and two of them are combined by connecting them in parallel. Then transmission line baluns are used at the input and output to match to 50 Ω. In this design each single-ended stage uses five cells and the device size is tapered to obtain the largest power bandwidth and PAE. In a DA topology, a major contribution to power is only given by the first two devices and the remaining devices help in realizing larger bandwidth. Thus large-size transistors are used at the input side to obtain high output power and small-size transistors are used at the output side to realize a large gate cutoff frequency. In the single-ended design example, the FETs have gate peripheries of 625, 625, 625, 470, and 300 μm. By using small-signal S-parameters, each stage was optimized for maximum gain, and good input and output VSWR. Figure 11.30 shows the physical layout of the 6–18-GHz distributed power amplifier.

The simulated output power and PAE of the 6–18-GHz 1-W DA are shown in Figure 11.31. The output power is about 31 dBm and PAE is in the range of 10–12%. The input and output return losses were better than 8 dB.

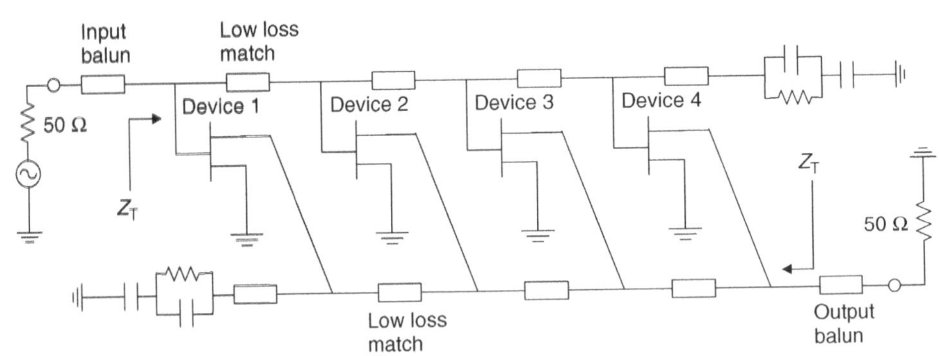

Figure 11.29 Configuration of a high-P_o and high-PAE ultra-broadband distributed amplifier.

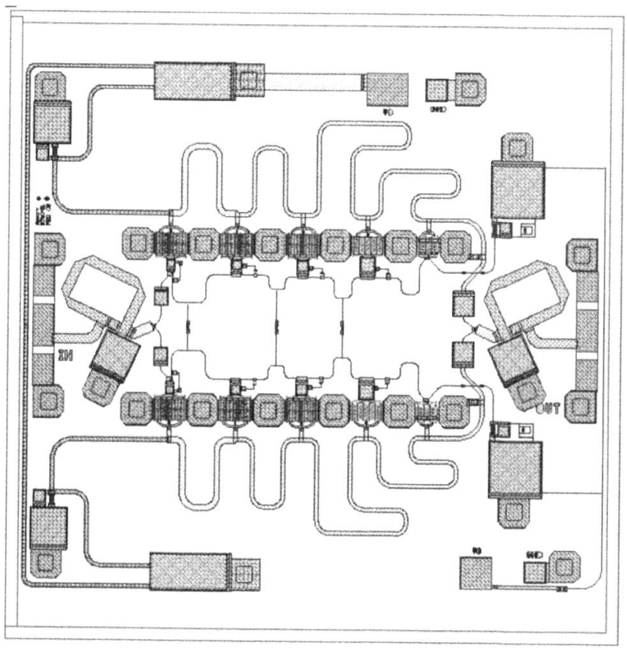

Figure 11.30 Layout of the 6–18-GHz 1-W distributed amplifier. Chip size is 3.0 × 3.0 mm^2.

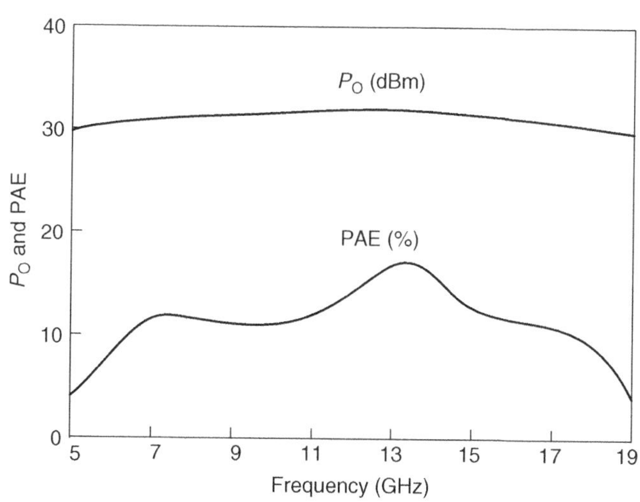

Figure 11.31 Simulated output power and PAE of the 6–18-GHz 1-W DA.

Two-Stage Power DA Example

Next, we describe an example of a 1-W two-stage DA working over 2–18 GHz. Like any other microwave circuit, DAs have limitations in terms of gain–bandwidth and power–bandwidth products. The bandwidth is limited by the input capacitance of the device, which is proportional to the output power level of the device. This means that if one wants to design a DA with double the output power, the device size becomes double and the gate cutoff frequency is reduced. Thus for a 2–18-GHz power amplifier, for a given device type, there is a limit for maximum achievable output power level. For MSAG power devices (see Chapter 5) operating at 10-V power supply, this power limit is about 0.5 W. In order to obtain higher power levels, one can further combine these circuits by using conventional hybrid combining techniques as described in Chapter 19. However, monolithic amplifier size and its cost preclude the design of amplifiers with power levels greater than 3–4 W.

The 1-W two-stage power amplifier topology was selected as shown in Figure 11.32. The first stage uses conventional DA configuration and in the second stage two DAs are combined using reactive combiners. In this design each stage uses five cells and the transistor size is tapered to obtain the largest power bandwidth and PAE. Device sizes for the first-stage FETs are 625, 625, 470, 470, and 300 μm. In the second-stage single-ended design, the FETs used are 625, 625, 625, 470, and 300 μm. By using small-signal *S*-parameters, each stage was optimized for maximum gain, and good input and output VSWR. A photograph of the two-stage 2–18-GHz distributed power amplifier is shown in Figure 11.33 and Figure 11.34 shows the measured saturated power at different drain biases.

11.2.5 Active Matching Broadband Technique

An active matching broadband technique suitable for RF and low microwave frequencies and low-power applications is shown in Figure 11.35. This is a multistage design that uses all three possible FET configurations as described in Chapter 4. The first stage uses a common-gate (CG) configuration for input match; the output stage is a common-drain (CD) configuration for output match and a conventional common-source (CS) configuration is sandwiched between these as a gain stage. Practically, there are

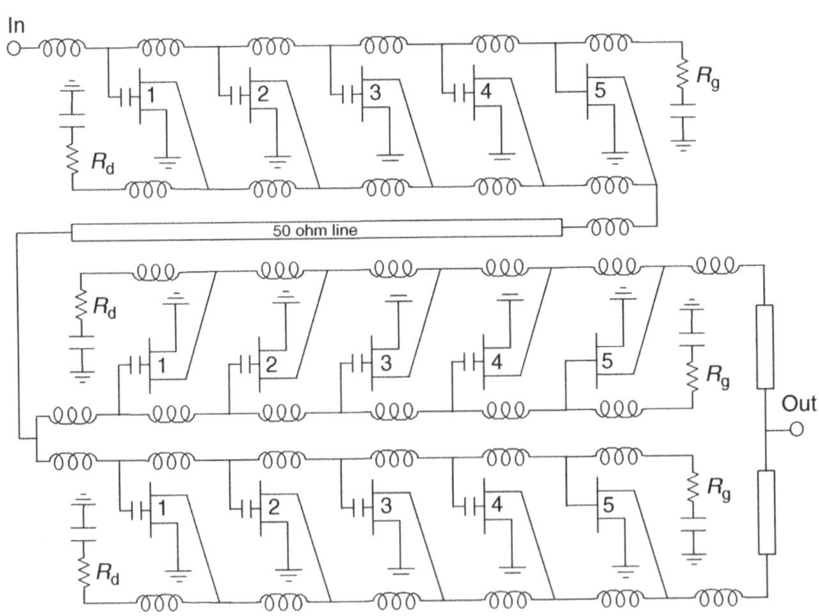

Figure 11.32 Schematic of a two-stage 1-W DA.

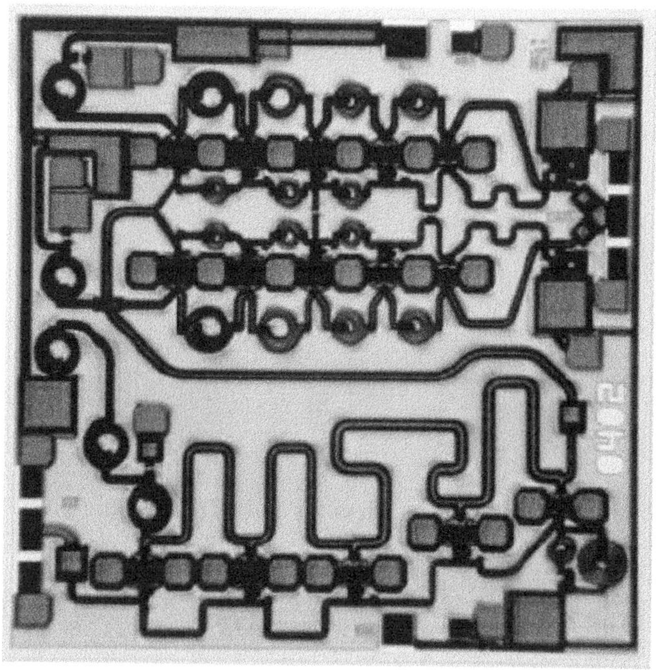

Figure 11.33 Photograph of the two-stage 2–18-GHz distributed power amplifier. Chip size is $3.0 \times 3.0\ mm^2$.

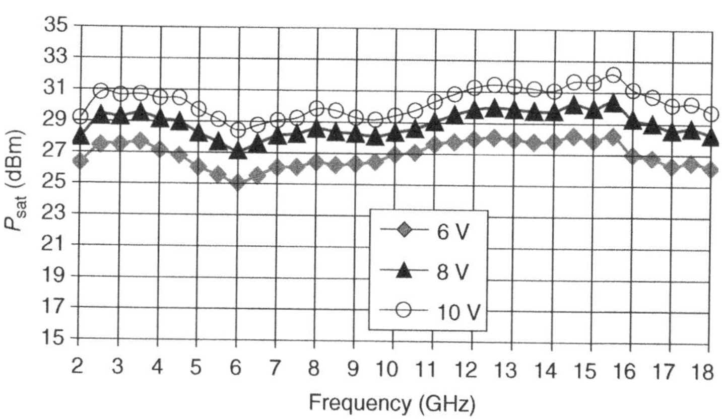

Figure 11.34 Saturated output power versus frequency and drain voltage.

no matching elements in the input and output. The interstage uses matching networks and the resistors R_1 and R_2 are the matching parts and also provide a required stable operation for the amplifier. The analysis for this topology is provided by Niclas [12]. It has great potential, especially in monolithic ICs [22].

Next, we discuss how the CG and CD configurations provide good input and output match, respectively. For a CG configuration, at low frequencies and ignoring

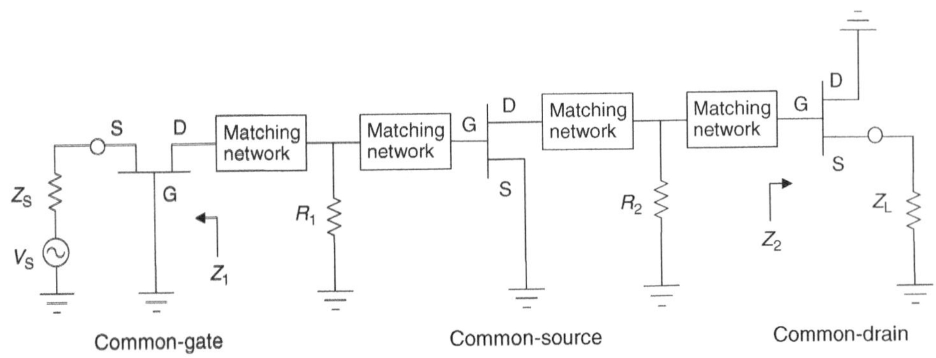

Figure 11.35 Schematic of an active broadband amplifier.

the device parasitic reactance, the *S*-parameters for an FET/HEMT are given by [12]

$$S_{11} = \frac{1 - g_m Z_S + (Z_1 - Z_S)/R_{ds}}{1 + g_m Z_S + (Z_1 + Z_S)/R_{ds}} \tag{11.28a}$$

$$S_{12} = \frac{2(Z_S Z_1)^{1/2}/R_{ds}}{1 + g_m Z_S + (Z_1 + Z_S)/R_{ds}} \tag{11.28b}$$

$$S_{21} = \frac{2(g_m + 1/R_{ds})(Z_S Z_1)^{1/2}}{1 + g_m Z_S + (Z_1 + Z_S)/R_{ds}} \tag{11.28c}$$

$$S_{22} = \frac{1 + g_m Z_S - (Z_1 - Z_S)/R_{ds}}{1 + g_m Z_S + (Z_1 + Z_S)/R_{ds}} \tag{11.28d}$$

where g_m and R_{ds} are the device's transconductance and output resistance, respectively. The impedances Z_S and Z_1 are shown in Figure 11.35. The input is perfectly matched when $S_{11} = 0$. From (11.28a),

$$g_m Z_S = 1 + (Z_1 - Z_S)/R_{ds} \tag{11.29}$$

Assuming $Z_1 = Z_S = Z_0 = 50\ \Omega$, from (11.28) and (11.29) we find

$$g_m = 1/Z_0 \tag{11.30a}$$

$$S_{12} = \frac{Z_0/R_{ds}}{1 + Z_0/R_{ds}} \tag{11.30b}$$

$$S_{21} = 1 \tag{11.30c}$$

$$S_{22} = \frac{1}{1 + Z_0/R_{ds}} \tag{11.30d}$$

Thus in a CG configuration, when the device size is selected such that the value of g_m is 20 mS, its input is matched to 50 Ω. This configuration has about unity gain.

Similarly, for a CD configuration, at low frequencies and ignoring the device parasitic reactance, the S-parameters for an FET/HEMT are given by

$$S_{11} = 1 \tag{11.31a}$$
$$S_{12} = 0 \tag{11.31b}$$
$$S_{21} = \frac{2g_m(Z_2 Z_L)^{1/2}}{1 + (g_m + 1/R_{ds})Z_L} \tag{11.31c}$$
$$S_{22} = \frac{1 - (g_m + 1/R_{ds})Z_L}{1 + (g_m + 1/R_{ds})Z_L} \tag{11.31d}$$

The output is perfectly matched when $S_{22} = 0$. From (11.31c),

$$(g_m + 1/R_{ds})Z_L = 1 \tag{11.32}$$

Assuming $Z_2 = Z_L = Z_0 = 50\ \Omega$, and $g_m \gg 1/R_{ds}$, from (11.31) and (11.32) we have

$$g_m = 1/Z_0 \tag{11.33a}$$
$$S_{21} = g_m Z_0 \tag{11.33b}$$

Thus in a CD configuration, when the device size is selected such that the value of g_m is 20 mS, its output is matched to 50 Ω. This configuration also has about unity gain.

EXAMPLE 11.7

Design a broadband low-power amplifier working from 0.1 to 5 GHz. The desired gain and VSWR values are 10 dB and 2:1. Select suitable devices and plot simulated performance.

SOLUTION A 150-μm 5-A FET biased at 3 V and 15% I_{dss} was selected with the following EC model parameters:

$$R_g = 1\Omega,\ R_i = 4\ \Omega,\ R_s = 2\ \Omega,\ R_d = 2\ \Omega,\ R_{ds} = 500\ \Omega$$
$$C_{gs} = 0.19\text{ pF},\ C_{gd} = 0.023\text{ pF},\ C_{ds} = 0.05\text{ pF}$$
$$g_m = 20.5\text{ mS},\ \tau = 2\text{ ps}$$
$$L_g = 0.02\text{ nH},\ L_s = 0.001\text{ nH},\ L_d = 0.02\text{ nH}$$

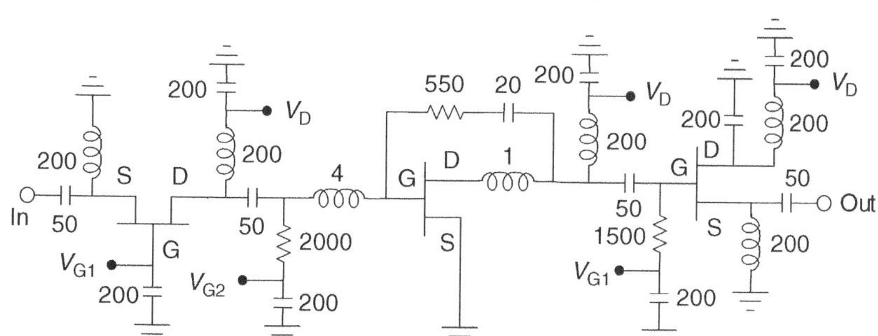

Inductors in nH, capacitors in pF, and resistors in ohm
V_D : 3 V, V_{G1} : −2.4 V (15%I_{dss}),V_{G2} : −1.2 V (45%I_{dss})

Figure 11.36 Schematic of an active broadband amplifier with biasing network.

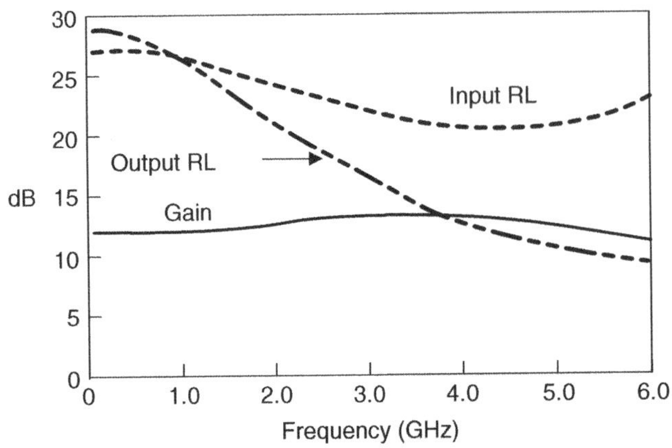

Figure 11.37 Simulated gain, input return loss (RL), and output return loss versus frequency of an ultra-broadband low-power amplifier.

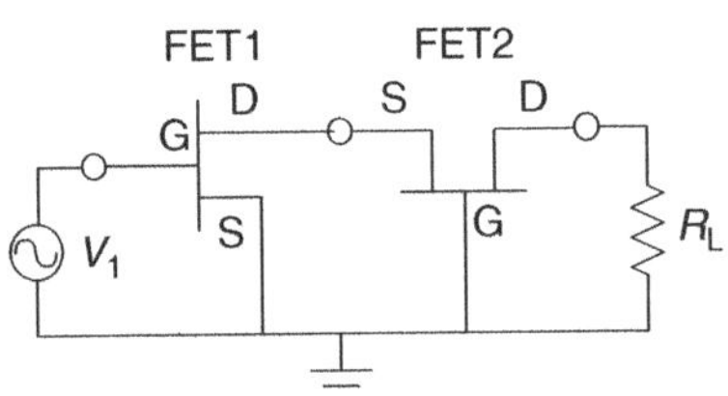

Figure 11.38 Basic cascode configuration.

The device has g_m of about 20 mS. The same FET but biased at 3 V and 45% I_{dss} was used for the CS configuration to obtain higher gain. The high-current version has g_m and C_{gs} values about 20% higher and C_{gd} value about 20% lower. Analyzing the data for the CG and CD configurations, it was found that the device isolation for the CD is not as good as for the CG. At 5 GHz, the isolation values for CG and CD configurations are about 19 dB and 11.4 dB, and these values are lower at higher frequencies. In order to flatten the gain, a feedback was used across the CS configuration, as shown in Figure 11.36. The value of the feedback resistor was selected to obtain flat gain within ±0.5 dB. Figure 11.37 shows the simulated data for the amplifier. ■

11.2.6 Cascode Configuration

When the CS and CG are connected in series, a cascode configuration is formed, as shown in Figure 11.38. In a CG configuration, if the gate is terminated in a resistor it makes the cascode cell a dual-gate FET and if the gate is connected to a capacitor it is equivalent to two FETs in series as discussed in Chapter 13. There are three main advantages of the cascode cell vis-à-vis CS: (a) high output impedance, (b) high reverse isolation due to low Miller feedback capacitance and therefore less sensitive to R_{ds}, and (c) high-voltage swing capability at the output. The g_m and C_{gs} for a cascode cell are approximately the same as for the CS case, but it has 3–4-dB higher gain due to low feedback capacitance and high output impedance. The cascode configuration has been used successfully in DAs to boost gain and bandwidth. The high-voltage swing capability feature has been used to improve output power as described in Chapter 13.

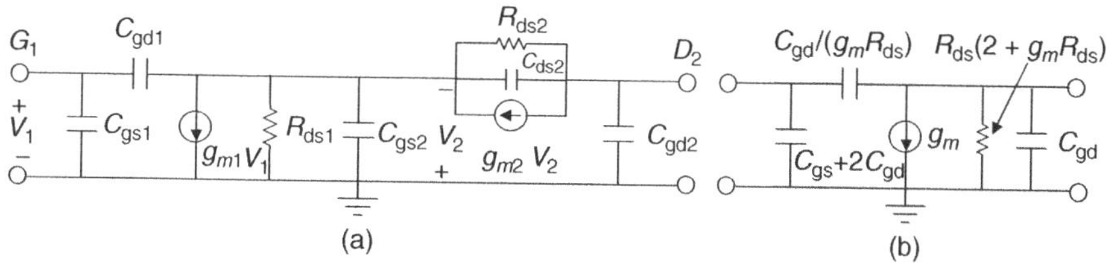

Figure 11.39 Low-frequency equivalent circuits for a cascode configuration: (a) two cascaded FETs and (b) simplified.

The common-source configuration is known as an inverting amplifier because it inverts the signals (i.e., multiplies the amplified signal by a negative constant factor). In this configuration, the input capacitance is larger for higher values of voltage gain due to the Miller effect. The C_{gd}, in addition to raising the input capacitance through the Miller effect, feeds back signals from the output to the input, contributing to possible instability, especially if the load itself is reactive. The Miller effect can be minimized by using the cascode configuration. Figure 11.39 shows equivalent circuits for a cascode cell.

The output impedance of the cascode cell is given by

$$R_{ds}^{cascode} = R_{ds1} + R_{ds2}(1 + g_{m2}R_{ds1}) \tag{11.34}$$

where the subscripts 1 and 2 designate FET1 (CS) and FET2 (CG), respectively. For the same FET widths, $g_{m1} = g_{m2} = g_m$ and $R_{ds1} = R_{ds2} = R_{ds}$,

$$R_{ds}^{cascode} = R_{ds}(2 + g_m R_{ds}) \tag{11.35}$$

For large $g_m R_{ds}$ values, the increase in the output impedance of a cascode cell is an order of magnitude higher in comparison to the CS case. Similarly, the decrease in feedback capacitance of a cascode cell is an order of magnitude lower in comparison to the CS case. Both these effects are responsible for the 3–4-dB gain increase in a cascode cell based amplifier.

11.2.7 Comparison of Broadband Techniques

A qualitative comparison of techniques for broadbanding of amplifiers previously described is given in Table 11.2. Broadband amplifiers can be designed using the impedance matching techniques described in Chapter 7 or CAD tools. CAD tools play a very important role in designing broadband amplifiers, as there are several required specifications such as gain, gain flatness, VSWR, noise figure, output power, PAE, and linearity that have to be met simultaneously. Analytical methods generally fall short in finalizing an amplifier design that meets all specifications.

11.3 BROADBAND POWER AMPLIFIER DESIGN CONSIDERATIONS

In the design of high-performance broadband power amplifiers, the selection of circuit topology, transistor aspect ratio, proper matching networks, and accurate thermal design become critical. These aspects are briefly discussed next.

Table 11.2 Comparison of Techniques Used for Broadband Amplifiers

Reactive/Resistive Matching	Balanced Circuit	Feedback	Distributed Approach
Good up to a two-octave bandwidth	Good up to a two-octave bandwidth	Good up to many-octave bandwidth	Good up to many-octave bandwidth
Requires large g_m devices	Requires matched devices pairs and quadrature couplers	Requires large g_m devices and uses negative as well as positive feedback	Requires many small devices
Moderate size	Size is relatively large	Size is relatively small	Size is moderate
Provides poor impedance match	Provides very good impedance match	Provides good impedance match	Provides good impedance match
Has low noise, high output power, and high PAE	Has high output power and moderate PAE	Has moderate output power and moderate PAE	Has low output power and poor PAE
Effect of fabrication tolerance is small	Effect of fabrication tolerance is small	Effect of fabrication tolerance is moderate	Effect of fabrication tolerance is moderate
Modular approach is not easy	Cascading of two or more gain modules is easy	Modular approach is easy	Modular approach is easy

11.3.1 Topology Selection

First and foremost, one must select a suitable amplifier topology to meet design requirements (available power supply, power output, and frequency range). For example, for bandwidths greater than two octaves, feedback is a preferred topology at low frequencies and DA is commonly used at high frequencies. However, for a bandwidth less than two octaves, the reactive/resistive technique provides the best output power and PAE.

11.3.2 Device Aspect Ratio

In a multistage amplifier design, a proper device aspect ratio plays a very important role in realizing high output power, high PAE, or a linear amplifier. Assuming that the output circuit loss can be kept to within 0.5 dB and another 0.5 dB due to mismatch between the desired and designed load impedance at the output of the device, the device must provide 5 W of output power to realize a 4-W power amplifier. In a two-stage amplifier design, the ratio of output to input stage device size depends on the device's compressed gain at the operating frequency, the PAE or linearity requirements, and the bandwidth. In broadband applications, a margin of 3–4-dB gain loss due to dissipative loss and mismatch loss between the input and output transistors is considered acceptable in the amplifier design. For example, FETs having 12 dB of compressed gain per stage at L- and S-bands require about 4:1 output to input stage FET size ratio for high-PAE applications. Therefore a two-stage 5-W HPA working over 1–4 GHz requires a 1.25-W device to drive a 5-W device. However, for high-linearity applications this ratio may be reduced to 3:1. At Ku- and K-bands, a 2:1 device aspect ratio is used due to lower transistor gain.

11.3.3 Low-Loss Matching Networks

It is very desirable to lower the dissipative loss in the power amplifier's output matching network using microstrip lines in order to improve the output power and PAE performance. Several techniques, discussed in Chapter 10 (Section 10.5.3), to improve PAE are applicable to broadband HPAs. The dissipation loss in the microstrip matching networks can be lowered by using relatively thick conductors [15] and using a low dielectric constant layer between the conductors and the substrate [20]. Thick conductors (8–10 μm) also improve their current and power handling capabilities. In the MMIC process, thick conductors are realized by combining two 4.5-μm thick conductors available in a multilevel plating (MLP) process [15]. The additional thick metallization layer available in the MLP process offers benefits in terms of lower loss at lower microwave frequencies and in the area of DC current routing.

11.3.4 Gain Flatness Technique

Most amplifiers require a flat gain response over the desired bandwidth. All amplifier design techniques described in Section 11.2 need some sort of gain compensation method depending on the required specifications. Gain compensation techniques described for reactive/resistive topologies can also be used in feedback and DAs. There is always a compromise between several critical requirements, including gain flatness, NF, input match, output power, PAE, or linearity. Any gain flatness technique used in broadband designs affects other parameters, depending on how stringent gain flatness requirements are. More demanding gain flatness requirements degrade NF and lower the output power and PAE.

11.3.5 Harmonic Termination

Since conventional harmonic terminations, described in Chapter 10, work over a narrow frequency (5–15%) band, an improved broadband harmonic termination technique is required. Another possible broadband harmonic termination technique uses the push–pull configuration described in Chapter 10. In this case, if the balun is broadband, the second harmonic is reactively terminated and results in higher PAE over a wider band. In this technique, a low-loss (0.3–0.5 dB) balun is needed. If the loss is about 1 dB, the push–pull scheme is not beneficial for gain and power, which are reduced by 2 dB and 1 dB, respectively. In this case, the net effect of broadband harmonic termination is diminished because of high balun loss and large size.

11.3.6 Thermal Design

Because of reduced PAE in broadband power amplifiers, the thermal design of such amplifiers needs special attention. In general, the PAE of multi-octave power amplifiers is about half of narrowband amplifiers; the temperature rise in broadband amplifiers is about twice. Thus the thermal design from the device level to the heat sink becomes important. Chapter 16 deals with all aspects of thermal design for power amplifiers.

REFERENCES

1. Y. Ayasi et al., A monolithic GaAs 1–13 GHz traveling-wave amplifier, *IEEE Trans. Microwave Theory Tech*., Vol. MTT-30, pp. 976–981, 1982.
2. J. P. Fraysse et al., A 2W, high efficiency, 2–8 GHz, cascode HBT MMIC power distributed amplifier, *IEEE Int. Microwave Symp. Dig*., pp. 529–532, 2000.
3. J. J. Xu et al., A 3–10-GHz GaN-based flip-chip integrated broad-band power amplifier, *IEEE Trans. Microwave Theory Tech*., Vol. 48, pp. 2573–2578, December 2000.
4. A. Sayed and G. Boeck, Two-stage ultrawide-band 5-W power amplifier using SiC MESFET, *IEEE Trans. Microwave Theory Tech*., Vol. 53, pp. 2441–2449, July 2005.
5. TriQuint Semiconductor, 13510 North Central Expressway, Dallas, TX.
6. M/A-COM, RF and Microwave Product Solutions, 1011 Pawtucket Blvd., Lowell, MA.
7. I. D. Robertson and I. J. Bahl Solid-state circuits, in *Electrical Engineering Handbook: Electronics, Power Electronics, Optoelectronics, Microwaves, Electromagnetics, and Radar* (R. C. Dorf, Editor), 3rd edition, CRC Press, Boca Raton, FL. 2006.
8. D. Conway, M. Fowler, and J. Redus, New process enables wideband high-power GHz amplifiers to deliver up to 20W, *Defense Electron*., pp. 8–11, February 2006.
9. I. J. Bahl, 0.7–2.7 GHz 12-watt power amplifier MMIC developed using MLP technology, *IEEE Trans. Microwave Theory Tech*., Vol. 55, pp. 222–229, February 2007.
10. I. J. Bahl, 2–8 GHz 8-watt power amplifier MMIC developed using MSAG MESFET technology, *IEEE Microwave Wireless Comp. Lett*., Vol. 18, pp. 52–54, January 2008.
11. K. B. Niclas et al., The matched feedback amplfier: ultrawide-band microwave amplification with GaAs MESFETs, *IEEE Trans. Microwave Theory Tech*., Vol. MTT-28, pp. 285–294, April 1980.
12. K. B. Niclas, Multi-octave performance of single-ended microwave solid-state amplifiers, *IEEE Trans. Microwave Theory Tech*., Vol. MTT-32, pp. 896–906, August 1984.
13. W. C. Peterson, A. Gupta, and D. R. Decker, A monolithic GaAs DC to 2-GHz feedback amplifier, *IEEE Trans. Microwave Theory Tech*., Vol. 31, pp. 27–29, January 1983.
14. K. Kurokawa, A wideband low noise L-band balanced transistor amplifier, *Proc. IEEE*, Vol. 53, p. 237, 1965.
15. I. Bahl, Ku-band MMIC power amplifiers developed using MSAG MESFET technology, *Microwave J*., Vol. 49, pp. 56–82, February 2006.
16. J. B. Beyer et al., MESFET distributed amplifier design guidelines, *IEEE Trans. Microwave Theory Tech*., Vol. 22, pp. 268–275, March 1984.
17. Y. Ayasli et al., Capacitively coupled traveling-wave power amplifier, *IEEE Trans. Microwave Theory Tech*., Vol. 32, pp. 1704–1709, December 1984.
18. S. N. Prasad, J. B. Beyer, and I. S. Chang, Power–bandwidth considerations in the design of MESFET distributed amplifiers, *IEEE Trans. Microwave Theory Tech*., Vol. 36, pp. 1117–1123, July 1988.
19. S. N. Prasad, S. Reddy, and S. Moghe, Cascaded-transistor cell distributed amplifiers, *Microwave Opt. Technol. Lett*., Vol. 12, No. 3, pp. 163–167, June 1996.
20. M. Ashman and I. Bahl, High performance wideband MSAG gain block/driver amplifier MMICs using MLP technology, *Microwave J*., Vol. 47, pp. 74–88, December 2004.
21. S. Olson, B. Thompson, and B. Stengel, Distributed amplifier with narrowband amplifier efficiency, *IEEE MTT-S Int. Microwave Symp. Dig*., pp. 155–158, 2007.
22. W. C. Peterson et al., A monolithic GaAs 0.1 to 10 GHz amplifier, *IEEE MTT-S Int. Microwave Symp. Dig*., pp. 354–357, 1981.

PROBLEMS

11.1 In a balanced amplifier S_{11A} and S_{11B} are the input reflection coefficients of single-ended A and B amplifiers. Show that at the input of the balanced amplifier the reflection coefficient

$$S_{11} = \frac{1}{2}[S_{11A} - S_{11B}].$$

11.2 Derive an expression for the noise figure F of a balanced amplifier in terms of the individual amplifier's noise figures F_1, F_2 and gains G_1, G_2 assuming the 3-dB 90° divider/combiner is ideal.

11.3 Prove that the noise figure of the balanced amplifier is equal to the loss in the coupler plus the noise figure of the single-ended amplifier.

11.4 A device load represented by a parallel RC is to be matched to 50 Ω over an octave bandwidth, 3–6 GHz. When $R = 100\ \Omega$ and $C = -0.5$ pF, what is the minimum reflection coefficient that can be obtained by using ideal matching elements?

11.5 A device input impedance represented by a series RC is to be matched to 50 Ω over an octave bandwidth, 3–6 GHz. When $R = 10\ \Omega$ and $C = 1.0$ pF, what is the minimum reflection coefficient that can be obtained by using ideal matching elements?

11.6 Determine the input return loss of the balanced amplifier when $S_{11A} = 0.5$ and $S_{11B} = 0.5e^{j\theta}$, where $\theta = 0, \pi/16$, and $\pi/8$.

11.7 The S-parameters for a transistor are given in Table 5.1. Calculate the maximum frequency of operation and gain for a five-section distributed amplifier using microstrip on 4-mil GaAs substrate.

11.8 Design a 2–8-GHz feedback amplifier with maximum stable gain using Table 5.1 and ideal elements. Calculate the amplifier's gain, S_{11}, and S_{22}.

Chapter 12

Linearization Techniques

An amplifier is called linear when the output power increases linearly with the input power. Linearity is defined based on generating a maximum level of acceptable distortion using a figure of merit such as TOI, ACPR, EVM or NPR. As input power increases, the amplifier transfer function becomes nonlinear and a stage is reached where the output power does not increase with the input power. This happens because of transistor current and voltage clipping and the variations of the transistor's transconductance and junction capacitances with input RF power. One of the measures of nonlinearity of amplifiers is intermodulation distortion (IMD). When more than one carrier frequency is present in a nonlinear amplifier, multiple sidebands will be generated as intermodulation products due to mixing in the nonlinear device. IMD can also arise from the combined effects of amplitude modulated (AM) signal introduced by a previous stage and AM to phase modulated (PM) conversion.

The reduction in gain (gain compression) and increase in gain (gain expansion) with input signal amplitude, referred to as AM–AM conversion, degrades the quality of amplitude modulated signals and increases adjacent channel leakage (ACL). This type of nonlinearity is known as AM–AM distortion. Gain expansion generally happens in class-B and class-C amplifiers. In amplifiers, in addition to gain compression at high input signal amplitudes, the transmission phase also varies with signal strength. Such nonlinearity is called phase distortion, also known as AM–PM distortion. This also degrades the signal quality and increases ACL. At high input signal amplitudes, the output signal also comprises harmonics of the fundamental frequency. This is known as harmonic distortion and is usually suppressed by incorporating filters in the amplifier matching networks or by using separate passive filters.

IMD prediction relies on either accurate simulation or measurements. IMD simulation requires accurate nonlinear models and appropriate computer simulation algorithms. This subject has been treated in detail in References 1–7. The purpose of this chapter is to provide an overview of amplifier linearization. Measurements of IMD will be treated in Chapter 22.

Amplitude modulated (AM) signals are strongly attenuated by saturated amplifiers, but frequency modulated (FM) signals are often passed through power amplifiers running at the point of maximum efficiency, which is usually above $P_{1\,\mathrm{dB}}$. Although an FM signal is less subject to distortion by gain compression than an AM signal, both are subject to distortion due to harmonic generation in nonlinear devices. The simplest form of distortion occurs when a sinusoidal input is converted to a square-wave output

Fundamentals of RF and Microwave Transistor Amplifiers. By Inder J. Bahl

by power-supply clipping. Under these conditions, the output signal will contain not only the fundamental frequency f but also the odd harmonics ($3f$, $5f, \ldots$) of the input signal. These harmonics will be more or less visible at the output, depending on the frequency response of the amplifier. If, for example, the output includes a lowpass filter that passes f but is cut off for $3f$, the output from a single-frequency input may be compressed but still purely sinusoidal. If clipping is asymmetrical, even and odd harmonics may be generated, but if filtering is applied to attenuate the $2f$ signal, the output will remain purely sinusoidal.

One of the largest users for power amplifiers is the wireless communications sector. Several applications include DECT, LAN systems, GSM, AMPS, WiMAX,CDMA, and WCDMA. These systems use different modulation schemes with significant amplitude and phase modulation to achieve higher spectral efficiency. These schemes mandate increased amplifier linearity as well as efficiency. Both requirements conflict with each other, as one can design either a linear or a high-efficiency amplifier. A power amplifier's efficiency is important to extend talk time for handsets or to reduce power dissipation in a base station transmitter. Thus suitable techniques to enhance the efficiency of linear amplifiers become important for modern power amplifiers.

12.1 NONLINEAR ANALYSIS

There are a number of different ways to measure the nonlinearity or distortion behavior of an amplifier. The simplest method is the measurement of the 1-dB gain compression power level, $P_{1\,\mathrm{dB}}$. For a single-carrier case, third-order intermodulation distortion measurement known as IP3 or TOI is performed, whereas for a multicarrier case ACPR, EVM, and NPR measurements are commonly used. Among these methods, the IP3 measurment is very popular and uses two closely spaced frequencies. When two signals at frequencies f_1 and f_2 are incident on an amplifier, the output of the amplifier contains these two signals, as well as intermodulation products at frequencies $mf_1 \pm nf_2$, where $m + n$ is known as the order of the intermodulation (IM) product.

Next, we discuss analytically the intermodulation products.

12.1.1 Single-Tone Analysis

Consider a weakly nonlinear two-port amplifier excited by an input sinusoidal signal $v_\mathrm{i}(t)$. The output voltage $v_\mathrm{o}(t)$ can be expressed as a power series of the input voltage as follows:

$$v_\mathrm{o} = k_1 v_\mathrm{i} + k_2 v_\mathrm{i}^2 + k_3 v_\mathrm{i}^3 + \cdots \tag{12.1}$$

where the k_i are complex coefficients. For weakly nonlinear amplifiers, $k_1 > k_2$, $k_2 > k_3, \ldots$. In a linear amplifier, $k_2 = k_3 = \cdots = 0$ and k_1 is known as the voltage gain. Neglecting phase characteristics, the k_i become real quantities. In a weakly nonlinear amplifier, the output voltage may be represented by only the first three terms in (12.1). For the input signal $v_\mathrm{i} = A\cos\omega t$, (12.1) becomes

$$\begin{aligned} v_\mathrm{o} &= k_1 A\cos\omega t + k_2 A^2\cos^2\omega t + k_3 A^3\cos^3\omega t \\ &= k_1 A\cos\omega t + k_2 A^2\left(\tfrac{1}{2} + \tfrac{1}{2}\cos 2\omega t\right) + k_3 A^3\left(\tfrac{3}{4}\cos\omega t + \tfrac{1}{4}\cos 3\omega t\right) \\ &= \tfrac{1}{2}k_2 A^2 + \left(k_1 A + \tfrac{3}{4}k_3 A^3\right)\cos\omega t + \tfrac{1}{2}k_2 A^2\cos 2\omega t + \tfrac{1}{4}k_3 A^3\cos 3\omega t \end{aligned} \tag{12.2}$$

Here, the sinusoidal signal is represented by the cos ωt function instead of sin ωt (as used in Chapter 8), because it allows a simple mathematical manipulation of mixing products. In Eq. (12.2), the first, second, third, and fourth terms represent DC, fundamental frequency (ω), second harmonic (2ω), and third harmonic (3ω) voltage components. The odd-order nonlinearities generate odd-order harmonics and even-order nonlinearities generate a DC component and even-order harmonics. This is the reason why a class-B amplifier operating over a half sinusoidal cycle results in even harmonics and DC current increased with v_i as nonlinearity increases. At the output of an amplifier the DC component is blocked by a blocking capacitor. The gain (G_{NL}) of the fundamental component under nonlinear operation is expressed as

$$G_{NL} = 20 \log \frac{k_1 A + \frac{3}{4} k_3 A^3}{A} = 20 \log \left(k_1 + \tfrac{3}{4} k_3 A^2\right) \tag{12.3}$$

For a linear amplifier, $v_o \propto v_i$ and $k_3 = 0$ and gain (G) becomes

$$G = 20 \log k_1 \tag{12.4}$$

For stable amplifiers $G > G_{NL}$, therefore $k_3 < 0$. When $G_{NL} < G$, the amplifier is in gain compression, and the 1-dB gain compression point is considered a figure of merit of an amplifier to evaluate its linear behavior. The 1-dB gain compression is defined as

$$G_{1dB} = \mathrm{G} - 1 \text{ dB} \tag{12.5}$$

Gain 1 dB lower is equivalent to a 0.8913 voltage ratio; therefore from (12.3) and (12.4),

$$0.8913 k_1 = k_1 + \tfrac{3}{4} k_3 A^2$$

or

$$A^2 = 0.145 \frac{k_1}{|k_3|} \tag{12.6}$$

Referring to Figure 12.1, when $Z_S = Z_L = Z_0$, the input power P_{in} and the output power P_{out} at ω are given (in dBm) as

$$P_{in} = 10 \log \left\{ \left(\frac{A}{\sqrt{2}} \right)^2 \frac{10^3}{Z_0} \right\} \tag{12.7}$$

$$P_{out} = 10 \log \left\{ \left(\frac{k_1 A + \frac{3}{4} k_3 A^3}{\sqrt{2}} \right)^2 \frac{10^3}{Z_0} \right\} \tag{12.8}$$

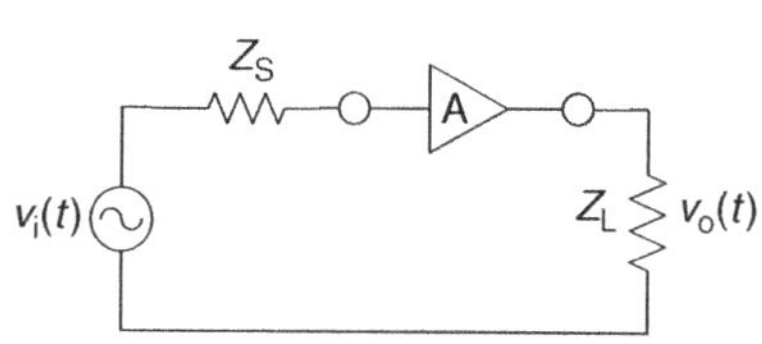

Figure 12.1 Schematic of an amplifier network.

From (12.3), (12.7), and (12.8), we find

$$P_{out} = G_{NL} + P_{in} \tag{12.9}$$

At 1-dB gain compression, P_{out} is given by

$$P_{1dB} = G - 1 + P_{in} \text{ dBm} \tag{12.10}$$

where G is power gain in dB. From (12.6), (12.7), and (12.10), when $Z_0 = 50\ \Omega$, we have

$$P_{1dB} = 10\ \log k_1^2 - 1 + 10\ \log\left[\frac{0.145k_1}{2|k_3|}20\right]\ \text{dBm} = 10\ \log\frac{k_1^3}{|k_3|} + 0.614\ \text{dBm} \tag{12.11}$$

This relationship will be used to establish a relation between single-tone P_{1dB} and the two-tone third-order intercept point. In (12.8), k_3 is a negative constant; under large conditions, it tends to lower P_{out}, giving rise to gain compression.

12.1.2 Two-Tone Analysis

Next, consider the input signal comprised of two signals of equal amplitude but different frequencies ω_1 and ω_2. In this v_i becomes

$$v_i = B[\cos\omega_1 t + \cos\omega_2 t] \tag{12.12}$$

where $\omega_1 = 2\pi f_1$ and $\omega_2 = 2\pi f_2$. Using (12.12) in (12.1) and retaining three terms, v_o becomes

$$\begin{aligned}
v_o &= k_1 B(\cos\omega_1 t + \cos\omega_2 t) + k_2 B^2(\cos\omega_1 t + \cos\omega_2 t)^2 + k_3 B^3(\cos\omega_1 t + \cos\omega_2 t)^3 \\
&= k_2 B^2 + k_2 B^2\cos(\omega_1 - \omega_2)t + \left(k_1 B + \tfrac{9}{4}k_3 B^3\right)\cos\omega_1 t \\
&\quad + \left(k_1 B + \tfrac{9}{4}k_3 B^3\right)\cos\omega_2 t + \tfrac{3}{4}k_3 B^3\cos(2\omega_1 - \omega_2)t \\
&\quad + \tfrac{3}{4}k_3 B^3\cos(2\omega_2 - \omega_1)t + k_2 B^2\cos(\omega_1 + \omega_2)t + \tfrac{1}{2}k_2 B^2\cos 2\omega_1 t \\
&\quad + \tfrac{1}{2}k_2 B^2\cos 2\omega_2 t + \tfrac{3}{4}k_3 B^3\cos(2\omega_1 + \omega_2)t + \tfrac{3}{4}k_3 B^3\cos(2\omega_2 + \omega_1)t \\
&\quad + \tfrac{1}{4}k_3 B^3\cos 3\omega_1 t + \tfrac{1}{4}k_3 B^3\cos 3\omega_2 t
\end{aligned} \tag{12.13}$$

Thus the output voltage, in addition to DC, fundamental frequencies, second harmonics, and third harmonics, also contains second-order intermodulation (SOI) products at $\omega_1 \pm \omega_2$ and third-order intermodulation (TOI) products at $2\omega_1 \pm \omega_2$ and $2\omega_2 \pm \omega_1$. In a narrowband amplifier and when ω_2 and ω_1 are very close to each other, all frequencies except ω_1, ω_2, $2\omega_1 - \omega_2$, and $2\omega_2 - \omega_1$ can be filtered out. The third-order products, which are very close to the fundamental frequency ω_1 or ω_2, will distort the desired signal. Again referring to Figure 12.1, the output powers (in dBm) for the fundamental and TOI product are given by

$$P_o = 10 \log \left\{ \left(\frac{k_1 B}{\sqrt{2}} \right)^2 \frac{10^3}{Z_0} \right\} \quad \text{(Linear)} \tag{12.14a}$$

$$P_{\omega_1} = 10 \log \left\{ \left(\frac{k_1 B + \frac{9}{4} k_3 B^3}{\sqrt{2}} \right)^2 \frac{10^3}{Z_0} \right\} \tag{12.14b}$$

$$P_{2\omega_1 - \omega_2} = 10 \log \left\{ \left(\frac{\frac{3}{4} k_3 B^3}{\sqrt{2}} \right)^2 \frac{10^3}{Z_0} \right\} \tag{12.14c}$$

By definition, linear $P_o = P_{2\omega_1 - \omega_2}$ at the third-order intercept point; from (12.14a) and (12.14c),

$$B^2 = \frac{4}{3} \frac{k_1}{|k_3|} \tag{12.15}$$

Therefore $P_{2\omega_1 - \omega_2}$ at the third-order intercept point denoted by P_I and expressed in dBm becomes

$$P_I = 10 \log \left(\frac{2}{3} \frac{k_1^3}{|k_3|} \right) \frac{10^3}{Z_0} = 10 \log \frac{k_1^3}{|k_3|} + 11.249 \tag{12.16}$$

From (12.11) and (12.16), we have

$$P_I - P_{1\,\text{dB}} = 10.635 \text{ dB} \tag{12.17}$$

This is a well-known relationship commonly used between the two-tone third-order intercept point (TOI or IP3) and the single-tone 1-dB gain compression power level ($P_{1\,\text{dB}}$). That is why IP3 is referred to 10 dB above $P_{1\,\text{dB}}$. From (12.14a), (12.14c), and (12.16), the power ratios are

$$P_o^3 = P_I^2 P_{2\omega_1 - \omega_2} \tag{12.18}$$

Expressed in dBm,

$$P_{2\omega_1 - \omega_2} = 3P_o - 2P_I \tag{12.19a}$$

In linear operation,

$$P_{2\omega_1 - \omega_2} \cong 3P_{\omega_1} - 2P_I \tag{12.19b}$$

Figure 12.2a illustrates the intermodulation (IM) products spectrum. The relative magnitude of the IM products depends on the details of how the amplifier saturates. For a bandwidth of an octave or less, only the third-order IM products that appear at

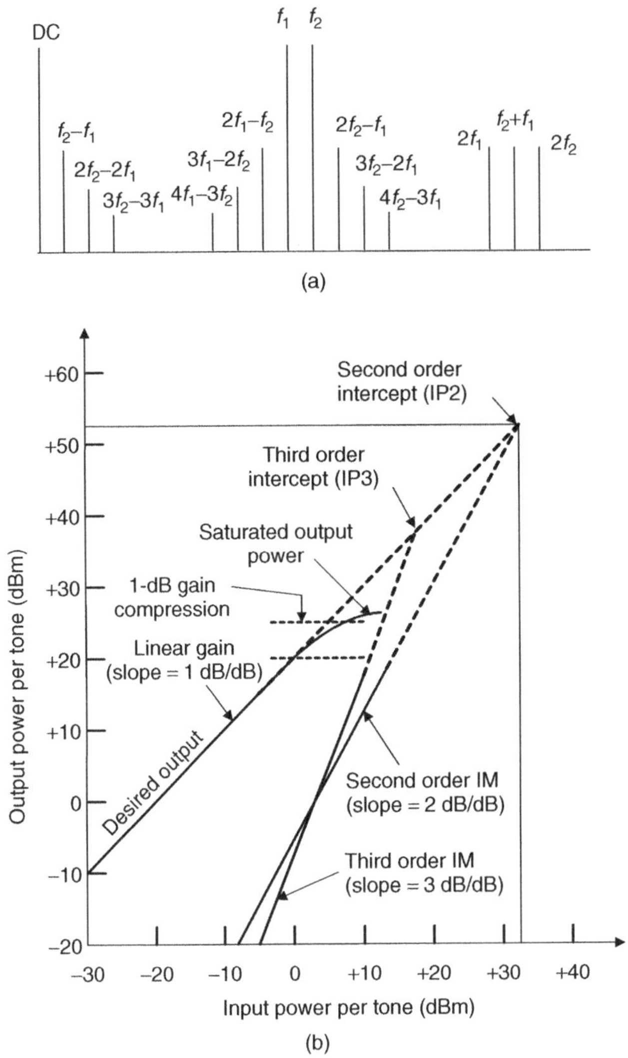

Figure 12.2 Spectrum showing two signals at frequencies f_1 and f_2 and their intermodulation products. (b) Variation of output power and intermodulation products with input power for a nonlinear amplifier.

$2f_1 - f_2$ and $2f_2 - f_1$ are the major sources of intermodulation distortion. However, second-order IM products appear in ultra-broadband amplifiers.

Figure 12.2b shows the transfer characteristics of a typical transistor amplifier having about 20 dB signal gain. With proper filtering, one of the fundamental frequencies and the distortion products are measured separately. Output powers of the second- and third-order IM products are also shown in Figure 12.2b. Since second- and third-order IM products correspond to square and cubic nonlinearities, respectively, the output power for these products increases by 2 dB/dB and 3 dB/dB as the input power is increased. If the fundamental linear transfer curve is extended, and the second- and third-order IM transfer curves are also extended, as shown in Figure 12.2b, these curves will meet at the point known as the intercept point. Intercept IP3 is the point where the power in the third-order product and fundamental tone are equal if the amplifier is assumed to be linear.

The third-order intercept power is typically 10 dB above the $P_{1\,\text{dB}}$ and is a very useful parameter for calculating low-level intermodulation effects.

If the intercept point is known, for any given fundamental output power the level of the second- or third-order IM products can easily be determined. Figure 12.2b also illustrates the 1-dB compression and the saturated output power characteristics. The third-order intercept point level and 1-dB compression point level for various amplifiers will be discussed later.

The ratio of third-order intermodulation products and the carrier is known as IM3 given by

$$IM3(\text{dBc}) = 10\ \log \frac{P_{2f_2-f_1}}{P_{f_2}} = 10\ \log \frac{P_{2f_1-f_2}}{P_{f_1}} \tag{12.20}$$

where P_{f_1}, P_{f_2}, $P_{2f_1-f_2}$, and $P_{2f_2-f_1}$ are the power outputs at frequencies f_1, f_2, $2f_1 - f_2$, and $2f_2 - f_1$, respectively. IM3 is described in dBc units, because it is specified at fundamental tone power output.

EXAMPLE 12.1

The output power transfer function of an amplifier designed in a 50-Ω system has $k_1 = 8$, $k_2 = 0$, and $k_3 = -0.5$. If the input power is 20 dBm, determine P_{out}. Also calculate $P_{1\text{dB}}$ and IP3 in dBm.

SOLUTION Gain $G = 20\ \log k_1 = 18.06$ dB. From (12.11), we find

$$P_{1\text{dB}} = 10\ \log(8^3/0.5) + 0.614 = 30.72\ \text{dBm}$$

At $P_{1\text{dB}}$, the input power is $P_{\text{in}} = 30.72 - 17.06 = 13.66$ dBm. At $P_{\text{in}} = 20$ dBm, from (12.7), we find

$$P_{\text{in}} = 20 = 10\ \log(0.5 \times A^2 \times 20), \quad A^2 = 10, \quad A = 3.1623\ \text{V}$$

From (12.8), the output power at $P_{\text{in}} = 20$ dBm is

$$P_{\text{out}} = 10\ \log\left[10\left(8 - \tfrac{3}{4} \times 0.5 \times 10\right)^2 \times 10\right] = 32.57\ \text{dBm}$$

From (12.16), IP3 $= 10\ \log(8^3/0.5) + 11.249 = 41.35$ dBm. ■

12.2 PHASE DISTORTION

In addition to the gain, power, and PAE of a power amplifier, phase and group delay characteristics are often very important. The transmission phase, ϕ_T (in radians), is given by

$$\phi_T = \text{ang}(S_{21}) \tag{12.21}$$

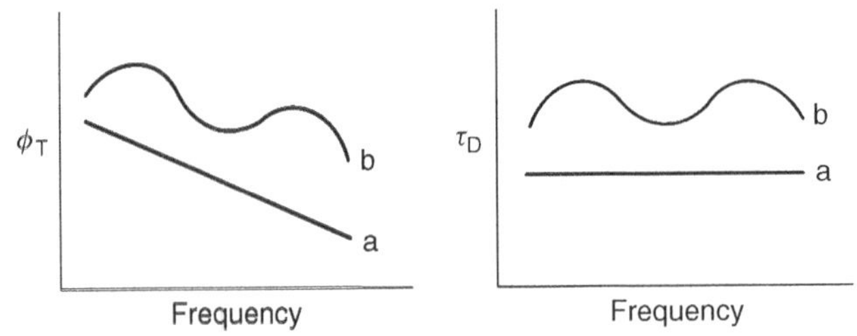

Figure 12.3 Transmission phase and group delay versus frequency. a: small signal and b: large signal.

and the group delay, τ_D (in seconds), is defined as

$$\tau_D = -\frac{d\phi_T}{d\omega} = -\frac{1}{2\pi}\frac{d\phi_T}{df} \tag{12.22}$$

where ω is in radians per second.

Group delay is important in several ways. It is a measure of how long a signal takes to propagate through the amplifier. Deviation from constant group delay over a given frequency band will cause FM signals to become distorted. With constant group delay, components of multifrequency signals will travel at the same velocity through the device with the result that there will be no frequency dispersion: sharp pulses will remain sharp and so on.

When the transmission phase of the amplified signal under power drive is a nonlinear function of frequency, the different frequency signals will have different group delays resulting in phase distortion. Examples of constant group delay (small-signal condition) and nonlinear group delay (2–3-dB power compression) are shown in Figure 12.3. The small-signal and large-signal data are represented by curves a and b, respectively.

There are two ways used to define the limits of phase distortion. The most common is to specify maximum permissible group delay variation over frequency. Alternatively, the maximum deviation from linear phase (DLP) may be specified. The DLP over a given frequency range is the maximum deviation between the device phase and a linear phase; that is,

$$DLP = \max[\phi_T - K\omega t] \tag{12.23}$$

where constant K has been chosen to minimize the deviation from linear phase. Note that if the group delay is constant, DLP is zero.

AM-to-PM Distortion

Power amplifiers have another type of phase distortion where the transmission phase is a function of input power drive level. In other words, at a given frequency, the transmission phase changes with input power drive level due to the change in C_{gs} and C_{gd} values in the transistor. This change is called AM-to-PM distortion and is expressed as degrees/dB at specified output power level. For example, in many applications, AM-to-PM specification is less than 5°/dB above $P_{1\,dB}$ level. For linear amplifiers AM-to-PM on the order of 1°/dB is desired as obtained in Example 10.1 in Chapter 10.

12.3 LINEARIZATION OF POWER AMPLIFIERS

Designing a power amplifier for systems that will have many closely spaced signals is very challenging. Intermodulation (IM) products typically fall on top of adjacent signals, potentially wrecking system performance. In this section we will introduce design concepts typically employed in power amplifiers for multicarrier systems.

The simplest and most widely used approach to countering IM is to run the power amplifier at power levels well below $P_{1\,\mathrm{dB}}$. There are many amplifiers in the field today running up to 10 dB below $P_{1\,\mathrm{dB}}$ (see Table 3.2 for different modulation schemes). This works, but at the cost of having a much larger amplifier than one might expect, running at low PAE. Backing down 10 dB might reduce a typical power amplifier from 50% to 5% PAE. However, recently developed transistors such as E-FET and E-pHEMT provide 20–25% PAE under a 10-dB power back-off condition but at lower power density.

Power amplifiers in communication satellites often counter IM by breaking the frequency band into narrow subbands, allowing the filtering of the IM products. The design trade of introducing much more hardware complexity for improved PAE is often chosen because prime power and cooling are major concerns for satellite operation. Cost constraints for terrestrial systems usually do not support this channelization scheme.

There is a great deal of activity underway in support of the telecommunication industry to develop alternative solutions to obtain high efficiency with excellent IM performance. High linearity in power amplifiers can also be achieved either by using highly linear devices such as pulse/spike doped devices [8, 9] along with optimum input and output match [10, 11] or by using predistortion techniques [12–29] at the circuit level or a cancellation technique such as feedforward [2, 30, 31]. These are briefly discussed next.

12.3.1 Pulsed-Doped Devices and Optimum Match

The IMD characteristics of GaAs FETs and their relation to doping profile have been studied, and it was found that pulsed doped FETs have 5–20-dB better IP3 performance compared to conventional FETs. Variation of transconductance (g_m) with gate–source voltage (V_{gs}) of two transistors is shown in Figure 12.4. In general, a transistor with constant g_m has better linear characteristics than for a transistor with

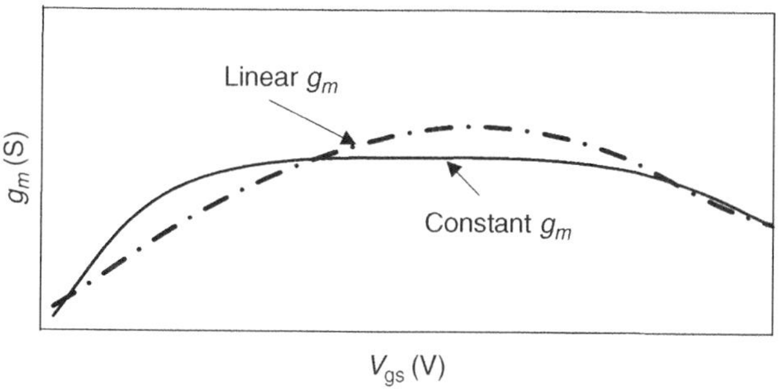

Figure 12.4 Variation of transconductance with gate–source voltage of two transistors.

linear g_m. Assuming only nonlinearity exists due to transconductance of the device, the output second-order intercept IP2 and third-order intercept point IP3 are given by [9]

$$IP2 = \frac{g_m^4 R_{\mathrm{ds}}}{2(g_m')^2} \tag{12.24}$$

$$IP3 = \frac{g_m^3 R_{\mathrm{ds}}}{g_m''} \tag{12.25}$$

where g_m' and g_m'' are the first and second derivatives of the transconductance and R_{ds} is the output resistance of the FET. Thus by minimizing g_m' and g_m'' values, one can obtain higher values of IP2 and IP3.

To characterize a linear device, the following figure of merit may be used:

$$IP3 = P_{\mathrm{DC}} + X \text{ (dBm)} \tag{12.26}$$

where X is expressed in dB. Desirable features are that the parameter X must be as large as possible and constant with FET gate periphery at the operating drain voltage. For a linear power transistor whose EC model parameter values are given in Table 5.5, the value of $X \geq 11$ dB when matched for high IP3. For example, a 600-μm transistor biased at 5 V and 120 mA, has an IP3 value of about 39 dBm. Such devices do not have good PAE under backed-off power conditions. For class-A and class-B depletion-mode transistors, the PAE under 4–10-dB power back-off, is on the order of 10–12%, whereas for enhanced-mode devices the PAE >20%. For example, both types as described in Chapter 5 have similar output power levels, that is, about 0.8 W/mm.

Power amplifiers can be designed for improved IM performance (typically 3–5 dB) by using either nonlinear modeling or source-pull and load-pull data to select optimum input match and output load impedance for the design of the output power level for best IM performance. An improvement of 3 dB may mean cutting prime power requirements by 50%.

The design of linear amplifiers does not depend on the number of carriers and the carrier modulation scheme. The linear power amplifiers can be classified into two categories, based on single carrier or multicarrier. The modulated input signal could have constant envelope or nonconstant envelope. Several papers have reported the design of both high-linearity and high-PAE power amplifiers under fixed 50-Ω load [32–36] and variable load condition [10, 11]. All these studies are based on predetermined device load conditions using the load-pull technique. The effect of input match and output match on the IP3 performance is very critical.

EXAMPLE 12.2

Design a linear power amplifier at a supply voltage of 5 V using the linear model given in Section 5.8.3 (Chapter 5) with an output IP3 = 38 dBm. The EC model for the device is given in Table 5.5 and the optimum source and load impedance model is given in Figure 5.17. The frequency of operation is 3.6 GHz and the output matching network loss is 0.5 dB. Calculate the input and output return losses, gain, and DC power (P_{DC}) and ratio of IP3 to P_{DC}. For the

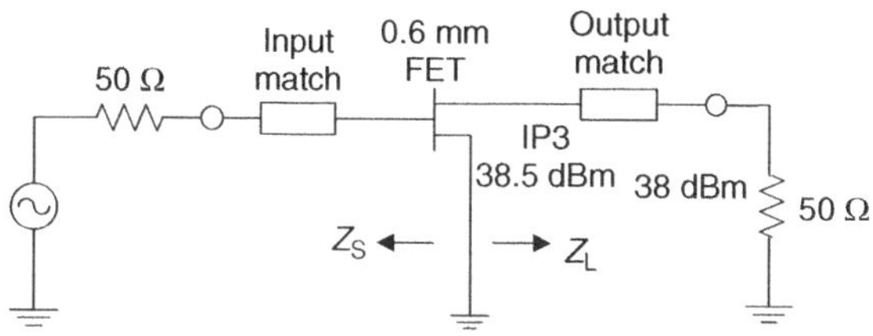

Figure 12.5 Schematic of the linear amplifier showing source and load impedances.

linear device, the IP3 load impedance, IP3 source impedance, IP3 and I_{ds} values are

$$R_S = 12\ \Omega \cdot \text{mm} \quad \text{and} \quad C_S = -1.5\ \text{pF/mm} \quad \text{(series)}$$
$$R_L = 23\ \Omega \cdot \text{mm} \quad \text{and} \quad C_L = -0.33\ \text{pF/mm} \quad \text{(shunt)}$$
$$IP3 = 41\ \text{dBm/mm} = 12.59\ \text{W/mm}$$
$$I_{ds} = 200\ \text{mA/mm}$$

SOLUTION Since the amplifier has 0.5-dB output matching network loss, the device must have 38.5 dBm = 7.08 W output IP3. Therefore the device size required is

$$7.08/12.59 = 0.5624\ \text{mm}$$

Here we selected the 0.6-mm device size whose EC model is given in Table 5.5. The IP3 and DC power for a 0.6-mm device are

$$IP3 = 7.08 \times 0.6 \div 0.5624 = 7.55\ \text{W} = 38.78\ \text{dBm}$$
$$P_{DC} = 5 \times 0.6 \times 200 = 600\ \text{mW} = 27.78\ \text{dBm}$$
$$IP3/P_{DC} = 7.55/0.6 = 12.58 = 11\ \text{dB} \quad \text{or} \quad 38.78\ \text{dBm} - 27.78\ \text{dBm} = 11\ \text{dB}$$

Figure 12.5 shows the schematic of the linear amplifier. For the 600-μm device,

$$R_S = 20\ \Omega \quad \text{and} \quad C_S = -0.9\ \text{pF} \quad \text{(series)}$$
$$Z_S = R_S - j/(\omega C_S) = 20 + j49.1\ \Omega$$
$$R_L = 38.3\ \Omega \quad \text{and} \quad C_L = -0.198\ \text{pF} \quad \text{(shunt)}$$
$$Z_L = R_L \,||- j/(\omega C_L) = 37.2 + j6.4\ \Omega$$

The circuit shown in Figure 12.6a was optimized to provide correct source and load impedances at the device terminals for the required IP3. The matching elements were assumed ideal components; bypass capacitors have very large values and their effects were ignored. The simulated input return loss and output return loss (RL), gain, and K factor (see Chapter 17 for details on stability factor K) are as follows

Frequency (GHz)	Input RL (dB)	Output RL (dB)	Gain (dB)	K
3.4	2.3	2.5	19.0	0.21
3.6	2.8	2.9	18.9	0.22
3.8	3.0	3.7	18.6	0.23

For this set of source and load conditions, the amplifier is potentially unstable. Next, a series resistance value of 14 Ω was added in series with a gate to stabilize the amplifier and the

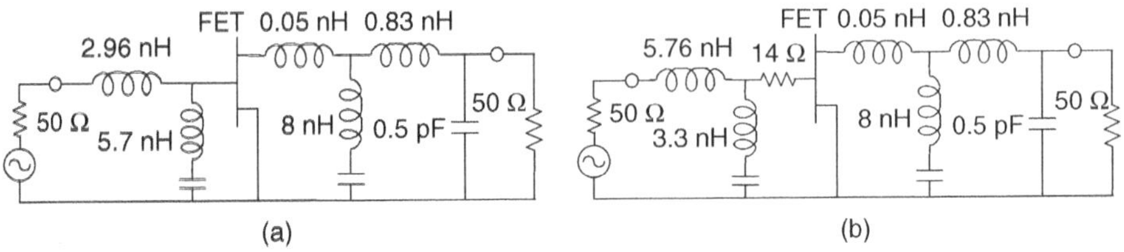

Figure 12.6 Schematic of the linear amplifier with matching element values.

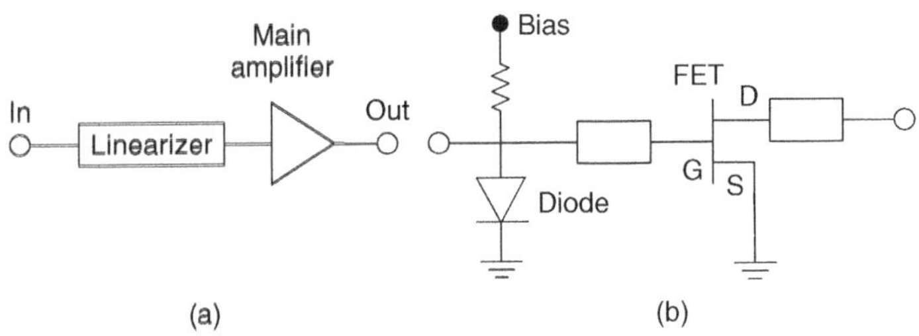

Figure 12.7 Configuration of a linearized amplifier using the predistortion technique and (b) schematic using a diode as a linearizer.

input matching circuit was reoptimized to get the K factor greater than 1 above 3.4 GHz. The recalculated performance of the amplifier, shown in Figure 12.6b, is as follows:

Frequency (GHz)	Input RL (dB)	Output RL (dB)	Gain (dB)	K
3.4	4.3	2.5	13.8	1.01
3.6	6.0	2.9	13.8	1.07
3.8	6.5	3.7	13.5	1.13

In this case, the amplifier's gain is reduced by about 5 dB and the input is still not matched to 50 Ω. In order to match the amplifier to 50 Ω (e.g., RL >20 dB) at the input and output, a balanced configuration may be used. Since the balanced amplifier has about 3-dB higher power/IP3 as compared to its single-ended version, the modified single-ended design will require about half the device size (300-μm FET instead of 600-μm FET). Power dissipation will be approximately the same. ■

12.3.2 Predistortion Techniques

The nonlinearity in an amplifier as previously discussed results in nonlinear behavior in amplitude (AM–AM) and phase (AM–PM) responses. Thus the distortion introduced in an amplifier can be explained in terms of AM–AM and AM–PM characteristics and strongly depends on the class of operation. In predistortion techniques, the amplitude and phase variation with input power of an amplifier are compensated for by adding extra elements in the circuit at the input of the amplifier, as shown in Figure 12.7. As an example, the diode as a predistortion circuit shown in Figure 12.7b has gain and transmission phase slopes as positive and negative, respectively, while the main amplifier has negative and positive slopes for the respective parameters. Figure 12.8 shows the power-dependent magnitude and phase and their cancellation in a linear

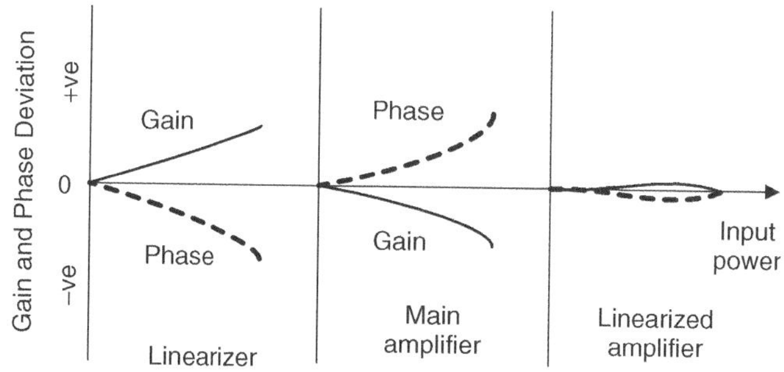

Figure 12.8 Basic power-dependent magnitude and phase cancellation technique for a linear amplifier.

amplifier using a predistortion scheme. This results in high PAE for digitally modulated signals.

A large number of linearization techniques have been used to linearize a nonlinear amplifier. These predistortion techniques include using dual-gate FET [12], series feedback [13], IF feedback [14], active feedback [15, 16], harmonic feedback [17–19], series [20, 21] and shunt diodes [22], passive FET [23], cascode [24], push–pull [25], second-order intermodulation component feedforwarding [26], current dumping [27], branch FET [28], and the Doherty technique [29]. Several of these techniques have shown drastic improvement in IP2/IP3 and ACPR. In most methods, accurate knowledge of amplitude and phase of the input signal is required along with accurate nonlinear device models predicting AM–AM and AM–PM characteristics. The predistortion technique is simple in concept but not so simple to implement in a stable reproducible way.

12.3.3 Feedforward Technique

A large amount of work on distortion cancellation has been published recently. Distortion cancellation techniques are used at the system level and are more complex and require sophisticated CAD. An overview of these techniques can be found in References 2, 30, and 31 and feedforward is the most popular. These techniques are very complicated for handset applications but are commonly used for base station applications. The basic idea in a feedforward design is described next.

A basic circuit schematic for the feedforward configuration is shown in Figure 12.9. For example, consider a two-tone signal at the input. The signal is split using coupler 1 between the main signal and the sample signal. The main signal is fed to the main power amplifier (path 1) and the sample signal enters the control circuitry (path 2). The amplified signal is again sampled using coupler 2 and combined with the path 2 signal using a combiner in such a way that the two-tone signal is cancelled and the intermodulation products are retained. Next, the intermodulation products are amplified using a relatively small amplifier, also known as a comparator or error amplifier, and combined with the main amplified signal in coupler 3. The amplitude and phase of the injected intermodulation products are so adjusted that they cancel out the intermodulation products present in the main signal and leave the output signal intermodulation products free. Feedforward amplifiers are very involved and mainly used in base station power transmission. Because it is not within the scope of this book to go into such detail, readers are referred to the aforementioned references.

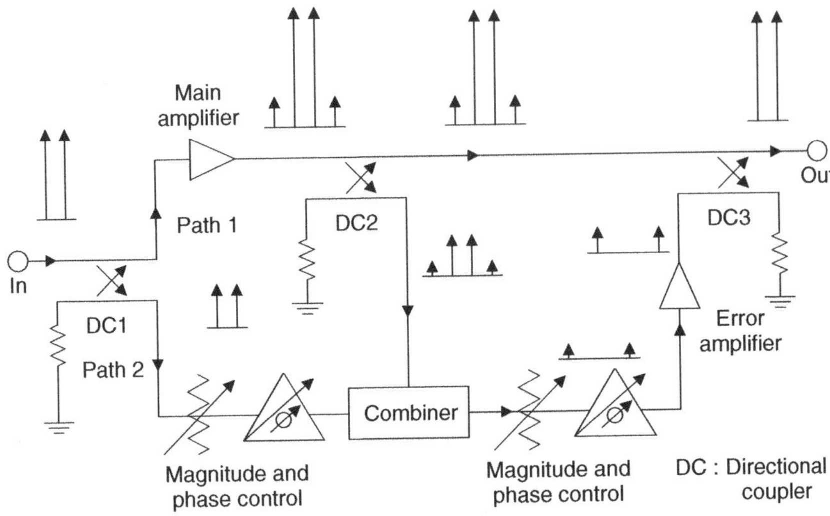

Figure 12.9 Schematic diagram of a simplified feedforward amplifier.

Among linearity boosting techniques, predistortion (PD) and feedforward (FF) methods are the most popular. Among these, the analog PD technique is the simplest and the FF technique is the best performing but has poor PAE.

12.4 EFFICIENCY ENHANCEMENT TECHNIQUES FOR LINEAR AMPLIFIERS

Base station transmitters used for modern wireless communications systems have stringent linearity and efficiency requirements for HPAs. Several methods at the system level are being used to improve both linearity and efficiency [7, 37–39]. Most efficiency enhancement techniques for linear amplifiers are quite old and use either complex circuits or multiple (mostly two) amplifiers connected in parallel. Some of the techniques are Chireix's outphasing [40–42], also known as linear amplification with nonlinear control (LINC), the Doherty technique [43–53], envelope elimination and restoration (EER) [54, 55], and bias adaptation [56, 57]. Except for the Doherty amplifier configuration, all other techniques require complex external circuitry. The Doherty topology with a predistortion technique meets the need of high PAE and linearity for base station amplifiers. These techniques are briefly discussed next. The Doherty technique is the most popular.

The Volterra-based adaptive digital predistortion (DPD) linearizer technique to improve linearity with high efficiency has been proposed recently [58]. This technique extends the PA's linear range with reduced peak-to-average power ratios (PARs) or crest factor. It provides a 4-dB improvement for orthogonal-frequency division multiplex (OFDM) signals while meeting ACPR and EVM specifications. This technique can be optimized for a variety of cellular standards including WCDMA, TD-SCDMA, MC-GSM, and Long-Term Evolution (LTE). This technique in combination with Doherty topology or class AB with dynamic envelope tracking improves PA efficiency from 5–15% to 15–50% [58].

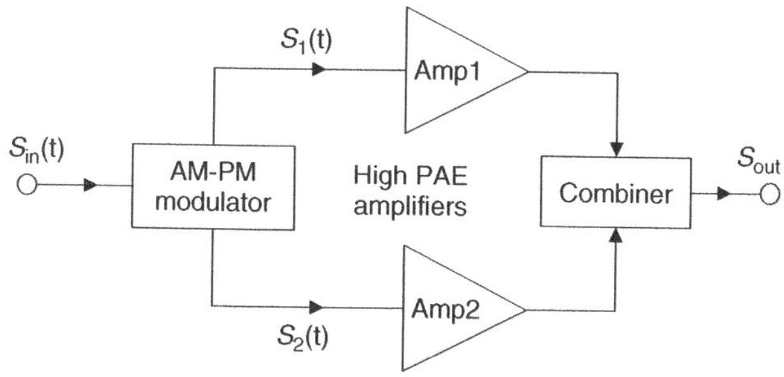

Figure 12.10 Schematic of an outphasing technique.

12.4.1 Chireix Outphasing

A simple outphasing scheme is shown in Figure 12.10. The amplitude modulated input signal $S_{\text{in}} = A(t)\cos(\omega t)$ is divided into two signals of constant envelopes with different phases, ϕ and $-\phi$. These two signals are amplified and added with peak output power when $\phi = 90^\circ$. The two signals are represented by

$$S_1(t) = \cos(\omega t + \phi) \quad (12.27a)$$

$$S_2(t) = \cos(\omega t - \phi) \quad (12.27b)$$

where

$$\phi = \cos^{-1}[A(t)] \quad (12.27c)$$

Using the trigonometric relationships described in Appendix E, we find

$$S_{\text{out}}(t) = G[S_1(t) + S_2(t)] = 2GA(t)\cos(\omega t) \quad (12.28)$$

where G is the gain of each amplifier. Thus the output signal of the system is the amplitude modulated input signal amplified by $2G$. Here, power amplifiers are matched and have high efficiency.

12.4.2 Doherty Amplifier

The design of Doherty amplifiers has been treated extensively in recent years [43–53] using LDMOS, GaN, pHEMT, CMOS, and HBT devices. The major application has been in base station transmitters. A basic topology for a Doherty amplifier is shown in Figure 12.11. It consists of a parallel combination of two different single-ended power amplifiers that combine their outputs reactively at the output. One single-ended branch is called the main/carrier amplifier and the other one is called the peaking/auxiliary amplifier.

The main amplifier devices are biased for class AB/B, while the peaking amplifier devices operate in class-C mode. During low output power operation, the peaking amplifier is turned off and the main amplifier is turned on. At low input power levels, the output of the peaking amplifier is considered an open circuit. As the input power is increased, the peaking amplifier starts turning on. The design of the Doherty amplifier

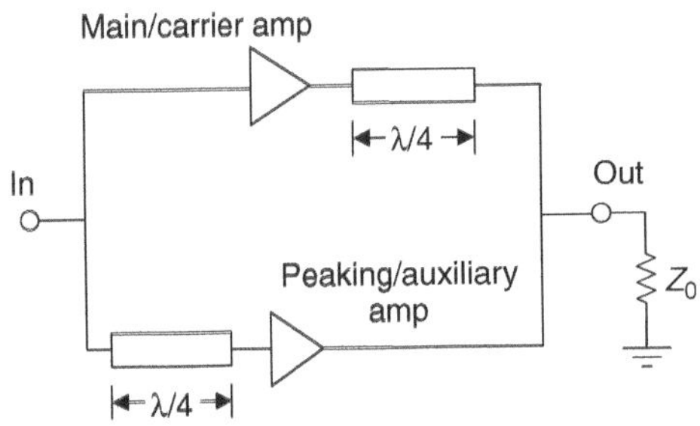

Figure 12.11 Basic Doherty amplifier topology.

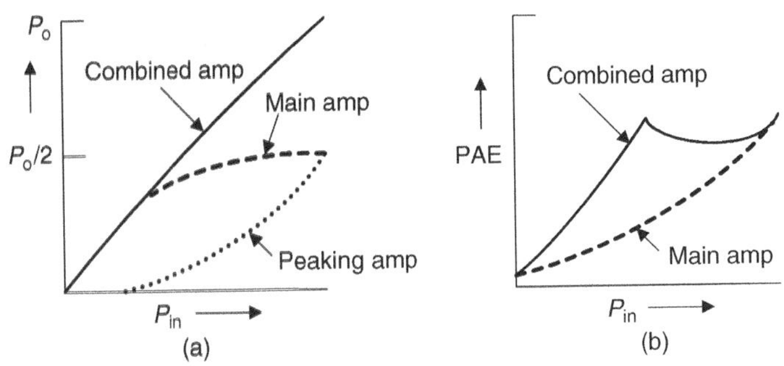

Figure 12.12 Operation of a Doherty amplifier versus input power: (a) output power and (b) PAE.

is done in such a way that maximum PAE of the amplifier is maintained over a large range of input power. Figure 12.12 shows the output power and PAE versus input power.

Figure 12.13 shows two implementations of a Doherty amplifier. The purpose of the input hybrid/coupler or power divider is to split the input signal and maintain good stability. In order to maintain good linearity in Doherty amplifiers, predistortion techniques are commonly used. The load impedances for the main (Z_m) and peaking (Z_p) amplifiers are given by [52]

$$Z_m = \frac{Z_0}{0.5[1 + I_p/I_m]} \tag{12.29a}$$

$$Z_p = 0.5Z_0[1 + I_m/I_p] \tag{12.29b}$$

where I_m and I_p are the main and peaking amplifier currents and they are strong functions of input power. Typical modulation range of Z_m is $2Z_0$ to Z_0 and for Z_p it is from infinity to Z_0.

A 35-W saturated power Doherty amplifier was developed at 2.14 GHz using two 25-W GaN HEMTs and the PCB technique [53]. The devices were biased at 28 V. The measured PAE and gain of the amplifier are shown in Figure 12.14. Measured saturated output power and PAE values were 35.5 W and 45%, respectively. At a 9-dB back-off power ($P_o = 36.5$ dBm), the measured PAE and ACLR values were 39.7% and −35.4 dBc, respectively.

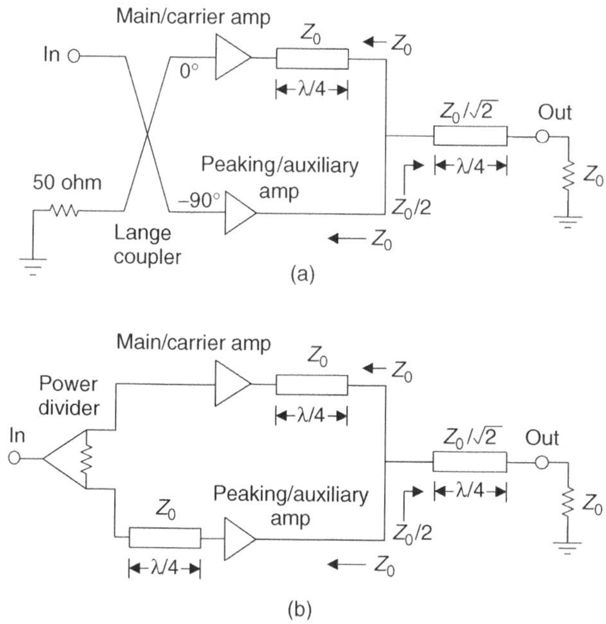

Figure 12.13 Implementation of a Doherty amplifier using (a) a Lange/hybrid coupler and (b) a power divider.

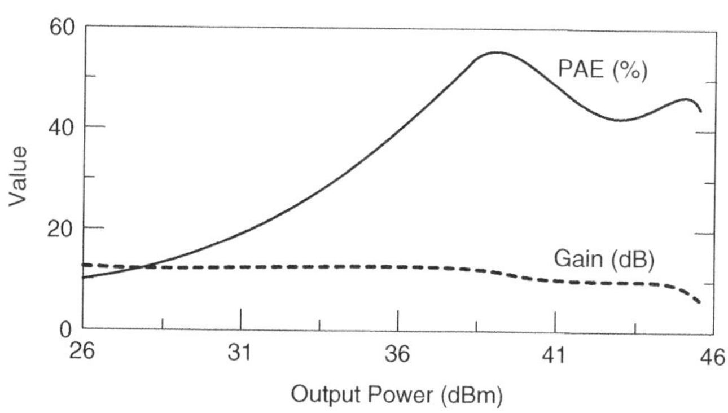

Figure 12.14 Measured gain and PAE of the 35-W Doherty amplifier.

12.4.3 Envelope Elimination and Restoration

The envelope elimination and restoration (EER) technique uses dynamic biasing for linearity and efficiency improvements. Figure 12.15 shows a schematic of an EER scheme. In this case the amplitude and phase information of the input signal are separated and the amplitude is kept constant using a limiter. The input signal is passed through a limiter, which provides constant amplitude to the high-efficiency amplifier and the envelope information is eliminated. The limiter also minimizes AM–PM distortion. In the other path, the detected input power is fed to the modulator. The amplitude is adjusted using the adaptive modulated bias control and the envelope of the output

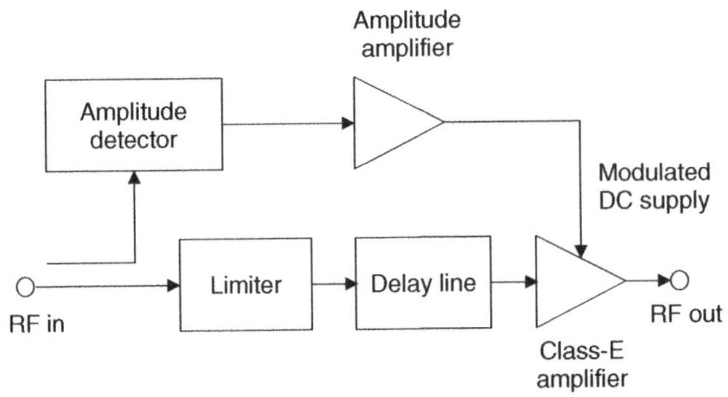

Figure 12.15 Schematic of an EER technique.

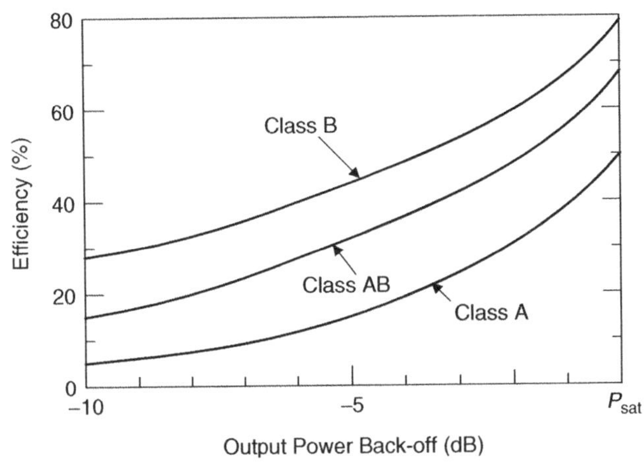

Figure 12.16 Amplifier efficiency versus output power back-off.

signal is restored. The delay line allows proper time alignment of the modulated supply with the power amplifier drive.

12.4.4 Bias Adaptation

The output power of an amplifier using transistors can be varied by input power level, controlling the drain/collector current and voltage. If the input power is lowered significantly so that the transistors are operating well below the saturation point, both output power and PAE are reduced with lowered input power. The amplifier efficiency versus output power back-off is shown in Figure 12.16. However, the efficiency is maintained high if the supply drain/collector voltage is reduced in the same ratio as the input power (see Section 20.2.2). This property of the transistor amplifiers has been used in the bias adaptation technique to maintain high efficiency at low output power levels.

Figure 12.17 shows a schematic for the bias adaptation technique. In this approach no limiter is used and the power amplifier is operated close to linear condition. In the other path the detected input power is fed to the modulator. The DC supply is

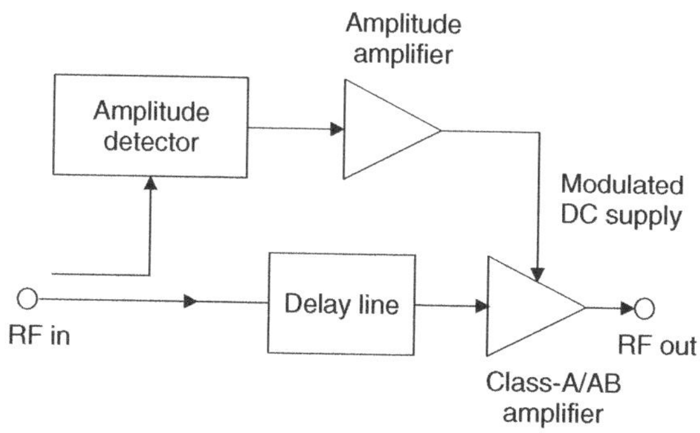

Figure 12.17 Schematic of a bias adaptation technique.

modulated with the input signal envelope. The output amplitude is adjusted using the adaptive modulated bias control. This scheme is simpler than EER but has lower efficiency. Power supply controlled methods such as EER and bias adaptation offer high-efficiency operation potentials for HPAs.

12.5 LINEAR AMPLIFIER DESIGN CONSIDERATIONS

The design factors that affect the linear operation of an amplifier are discussed next.

12.5.1 Amplifier Gain

In a multistage amplifier the gain of each stage affects the output as well as the input IP3. There are trade-offs between noise figure, gain, DC power, and IP3. As an example, for a given high output IP3, higher gain for each stage is required, whereas for higher input IP3, lower gain for each stage is required. The latter is quite a contradictory requirement for a low-noise multistage amplifier. Table 12.1 provides a cascaded analysis of three-stage amplifiers with different gain and IP3 values. The analysis was carried out using Eq. (3.56). Figure 12.18 shows output TOI or IP3 versus gain for a multistage amplifier. In this case, the output stage TOI is 33 dBm and the device aspect ratio is 2:1. The gain range is 20–27 dB. Thus lower gain per stage lowers the output TOI. When the gain/stage is 5 dB, the output TOI is only 29 dBm as compared to 32 dBm for a 9-dB/stage case.

Next, the effect of gain/stage on the required DC supply power, P_{DC}, for a multistage amplifier is considered. Assumptions are that an amplifier has three stages and TOI is a linear function of device size. Here, the X value in (12.26) is assumed to be 10 dB; that is, output TOI for a device is 30 dBm when $P_{DC} = 20$ dBm. Each stage has the same gain and the lower the gain, the higher will be the DC supply power required in each stage. The targeted output TOI is 32 dBm. The DC power required to achieve an output TOI of 32 dBm for various device aspect ratios is shown in Figure 12.19. Thus P_{DC} is a very strong function of device gain and the device aspect ratio. P_{DC} is increased from 323 mW to 492 mW when the device gain is dropped from 10 dB to 6 dB per stage. Optimum device ratios are about 2:1, 2.5:1, and 3.5:1 for 6-, 8-, and 10-dB gain, respectively.

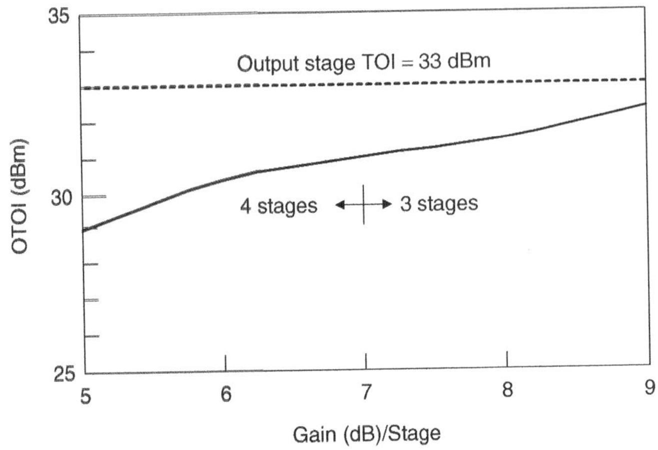

Figure 12.18 Output TOI versus gain for a multistage amplifier. The output stage TOI is 33 dBm and the device aspect ratio is 2:1. The gain range is 20–27 dB.

Table 12.1 Cascaded Input and Output IP3 Values for a Three-Stage Amplifier

Stage 1		Stage 2		Stage 3			
G	IP3	*G*	IP3	*G*	IP3	Input IP3	Output IP3
10	30	10	35	10	40	8.5	38.5
8	30	8	35	8	40	13.6	37.6
6	30	6	35	6	40	18.2	36.2
10	34	10	37	10	40	9.1	39.1
8	34	8	37	8	40	14.5	38.5
6	34	6	37	6	40	19.6	37.6

Gain: *G* in dB and IP3 in dBm. IP3 for each stage is for output.

12.5.2 Minimum Source and Load Mismatch

When an amplifier is used in a system, non 50-Ω impedances provided at the input and output affect the amplifier's TOI. Also, in a linear amplifier, the device's input and output are matched for an optimum TOI not for good match to 50 Ω as described in Chapter 5. In such cases, a balanced configuration can be used which allows one to design the input and output match of the single-ended amplifier for optimum TOI, and the balanced topology provides a good match and stable operation and minimum interaction effects in the system. A balanced configuration also allows one to design a broadband amplifier. A balanced amplifier configuration is shown in Figure 12.20 and discussed in detail in Chapter 11.

12.6 LINEAR AMPLIFIER DESIGN EXAMPLES

Load-pull contours for ACPR and PAE for a power FET (8-mm gate periphery biased at 3.5 V) were obtained at 1.9 GHz to determine a load that is more tolerant to mismatch at the output [11]. In a 50-Ω system, the load $Z_{OL1} = 10 + j8\ \Omega$ results in $P_o \cong 0.16$

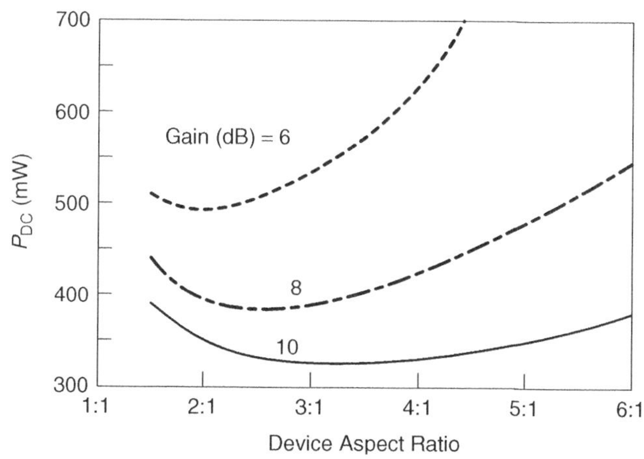

Figure 12.19 DC power versus device aspect ratio and gain/stage for a multistage amplifier.

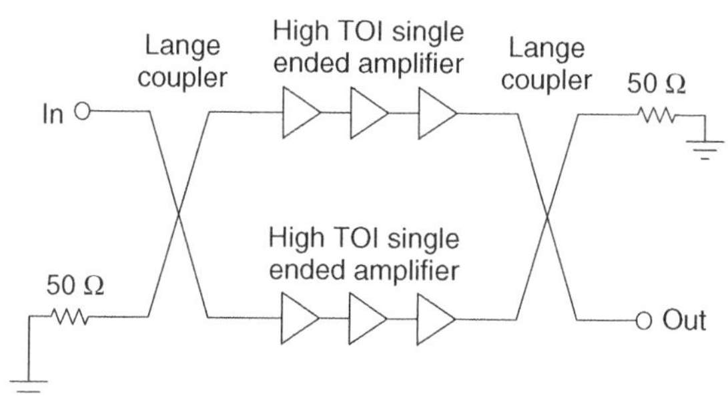

Figure 12.20 A balanced amplifier configuration for high-TOI applications and allowing the single-ended amplifier's S_{11} and S_{22} to be poor for optimal TOI.

W, PAE $\cong$ 40%, and ACPR ≤ -60 dBc, where the load $Z_{0L2} = 18 + j2\ \Omega$ achieved $P_o \cong 0.12$ W, PAE $\cong$ 35%, and ACPR ≤ -55 dBc when the system's impedance represents 3:1 VSWR ($\rho = 0.5$) as a load. As an example, Figure 12.21 shows a typical implementation of an output match for $Z_{0L2} = 18 + j2\ \Omega$.

EXAMPLE 12.3

Next, we describe an example of an MMIC linear C-band driver amplifier based on MESFET technology with the following specifications:

$$
\begin{aligned}
\text{Frequency range} &= 5.7 - 8.5 \text{ GHz} \\
\text{Gain} &= 22 \text{ dB} \\
\text{OIP3} &= 32 \text{ dBm} \\
P_{1dB} &= 20 \text{ dBm} \\
\text{NF} &= 5 \text{ dB, maximum} \\
\text{Return loss} &= 10 \text{ dB, typical} \\
V_D &= 5 \text{ V} \\
I_d &= 165 \text{ mA}
\end{aligned}
$$

The MMIC chip is packaged into a 4-mm 16-lead PQFN plastic package.

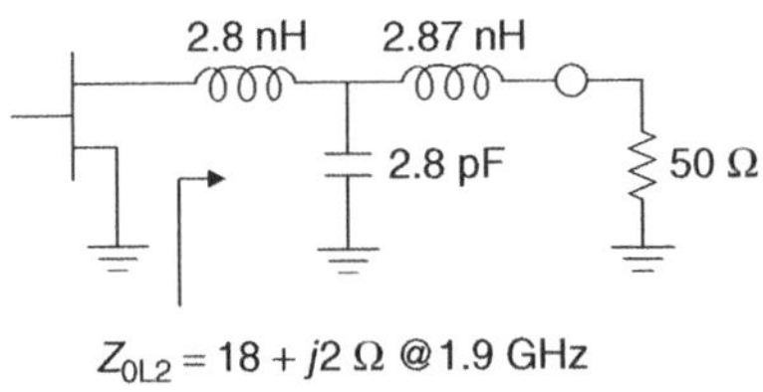

Figure 12.21 Load realization by using lumped components.

SOLUTION The design of an MMIC multistage amplifier starts with the selection of suitable device types and their sizes to meet the required specifications with margins. Here, we selected MESFET MMIC technology. The number of stages is based on the gain requirements and a three-stage topology is selected. In the first stage a low-noise FET is used to keep the noise figure less than 5 dB and is biased at 25% I_{dss}. The second and third stages use linear FETs and are biased at about 50% I_{dss}. The gate peripheries for first-, second-, and third-stage FETs are 300, 300, and 450 μm, respectively. Typical measured NF and output IP3 data at 10 GHz for these FETs are as follows:

Parameter	Units	FET1	FET2	FET3
V_{ds}	V	3	3.8	3.8
V_{gs}	V	−0.7	−1.2	−1.2
I_{ds}	mA	15	60	90
NF	dB	0.81	3.06	3.66
IP3	dBm	24	34	36

Assuming 8-dB gain per stage, the calculated output IP3 using (3.57) is approximately 33.8 dBm and the NF calculated using (3.62) is 1.46 dB.

The transistors are biased with a single power supply using a self-bias scheme as discussed in Chapter 18. The bias stability is improved by $1/[1 + g_m R_S]$ in the self-bias case and is independent of temperature. Here, g_m and R_S are the device transconductance and the self-bias resistor, respectively.

Thermal modeling of FETs is performed by using the Cooke model as described in Chapter 16. The first step is to calculate the thermal resistance (R_{th}) of each FET used in the design. Based on the FET structure (gate–gate pitch, unit gate width, and FET size), the substrate properties, and the maximum channel temperature, the thermal resistance is calculated. Table 12.2 summarizes the thermal resistance calculation for the FETs used in the design. The GaAs substrate thickness and thermal conductivity at room temperature are 75 μm and 0.46 W/cm·°C, respectively. The difference in the temperature, ΔT, from the bottom surface (carrier) to the top surface (channel) of the MMIC chip is calculated using $\Delta T = R_{th} \times P_D$, where P_D is the net power dissipated in the device. For these calculations, the GaAs thermal conductivity is chosen based on the maximum allowed junction temperature of 150°C. However, for small devices (gate periphery less than 0.6 mm), the Cooke model overestimates the thermal resistance because it does not include the extra heat dissipation due to metal filled via holes and the heat spreading at the gate finger ends.

The amplifier circuit is optimized for minimum noise figure and high gain using *S*-parameters and EC models. The first-stage FET has a series feedback as well as the parallel feedback in order to keep the NF below 3 dB, achieve flat gain, and maintain good match over the desired frequency band. The lead frame with bond-wire model used in the design is shown in Figure 12.22. Single drain pad supply operation was used. The first two stages use small resistors in the drain bias lines for stabilization (for more discussions see Chapter 18). The values of resistors and their sizes were selected so that the voltage drop across them is

Table 12.2 Thermal Analysis of the Three FETs[a]

Parameter	Unit	Values		
FET size	μm	300	300	450
Number of fingers	—	4	4	4
Gate–gate pitch	μm	20	30	30
R_{th}	°C/W	293.5	267.9	178.6
R_{th} die attach (epoxy)	°C/W	21.5	19.1	16.6
R_{th} carrier/lead frame	°C/W	20.0	20.0	20.0
Total R_{th}	°C/W	335.0	307	215.2
P_D	W	0.045	0.228	0.342
ΔT	°C	15.1	70.0	73.6

[a]Substrate: 75-μm thick GaAs.

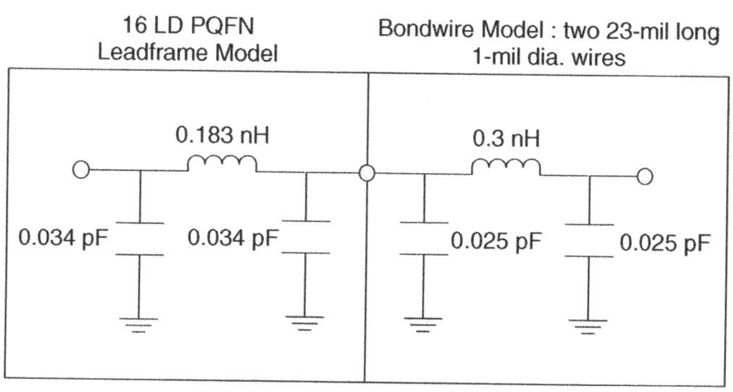

Figure 12.22 Lead frame with bond-wire model used in the designs.

less than 0.5 V and they also meet electromigration requirements. The nominal drain supply voltage is 5 V. However, if the supply voltage available is only 8 V, it can be applied through a suitable high-power resistor. Figures 12.23 and 12.24 show the schematic and physical layout of the MMIC amplifier, respectively.

The simulated gain and return loss of the linear amplifier are shown in Figure 12.25. The gain is flat 25–26 dB and the return loss is better than 15 dB over the design frequency band of 5.7–8.5 GHz. The simulated noise figure is shown in Figure 12.26 and is less than 3.5 dB over the frequency band. Each stage is unconditionally stable and the K factor for each stage is shown in Figure 12.27. For the three-stage amplifier, the K factor is much greater than 1.

The MESFET amplifier chip is epoxied into a 16-lead 4 mm plastic package (see Chapter 21 for more details). A high thermal conductivity epoxy is used to bind the chip to the lead frame paddle. The chip is wire bonded to the lead frame with 1-mil diameter gold wire. The lead frame is soldered to an FR-4 prototyping board. Each lead frame pin sits on an 18-mil wide by 50-mil pad from which microstrip or coplanar transmission lines are connected. Figure 12.28 shows the wiring of the MMIC chip into the plastic package.

Typical measured gain, noise figure, and output IP3 are given in Figure 12.29. Their values are 24 dB, 3.5 dB, and 32 dBm, respectively. Measured P_{1dB} and return loss are better than 23 dBm and 10 dB, respectively. ■

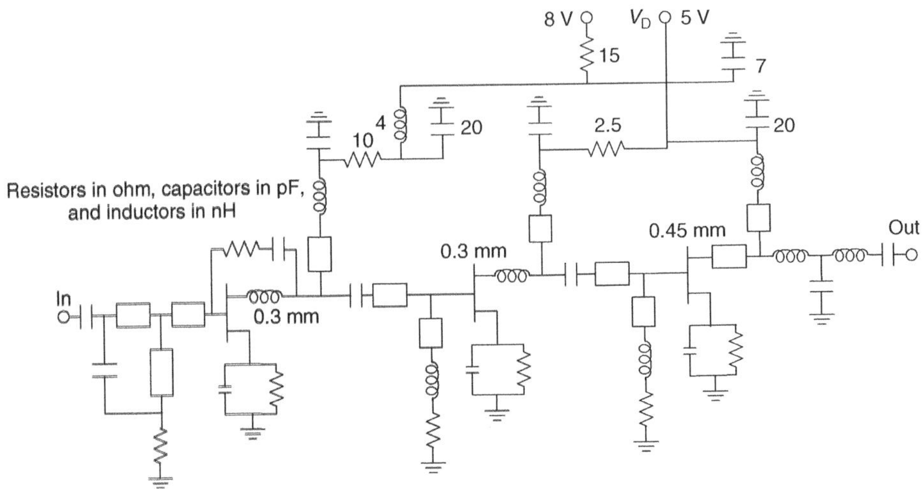

Figure 12.23 Schematic of the three-stage linear amplifier using a single power supply.

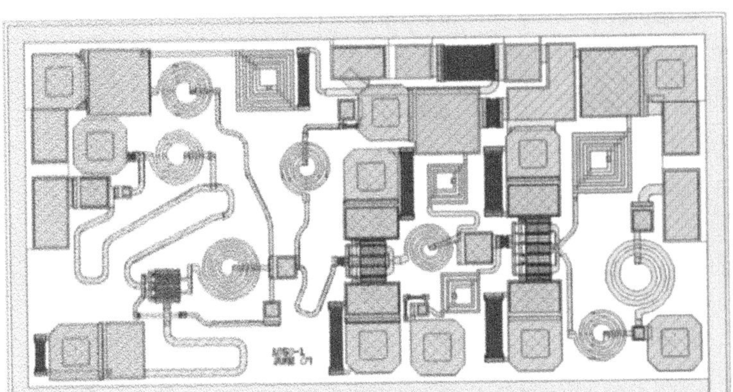

Figure 12.24 Layout of the three-stage linear amplifier. Chip size is 2.3 mm^2.

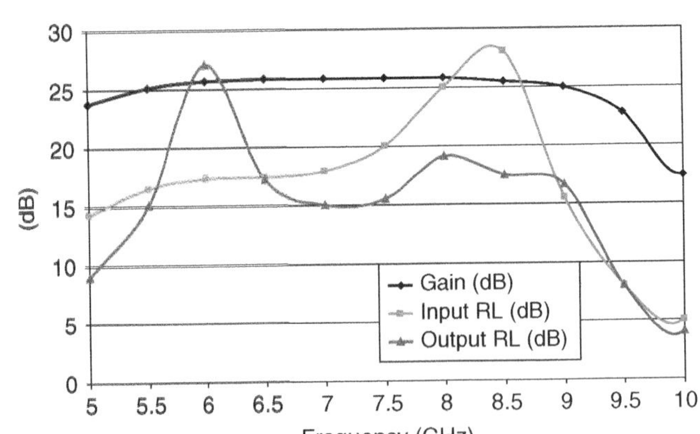

Figure 12.25 Simulated gain and return loss of the of the three-stage linear amplifier.

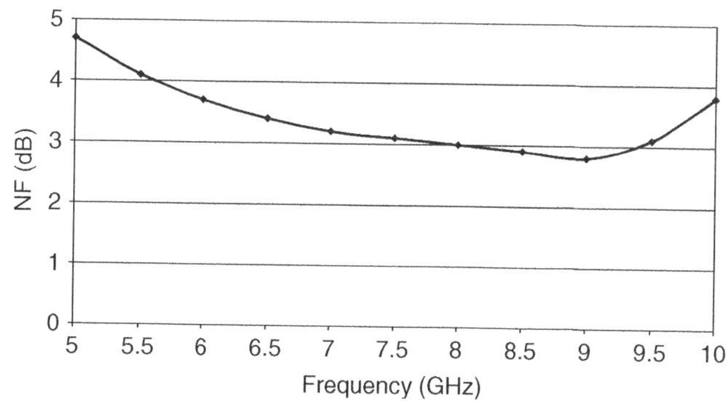

Figure 12.26 Simulated noise figure of the three-stage linear amplifier.

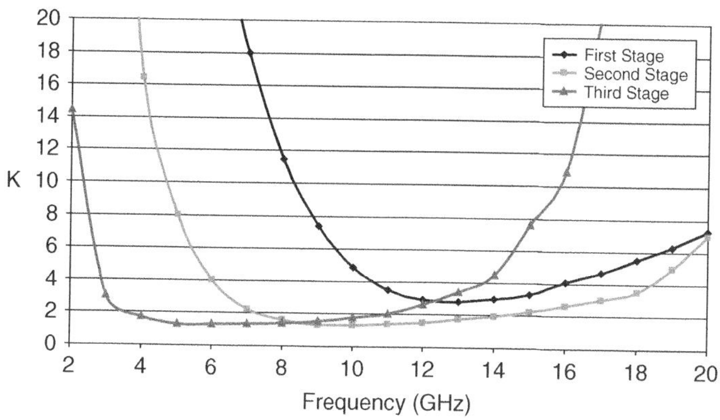

Figure 12.27 Simulated K factor for each stage of the three-stage linear amplifier.

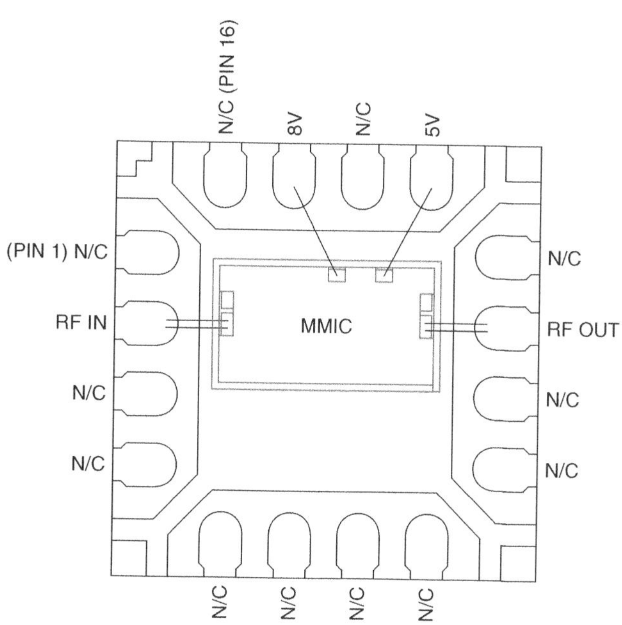

Figure 12.28 Bonding diagram for the linear amplifier into a 16-lead plastic package.

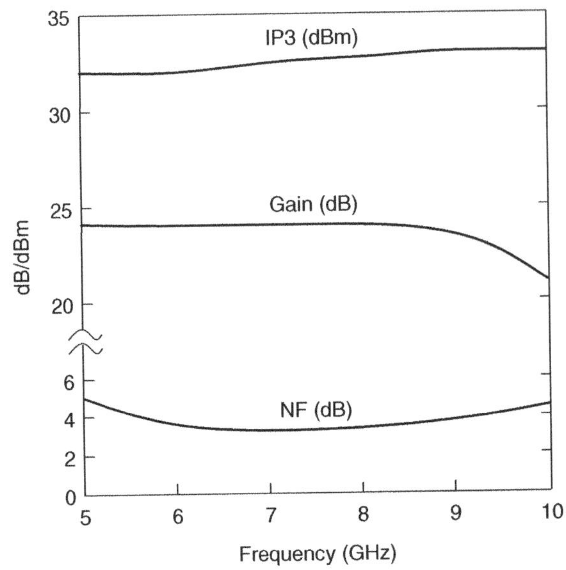

Figure 12.29 Typical measured gain, noise figure, and IP3 versus frequency of the linear amplifier.

EXAMPLE 12.4

Now we describe an example of a S-band 10-W linear HPA based on 0.5-gate length pHEMT technology with the following target specifications:

$$\begin{aligned}\text{Frequency} &= 3.55\ \text{GHz}\\ \text{Gain} &= 10\ \text{dB}\\ \text{ACPR} &= -35\ \text{dBc @ } P_o = 30\ \text{dBm}\\ P_{1dB} &= 40\ \text{dBm}\\ \text{PAE} &= 20\%\ \text{@ } P_o = 30\ \text{dBm}\\ V_D &= 12\ \text{V}\end{aligned}$$

SOLUTION The unmatched 10-W pHEMT device was mounted in a 3-mm 16-lead PQFN surface mount plastic package, as shown in Figure 12.30. The measured optimum source and load reflection coefficients for the device at 3.55 GHz were $\Gamma_S = 0.946\angle - 169.5^\circ$ and $\Gamma_L = 0.851\angle - 179.3^\circ$, respectively. The amplifier was designed using multisection microstrip matching networks on an evaluation PCB, as shown in Figure 12.31. Typical measured gain, PAE, input return loss (IRL), and ACPR as a function of output power are shown in Figure 12.32. At $P_o = 30$ dBm, the values of gain, PAE, IRL, and ACPR are 10 dB, 20%, 12 dB, and −38 dBc, respectively. ■

There are several techniques to measure intermodulation distortion (IMD) in amplifiers [46–50]. The most popular methods are third-order intermodulation (TOI or IP3), adjacent channel power ratio (ACPR), noise power ratio (NPR), and error vector magnitude (EVM). These techniques are briefly described in Chapter 22.

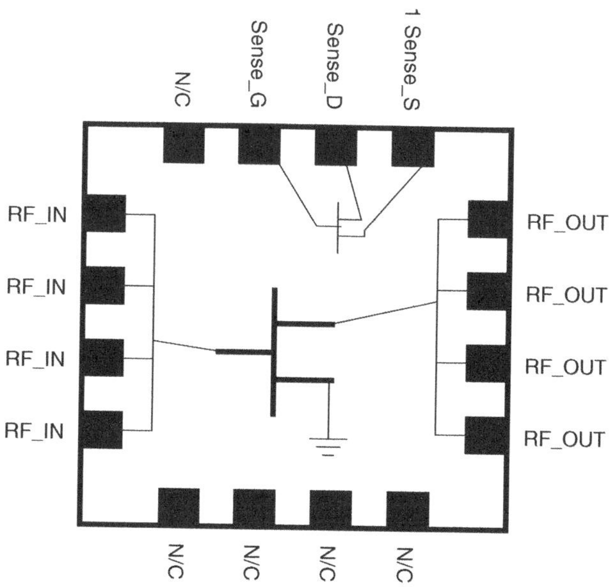

Figure 12.30 A 10-W pHEMT in a 16-lead plastic package.

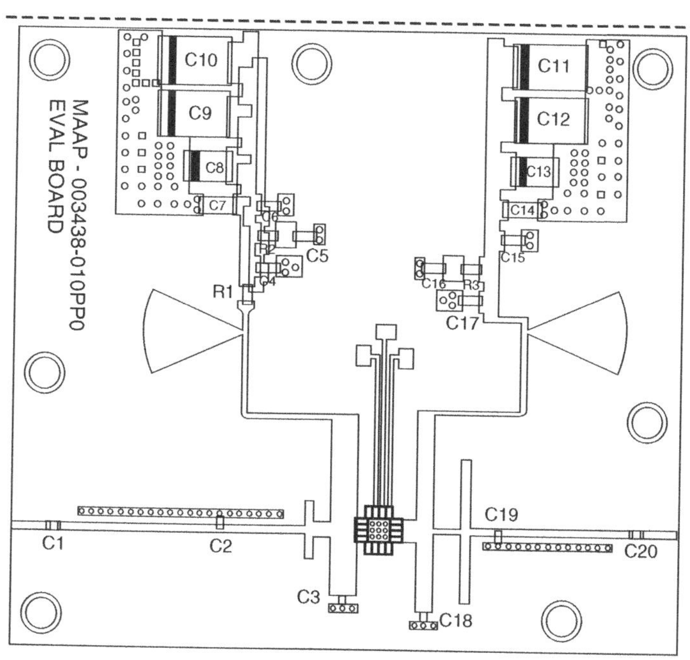

Figure 12.31 Test board for 10-W pHEMT in a 16-lead plastic package.

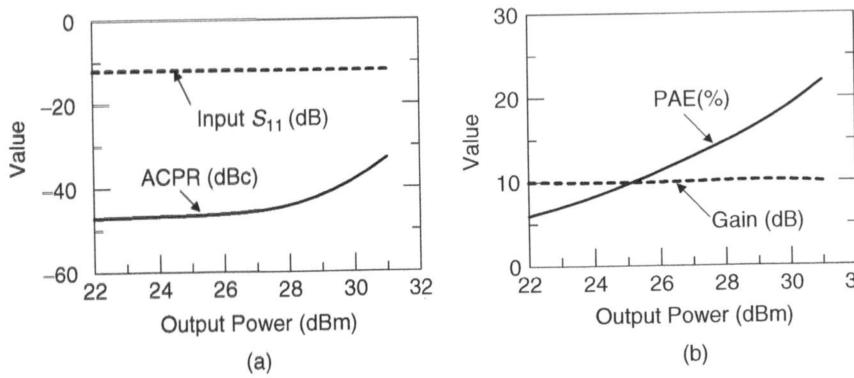

Figure 12.32 Measured performance of the 10-W pHEMT amplifier in a 16-lead plastic package. (a) ACPR and input match versus output power, and (b) gain and PAE versus output power.

REFERENCES

1. N. Pothecary, *Feedforward Linear Power Amplifiers*, Artech House, Norwood, MA, 1999.
2. J. Vuolevi and T. Rahkonen, *Distortion in RF Power Amplifiers*, Artech House, Norwood, MA, 1999.
3. J. Pedro and N. B. Carvalho, *Introduction to Distortion in Microwave and Wireless Circuits*, Artech House, Norwood, MA, 2003.
4. F. Giannini and G. Leuzzi, *Nonlinear Microwave Circuit Design*, Wiley-Interscience, Hoboken, NJ, 2004.
5. G. D. Vendelin et al., *Microwave Circuit Design Using Linear and Nonlinear Techniques*, 2nd edition, Wiley-Interscience, Hoboken, NJ, 2005.
6. J. C. Pedro, Intermodulation, in *Encyclopedia RF and Microwave Engineering*, K. Chang (Ed.), Vol. 3, John Wiley & Sons, Hoboken, NJ, 2005, pp. 2189–2215.
7. S. C. Cripps, *RF Power Amplifiers for Wireless Communications*, 2nd edition, Artech House, Norwood, MA, 2006.
8. P. K. Ikalainen, L. C. Witkowski, and K. R. Varian, Low-noise, low DC power linear amplifiers, *IEEE Microwave Millimeter-Wave Monolithic Circuits Symp. Dig*., pp. 159–162, 1993.
9. J. C. Pedro, Evaluation of MESFET nonlinear intermodulation distortion reduction by channel-doping control, *IEEE Trans. Microwave Theory Tech*., Vol. 45, pp. 1989–1997, November 1997.
10. H. Ikeda, H. Kosugi, and T. Uwano, A low distortion and high efficiency parallel-operation power amplifier combined in different phases in wide range of load impedances, *IEEE MTT-S Int. Microwave Symp. Dig*., pp. 535–538, 1996.
11. K. Ishida et al., A high efficiency and low distortion GaAs power MMIC design in the wide load impedance range by extended use of load-pull method, *IEEE MTT-S Int. Microwave Symp. Dig*., pp. 775–778, 1999.
12. M. Kumar, J. C. Whartenby, and H. J. Wolkstein, Predistortion linearizer using GaAs dual-gate MESFET for TWTA and SSPA used in satellite transponders, *IEEE Trans. Microwave Theory Tech*, Vol. MTT-33, pp. 1479–1488, December 1985.
13. M. Nakayama et al., A novel amplitude and phase linearizing technique for microwave power amplifiers, *IEEE MTT-S Int. Microwave Symp. Dig*., pp. 1951–1954, 1995.
14. K. G. Voyce and J. H. McCandless, Power amplifier linearization using IF feedback, *IEEE MTT-S Int. Microwave Symp. Dig*., pp. 863–866, 1989.

15. J. C. Pedro and J. Perez, An MMIC linearized amplifier using active feedback, *IEEE Microwave Millimeter-Wave Monolithic Circuits Symp. Dig.*, pp. 113–116, 1993.
16. K. Nishikawa and T. Tokumitsu, An MMIC low-distortion variable-gain amplifier using active feedback, *IEEE Trans. Microwave Theory Tech.*, Vol. 43, pp. 2812–2816, December 1995.
17. M. R. Moazzam and C. S. Aitchison, A low third order intermodulation amplifier with harmonic feedback circuitry, *IEEE MTT-S Int. Microwave Symp. Dig.*, pp. 827–830, 1996.
18. F. Ali, M. R. Moazzam, and C. Aitchison, IMD elimination and ACPR improvement for an 800 MHz HBT MMIC power amplifier, *IEEE RFIC Symp. Dig.*, pp. 69–71, 1998.
19. D. Jing et al., New linearization method using interstage second harmonic enhancement, *IEEE Microwave Guided Wave Lett.*, Vol. 8, pp. 402–404, November 1998.
20. K. Yamauchi et al., A novel series diode linearizer for mobile radio power amplifiers, *IEEE MTT-S Int. Microwave Symp. Dig.*, pp. 831–834, 1996.
21. C. Haskins, T. Winslow, and S. Raman, FET diode linearizer optimization for amplifier predistortion in digital radios, *IEEE Microwave Guided Wave Lett.*, Vol. 10, pp. 21–23, January 2000.
22. K. Yamauchi et al., A microwave miniaturized linearizer using a parallel diode with a bias feed resistance, *IEEE Trans. Microwave Theory Tech.*, Vol. 45, pp. 2431–2434, December 1997.
23. A. Katz, S. Moochalla, and J. Klatskin, Passive FET MMIC linearizers for C, X and Ku-band satellite applications, *IEEE Microwave Millimeter-Wave Monolithic Circuits Symp. Dig.*, pp. 155–158, 1993.
24. H. Hayashi, M. Nakatsugawa, and M. Muraguchi, Quasi-linear amplification using self phase distortion compensation technique, *IEEE Trans. Microwave Theory Tech.*, Vol. 43, pp. 2557–2564, November 1995.
25. D. E. Meharry, Multi-octave transformer coupled differential amplifier for high dynamic range, *IEEE GaAs IC Symp. Dig.*, pp. 225–228, 1998.
26. Y. Yang and B. Kim, A new linear amplifier using low-frequency second-order intermodulation component feedforwarding, *IEEE Microwave Guided Lett.*, Vol. 9, pp. 419–421, October 1999.
27. P. Sandhiya et al., Current dumping: a novel linearization technique for RF power transistors, *IEEE MTT-S Int. Microwave Symp. Dig.*, pp. 815–818, 1999.
28. M. G. Kim et al., An FET-level linearization method using a predistortion branch FET, *IEEE Microwave Guided Wave Lett.*, Vol. 9, pp. 233–235, June 1999.
29. C. F. Campbell, A fully integrated Ku-band Doherty amplifier MMIC, *IEEE Microwave Guided Wave Lett.*, Vol. 9, pp. 114–116, March 1999.
30. P. B. Kenington, Methods linearize RF transmitters and power amps, Part 1, *Microwaves RF*, Vol. 37, pp. 102–116, December 1998.
31. P. B. Kenington, Methods linearize RF transmitters and power amps, Part 2, *Microwaves RF*, Vol. 38, pp. 79–89, January 1999.
32. H. Ikeda et al., Phase distortion mechanism of a GaAs FET power amplifier for digital cellular application, *IEEE MTT-S Int. Microwave Symp. Dig.*, pp. 541–544, 1992.
33. T. Yokoyama et al., High-efficiency low adjacent channel leakage GaAs power MMIC for 1.9GHz digital cordless phones, *IEEE Trans. Microwave Theory Tech.*, Vol. 42, pp. 2623–2628, December 1994.
34. K. Tateoka et al., A GaAs MCM power amplifier of 3.6V operation with high efficiency of 49% for 0.9GHz digital cellular phone system, *IEEE MTT-S Int. Microwave Symp. Dig.*, pp. 569–572, 1994.
35. Y. Hasegawa et al., 3.4V Operation power amplifier multi-chip ICs for digital cellular phone, *IEEE GaAs IC Symp. Dig.*, pp. 63–66, 1995.
36. M. Nakamura et al., A buried p-gate heterojunction field effect transistor for a power amplifier of digital wireless communication systems, *IEEE MTT-S Int. Microwave Symp. Dig.*, pp. 1095–1098, 1999.

37. A. Grebennikov, *RF and Microwave Power Amplifier Design*, McGraw-Hill, New York, 2004.
38. T. H. Lee, *The Design of CMOS Radio-Frequency Integrated Circuits*, 2nd edition, Cambridge Press, Cambridge, UK, 2004.
39. A. Raghavan, N. Srirattana, and J. Laskar, *Modeling and Design Techniques for RF Power Amplifiers*, John Wiley & Sons, Hoboken, NJ, 2008.
40. H. Chireix, High power outphasing modulation, *Proc. IRE*, Vol. 23, pp. 1370–1392, November 1935.
41. X. Zhang, L. E. Larson, and P. M. Asbeck, *Design of Linear RF Outphasing Power Amplifiers*, Artech House, Norwood, MA, 2003.
42. A. Huttunen and R. Kaunisto, A 20-W Chireix outphasing transmitter for WCDMA base stations, *IEEE Trans. Microwave Theory Tech*. Vol. 55, pp. 2709–2718, December 2007.
43. W. H. Doherty, A new high efficiency power amplifier for modulated waves, *Proc. IRE*, Vol. 24, pp. 1163–1182, September 1936.
44. N. Srirattana et al., Analysis and design of a high-efficiency multistage Doherty power amplifier for wireless communications, *IEEE Trans. Microwave Theory Tech*., Vol. 53, pp. 852–859, March 2005.
45. K.-J. Cho et al., Linear optimization of a high power Doherty amplifier, *IEEE MTT-S Int. Microwave Symp. Dig*., TH1B-5, 2005.
46. Y. Zhao et al., Linearity improvement of HBT-based Doherty power amplifiers based on a simple analytical model, *IEEE Trans. Microwave Theory Tech*., Vol. 54, pp. 4479–4488, December 2006.
47. I. Takenaka et al., A distortion-cancelled Doherty high-power amplifier using 28-V GaAs heterojunction FETs for W-CDMA base stations, *IEEE Trans. Microwave Theory Tech*., Vol. 54, pp. 4513–4529, December 2006.
48. N. Wongkomet, L. Tee, and P. R. Gray, A +31.5 dBm CMOS RF Doherty power amplifier for wireless communications, *IEEE J. Solid State Circuits*., Vol. 41, pp. 2852–2858, December 2006.
49. J. Kim, B. Kim, and Y. Y. Woo, Advanced design of linear Doherty amplifier for high efficiency using saturation amplifier, *IEEE MTT-S Int. Microwave Symp. Dig*., pp. 1573–1576, 2007.
50. C. T. Burns, A. Chang, and D. W. Runton, A 900MHz, 500W Doherty power amplifier using optimized output matched Si LDMOS power transistors, *IEEE MTT-S Int. Microwave Symp. Dig*., pp. 1577–1580, 2007.
51. K.-J. Cho et al., *N*-way distributed Doherty amplifier with an extended efficiency range, *IEEE MTT-S Int. Microwave Symp. Dig*., pp. 1581–1584, 2007.
52. J. Kim et al., Analysis of a fully matched saturated Doherty amplifier with excellent efficiency, *IEEE Trans. Microwave Theory Tech*., Vol. 56, pp. 328–338, February 2008.
53. Y.-S. Lee et al., Highly efficient class-F GaN HEMT Doherty amplifier for WCDMA applications, *Microwave Opt. Technol. Lett*., Vol. 50, pp. 2328–2331, September 2008.
54. L. R. Kahn, Single sideband transmission by envelope elimination and restoration, *Proc. IRE*, Vol. 40, pp. 803–806, July 1952.
55. N. Wang et al., Linearity of X-band class-E power amplifiers in EER operation, *IEEE Trans. Microwave Theory Tech*., Vol. 53, pp. 1096–1102, March 2005.
56. F. Wang et al., Design of wide-bandwidth envelope-tracking power amplifiers for OFDM applications, *IEEE Trans. Microwave Theory Tech*., Vol. 53, pp. 1244–1255, April 2005.
57. G. Hanington et al., High-efficiency power amplifier using dynamic power-supply voltage for CDMA applications, *IEEE Trans. Microwave Theory Tech*., Vol. 47, pp. 1471–1476, August 1999.
58. H. Gandhi, Digital predistortion linearizers broadband PAs, *Microwaves RF*, Vol. 47, pp. 66–76, July 2008.

PROBLEMS

12.1 Describe various sources of distortion in amplifiers.

12.2 If the RF output voltage, v_o, of an FET amplifier is represented by

$$v_o = C_1 v_i + C_2 v_i^2 + C_3 v_i^3$$

where the input voltage $v_i = A \cos \omega_1 t$, show that the gain of the amplifier, G, in decibels is given by

$$G = 20 \log(C_1 + 0.75 C_3 A^2)$$

12.3 In Problem 12.2 if $C_1 = 10$, $C_2 = 0$, and $C_3 = -1$, determine the 1-dB compression point for the amplifier having input and output impedances of 50 Ω.

12.4 Consider an amplifier designed for a 50-Ω source and load. The transfer characteristics are given by

$$v_o = 10 v_i - 0.5 v_i^3$$

Calculate the output P_{1dB} level.

12.5 An amplifier has output P_{1dB} and IP3 values of 30 and 40 dBm, respectively. The P_{1dB} level was measured using a single tone. Determine IM3 if the output power is 25 dBm.

12.6 Design a two-stage linear amplifier at a supply voltage of 5 V using the linear model given in Section 5.8.3 (Chapter 5) with an output IP3 = 42 dBm. The EC model for the device is given in Table 5.5 and the optimum source and load impedance model is given in Figure 5.17. The frequency of operation is 10 GHz and the matching networks loss may be considered negligible. Calculate the DC power (P_{DC}) and ratio of IP3 to P_{DC}.

Chapter 13

High-Voltage Power Amplifier Design

Today, solid state amplifier designers have a wide range of existing and emerging transistor technologies at their disposal to design products to meet their goals. One of the applications is mobile wireless communication, which has the highest potential for volume. Improved linearity is one of the important requirements for such amplifiers. Another equally important application is for the base station, where one needs high power, efficiency, and linearity for power amplifiers operated at high voltages. For such applications high-voltage devices such as GaAs FETs, pHEMTs and HBTs, GaN HEMTs, and Si LDMOS are competing against each other. The GaAs transistors have high performance, LDMOS transistors are cost effective, and GaN HEMTs have the highest power density capability but are the most expensive.

For high-power transmitters in L- and S-band radars and base stations, discrete Si bipolar and LDMOS devices are commonly used. Si is preferred to GaAs because of its lower fabrication cost and a three times higher thermal conductivity. Below 3.5 GHz, bipolar junction transistors (BJTs) and LDMOS transistors with an output power of hundreds of watts have been established as cost-effective and high-performance devices for the aforementioned applications. Table 13.1 provides a comparison of Si bipolar and LDMOS transistors.

13.1 PERFORMANCE OVERVIEW OF HIGH-VOLTAGE TRANSISTORS

The BJTs have poor gain and gain flatness above 2 GHz and are also thermally limited for longer pulses, whereas LDMOS [1–9] devices have reduced power density above 2 GHz. Both of these devices have degraded performance above 4 GHz. Above S-band frequencies the high-power amplifier (HPA) function in microwave transmit/receive modules is typically implemented using GaAs based internally matched transistors and microwave monolithic integrated circuit (MMIC) HPAs generating on the order of 20–30 W of output power operating from a CW/pulsed 8–10-V power supply. Future radars are expected to require much higher transmit/receive (T/R) module output power levels. These increased power levels cannot be achieved using 10-V GaAs technology.

Fundamentals of RF and Microwave Transistor Amplifiers. By Inder J. Bahl

Table 13.1 Comparison of Si Bipolar and LDMOS High-Power Solid State Amplifiers

Typical Frequency (GHz)	Power (W)	Gain (dB)	PAE (%)	Supply Voltage (V)	Device	Technology	Applications	Advantage	Disadvantage
0.93–1.03	200	12	40	28	LDMOS	Hybrid	Base station	High supply voltage Lower current	Needs matching network
1.2–1.4	300	11	40	36	BJT	Hybrid	Radar	High supply voltage Lower current	Needs matching network
2.11–2.17	200	22	34	28	LDMOS	Hybrid	Base station	High supply voltage Lower current	Limited power density; limited to lower frequencies
2.7–3.1	150	9	40	36	BJT	Hybrid	Radar	Highest power density	Limited gain and flatness, thermally limited to below 200-μs pulse width and 10% duty cycle
3.1–3.5	90	7.5	35	36	BJT	Hybrid	Radar	Highest power density	Limited gain and flatness, thermally limited to below 200-μs pulse width and 10% duty cycle

During the past decade there has been significant progress made in high-voltage (HV) transistor based microwave power amplifiers operating over both narrowbands and broadbands. The focus has been to develop efficient, reliable, and low-cost HV, high-power amplifiers for communications and active aperture array radar applications.

In order to meet the challenge, several different device technologies are being pursued including field plate (FP) FET on GaAs [10–14], MSAG FP-FET on GaAs [15–19], FET on SiC [20–25], pHEMT on GaAs [26–28], FP-pHEMT on GaAs [29–32], GaN HEMT on SiC/Si [33–57], GaN FP-HEMT on SiC/Si [58, 59], and HBT on GaAs [60–62]. Progress in this area has been summarized and given in Table 13.2 for discrete devices and in Table 13.3 for MMICs. It may be noted from these two tables that the power combining using hybrid technology is much more aggressive (an order of magnitude higher) than on MMIC semiconductor substrates.

Wide bandgap (WBG) semiconductors, such as SiC or GaN, have basic material properties that are more favorable to very-high-power amplifier realization than GaAs. Considerable research effort is being invested in WBG HPA development and has resulted in impressive progress, as illustrated in Tables 13.2 and 13.3.

Table 13.2 High-Voltage Hybrid HPA Examples

Frequency (GHz)	Supply Voltage (V)	Output Power (W)	Power Density (W/mm)	PAE (%)	Technology	Year [Reference]
1.5	65	500	3.5	48	GaN HEMT, SiC	2006 [50]
2.0	28	20	12.5	62	HBT, GaAs	1997 [60]
2.0	40	56	1.9	53	FET, SiC	2005 [24]
2.0	47	149	4.7	64	GaN FP-HEMT, SiC	2004 [59]
2.0	53	230	4.8	67	GaN FP-HEMT, SiC	2004 [38]
2.1	28	250	0.7	27	FP-FET, GaAs	2004 [13]
2.1	12	300	0.29	25	FP-FET, GaAs	2001 [10]
2.1	50	250	3.5	46	GaN HEMT, SiC	2004 [34]
2.14	26	26	1.8	61	pHEMT, GaAs	2004 [26]
2.14	26	45	1.4	30	FP-pHEMT, GaAs	2004 [28]
2.14	28	330	0.67	42	FP-pHEMT, GaAs	2006 [32]
5.0	60	220	4.6	38	GaN HEMT, SiC	2007 [57]
5.5	26	83	0.67	34.7	FP-FET, GaAs	2004 [14]
30.0	28	5.8	5.8	43	GaN HEMT, SiC	2004 [35]

Table 13.3 High-Voltage MMIC HPA Examples

Frequency (GHz)	Supply Voltage (V)	Output Power (W)	Power Density (W/mm)	PAE (%)	Technology	Year [Reference]
0.9	28	25	—	60	HBT, GaAs	2004 [61]
2.0	12	50	—	45	pHEMT, GaAs	2004 [27]
3.3	24	50	1	40	FP-FET, GaAS	2007
3.5	55	36.3	3	20.6	MESFET, SiC	2002 [21]
10.0	20	8.0	5	36.7	GaN HEMT, SiC	2004 [37]
10.0	40	20.0	3.3	25	GaN HEMT, SiC	2006 [55]
16.0	31	24.2	—	22.2	GaN HEMT, SiC	2002 [21]
31.0	20	11	—	—	GaN HEMT, SiC	2006 [63]
33.0	13	2.2	2.3	18.6	GaN HEMT, SiC	2004 [36]
35.0	24	4.0	3.3	23	GaN HEMT, SiC	2006 [52]

13.1.1 Advantages

To achieve output power levels far higher and at greater bandwidths than are available today, future solid state device based systems are expected to operate transistors at much higher voltages than the normal 8–10-V operation. High-voltage operation above 20 V lowers the DC–DC conversion power loss and the bias interconnect I^2R power loss. The design of input and output matching networks becomes simpler because of higher power density and higher load impedance, respectively. This also leads to greater bandwidths. GaN HEMT and other transistors such as GaAs FET and HEMT have similar input impedance per unit gate width. However, the output impedance is about twice for the same unit gate width. Since the power density for GaN HEMTs is about 10 times greater than the power density for GaAs FETs and HEMTs, HV GaN HEMTs for the same output power require 10 times smaller input and 20 times smaller output impedance transformation ratios.

Because of higher power density, one needs fewer transistors for a given power level, leading to lower cost, higher yield, and more reliable power amplifiers. Replacing conventional transistors with high-voltage transistors can reduce amplifier cost drastically by lowering the packaging and integration costs.

At a nominal 10-V supply there is a limit to how many transistors can be combined in parallel in a single transistor or MMIC. At the low end of the S-band for discrete transistors, this limit is 150–200 W, while for a single MMIC at 10 GHz, this limit is about 20 W. This limit is caused by transverse resonance along the width and the inability to match a very low impedance device. With HV transistors this limit is extended by a factor of 5–10.

13.1.2 Applications

Many commercial and military applications could benefit using high-voltage power amplifiers. Potential applications are active phased array (APA) radars, base station transmitters, satellite communications, broadband wireless technology (WiMAX), electronic warfare (EW), multiple octave band power amplifiers for JTRS, and secure communication for homeland security.

13.2 HIGH-VOLTAGE TRANSISTORS

In recent years, rapid progress has been witnessed in high-voltage transistors such as LDMOS, high-voltage MESFETs, HBTs, and AlGaN/GaN HEMTs. In comparison to a conventional gate, a field plate (FP) MESFET/HEMT has increased breakdown voltage and increased gate capacitance. Therefore the power density and operating voltage of FP transistors are higher, whereas the gain and operating frequency are lower. The high breakdown voltage combined with high electron velocity of GaN based power transistors has enabled designers to achieve an order of magnitude higher power densities when compared with conventional 10-V transistors. Basic operation of several devices was discussed in Chapter 4, while salient features of HV transistors are given in this section.

A comparison of device and substrate properties of various HV transistors is given in Table 13.4a. Typical measured performance of HV transistors at 2.1 GHz is summarized in Table 13.4b.

Table 13.4 Comparison of Devices and Performance

(a) Transistors and Substrate Properties at Room Temperature

Device	Substrate	Bandgap (eV)	Dielectric Constant	Thermal Conductivity W/cm·°C
LDMOS	Si	1.12	11.7	1.45
MESFET	GaAs	1.42	12.9	0.46
MESFET	SiC	2.86	9.7	3.5
GaN HEMT	SiC	3.39	9.7	3.5
GaN HEMT	GaN	3.39	8.9	1.3

(b) Ceramic Packaged Transistor's Performance at 2.1 GHz

Device	Typical Breakdown Voltage (V)	Operating Voltage (V)	Output Power (W)	Gain (dB)	Efficiency (%)
Si BJT	63	36	110	7.4	40
Si LDMOS	70	28	100	13	55
GaAs FP-MESFET	60	28	300	14	63
SiC MESFET	90	40	56	10	55
GaAs FP-pHEMT	50	26	43	11.5	56
SiC GaN HEMT	80	48	370	10	50
GaAs HBT	70	28	20	11	70

13.2.1 Si Bipolar Junction Transistors

A high-voltage power version of a microwave BJT consists of a base–emitter multifinger interdigital configuration. The multicell is comprised of narrow and long emitter electrodes with emitter ballasting resistors to minimize thermal runaway. The breakdown voltage of such transistors is about 50–70 V. The reliability of linear and class-C power BJTs has been well established, with demonstrated MTTF greater than 10^7 hours at 200 °C junction temperature. The devices are low in cost, immune to static discharge, and rugged enough to withstand very large VSWR mismatch conditions. The commercial HV BJTs have power levels over 1 kW and 150 W (under pulsed operation) at 1 GHz and 3 GHz, respectively.

13.2.2 Si LDMOS Transistors

A power version of the MOSFET, commonly referred to as the laterally diffused metal oxide semiconductor (LDMOS) field effect transistor, is preferred to the bipolar transistor because it has better temperature stability, greater reliability, and a more robust structure. Since the channel region on the drain side is lightly doped and is fully depleted under large drain voltage, LDMOS transistors have very high breakdown voltages.

Typical values of f_T and f_{max} for LDMOS FETs are 5 and 15 GHz, respectively. The drain–source breakdown voltage BV_{DS} is in excess of 70 V. Thus the LDMOS FETs are normally operated at 24–28 V and some devices are operated at as high as 48 V. Although, the f_T for LDMOS FETs is much lower than for the small-signal MOSFETs, their usage for power performance is suitable below 3 GHz. At 2 GHz, LDMOS FETs have about 200–300-W output power with associated PAE and gain values of 50% and 10 dB, respectively. In high-power applications, especially for base stations, where one needs power levels in excess of 100 W, Si LDMOS FETs are commonly used.

The LDMOS transistor's cross section is shown in Figure 13.1a. The starting material is a lightly doped p^- epitaxial layer on 8-inch p-type Si substrate. Multiple ion implantations are performed—typically three n-type and three p-type. The nominal gate length is 0.5 μm and is fabricated with WSi or CoSi on polysilicon. The processing is similar to the Si CMOS with the addition of thick interconnect metallization for higher current handling capacity.

LDMOS transistors are available in different packages and frequencies. Power levels are 25–50 W. Most devices are housed in copper–tungsten (CuW) flange based ceramic packages. The electrical and thermal properties of these devices are improved by using a copper flange based plastic open-cavity package. Such packages have lower thermal resistance. Improved versions of LDMOS transistors are under development to improve power density, reduce thermal resistance, and achieve high efficiency. Plastic packages have emerged recently as an alternative low-cost solution. Nominal operating voltage is 28 V. Figure 13.1b shows current–voltage (I–V) characteristics for a 1.5-mm gate width LDMOS transistor.

The lumped-element model for an LDMOS is shown in Figure 13.1c. The EC model parameter values for a device having gate length of 1.25 μm and gate width of 1.44 mm, and biased at $V_{ds} = 28$ V and $I_{ds} = 15$ mA, are as follows [7]:

$$R_g = 13.7\ \Omega, \quad R_i = 1.2\ \Omega, \quad R_s = 0.87\Omega, \quad R_d = 9.1\ \Omega, \quad R_{ds} = 4.37\ \text{k}\Omega$$
$$C_{gs} = 0.73\ \text{pF}, \quad C_{gd} = 0.029\ \text{pF}, \quad C_{ds} = 0.31\ \text{pF}, \quad C_{gp} = 0.28\ \text{pF}, \quad C_{dp} = 0.26\ \text{pF}$$
$$g_m = 21.1\ \text{mS}, \quad \tau = 14.1\ \text{ps}$$
$$L_g = 0.059\ \text{nH}, \quad L_s = 0.001\ \text{nH}, \quad L_d = 0.057\ \text{nH}$$

The calculated $f_T (= g_m/2\pi C_{gs})$ for this device is 4.6 GHz.

The load impedance at 2.1 GHz of a ceramic packaged 60-W LDMOS transistor is about $1.05 - j3.0\ \Omega$. The reactive part of the load is capacitive due to the inductive package lead reactance.

13.2.3 GaAs Field Plate MESFETs

The fabrication of field plate (FP) MESFETs is similar to conventional FETs. The process is relatively simple, using four to five mask layers to form discrete FETs and 10 or more for MMICs. To achieve high breakdown voltage and high-voltage operation, a field-modulating plate (or simply a field plate) FET has been developed [10, 29].

In this structure, as shown in Figure 13.2a, the gate is extended and the metal–insulator–semiconductor (MIS) type configuration between the gate and drain is used, which results in breakdown voltages on the order of 50–60 V. The average field plate length is on the order of 1 μm. Both recessed gate [10–14] and self-aligned gate (SAG) [15–19] fabrication processes have been used to manufacture high-voltage FETs (or simply HVFETs). An improved version of the FP-FET is shown in Figure 13.2b. In this configuration, the second FP is connected to the source and is known as the source-connected FP-FET. The cross-sectional view of the HVMSAG MESFET is shown in Figure 13.3. An addition of a field plate on the gate increases the HVMSAG breakdown voltage while maintaining high peak current capability. Such devices can be operated up to 6 GHz.

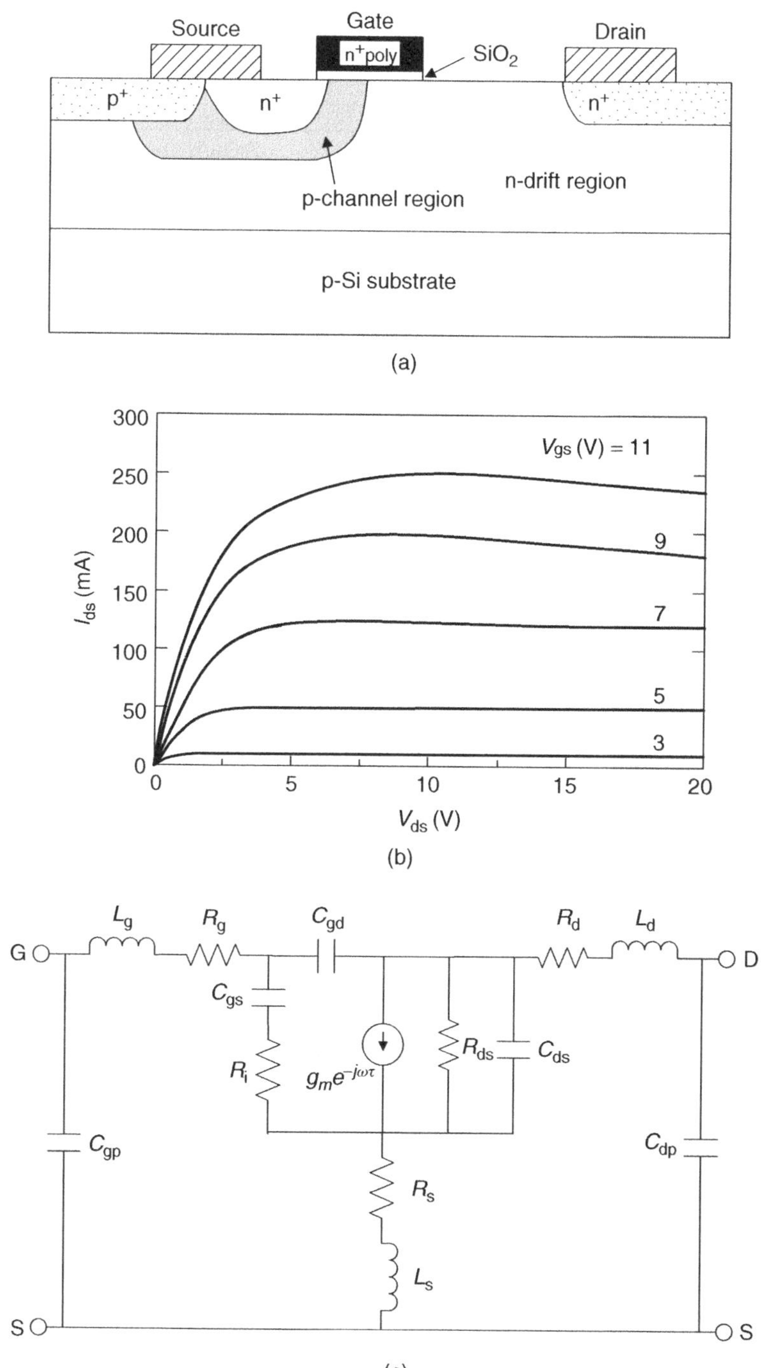

Figure 13.1 Physical cross section of an LDMOS transistor. (b) $I-V$ representation of a 1.5-mm gate periphery LDMOS transistor. (c) Equivalent circuit model of a packaged LDMOS.

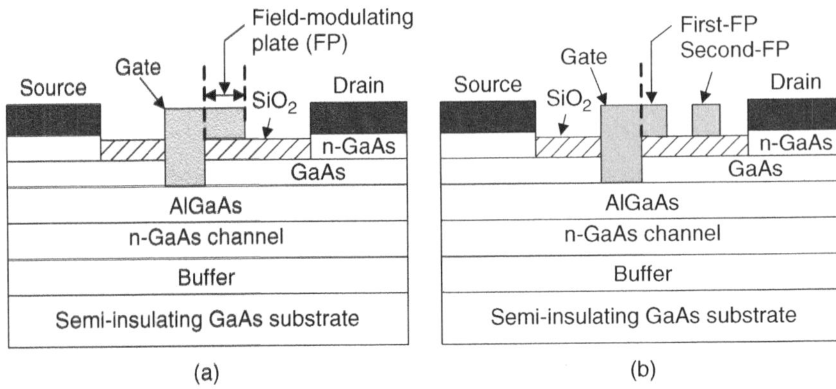

Figure 13.2 Cross-sectional views of field plate MESFETs: (a) single field plate and (b) double field plate.

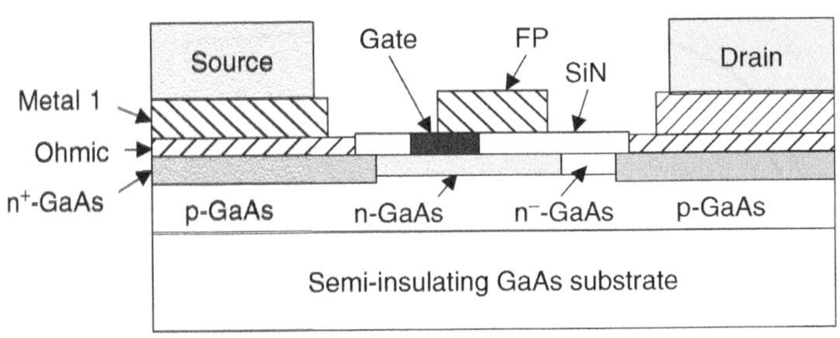

Figure 13.3 Cross section of the HVMSAG field plate MESFET.

13.2.4 GaAs Field Plate pHEMTs

The fabrication of GaAs FP-pHEMTs is similar to that for GaAs FP-FETs, except the starting material has epitaxial layers on it. The devices consist of a double-recessed AlGaAs/InGaAs double heterojunction pHEMT structure. The starting wafers having the required epitaxial layers are commercially available. Typical values of I_{max}, I_{dss}, g_m, f_T, f_{max}, and breakdown voltage for a GaAs FP-pHEMT are 355 mA/mm, 204 mA/mm, 180 mS/mm, 6 GHz, 28 GHz, and 50 V, respectively. For a 13.4-mm device at 2.14 GHz and 26 V, measured values for P_{1dB} and PAE are 26 W (1.9 W/mm) and 61%, respectively. For a 32.4-mm device, these values are 43 W and 56%. Typical small-signal gain is about 11.5 dB. At X-band frequencies, 20–25-W power levels have been demonstrated; however, their usage is limited to below the Ku-band.

13.2.5 GaAs HBTs

GaAs HBTs can also be operated at 24–28 V for high-power amplifiers [60–62]. These devices use a relatively thicker collector layer in comparison to their low-voltage version. AlGaAs/GaAs and InGaP/GaAs HBTs, biased at 24–28 V, have demonstrated over 20-W output power. In an AlGaAs/GaAs HBT epitaxial structure, the active layers are grown on GaAs using the MOCVD technique. For the 2.8-μm thick collector layers, the measured base–collector breakdown voltage was about 70 V [63]. Application of HV HBTs is limited to below the C-band.

13.2.6 SiC MESFET

The SiC MESFET is a potential candidate for high-voltage operation by virtue of silicon carbide's high-field breakdown. The devices are fabricated by growing an n-type epilayer on SiC wafers. Ni/Pt/Au is evaporated to form the gates. Despite high saturation velocity, SiC MESFETs have low electron mobility, low g_m and gain, high knee voltage, high pinch-off voltage, and low frequency of operation. Although the fabrication technology is more mature than that for GaN HEMTs, SiC MESFETs have a high cost because of expensive 3–4-inch SiC substrates. Typical measured breakdown voltages are in the 80–100-V range and their usage is limited to below the C-band.

13.2.7 SiC GaN HEMTs

The fabrication of GaN HEMTs is similar to that for GaAs pHEMTs. A major difference is in the AlGaN/GaN materials, which have high-field breakdown, giving rise to high-voltage operation of GaN HEMTs. GaN HEMT layers can be grown on sapphire, silicon, or SiC substrates. SiC substrates are generally preferred because of their very high thermal conductivity, which greatly helps these devices for high-power density operation. Although 3-inch diameter SiC substrates are available, they are very expensive and have high defect densities. The high electron mobility of GaN layers allows GaN HEMTs to be used at millimeter-wave frequencies. SiC based GaN HEMTs have the highest power density; the next closest are Si based GaN HEMTs. Because SiC is a good electrical insulator, the SiC based GaN HEMT process is very suitable for IC production due to the low-loss matching circuitry.

The GaN transistors can also be used as front-end LNAs. The devices have excellent NF and input power limiting/handling capability. The NF values of a 0.5-μm gate GaN HEMT are about 0.5 dB at 2 GHz and 1.2 dB at 20 GHz. These NF values are about two times higher than the state-of-the-art NF for GaAs based pHEMT devices. Appropriately sized transistors can handle several watts of input power without sustaining damage to the device. These power levels are an order of magnitude higher than the low-noise equivalent GaAs transistors can usually handle. Thus GaN devices can be used as limiter/LNAs.

Typical values of I_{max}, I_{dss}, g_m, f_T, and f_{max} for a power GaN HEMT are 900 mA/mm, 600 mA/mm, 290 mS/mm, 50 GHz, and 80 GHz, respectively. For a 1-mm device at 4 GHz and 28 V, measured values for P_o and PAE are 5 W and 70%, respectively. Typical small-signal gain is about 20 dB. Power densities of 10 W/mm and 5 W/mm at 4 GHz and 35 GHz, respectively, have been demonstrated. Measured breakdown voltage is 60–130 V for GaN HEMTs and 100–250 V for GaN FP-HEMTs. The drain voltage operation of GaN HEMTs has been reported as high as 65 V and for GaN FP-HEMTs as high as 118 V.

A theoretical limit for the power density (W/mm) of GaN FP-HEMTs at 4 GHz is approximately given by [42]

$$P_o \cong 7\frac{V_{ds}}{28} \tag{13.1}$$

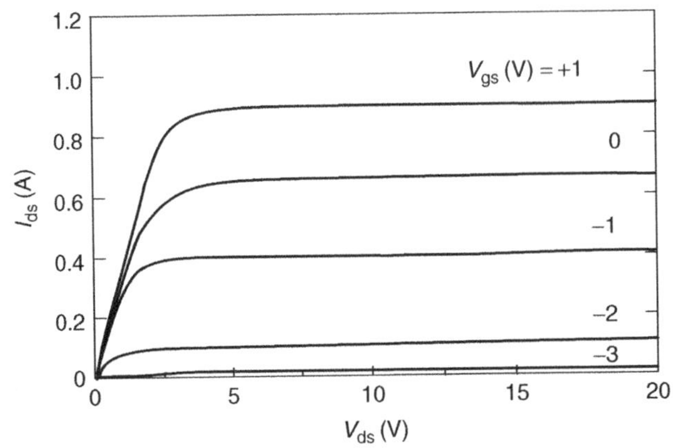

Figure 13.4 Pulsed $I-V$ characteristics of a 1-mm gate GaN HEMT on SiC.

where V_{ds} is the applied drain–source voltage. At $V_{ds} = 100$ V, $P_o \cong 25$ W/mm. Thus high-voltage operation of GaN devices results in very high power density, which enables the development of very-high-power and high-efficiency microwave and millimeter-wave power amplifiers. The future success of GaN devices depends on their performance, reliability, and cost. Figure 13.4 shows pulsed $I-V$ data for a 1-mm GaN HEMT on SiC.

Most recently, GaN HEMT devices have received considerable attention for high power density applications because of their wide-bandgap property. The HEMT structure uses AlN as a barrier layer and GaN as the conducting channel. GaN on SiC has very high power density (PD) capability—as high as 10 W/mm. These values are significantly higher than GaAs FETs and the LDMOS. For high-voltage GaAs transistors, the values for PD are 1.5–2 W/mm, while for the LDMOS their number is about 1 W/mm. Although 10 W/mm is a very attractive power density, the associated heat flux density is very high. Thus more advanced techniques are needed to extract the heat from the package to maintain a reasonable device junction temperature. More realistic power density values may be below 4–5 W/mm. The GaN devices have great potential for cellular and WiMAX base station, broadband EW, and high-power radar applications.

GaN transistors are available in different packages and frequencies. Power levels are 15–90 W. Most devices are assembled in CuW flange based ceramic packages. The operating voltage range is 24–65 V. Discrete transistors are available from TriQuint, Cree, Nitronex, and Eudyna. The available discrete transistors can be operated up to 18 GHz. Applications include WiMAX and WCDMA; a pulsed peak power as high as 650 W for GaN HEMTs for WCDMA base station applications has been reported. For GaN transistors, the base substrates Si and SiC are being used. However, SiC is preferred due to its very high thermal conductivity.

13.3 HIGH-POWER AMPLIFIER DESIGN CONSIDERATIONS

High-power and high-voltage operation of very-high-power amplifiers (VHPAs) require different trade-offs that affect performance, reliability, and cost. Achieving reliable and cost-effective output power substantially higher than that available today requires a new set of design rules for passive components and thermal models for devices as well as for amplifier assemblies. In the new arena of VHP it is also very important to pay more attention to the effect of voltage, current, power, and temperature on passive components, including microstrip, CPW, MIM capacitors, inductors, resistors, crossovers, and via holes. In such amplifiers, more margins in thermal design and operating voltage and current limits are needed because of the imperfect assembly interfaces and voltage and current transients in the circuits.

13.3.1 Thermal Design of Active Devices

WBG transistors can generate higher output power per millimeter of periphery and dissipate more efficiently the additional generated heat. A comparison of substrate thermal properties for various devices is given in Table 13.4a. The power dissipation in HV devices is much higher than in conventional power transistors; therefore the thermal design of HV based power amplifiers and their assemblies is the most critical aspect of their success. Minimal thermal resistance of the die attach and housing is essential. The die attach must be of high quality with minimal voiding ($<5\%$). Due to the high operating power density of HV devices, microvoids below a transistor must be avoided.

The thermal modeling of HVMSAG FETs was performed using the Cooke model [64] described in Chapter 16. For larger HV FETs (gate periphery greater than 1.5 mm), measurements using infrared (IR) and liquid crystal technique (LCT) have shown a close agreement with the Cooke model predictions. However, for smaller FET sizes, the Cooke model can overestimate the channel temperature rise by a factor of 1.5 because it does not include the cooling due to via sources and heat spreading in the lateral direction.

As an example, Table 13.5 [64] summarizes the thermal resistance calculation for 2-mm and 3.2-mm FETs. The GaAs substrate thickness and thermal conductivity at room temperature are 50 μm and 0.46 W/cm·°C, respectively. Since SiC has about 10 times higher thermal conductivity than GaAs, GaN HEMTs on SiC with similar dimensions will have about 10 times smaller thermal resistance values, that is, about 6 °C/W·mm.

The thermal design of power transistors is treated in Chapter 16 and some general observations for HV devices are as follows:

1. The thermal resistance of the HV device depends on the electrode configuration, substrate thickness, via hole locations, and backside metallization thickness. Larger gate–gate pitch, thinner substrate, proper via hole locations, and thick backside metallization are required to achieve the lowest thermal resistance.
2. A conventional stack-up of various layers consisting of shim, solder, carrier, epoxy, and fixture/heat sink can give rise to excessive junction temperature.

Table 13.5 Thermal Calculations Using the Cooke Model[a]

FET Size (mm)	Gate Pitch (μm)	Number of Fingers	Unit Gate Width (μm)	R_{th}(°C/W)
2.0	32	12	167	31.5
3.2	32	18	178	19.8

[a]GaAs substrate thickness is 50 μm

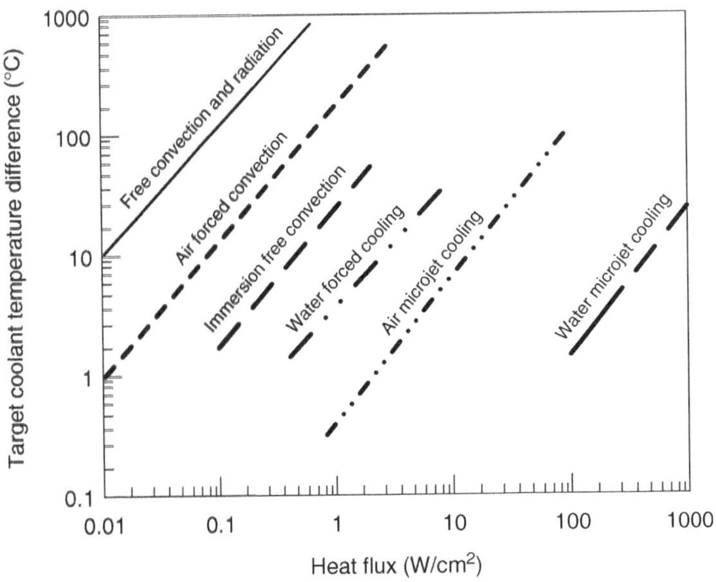

Figure 13.5 Various cooling technique requirements versus device heat flux.

Devices with minimal interlayers between the device and heat sink are desirable. For example, carriers on which transistors or amplifiers are soldered must be bolted directly to heat sinks.

Figure 13.4 shows various cooling technology requirements versus heat flux passing through a chip. For high power density devices, assuming good die attach, high heat flux removal capacity heat spreaders will be required to maintain reliable temperatures within the amplifier. For example, a 25-W GaN SiC amplifier chip has 50% efficiency and 4-mm^2 chip size. An estimated heat flux at the back of the chip (BOC) would be 25 W/0.04 cm^2 = 625 W/cm^2. If the temperature difference between the channel and BOC is allowed to be 100 °C, by referring to Figure 13.5, one requires somewhere in-between air microjet and water microjet cooling techniques.

Diamond heat spreaders and spray cooling can also be used for thermal management of HV devices. However, all these heat removal techniques add various degrees of complexity and cost to the amplifier product [65].

13.3.2 Power Handling of Passive Components

To ensure reliable operation of HV amplifiers, using multilayered passive components such as microstrips, inductors, capacitors, resistors, crossovers, and inductor transformers for high-power applications, power handling capability models are needed for these structures. Power handling capability of multilayer microstrip lines used in MICs and MMICs has been described in the literature [66–68]. The power handling capacity of passive components is limited by heating, the result of ohmic and dielectric losses, and by dielectric breakdown. An increase in temperature due to conductor and dielectric losses limits the average power, while the breakdown between the strip conductor and ground plane limits the peak power. In general, the peak power handling for microwave circuits is much higher than the average power handling capability.

Microstrip

Microstrip lines are well suited for MIC and MMIC applications and have been extensively used in power amplifiers. The average power handling capability (APHC) of microstrip lines has been discussed in References [66 and 67]. Recent advance in multilayer microstrip line technologies have made it possible to realize compact monolithic microwave integrated circuits (MMICs), compact modules, low-loss microstrip lines [69], and high-Q inductors [70]. In multilayered components, along with substrate materials, low dielectric constant materials such as polyimide or benzocyclobutene (BCB) are used as a multilayer dielectric. The thermal resistance of polyimide or BCB is about 200 times the thermal resistance of GaAs or alumina; therefore proper attention has to be paid to the thermal properties of these layers.

The APHC of a multilayer microstrip is determined by the temperature rise of the strip conductor and the supporting dielectric layers and the substrate. The parameters that play major roles in the calculation of average power capability are (a) transmission line losses, (b) thermal conductivity of dielectric layers and the substrate material, (c) surface area of the strip conductor, (d) maximum allowed operating temperature of the microstrip structure, and (e) ambient temperature—that is, temperature of the medium surrounding the microstrip. Therefore dielectric layers and substrates with low-loss tangents and large thermal conductivities will increase the average power handling capability of microstrip lines.

Typically, a procedure for APHC calculation consists of the calculation of conductor and dielectric losses, heat flow due to power dissipation, and the temperature rise. The temperature rise of the strip conductor can be calculated from the heat flow field in the microstrip cross section using an analogy between the heat flow field and the electric field. As a first-order approximation, the heat flow from the microstrip conductor may be considered to follow the rule of 45° thermal spread angle, as shown in Figure 13.6 for a two-layered microstrip configuration. This means that the heat generated in the microstrip conductor (assuming there are no other heat sources and heat flow is mainly by conduction) flows down through the dielectric materials through areas larger than the strip conductor as it approaches the ground plane, where the ground plane acts as a heat sink. However, to account accurately for the increase in area normal to heat flow lines, the parallel plate waveguide model of a microstrip has been used. For $T_{max} = 150\,°C$, $T_{amb} = 25\,°C$, and $Z_0 = 50\ \Omega$, values of the APHC for various substrates at 10 GHz are calculated and given in Table 13.6. Among the dielectrics considered, the APHC is the lowest for Duroid (0.144 kW) and a maximum for BeO (52.774 kW). For commonly used alumina (or sapphire) substrates, a 50-Ω microstrip can carry about 4.63 kW of CW power at 10 GHz. Table 13.7 shows the APHC values of several multilayer 50-Ω microstrip lines on 75-μm thick GaAs at several frequencies. The APHC decreases with increasing frequency. Lines having characteristic impedance higher than 50 Ω will have lower APHC values as given in Table 13.6 due to the higher loss and narrower line widths. The APHC of microstrip lines on SiC is much higher than on GaAs or alumina.

Capacitors

For high-voltage and high-power applications, circuit designers must have maximum voltage, current, and power handling ratings for the components they intend to use in their circuit designs. For chip capacitors, such ratings depend on the materials used in the fabrication, capacitance value, and the area.

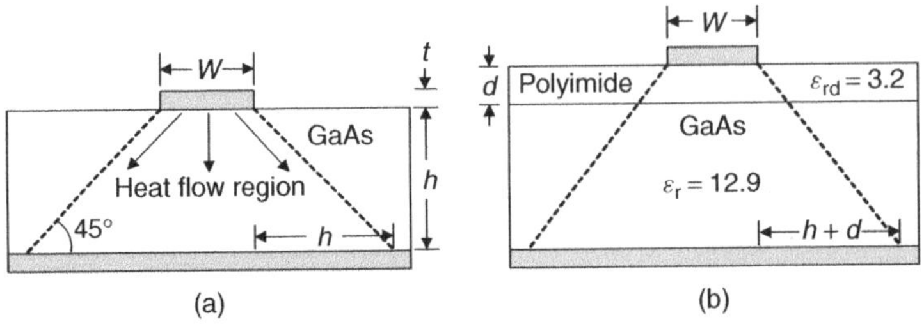

Figure 13.6 Schematic of heat flow based on 45° thermal spread angle rule: (a) microstrip line and (b) multilayer microstrip line.

Table 13.6 Comparison of APHC of 50-Ω Microstrip Lines on Various Substrates[a] at 10 GHz

Substrate	ε_r	tan δ	h (μm)	W (μm)	ΔT (°C/W)	Maximum Average Power (kW)
Duroid	2.2	0.0009	250	760	0.8682	0.144
Si	11.7	0.1540	100	75	0.126	0.992
GaAs	12.9	0.0010	75	50	0.0865	1.445
Al_2O_3	9.8	0.0002	250	235	0.027	4.630
BeO	6.4	0.0003	250	352	0.00237	52.774
SiC	9.7	0.005	100	95	0.0051	24.510

[a]Gold Conductors are 4.5 μm thick; $d = 0$ and $T_{amb} = 25\,^{\circ}$C.

Table 13.7 Comparison of APHC of 50-Ω Multilayer Microstrip Lines[a] on 75-μm Thick GaAs

Polyimide Thickness, d (μm)	Maximum Average Power (W) 5 GHz	10 GHz	20 GHz	40 GHz
0	2049	1445	1020	720
1	260	181	129	91
3	107	76	53	38
7	71	51	36	25
10	63	44	31	22

[a]Gold Conductors are 4.5 μm thick except in the 3-μm polyimide ($\varepsilon_{rd} = 3.2$) case, where thickness is 9 μm.

Maximum Voltage Rating For chip capacitors, the maximum voltage rating depends on the breakdown voltage (V_B) between the plates. Typical values of V_B for chip capacitors are 100–200 V, while for GaAs MMIC capacitors the values are 40–60 V. Many times, the surface breakdown due to sharp edges, contamination, and surrounding humidity conditions sets the working voltage ratings rather than the internal breakdown voltage between plates. Surface breakdown occurs outside the device package and is also known as the *flashover voltage*.

Maximum RF Current Rating If V_B is the capacitor's breakdown voltage and V_a is the DC applied voltage, the maximum (rms) current I_{mv} allowed to pass through a capacitor C is given by

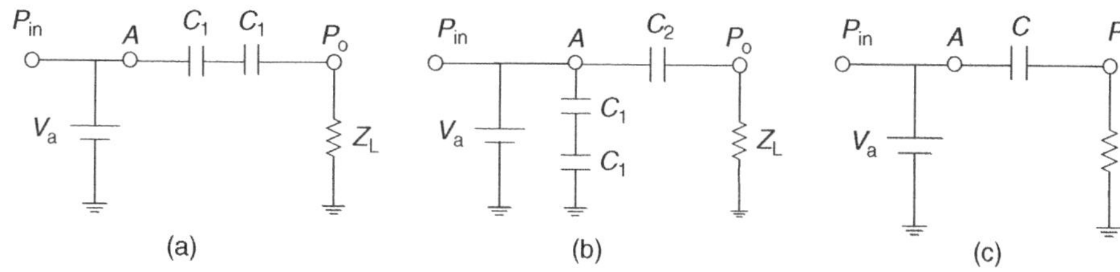

Figure 13.7 Series connections of MIM capacitors to improve current ratings: (a) two series capacitors in series configuration, (b) two series capacitors in shunt configuration, and (c) DC block MIM capacitor at the output of an amplifier in series configuration.

$$I_{mv} = \frac{1}{\sqrt{2}}(V_B - V_a)\omega C \tag{13.2a}$$

Thus for given values of V_B, V_a, and C, the maximum current is proportional to the operating frequency. For example, for $V_B = 200$ V, $V_a = 28$ V, and $C = 5$ pF, the values of I_{mv} are 3.82 A and 19.1 A at 1 GHz and 5 GHz, respectively. For a capacitor operating at 28 V and 3 GHz, with a breakdown voltage of 50 V,

$$I_{mv} = 0.293C \tag{13.2b}$$

where I_{mv} is in amperes and C is in picofarads. The current rating of MIM capacitors can be increased by connecting more capacitors in series. For example, when two capacitors of double capacitance value are used in series, as shown in Figure 13.7a, the current rating is increased by more than four times. Here, for RF design, $C_1 = 2C$ and the breakdown voltage across a series combination of two capacitors is doubled. Equation (13.2) becomes

$$I_{mv} = 1.172C \tag{13.3}$$

Maximum RF Voltage Rating The maximum RF voltage rating of a capacitor can be calculated by referring to Figure 13.7c. If V_a is the DC applied voltage and P_o is the output power across the load, the maximum voltage across C is

$$V_{mc} = V_a + \sqrt{2P_oR_L} \tag{13.4}$$

which should be less than V_B. If $V_{mc} > V_B$, the maximum working voltage of the capacitor can be increased by connecting more capacitors in series, as shown in Figure 13.7b. Equation (13.4) does not include any voltage standing wave effects, as discussed in References 67 and 68.

Maximum Power Dissipation The maximum current rating of a capacitor is also determined from the maximum allowed power dissipation to maintain the capacitor's temperature below the rated value, for example, 150 °C for a monolithic MIM capacitor on GaAs substrate. If P_{diss} is the maximum allowed power dissipation, the maximum current is given by

$$I_{mp} = \sqrt{\frac{P_{diss}}{ESR}} \tag{13.5}$$

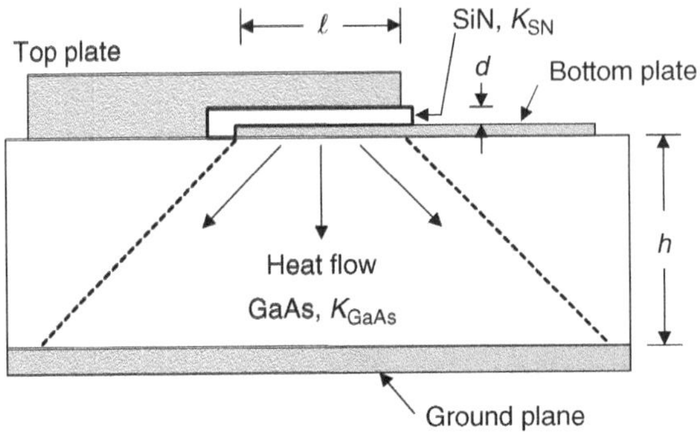

Figure 13.8 Heat flow schematic of a MIM capacitor.

where *ESR* is the effective series resistance of the capacitor. The maximum allowed power dissipation is calculated in the following.

The calculation of thermal resistance for an MIM capacitor is based on vertical heat flow, as shown in Figure 13.8. Consider that the RF power is dissipated in both the top and bottom conductors having surface resistances of R_{ST} and R_{SB}, respectively. Assuming the MIM capacitor as a square structure and ignoring dielectric losses, the temperature rise is given by

$$\Delta T = P_{DCT} R_{thc} + (P_{DCT} + P_{DCB}) R_{thg} \tag{13.6}$$

where P_{DCT} and P_{DCB} are the power dissipated in the top and bottom plates, respectively. R_{thc} and R_{thg} are the thermal resistances of the capacitor dielectric and GaAs substrate, respectively. Assuming the 45° heat spreading rule for the bottom conductor along its width only and the capacitor's dielectric is Si_3N_4 of thickness d,

$$\Delta T = I_{mp}^2 \left[R_{ST} \frac{d}{K_{SN} \ell W} + (R_{ST} + R_{SB}) \frac{h}{K_{GaAs} \ell (W + 2h)} \right] \tag{13.7}$$

where K_{SN} and K_{GaAs} are the thermal conductivities of the Si_3N_4 and GaAs, respectively. The width and length of the capacitor are W and l, respectively. Table 13.8 lists the maximum power dissipation and I_{mp} for various capacitor values. The parameters used in the calculations are

$$\ell = W = 0.5774 \times 10^{-2} \sqrt{C} \text{ cm}, \text{ where } C \text{ is in pF}$$
$$d = 0.2\ \mu\text{m}, \quad h = 75\ \mu\text{m}$$
$$K_{SN} = 0.1 \text{ W/cm} \cdot {}^\circ\text{C} \quad \text{and} \quad K_{GaAs} = 0.294 \text{ W/cm} \cdot {}^\circ\text{C at } 150\,^\circ\text{C}$$
$$T_m = 150\,^\circ\text{C}, \quad T_a = 25\,^\circ\text{C}, \quad \Delta T = 125\,^\circ\text{C}$$
$$R_{ST} = 0.007\ \Omega, \quad R_{SB} = 0.07\ \Omega, \quad \text{and} \quad ESR = 0.13\ \Omega$$

Resistors

The power handling capacity of resistors is limited due to burnout of the thin film by overheating. The power handling capacity of monolithic resistors can be determined in a way similar to that for microstrip lines. In this case, the resistor strip is considered as a lossy microstrip line. Since the loss in the resistor conductor is much higher than the dielectric loss, only conductor loss is considered in the calculation of dissipated

Table 13.8 Calculated Maximum Power Dissipation and I_{mp} for MIM Capacitors on 75-μm Thick GaAs Substrate

C (pF)	P_{diss} (W)	I_{mp} (A)
0.5	0.27	1.44
1.0	0.44	1.84
2.0	0.71	2.34
5.0	1.43	3.32
10.0	2.50	4.38
20.0	4.48	5.87
50.0	10.07	8.80

power. The temperature difference ΔT (°C) between the resistor film (of length $\ell$and width W) and the back side of the substrate due to P_{diss} (watt) power dissipated in the resistor is given by

$$\Delta T = P_{diss} R_{th} = P_{diss} \frac{h}{KA} \tag{13.8}$$

where R_{th} is the thermal resistance, A is the equivalent area of the resistor, h is the thickness, and K is the thermal conductivity of the substrate. Dimensions of h, A, and K are cm, cm^2, and W/cm·°C. The resistor area is given by

$$A = W_e(\ell + 2\ell') \tag{13.9}$$

where ℓ' is the length of the ohmic contact used to connect the resistor film to other circuitry and W_e is the effective width calculated from the parallel plate waveguide model [66]. Approximately, $W_e = W + 2h$. If R_s is the sheet resistance of the film, the total resistance R is given by

$$R = R_s \frac{\ell}{W} \tag{13.10}$$

From (13.9) and (13.10), we find

$$A = (W + 2h)\left(\frac{R}{R_s} W + 2\ell'\right) \tag{13.11}$$

From (13.8) and (13.11), we have

$$W^2 + pW - q = 0 \tag{13.12}$$

where

$$p = 2h + \frac{2\ell' R_s}{R} \tag{13.13a}$$

$$q = \frac{R_s h}{R}\left[\frac{P_{diss}}{K\Delta T} - 4\ell'\right] \tag{13.13b}$$

Equation (13.13b) is valid for positive values of q.

Table 13.9 Calculated Required Minimum Width (in μm) for Thin-Film Resistors on 100-μm Thick GaAs Substrate for Several Dissipated Power Values[a]

Power Dissipated,	Resistor Value (Ω)		
P_{diss} (W)	10	20	50
0.5	22.7	12.9	5.6
1.0	66.8	39.7	18.3
2.0	133.2	82.4	40.3
5.0	269.6	174.5	92.0
10.0	426.0	282.7	156.3
20.0	648.9	438.5	251.5

[a] $T_m = 150\,^\circ$C and $T_a = 25\,^\circ$C

The solution of (13.12) for a positive value of W is

$$W = \frac{-p + \sqrt{p^2 + 4q}}{2} \tag{13.14}$$

Thus the required resistor width for given power dissipation and resistance value can be calculated from (13.14). Table 13.9 summarizes W values for several power dissipation and resistor values. This does not include the effect of resistor film thickness.

EXAMPLE 13.1

Consider an HPA with a 50-Ω resistor to dissipate 1 W with the top surface's maximum allowed operating temperature of $T_m = 150\,^\circ$C. The base plate temperature is 25 °C. Other parameters are

$$h = 100\ \mu\text{m}$$
$$\ell' = 20\ \mu\text{m}$$
$$R_s = 10.8\ \Omega/\square$$
$$K = 0.294\ \text{W/cm}\cdot{}^\circ\text{C (GaAs at 150}\,^\circ\text{C)}$$

Determine the width and length of the resistor.

SOLUTION Here, $P_{diss} = 1$ W. The calculated values of ΔT and p and q from (13.13) are

$$\Delta T = 150 - 25\,^\circ\text{C} = 125\,^\circ\text{C}$$
$$p = 200 + 40 \times 10.8/50\ \mu\text{m} = 208.64 \times 10^{-4}\ \text{cm}$$
$$q = \frac{10.8 \times 100}{50}\left[\frac{1}{0.294 \times 10^{-4} \times 125} - 80\right]\mu\text{m}^2 = 0.4150 \times 10^{-4}\ \text{cm}^2$$

From (13.14), we find

$$W = \frac{-208.64 \times 10^{-4} + \sqrt{208.64^2 \times 10^{-8} + 4 \times 0.415 \times 10^{-4}}}{2}\ \text{cm} = 18.3\ \mu\text{m}$$
$$\ell = 50 \times 18.3/10.8 = 84.7\ \mu\text{m}$$

The following design notes are applicable for monolithic resistors:

- Select resistor width to meet electromigration rules as provided by the manufacturer.
- Maximize resistor area for best power dissipation capability.
- Make the resistor width as close to the connecting transmission line width as possible to minimize discontinuity effects. Discontinuity effects can usually be ignored if $\frac{1}{3} < W_M/W_R < 3$, where W_M and W_R are the line widths of the microstrip and resistor, respectively. Resistors shorter than λ/50 can simply be treated as a lumped resistance in series with a microstrip line of equivalent line dimension. ■

High-Power Resistors

High-power chip resistors and terminations are required to absorb unwanted power in RF and microwave couplers/hybrids, and power dividers/combiners used in HPAs. Reliable design of such components requires substrate materials having the following electrical and mechanical characteristics:

- Low dielectric constant (ε_r)
- High thermal conductivity (K) at high operating temperatures
- Coefficient of thermal expansion (CTE) close to that of resistive film and of copper used as a heat sink
- Good insulator having high electrical resistivity
- Good adhesion to resistive film and bonding contact metal
- Good machinability for batch production
- Low cost

Table 13.10 lists several potential candidates for chip resistors. Beryllium oxide (BeO) has been widely used for such applications as a substrate of choice. However, BeO powder and dust require special handling as they are known to be hazardous materials. A comparable material to BeO is aluminum nitride (AlN) and its usage is steadily increasing because of the international community's concern regarding the hazards of BeO. Other possible replacement substrate materials are silicon carbide (SiC), boron nitride (BN), alumina (Al_2O_3), and diamond. As compared in Table 13.10, diamond has the highest thermal conductivity and is used where good heat sinks are required.

Both thin-film and thick-film technologies are being used to manufacture high-power chip resistors. Nichrome and tantalum nitride are commonly used resistor thin films, TaN is preferred to NiCr, due to the presence of undesirable nickel in NiCr,

Table 13.10 Comparison of Dielectric Substrates for High-Power Chip Resistors

Property	Alumina	BN	BeO	AlN	SiC	Diamond
ε_r	9.9	4.2	6.7	8.5	9.7	5.7
K (W/m·°C) @ 25°C	37	70	280	230	430	689
@ 100°C	31	20	200	203	310	—
@ 200°C	—		150	170	217	—
CTE (ppm/°C)	6.9	5.0	6.4	4.6	3.8	1.2
Shunt capacitance	Medium	Small	Small	Medium	Large	Small
Film adhesion	Excellent	Poor	Excellent	Good	Good	Poor
Machinability	Good	Good	Good	Good	Good	Poor
Cost	Low	Low	Medium	Medium	High	Very high

Table 13.11 Chip Resistor Film Dimensions for Various Power Handling Levels Using 25-mil BeO Substrate[a]

Dissipated Power (W)	Area (mm^2)	$W = \ell$ (mm)
10	0.29	0.539
20	0.58	0.762
50	1.45	1.204
100	2.90	1.703
200	5.80	2.408

[a] $T_m = 150\,^\circ$C and $T_a = 25\,^\circ$C

which is believed to introduce unwanted intermodulation products in multicarrier communication system application apparatus.

The substrate thickness is selected based on mechanical strength, handling, parasitic capacitance, and cost. Maximum allowed film temperature is usually 150 °C and the ambient temperature with chip resistors attached to heat sinks with additional cooling is about 25 °C. Let us assume that the BeO substrate is 25 mils (0.635×10^{-3}m) thick and K at 150 °C is about 175 W/m·°C. The calculated resistive film areas for 10-, 20-, 50-, 100-, and 200-W power handling are summarized in Table 13.11. If the substrate selected is 50 mils thick, the resistor width and length for a square area given in Table 13.11 are $\sqrt{2}$ times more.

13.4 POWER AMPLIFIER DESIGN EXAMPLES

The high-voltage operation of power amplifiers can be achieved by using wide-bandgap devices such as GaN or SiC MESFET or field plate MESFETs or pHEMTs, or by arranging low-voltage conventional devices in series. In this section we describe examples dealing with high-voltage operation.

Let us look at the practical aspect of high-PAE HPAs. At UHF and RF, the device and matching networks parasitic reactances are negligible and transistors have very high gain (20 dB or higher). At these frequencies, the transistors can be operated at class B and still have high gain (15 dB or higher). By terminating even harmonics, and third and fifth harmonics, a drain efficiency on the order of 90% may be obtained. Similar efficiency value is possible by using class-E operation, but at lower output power levels. Table 8.3 shows a comparison of RF performance measured using LDMOS transistors at 0.5 GHz. It may be noted that both class-E and class–F amplifiers have similar performance. The PAE and power level capabilities for class-E amplifiers are adversely affected at higher voltage and frequency.

13.4.1 HV Hybrid Amplifiers

The design of HV hybrid power amplifiers is similar to that of conventional hybrid power amplifiers described in the next chapter. The main advantage of using HV transistors is that one needs fewer devices to achieve high amplifier power levels. On the other hand, the thermal design of the amplifier becomes as important as the electrical design due to the much higher device power density. For GaN HEMTs on SiC, the limiting factor in thermal design is not the device's thermal resistance but the stacked layers under the transistor chip.

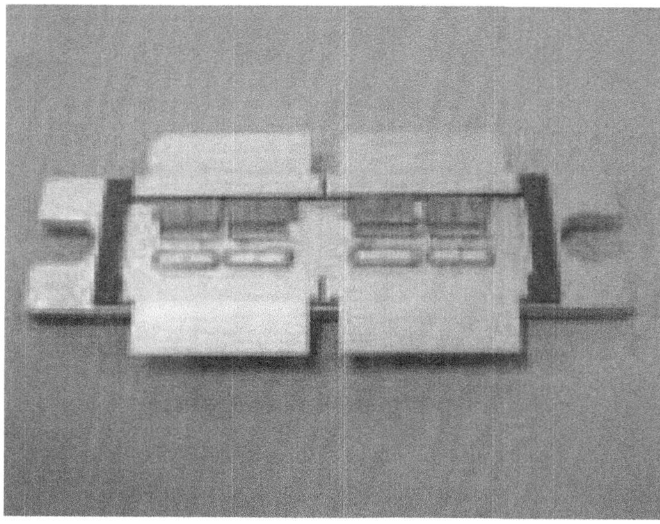

Figure 13.9 Photograph of the 250-W GaAs FP-FET amplifier in a ceramic package. (From Nagahara et al. [13]. Reprinted with permission of IEEE.)

FP MESFET Amplifier

Figure 13.9 shows a photograph of a 2.1-GHz 250-W amplifier. The drain bias voltage was 28 V. Four 90-mm GaAs FP-FETs were combined in a package in a push–pull configuration. The input and output matching networks consist of lowpass topology realized by employing chip capacitors and wires. Using external baluns, at 2.1 GHz the measured saturated power, power linear gain, drain efficiency, and IM3 values (under pulsed conditions of 5-μs pulse width and 0.5% duty cycle) were 250 W, 15.5 dB, 25%, and −33 dBc, respectively.

GaN HEMT on SiC Amplifiers

Figure 13.10 shows a photograph of a 1.5-GHz 500-W GaN HEMT hybrid amplifier. The devices were operated near class B and the drain bias voltage was at 65 V. Four 36-mm GaN HEMT chips were combined in a package in a push–pull configuration. The input and output matching networks were designed using a high dielectric constant substrate. The input and output impedances were 25 Ω. Using external baluns, at 1.5 GHz, the measured pulsed saturated power, power gain, and drain efficiency values were 500 W, 17.8 dB, and 49%, respectively. For pulsed measurements, the pulse width and duty cycle were 100 μs and 10%, respectively. This HPA has 2.5 times less transistor size and double the output power as compared to the previous example.

Figure 13.11 shows a photograph of another GaN HEMT hybrid amplifier operated at the C-band. Two devices, each having a gate periphery of 8 mm, were combined using input and output matching networks. The output matching network was comprised of a short-circuited stub, open-circuited stub, and quarter-wave impedance transformer as described in Chapter 10 for second-harmonic termination. The amplifier was fabricated on an alumina substrate and its size was 8×7 mm^2. The drain bias voltage was 40 V. At C-band over narrowband, the measured pulsed saturated power and PAE values were 80 W and 46%, respectively.

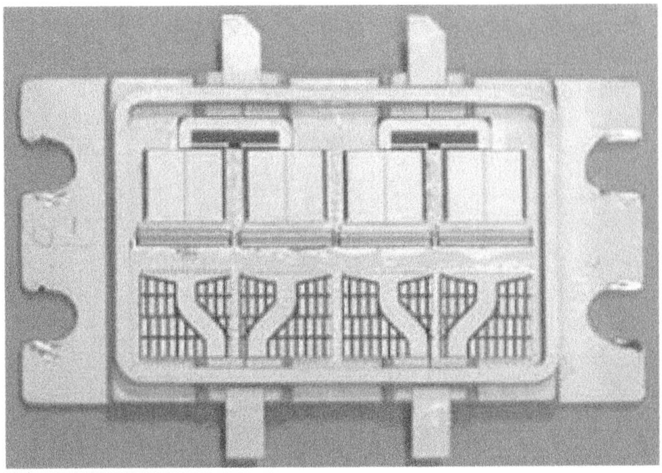

Figure 13.10 Photograph of the 500-W GaN HEMT amplifier in a ceramic package. (From Maekawa et al. [50]. Reprinted with permission of IEEE.)

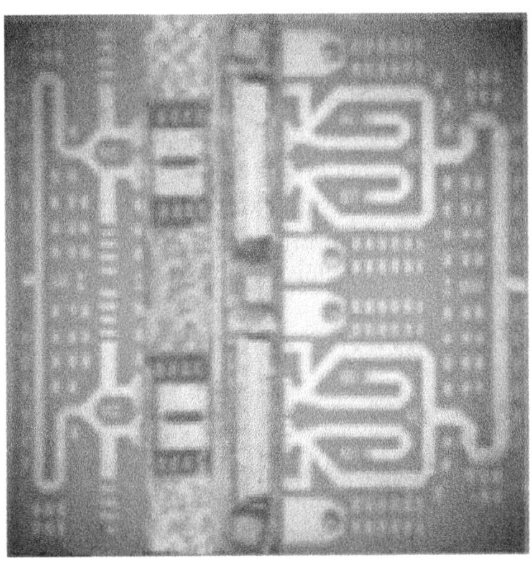

Figure 13.11 Photograph of the 80-W GaN HEMT amplifier. (From Iyomasa et al. [58]. Reprinted with permission of IEEE.)

A single-stage 50-W amplifier using 0.6-μm gate GaN HEMT operated at 6 GHz and 39 V was reported [71, 72]. The transistor's I_{peak} was 1 A/mm and breakdown voltage was 80 V. The total gate periphery used was 8 mm. The measured output power for the amplifier was about 51 W.

EXAMPLE 13.2

Design a power amplifier at 3.5 GHz using a GaN HEMT that has 8-W output power and 60% drain efficiency (η_D) when operated at 28 V. The transistor's f_T is 11.37 GHz and load impedance (parallel combination of R_L and C_L) at 28 V is $R_L = 41\ \Omega$ and $C_L = -1.1$ pF. The transistor's lumped-element EC model (Fig. 5.2) values are

$$R_g = 2\ \Omega, \quad R_i = 0.5\ \Omega, \quad R_s = 0.3\ \Omega, \quad R_d = 0.5\ \Omega, \quad R_{ds} = 150\ \Omega$$
$$C_{gs} = 4.2\ \text{pF}, \quad C_{gd} = 0.18\ \text{pF}, \quad C_{ds} = 0.8\ \text{pF},$$
$$g_m = 300\ \text{mS}, \quad \tau = 6\ \text{ps}$$
$$L_g = 0.02\ \text{nH}, \quad L_s = 0.001\ \text{nH}, \quad L_d = 0.02\ \text{nH}$$

The amplifier is fabricated using a microstrip on a 15-mil thick alumina substrate ($\varepsilon_r = 9.9$). Determine the matching network's dimensions and the amplifier's performance including gain, input return loss, K factor, output power, and PAE. The amplifier must be unconditionally stable.

SOLUTION At 3.5 GHz,

$$Z_L = 20.7 + j20.5\ \Omega$$

and a schematic selected for the amplifier is shown in Figure 13.12a. For the transistor, at 3.5 GHz, $K = 0.727$. Here the resistor in the input matching network is used to stabilize the amplifier. The circuit is optimized in terms of microstrip line length as well as width. The output is matched for correct load impedance. Physical dimensions for optimized matching elements on alumina substrate ($\varepsilon_r = 9.9$, $h = 15$ mils) are given in Figure 13.12a. The effect of T-junction discontinuity, bond wires, parasitic reactance of DC blocking, and bypass capacitors have been ignored in the design. The calculated small-signal performance of this amplifier is gain = 14.9 dB, input return loss = 16 dB, output return loss = 9.4 dB, and $K = 1.06$.

The output matching network loss is about 0.2 dB. Let us assume that the transistor's gain is compressed by 3 dB at maximum output power and PAE. Therefore

$$\text{Power gain, } G = 14.9 - 3.0 = 11.9\ \text{dB} = 15.5$$
$$\text{Transistor's } P_o = 8\ \text{W} = 39\ \text{dBm}$$
$$\text{Amplifier's } P_o = (39 - 0.2)\ \text{dBm} = 38.8\ \text{dBm} = 7.6\ \text{W}$$
$$\text{Transistor's } \eta_D = 60\%$$
$$\text{Amplifier's } \eta_D = 60 \times \frac{7.6}{8.0}\% = 57\%$$
$$\text{Amplifier's PAE} = 57 \times (1 - 1/G)\% = 57 \times (1 - 1/15.5)\% = 53\%$$

Thus both the gain and the output matching network's loss affect the PAE of the amplifier. ■

A very high-efficiency 1.2-GHz HV GaN power amplifier was designed, fabricated, and tested [73]. The transistor used in the design was 10-W GaN HEMT from Cree (CGH40010) in a screw-down package configuration. The input and output matching network topologies were selected so that up to fifth harmonics were reactively terminated. A simple single-stage circuit was optimized using a nonlinear model with loose convergence tolerance setting in the harmonic balance simulator. The circuit was optimized for PAE > 85% and output power greater than 6 W. Figure 13.12b shows the schematic of the high-efficiency HPA. The amplifier was fabricated using Rogers 4003 PCB material. The measured output power and PAE values at 1.2 GHz were 8 W and 90%, respectively.

At RF and low microwave frequencies, using the aforementioned optimization technique, it is not easy to classify the amplifier based on impedance termination conditions. The effect of parasitic reactance is not accurately determined. However,

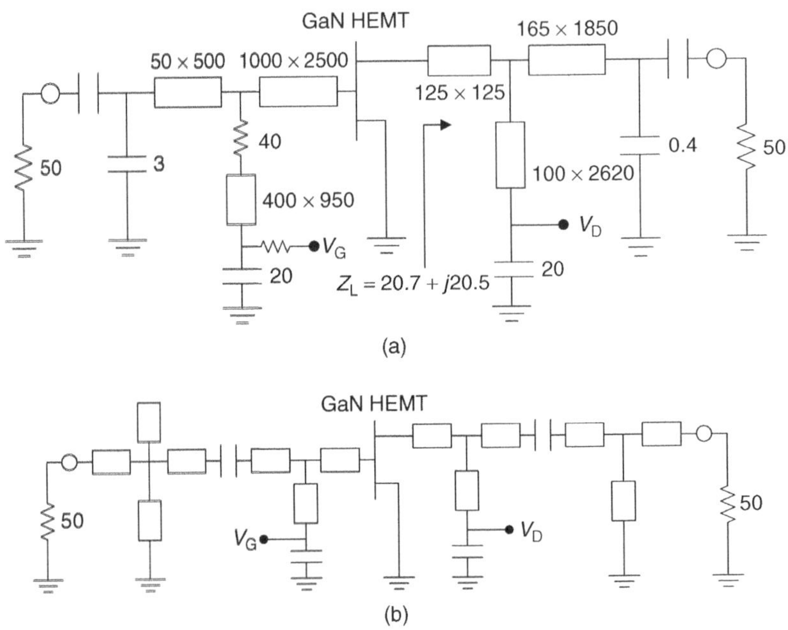

Figure 13.12 Schematic of the single-stage HPA. Dimensions (width × length) are in μm, capacitors in pF and resistors in Ω. (b) Schematic of the single-stage high-efficiency HPA.

one can measure output current and voltage waveforms using a superfast oscilloscope and approximately determine the class of operation. In general, an amplifier tuned for maximum PAE may be a blended version of class E, class F, or class F^{-1}.

13.4.2 HV Monolithic Amplifiers

Achieving reliable and cost-effective MMIC HPAs having output power substantially higher than is available today involves quite a range of design considerations, many of which require trade-offs. Some of the top-level design parameters are given in Table 13.12 along with their expected impacts on the performance, reliability, and cost. Fundamentally, an optimum MMIC design provides higher output power density and high PAE at much higher drain voltage, while simultaneously reducing MMIC chip size and keeping channel temperatures below 150 °C.

High-voltage operation for a given output power simplifies the MMIC chip and system current routing. Higher voltage operation will increase the load impedance, making it easier to achieve the necessary matching for output power and PAE over the operating frequency band. Wider gate–gate pitch and evenly distributed active devices on the chip will keep the transistor channel temperatures to reasonable levels for reliable HPA operation. Transmission lines and inductors have substantially higher dissipative loss on thin semiconductor substrates than on alumina substrates. The low-loss matching networks, as discussed in Chapter 10, are realized by adding a polyimide layer between the transmission line conductors and the GaAs substrate, lowering the transmission loss by a factor of 2.

Table 13.12 Design Trades to Be Considered in the Development of Very High Power MMIC Amplifiers

Design Parameter	Performance Impact	Reliability Impact	Cost Impact
Higher drain voltage	+ Reduced current + Higher load impedance + Reduced transistor periphery and easier to match + Reduced I^2R loss	+ Reduced current − Increased channel temperature due to increased dissipated power density − Higher voltage stress of capacitors − Higher voltage spikes	+ Smaller chip + System costs less for higher voltage and lower current
Wider gate–gate pitch	− Lower gain due to higher drain capacitance	+ Reduces channel temperature	− Increases chip width
Lower loss output match	+ Higher electrical performance	+ Lower power dissipation in the chip	− Higher fabrication steps
Harmonic termination	+ Higher PAE	+ Cooler channel	− Slightly larger chip

Reliable HV operation requires the transistor channel temperatures to be kept at reasonable levels by designing high-PAE HPAs. Harmonic terminations and class-E operation are important design parameters in an HPA next to a low-loss output matching network in order to obtain high PAE. HV device based MMIC power amplifiers have been developed working in the S- through Ka-bands. Below 4 GHz, several devices including FP MESFETs, HBTs, pHEMTs, and GaN HEMTs have been used to develop MMICs, whereas above 4 GHz, MMICs are primarily based on GaN HEMTs. Table 13.3 summarized progress in HV transistor based MMICs.

HV FET MMIC Amplifier

A 10-W output power, an MMIC amplifier was reported using HV MSAG FET [16]. The three-stage design used a 9.6-mm FET periphery at the output stage to achieve 10 W of output power at P_{1dB} gain compression and a bias supply of 24 V. Figure 13.13a shows the photograph of the MMIC. The chip size is 3 mm^2. Typical measured CW output power and PAE of the packaged chip is shown in Figure 13.13b. The PAE was greater than 30% over the 3–3.8-GHz band.

GaN on SiC HEMT MMIC Amplifiers

A 24-W MMIC power amplifier was reported using GaN 0.4-μm gate HEMT at 16 GHz and biased at 31 V. The two-stage amplifier was reactively matched to 50 Ω and 25 Ω at the input and output, respectively. The input and output stages used 3-mm and 6-mm transistors, respectively. The measured gain, output power, and PAE, at 16 GHz, were 12.8 dB, 24.2 W, and 22%, respectively [21].

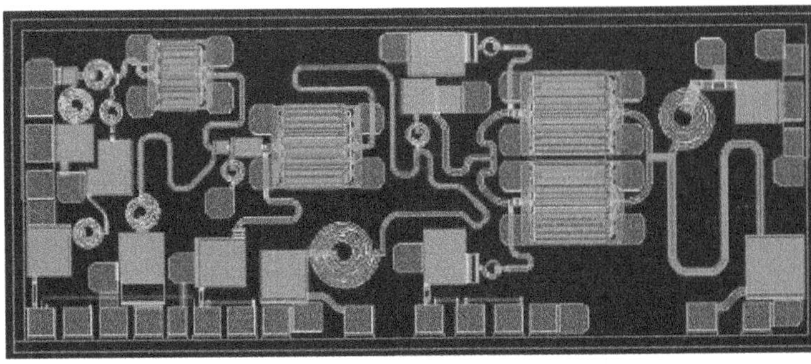

(a)

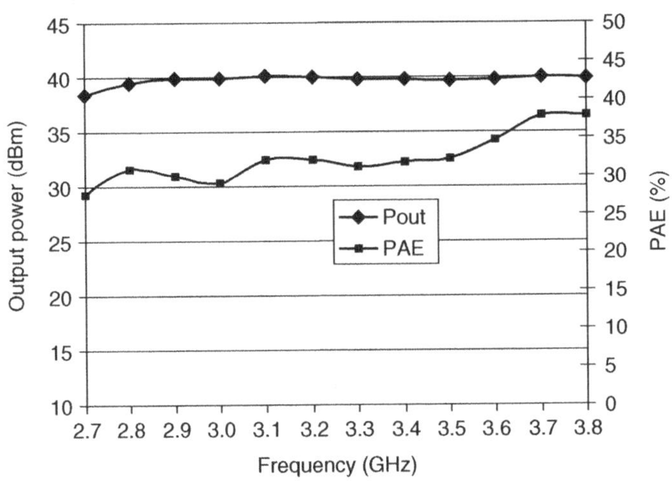

(b)

Figure 13.13 (a) Photograph and (b) output power and PAE of the 10-W three-stage HVMSAG MMIC.

A Ka-band, a 4-W two-stage power amplifier MMIC was developed using Al/GaN/GaN HEMT on SiC. The output stage used a 0.18-μm gate length and two 0.6-mm gate periphery transistors. The device aspect ratio in the two-stage design used was 2:1 [52]. For a 400-μm gate width device, the measured I_{dss}, g_m, f_T and f_{max} values were 1 A/mm, 290 mS/mm, 84 GHz, and 114 GHz, respectively. The MMIC design was performed using measured load-pull data and S-parameters. The optimum power load for the 0.6-mm device was close to 50 Ω. Figure 13.14 shows the photograph of the 4-W PA MMIC. The measured small-signal gain, saturated output power, and PAE values, at 35 GHz and drain bias of 24 V, were 12 dB, 4 W, and 23%, respectively. Thus a power density of 3.2 W/mm was demonstrated at the Ka-band.

13.5 BROADBAND HV AMPLIFIERS

Achieving reliable and cost-effective broadband HV HPA MMICs having output power substantially higher than is available today involves a wide range of design considerations, many of which require trade-offs in terms of bandwidth, performance, reliability,

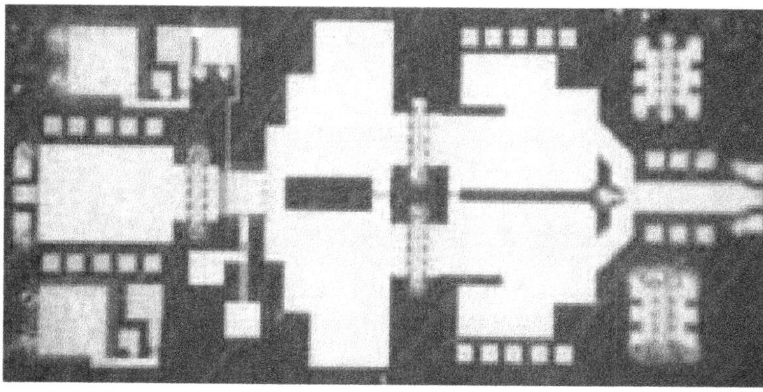

Figure 13.14 Photograph of the 4-W two-stage GaN HEMT MMIC. (From Darwish et al. [52]. Reprinted with permission of IEEE.)

and cost. Some of the broadband top-level design methodologies are (a) feedback topology for low band (30 MHz to 2 GHz), discussed in Chapter 11; (b) reactive/resistive topology for midband (0.5–4 GHz or 1–8 GHz), discussed in Chapters 7 and 9; and (c) distributed configuration for high band (2–20 GHz), discussed in Chapter 11.

For the low band (30 MHz to 2 GHz), a major problem is in the design of the output matching network over such a wide band. In addition to feedback, an amplifier topology having no output match can be used. Since, at low frequencies, the transistor's output reactance is much lower than the load impedance required at the device output, the transistor size is selected such that the load impedance value is 50 Ω. Approximate values of load impedance R_L and power density P_L, normalized to a gate periphery of 1 mm, for various devices are

$$R_L = 5.0V_{ds} \quad \Omega \cdot \text{mm for MSAG@10V} \tag{13.15a}$$
$$R_L = 7.5V_{ds} \quad \Omega \cdot \text{mm for HVMSAG@30V} \tag{13.15b}$$
$$R_L = 7.5V_{ds} \quad \Omega \cdot \text{mm for GaN HEMTs on SiC} \tag{13.15c}$$

$$P_L = 0.8 \text{ W/mm} \quad \text{for MSAG} \tag{13.16a}$$
$$P_L = 1.5 \text{ W/mm} \quad \text{for HVMSAG} \tag{13.16b}$$
$$P_L = 5.0 \text{ W/mm} \quad \text{for GaN HEMTs} \tag{13.16c}$$

In this case, the device sizes required to have an R_L value of 50 Ω are 1 mm, 4.5 mm, and 7.5 mm for MSAG @ 10 V, HVMSAG @ 30 V, and GaN HEMT @ 50 V, respectively. Using power densities given in (13.16), the maximum output power P_o possible for HPAs without any output match are shown in Figure 13.15. To get high power levels over ultra-wideband, high power density devices, such as GaN HEMTs operating at very high voltage, are essential.

The operating voltage of power devices can also be increased by connecting them in series. For example, two 28-V HVMSAG devices connected in series can be operated at 56 V. With this technique, one can increase the required load value as well as the power density manyfold. Series connection of transistors is discussed in the next section.

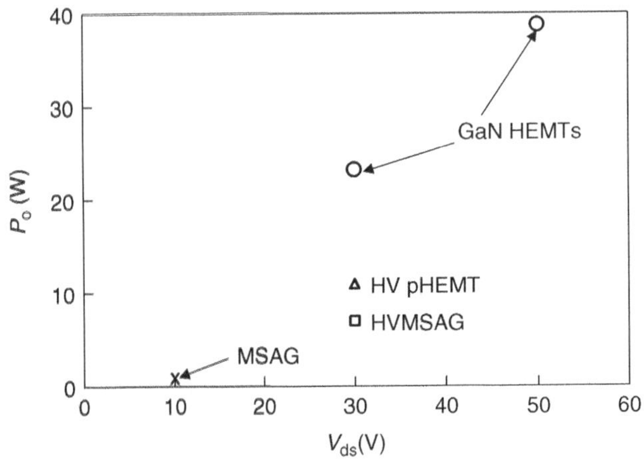

Figure 13.15 A plot of maximum output power versus drain voltage for HPAs using no output match.

The design of high-power amplifiers for midband, three-octave bandwidth (0.5–4 GHz, 1–8 GHz) can be executed using reactive/resistive matching techniques employing broadband matching networks, as discussed in Chapter 7. In this configuration, the baluns are used for impedance matching as well as for biasing chokes. This allows a reduction in chip size as well as transforming the high-power device's very low impedances to high impedance over 2–3-octave bandwidths. This technique has been applied successfully over two-octave bandwidths to achieve 13-W output power using MSAG FETS operating at 10 V.

The design of high-power amplifiers for high-band, 2–3-octave bandwidth (4–20 GHz, 6–20 GHz) can be implemented using modified distributed topology, as discussed in Chapter 11. In this configuration the distributed power amplifier (DPA) is matched to a 3–12-Ω source and load impedance and an ultra-broadband balun is used to transform the 3–12-Ω impedance to 50 Ω. A high PAE, DPA design requires accurate nonlinear device models, low-loss matching networks, and low-loss ultra-broadband baluns. The LLM technique described in Chapter 9 is a very suitable method in designing broadband HPAs. The total loss required for each stage to compensate device gain slope for flat amplifier gain is adjusted by maintaining the power and PAE at optimum value.

A 2.5-W broadband MMIC power amplifier was reported using 0.15-μm GaN HEMT at 35 V. The single-stage balanced amplifier used a four-cell distributed approach. Th MMIC size was 4.2×4.9 mm^2. The measured gain, output power, and PAE were 8 dB, 2.5 W, and 8%, respectively, over the 3–18-GHz frequency range [74].

13.6 SERIES FET AMPLIFIERS

The operation of low supply voltage devices ($\leq$ 10 V) or high-voltage devices ($\geq$ 20 V) can be extended to much higher supply voltages [75–78]. In this case, the operating voltage is much higher than the device breakdown voltage. This has been achieved by connecting devices in series. When the devices are connected for DC bias in series and for RF in parallel [75, 76], the circuit is biased at higher voltages; however, the RF load impedance of parallel devices becomes very low. In this case, the design of matching networks becomes difficult for broadband applications. Recently, an improved version

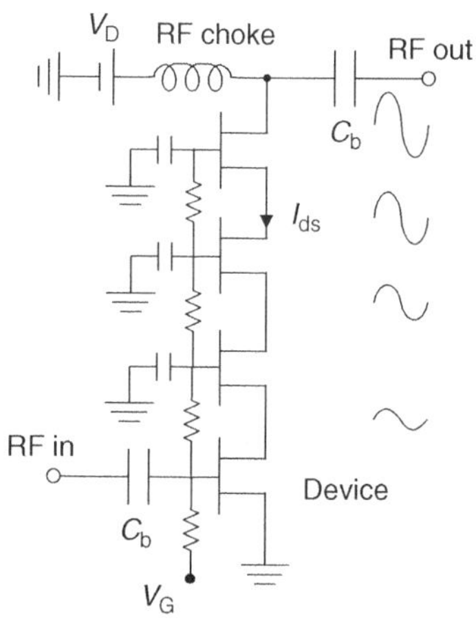

Figure 13.16 Configuration of four FETs in series for high-voltage operation.

of this topology has been introduced, where devices are connected in series for both DC bias and RF load [78]. This configuration using four devices in series, as shown in Figure 13.16. In this case, the supply voltage (V_D) and the load impedance (Z_L) for FETs are

$$V_D = NV_{ds} \tag{13.17}$$

$$Z_L \cong NV_{ds}/I_{ds} \tag{13.18}$$

where $N = 4$ is the number of FETs in series, and V_{ds} and I_{ds} are the operating voltage and current, respectively, for each FET. The power gain (G) and output power (P_o) are increased by the number of FETs:

$$G = NG_S \tag{13.19}$$

$$P_o = NP_S \tag{13.20}$$

where G_S and P_S are the gain and power output for a single FET.

Figure 13.17 shows an example of a four-FET high-voltage configuration with both DC and RF connections in series. A two-stage MMIC was designed using a low-voltage M/A-COM foundry. The measured data achieved power gain = 21 ± 1 dB, $P_{1dB} = 2$ W, and PAE = 20% over the 30-MHz to 2.5-GHz frequency band, at a bias voltage of +20 V. This configuration can be used for other devices including HVFETs and HEMTs.

EXAMPLE 13.3

Consider a low-frequency broadband HPA using HV transistors whose input is matched. The transistor's power density and load (parallel combination of R_L and C_L) at 28 V are

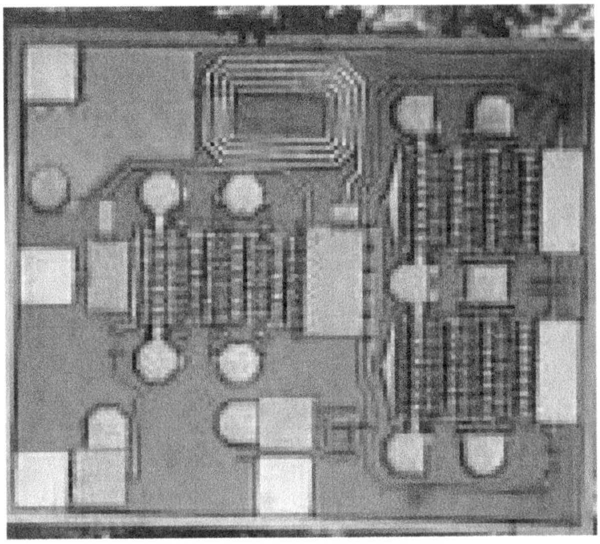

Figure 13.17 Photograph of MMIC using four FETs in series for high-voltage operation. (From Ezzeddine and Huang [78]. Reprinted with permission of IEEE.)

$$P_o = 4\ \text{W/mm}$$
$$R_L = 90\ \Omega \cdot \text{mm}$$
$$C_L = -0.5\ \text{pF/mm}$$

When there is no output matching network at the output terminal of the device, determine its maximum output power capability and operating frequency for a single transistor as well as for two transistors connected in series (similar to Fig. 13.16). In this case, it is assumed that the input is matched and the maximum operating frequency is limited by the output reflection coefficient ρ due to C_L. Here, $\rho = 0.3$ and the system impedance $Z_0 = 50\ \Omega$.

SOLUTION For the single-transistor case, for maximum power transfer $R_L = 50\ \Omega$, and the maximum output power for a single transistor is given by

$$P_{o\,\max} = 4 \times 90/50 = 7.2\ \text{W}$$

In this case, the transistor size is 1.8 mm. The output reflection coefficient ρ is given by

$$|\rho| = \left|\frac{Z_L - Z_0}{Z_L + Z_0}\right| = \left|\frac{Y_0 - Y_L}{Y_0 + Y_L}\right| = \frac{\omega|C_L|}{\sqrt{0.04^2 + (\omega C_L)^2}} \tag{13.21}$$

At $\rho = 0.3$ the maximum operating frequency is $f_{\max}$; rearranging (13.21), we find

$$f_{\max}|C_L| = 2 \tag{13.22}$$

where $f_{\max}$ is in GHz and C_L is in pF. For a 1.8-mm device, $C_L = -0.9$ pF and $f_{\max} = 2.2$ GHz. For two transistors in series, using (13.18) and (13.20), we have

$$P_o = 8\ \text{W/mm}$$
$$R_L = 180\ \Omega \cdot \text{mm}$$
$$C_L = -0.25\ \text{pF/mm}$$

For maximum power transfer $R_L = 50\ \Omega$ and the maximum output power is given by

$$P_{o\,max} = 8 \times 180/50 = 28.8 \text{ W}$$

Here, two 3.6-mm transistors are in series and $C_L = -0.9$ pF. Thus the output power is increased by a factor of 4 but the value of C_L remains the same as for a single transistor.

Therefore $f_{max} = 2.2$ GHz in this case also. ■

EXAMPLE 13.4

Consider an ideal HV transistor with a maximum current of I_{max}, drain–source voltage of V_{ds}, and breakdown voltage of 50 V. Determine its maximum output power when the supply voltages V_D are 28, 48, 100, and 200 V and transistor's output load R_L is 50 Ω. The effect of C_L is neglected.

SOLUTION Since the device's breakdown voltage is greater the $2V_{ds}$,

$$V_D = nV_{ds}$$

were n is number of transistors connected in series. Here, for $V_D = 28$, 48, 100, and 200 V, the values of n are 1, 2, 4, and 8, respectively. The output power and the transistor load may be expressed as follows:

$$P_o = \frac{I_{max} V_D}{4}$$

$$R_L = \frac{2V_D}{I_{max}}$$

or

$$P_o = \frac{V_D^2}{2R_L}$$

For

$$R_L = 50\ \Omega,\ P_o = \frac{V_D^2}{100}\ W$$

For $V_D = 28$, 48, 100, and 200 V, the values of P_o are 7.8, 23, 100, and 400 W, respectively. ■

The HV devices and HV amplifier design techniques including several hybrid and monolithic amplifier examples have been described. The discussion on realization of the amplifier circuits in the hybrid and monolithic technologies will continue in the next two chapters.

REFERENCES

1. J. R. Gajadharsing, O. Bosma, and P. Westen, Analysis and design of a 200 W LDMOS based Doherty amplifier for 3 G base stations, *IEEE MTT-S Dig.*, pp. 529–532, 2004.
2. Y. Chung et al., Efficiency-enhancing technique: LDMOS power amplifier using dual-mode operation design approach, *IEEE MTT-S Dig.*, pp. 859–862, 2004.
3. K. Goverdhanam et al., Modeling high power RF LDMOS amplifiers, *IEEE MTT-S Dig.*, pp. 1069–1072, 2004.
4. D. W. Beishline et al., Highly efficient 60 watt W-CDMA LDMOS power amplifier using the modified Doherty configuration, *34th Eur. Microwave Conf. Dig.*, pp. 1169–1172, 2004.
5. Z. Xie et al., High performance and low cost LDMOS devices for 3 G base stations, *34th Eur. Microwave Conf. Dig.*, pp. 1177–1180, 2004.
6. F. Lepine, A. Adahl, and H. Zirath, A high efficient LDMOS power amplifier based on an inverse class F architecture, *34th Eur. Microwave Conf. Dig.*, pp. 1181–1184, 2004.
7. A. Grebennikov and N. O. Sokal, *Switchmode RF Power Amplifiers*, Elsevier, Burlington, MA, 2007.
8. P. H. Aaen, J. A. Pla, and J. Wood, *Modeling and Characterization of RF and Microwave Power FETs*, Cambridge University Press, Cambridge, UK, 2007.
9. K. Horiguchi et al., Efficiency enhancement of 250 W Doherty power amplifiers using virtual open stub techniques for UHF-band OFDM applications, *IEEE MTT-S Dig.*, pp. 1356–1359, 2006.
10. K. Ebihara et al., An ultra broadband 300 W GaAs power FET for W-CDMA base stations, *IEEE MTT-S Dig.*, pp. 649–652, 2001.
11. K. Inoue et al., A high gain L-band GaAs FET technology for 28 V operation, *IEEE MTT-S Dig.*, pp. 821–824, 2004.
12. R. Hajji, J. Shumaker, and E. Camargo., 28 V High-voltage GaAs FET large-signal modeling achieves power and linearity prediction, *IEEE MTT-S Dig.*, pp. 1065–1068, 2004.
13. M. Nagahara et al., A 28 V 250 W GaAs power FET with high gain of 15.5dB for W-CDMA base stations, *IEEE MTT-S Dig.*, pp. 1359–1362, 2004.
14. T. Yamamoto et al., A 5.5GHz, 25 W GaAs power-FET chip at 26 V operation, *34th Eur. Microwave Conf. Dig.*, pp. 1313–1316, 2004.
15. D. Miller and M. Drinkwine, High voltage microwave devices: an overview, in *GaAs MANTECH Conference*, May 19–22, 2003, paper #7-1.
16. High voltage low cost FETs technology for HPA MMIC applications, *Microwave J.*, Vol. 47, pp. 16–128, December 2004.
17. B. Yang et al., 28 V Planar GaAs MESFETS for wireless base-station power amplifiers, *IEEE Compound Semiconductor Integrated Circuit Dig.*, pp. 244–247, 2005.
18. T. A. Winslow, Power dependent input impedance of field plate MESFETs, *IEEE Compound Semiconductor Integrated Circuit Dig.*, pp. 240–243, 2005.
19. T. A. Winslow, D. M. Conway, and E. L. Griffin, High voltage MESFET technology for MMIC applications, *GOMAC Dig.*, pp. 56–59, 2005.
20. J. W. Palmour et al., Wide bandgap semiconductor devices and MMICs for RF power applications, *Int. Electron Device Meeting Proc.*, pp. 17.4.1–17.4.4, 2001.
21. W. L. Pribble et al., Applications of SiC MESFETs and GaN HEMTs in power amplifier design, *IEEE MTT-S Dig.*, pp. 1819–1822, 2002.
22. A. Sayed et al., An ultra wideband 5 W power amplifier using SiC MESFETs, *34th Eur. Microwave Conf. Dig.*, pp. 57–60, 2004.
23. J. W. Palmour et al., SiC and GaN based transistor and circuit advances, *12th GAAS Symp. Dig.*, pp. 555–558, 2004.
24. B. Luo et al., 56 W SiC MESFET transistors with >50% PAE for L-band applications, *GOMAC Dig.*, pp. 49–51, 2005.
25. M. Sudow et al., A SiC MESFET-based MMIC process, *IEEE Trans. Microwave Theory Tech.*, Vol. 54, pp. 4072–4077, December 2006.

26. B. M. Green et al., A high power density 26 V GaAs pHEMT technology, *IEEE MTT-S Dig.*, pp. 817–820, 2004.
27. M. Akkul et al., 50 W MMIC power amplifier design for 2GHz applications, *IEEE MTT-S Dig.*, pp. 1355–1358, 2004.
28. P.-M. Piel, M. Miller, and B. Green, A 26V, 45 watt GaAs pHEMT for 2GHz WCDMA applications, *IEEE MTT-S Dig.*, pp. 1363–1366, 2004.
29. A. Wakejima et al., A GaAs-based field-modulating plate HFET with improved WCDMA peak-output-power characteristics, *IEEE Trans. Electron Devices*, Vol. 50, pp. 1983–1987, September 2003
30. M. Miller, Design, performance and applications of high voltage GaAs FETs, *IEEE Compound Semiconductor Integrated Circuit Dig.*, pp. 236–239, 2005.
31. K. Ishikura et al., A 28V, over 300 W GaAs heterojunction FET with dual field-modulating-plates for W-CDMA base stations, *IEEE MTT-S Dig.*, Paper # WE3E-1, 2005.
32. I. Takenaka et al., A 330 W distortion-cancelled Doherty 28 V GaAs HJFET amplifier with 42% efficiency for W-CDMA base stations, *IEEE MTT-S Int. Microwave Symp. Dig.*, pp. 1344–1347, 2006.
33. L. F. Eastman and U. K. Mishra, The toughest transistor yet, *IEEE Spectrum*, pp. 28–33, May 2002.
34. T. Kikkawa et al., An over 200-W output power GaN HEMT push–pull amplifier with high reliability, *IEEE MTT-S Dig.*, pp. 1347–1350, 2004.
35. T. Inoue et al., 30GHz-band 5.8 W high-power AlGaN/GaN heterojunction-FET, *IEEE MTT-S Dig.*, pp. 1649–1652, 2004.
36. M. Micovic et al., Ka-band MMIC power amplifier in GaN HFET technology, *IEEE MTT-S Dig.*, pp. 1653–1656, 2004.
37. R. Behtash et al., Coplanar AlGaN HEMT power amplifier MMIC at X-band, *IEEE MTT-S Dig.*, pp. 1657–1660, 2004.
38. Y. Okamoto et al., Improved power performance for recessed-gate AlGaN–GaN heterojunction FET with a field-modulating plate, *IEEE Trans. Microwave Theory Tech.*, Vol. 52, pp. 2536–2540, November 2004.
39. P. Saunier, GaN technology overview: accomplishments and challenges, *12th GAAS Symp. Dig.*, pp. 543–546, 2004.
40. D. C. Streit et al., The future of compound semiconductors for aerospace and defense applications, *IEEE Compound Semiconductor Integrated Circuit Dig.*, pp. 5–8, 2005.
41. D. Kimball et al., 50% PAE WCDMA base station amplifier implemented with GaN HFETs, *IEEE Compound Semiconductor Integrated Circuit Dig.*, pp. 89–92, 2005.
42. Y.-F. Wu et al., Field-plated GaN HEMTs and amplifiers, IEEE Compound Semiconductor Integrated Circuit Dig., pp. 170–172, 2005.
43. M. Micovic et al., GaN MMIC technology for microwave and millimeter-wave applications, *IEEE Compound Semiconductor Integrated Circuit Dig.*, pp. 173–176, 2005.
44. C. Lee, P. Saunier, and H.-Q. Tserng, High-temperature power performance of X-band recessed-gate AlGaN/GaN HEMTs, *IEEE Compound Semiconductor Integrated Circuit Dig.*, pp. 177–180, 2005.
45. P. Saunier et al., AlGaN/GaN on SiC and on Si for RF power applications, *GOMAC Dig.*, pp. 464–467, 2005.
46. W. Nagy et al., 150 W GaN-on-Si RF power transistor, *IEEE MTT-S Int. Microwave Symp. Dig.*, WE1E-1, 2005.
47. R. J. Trew et al., RF breakdown and large-signal modeling of AlGaN/GaN HFETs, *IEEE MTT-S Int. Microwave Symp. Dig.*, pp. 643–646, 2006.
48. B. M. Green et al., A GaN HFET device technology on 3″ SiC substrates for wireless infrastructure applications, *IEEE MTT-S Int. Microwave Symp. Dig.*, pp. 706–709, 2006.
49. R. Therrien et al., AlGaN/GaN HFETs on Si substrates for WiMA applications, *IEEE MTT-S Int. Microwave Symp. Dig.*, pp. 710–713, 2006.

50. A. Maekawa et al., A 500 W push–pull AlGan/GaN HEMT amplifier for L-band high power application, *IEEE MTT-S Int. Microwave Symp. Dig.*, pp. 722–725, 2006.
51. P. Schuh et al., 20 W GaN HPAs for next generation X-band T/R-modules, *IEEE MTT-S Int. Microwave Symp. Dig.*, pp. 726–729, 2006.
52. A. M. Darwish et al., AlGaN/GaN Ka-band 5-W MMIC amplifier, *IEEE Trans. Microwave Theory Tech.*, Vol. 54, pp. 4456–4463, December 2006.
53. P. Schuh et al., Linear broadband GaN MMICs for Ku-band applications, *IEEE MTT-S Int. Microwave Symp. Dig.*, pp. 1324–1326, 2006.
54. A. Wakejima et al., 370-W output power GaN-FET amplifier with low distortion for W-CDMA base stations, *IEEE MTT-S Int. Microwave Symp. Dig.*, pp. 1360–1363, 2006.
55. F. van Raay et al., X-band high-power microstrip AlGaN/GaN HEMT amplifier MMICs, *IEEE MTT-S Int. Microwave Symp. Dig.*, pp. 1368–1371, 2006.
56. M.-Y. Kao et al., AlGaN/GaN HEMTs with PAE of 53% at 35GHz, *IEEE MTT-S Dig.*, pp. 627–630, 2007.
57. K. Yamanaka et al., C-band GaN HEMT power amplifier with 220 W output power, *IEEE MTT-S Dig.*, pp. 1251–1254, 2007.
58. K. Iyomasa et al., GaN HEMT 60 W output power amplifier with over 50% efficiency at C-band 15% relative bandwidth using combined short and open circuited stubs, *IEEE MTT-S Dig.*, pp. 1255–1258, 2007.
59. Y. Okamoto et al., A 149 W recessed-gate AlGaN/GaN FP-FET, *IEEE MTT-S Dig.*, pp. 1351–1354, 2004.
60. D. Hill and T. S. Kim, 28-V low thermal impedance HBT with 20-W CW output power, *IEEE Trans. Microwave Theory Tech.*, Vol. 45, pp. 2224–2228, December 1997.
61. N. L. Wang et al., Linearity improvement of multi-watts 24–28 V InGaP/GaAs HBT by low frequency low source impedance matching, *IEEE MTT-S Dig.*, pp. 541–544, 2004.
62. S. Piotrowicz et al., High power and high efficiency 30 W compact S-band HBT power chips with gold or diamond heat spreaders, *IEEE MTT-S Dig.*, pp. 1527–1530, 2004.
63. A. Oki et al., HEMT and HBT MMIC power amplifiers, in *IEEE MTT-S Int. Microwave Symp. Workshop #WMH*, 2006.
64. H. F. Cooke, Precise technique finds FET thermal resistance, *Microwaves RF*, pp. 85–87, August 1986. Correction of this paper in *Microwaves RF*, p. 13–, February 1987.
65. B. A. Kopp, A. J. Billups, and M. H. Leusse, Thermal analysis and considerations for gallium nitride microwave power amplifier packaging, *Microwave J.*, Vol. 44, pp. 72–82, December 2001.
66. K. C. Gupta, R. Garg, I. Bahl, and P. Bhartia, *Microstrip Lines and Slotlines*, 2nd edition, Artech House, Norwood, MA, 1996.
67. I. J. Bahl, Average power handling capability of multilayer microstrip lines, *Int. J. RF Microwave Computer-Aided Eng.*, Vol. 11, pp. 385–395, November 2001.
68. I. Bahl, *Lumped Elements for RF and Microwave Circuits*, Artech House, Norwood, MA, 2003.
69. I. J. Bahl, E. L. Griffin, J. Dilley, and M. Balzan, Low loss multilayer microstrip line for monolithic microwave integrated circuits applications, *Int. J. RF Microwave Computer-Aided Eng.*, Vol. 8, pp. 441–454, November 1998.
70. I. J. Bahl, High current capacity multilayer inductors for RF and microwave circuits, *Int. J. RF Microwave Computer-Aided Eng.*, Vol. 10, pp. 139–146, March 2000.
71. J.-W. Lee and I. Adesida, Gallium nitride for electronics, in *Encyclopedia RF and Microwave Engineering*, K. Chang (Ed.), Vol. 2, John Wiley &, Hoboken, NJ, 2005, pp. 1758–1767.
72. Y.-F. Wu et al., A 50-W AlGaN/GaN HEMT amplifier, *IEEE IEDM Tech. Dig.*, pp. 375–376, 2000.
73. M. Boers, A. Parker, and N. Weste, A GaN HEMT amplifier with 6-W output power and $>85\%$ power-added efficiency, *IEEE Microwave Mag.*, Vol. 9, pp. 106–110, April 2008.
74. D. E. Meharry et al., Multi-watt wideband MMICs in GaN and GaAs, *IEEE MTT-S Int. Microwave Symp. Dig.*, pp. 631–634, 2007.

75. A. K. Ezzeddine, H.-L. Hung, and H. C. Huang, High-voltage FET amplifiers for satellite and phased array applications, *IEEE MTT-S Int. Microwave Symp. Dig.*, pp. 336–339, 1985.
76. K. E. Peterson et al., Monolithic high-voltage FET power amplifiers, *IEEE MTT-S Int. Microwave Symp. Dig.*, pp. 945–948, 1989.
77. D. Willems, I. Bahl, and K. Wirsing, High voltage distributed amplifier, *Electron. Lett.*, Vol. 27, pp. 2368–2370, December 1991.
78. A. K. Ezzeddine and H. C. Huang, Ultra-broadband GaAs HIFET MMIC PA, *IEEE MTT-S Int. Microwave Symp. Dig.*, pp. 1320–1323, 2006.

PROBLEMS

13.1 Describe five major advantages and two major disadvantages of HV HPAs.

13.2 Compare qualitatively GaAs MESFET, pHEMT, Si BJT, LDMOS and SiC MESFET, and GaN HEMT for very-high-power amplifier applications. What is the major issue with a GaN transistor?

13.3 Describe various types of device cooling techniques. If an amplifier has 25-mm^2 size and dissipates 100 W of power, what type of heat sink or cooling is required?

13.4 A 50-Ω microstrip, a 10-pF MIM capacitor, and a 50-Ω resistor are fabricated on a 3-mil GaAs substrate. The capacitor density is 300 pF/mm^2 and the resistor size is $100 \times 500\ \mu m^2$. Calculate the average power handling capability of each component at 10 GHz. The chip is mounted on an ideal heat sink at 25 °C, the maximum GaAs surface operating temperature is 150 °C, and conductors are 5 μm thick.

13.5 A high-power discrete chip resistor is to be designed to dissipate 100 W. Determine the resistor type and its dimensions. The operating frequency is 2 GHz.

13.6 Design a 16-W power amplifier at 1 GHz using an LDMOS whose parameters are given in Section 13.2.2. Assume a device power density of 1 W/mm. Use ideal LC components.

13.7 Design a 16-W power amplifier at 2 GHz using a GaN HEMT whose parameters are given in Example 13.2. Use ideal LC components.

13.8 Design an amplifier for maximum power and bandwidth by connecting devices in series as described in Section 13.6. Use a GaN HEMT whose parameters are given in Example 13.2. Determine the maximum achievable power over wideband at RF. Assume the input matching network is not limiting the amplifier performance and the maximum supply voltage in 56 V.

Chapter 14

Hybrid Amplifiers

Amplifier circuits are fabricated using either hybrid technologies or monolithic technologies. Monolithic based amplifiers are treated in Chapter 15. Packages and assembly techniques are described in Chapter 21. The objective of this chapter is to provide a brief overview of hybrid amplifier manufacturing technologies. The reader can find detailed information available in published literature [1–22]. Finally, design examples of small-signal and power hybrid amplifiers are described.

14.1 HYBRID AMPLIFIER TECHNOLOGIES

At RF and microwave frequencies the hybrid technologies are employed for matching circuit elements for amplifiers, internally matched amplifiers, the assembly of monolithic microwave integrated circuit (MMIC) amplifiers, and high power combining. The hybrid fabrication of amplifiers employs one or more technologies including printed circuit board, thin, thick, and cofired ceramic, and multichip module. At radiofrequencies discrete matching components such as inductors, capacitors, and resistors are added on printed circuit board to design amplifiers. At microwave frequencies thin-film technology is used to design hybrid amplifiers. These fabrication techniques for hybrid amplifiers are described next.

14.2 PRINTED CIRCUIT BOARDS

Printed circuit boards (PCBs) or printed wiring boards (PWBs) are used extensively for electronic packaging and RF front-end circuit boards [12, 13]. In these applications, the primary function of PCBs is to provide mechanical support and multilevel electrical interconnections for packaged solid state devices, resistors, capacitors, and inductors. For RF/microwave applications, there is a need for high-performance and low-cost PCB materials that can provide low-loss finer lines ($\cong$5 mils wide) and narrower spacing ($\cong$5 mils) for high-density circuits as well as provide limited impedance matching capability. All these materials have low-loss copper conductors capable of carrying high current densities. The PCB can be single sided or double sided or can consist of multilayer substrates. Multilayer PCBs have two or more layers of dielectric and metallization layers, with the latter being interconnected by plated-through via holes. Substrates may be rigid or flexible.

Fundamentals of RF and Microwave Transistor Amplifiers. By Inder J. Bahl

Table 14.1 Electrical Properties and the Thermal Expansion Characteristics of a Wide Range of Dielectric Materials

Material	Dielectric Constant	Dissipation Loss	CTE xy (ppm/°C)	CTE z (ppm/°C)
FR4/glass	4.5	0.03	16–20	50–70
Driclad/glass	4.1	0.01	16–18	55–65
BT/epoxy/glass	4.0	0.01	17	55–65
Epoxy/PPO/glass	3.9	0.01	12–18	150–170
Cyanate ester/glass	3.5	0.01	16–20	50–60
Polyimide/glass	4.5	0.02	12–16	65–75
Ceramic fill thermoset	3.3	0.0025	15	50
EPTFE w/thermoset	2.8	0.004	50–70	50–70
Silica filled PTFE	2.9	0.003	16	24–30
PTFE/glass	2.4	0.001	12–20	140–280
PTFE	2.1	0.0004	70–90	70–90

Substrate manufacturers have tried to combine the characteristics of various basic materials to obtain desired electrical and mechanical properties. The resulting material is called a composite. By adding fiberglass, quartz, or ceramic in suitable proportions to organic or synthetic materials, the mechanical properties are modified and the dielectric constant value is adjusted. A very wide variety of products are now available with a dielectric constant range of 2.1–10 and tan δ values from 0.0004 to 0.01. Table14.1 shows important electrical and thermal parameters of several PCB materials that are currently in use. FR-4 (fire retardant) is an epoxy based glass woven substrate that is widely used and has the lowest cost, while PTFE (polytetrafluoroethylene) gives the highest performance and can be operated above 300 °C. FR-4, BT/epoxy, and polyimide, called thermoset materials, are hard and elastic. These materials become soft above their glass transition temperature (T_g). The glass transition temperatures of FR-4, BT/epoxy, and polyimide are about 150, 210, and 250 °C, respectively. Materials such as PTFE/glass, known as thermoplastics, become soft and melt if heated. The melting temperature (T_m) of PTFE/glass is about 325 °C. The coefficient of thermal expansion (CTE) as given for several materials in Table 14.1 is a measure of the dimensional stability with temperature. The thermal conductivity of these materials is quite poor, and a typical value is about 0.2 W/m·°C. Glass-reinforced epoxy laminates offer the lowest cost while PTFE based laminates have the lowest dielectric constant and loss. PTFE substrates also provide better protection from moisture and have ultrahigh adhesion strength. The high loss tangent of FR-4 and relatively variable ε_r limit its use to below 3 GHz. The values of parameters of composite materials vary slightly from manufacturer to manufacturer.

In a basic multilayer PCB fabrication process, first a copper foil is laminated to the dielectric sheets and the required interconnect/wiring patterns are etched on all substrates by using a photolithography technique. The substrates are then stacked and laminated under heat and pressure to make a monolithic board. Next, via holes are drilled in the board to make interlayer metallic connections and are catalyzed, and the whole board is plated with electroless copper. This increases the thickness of the surface conductor pattern and provides the copper layer in the via holes. The board is then tinned for soldering or nickel or gold plated for gold wire bonding. Finally, the board is cut into the required small sizes.

Figure 14.1 A prototyping printed circuit board configuration for testing a plastic packaged PA.

The RF prototyping printed circuit board is generally made from multilayer FR-4. The top dielectric layer is 10 mils thick. The top metal layer is made from 1 oz Cu (1.4 mils thick). The ground plane is made from 2 oz Cu (2.8 mils). The 10-mil thickness (between the RF layer and the ground layer) sets the width of a 50-Ω microstrip line to 17.5 mils. The total board thickness is set to 62 mils to make it compatible with standard RF connectors.

14.3 HYBRID INTEGRATED CIRCUITS

Hybrid microwave integrated circuits (MICs) have been used almost exclusively in the frequency range of 1–20 GHz for wireless, space, and military applications, because they meet the requirements for shock, temperature conditions, and severe vibration. This section is intended to provide a brief introduction to several hybrid technologies such as thin-film, thick-film, and cofired ceramic. The most commonly used ceramic for MICs is alumina (Al_2O_3). There are a number of other ceramic materials available, with ε_r ranging between 6 and 150. High dielectric constant materials are useful for important circuit size reduction at RF and low microwave frequencies. Hybrid integrated technologies are used exclusively to manufacture discrete lumped inductors, capacitors, and resistors as well as lumped-element based passive components. The lumped elements are realized by using multilevel sputtering of different materials.

The basic materials for fabricating printed circuits and MICs, in general, are divided into four categories:

1. Substrate materials—for example, sapphire, alumina, beryllia, ferrite/garnet, RT/duroid, quartz, silicon, GaAs, InP
2. Conductor materials—for example, copper, gold, silver, aluminum
3. Dielectric films—for example, SiO, SiO_2, Si_3N_4, Ta_2O_5
4. Resistive films—for example, NiCr, Ta, Ti, TaN, WN, Cermet, GaAs

Substrate Materials

The substrate selected for MICs must have the following general characteristics [5–7]:

1. The cost of the substrate must be justifiable for the application.
2. The choice of thickness and permittivity determines the achievable impedance range as well as the usable frequency range.
3. The loss tangent should be low, for negligible dielectric loss.
4. The substrate surface finish should be good (~0.05-0.1 μm finish), with relative freedom from voids, to keep conductor loss low and maintain good metal—film adhesion.
5. There should be good mechanical strength and thermal conductivity.
6. No deformation should occur during processing of the circuit.
7. Matching coefficient of thermal expansion (CTE) with solid state devices or package materials is important if they are attached to such substrates to avoid susceptibility to large temperature variations for improved reliability.

The dielectric constant of the substrate should be high while meeting the other criteria to keep the circuit size small. A variety of substrate materials with their properties are listed in Table 14.2. The data provided here for materials is of a general nature and must be used with care. For accurate data, refer to the manufacturer's data sheets.

Use of a high dielectric constant substrate $\varepsilon_r \cong 10$ is highly desirable. However, the substrate thickness is limited by the presence of higher order modes. High-impedance lines on thin substrates require very narrow conductors, which become lossy, and the definition of these narrow conductors can be difficult. Alumina (Al_2O_3) is one of the most suitable substrate materials for use up to 20 GHz. Quartz with a dielectric constant of 4 is more suitable and widely used for high-frequency (>20 GHz) microwave and millimeter-wave integrated circuits. Beryllia and aluminum nitride (AlN) have excellent thermal conductivity and are suitable for power applications where heat dissipation from active devices mounted on the substrate is large and a low thermal resistance is required.

Conductor Materials

Conductor materials used for MICs should have high conductivity, a low temperature coefficient of resistance, low RF resistance, good adhesion to the substrate, and good etchability and solderability and should be easy to deposit or electroplate [5, 7]. The resistance is determined by the RF surface resistivity and skin depth; thus the skin depth determines the thickness required. The conductor thickness should be at least three to four times the skin depth, to include 98% of the maximum possible current density within the conductor. Table 14.3 shows the properties of some widely used conductor materials for MICs.

Dielectric Film Materials

Dielectric films in MICs are used as insulators for capacitors, protective layers for active devices, and insulating layers for passive circuits. The desirable properties of these dielectric materials are reproducibility, high breakdown field, low loss tangent, and the ability to undergo processing without developing pin holes [7]. Table 14.4

Table 14.2 Properties of Substrates for MICs

Material	Surface Roughness (μm)	Loss Tangent (tan δ) at 10 GHz (10^{-4})	Relative Dielectric Constant (ε_r)	Thermal Conductivity (W/cm·°C)	Dielectric Strength (kV/cm)	MIC Applications
Alumina 99.5%	2–8	1–2	10	0.37	4×10^3	Microstrip, suspended substrate
96%	20	6	9	0.28	4×10^3	
85%	50	15	8	0.2	4×10^3	
Sapphire	1	1	9.3–11.7	0.4	4×10^3	Microstrip, lumped element
AlN	1–2	5	8.8	2.3	—	Compound substrate, package
Glass	1	20	5	0.01	—	Lumped element
Beryllia (BeO)	2–50	1	6.6	2.5	—	Compound substrate, package
Rutile	10–100	4	100	0.02	—	Microstrip
Ferrite/garnet	10	2	13–16	0.03	4×10^3	Microstrip, coplanar
GaAs (high resistivity)	1	6	12.9	0.46	350	High frequency, microstrip, monolithic MMICs
Si (high resistivity)	1	10–100	11.7	1.45	300	MMICs, RFICs
Quartz	1	1	3.8	0.01	10×10^3	Microstrip, high frequency
Polyolefin	1	1	2.3	0.001	~300	
InP	—	—	14	0.68	—	MMICs

shows some of the properties of commonly used dielectric films in MICs. SiO is not very stable and can be used in noncritical applications, such as bypass and DC blocking capacitors. A quality factor Q of more than 100 can be obtained for capacitors using SiO_2, Si_3N_4, and Ta_2O_5 materials.

Resistive Films

Resistive films in MICs are required for fabricating resistors for terminations, for attenuators, for amplifier stabilization, and for bias networks. The properties required for a resistive material are good stability, low temperature coefficient of resistance (TCR), and sheet resistance in the range of 10–2000 Ω/□ [7, 8]. Table 14.5 lists some of the thin-film resistive materials used in MICs. Evaporated nichrome and tantalum nitride are the most commonly used materials.

Table 14.3 Properties of Conductors for MICs

Material	Surface Resistivity [(Ω/□) × 10^{-7}]	Skin Depth δ at 2 GHz (μm)	Coefficient of Thermal Expansion [(α_t/°C) × 10^6]	Adherence to Dielectrics	Deposition Technique
Ag	2.5	1.4	21	Poor	Evaporation
Cu	2.6	1.5	18	Poor	Evaporation plating
Au	3.0	1.7	15	Poor	Evaporation plating
Al	3.3	1.9	26	Poor	Evaporation
Cr	4.7	2.7	9.0	Good	Evaporation
Ta	7.2	4.0	6.6	Good	Electron-beam evaporation, sputtering
Ti	—	—	—	Good	Evaporation, sputtering
Mo	4.7	2.7	6	Fair	Electron-beam evaporation, sputtering
W	4.7	2.6	4.6	Fair	Sputtering, vapor phase, electron-beam evaporation
Pt	—	3.6	9	—	Sputtering, electron beam evaporation
Pd	—	3.6	11	—	Evaporation, sputtering, electroplating

Table 14.4 Properties of Dielectric Films for MICs

Material	Method of Deposition	Relative Dielectric Constant (ε_r)	Dielectric Strength (V/cm)	Microwave Q
SiO	Evaporation	6–8	4×10^5	30
SiO_2	Deposition	4	10^7	100–1000
Si_3N_4	Vapor-phase sputtering	7.6	10^7	
Al_2O_3	Anodization evaporation	7–10	4×10^6	
Ta_2O_5	Anodization evaporation	22–25	6×10^6	100

Table 14.5 Properties of Resistive Films for MMICs

Material	Method of Deposition	Resistivity (Ω/□)	TCR (%/°C)	Stability
Cr	Evaporation	10–1000	−0.100 to +0.10	Poor
NiCr	Evaporation	40–400	+0.001 to +0.10	Good
Ta	Sputtering	5–100	−0.010 to +0.01	Excellent
Cr-SiO	Evaporation or cement	Up to 600	−0.005 to −0.02	Fair
Ti	Evaporation	5–2000	−0.100 to +0.10	Fair

Mask Layouts

Any MIC design starts with a schematic diagram for the circuit. After the circuit is finalized, a rough layout is drawn. The next step is to obtain an accurate mask layout for producing a single mask layer for hybrid MICs or a set of masks for multilevel miniature MICs and MMICs. Finally, hybrid MIC substrates are etched using these masks for the required pattern, and for miniature and monolithic MICs various photolithographic steps are carried out using a set of masks.

14.3.1 Thin-Film MIC Technology

Thin-film fabrication technology used for MICs is continuously developing to meet the requirements of increasing frequency of operation, higher yield, and reduced costs. A thin-flim MIC consists of an assembly that combines different circuit functions formed by strip or microstrip conductors and incorporates discrete semiconductor devices and lumped elements. This can be achieved by a thin-film manufacturing process that is carefully controlled and repeatable in a clean room environment. The first step in the fabrication process is the deposition of a first layer (seed layer) of metal film on the substrate. The selection of the film is made based on the criterion of good adhesion to the substrate and is one of the most important factors in selection of a conductor material for the first layer of metal film. Some precautions specific to MIC conductors should be mentioned with regard to the deposition techniques. At radiofrequencies the electromagnetic fields are confined to several skin depths of the conductors. In order to achieve low loss, the layer of high-resistivity material such as chromium used for adhesion must be extremely thin. The main conductor must have a low bulk DC resistivity for low-loss propagation. Improper processing techniques can result in high RF loss for a low sheet resistance material made of thin chromium and a thicker gold structure. As a result of very high substrate temperatures ($>300\,^\circ$C) sometimes encountered during sputtering, this thin sputtered chromium layer will diffuse into an overlaying gold film. This results in a high RF loss, even though the sheet resistance may be low with a thick gold layer. Therefore techniques such as sputtering must be used with care for MIC materials. Metal films are deposited on substrates by three methods: vacuum evaporation, electron-beam evaporation, and sputtering.

A typical metal combination for alumina substrate is Cr/Cu/Au or NiCr/Ni/Au. A very thin seed layer of suitable metal is deposited on the substrate by one of the preceding techniques and then the bulk conductor metal is deposited by electroplating techniques. The seed layers of metal provide mechanical and electrical foundation layers on which to electroplate a good-quality bulk conductor metal. The circuit definition can be accomplished by a plate-through technique or by an etchback technique [20]. The techniques that are used to define patterns in metal layers can influence the deposition choice. The plate-through technique begins with a substrate coated with a thin layer of evaporated metal followed by an application of a thick photoresist. The thickness of this photoresist is similar to the thickness of the final metal film required. After defining a pattern in the photoresist, the second metal layer is plated up to the desired thickness with precise definition, and only in the areas where metal is required. The photoresist layer is then washed away and the thick seed-metal is etched with very little undercut from the undesired areas. This technique is also suitable for fabricating lines that are 25–50 μm wide and/or when the separation between them is 25–50 μm.

The second technique is the etchback technique that utilizes a thick metal layer defined either completely by evaporation or by a combination of a thin evaporated layer and a thicker plated layer. A thin photoresist layer is used as a mask to define the circuit pattern. The undesired areas of metal are then removed by etching. This technique results in undercutting the metal film by about twice the line conductor thickness. The plate-through technique not only permits better definition for thick conductors, but also saves on cost in that only the required material is deposited.

Traditionally, single-layer discrete capacitors are manufactured by firing ceramic substrates having typically 5-mil thickness [16]. Then both sides are metallized using a thin- or thick-film process. Finally, the substrate is sawed into chip capacitors. Thin-film resistors are realized by depositing nichrome or tantalum nitride films on alumina substrates, similar to the process described for conductor films. A laser trimming technique is used to achieve $\pm 1\%$ tolerances in the resistor values. Termination or connecting pads are of metallized chromium–gold. For high-power resistors, substrates used are BeO and AlN. Finally, the substrate is sawed into chip resistors.

In the early 1980s, a thin-film technology variant was introduced called miniature hybrid [15]. Miniature hybrid MIC technology is based on thin-film technology in which the multilevel passive circuits including lumped resistors and capacitors are batch fabricated on the substrate and solid state devices are externally attached to these circuits. The advantages of this circuit technology are small size, light weight, excellent heat dissipation, and broadband performance.

14.3.2 Thick-Film MIC Technology

Thick-film MICs are manufactured using various inks pressed through patterned silk screens. Thick-film MICs are inexpensive and are generally limited to the lower end of the microwave spectrum. In a conventional thick-film technology, the multilayer interconnects are formed by successive screen printing of conductors, dielectric layers, and resistor patterns on a base substrate. The materials are in the form of inks or pastes. After screen printing, each layer is dried at about 150 °C for 15 minutes and fired at about 850 °C for 30–60 minutes. The printing, drying, and firing steps are repeated to fabricate the multilayer circuitry in a fully automated way to produce high-volume, cost-effective components.

The commonly used base substrate materials are alumina (Al_2O_3), beryllia (BeO), and aluminum nitride (AlN). The dielectric pastes are typically glass–ceramic compositions having low dielectric constant and loss tangent, high breakdown voltage, and a CTE matched to the substrate material. The conductors may be gold, copper, silver, palladium–silver/gold, and platinum–silver/gold. Properties of various conductor materials are given in Table 14.2. The commonly used resistor material is ruthenium doped glass (RuO_2).

Recently, this technology [17] has been improved by using a photoimageable thick film process that is capable of producing 1-mil lines and gaps and 3-mil vias. In this process, both Cu and Au conductors up to 10 layers can be used. Earlier, thick-film technology was used to interconnect discrete components; however, improved technology is also capable of printing conductor patterns for low-loss passive circuits at RF and low microwave frequencies.

14.3.3 Cofired Ceramic and Glass–Ceramic Technology

Around the time of the introduction of hybrid miniature MICs, a thick-film variant known as low-temperature cofired ceramic (LTCC) was also introduced [18]. The LTCC manufacturing process is similar to the thick-film process except that it does not use a base substrate. Dielectric layers are in the form of unfired ceramic tapes (also called green tapes) instead of paste. This technology also enables the printing of reliable capacitors and resistors. The process consists of blanking, punching vias, conductor

Table 14.6 Typical Electrical and Thermal Properties of Ceramic Materials

Property	Materials				
	Al_2O_3	HTCC	LTCC	BeO	AlN
Relative dielectric constant at 1 MHz	9.8	9.5	5.0	6.6	8.8
Loss tangent at 1 MHz	0.0002	0.0004	0.0002	0.0003	<0.001
Coefficient of thermal expansion, $10^{-6}/°C$	6.5	7.1	3.0	7.2	4.4
Thermal conductivity (W/m · °C)	37	25	2	250	230
Dielectric strength (kV/m)	25	23	1.5	26	14
Density (g/cm^3)	3.8	3.9	2.6	2.8	3.3

screen printing, collating, laminating, and firing. The vias are punched in the green tape and filled with conducting paste. At the same time conductor patterns are screen printed. This process is carried over for each dielectric layer, and finally the composite structure is fired to obtain a monolithic substrate. The firing temperature for glass–ceramic substrates is 850–900 °C and this technology is known as LTCC technology. Because of the low-temperature firing, LTCC allows one to use high-conductivity metals such as Ag, Cu, and Au. The dielectric tapes use a glass–ceramic composite optimized for a better CTE match with base metal and the semiconductor chips. As many as 50 layers can be combined in a single LTCC substrate measuring 6 × 6 inches. When ceramic tapes are used they are fired at 1500–1600 °C and the technology is known as high-temperature cofired ceramic (HTCC) technology. Commonly used conductors in this case are tungsten (W) and molybdenum (Mo). Dielectric properties of cofired glass–ceramic are compared with cofired alumina–ceramic, alumina, BeO, and AlN in Table 14.6.

LTCC technology, due to its multilayer process, offers several advantages over conventional thin-film, thick-film, and HTCC technologies. These advantages include a higher level of integration of components (e.g., capacitors, resistors, inductors, inductor transformers, transmission lines, and bias lines) and greater design flexibility by enabling the realization of different types of transmission line media such as microstrip, strip line, coplanar waveguide, and rectangular coax. Passive components, matching networks, bias lines, and shielding of RF lines can be combined in LTCC technology using several available ceramic and metal layers. Finally, solid state low-power devices are attached on the top surface to realize active or passive circuits. High-power devices can be integrated with LTCC by attaching the devices directly to the next level assembly chassis through holes fabricated in the LTCC MIC.

The conductors for inductors, transformers, capacitors, interconnects, and other passive components are screen printed by using conductive pastes of gold, silver, or copper. Their sheet resistances measured in milliohms/square (mΩ/□) are in the range of 4–10, 2–8, and 3–4, respectively. A paste of glass frit and conductive powder is used to screen print thick-film resistors. The ratio of glass frit content to conductive powder is adjusted to vary the sheet resistance from about 5 Ω/□ to 10 MΩ/□. The surface resistors are trimmed to achieve ±1% tolerances in the resistor values, while for buried resistors the tolerance is generally ±25%. The materials for MIM capacitors are available in both paste and tape forms. The dielectric constant (also referred to as K) value varies from 5 to 200. The capacitance range and dissipation factor for low dielectric constant materials ($\varepsilon_r = 5 - 10$) are 1–200 pF and <0.3%, while for high-K materials ($\varepsilon_r = 100 - 200$) these values are 10–3000 pF and <2%, respectively. The breakdown voltage and capacitance tolerance are 500 V and ±10%, respectively, for

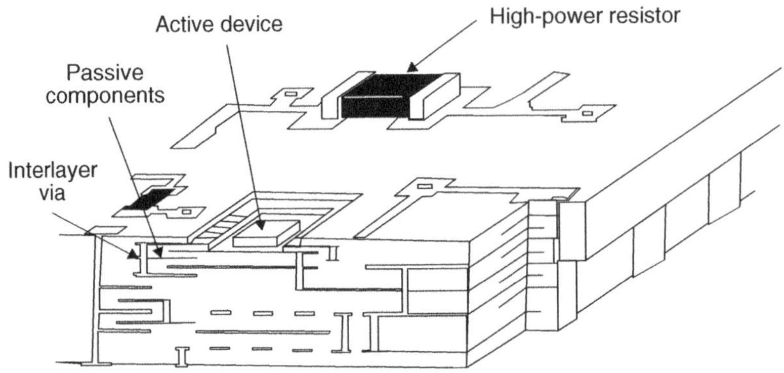

Figure 14.2 A three-dimensional view of the LTCC module.

low-*K* materials, and 200 V and ±20%, respectively, for high-*K* materials. Figure 14.2 shows a three-dimensional view of an LTCC module with embedded passive components and bias lines. Wire bonding of a solid state device and surface mounting of a bypass capacitor are also shown.

MIC technology is very diverse in its application of materials and processes to implement a broad array of functions. Table 14.7 lists some of these materials and processes.

14.4 DESIGN OF INTERNALLY MATCHED POWER AMPLIFIERS

Since the total gate periphery/width is very large in the case of high-power transistors, the impedance of device chips becomes so low that input and output impedances are affected by parasitic capacitance and inductance of a package. It is difficult to match the

Table 14.7 Summary of Typical Materials and Processes Used to Fabricate MICs

Materials/ Processes	Microwave Printed Circuit	Thin Film	Cofired Glass–Ceramic (LTCC)
Base substrates	PTFE glass fiber, PTFE ceramic, hydrocarbon ceramic, polyester glass	Al_2O_3,AIN, BeO, quartz, glass–ceramic	N/A
Conductors	Cu	Au, Al, Cu	Au, Ag, PdAgCu
Dielectrics	N/A	SiO_2, polyimide, benzocyclobutene (BCB)	Glass–ceramic tape
Resistors	N/A	NiCr, TaN	RuO_2 doped glass
Processes	Photolithography, etch, collate sheets, bonding	Sequentially vacuum deposit, spin coat, and/or plate conductors, dielectrics, and resistors; photolithography; etch	Punch vias, print and dry conductors on tape, collate layers, laminate, cofire

transistor out of a package for an amplifier operation, especially at higher frequencies. Under prematch conditions, added reactance of the package's lead frame also significantly affects the amplifier's performance in terms of power, PAE, and bandwidth. One of the most practical methods of designing hybrid high-power amplifiers is to use internal matching within a microwave package to deal with the low input and output impedances of the device.

Lumped and/or distributed circuit elements for matching networks can be used. For broadband and high power levels, lumped inductors and capacitors are generally preferred for input matching of power transistors. Lumped inductors are realized by using bond wires and capacitors are of the MIM type that use high dielectric constant ceramics. Capacitors must have small parasitic inductance and resistance, sufficient thermal and mechanical strength, a small temperature coefficient, a breakdown voltage 3–4 times the amplifier's operating voltage, and low cost. Since the transistor's output impedance is much higher than the input impedance, the output matching networks are realized using both lumped and distributed circuit elements. Microstrip lines on ceramic substrates are generally used for distributed circuit elements.

The schematic of an internally matched power FET/pHEMT is shown in Figure 14.3. The input match has more matching elements because high impedance transformation is required. A 2.2-GHz 140-W internally matched heterojunction FET used in a push–pull amplifier for cellular base station applications is depicted in Figure 14.4. This particular device has a total gate periphery of 332 mm, which gives a power output density of about 0.42 W/mm. The FET has two input and output leads for push–pull combining using external baluns. The design and fabrication of this circuit are described in Reference 23.

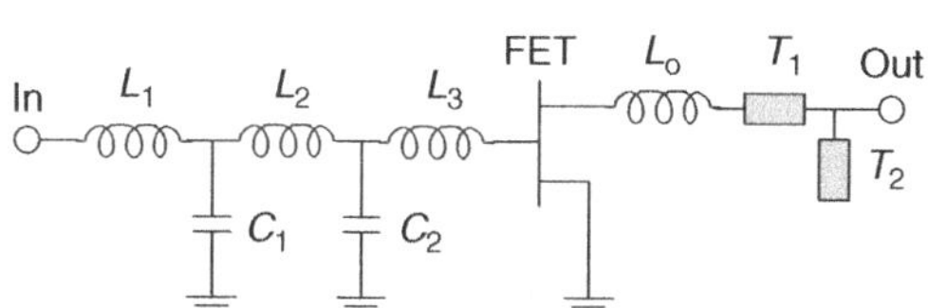

Figure 14.3 Schematic of internally matched power FET/pHEMT.

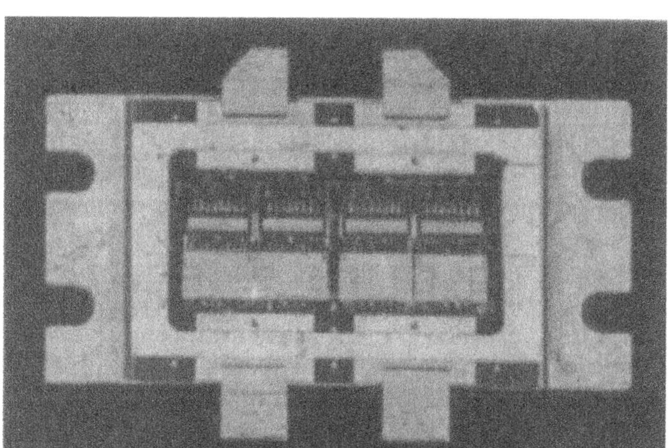

Figure 14.4 Photograph of a 2.2-GHz, 140-W internally matched power FET. (From Takenaka et al. [23]. Reprinted with permission of IEEE.)

14.5 LOW-NOISE AMPLIFIERS

In this section we describe narrowband, ultra-wideband, and broadband distributed low-noise amplifier examples using hybrid MIC technology.

14.5.1 Narrowband Low-Noise Amplifier

Design an X-band low-noise amplifier using *S*-parameters and noise parameters. The amplifier is required to be unconditionally stable up to 50 GHz. The specifications for the LNA to be fabricated using a microstrip on a 15-mil thick alumina substrate ($\varepsilon_r = 9.9$) are the following:

Frequency	10 GHz
Bandwidth	5%
Gain (min)	20 dB
Noise figure (max)	1.0 dB
VSWR (max)	1.5:1

The pHEMT listed in Table 5.9 satisfies the NF requirements and the amplifier needs two stages to obtain 20-dB gain. For this device the stability factor $K = 0.647$ at 10 GHz. Thus the amplifier topology must provide simultaneously low noise figure and good input match. Since $K < 1$, each stage as well as the complete amplifier must be designed for unconditional operation. A simple amplifier configuration consisting of input, interstage, and output matching networks is shown in Figure 14.5. The square/rectangular isolated pads are used for circuit tuning via wire bonds. In a two-stage low-noise amplifier, the input port is matched for optimum noise figure, and the interstage and output port are matched for maximum and flat gain. In this case, each device is biased approximately at 10 mA (an optimum bias current, 20–25% I_{dss}, for minimum NF); however, biasing the output stage for higher current increases device gain and output TOI/IP3. In both devices inductive source feedback is used to obtain optimum NF, good input match, and circuit stabilization. Additional circuit stabilization at 10 GHz and up to 50 GHz was provided by resistors used in the interstage and output matching networks. The matching circuit elements are determined using a circuit optimization technique by employing a commercial CAD tool. Here 10-mil wide (60 Ω as the characteristic impedance) microstrip lines have been used. Physical dimensions for matching elements on alumina substrate ($\varepsilon_r = 9.9$, $h = 15$ mils) are also given in Figure 14.5. It is a common practice to bond wire in or out the rectangular/square patches to tune the circuit for improved performance. The effect of T-junction discontinuity and bond wires has been included in the design. However, the effect of associated parasitic reactance of DC blocking and bypass capacitors has been ignored. The resistors are assumed to be of printed circuit type. The discrete components are connected using two 10–12-mil long and 1-mil diameter wires. The calculated K factor is greater than 1 from 0.1 to 50 GHz and is calculated using an EC model for the device. The simulated NF over 9.5–10.5 GHz is less than 0.7 dB. The calculated gain and return loss (RL) of this amplifier are shown in Figure 14.6 and are greater than 20 dB and 15 dB, respectively.

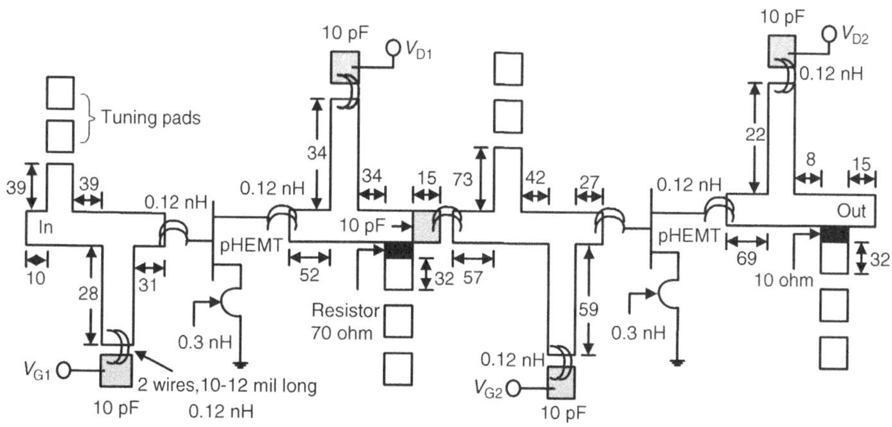

Figure 14.5 Schematic of a two-stage narrowband LNA using a pHEMT. All dimensions are in mils.

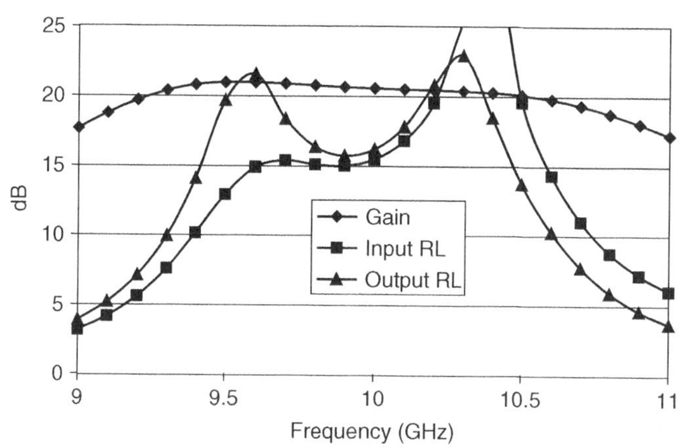

Figure 14.6 Simulated performance of the narrowband LNA.

14.5.2 Ultra-wideband Low-Noise Amplifier

Next, a design for an ultra-wideband (UWB) LNA is described. Over 3–11 GHz the target value of gain and NF are 9 dB and 3 dB, respectively. The amplifier is required to have good input and output match (VSWR = 2:1) and also must be unconditionally stable up to 50 GHz.

The pHEMT listed in Table 5.9 satisfies the NF requirements and a single stage meets the 9–10-dB gain target value. For this device, $K < 1$. Thus the broadband amplifier topology must provide simultaneously low noise figure and good input match. Since $K < 1$, the amplifier must be designed for unconditional operation. A simple feedback amplifier configuration consisting of reactive input and reactive/resistive output matching networks is shown in Figure 14.7. The low-noise amplifier is matched for optimum noise figure, good input and output match, and flat gain. The device is biased approximately at 10 mA (an optimum bias current for minimum NF). The matching circuit elements are determined using a commercial CAD tool. The circuit is optimized in terms of microstrip line length as well as width. Physical dimensions

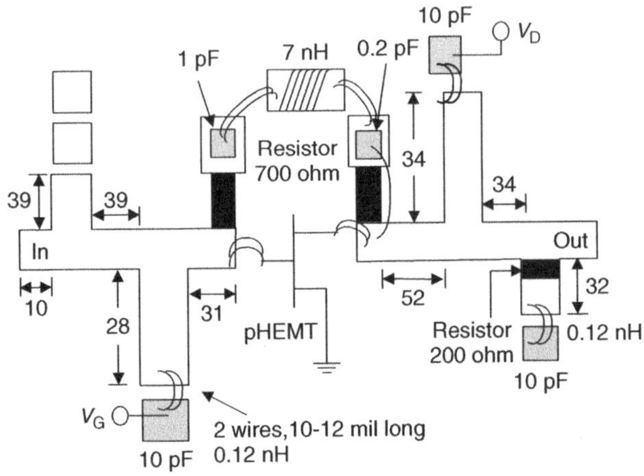

Figure 14.7 Schematic of a single-stage UWB LNA using a pHEMT. All dimensions are in mils.

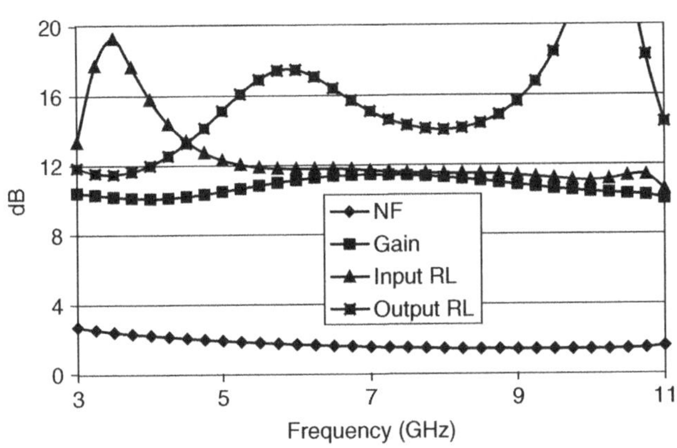

Figure 14.8 Simulated performance of the UWB LNA.

for matching elements on alumina substrate ($\varepsilon_r = 9.9$, $h = 15$ mils) are also given in Figure 14.7 Here 10-mil wide microstrip lines have been used. The effect of T-junction discontinuity and bond wires has been included in the design. However, the effect of associated parasitic reactance of DC blocking and bypass capacitors has been ignored. The resistors are assumed to be of printed circuit type. The discrete components are connected using two 10–12-mil long and 1-mil diameter wires. The device is assumed to have source via grounding. The calculated K was greater than 1. The simulated performance of this amplifier is shown in Figure 14.8. The NF is less than 3 dB, gain is greater than 10 dB, and return loss is greater than 10 dB.

14.5.3 Broadband Distributed LNA

Next, we describe an example of a single-stage distributed LNA designed using the S-parameters and noise parameters given in Table 5.9. Over 2–12 GHz the target value of gain and NF are 12 dB and 3.5 dB, respectively. The amplifier is required to have good input and output match, and also must be unconditionally stable up to 50 GHz.

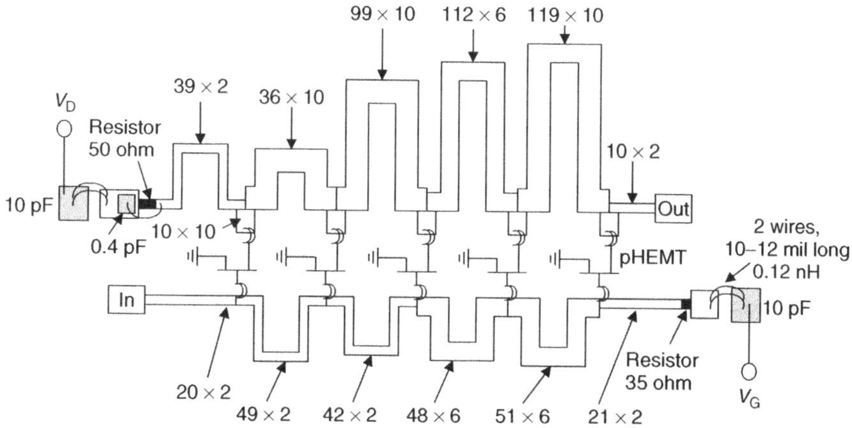

Figure 14.9 Schematic of a single-stage broadband distributed LNA using a pHEMT.

The pHEMT listed in Table 5.9 meets the NF requirements and the value of $K < 1$. Thus the broadband amplifier topology must provide simultaneously low noise figure and good input match. Since $K < 1$, the amplifier must be designed for unconditional operation. A simple distributed topology having five cells was selected and is shown in Figure 14.9. The low-noise amplifier is matched for optimum noise figure, good input and output match, and flat gain. The device is biased approximately at 10 mA and the matching circuit elements are optimized in terms of microstrip line length as well as width. Physical dimensions (length × width) for matching elements on alumina substrate ($\varepsilon_r = 9.9$, $h = 15$ mils) are given in Figure 14.9. The effect of T-junction discontinuity and bond wires has been included in the design. However, the effect of associated parasitic reactance of DC blocking and bypass capacitors has been ignored. The resistors are again of the printed circuit type and the discrete components are connected using two 10–12-mil long wires. The device is assumed to have source via grounding. The calculated K was greater than 1. The simulated performance of this amplifier is shown in Figure 14.10. The NF is less than 3.5 dB, gain is greater than 12 dB, and return loss is better than 11.5 dB. This topology uses five devices as compared to one device in the feedback configuration; however, a distributed approach has much higher single-stage gain, P_{1dB} and TOI values.

14.6 POWER AMPLIFIERS

In this section, narrowband and broadband power amplifier examples using hybrid MIC technology are described.

14.6.1 Narrowband Power Amplifier

A two-stage narrowband (1626.5–1660.5 MHz) 10-W power amplifier was designed and tested using packaged devices [24]. The design example selected is based on the following two commercially available devices:

Q1 Motorola GaAs Integrated Power Amplifier MRFIC1818

Q2 Ericsson GOLDMOS FET PTF10053

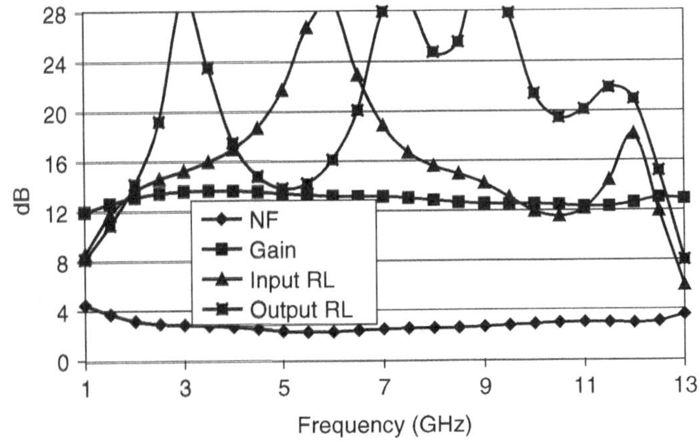

Figure 14.10 Simulated performance of the broadband distributed LNA.

Table 14.8 Performance Parameters of MRFIC1818 and PTF10053 Transistors

Device	Frequency Range (MHz)	Input Power (dBm)	Output Power (dBm)	Nominal Operating Voltage (V)	Nominal RF Gain (dB)	Input Impedance $Z_{in}(\Omega)$	Output Impedance (Conjugate) Z^*_{OL} (Ω)
Q1 (MRFIC1818)	1700–1900	3	33	+5	30	$9.19 - j30.10$ at 1.71-GHz	$6.00 + j3.80$ at 1.71-GHz
Q2 (PTF10053)	1000–2000	30	40.8	+26	11	$2.70 - j4.0$ at 1.75-GHz	$3.3 + j2.5$ at 1.75-GHz

The electrical performance of these two devices is given in Table 14.8. The design principle used is quite generic and is used for multistage hybrid and MMIC power amplifiers, shown schematically in Figure 14.11. Note that the input and output impedance data presented in Table 14.8 are not in the frequency band of interest (1626.5–1660.5 MHz). Both input and output impedances were modified using extrapolation techniques. The two-stage design was optimized using a commercial CAD and the substrate selected was a 30-mil thick GETEK material. A single-layer PCB layout of the amplifier is shown in Figure 14.12. A solid rectangular metal plate was attached to the PCB ground to provide thermal and mechanical stability. The PCB size was 2.5 in. by 1.37 in.

The amplifier was tested under CW conditions and a summary of measured data is provided in Table 14.9.

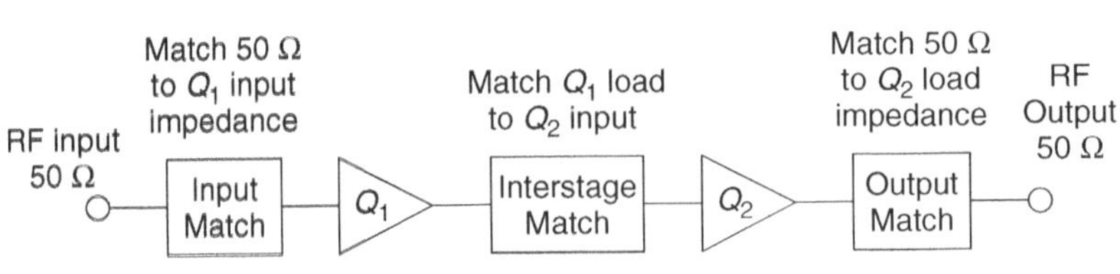

Figure 14.11 Schematic of a two-stage hybrid amplifier.

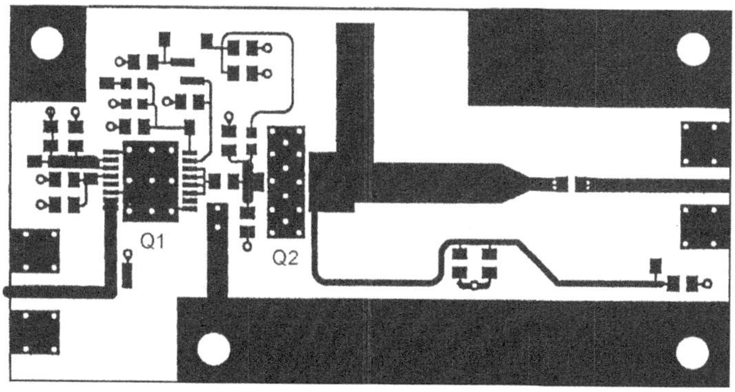

Figure 14.12 PCB layout of the two-stage 10-W hybrid amplifier.

Table 14.9 Summary of Measured Data for the 10-W Hybrid Power Amplifier at 1.6605 GHz

Parameter	Performance
1-dB Compression point	40.9 dBm (12.3 W)
Gain flatness within 34-MHz bandwidth from center	±0.3 dB
Input return loss	Better than 14 dB
Output load mismatch	VSWR 2:1 with no oscillations
Oscillations	None up to 18 GHz
Spurious excluding harmonics	−50 dBc or better, up to 18 GHz
Second harmonic	−35 dBc or better
Third harmonic	−55 dBc or better
MRFIC1818 peak RF current (+5-V operation with full RF output)	1900 mA
PTF10053 peak RF current (+26-V operation with full RF output)	800 mA
PAE	35%

Figure 14.13 Example of a broadband 24-V MESFET amplifier utilizing broadband conical inductors as multidecade bias chokes.

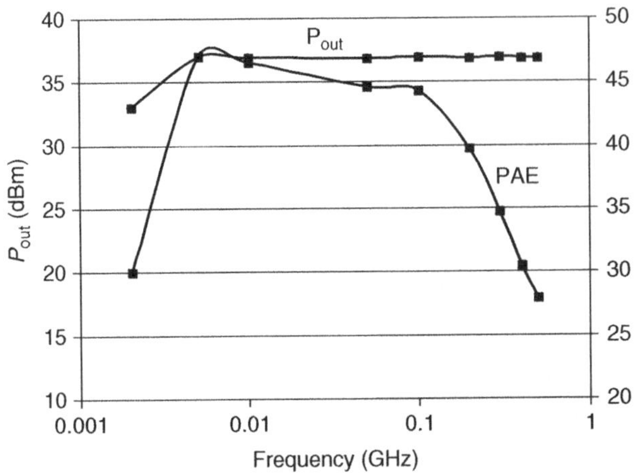

Figure 14.14 RF power performance of a broadband 24-V MESFET amplifier utilizing broadband conical inductors as multidecade bias chokes.

14.6.2 Broadband Power Amplifier

An application using the CC75T36K240G5 conical inductor (described in Chapter 6) in an ultra-broadband power amplifier is shown in Figure 14.13. A low-frequency broadband MMIC power amplifier utilizing a new 24-V high power density MESFET process and feedback topology was developed [25]. The amplifier chip was mounted in a ceramic package and conical inductors were placed on the printed circuit board. Conical inductors were used for broadband low-frequency bias choking in this power amplifier application that is designed to operate down to 10 MHz. The measured power and PAE of the amplifier are shown in Figure 14.14. Output power of 5 W and 30% PAE across nearly two decades of bandwidth were achieved. The conical inductor is capable of much more bandwidth than this particular application requires.

REFERENCES

1. R. R. Tummala and E. J. Rayaszewski (Eds.), *Microelectronic Packaging Handbook*, Van Nostrand Reinhold, New York, 1989.
2. L. T. Manzione, *Plastic Packing of Microelectronic Devices*, Van Nostrand Reinhold, New York, 1990.
3. J. E. Sergent and C. A. Harper (Eds.), *Hybrid Microelectronic Handbook*, 2nd edition, McGraw-Hill, New York, 1995.
4. P. E. Garrou, and I. Turlik, *Multichip Module Technology Handbook*, McGraw-Hill, New York, 1998.
5. F. Z. Kesiter, An evaluation of materials and processes for integrated microwave circuits, *IEEE Trans. Microwave Theory Tech.*, Vol. MTT-16, pp. 469–475, July 1978.
6. M. Caulton and H. Sobol, Microwave integrated circuit technology—a survey, *IEEE J. Solid-State Circuits*, Vol. SC-5, pp. 292–303, December 1970.
7. H. Sobol, Applications of integrated circuit technology to microwave frequencies, *Proc. IEEE*, Vol. 59, pp. 1200–1211, August 1971.
8. H. Sobol, Technology and design of hybrid integrated circuits, *Solid State Technol.*, Vol. 13, pp. 49–57, February 1970.

9. T. S. Laverghetta, *Microwave Materials and Fabrication Techniques*, Artech House, Norwood, MA, 1984.
10. R. D. Jones, *Hybrid Circuit Design and Manufacture*, Marcel Dekker, New York, 1982.
11. D. J. Elliott, *Integrated Circuit Mask Technology*, McGraw-Hill, New York, 1985.
12. D. S. Farquhar, A. M. Seman, and M. D. Paliks, Manufacturing experience with high performance mixed dielectric circuit board, *IEEE Trans. Advanced Packaging*, Vol. 22, pp. 153–159, May 1999.
13. A. Aguayo, PTFE and low cost, high performance materials for high speed and high frequency applications, *Proc. 6th Annual Wireless Symp.*, pp. 243–267, 1998.
14. L. M. Walpita, et al. Temperature-compensated thermoplastic high dielectric-constant microwave laminates, *IEEE Trans. Microwave Theory Tech.*, Vol. 47, pp. 1577–1583, August 1999.
15. F. Sechi, et al. Miniature beryllia circuits—a new technology for microwave power circuits, *RCA Rev.*, Vol. 43, p. 363, 1982.
16. L. Devoe and A. Devoe, Technology and innovation in single layer capacitors, *Microwave J.*, Vol. 37, pp. 144–152, February 2002.
17. P. Barnwell and J. Wood, A novel thick film on ceramic MCM technology offering MCM-D performance, Presented at *6th International Conference Multichip Modules*, Denver (CO), April 1997.
18. R. L. Brown, P. W. Polinski, and A. S. Shaikh, Manufacturing of microwave modules using low-temperature cofired ceramics, *IEEE MTT-S Int. Microwave Symp. Dig.*, pp. 1727–1730, 1994.
19. A. A. Mohammed, LTCC for high-power RF applications, *Advanced Packaging*, pp. 46–50, October 1999.
20. I. Bahl and P. Bhartia, *Microwave Solid State Circuit Design*, 2nd edition, John Wiley & Sons, Hoboken, NJ, 2003, Chapter chap10.
21. I. Bahl, *Lumped Elements for RF and Microwave Circuits*, Artech House, Norwood, MA, 2003.
22. K. C. Gupta, R. Garg, I. Bahl, and P. Bhartia, *Microstrip Lines and Slotlines*, 2nd edition, Artech House, Norwood, MA, 1996.
23. I. Takenaka, et al. Low distortion high power GaAs pseudomorphic heterojunction FETs for L/S-band digital cellular base stations, *IEEE MTT-S Int. Microwave Symp. Dig.*, pp. 1711–1715, 2000.
24. R. S. Tomar and P. Bhartia, A simple interstage matching technique for designing hybrid microwave power amplifiers, *IEEE Microwave Mag.*, Vol. 4, pp. 67–84, September 2003.
25. T. A. Winslow, Conical inductors for broadband applications, *IEEE Microwave Mag.*, Vol. 6, pp. 68–72, March 2005.

PROBLEMS

14.1 Compare qualitatively and quantitatively microstrip, CPW, and lumped elements for hybrid MIC amplifier applications.

14.2 A transmission line has an attenuation constant of α (nepers/unit length) and length ℓ. Show that the noise figure F (power ratio) of the line is given by

$$F = e^{2\alpha\ell}$$

When the attenuation is 0.2 dB/cm and the length is 0.5 cm, determine its noise figure F (also express in dB).

14.3 Input match of an LNA needs a $\lambda/4$ transformer from 50 Ω. Determine its dimensions at 2 GHz using FR-4 for minimum noise contribution as well as its contribution to the LNA's noise figure in decibels.

14.4 Design a single-stage LNA at 5 GHz using the *S*-parameters and noise parameters given in Table 5.9. The substrate is 15-mil thick alumina. Use microstrip in series and open-circuited stub configurations. The design must be unconditionally stable (if required use source series inductor) and output return loss must be greater than 15 dB. Calculate the NF, gain, and input and output return losses. The transistor is connected using 1-mil diameter and 10-mil long wires.

14.5 Discuss the pros and cons of PCB, thin-film, thick-film, LTCC, and HTCC technologies for amplifier applications at 2 GHz and 15 GHz.

14.6 Describe the advantages of microwave integrated circuits over conventional circuits.

Chapter 15

Monolithic Amplifiers

Over the past twenty years, microwave technology has gone through a significant evolution to meet necessary requirements for lower-cost solutions, circuit miniaturization, improved reliability, lower power consumption, low-voltage operation, high-voltage operation, and high-volume applications. Component size and weight are prime factors in the design of electronic systems for satellite communications, phased-array radar (PAR), electronic warfare, and other airborne applications, whereas high volume and low cost drive the phased array radar (PAR) and consumer electronics market. Building upon the success of MIC technology, a new microwave GaAs semiconductor based technology known as monolithic microwave integrated circuit (MMIC) was introduced in the mid-1970s. Unlike MICs, in MMICs all active and passive circuit elements are fabricated together on a semi-insulating substrate. MMIC amplifiers are integral parts of most commercial and military systems.

For RF wireless applications, several Si based device technologies including bipolar, CMOS, BiCMOS, and SiGe HBT are being pursued to obtain an optimum solution. In the Si based processes, Si wafers are larger and cheaper than GaAs wafers but the fabrication involves a relatively larger number of process steps. Salient features of a commonly used n-type or n-well CMOS fabrication process are discussed at the end of this chapter. Both RFICs and MMICs have low Q passives, no postmanufacture tuning or "tweaking" to obtain the optimum performance, expensive nonrecurring engineering cost and long development cycle time.

15.1 ADVANTAGES OF MONOLITHIC AMPLIFIERS

Whereas most MMIC amplifiers currently in production operate in the 0.5–40-GHz microwave range, applications covering the millimeter-wave (mmW) spectrum from 30 to 300 GHz are increasing. Monolithic technology is particularly suited for millimeter-wave applications through the elimination of the parasitic effects of bond wires, which connect discrete components in conventional hybrid microwave integrated circuits (HMICs). In MMIC based mmW subsystems, the cost can be lowered by a factor of 10 or more as compared to hybrid solutions. Advantages of MMIC amplifiers include low cost, small size, light weight, circuit design flexibility, broadband performance, elimination of circuit tweaking, high-volume manufacturing capability, package simplification, improved reproducibility, radiation hardness, and improved reliability.

Fundamentals of RF and Microwave Transistor Amplifiers. By Inder J. Bahl

MMIC power amplifiers have the following potential advantages as compared to commonly available internally matched power amplifiers:

- Multistage designs with higher gain (15–25 dB)
- Higher overall power added efficiency (PAE)
- Better unit-to-unit amplitude and phase tracking
- Compact in size and light in weight
- Lower parts count, higher reliability, and lower cost
- No external biasing chokes required

15.2 MONOLITHIC IC TECHNOLOGY

In fabricating MMICs, all active and passive circuit elements and interconnections are formed together on the surface of a semi-insulating substrate (usually gallium arsenide). Typically, MMICs use microstrip and metal–insulator–metal (MIM) capacitors for the matching networks, whereas at low microwave frequencies, lumped inductors and MIM capacitors are commonly used. Via holes, metal-filled holes from the bottom of the substrate (ground plane) to the top surface of MMICs, provide low-loss and low-inductance ground connections. Figure 15.1 shows a three-dimensional view of an MMIC.

15.2.1 MMIC Fabrication

There are many ways to fabricate MMIC amplifiers. MMICs using MESFETs, HBTs, and HEMTs are most commonly fabricated by a recessed-gate process, but the self-aligned gate (SAG) FET process is gaining popularity because it permits the efficient fabrication of devices optimized for different functions (e.g., microwave small signal, microwave power, and digital) on the same wafer at the same time. The SAG process has demonstrated superior performance uniformity in a manufacturing environment.

In order to give the reader an understanding of the relative complexity of GaAs MMIC manufacturing, a process flowchart for the SAG process is given in Figure 15.2. The process for recessed-gate MMICs has many similarities. The process includes

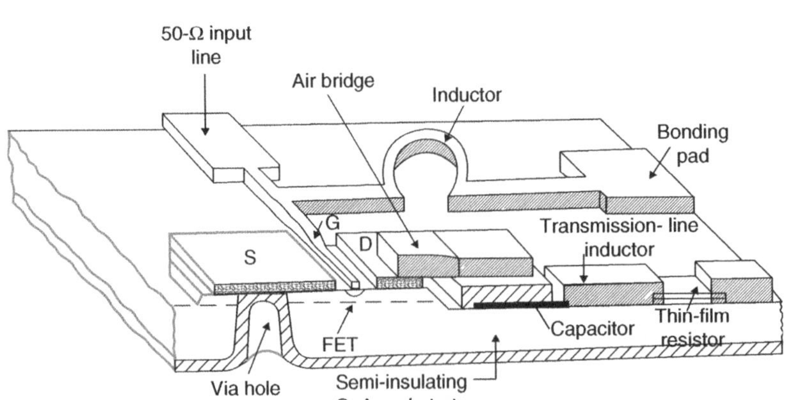

Figure 15.1 Three-dimensional view of an MMIC amplifier.

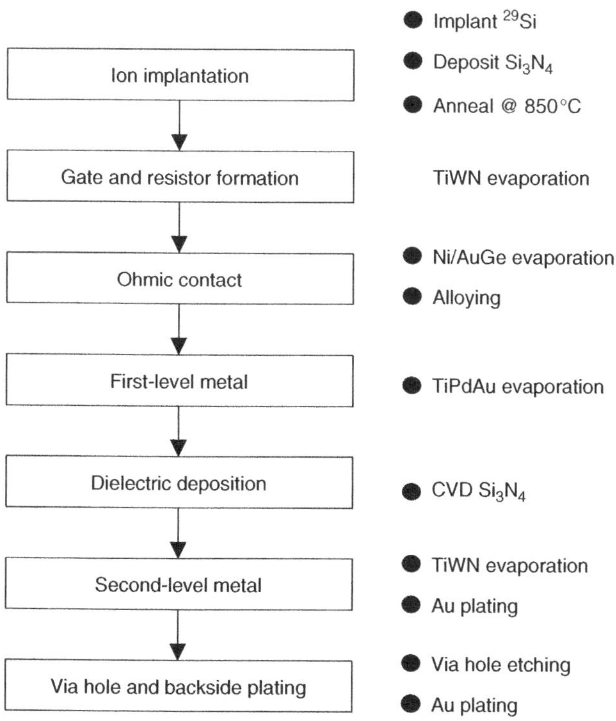

Figure 15.2 MMIC process flowchart for multifunction self-aligned gate (MSAG) process.

the fabrication of active devices, resistors, capacitors, inductors, distributed matching networks, air bridges, and via holes for ground connections through the substrate. Basic process steps are similar for any MMIC technology.

It should be noted that GaAs MMIC processing is less complex than silicon processing for devices operating at the low end of the microwave spectrum. Because silicon has inherently lower frequency capability and poorer isolation properties for integration purposes, more exotic processing is required to compete in the frequency region of overlap with GaAs applicability. For example, a silicon bipolar complementary metal oxide semiconductor (BiCMOS) process for such IC applications may require 2–3 times as many mask layers, adding significantly to the cost.

IEEE Microwave and Millimeter-Wave Monolithic Circuits Symposium Digest, published from 1982 to 1996, *IEEE RFIC Symposium Digest*, published since 1997, and *IEEE GaAs IC Symposium Digest*, published since 1980 include comprehensive information on the design, fabrication, and performance of monolithic microwave and millimeter-wave integrated circuits as well as their applications. Several other books [1–22] deal with this subject either partially or exclusively.

15.2.2 MMIC Substrates

Any assessment of MMIC technology options available to the microwave designer will generally be in terms of chip size, weight, reliability, reproducibility, cost, maximum frequency of operation, and availability of a wide range of active devices for design flexibility. Various substrate materials used for MMICs are bulk silicon,

silicon carbide, GaAs, InP, and GaN. Their electrical and physical properties are compared in Table 4.1, Chapter 4. The semi-insulating and high thermal conductivity property of the substrate material is crucial to providing higher device isolation and lower dielectric loss, and good heat dissipation path for power MMICs. Silicon dominates the marketplace and GaAs is used widely at RF, microwave, and millimeter-wave frequencies. For high-voltage, high-power, and high-temperature applications, wide-bandgap materials with relatively high thermal conductivity, such as SiC and GaN, play a significant role as substrate materials. Recent development of high-voltage active devices with very high power densities on a SiC substrate, as described in Chapter 13, was only possible due to its high thermal conductivity, which is a prime requirement for any semiconductor material to be used as a substrate for high-voltage and high-power-density devices and MMICs.

15.2.3 MMIC Active Devices

Since the first reported GaAs MMIC, the MESFET and the Schottky diode have been the workhorses for analog integrated circuits (ICs). MESFET technology commonly uses 0.25–1.0-μm gate lengths for microwave applications. MESFET power MMICs demonstrate excellent performance at microwave frequencies. However, increasing emphasis is being placed on new devices for better performance and higher frequency operation. HEMT and HBT devices offer potential advantages in microwave and millimeter-wave IC applications, arising from the use of heterojunctions to improve charge transport properties (as in HEMTs) or pn-junction injection characteristics (as in HBTs). HEMTs appear to have a niche in ultra-low-noise and high-frequency (mmW) applications. The MMICs produced using novel structures such as pseudomorphic, lattice-matched HEMTs, also known as pHEMTs, have significantly improved the noise performance and high-frequency (up to 280 GHz) operation. AlGaN/GaN HEMT devices have demonstrated power densities greater than five times higher than that of conventional GaAs based transistors [23, 24]. HBTs are vertically oriented heterostructure devices and are gaining popularity as power devices. GaAs HBTs are extensively used as power devices for high-volume wireless applications because of their high-gain, good efficiency, and single power supply low-voltage operation. They also offer better linearity and lower phase noise than do FETs and HEMTs. In a pHEMT structure there is another InGaAs active layer between the AlGaAs spacer and GaAs buffer that provides better carrier concentration in the channel than does a conventional HEMT structure. MESFETs, HEMTs, and HBTs have been described in detail in Chapter 5. These devices can be used for low-noise, switching, mixing, and power amplification depending on application requirements. For power circuits, where one needs much higher current, either a large number of cells are employed or larger gate periphery devices are used.

The upper frequency limit of MMICs is generally dictated by the active devices used. The performance of microwave transistors in MMIC technologies is improving every year. The performance of these devices (FETs, HEMTs, and HBTs) depends on the substrate material, process type, and channel physical dimensions. A commonly used figure of merit for devices is known as the maximum frequency of oscillation and denoted by $f_{\max}$. Generally, for amplifiers the maximum frequency of operation is about $\frac{1}{2}f_{\max}$. For FETs on a GaAs substrate, a simplified expression for $f_{\max}$ is given by [25]

$$f_{\max} = 38.05L^{-0.953} \tag{15.1}$$

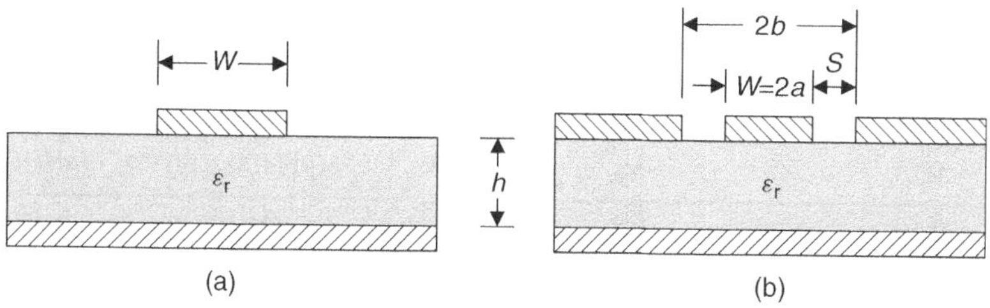

Figure 15.3 Transmission lines for MMICs: (a) microstrip and (b) coplanar waveguide.

where L is the gate length in micrometers. Thus for FETs having gate length of 0.25 μm, the f_{max} value is about 140 GHz. As reported in the literature, the f_{max} values for a 0.1-μm gate length pHEMT on an InP substrate is about 600 GHz, and for a 1-μm emitter HBT it is about 170 GHz. A three-stage amplifier fabricated using a 0.1-μm pHEMT on an InP substrate has exhibited about 12-dB gain at 153–155 GHz [26].

15.2.4 MMIC Matching Elements

Similar to hybrid ICs, monolithic circuits use distributed as well as lumped matching elements. The microstrip line and coplanar waveguide (CPW) are the two most commonly used transmission media in MMICs. The microstrip is more popular because of its quasi-TEM nature and excellent layout flexibility. Cross-sectional views of these lines with physical parameters are shown in Figure 15.3. Sections of microstrip lines and coplanar waveguides constitute the basic passive component building blocks of MMICs. When the size of the microstrip section is reduced to dimensions much smaller than the wavelength, the section can be used as a lumped element. Examples of lumped microstrip elements are spiral inductors, thin-film resistors, and interdigital capacitors. Microstrip sections in lumped and distributed forms are commonly used in passive and active MMICs. To realize compact circuits, lumped-element matching networks or lumped-distributed circuit elements are utilized to transform device impedance to 50 Ω. An overview of these circuits elements is given next.

Microstrip

The microstrip line is used exclusively as a transmission medium in MMIC amplifiers due to high power handling capability. Several methods used to determine microstrip parameters are summarized in Chapter 6 [27]. The microstrip propagation properties are controlled by conductor width W and substrate height h for a given dielectric constant value ($\varepsilon_r = 12.9$ for GaAs). Table 15.1 summarizes Z_0, ε_{re}, α, line capacitance, and line inductance for GaAs and Si substrates. As an example, for a 50-Ω line on a GaAs substrate, the value of W/h is about 0.7. As shown in Figure 15.4, the characteristic impedance Z_0 decreases and the effective dielectric constant ε_{re} increases when the strip width-to-height ratio W/h of the line is increased. The measured attenuation constant of microstrip on 100-μm thick GaAs at 1, 10, 20, and 30 GHz is shown in Figure 15.5.

Wavelength in the microstrip, λ, is related to ε_{re} by

$$\lambda = \lambda_0/\sqrt{\varepsilon_{re}} \tag{15.2}$$

Table 15.1 Microstrip Data Summary

GaAs Substrate $h = 100\ \mu m$, $t = 5\ \mu m$, $\tan\delta = 0.0005$, $f = 10$ GHz, and $\varepsilon_r = 12.9$

W (μm)	W/h	Z_0	ε_{re}	α (dB/cm)	Line Capacitance (pF/100 μm)	Line Inductance (nH/100 μm)
10	0.10	87.8	6.89	0.716	0.010	0.077
20	0.20	75.1	7.23	0.541	0.012	0.067
30	0.30	67.2	7.45	0.468	0.014	0.061
40	0.40	61.4	7.62	0.422	0.015	0.056
50	0.50	56.8	7.76	0.390	0.016	0.053
75	0.75	48.4	8.06	0.342	0.020	0.046
100	1.00	42.5	8.31	0.315	0.023	0.041
125	1.25	38.1	8.52	0.301	0.026	0.037
150	1.50	34.5	8.71	0.293	0.028	0.034
200	2.00	29.2	9.03	0.282	0.034	0.029
250	2.50	25.4	9.30	0.276	0.040	0.026
300	3.00	22.5	9.52	0.271	0.046	0.023
400	4.00	18.3	9.89	0.265	0.057	0.019
500	5.00	15.5	10.18	0.262	0.069	0.016

(b) Si Substrate
$h = 100\ \mu m$, $t = 5\ \mu m$, $\tan\delta = 0.0005$, $f = 10$ GHz, and $\varepsilon_r = 11.7$

W (μm)	W/h	Z_0	ε_{re}	α (dB/cm)	Line Capacitance (pF/100 μm)	Line Inductance (nH/100 μm)
10	0.10	91.8	6.30	0.685	0.009	0.077
20	0.20	78.6	6.60	0.517	0.011	0.067
30	0.30	70.3	6.80	0.447	0.012	0.061
40	0.40	64.2	6.95	0.403	0.014	0.056
50	0.50	59.4	7.08	0.373	0.015	0.053
75	0.75	50.7	7.35	0.327	0.018	0.046
100	1.00	44.5	7.57	0.301	0.021	0.041
125	1.25	39.9	7.76	0.287	0.023	0.037
150	1.50	36.2	7.93	0.280	0.026	0.034
200	2.00	30.6	8.22	0.269	0.031	0.029
250	2.50	26.6	8.46	0.263	0.036	0.026
300	3.00	23.5	8.66	0.258	0.042	0.023
400	4.00	19.2	8.99	0.253	0.052	0.019
500	5.00	16.2	9.25	0.250	0.062	0.016

where λ_0 is the free-space wavelength. The maximum frequency of operation of a microstrip transmission line is given by (6.5), Chapter 6. The calculated value for the maximum thickness of the GaAs substrate for microstrip circuits designed at 100 GHz is less than 0.3 mm.

Since it is impossible to do tuning on GaAs MMICs, an accurate and comprehensive modeling of microstrip discontinuities is required to save expensive and time-consuming iteration of mask and wafer fabrication and evaluation. As the yield of MMICs depends on the size (the smaller the chip, the higher the yield) and acceptable circuit electrical performance, discontinuities play an important part in the development of MMICs. The effect of discontinuities becomes more critical at higher frequencies. The discontinuities should be either taken into account or compensated for at the final stage of design. In most cases, discontinuities are basically undesirable circuit reactances, and in a good circuit design, efforts are made to reduce or compensate for these reactances shown in Figure 6.5, Chapter 6.

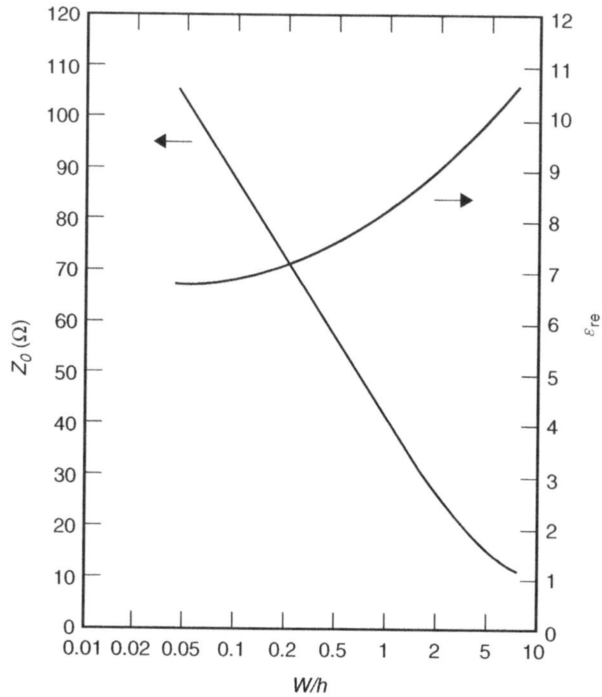

Figure 15.4 Characteristic impedance and effective dielectric constant of microstrip on 100-μm thick GaAs.

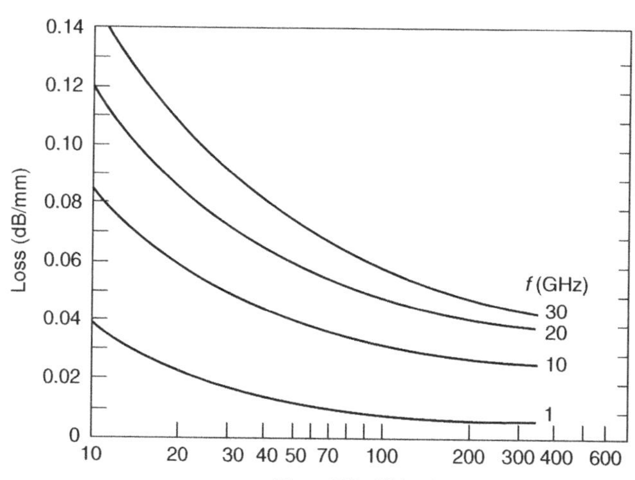

Figure 15.5 Measured attenuation constant of microstrip on 100-μm thick GaAs at 1, 10, 20 and 30 GHz.

CPW

CPW properties are controlled by the center conductor width W and the spacing between the strip and the ground-plane conductor denoted by S in Figure 15.3b. In a CPW, the substrate thickness generally used is large so that if the substrate has a

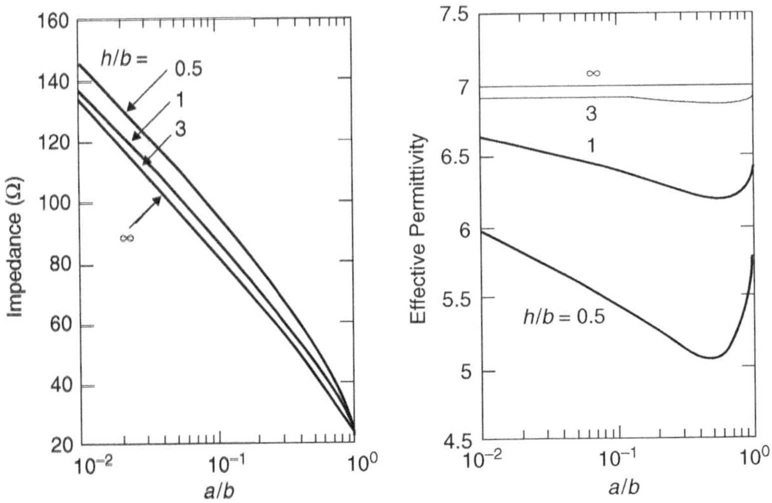

Figure 15.6 Characteristic impedance and effective dielectric constant of CPW on GaAs.

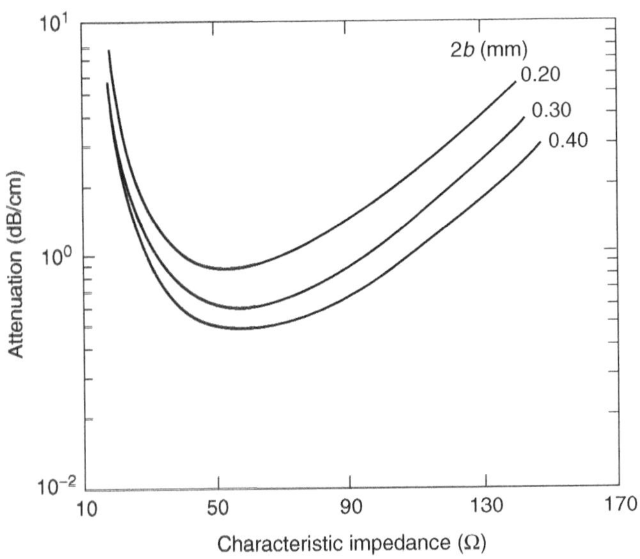

Figure 15.7 Measured attenuation constant of CPW on 100-μm thick GaAs at 60 GHz.

conductor backing to improve the mechanical strength, it does not affect the electrical characteristics of the CPW. Figure 15.6 shows the variation of Z_0 and ε_{re} as a function of the conductor width-to-gap separation ratio. The measured attenuation versus characteristic impedance Z_0 for CPW is shown in Figure 15.7.

In addition to dielectric and ohmic losses, coupling of power to surface waves and radiation from unwanted (parasitic) modes contribute to the total loss of the coplanar lines. The parasitic mode in a coplanar waveguide is the odd mode with antiphase voltages in the two slots. This mode can be excited at discontinuities, and radiation may occur. Radiation from this mode can be minimized by maintaining symmetry of the circuits (and thus avoiding its excitation) or by using air bridges connecting the ground planes at regular intervals to short circuit it out. In a conductor-backed coplanar waveguide, the parallel-plate waveguide modes are other parasitic modes. Surface waves or the substrate modes are the TM and TE modes supported by the substrate. Excitation of these modes can be avoided if a thin substrate is used such that the cutoff frequency of the surface modes is pushed above the operating frequency. This is achieved if the substrate thickness h is chosen such that

$$h \leq 0.12\lambda_0/\sqrt{\varepsilon_r} \tag{15.3}$$

where λ_0 and ε_r are, respectively, the free-space wavelength and dielectric constant of the substrate.

CPW MMICs, compared with microstrip based MMICs, can have lower loss at millimeter-wave frequencies by properly designing matching networks, require no via-hole technology for RF ground connections, and are more suitable for flip-chip mounting. Similar to microstrip discontinuities, CPW discontinuity effects must also be taken into consideration.

Lumped Elements

At radiofrequencies, lumped inductors, MIM capacitors, and thin-film resistors are mostly used in MMIC matching networks. Lumped-element amplifier circuits have the advantage of smaller size, lower cost, and wider band characteristics. These are especially suitable for MMICs where real-estate requirements are of prime importance and applications where broadband is required. Currently, MMIC technologies have reached a mature stage; lumped elements working even up to 30 GHz are more suitable for low-cost circuit solutions. At frequencies below C-band, MMICs using lumped inductors and capacitors are an order of magnitude smaller than ICs using distributed elements fabricated in microstrip or CPW. At RF and the low end of the microwave band, the use of lumped elements makes the chip size significantly smaller without affecting the RF performance or increasing the number of chips per wafer, and gives improved visual and RF yields. All these factors can reduce the chip cost drastically.

Electromigration Requirements

Electromigration is the transport of material caused by the gradual movement of the ions in a conductor due to high current densities flowing through it. When the current density in the conductors is on the order of 10^6 A/cm^2 or higher, electromigration has a continuous impact on the metal grain, causing the metal to pile up in the direction of current flow. In thin conductors, electromigration-induced damage usually occurs in the form of voids and hillocks in the metal. Depletion and accumulation of metal grains due to heavy flow of electrons give rise to voids and hillocks. This also occurs in transistor gates and drain and source pads, as well as in ohmic contacts. Mostly voids result in higher ohmic contact resistance. The effect of electromigration becomes more pronounced at elevated temperatures. The voids and hillocks due to electromigration grow over extended periods, resulting in open circuits in conductors and short circuits between closely spaced conductors. In FETs, open circuits in gates give rise to limited control on drain–source current and higher drain–source current values. Growing hillock formations short circuit the fingers.

Since in MMICs the conductors are much thinner than in MICs, special attention to current handling of such conductors becomes important. Thus the current handling capability of a conductor is limited by the onset of electromigration [19]. The conductor conductivity, thickness, and line width determine the current carrying capacity of the inductor. A safe value of maximum current density of gold conductors on a flat GaAs surface is $2.22 \times 10^5 A/cm^2$. Thus electromigration requirements dictate the microstrip and inductor line widths carrying direct current (DC). The current density of $2.22 \times 10^5 A/cm^2$ translates to a maximum allowed current per unit line width of 10 mA/μm for 4.5-μm thick gold conductors and 20 mA/μm for 9-μm thick gold conductors.

15.3 MMIC DESIGN

The design of MMICs requires state-of-the-art computer-aided design (CAD) tools. The need for increased design sophistication arises from the fact that the postfabrication tuning flexibility available in conventional hybrid microwave circuits is no longer present in the monolithically fabricated circuits. Consequently, a new design methodology is required. This includes development of accurately characterized standard library cells as well as subcircuits, accurate models for linear and nonlinear active devices, accurate passive component models, use of circuit topology and circuit elements that are more tolerant to process variations, tolerance centering of designs, proximity effect models, comprehensive simulation of complete circuits, and automatic RF testing of ICs on wafer. The latter is needed in order to obtain sufficient statistical characterization data without having to do expensive mounting or packaging.

15.3.1 CAD Tools

Figure 15.8 shows [28] a comprehensive CAD tool that consists of device, circuit, system simulators, and their accurate models (including physics based and electromagnetic), and statistical design feature. Several microwave CAD tools are available to designers, including Agilent's ADS, Ansoft's Ansoft Designer, and Applied Wave Research's Microwave Office and Cadence. A workstation based MMIC CAD tool [29] is conceptually shown in Figure 15.9. This interactive system will provide efficient coupling between the circuit simulation, the schematic captive/text editor, and the layout generator, greatly improving overall accuracy and reducing design cycle time. With such a system, first-pass design success for simple microwave functions should be achievable.

15.3.2 Design Procedure

The evolution of a typical MMIC power amplifier design generally follows the flow diagram depicted in Figure 15.10. The design starts with the circuit specifications, which derive from the system requirements. System requirements also dictate the circuit topology along with the types of passive elements and active devices to be used (e.g., distributed or lumped passive elements, single- or dual-gate FETs, and low-noise or power FETs). Comprehensive passive element and active device models developed by a foundry or by users are used to simulate circuit functions. The final design is completed by taking into account layout discontinuities, interaction between the components, stability analysis in the case of amplifiers, and circuit yield analysis by considering process variations. In the case of nonlinear circuit design, an accurate nonlinear model [30–37] for each device used is essential in order to design the circuit accurately. An overview of MMICs is given in References 15 and 38.

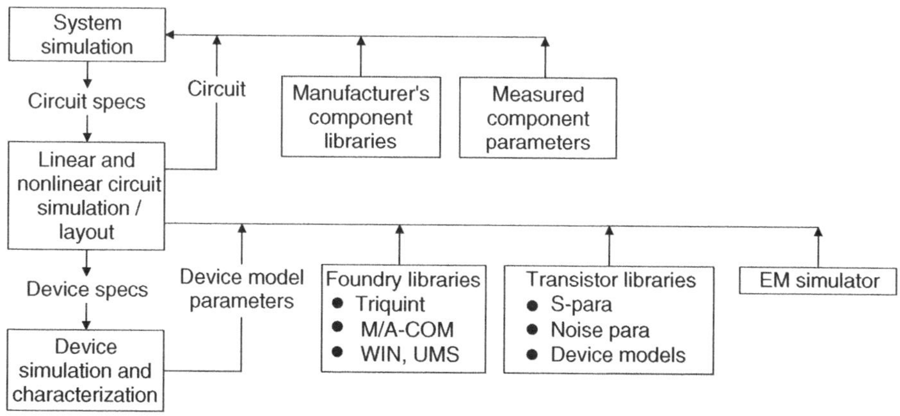

Figure 15.8 MMIC design system.

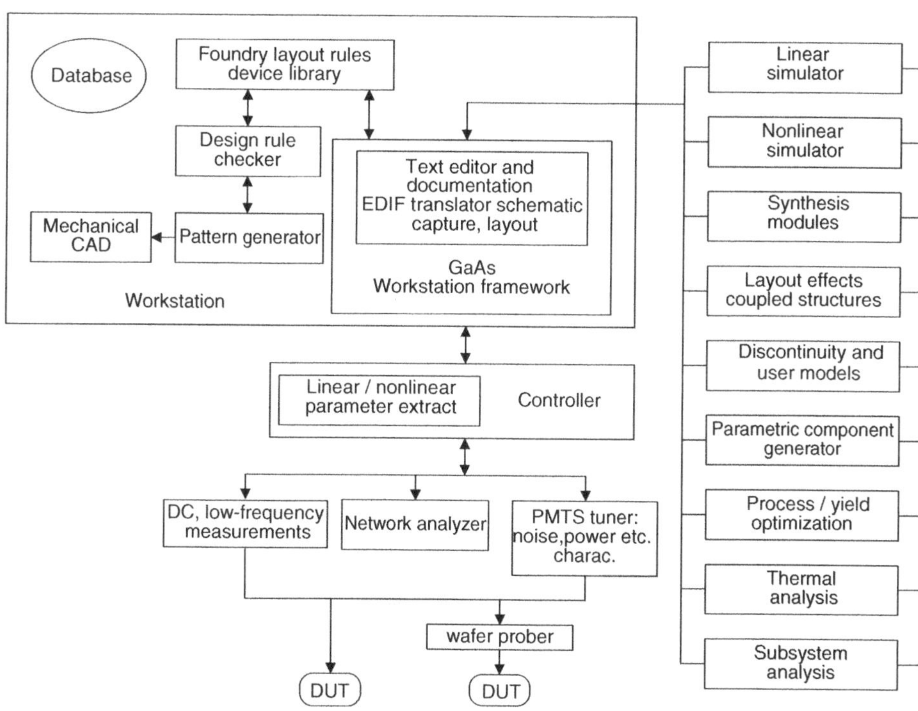

Figure 15.9 Next generation MMIC CAD workstation concept.

15.3.3 EM Simulators

The main contribution of electromagnetic (EM) simulators to MMIC CAD tools has been in the area of accurate modeling of passive circuit elements and components. These simulators, also known as field solvers, are commonly used to model circuit elements such as microstrip and coplanar waveguide structures, discontinuities, and coupling between transmission-line sections and discontinuities, structures using

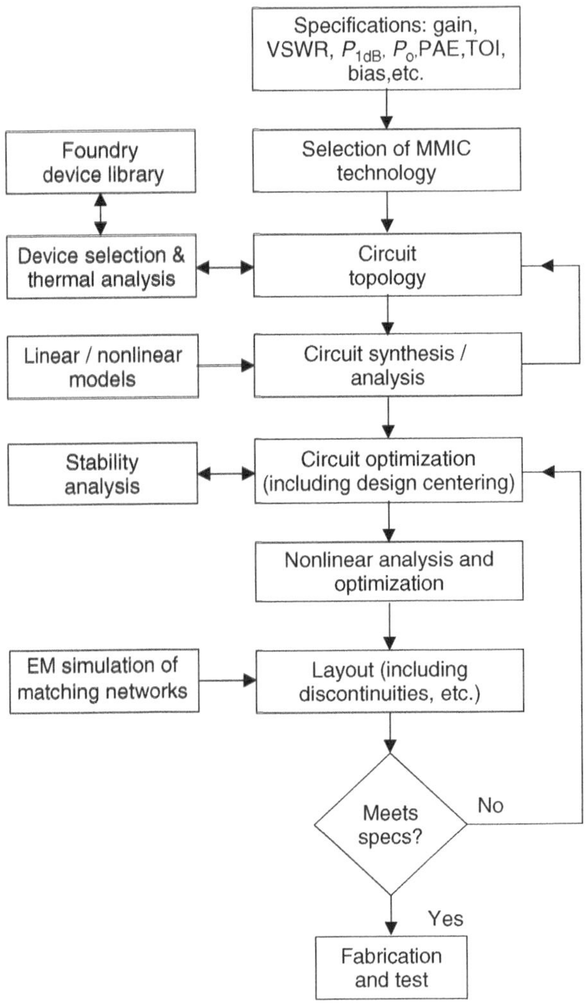

Figure 15.10 Typical flowchart for an MMIC power amplifier design.

multilayer dielectric and plating, inductors, capacitors, via holes, and crossovers. Passive components, such as filters, couplers, resonators, power dividers/combiners, baluns, matching impedance transformers, and several types of interconnects and packages, are accurately simulated using EM simulators. Accurate characterization of active device parasitic reactance also requires EM simulation. Another key and important role of EM simulators in successful MMIC design is the capability of incorporation of parasitic coupling effects among various parts of the circuit layout. Accurate evaluation of radiation and surface waves can be performed using EM simulators only. These effects become increasingly important as MMIC designs become more compact and are not easily incorporated using conventional network theory based CAD tools. However, due to the very large computation time, only a small portion of a circuit is analyzed using EM simulators, and the numerical results are combined with conventional CAD tools to obtain the response of the complete circuit. Most EM simulators work in the integrated simulation environment (i.e., they can be interfaced with microwave computer-aided design and engineering tools). In the past decade, outstanding progress made on personal computers and workstations led to commercial

Table 15.2 Overview of Some Electromagnetic Simulators Being Used for MMICs

Company	Software Name	Type of Structure Three-Dimensional	Method of Analysis	Domain of Analysis
HP-EEsof	Momentum	Planar	FEM	Frequency
	HFSS	Arbitrary	FEM	Frequency
Sonnet Software	EM	Planar	MoM	Frequency
Jansen Microwave	Unisim	Planar	Spectral domain	Frequency
	SFMIC	Planar	MoM	Frequency
Ansoft Corporation	Maxwell® 2D	Planar	MoM	Frequency
	Maxwell® SI Eminence 3D	Arbitrary	FEM	Frequency
Compact Software	Microwave Explorer	Planar	MoM	Frequency
MacNeal–Schwendler Corp.	MSC/EMAS	Arbitrary	FEM	Frequency
Zeland Software	IE3D	Arbitrary	MoM	Frequency
Kimberly Communications Consultants	Micro-Stripes	Arbitrary	TLM	Time
Remco	XFDTD	Arbitrary	FDTD	Time

EM simulators. More comprehensive information on these tools can be found in available publications [39–44]. An overview of commercially available EM simulators is given in Table 15.2.

15.4 DESIGN EXAMPLES

Since the early 1980s tremendous progress has been made in amplifiers including low noise, drivers, and power. These components are integral parts of most RF and microwave front ends. In this section we describe various types of MMIC amplifiers designed for low noise figure, narrowband, broadband, high-power, and high-efficiency applications. Salient features of each design are discussed briefly.

15.4.1 Low-Noise Amplifier

A low-noise amplifier (LNA) at a receiver front end sets the system noise figure. Narrowband and broadband LNAs are required depending on the system application. Table 15.3 compares the state of the art in narrowband MMIC multistage LNAs developed using pHEMTs [45]. Noise figures of about 1 dB and 3.3 dB have been achieved at 10 GHz and 94 GHz, respectively. Figure 15.11 shows the photograph of a 20–40-GHz two-stage LNA fabricated using pHEMT GaAs MMIC technology.

15.4.2 High-Power Limiter/LNA

Due to the fine geometry used in amplifier transistors, these circuits are susceptible to damage from high-power spurious EM radiation. To protect the front-end receiver and maintain a low noise figure, a high-power and low-loss limiter in front of an LNA

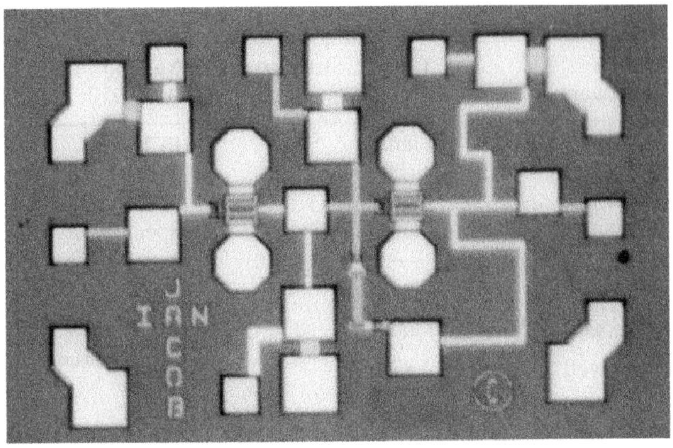

Figure 15.11 A 20–40-GHz two-stage LNA using pHEMT GaAs MMIC technology.

Table 15.3 Best Reported InP HEMT MMIC LNA Results

Frequency (GHz)	Number of Stages	Minimum NF (dB)	Gain (dB)	NF over Band (dB)	Year Reported
2.3–2.5	3	0.4	35	0.5 max	1993
7–11	2	1.0	21	1.2 max	1993
19–22	3	1.1	38	1.2 max	1995
40–50	3	1.9	25	2.3 ave	2004
43–46	3	1.9	22	2.0 ave	1995
50	2	2.8	9	—	1994
63	2	3.0	18	—	1990
56–60	2	3.2	15	4.2 ave	1992
56–64	3	2.7	25	3.0 ave	1993
58–62	2	2.2	16	2.3 ave	1995
75–110	3	3.3	11	5.0 max	1993
75–110	4	6.0	23	—	1993
92–96	3	3.3	20	4.4 max	1995
112–120	3	3.8	15	5.1 max	1999
120–124	2	—	11	—	1994
142	2	—	9	—	1995
155	3	5.1	12	—	1998

is required. A novel high-power three-stage limiter/LNA, which can handle greater than 10-W CW power at X-band, was developed. The design was similar to one reported in References 46 and 47 and its photograph is shown in Figure 15.12. The limiter/LNA design is a balanced configuration using Lange couplers with a Schottky diode limiter circuit followed by a three-stage LNA. A high-power termination resistor is also included on the chip. Typical measured performance shows gain greater than 19 dB, NF $<$ 2.7 dB, return loss better than 18 dB, and output third-order intercept greater than 28 dBm over the 8.5–12-GHz frequency range. The measured recovery time was about 35 ns.

15.4.3 Narrowband PA

A three-stage Ku-band 7-W MMIC power amplifier was reported in the literature [48]. The 7-W HPA design uses two 1.8-mm FETs driving four 1.8-mm FETs and driving

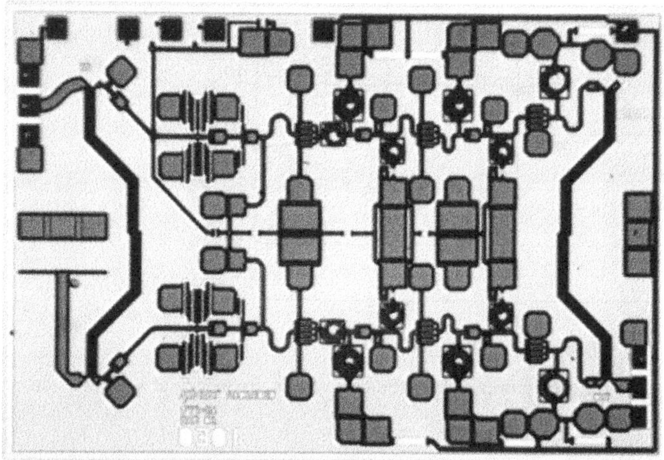

Figure 15.12 Photograph of an X-band three-stage integrated MMIC LNA/limiter. Chip size is $4.6 \times 3.1\ \text{mm}^2$.

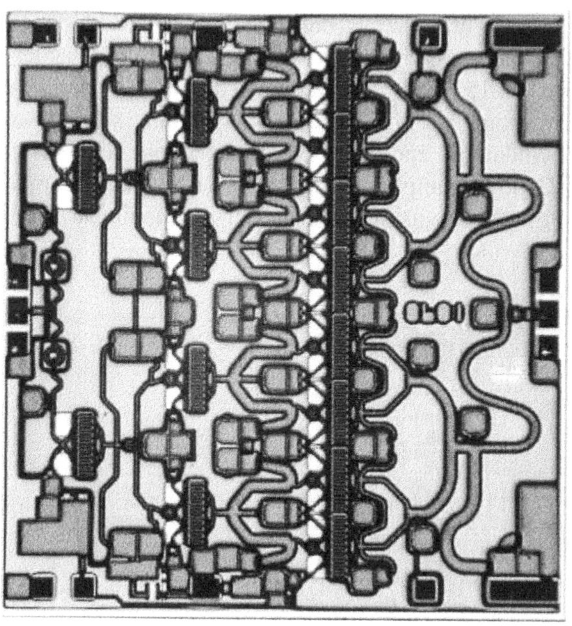

Figure 15.13 Photograph of the Ku-band 7-W HPA. Chip size is $4.2 \times 4.4\ \text{mm}^2$.

eight 1.8-mm FETs. Thermal analysis of these FETs is also given. Here, the FET aspect ratio is 2:1 and the circuit is designed for maximum output power and PAE under saturation. The matching circuit microstrip lines are on 10-μm polyimide [49]. Figure 15.13 shows the photograph of the 7-W MMIC HPA. This design requires a bias supply from both sides. Typical measured CW P_{out} and PAE for the 7-W MMIC power amplifier at $V_{ds} = 8$ V and $P_{in} = 23$ dBm are shown in Figure 15.14. The amplifier has greater than 38.5-dBm output power and better than 27% PAE over the 12.5–14.5-GHz frequency range. The measured second- and third-harmonic power levels at $V_{ds} = 8$ V and $P_{in} = 23$ dBm were below −40 dBc and −75 dBc, respectively.

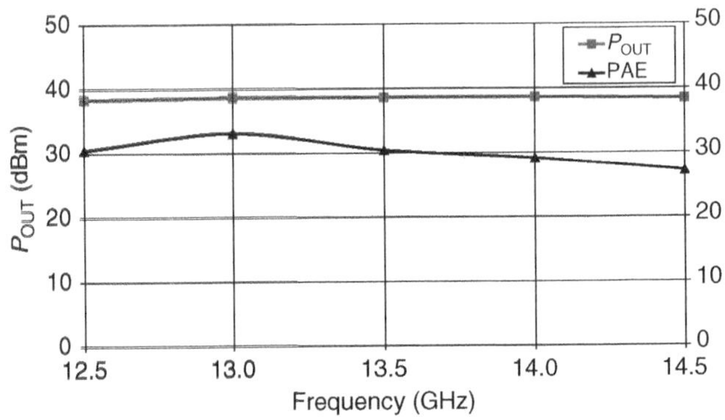

Figure 15.14 Output power and power added efficiency versus frequency at $V_{ds} = 8$ V and $P_{in} = 23$ dBm.

15.4.4 Broadband PA

C-Band PA

Next, a broadband MMIC power amplifier is described. In the IC design, a binary matching scheme employing lowpass networks, as described in Chapter 9, was used. As shown in Figure 15.15, both lumped and distributed circuit elements were used for impedance matching networks. Lowpass matching sections consist of series high-impedance lines/inductors and shunt MIM capacitors. The design of the two-stage broadband MMIC power amplifier was based on the conventional power amplifier design methodology, that is, using small-signal and nonlinear FET models,

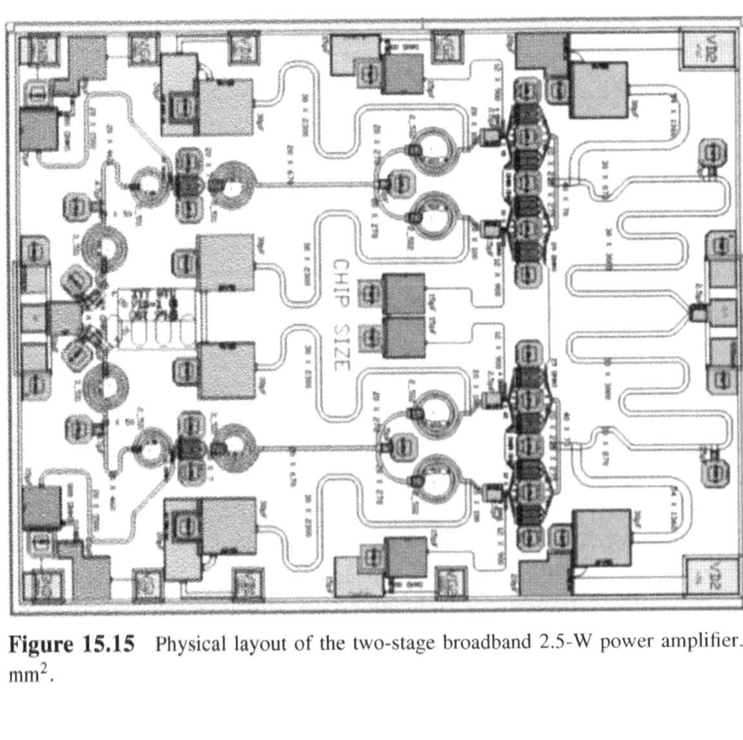

Figure 15.15 Physical layout of the two-stage broadband 2.5-W power amplifier. Chip size is 3.4×4.4 mm^2.

and load-pull data obtained at the operating bias point. The Q-point was selected for class-AB operation (25% I_{dss}) of the device in order to obtain the best compromise of power output, gain, PAE, linearity, and variable power supply operation over the C-band. The power output goal for the IC was 2.5 W.

Based on the output matching network's dissipative and mismatch loss and 0.8-W/mm power output at $V_{ds} = 8$ V for the FETs, a total of 5-mm gate periphery for the output stage FETs was used. The first stage uses two 0.625-mm gate periphery FETs, resulting in an FET ratio of 4:1. The 4:1 FET aspect ratio was chosen in order to have the best compromise between the PAE and linearity as well as circuit operation over a large drain voltage range. The thermal resistance measured between the die and the carrier is 16°C/W. Typical CW measured P_{out}, PAE, and input VSWR for MMIC chips packaged using standard carriers at $V_{ds} = 8$ V, $I_{DQ} = 360$ mA, and $P_{in} = 18$ dBm are shown in Figure 15.16 as a function of frequency. The amplifier has greater than 34-dBm output power, 16-dB gain, and better than 45% PAE over the 4.5–7.4-GHz frequency range. The typical measured second- and third-harmonic levels at $V_{ds} = 8$ V and $P_{in} = 18$ dBm were below −25 dBc and −28 dBc. Typical measured noise figure and output third-order intercept point values over the 4.5–7.1-GHz frequency range were 8.8 dB and 43 dBm, respectively. All nonharmonically related outputs were below −70 dBc for $P_{in} = 5 - 20$ dBm, $V_{ds} = 3 - 10$ V, and load VSWR = 3:1 [50].

X-Band PA

The design of the three-stage 10-W high-efficiency MMIC power amplifier was based on small-signal and large-signal FET models, and load-pull data obtained at the operating bias point. The Q-point was selected for class-AB operation (25% I_{dss}) of the device in order to obtain the best compromise of power output, gain, and PAE over the X-band. Here, a reactive binary matching topology employing lowpass and highpass networks provided high power output and PAE. Both lumped elements and distributed circuit elements were used for impedance matching networks. In the design optimization using a load-line technique, four sets of S-parameter data, corresponding to low

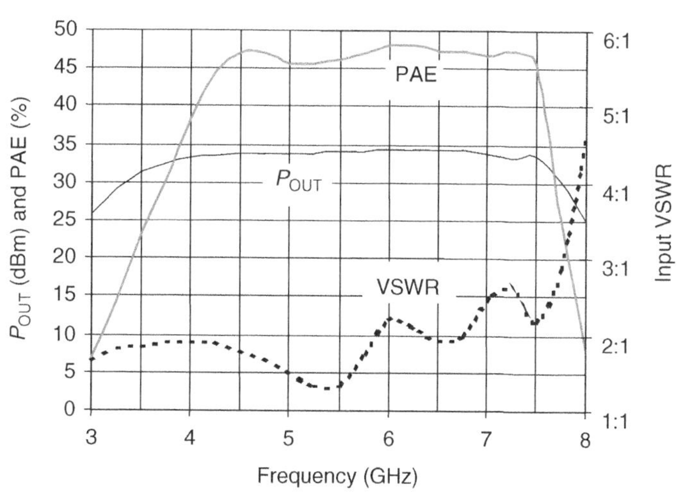

Figure 15.16 Output power, efficiency, and input VSWR versus frequency.

gain, high gain, low current, and high current, were used. These data files represent the possible fabrication changes and allow the realization of more process-tolerant design.

In the second iteration the circuit was further optimized using the Taguchi technique to improve the bandwidth and output power as described in Chapter 9. The design used an FET periphery of two 0.94-mm (10 fingers each) in the input stage, four 1.5-mm (14 fingers each) in the interstage, and eight 2.5-mm (24 fingers each) in the output stage. The GaAs substrate thickness was 75 μm. Figure 15.17 shows the photograph of the 10-W HPA [51].

Figure 15.18 shows typical measured CW output power, PAE, and gain versus frequency for a packaged die. As can be seen, efficiency is in the 34–43% range and an output power of 10 W was achieved across the 8.5–11-GHz frequency band. Over the narrowband (9.5–10.5 GHz), the output power was 12 W with over 40% PAE.

Figure 15.17 Photograph of the 10-W HPA. Chip size is 4.6 × 4.6 mm^2.

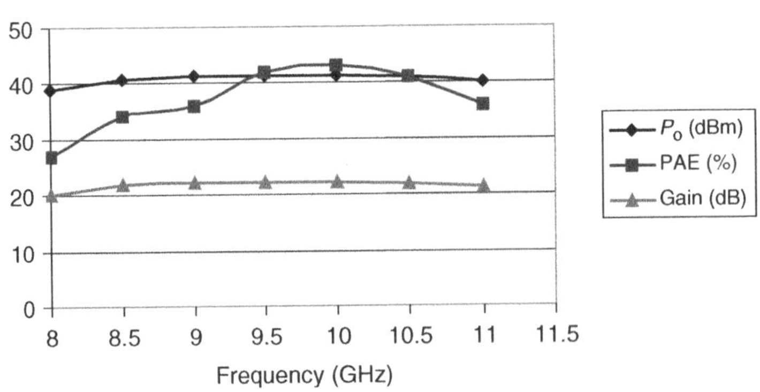

Figure 15.18 Typical measured CW output power, PAE, and gain of a 10-W HPA. $V_{ds} = 10$ V and $P_{in} = 19$ dBm. Baseplate temperature was 60 °C.

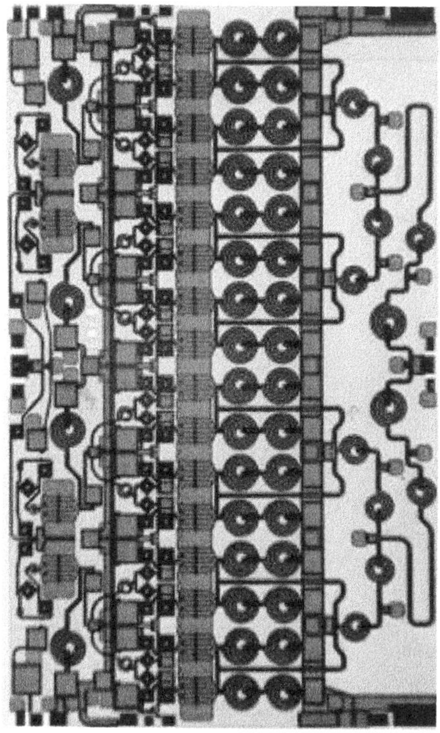

Figure 15.19 Photograph of the L/S-band 15-W MMIC power amplifier. Chip size is $5.0 \times 8.0\ mm^2$.

15.4.5 Ultra-Wideband PA

L- to S-Band

The design and test results of a monolithic class-AB broadband 12-W MMIC power amplifier to cover the frequency range from 0.7 to 2.7 GHz have been reported recently [52]. The design methodology for this HPA has been described in Chapter 11. Another high-power amplifier working over 1.2–2.4 GHz with a target output power of 15 W was designed using a reactive/resistive matching technique and 0.4-μm MSAG FET. The two-stage HPA consists of four 2.0-mm FETs at the input driving sixteen 2.0-mm FETs at the output. The amplifier was operated at a nominal power supply voltage of 10 V.

Figure 15.19 shows the photograph of the 15-W broadband HPA. Average measured output power and PAE of the two-stage broadband power amplifier are shown in Figure 15.20. The saturated output power was greater than 15 W and the PAE better than 29%. Over 1.25–2.75 GHz, the small-signal gain was better than 20 dB and input and output VSWR values were less than 2:1.

2–8 GHz PA

The design approach and test data for a two-octave bandwidth HPA developed using GaAs based MSAG MESFET with multilevel plating MMIC technology have been reported recently [53]. A low-loss matching design technique, as discussed in Chapter 9, was used in the design of the two-stage power amplifier. In this case, eight 0.94-mm

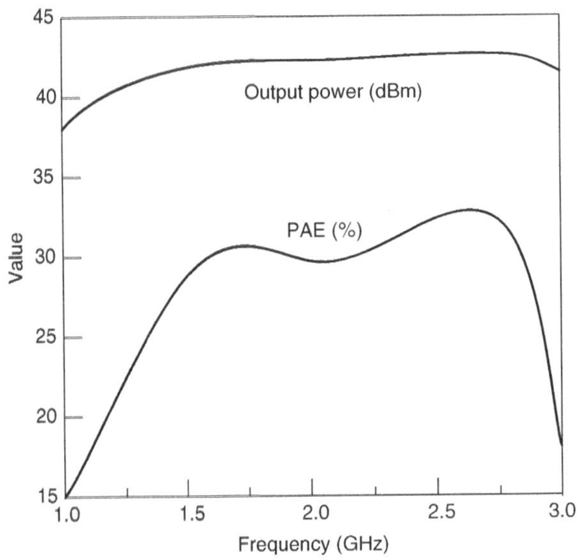

Figure 15.20 Output power and power added efficiency at $V_{ds} = 10$ V.

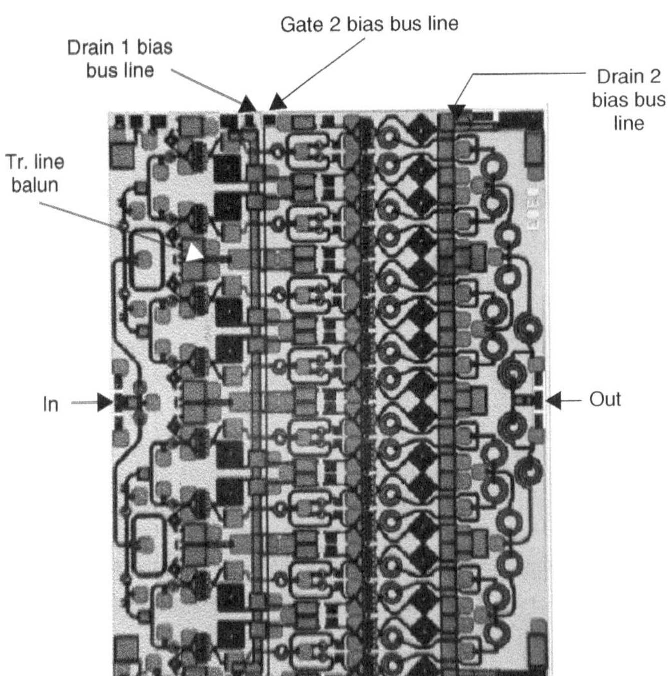

Figure 15.21 Photograph of the S/C/X-band 8-W HPA. Chip size is 5.0 × 6.3 mm^2.

FETs in the input stage drive sixteen 0.94-mm FETs in the output stage. The output stage matching network used a 16-way binary reactive combining topology to obtain two-octave bandwidth. Figure 15.21 shows the photograph of the 8-W broadband MMIC HPA.

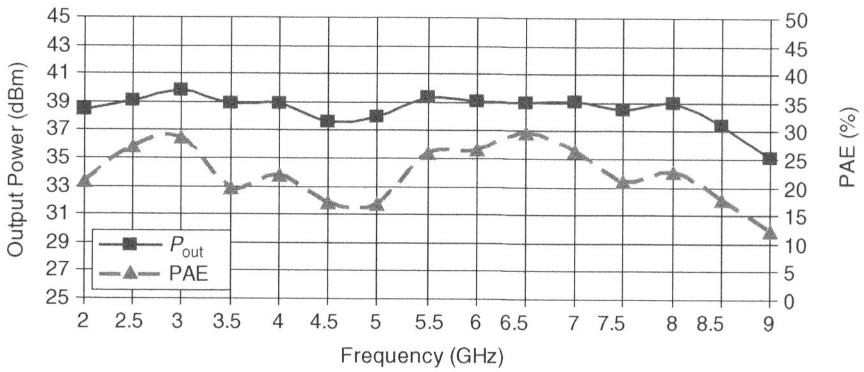

Figure 15.22 Output power and power added efficiency versus frequency of the 8-W S/C/X-band MMIC HPA.

The typical measured P_{out} and PAE for the power amplifier MMIC are shown in Figure 15.22. The amplifier has greater than 37.6-dBm output power and better than 16% PAE over the 2–8.5-GHz frequency range.

2–18 GHz PA

Next, we describe a 0.5-W 2–18-GHz two-stage distributed amplifier (DA). Each stage is matched to 50 Ω and uses a conventional DA configuration as shown in Figure 11.17, Chapter 11. In this design each stage uses five cells and the device size is tapered to obtain the largest power bandwidth and PAE. Device sizes for the first-stage FETs are 300, 300, 300, 300, and 300 μm. In the second-stage single-ended design, the FETs used are 625, 625, 470, 470, and 300 μm. By using small-signal S-parameters, each stage was optimized for maximum gain, and good input and output VSWR values. Figure 15.23 shows the physical layout of the broadband DA. The simulated gain and output power are shown in Figure 15.24.

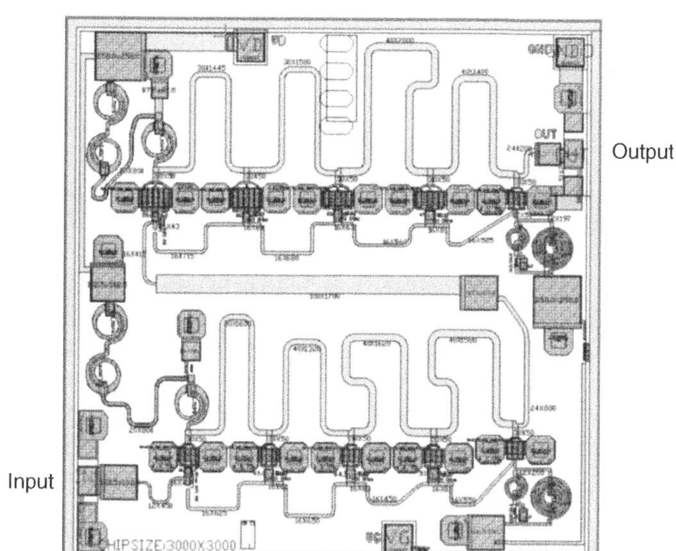

Figure 15.23 Layout of the 2–18-GHz 0.5-W HPA. Chip size is 3 × 3 mm^2.

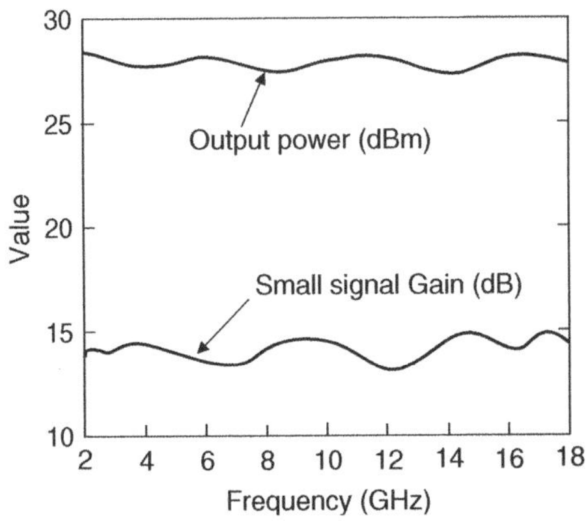

Figure 15.24 Simulated gain and output power versus frequency of the 2–18-GHz MMIC HPA.

15.4.6 High-Power Amplifier

Monolithic high-power amplifier design involves power combining as many devices as is practical in order to achieve increased power levels. A single large device is impractical on an MMIC because of matching of the very low device input impedance. The cluster matching technique, as described in Chapter 9, has emerged as the optimum means of integrating the matching network into the splitting and combining manifolds. The insertion loss of this output manifold imposes a practical limit on the number of devices that can be combined—both for economical reasons (wasteful use of expensive chip space) and because the efficiency drops quickly as the combining loss increases.

S-Band 50-W HPA

The 50-W MMIC amplifier [54] shown in Figure 15.25 illustrates very clearly the output manifold. The two-stage 2-GHz design is based on a 0.5-μm gate GaAs pHEMT and used 128-mm total gate periphery at the output stage. The measured CW power, at a nominal 12-V supply voltage, was 50 W with an efficiency of 45% over 10% bandwidth.

X-Band 20-W HPA

The 20-W X-band HPA consists of two 10-W power amplifiers fully matched to 50 Ω on both input and output, and combined "on-chip" using a two-way Wilkinson type power splitter/combiner. The 10-W high-power amplifier design consists of three stages [55]. The choice of FET cell size impacts the matching networks, combining topology, chip size, and electrical performance. A binary corporate feed combining was used consisting of two 0.625-mm FETs driving four 1.1-mm FETs, which finally drive eight 2.0- mm FETs. Here the FET aspect ratio is 3.6:1. Each FET had 20-μm gate-to-gate pitch. Figure 15.26 shows the photograph of the 20-W HPA. The test fixture used for the characterization of 20-W amplifier chips is shown in Figure 15.27. The test fixture was fitted with high-performance microstrip to coaxial connectors having return loss better than 20 dB up to 18 GHz.

Figure 15.28 shows the typical measured CW output power and PAE data for a packaged die. The measured output power and PAE are at $V_{ds} = 10$ V and $P_{in} = 18$ dBm. The measurements were taken at 25 °C baseplate temperature. The amplifier has

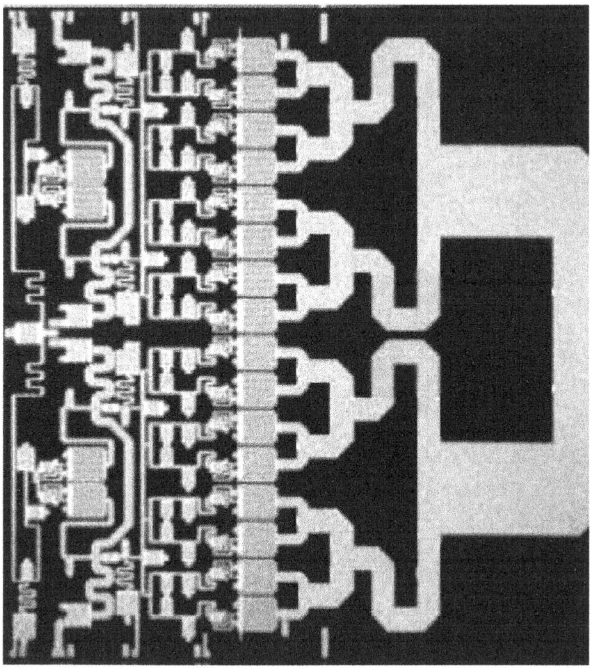

Figure 15.25 A 50-W 2.1–2.2-GHz power amplifier MMIC (10×10 mm^2). (From Akkul et al. [54]. Reprinted with permission of IEEE.)

greater than 43-dBm output power, 25-dB gain, and better than 33% PAE over the 8.0–10.0-GHz frequency range. Over 7.8–10.5 GHz, the output power and PAE are greater than 16 W and 27%, respectively.

15.4.7 High-Efficiency PA

A single-stage 15-W MMIC power amplifier using MESFET technology is shown in Figure 15.29. This C-band amplifier used 28-mm gate periphery and matched to the 25-Ω input and output system impedance [56]. Typical measured output power and PAE at $V_{ds} = 10$ V are shown in Figure 15.30. The circuit was optimized using the Taguchi technique and described in Chapter 9. The measurements were taken at 25 °C baseplate temperature. The amplifier has greater than 14-W output power and better than 55% PAE over the 4.7–5.5-GHz frequency range. Over a narrower band, the PAE was close to 60%.

15.4.8 Millimeter-Wave PA

A monolithic high-power and high-PAE amplifier operating from 27.5 to 29.5 GHz was reported for local multipoint distribution service. The amplifier consists of two stages and each stage is comprised of a balanced topology. The design was based on 0.15-μm gate length pHEMTs. Figure 15.31 shows the photo of this amplifier [57]. The measured values for small-signal gain, output power, and PAE were 16 dB, 1.6 W, and 35%, respectively.

Table 15.4 compares some MMICs that have achieved the highest powers for various frequency ranges. To further increase the achievable power, it is necessary to

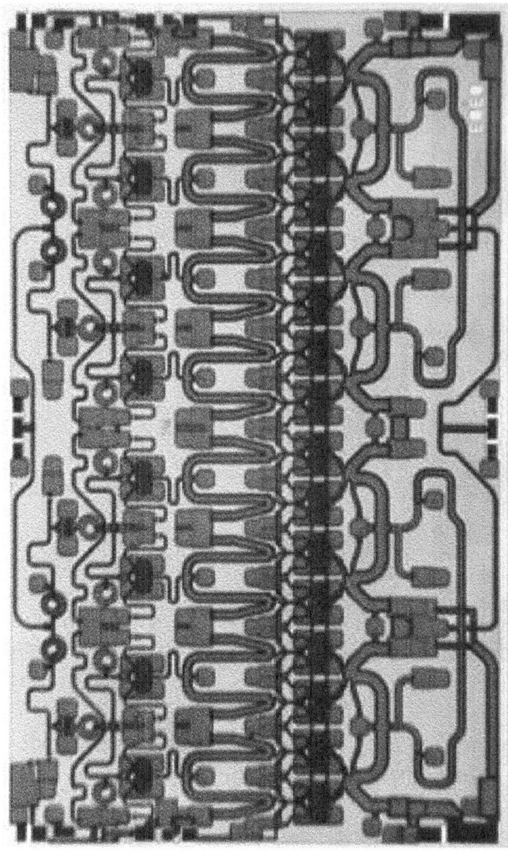

Figure 15.26 Photograph of the 20-W X-band HPA. Chip size is $5 \times 8\ \text{mm}^2$.

power combine multiple MMICs. Depending on the frequency, this might be done at the package level or by using very-low-loss combining in a waveguide, or even quasioptically as discussed later in Chapter 19.

15.4.9 Wireless Power Amplifier Design Example

The example provided here is for a GSM power amplifier (PA), whose typical performance specifications are the following:

Frequency range	880–915 MHz
Power gain	30 dB
Output power	3 W
PAE	50%
Input VSWR	2:1
Supply voltage	3.5 V
Control current	5 mA @ 2.7 V

To design such an amplifier, a 3-μm emitter width HBT based monolithic technology was selected. HBT technology has gained acceptance as a cost-effective alternative to MESFET based power amplifiers and has several advantages over MESFET power

Figure 15.27 Photograph of 20-W MMIC amplifier test fixture.

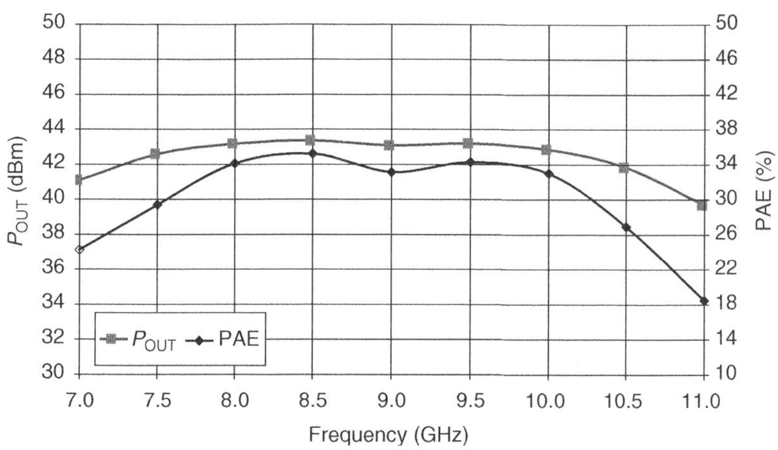

Figure 15.28 Output power and power added efficiency versus frequency at $V_{ds} = 10$ V and $P_{in} = 18$ dBm.

amplifiers. An HBT power amplifier operates from a single positive DC power supply resulting in reduced overall amplifier design complexity. HBT power amplifiers are capable of very high power densities, which reduces the overall chip size and cost compared to MESFET power amplifiers. However, the thermal design of HBT power amplifiers is much more critical than with MESFET amplifiers. Great care must be taken to prevent thermal runaway in HBT power amplifier designs. The process of designing an HBT power amplifier can be broken down into two important areas—electrical design and thermal design.

Electrical Design

To keep the device operational current density at $\sim 0.1 - 0.2$ mA/μm^2, the load line and class of operation of the output stage must be properly designed. Typically, the first

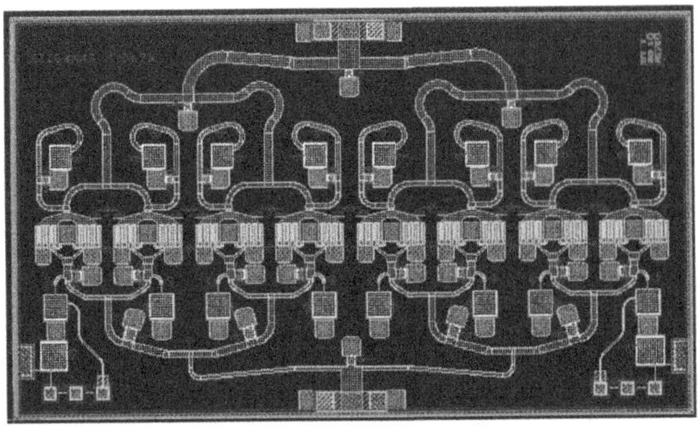

Figure 15.29 Physical layout of the 15-W C-band HPA. Chip size is 3.7 × 6.4 mm^2.

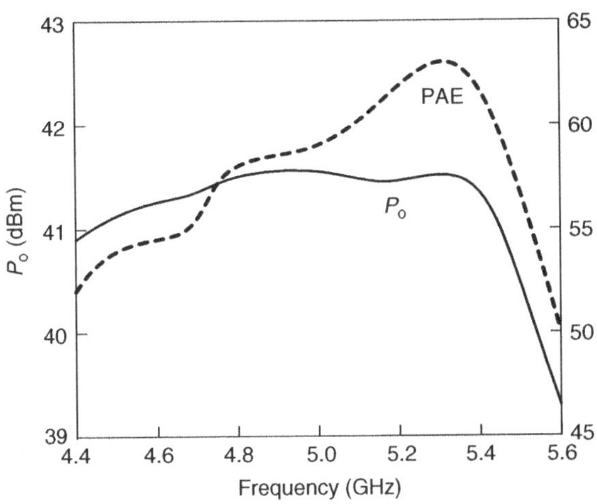

Figure 15.30 Output power and PAE versus frequency at $V_{ds} = 10$ V and $P_{in} = 32$ dBm.

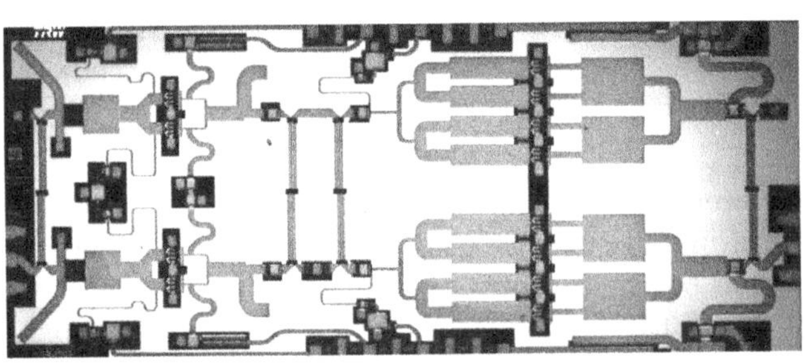

Figure 15.31 Photograph of the 1.6-W Ka-band HPA. Chip size is 6.0 × 2.5 mm^2.

stage is biased class A, 0.1-mA/μm^2 current density, but the last two stages are biased deep class AB or class B, resulting in a quiescent current density of approximately 0.008 mA/μm^2. Utilizing thick emitter metallization and proper via placement for minimum unit cell thermal resistance, the output stage was designed to operate at ~ 0.14-mA/μm^2 current density under RF power. The output stage is sized at 11880 μm^2, which requires

Table 15.4 Summary of Narrowband GaAs MMIC Power Amplifier Performance

Frequency (GHz)	Power (W)	Number of Stages	Gain (dB)	PAE (%)	Device	Reference
2.1–2.2	50	2	21	50%	pHEMT	54
8–10	20	3	25	33%	MESFET	55
12.5–14.5	7	3	18	27	MESFET	48
29–31	4	3	20	25%	pHEMT	58
42–46	2.8	2	17	24%	pHEMT	59
95	0.427	2	15	19%	InP pHEMT	60

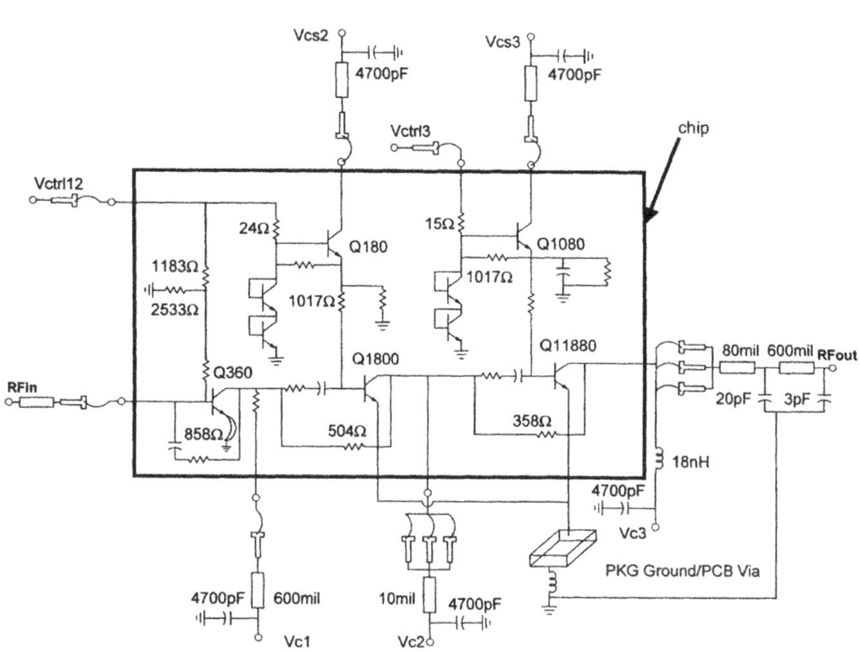

Figure 15.32 Schematic of GSM power amplifier for handset applications. Package grounding effects must be considered to maximize performance and stability.

six parallel arms with 11 cells per arm. Each cell consists of two fingers having a total area of $2 \times 3 \times 30 = 180\ \mu m^2$. To meet the target gain specification, the power amplifier requires three stages. The device size ratio of the last two stages is 6.6:1 and the device size ratio of the first two stages is 5:1. The fundamental load impedance of the output stage as represented by a parallel *RC* combination is approximately 2 Ω and −26 pF. The fundamental load impedance of the second stage is approximately 13 Ω and −3.9 pF. The first-stage load impedance is easier to match than the last stage and is approximately 65 Ω with less than −1-pF capacitance.

The output match is formed using the output wire bonds and package inductance in series with an 80-mil long 50-Ω transmission line segment followed by two shunt 10-pF high-*Q* adjacent capacitors in parallel forming 20 pF. A secondary output capacitance of 3 pF is located 600 mils beyond a 20-pF capacitor by using another 50-Ω transmission line section. The transmission lines can be formed on standard 10-mil thick FR-4 dielectric using 1-oz copper traces.

The collector bias can be fed into the output stage through a quarter-wave line, which can also act as a second-harmonic short, or a low DC resistance 18-nH choke. The load impedance for the second stage is realized using the input capacitance of the third stage in parallel with a shunt inductance formed by series parallel wire bonds directly off the collector of the second stage followed by a large-valued shunt RF bypassing capacitor. The collector bias of the second stage can be fed through this low-impedance point. The shunt inductive bias path is in parallel with the capacitive input of the last stage, forming a parallel resonance slightly offset from the center frequency of the band. The second- to third-stage interstage match is highly frequency tunable and is very sensitive, but allows for a simple and very compact layout, requiring non on-chip inductors. The interstage match between the first and second stages of the amplifier is more forgiving and can be formed by using either a shunt transmission line or shunt inductance connected to the collector of the first stage. Bias for the first stage is supplied through bypassing on this line, as seen in the schematic shown in Figure 15.32. All three stage devices have parallel feedback resistors for stabilization. The current design also includes a control circuit on the chip for stable operation of the PA over a range of temperatures.

Thermal Design Considerations

Two important thermally related issues must be addressed when designing an HBT PA. The first issue is ballasting. When large numbers of small unit HBT cells (e.g., 2 fingers by 3 μm $\times$ 30 μm $= 180$ μm^2) are joined together in parallel to form an output stage necessary for generating the desired target output power, small variations between unit devices can cause current hogging during operation. GaAs HBTs have a positive temperature coefficient, which means that as an HBT cell draws more current as it heats up, it further increases the operational junction temperature. Current hogging is a phenomenon that results from the slight variation between unit HBT cells, causing a few to draw more current than their neighbors. Ultimately, a few cells will tend to draw excessive amounts of current (hogging) and reach a junction temperature that can cause serious performance degradation with serious lifetime degradation or even catastrophic junction failure. The simplest method to prevent this is base or emitter ballasting. Ballasting involves the placement of a small resistor in the emitter of each unit cell or in the base bias path of each unit cell to prevent current hogging. When a "hot" HBT cell within an array starts to draw slightly more current than the surrounding cells, the voltage across the ballasting resistance will increase, which will reduce the voltage applied across the base–emitter junction of the current-hogging cell. This breaks the positive thermal feedback within the unit cell and will allow an even distribution of heat across the output array of an HBT PA as it draws current under operation.

Another important aspect of HBT PA thermal design is the maximum operating current density. Since HBT power amplifiers are not as current limited as MESFET PAs, they can be designed to operate at very high current densities (and thus junction temperature) by a proper selection of load line and bias. Care must be taken to design the HBT PA to operate at a current density that will keep the junction temperature of the unit cells below 150 °C. This will depend on process and layout specific factors, such as emitter metallization thickness, wafer thickness, via location, cell pitch, and emitter dimensions. Excessive junction temperature can cause degradation in operational lifetime, degradation in RF performance, or even destructive failure.

Backside vias are useful for decreasing the effective thermal resistance of the last stage of a PA because they help to transfer heat from the emitters to the backside of the chip. The thermal resistance of the HBT PA is approximately 18 °C/W. The layout symmetry of the output stage can help to improve stability, thermal uniformity, and power gain of the overall amplifier. See Figure 15.33 for a layout of the example amplifier. The typical measured performance of this HBT PA design is shown in Figure 15.34. Peak power added efficiency (PAE) is 54% at an output power of 35.5 dBm and 30-dB gain.

In this section, we have provided an overview of MMIC amplifier performance. Several other MMIC amplifiers have been discussed in previous chapters. High-voltage MMIC power amplifiers are described in Chapter 14.

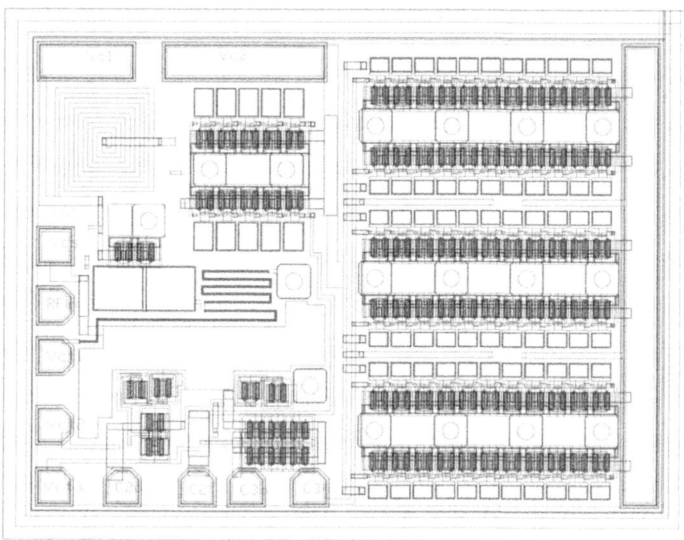

Figure 15.33 GSM power amplifier chip layout.

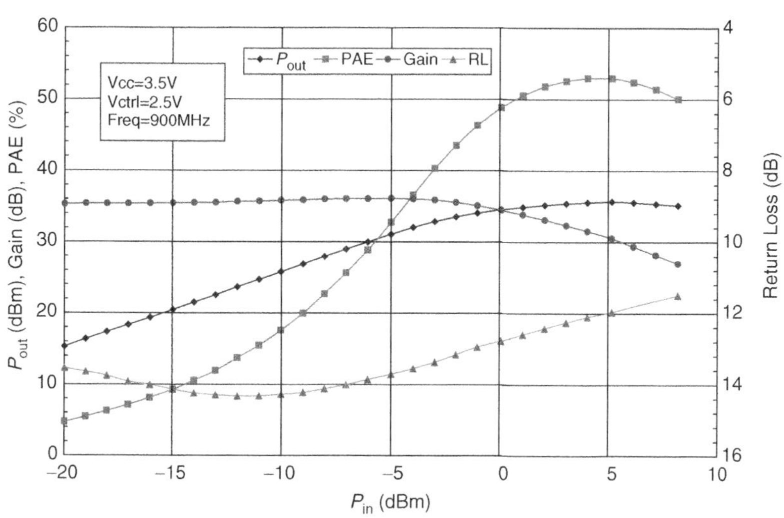

Figure 15.34 Measured performance of GSM power amplifier design.

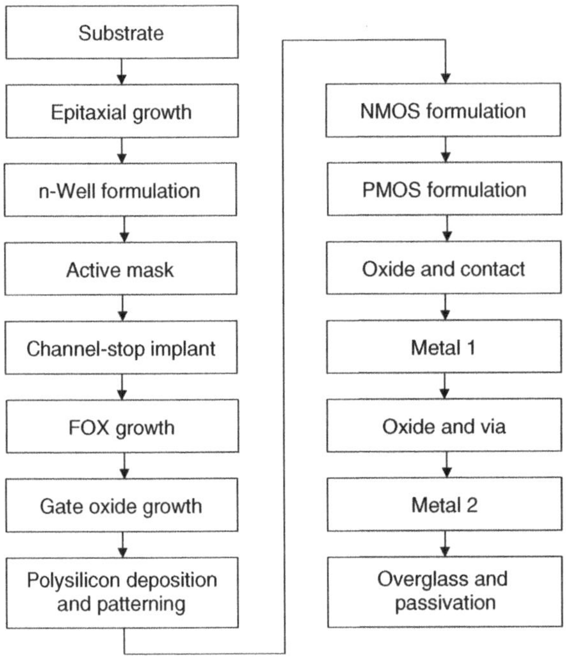

Figure 15.35 Basic steps for an n-well CMOS process.

15.5 CMOS FABRICATION

CMOS integrated circuit fabrication starts with thin circular (6–12-in.) silicon wafers, also known as substrates. Salient features of a commonly used n-type or n-well CMOS fabrication process [61–63] are discussed next. The wafers are doped with phosphorus (donor atoms) or boron (acceptor atoms) for an n-type or p-type wafer, respectively. For n-channel MOSFETs or NMOS, a p-type wafer is used. Basic steps for an n-well CMOS fabrication are shown in Figure 15.35. The process starts with the epitaxial growth of a p-type layer on a wafer/substrate. This layer provides a controlled thickness (1–20 μm) of high-purity silicon. The next step is n-well formulation using ion implantation. Following that, a layer of SiO_2 is grown and covered with Si_3N_4 to define the active device areas, followed by a p^+ field implant (known as channel-stop implant) outside the active device areas to isolate the active devices. After the channel-stop implant, a thick layer of field oxide (FOX) is grown in areas where the Si_3N_4 layer is absent, and then the Si_3N_4 and SiO_2 layers are removed. Following these steps, a thin layer of gate oxide ($\cong$80–200 Å) is grown in active device areas.

Next, a layer of polysilicon is deposited and transistor gates, resistor connections, capacitor electrodes, and interconnections are defined. Oxide areas at the drain and source locations of the n-type channel device are removed and donor atoms are introduced by diffusion or ion implantation to make n^+ regions for connections. Similarly, acceptor atoms are introduced for p-type channel devices. This is followed by depositions of thick layers of oxide and metal 1 and metal 2. Connections between polysilicon and various metals are made through etched vias. At the end of the process, the wafers are passivated by another SiO_2 layer, leaving opened bonding pad areas for wire connections.

Lumped-element passive components such as resistors, capacitors, and inductors are realized in many different ways. Resistors include well resistors, diffused,

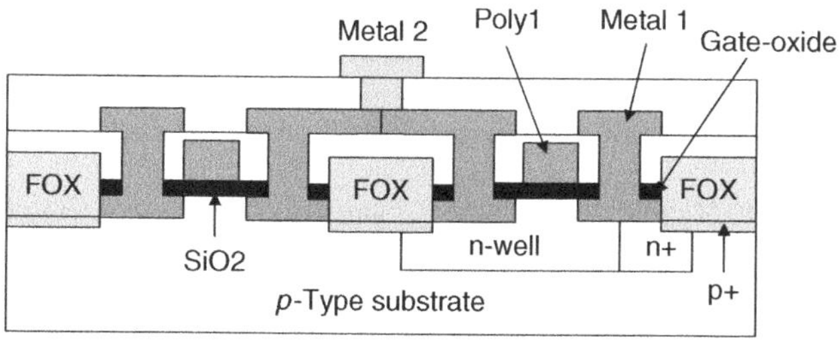

Figure 15.36 Cross section of a finished CMOS device.

polysilicon, and thin film. Capacitors are comprised of poly–poly, metal–poly, metal–silicon, and silicon–silicon. Inductors include metal 1–metal 2 and many other metal layers.

The well and diffused resistors are formed by connecting a well region and n-channel or p-channel regions through source and drain connections. The well region's sheet resistivity is on the order of 10 kΩ/□ while diffused resistors have resistivity in the range of 100–200 Ω/□. Polysilicon gate layers are also used as resistors. The sheet resistivity of polysilicon layers used in CMOS fabrication is on the order of 20–80 Ω/□. Thin-film resistors use nichrome and tantalum and are deposited on an SiO_2 layer. Depending on their thickness and composition, the range of sheet resistivity is 100–1000 Ω/□.

An SiO_2 layer sandwiched between two polysilicon electrodes comprises poly–poly capacitors. The tolerance on the value of the capacitor is 10–20%. In the case of a poly–metal capacitor, one can adjust the thickness of SiO_2 between the polysilicon and metal 1 electrodes. Properties of such capacitors are similar to poly–poly capacitors. Figure 15.36 shows a cross-sectional view of a typical CMOS circuit.

For RF applications, inductors are also becoming a part of the CMOS standard process. In this case the inductors are realized by using many available metal layers, to improve the quality factor. In recent years, Si based CMOS technology circuits have advanced dramatically and are competing with MMIC technologies in the microwave as well as in the millimeter-wave frequency range.

REFERENCES

1. J. V. Dilorenzo and D. D. Khandelwal (Eds.), *GaAs FET Principles and Technology*, Artech House, Norwood, MA, 1982.
2. R. S. Pengelly, *Microwave Field-Effect Transistors—Theory, Design and Applications*, John Wiley & Sons, Hoboken, NJ, 1982.
3. R. Soares, J. Graffeuil, and J. Obregon (Eds.), *Applications of GaAs MESFETs*, Artech House, Norwood, MA, 1983.
4. R. E. Williams, *Gallium Arsenide Processing Techniques*, Artech House, Norwood, MA, 1984.
5. R. A. Pucel (Ed.), *Monolithic Microwave Integrated Circuits*, IEEE Press, Piscataway, NJ, 1985.
6. D. K. Ferry (Ed.), *Gallium Arsenide Technology*, Howard Sams, Indianapolis, IN, 1985.
7. N. G. Einspruch and W. R. Wisseman, *GaAs Microelectronics*, Academic Press, San Diego, CA, 1985.
8. I. J. Bahl and P. Bhartia, *Microwave Solid State Circuit Design*, 2nd edition, John Wiley & Sons, Hoboken, NJ, 2003, Chapter 15.

9. R. Soares (Ed.), *GaAs MESFET Circuit Design*, Artech House, Norwood, MA, 1989.
10. J. Mun (Ed.), *GaAs Integrated Circuits: Design and Technology*, Macmillan, New York, 1988.
11. P. H. Ladbrooke, *MMIC Design: GaAs FETs and HEMTs*, Artech House, Norwood, MA, 1989.
12. R. Goyal (Ed.), *Monolithic Microwave Integrated Circuits: Technology and Design*, Artech House, Norwood, MA, 1989.
13. F. Ali, I. Bahl, and A. Gupta (Eds.), *Microwave and Millimeter-Wave Heterostructure Transistors and Their Applications*, Artech House, Norwood, MA, 1989.
14. F. Ali and A. Gupta (Eds.), *HEMTs and HBts: Devices, Fabrication and Circuits*, Artech House, Norwood, MA, 1991.
15. D. Fisher and I. Bahl, *Gallium Arsenide IC Applications Handbook*, Academic Press, San Diego, CA, 1995.
16. R. Goyal (Ed.), *High Frequency Analog Integrated Circuit Design*, John Wiley & Sons, Hoboken, NJ, 1995.
17. W. R. Deal et al., Demonstration of a 270-GHz MMIC amplifier using 35-nm InP HEMT technology, *IEEE Microwave Wireless Components Lett.*, Vol. 17, pp. 391–393, May 2007.
18. K. Chang, I. Bahl, and V. Nair, *RF and Microwave Circuit and Component Design for Wireless Systems*, John Wiley & Sons, Hoboken, NJ, 2002.
19. I. J. Bahl, *Lumped Elements for RF and Microwave Circuits*, Artech House, Norwood, MA, 2003.
20. I. J. Bahl, Solid state circuits, in *Electrical Engineering Handbook*, 2nd edition (R. C. Dorf, Ed.), CRC Press, Boca Raton, FL, 1997.
21. I. D. Robertson and I. J. Bahl, Solid state circuits, in *Electrical Engineering Handbook*, 3rd edition (R. C. Dorf, Ed.), CRC Press, Boca Raton, FL, 2006.
22. I. J. Bahl, Monolithic microwave integrated circuits (MMICs), in *Encyclopedia of RF and Microwave Engineering*, John Wiley & Sons, Hoboken, NJ, 2005.
23. W. L. Pribble et al., Applications of SiC MESFETs and GaN HEMTs in power amplifier design, *IEEE MTT-S Int. Microwave Symp. Dig.*, pp. 1819–1822, 2002.
24. R. S. Pengelly, Improving the linearity and efficiency of RF power amplifiers, *High Frequency Electronics*, pp. 26–34, September 2002.
25. J. M. Golio, *Microwave MESFETs and HEMTs*, Artech House, Norwood, MA, 1991.
26. H. Wang et al., A 155-GHz monolithic InP-based HEMT amplifier, *IEEE MTT-S Int. Microwave Symp. Dig.*, pp. 1275–1278, 1997.
27. K. C. Gupta, R. Garg, I. Bahl, and P. Bhartia, *Microstrip Lines and Slotlines*, 2nd edition, Artech House, Norwood, MA, 1996.
28. M. B. Steer, J. W. Bandler, and C. M. Snowden, Computer-aided design of RF and microwave circuits and systems, *IEEE Trans. Microwave Theory Tech.*, 50th Anniversary Issue, Vol. 50, pp. 996–1005, March 2002.
29. U. L. Rohde et al., MMIC workstations for the 1990s, *Microwave J.*, State of the Art Reference, Vol. 32, pp. 51–77, 1989.
30. R. Anholt, *Electrical and Thermal Characterization of MESFETs, HEMTs and HBTs*, Artech House, Norwood, MA, 1995.
31. Special Issue on Process-Oriented Microwave CAD and Modeling, *IEEE Trans. Microwave Theory Tech.*, Vol. 40, July 1992.
32. Special Issue on Computer-Aided Design of Nonlinear Microwave Circuits, *Int. J. Microwave Millimeter-Wave Computer-Aided Eng.*, Vol. 6, January 1996.
33. Special Issue on Optimization-Oriented Microwave Computer-Aided Design, *Int. J. Microwave Millimeter-Wave Computer-Aided Eng.*, Vol. 7, January 1997.
34. F. Bonani et al., Physics-based large-signal sensitivity analysis of microwave circuits using technological parametric sensitivity from multidimensional semiconductor device model, *IEEE Trans. Microwave Theory Tech.*, Vol. 45, pp. 846–854, May 1997.
35. D. Estreich, Nonlinear modeling for MMICs, *IEEE Microwave Millimeter-Wave Monolithic Circuits Symp. Dig.*, pp. 93–96, 1987.

36. R. J. Trew, MESFET models for microwave CAD applications, *Int. J. Microwave Millimeter-Wave Computer-Aided Eng*., Vol. 1, pp. 143–158, April 1991.
37. J. L. B. Walker (Ed.), *High-Power GaAs FET Amplifiers*, Artech House, Norwood, MA, 1993.
38. E. C. Niehenke, R. A Pucel, and I. J. Bahl, Microwave and millimeter-wave integrated circuits, *IEEE Trans. Microwave Theory Tech*., 50th Anniversary Issue, Vol. 50, pp. 846–857, March 2002.
39. T. Itoh (Ed.), *Numerical Techniques for Microwave and Millimeter-Wave Passive Structures*, John Wiley & Sons, Hoboken, NJ, 1989.
40. R. Sorrentino (Ed.), *Numerical Methods for Passive Microwave and Millimeter-Wave Structures*, John Wiley & Sons, Hoboken, NJ, 1989.
41. Special Issue on Engineering Applications of Electromagnetic Field Solvers, *Int. J. Microwave Millimeter-Wave Computer-Aided Eng*., Vol. 5, September 1995.
42. Special Issue on Automated Circuit Design Using Electromagnetic Simulators, *IEEE Trans. Microwave Theory Tech*., Vol. 45, November 1997.
43. A. Conrad and J. Browne, EM tools enhance simulation accuracy, *Microwaves RF*, Vol. 36, pp. 133–136, November 1997.
44. D. Swanson and W. Hoefer, *Electromagnetic Simulators*, Artech House, Norwood, MA, 2003.
45. P. M. Smith, Status of InP HEMT technology for microwave receiver applications, *IEEE Trans. Microwave Theory Tech*., Vol. 44, pp. 2328–2333, December 1996.
46. I. J. Bahl, 10W CW broadband balanced limiter/LNA fabricated using MSAG MESFET process, *Int. J. RF Microwave Computer-Aided Eng*., Vol. 13, pp. 118–127, March 2003.
47. J. Looney, D. Conway, and I. Bahl, An examination of recovery time of an integrated limiter/LNA, *IEEE Microwave Mag*., Vol. 5, pp. 83–86, March 2004.
48. I. Bahl, Ku-band MMIC power amplifiers developed using MSAG MESFET technology, *Microwave J*., Vol. 49, pp. 56–82, February 2006.
49. I. J. Bahl et al., Low loss multilayer microstrip line for monolithic microwave integrated circuits applications, *Int. J RF Microwave Computer-Aided Eng*., Vol. 8, pp. 441–454, November 1998.
50. I. J. Bahl, Design of a generic 2.5W, 60 percent bandwidth, C-band MMIC amplifier, *Microwave J*., Vol. 45, pp. 54–70, August 2002.
51. E. L. Griffin, X-band GaAs MMIC size reduction and integration, *IEEE MTT-S Int. Microwave Symp. Dig*., pp. 709–712, 2000.
52. I. J. Bahl, 0.7–2.7GHz 12-watt power amplifier MMIC developed using MLP technology, *IEEE Trans. Microwave Theory Tech*., Vol. 55, pp. 222–229, February 2007.
53. I. J. Bahl, 2–8GHz 8-watt power amplifier MMIC developed using MSAG MESFET technology, *IEEE Microwave Wireless Comp. Lett*., Vol. 18, pp. 52–54, January 2008.
54. M. Akkul et al., 50 Watt MMIC power amplifier design for 2GHz applications, *IEEE MTT-S Int. Microwave Symp. Dig*., pp. 1355–1358, 2004.
55. D. Conway, M. Fowler, and J. Redus, New process enables wideband high-power GHz amplifiers to deliver up to 20W, *Defense Electron*., pp. 8–11, February 2006.
56. W. L. Pribble and E. L. Griffin, An ion-implanted 13 watt C-band MMIC with 60% peak power added efficiency, *IEEE Microwave Millimeter-Wave Monolithic Circuits Symp. Dig*., pp. 25–28, 1996.
57. M. K. Siddiqui et al., A high-power and high-efficiency monolithic power amplifier at 28GHz for LMDS applications, *IEEE Trans. Microwave Theory Tech*., Vol. 46, pp. 2226–2232, December 1998.
58. K. Kong et al., Ka-band MMIC power amplifier (4W at 30GHz) with record compact size, *IEEE Compound Semiconductor IC Symp. Dig*., pp. 232–235, 2005.
59. M. V. Aust et al., A 2.8-W Q-band high-efficiency power amplifier, *IEEE J. Solid State Circuits*, Vol. 41, pp. 2241–2247, October 2006.
60. Y. C. Chen et al., A 95-GHz InP HEMT MMIC amplifier with 427-mW power output, *IEEE Microwave Guided Wave Lett*., Vol. 8, pp. 399–401, November 1998.

61. R. J. Baker, H. W. Li, and D. E. Boyce, *CMOS Circuit Design, Layout, and Simulation*, IEEE Press, Piscataway, NJ, 1997.
62. B. Kang, S. Lee, and J. Park, *CMOS Layout Design*, MyCAD, Sunnyvale, CA, 2000.
63. P. R. Gray and R. G. Meyer, *Analysis and Design of Analog Integrated Circuits*, 3rd edition, John Wiley & Sons, Hoboken, NJ, 1993.

PROBLEMS

15.1 Compare qualitatively and quantitatively microstrip, CPW, and lumped elements for monolithic MIC amplifier applications.

15.2 Compare qualitatively hybrid and monolithic MIC technologies.

15.3 A monolithic amplifier is to be designed on InP substrate at 300 GHz. Determine the substrate thickness, maximum microstrip conductor width, and gold conductor thickness.

15.4 Design a single-stage MMIC LNA at 5 GHz using the S-parameters and noise parameters given in Table 5.1. The substrate is 5-mil thick GaAs. Use microstrip in series and open-circuited stub configurations. The design must be unconditionally stable (if required use a source series inductor) and output return loss must be greater than 15 dB. Calculate the NF, gain, and input and output return losses. Compare MMIC LNA performance with hybrid LNA in Chapter 14, Problem 14.4.

15.5 Compare qualitatively the GaAs MMIC and Si IC processes.

15.6 Discuss the pros and cons of hybrid and MMIC amplifiers. What are the three major limitations in the development of MMIC very-high-power amplifiers?

15.7 Describe the advantages of monolithic millimeter wave monolithic integrated circuits over hybird integrated circuits.

15.8 (a) List the basic properties required for an ideal substrate material, conductor material, resistive film, and capacitor dielectric film used in integrated circuits.

(b) Compare the advantages and disadvantages of thin-film versus active-layer resistors.

Chapter 16

Thermal Design

Both low-noise and power transistors have a maximum allowed operating channel/junction temperature, which if exceeded will reduce operating life to less than required. In addition, by operating at lower channel temperature the electrical performance will improve. The channel temperature is a function of both power dissipation and heat removal. Power dissipation, in general, in low noise and small-signal devices is much lower than in power devices. Thus in the design of a power amplifier product, it is very important to effectively take out the excessive heat generated in the device. The thermal design of power amplifiers is performed at two levels: active device and amplifier assembly.

Thermal modeling of semiconductor transistors can be performed by using numerical techniques based ANSYS Multiphysics software or by using simple analytic methods such as the Cooke model [1] to calculate the thermal resistance (R_{th}) for the transistor. For larger FETs/HEMTs (gate periphery greater than 1.5 mm), measurements using IR and liquid crystal techniques have shown a close agreement with the Cooke model predictions. However, for small-size transistors (gate periphery less than 0.6 mm), the Cooke model overestimates the thermal resistance because it does not include the extra heat dissipation due to metal-filled via holes and the heat spreading at the gate finger ends. Thus for small devices, the Cooke model seems to predict overly high FET channel temperature. Based on the FET structure (gate–gate pitch, unit gate width, and FET size), the substrate properties, and the maximum channel temperature, the thermal resistance is calculated. This method is discussed in detail in Section 16.2.

Several methods of thermal monitoring of power transistor/MMIC chips are being used. Among them the most popular are:

- Infrared imaging
- Liquid crystal measurement
- Electrical characterization of the Schottky diode

These methods use different techniques to measure the device active area temperatures. For a complete discussion of the advantages and capabilities of each method, refer to References [2–4]. Salient features of theses methods are discussed in Section 16.6.

Fundamentals of RF and Microwave Transistor Amplifiers. By Inder J. Bahl

16.1 THERMAL BASICS

Spreading or dissipation of heat occurs through conduction, convection, and radiation. Since in solid state devices, removal of heat is predominantly through conduction, devices must be mounted on good heat sinks to conduct the heat away. First consider that the chip/die thickness is small compared to its heat-generating surface dimensions. In this case, the heat flows in a vertical column from the top surface to the bottom surface, as shown in Figure 16.1. The temperature difference between the top and bottom surfaces, ΔT, is approximately given by

$$\Delta T = T_s - T_a = \frac{Ph}{K_{th}A} = \rho\frac{Ph}{A} \tag{16.1}$$

where P is the power dissipated, h is the die thickness, K_{th} is the thermal conductivity of the die/semiconductor material, ρ is the specific thermal resistance ($= 1/K_{th}$), and A is the die area ($A = W \times L$). Here, T_s is the surface or channel or junction temperature and T_a is the ambient or room or heat sink temperature. Thus for smaller values of ΔT, one needs thinner and larger chips with larger values of thermal conductivity. Table 16.1 gives the thermal conductivities of several ceramics and semiconductors. The thermal conductivity of copper is also included for comparison. The thermal conductivity for most semiconductors is temperature dependent. The higher the temperature, the lower will be its value.

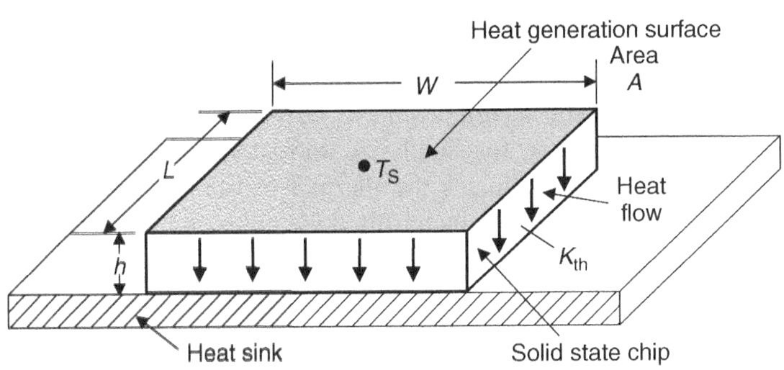

Figure 16.1 Die with vertical heat flow configuration.

Table 16.1 Typical Values for Thermal Conductivity of Ceramics and Semiconductors

Material	K_{th}(W/m·°C)
Alumina	37
AlN	230
BeO	250
GaAs	46
GaN	130
Si	145
SiC	350
Diamond	689
Copper	393

Equation (16.1) may also be expressed as

$$\Delta T = P R_{\text{th}} \ ^\circ\text{C} \tag{16.2}$$

where

$$R_{\text{th}} = \frac{h}{K_{\text{th}} W L} \ ^\circ\text{C/W} \tag{16.3}$$

and is defined as the thermal resistance analogous to the electrical resistance. When ΔT is expressed in °C, the units of P, h, A, and K_{th} are W, m, m^2, and W/m·°C, respectively, and R_{th} is expressed in °C/W. Units of h, A, and K_{th} may also be used as cm, cm^2, and W/cm·°C, respectively.

When the width of the heat-generating area of the die is comparable to the thickness, the heat spreading occurs laterally as well as vertically, as shown in Figure 16.2. As a first-order approximation, the heat flow from the surface may be considered to follow the rule of 45° thermal spreading angle. This means that the heat generated in the surface (assuming no other heat sources and heat flow is mainly due to conduction) flows down through the dielectric materials through expanding areas larger than the heat-generating surface as it approaches the ground plane, where the ground plane acts as a heat sink. In this case, heat spreading takes place on all four sides and (16.3) becomes

$$R_{\text{th}} = \frac{h}{K_{\text{th}}(W + 2h)(L + 2h)} \tag{16.4}$$

EXAMPLE 16.1

A GaAs chip has a thickness of 75 μm and the transistor area is 100 × 100 μm^2. When the thermal conductivity is 46 W/m·°C, determine the thermal resistance for chip areas of (a) 100 × 100 μm^2 and (b) 300 × 300 μm^2.

SOLUTION (a) In this case,

$$W = 100 \ \mu\text{m}$$
$$L = 100 \ \mu\text{m}$$
$$R_{\text{th}} = \frac{75 \times 10^{-6}}{46 \times 100 \times 10^{-6} \times 100 \times 10^{-6}} = 163 \ ^\circ\text{C/W}$$

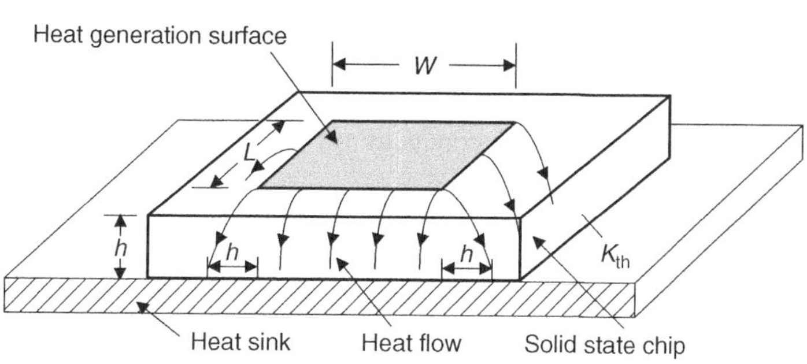

Figure 16.2 Heat flow fields from a heat source.

(b) Here,

$$W + 2h = 100 + 150 = 250\ \mu\text{m}$$
$$L + 2h = 100 + 150 = 250\ \mu\text{m}$$
$$R_{\text{th}} = \frac{75 \times 10^{-6}}{46 \times 250 \times 10^{-6} \times 250 \times 10^{-6}} = 26.1\ ^\circ\text{C/W}$$

Thus, due to higher heat spreading, the transistor with larger chip area has much lower thermal resistance. ■

16.2 TRANSISTOR THERMAL DESIGN

Power transistor junction/channel temperature is an important parameter in the device design since it not only affects its performance but high channel temperatures degrade the reliability. Thus it is critical to keep the device's thermal resistance as small as possible.

Accurate knowledge of transistor operation as a function of channel temperature is required to determine the performance of the device. As DC power is dissipated in the device conducting channel, a temperature rise occurs. The increased temperature causes a degradation of the electron mobility and saturation velocity due to increased lattice scattering. Increased parasitic resistances and reduced channel current result. Also, the gate–drain leakage current is sensitive to temperature and increases with increasing temperature. These effects, when coupled together, result in significant degradation of the RF performance of the transistor as temperature increases.

The thermal design of power transistors is as important as the electrical design. The power devices must be carefully designed for minimum thermal resistance to optimize the performance as well as to increase its reliability. The parameters that most significantly affect the thermal resistance of the FET/HEMT are the substrate thickness, gate-to-gate separation, and effective gate length acting as a heat source.

16.2.1 Cooke Model

Thermal modeling of semiconductor devices can be performed by using a simple analytic method such as the Cooke model. The Cooke model uses the analogy between the heat flow and the electric field. In an FET the heat source is primarily located in the channel under the gate, where most of the drain–source voltage drop occurs. The temperature rise of the channel is calculated from the thermal resistance (inverse of thermal conductance), which is obtained by calculating the electric capacitance because both these quantities are mathematically analogous to each other [5, 6]. Table 16.2 summarizes the analogy between the heat flow and electric field quantities [7, 8]. For a structure where C is the electric capacitance, G_{th} is the thermal conductance, ε is the permittivity, and K_{th} is the thermal conductivity, we have

$$\frac{C}{G_{\text{th}}} = \frac{\varepsilon}{K_{\text{th}}} \tag{16.5}$$

In this case the thermal resistance, $R_{\text{th}} = 1/G_{\text{th}}$, is calculated from

$$R_{\text{th}} = \frac{\varepsilon}{C K_{\text{th}}} \tag{16.6}$$

Table 16.2 Analogy Between Heat Flow and Electric Field

Heat Flow Field	Electric Field
1. Temperature, T (°C)	Potential, V(V)
2. Temperature gradient, T_g (°C/m)	Electric Field, E (V/m)
3. Heat flow rate, Q (W)	Flux, ϕ (coulomb)
4. Density of heat flow, q (W/m^2)	Flux density, D (coulomb/m^2)
5. Thermal conductivity, K (W/m·°C)	Permittivity, ε (coulomb/m/V)
6. Density of heat generated, ρ_h (W/m^3)	Charge density, ρ (coulomb/m^3)
7. $q = -K\nabla T$	$D = -\varepsilon \nabla V$
8. $\nabla \cdot \mathbf{q} = \rho_h$	$\nabla \cdot \mathbf{D} = \rho$
9. $\nabla^2 T = 0$	$\nabla^2 V = 0$

Next, single- and multiple-gate FET configurations are considered. When the current starts to flow through the devices, the heat is generated due to power dissipation under the gate in the FET/HEMT and under the emitter in a BJT. The active areas become heat sources and the back of the chip (i.e., ground plane and package) becomes the heat sink. Thus the maximum and minimum temperatures occur at the top and bottom surfaces, respectively, of the substrate. The heat transfer takes place due to the temperature gradient between the heat source and heat sink. Larger heat dissipation areas and thinner substrate thicknesses lead to a lower rise in channel temperatures.

In a single heat source, the heat spreading occurs vertically down as well a sideways at a 45° angle, as shown in Figure 16.3. However, in multiple heat sources on a surface, sideways heat spreading depends on the separation between sources, Figure 16.4. Separation between the gates (or gate–gate pitch) that is smaller than the substrate thickness gives rise to higher channel temperatures due to mutual heating between the sources.

16.2.2 Single-Gate Thermal Model

A single-finger FET thermal model is based on Cohn's calculations [9] of the microstrip line capacitance. The thermal resistance, R_{th}, is given as

$$R_{th} = \frac{1}{2K_{th}W_g}\frac{K(k)}{K(k')} \tag{16.7}$$

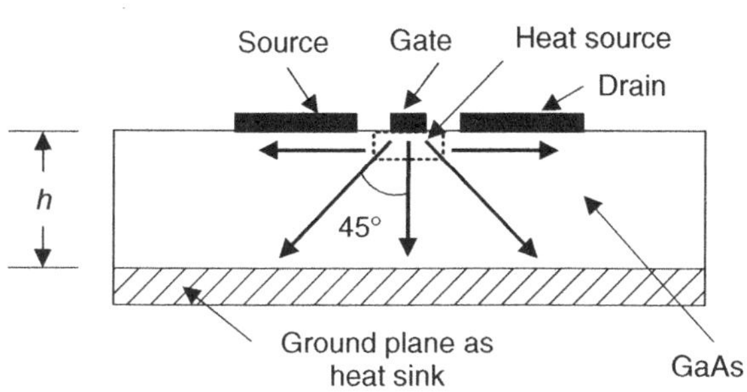

Figure 16.3 Heat flow fields in a single-gate FET/HEMT.

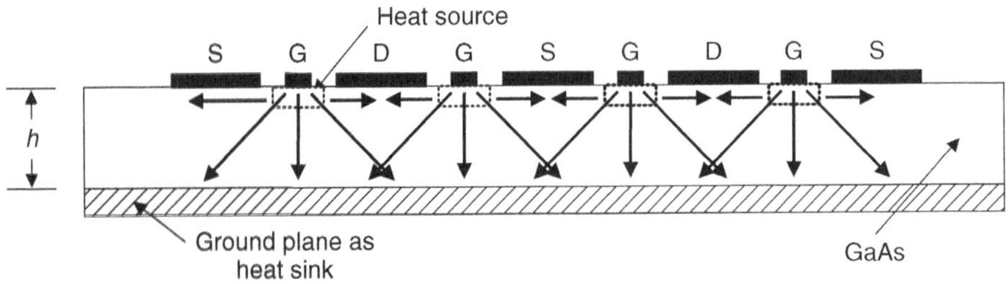

Figure 16.4 Heat flow fields in a multigate FET/HEMT. For close heat sources (i.e., gates), heat flow will be impaired as fluxes combine before reaching the heat sink.

where W_g is the unit gate width, K_{th} is the thermal conductivity of the substrate material, and K is the complete elliptical integral of the first kind:

$$k = \operatorname{sech}(\pi L'/4h)$$
$$k' = \tanh(\pi L'/4h), \quad L' = \text{effective gate length}, h = \text{die thickness}$$

A simplified expression for $K(k)/K(k')$ is given by (6.7) in Chapter 6.

16.2.3 Multiple-Gate Thermal Model

A multigate thermal model can be derived by treating heat flow in a multigate FET, as shown in Figure 16.5, as analogous to the capacitances of multiple coupled transmission lines [10]. The expression for the thermal resistance is given by Cooke [1]:

$$R_{th} = \frac{n/(ZK_{th})}{\pi\left[\frac{2(n-1)}{\ell n\ (M)} - \frac{(n-2)}{\ell n\ (P)}\right]}\ (^{\circ}\mathrm{C/W}) \tag{16.8}$$

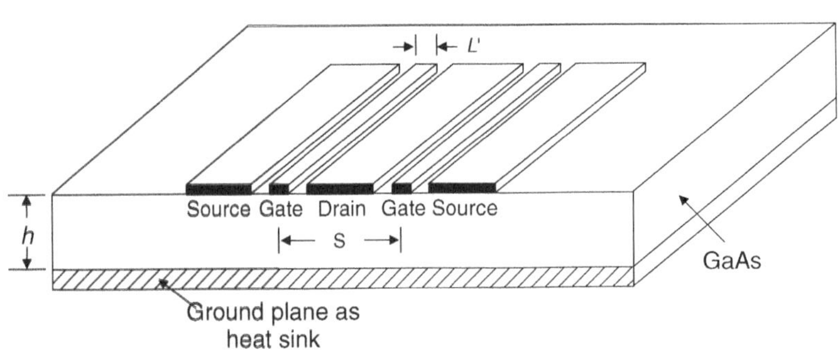

Figure 16.5 Multifinger FET analogous to multiconductor microstrip.

$$P = 2\sqrt{\frac{1 + \operatorname{sech}\left(\frac{\pi L'}{4h}\right)}{1 - \operatorname{sech}\left(\frac{\pi L'}{4h}\right)}} \tag{16.9a}$$

$$M = \frac{2\left[\cosh \pi \left(\frac{S+L'}{4h}\right) / \cosh \pi \left(\frac{S-L'}{4h}\right)\right]^{1/2} + 1}{\left[\cosh \pi \left(\frac{S+L'}{4h}\right) / \cosh \pi \left(\frac{S-L'}{4h}\right)\right]^{1/2} - 1} \tag{16.9b}$$

where the thermal conductivity (K_{th}) of GaAs and SiC as a function of temperature are given by

$$K_{\text{th}}(T) = 1.08T^{-0.26}(\text{W/cm} \cdot {}^\circ\text{C}) \quad \text{GaAs} \tag{16.10a}$$

$$K_{\text{th}}(T) = 3.47\left[\frac{300}{273 + T}\right]^{1.5} (\text{W/cm} \cdot {}^\circ\text{C}) \quad \text{SiC} \tag{16.10b}$$

The parameters are defined as follows:

L' = Effective gate length in microns
S = Gate-to-gate spacing or pitch in microns
h = GaAs thickness in microns
Z = Total gate periphery in cm
n = Number of gate fingers
T = Maximum channel temperature in °C

Thermal resistance of a 2.5-mm FET with a maximum operating channel temperature of 150 °C was calculated using the above equations as a function of gate length for three different gate–gate pitches. The other parameters are $n = 10$, $h = 75$ μm, unit gate width $Z/n = W_g = 250$ μm, and GaAs substrate. Variation of thermal resistance as a function of gate length is shown in Figure 16.6. As expected, the thermal resistance increases as gate length is reduced. A 0.25-μm gate length and 20 μm gate–gate pitch device has a thermal resistance of 43.5 °C/W. Increasing the gate length to 2 μm results in a reduction in the thermal resistance to about 35.3 °C/W; increasing the number of gate fingers increases the thermal resistance, as shown in Figure 16.7. In this case the gate length is 1 μm, gate–gate pitch is 20 μm, and the gate width is held constant at 2.5 mm. The unit gate width is reduced as the number of gate fingers is increased to maintain the 2.5 mm. Increasing the number of fingers from 1 to 16 results in an increase in thermal resistance from 25.8 °C/W to about 38.7 °C/W, which is about 50% higher. This means that the R_{th} value for a group of closely spaced heat sources is about 50% higher than the isolated heat source.

Consider an FET having 100-μm unit gate width and 24-μm gate–gate spacing. According to Cooke's model, the incremental increase in R_{th} value per unit width of an FET from fingers 1 to 2, 2 to 4, 4 to 18, and 18 to 144 fingers is 19%, 14%, 9%, and 2%, respectively. This means for a large size transistor with many fingers, the maximum increase in the R_{th} value occurs due to the first 10–12 fingers. Thus using

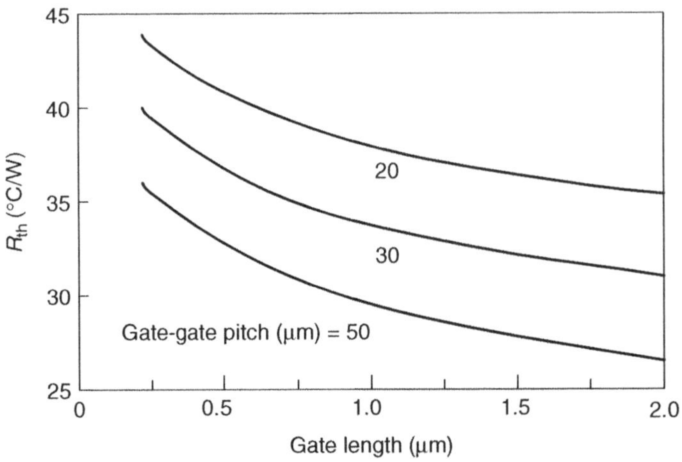

Figure 16.6 Thermal resistance versus gate length for different gate–gate pitch values for a GaAs 2.5-mm MESFET. Number of fingers is 10 and the maximum channel temperature is 150 °C.

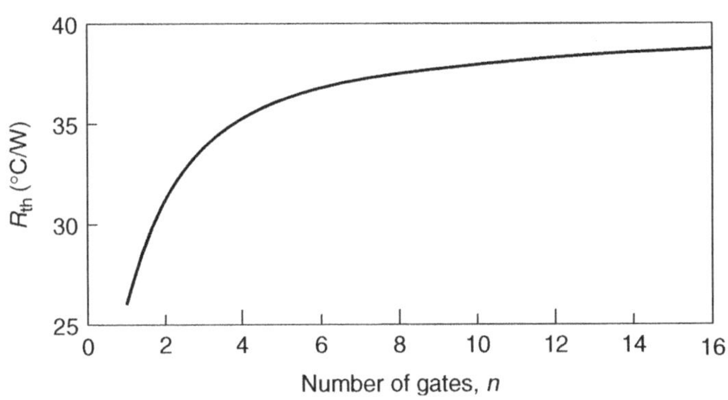

Figure 16.7 Thermal resistance versus number of gates for a 2.5-mm GaAs MESFET. The gate length is 1 μm and the maximum channel temperature is 150 °C.

small FET size and adequate physical separation between the FETs helps in lowering the R_{th} value.

The difference in the temperature, ΔT, from the bottom surface (carrier) to the top surface (channel) of the device is calculated using $\Delta T = R_{th} \times P_D$, where P_D is the net or total power dissipated in the device. For these calculations, the GaAs thermal conductivity is chosen based on the maximum allowed junction temperature of 150 °C. The next step is to calculate the net power dissipation in the device under RF drive. The net power dissipated is calculated from the power delivered to the device and the power delivered out of the device.

If T_C is the case temperature, in this case the back of the chip, the channel or junction temperature T_{ch} under RF conditions is given by

$$T_{ch} = P_D R_{th} + T_C$$

or

$$\Delta T = T_{ch} - T_C = P_D R_{th} \quad (16.11)$$

$$T_{ch} = [P_{DC} + P_{in}(RF) - P_o(RF)]R_{th} + T_C \quad (16.12)$$
$$= [P_{DC}(1 - PAE)]R_{th} + T_C$$

where $P_D = P_{DC} + P_{in} - P_o$; P_{DC} is the DC power, P_{in}(RF) is the input signal power level, and P_o(RF) is the output power delivered to the load.

EXAMPLE 16.2

The parameters of a single-stage GaAs HPA are $P_o = 2$ W, $G = 10$ dB, PAE = 50%, and $R_{th} = 35\,^\circ$C/W. Determine the channel temperature of the chip when the case temperature is 50 °C.

SOLUTION Using PAE $= P_o(1 - 1/G)/P_{DC}$, $G = 10\text{dB} = 10$, PAE = 50% = 0.5, and $P_{in} =$ 0.2 W, we find

$$P_{DC} = P_o(1 - 1/G)/\text{PAE} = 2 \times (1 - 1/10)/0.5 = 3.6 \text{ W}$$

From (16.12), we have

$$T_{ch} = [3.6 + 0.2 - 2] \times 35 + 50 = 113\,^\circ\text{C}$$

■

16.3 AMPLIFIER THERMAL DESIGN

When a power transistor/MMIC power amplifier is bonded into a package or on a shim (pallet) or a carrier, several thermal resistance barriers may exist between the die and the final assembly used for testing or for the subsystem. These barriers may be a thin layer of die attachment solder or silver (Ag) filled epoxy, package base plate or carrier or shim, air between the package or carrier and fixture, and the fixture's base and heat sink. All these thermal barriers must be taken into account in the theoretical analysis of the device's junction temperature. In some cases, the ΔT rise due to solder/package/fixture/heat sink may be higher than the ΔT due to the die itself.

A power amplifier assembly using a heat sink is shown in Figure 16.8. The power dissipated by an FET or MMIC/HMIC amplifier is related to the device junction through air or ambient temperature by the thermal resistances, as shown in Figure 16.9, according to the following equation:

$$P_D = \frac{T_{ch} - T_a}{R_{JC} + R_{CS} + R_{SA}} \quad (16.13)$$

where P_D is the power dissipated, T_{ch} is the junction temperature, T_a is the ambient temperature (usually 25 °C), and R_{JC}, R_{CS}, and R_{SA} are the thermal resistances between the device junction–case, case–heat sink, and heat sink–ambient, respectively. Power is measured in watts, temperature in °C, and resistance in °C/W. R_{JC} includes the GaAs chip, die attach solder material, and case or carrier; R_{CS} includes the package or fixture and silicon grease. Since the device carrier must make good thermal contact

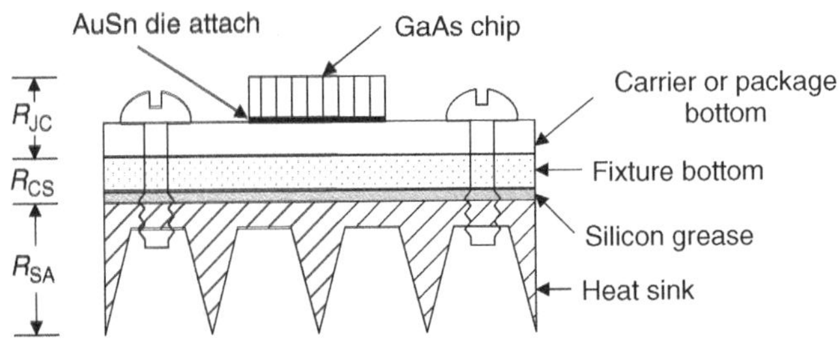

Figure 16.8 An MMIC power amplifier assembly using a heat sink.

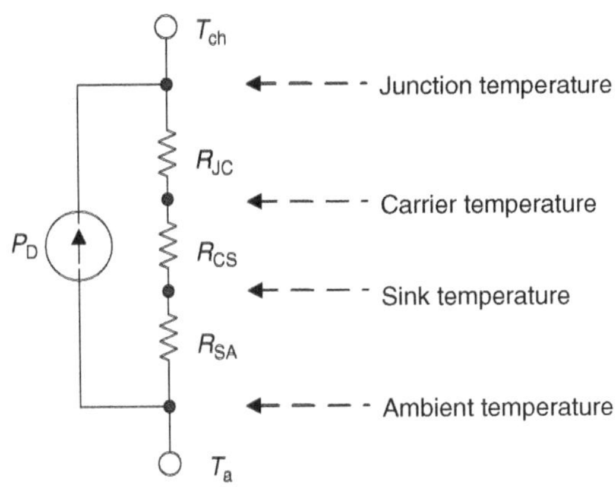

Figure 16.9 Thermal equivalent model of a power amplifier assembly.

with the heat sink, silicon grease is generally used between the package or carrier and the heat sink to fill up the gaps, which reduces R_{CS} by a factor of 2. From (16.13), the junction temperature is given by

$$T_{ch} = (R_{JC} + R_{CS} + R_{SA})P_D + T_a \tag{16.14}$$

For given T_{ch}, R_{JC}, R_{CS}, P_D, and T_a, the smaller value of R_{SA} becomes very important in order to lower T_{ch}. Thus the design of the heat sink is critical in packaging power amplifiers. In high voltage HPAs, R_{SA} is the main limiting factor in achieving very high power levels.

The power amplifier chips are generally soldered to carriers using 80–20 gold–tin preforms, which are cut from sheets. The carrier material may be copper or CuW or CuMo. Copper has the best thermal conductivity, whereas CuW and CuMo have better thermal expansion coefficient match with GaAs and alumina. Generally, materials having better matched thermal expansion coefficients have lower thermal conductivity, but higher material, machining, and gold plating costs. All carriers or packages are plated with nickel and gold. In the case of copper carriers, one is limited to smaller chip sizes. Also, copper carriers need thicker nickel so that there are no cracks due to mismatch in the thermal expansion coefficients; and gold does not leach into copper at solder temperatures.

The thermal conductivity of commonly used gold–tin solder is 57 W/m·°C, whereas the thermal conductivity of silver epoxy materials Kidd AG-03HTL, Std Ablebond 84-1LMISR4 (~Ablebond 8360), and Ablebond RP-316-1 are 300, 2.8, and 10 W/m·°C, respectively. Among these materials, Kidd AG-03HTL appears to be the best for packaging active devices or MMIC chips.

EXAMPLE 16.3

Consider a two-stage driver amplifier using a 0.3-mm FET in the input and a 1.25-mm FET at the output. The substrate is 75-μm thick GaAs and the maximum channel temperature assumed is 150 °C. The devices are biased at a Q-point current of 72 mA/mm and 8.5 V. Calculate the maximum allowed case temperature under each FET.

SOLUTION The thermal resistance for each device is calculated using (16.8)–(16.10). The calculated R_{th} values are 231 °C/W and 58.8 °C/W for the 0.3-mm and 1.25-mm FETs, respectively. At the back of the chip, the maximum allowed case temperature under the 0.3-mm FET is $150 - 231 \times 0.187 = 106.8\,^{\circ}$C and under the 1.25-mm FET it is $150 - 58.8 \times 0.765 = 105\,^{\circ}$C. The thermal calculations for the two-stage amplifier are summarized in Figure 16.10. ■

EXAMPLE 16.4

Next, consider a three-stage 12-W power amplifier using two 0.94-mm FETs in the input, four 1.6675-mm FETs in the interstage, and eight 2.625-mm FETs at the output. The FETs are isolated from each other. The substrate is 75-μm thick GaAs and the maximum channel temperature assumed is 150 °C. Calculate the temperature rise for each FET under the applied RF. Also, calculate the temperature rise for each FET when the PAE is reduced by 20%.

SOLUTION The first step is to calculate the thermal resistance (R_{th}) for each FET using the Cooke model. Based on the FET structure and substrate properties (FET size, gate–gate pitch, number of fingers or unit gate width, and substrate thickness), the value of R_{th} is calculated. These are given below.

First Stage

FET size	= 0.94 mm
Gate–gate pitch	= 30 μm
Number of fingers	= 10
R_{th}	= 89.94 °C/W

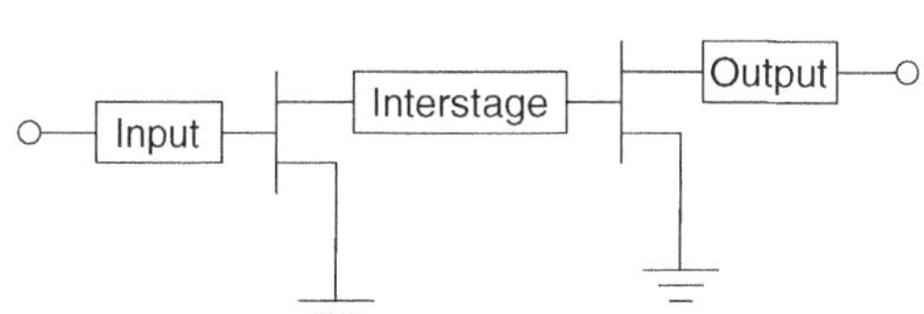

FET size (mm)	0.3	1.25
Gate–gate pitch (μm)	50	50
Number of fingers	2	8
V_{ds} (V) @ Q-point	8.5	8.5
I_{ds} (mA) @ Q-point	22	90
P_{DC} (mW) @ Q-point	187	765
R_{th}(°C/W)	231	58.8
ΔT (°C)	43.2	45

Figure 16.10 Thermal analysis of a two-stage driver amplifier. Substrate is 75-μm thick GaAs and maximum channel temperature is 150 °C.

Second Stage

FET size	= 1.6675 mm
Gate–gate pitch	= 30 μm
Number of fingers	= 14
R_{th}	= 51.08 °C/W

Third Stage

FET size	= 2.625 mm
Gate–gate pitch	= 20 μm
Number of fingers	= 24
R_{th}	= 37.25 °C/W

The difference in the temperature from the bottom surface (carrier) to the top surface (channel) of the MMIC power amplifier chip, ΔT, is calculated using (16.11). For these calculations, the GaAs thermal conductivity is chosen to be at 150 °C. Using (16.10a), $K_{th} = 0.2935$ W/cm·°C at 150 °C.

The next step is to calculate the channel temperature under RF conditions. In this case one needs the net power dissipated in the devices. Based on the design of the amplifier, using a CAD, the dissipative and mismatch losses for each matching network, the power delivered to each FET and the power delivered out of each FET were calculated and are shown in Figure 16.11. Over the desired frequency range, these calculations are done at each frequency and averaged. After calculating the net dissipated power, the ΔT value for each stage is calculated. For the reduced value of PAE, the net dissipated power is increased by the same ratio (i.e., 20%), and the ΔT values are recalculated and are shown in Figure 16.11. ■

16.4 PULSED OPERATION

Solid state pulsed power amplifiers are used in phased array radars, mobile communication, distance measuring equipment, portable-tactical-approaching guidance equipment,

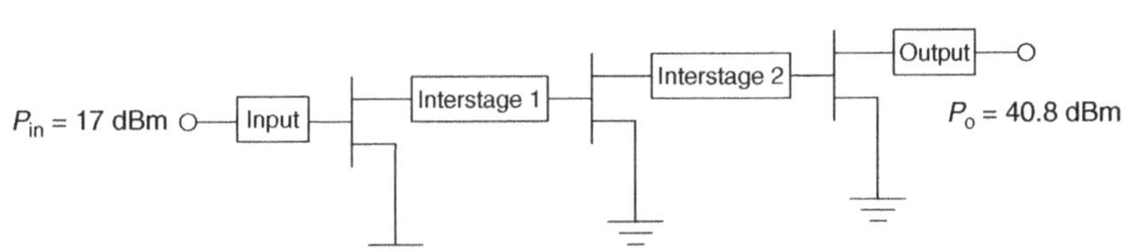

Total FET size (mm)		1.88		6.67		21.0	
Total DC power (W)		1.632		7.6		23.7	
Matching network loss (dB)	1.6		1.8		1.4		0.9
Power delivered to FETs (W)		0.035		0.25		2.5	
Power delivered out of FETs (W)		0.38		3.5		14.8	
Net power dissipated (W)		1.287		4.35		12.4	
R_{th}(°C/W)		44.97		12.77		4.66	
ΔT (°C)		57.9		55.5		57.8	
ΔT (°C) @ 20% reduced PAE		69.4		66.6		69.4	

Figure 16.11 Thermal analysis of a three-stage power amplifier. Average power level and ΔT calculations are given at each stage of the 12-W HPA. Substrate is 75-μm thick GaAs and maximum channel temperature is 150 °C.

and pulsed-width modulators. The design of pulsed power amplifiers is very similar to CW power amplifiers except that the thermal requirements in CW operation are more stringent than in pulsed operation. Under pulsed operation, the electrical performance and reliability of power amplifiers are enhanced.

The operation of solid state devices under pulsed conditions does not improve the output power significantly because the P_o is limited by BV_{gd} and I_p. However, depending on the pulse width or length and the thermal time constant of the devices, the case temperature may be raised during pulsed operation without exceeding the device's maximum junction or channel temperature, thus enhancing the reliability of the devices. For shorter pulses (≤ 30 μs), the GaAs MMIC HPAs have about 0.5–1 dB higher P_o and 3–5% better PAE as compared to a CW.

When a power transistor is operating under pulsed conditions, the channel temperature rise after applying DC power is approximately given by [3]

$$T_{ch} = P_D R_{th} \left[1 - \frac{8}{\pi^2} \sum_{n=1,3,\ldots}^{\infty} (e^{-n^2 t/\tau}/n^2) \right] + T_a \quad (16.15)$$

where τ is the thermal time constant, P_D is the net power dissipated in the device, R_{th} is the thermal resistance, and t is the time after DC power is applied to the device. For $t < \tau$, an approximate expression for T_{ch} may be written [3, 11]:

$$T_{ch} = P_D R_{th} \left(\frac{4}{\pi^{3/2}} \right) \left(\frac{t}{\tau} \right)^{1/2} + T_a \quad (16.16a)$$

or

$$T_{ch} \cong P_D R_{th} \left(\frac{t}{2\tau} \right)^{1/2} + T_a \quad (16.16b)$$

An approximate expression for the thermal time constant is given by [11]

$$\tau = \left(\frac{2h}{\pi} \right)^2 \left(\frac{\rho C}{K_{th}} \right) \quad (16.17)$$

where

h = die thickness
ρ = density of semiconductor
K_{th} = thermal conductivity
C = specific heat of semiconductor

For gallium arsenide (GaAs),

$$\rho = 5.31 \text{g/cm}^3$$
$$C = 0.35 \text{J/g} \cdot {}^\circ\text{C} = 0.35 \text{W} \cdot \text{s/g} \cdot {}^\circ\text{C}$$
$$K_{th} = 0.46 \text{W/cm} \cdot {}^\circ\text{C}$$

Then

$$\frac{\rho C}{K_{\text{th}}} = \frac{5.31 \times 0.35}{0.46} = 4.04\ \text{s/cm}^2$$

For silicon (Si),

$$\rho = 2.33\text{g/cm}^3$$
$$C = 0.7\text{J/g} \cdot {}^\circ\text{C} = 0.7\text{W} \cdot \text{s/g} \cdot {}^\circ\text{C}$$
$$K_{\text{th}} = 1.45\text{W/cm} \cdot {}^\circ\text{C}$$

Then

$$\frac{\rho C}{K_{\text{th}}} = \frac{2.33 \times 0.7}{1.45} = 1.13\ \text{s/cm}^2$$

For silicon carbide (SiC),

$$\rho = 3.1\text{g/cm}^3$$
$$C = 0.75\text{J/g} \cdot {}^\circ\text{C} = 0.75\text{W} \cdot \text{s/g} \cdot {}^\circ\text{C}$$
$$K_{\text{th}} = 3.5\text{W/cm} \cdot {}^\circ\text{C}$$

Then

$$\frac{\rho C}{K_{\text{th}}} = \frac{3.1 \times .75}{3.5} = 0.66\ \text{s/cm}^2$$

EXAMPLE 16.5

Consider a 12-W MMIC HPA on 75-μm thick GaAs. Under RF conditions, $G = 10$ dB, PAE = 45%, $P_o = 12$ W, and $I_{ds} = 2.4$ A at 10 V. The thermal resistance of the chip $R_{th} = 4\,^\circ$C/W. Assuming a 25 °C temperature at the back of the chip, calculate the temperature rise of the channel for (a) 10 μs, (b) 50 μs, and (c) CW operation.

SOLUTION Here,

$$\tau = \left(\frac{2h}{\pi}\right)^{24} .04\,\text{s} = \left(\frac{2 \times 0.0075}{\pi}\right)^{24} .04\ \text{s} = 92.1\ \mu\text{s}$$

$G = 10\text{dB} = 10$, $P_{\text{in}} = P_o/G = 12/10 = 1.2\text{W}$.

Net power dissipation, P_D, in the chip is given by

$$P_D = P_{DC} + P_{\text{in}}(\text{RF}) - P_o(\text{RF}) = 2.4 \times 10 + 1.2 - 12\ W = 13.2\ \text{W}$$

(a) For 10-μs pulse width,

$$T_{ch} = P_D\ R_{th}\left(\frac{4}{\pi^{3/2}}\right)\left(\frac{t}{\tau}\right)^{1/2} + T_a$$

$$T_{ch} = 13.2 \times 4\left(\frac{4}{\pi^{3/2}}\right)\left(\frac{10}{92.1}\right)^{1/2} + 25\,^\circ\text{C} = 37.5\,^\circ\text{C}$$

(b) For 50-μs pulse width,

$$T_{ch} = 13.2 \times 4 \left(\frac{4}{\pi^{3/2}}\right)\left(\frac{50}{92.1}\right)^{1/2} + 25\,^\circ\mathrm{C} = 52.95\,^\circ\mathrm{C}$$

(c) For CW operation, using (16.11), we have

$$T_{ch} = 13.2 \times 4 + 25\,^\circ\mathrm{C} = 77.8\,^\circ\mathrm{C}$$

Thus the channel temperature is decreased by 40.3 °C from CW to 10-μs pulsed operation; correspondingly, using (3.68) the power is increased by about 0.4 dB. ■

For long pulses or pulse lengths greater than 2τ, the channel temperature will correspond to CW operation. For $T_{ch} - T_a = \Delta T$ to be small, $t < \tau$, that is, pulse width must be smaller than τ. When a device is subjected to pulsed operation, the channel temperature after t_o seconds of pulsed power is given by

$$T_{ch} = P_D\, R_{th} \left(\frac{t - t_o}{2\tau}\right)^{1/2} + T_a \tag{16.18}$$

If $t - t_o$ corresponds to the pulse width (PW), the maximum channel temperature is

$$T_{max} = P_D\, R_{th} \left(\frac{PW}{2\tau}\right)^{1/2} + T_a \tag{16.19}$$

and occurs while the pulse is on. When the pulse goes off, the die starts to cool down. After tp seconds the device channel/junction temperature becomes

$$T_{ch}(t) = T_{max}\left[1 - \left(\frac{PW - tp}{2\tau}\right)^{1/2}\right] + T_a \tag{16.20}$$

or when $tp = PW$, the channel temperature becomes T_a. Figure 16.12a shows the device channel temperature as a function of time. Here, PRR is the pulse repetition rate or the duty cycle. In order to maintain a low channel temperature, PW $\ll \tau$ and PRR<50%.

As an example, a packaged 12-W HPA was power tested under various pulsed conditions and its power response is shown in Figure 16.12b. The duty cycle was 10%. When PW = 10 μs, no droop in power with time was measured; however, for PW = 100 μs, the power was lowered by 0.3 dB over the pulsed width.

16.5 HEAT SINK DESIGN

A heat sink [12, 13] is a good conductor structure, generally coated with black paint (for good heat radiation) and connected to the base of the semiconductor device. It provides a path for heat generated in the device to be drained out by the radiation mechanism. Heat sinks with very low thermal resistance are designed with large surface areas, which usually are realized by using many fins. Additionally, their efficiency can be improved using cooling fans.

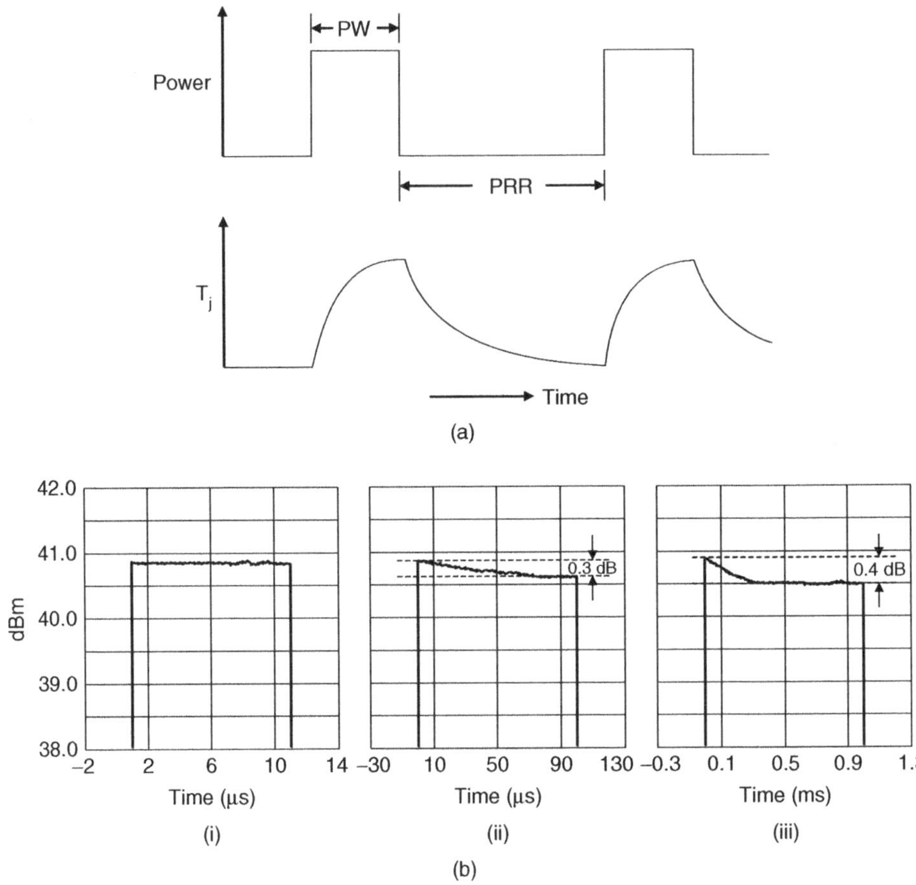

Figure 16.12 (a) Transistor's channel temperature versus time under pulsed operation. (b) Effect of pulsed width (PW) on the power performance of a 12-W HPA: (i) PW = 10 μs, (ii) PW = 100 μs, and (iii) PW = 1.0 ms.

The main function of a heat sink is to provide very low thermal resistance from the case to the air and maintain the junction temperature of the device below the maximum operating temperature specified by the semiconductor manufacturer. The thermal resistance (R_{SA}) plays a significant role in the selection of the proper heat sink. The smaller the value for R_{SA}, the more power the device can handle without exceeding its maximum junction temperature. The thermal resistance from the heat sink to the air consists of three parts: conduction, convection, and radiation, as given [13] by

$$R_{SA} = \underbrace{\rho \frac{d}{A_1}}_{\text{Conduction}} + \frac{1}{\underbrace{h_c A_2}_{\text{Convection}} + \underbrace{h_r A_3}_{\text{Radiation}}} \tag{16.21}$$

where ρ is the specific thermal resistance of the heat sink material, d is the average thickness of the surface having area A_1, h_c and h_r are the convection and radiation heat transfer coefficients, respectively, and A_2 and A_3 are the heat sink areas in contact with the moving ambient fluid and the effective area available for radiation heat transfer. The properties of various conductor materials are given in Table 16.3. For aluminum having $d = 0.005$ m and area $A_1 = 0.05 \times 0.05\ m^2$, the value of R_{SA} due to conduction is only about 0.0091 °C/W. This can be further reduced by minimizing d and maximizing A_1. The heat transfer coefficient, h_c, for both natural and forced convection is a complex quality and cannot easily be determined accurately. In forced convection, h_c is significantly larger (3–4 times) than the value in natural convection. In order to reduce R_{SA} due to convection, usually it is easier to increase A_2 instead of increasing h_c. This is why heat sinks are big structures. For a given temperature difference, the heat transfer coefficient due to radiation, h_r, is a function of the emissivity of the heat sink surface (perfect black body has emissivity of 1.00). An emissivity value for a clean, smooth aluminum surface is about 0.1, while a black anodized or painted surface has a value of about 0.9; thus it is obvious that a heat sink's performance for natural radiation can be significantly improved (10–15%) by just blackening the surface. For forced convection, the amount of heat being dissipated by radiation is so small (3%) that the added cost of blackening is not justified [13]. It should be noted again that A_3 is not the same as A_2, though in both cases, the term heat sink surface area is used. In this case, A_3 represents the effective radiation surface area, and with normal extruded heat sink designs (i.e., fins that are four times higher than the spacing between them), it is only about 20% of A_2. Generally, an increase in surface area A_2 will result in the same proportional improvement of A_3 [13].

Table 16.4 gives typical parameters for EG & G Wakefield Engineering heat sinks made from aluminum. The heat-dissipating surface area (HDS) in this table is the total area in contact with air, assuming 1-inch width of the heat sink and a surface with black anodized finish. The flatness of the heat sink mounting surface must be 0.01 mm with a surface roughness less than 0.5 μm, which is usually obtained by grinding or lapping. In order to achieve minimum possible thermal resistance between the carrier and the heat sink, the carrier must be tightly bolted to it. After each test, grind the heat sink before reusing it.

Table 16.3 Properties of Various Conductor Materials

Metal	Melting Point (°C)	Electrical Resistivity (10^{-6} Ω·cm)	Thermal Expansion Coefficient (10^{-6}/°C)	Thermal Conductivity, K_{th} (W/m·°C)
Copper	1093	1.7	17.0	393
Silver	960	1.6	19.7	418
Gold	1063	2.2	14.2	297
Tungsten	3415	5.5	4.5	200
Molybdenum	2625	5.2	5.0	146
Platinum	1774	10.6	9.0	71
Palladium	1552	10.8	11.0	70
Nickel	1455	6.8	13.3	92
Chromium	1900	20.0	6.3	66
Kovar	1450	50.0	5.3	17
Aluminum	660	4.3	23.0	220
Au–20% Sn	280	16.0	15.9	57
Pb–5% Sn	310	19.0	29.0	63
Cu–W(20% Cu)	1083	2.5	7.0	248
Cu–Mo(20% Cu)	1083	2.4	7.2	197

Table 16.4 Typical Parameters for EG&G Wakefield Engineering Flatback Extruded Heat Sinks

Model Number	HDS (in.^{2}/in.)	Weight (lb/ft)	Extrusion Length (cm)	Height (in.)	R_{SA} (°C/W)
4242	5.16	0.25	1.106	0.375	14.0
5947	13.0	0.57	2.025	1.180	5.7
5596	30.93	2.20	3.136	2.25	2.4
5079	39.84	2.66	3.766	1.976	1.5
5271	60.70	4.34	7.03	1.312	1.3
5584	63.5	7.17	11.4	1.5	1.1
6274	101.0	9.15	6.961	2.79	0.8
4559	87.08	16.81	10.71	3.74	0.5

16.5.1 Convectional and Forced Cooling

Basically, there are three methods of cooling: natural convection, forced convection, and liquid cooling [14]. The data for heat sinks given in Table 16.4 is valid for natural convection. Given volume and heat dissipation requirements, one can select heat sinks from manufacturer catalogs to meet the requirements. However, for large power dissipation, the heat sink size becomes bulky. Figure 16.13 shows that for natural convection, the heat sink volume has to be approximately quadrupled to reduce the thermal resistance by half. Whenever a large amount of heat is removed from a small volume, either forced convection or liquid cooling techniques are preferred. Forced convection cooling is generally used up to a few hundred watts of dissipation and liquid cooling systems are required to remove hundreds of watts from a compact volume.

Forced convection typically uses a fan or a blower to move air along the fins of the heat sink. In this case, the cooling factor becomes a function of air flow rate. Figure 16.13 shows how thermal resistance drops with forced convection cooling from the originally designed value under natural convection. Forced convection cooling

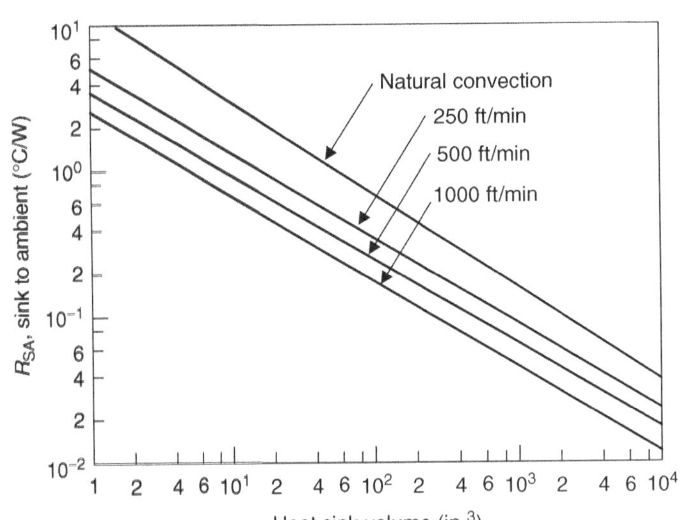

Figure 16.13 Thermal resistance versus heat sink volume for various cooling methods.

allows the use of small size for the same power dissipation. For example, 1 °C/W thermal resistance requires about a 56-in.^{3} volume for a heat sink under natural convection, whereas with a 500-ft/min air flow rate, the volume is only 9 in.^{3}, which gives a volume reduction factor greater than 6.

In the liquid cooling method, the calorific capacity of water is about 4000 times greater than that of air [14], which allows the extraction of several kilowatts of heat with a flow rate of a few liters/minute through pipes of small cross section. A more promising method of heat transfer is that offered by the heat pipe, which has been used successfully to cool a compact 40-W C-band power amplifier that dissipated about 300 W [15]. The thermal resistance of the finned heat sink was 0.092 °C/W.

16.5.2 Design Example

An example of a thermal design of a power amplifier using a heat sink with natural convection is shown in Figure 16.14. A power FET chip capable of delivering 12-W RF power (P_o) at 5 GHz and 10-dB gain (G) is dissipating 30 W of DC power. The thermal resistance of this device is 2 °C/W and the maximum allowed channel and ambient temperatures are 150 °C and 50 °C, respectively.

The channel junction temperature is given by

$$T_{ch} = \left[P_{DC} - P_o\left(1 - \frac{1}{G}\right)\right] R_{th} + T_a \qquad (16.22)$$

where $R_{th} = R_{JC} + R_{CS} + R_{SA}$. The highest channel temperature occurs when no RF signal is applied; that is,

$$T_{ch} = P_{DC} R_{th} + T_a \quad \text{or} \quad R_{th} = (150 - 50)/30 = 3.33\,^\circ\text{C/W}$$

In many instances, GaAs chip carriers or packages are mounted into fixtures and then onto heat sinks. Good polished surfaces are required between these junctions to minimize microscopic air pockets. Since obtaining such polished surfaces is very expensive, usually silicone grease or thin conformal shims with good thermal conductivity are used between the surfaces. Table 16.5 gives typical values for specific thermal resistance (ρ) of various materials. The thermal resistance is obtained using

$$R_{th} = \rho \frac{h}{A} \quad \text{and} \quad \rho = \frac{1}{K_{th}} \qquad (16.23)$$

where h and A are the average thickness and contact area of the material, respectively.

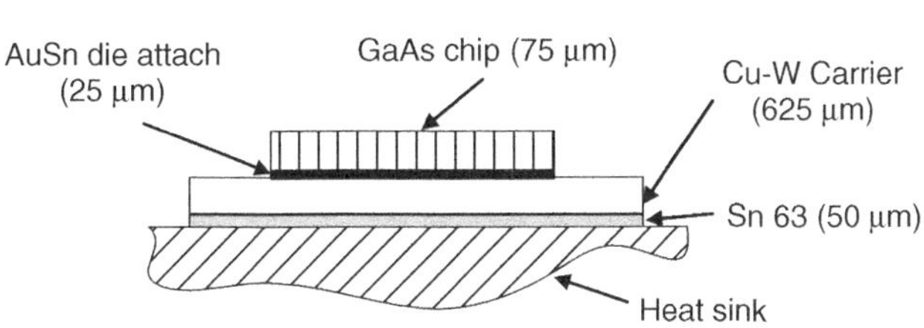

Figure 16.14 Typical MMIC power amplifier assembly using a heat sink.

Table 16.5 Typical Values for Specific Thermal Resistance (ρ) of Various Materials

Material	ρ(°C·m/W)
Still air	30.48
Mylar film	5.99
Silicone grease	5.18
Filled silicone rubber	2.06
Mica	1.68
Aluminum	0.00450
Copper	0.00254

If the aluminum fixture and silicon grease dimensions (length × width × height) are $0.025 \times 0.0125 \times 0.002$ m^3 and $0.025 \times 0.0125 \times 10^{-5}$ m^3, respectively, using (16.3) we find

$$R_{CS} = \frac{0.0045 \times 0.002}{0.025 \times 0.0125} + \frac{5.18 \times 10^{-5}}{0.025 \times 0.0125}\ {}^\circ\text{C/W} = 0.029 + 0.17 = 0.2\ {}^\circ\text{C/W}$$

Thus the heat sink thermal resistance required is

$$R_{SA} = R_{th} - R_{JC} - R_{CS} = 3.33 - 2.0 - 0.2 = 1.13\ {}^\circ\text{C/W}$$

From Table 16.4, heat sink model No. 5584 is good enough. Under RF conditions, the net power dissipated is

$$P_D = 30 - 12 \times (1 - 1/10) = 19.2\text{W}$$
$$R_{th} = (150 - 50)/19.2 = 5.21\ {}^\circ\text{C/W}$$
$$R_{SA} = R_{th} - R_{JC} - R_{CS} = 5.21 - 2.0 - 0.2 = 3.01\ {}^\circ\text{C/W}$$

From Table 16.2, heat sink model No. 5596 or slightly smaller is required. Thus, under RF conditions, the required size for the heat sink is much smaller.

16.6 THERMAL RESISTANCE MEASUREMENT

The thermal resistance of the transistor or amplifier can be calculated from the measurement of channel temperature using the infrared (IR) technique [2–4, 16, 17], the liquid crystal technique [2–4, 18], and the electrical measurement [2–4, 19, 20] technique.

16.6.1 IR Image Measurement

In the IR image method, an infrared radiometric microscope is used to measure the channel temperature. The principle of this instrument is based on measurement of the IR energy radiated from the device channel. The measured device's surface radiance is proportional to the fourth power of its temperature. A recent infrared radiometric microscope has a spot size resolution on the order of 2–3 μm. After calibrating the instrument carefully, one can make very accurate channel temperature measurements.

However, if the features of the active area are smaller than this spot size and have varying emissivity (e.g., Au $\cong$ 0.1,GaAs $\cong$ 0.7), significant errors can be introduced.

An MMIC power amplifier assembly using a heat sink and showing the position of the miniature thermocouples is shown in Figure 16.15. The thermocouples are used to measure the carrier or base plate and heat sink temperatures. From the measured channel temperature (T_{ch}), the power dissipated (P_D), and the carrier temperature measured by a thermocouple, the GaAs chip thermal resistance is readily calculated using (16.11).

An illustration of good agreement between the IR measured and the Cooke model calculated temperatures is shown in Figure 16.16. This figure shows the measured maximum allowed base plate temperature of a 3-W single-stage GaAs power amplifier chip. The amplifier used a 5-mm FET with 20 μm gate–gate pitch. The number of fingers were 20 and the GaAs substrate was 75 μm thick. The channel temperature was kept at 150 °C, as determined by an IR scan, by heating the carrier for different DC power dissipation values in the chip. The measured R_{th} value was 19.3 °C/W, while the calculated R_{th} value using the Cooke model is 19.5 °C/W. As an example, an IR image of a 2-W Ku-band four-stage driver power amplifier in a plastic package, taken with 5-μm spot resolution system, is shown in Figure 16.17. The thermal resistance was 8.4 °C/W. For 8 W of power dissipated, the peak channel temperature was 97.4 °C.

16.6.2 Liquid Crystal Measurement

In this method, nematic liquid crystal materials with different transition temperatures are used to measure the channel temperature. This technique is able to identify hot spots within the device when it is viewed under a polarizing microscope. The device chip is mounted on a heat sink and a thermocouple is placed at the back of the chip for temperature measurement. The device is DC biased and a liquid crystal compound with a known transition temperature is placed on the chip. The DC power is adjusted to observe the transition in the liquid crystal. This corresponds to the channel temperature. At this point, both the DC power dissipated and the back of the chip temperature are measured.

Consider a liquid crystal with a transition temperature of 120 °C. At the transition point, 5 W of power is dissipated in the device and at the back of the chip the measured temperature is 50 °C. In this case the channel temperature is 120 °C. The thermal resistance of the device calculated is $R_{th} = (120 - 50)/5 = 14\,°C/W$.

Using a liquid crystal method, the channel temperatures of the FETs of a 12-W HPA were measured. In this method, DC power of 17.3 W was dissipated in the output

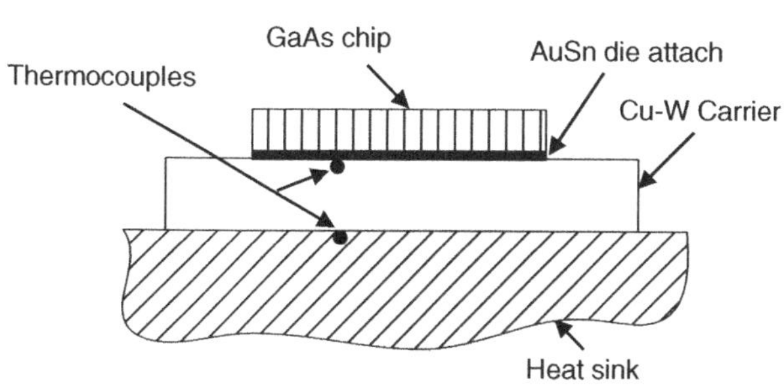

Figure 16.15 MMIC power amplifier assembly using a heat sink and showing the position of the miniature thermocouples used to measure the carrier or base plate and heat sink temperatures.

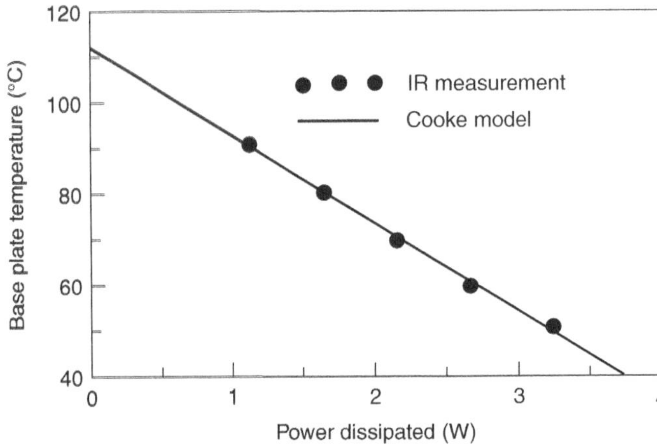

Figure 16.16 Measured maximum allowed base plate temperature of the 3-W power amplifier chip versus power dissipation.

Figure 16.17 IR image of a 2-W Ku-band driver power amplifier.

stage of the amplifier and the liquid crystal's transition temperature was measured. The liquid crystal used in the measurements has a transition temperature of 120 °C. The base of the carrier was cooled so that the maximum FET channel temperature is 120 °C by observing the transition in the liquid crystal. After measuring the temperature, the temperature of the base of the carrier is noted. The calculated thermal resistance including the die attach was 5.205 °C/W. In this case, the value of the die attach thermal resistance is about 0.7 °C/W. The value of ΔT calculated for the output stage FETs is about 78 °C, which gives $R_{th} = 4.505$ °C/W for the die only. From the 12-W

HPA example, when the output stage uses 8 FETs, $R_{th} = 4.66\,^{\circ}$C/W. Therefore both numbers are very close to each other and confirm Cooke's prediction.

16.6.3 Electrical Measurement Technique

The electrical measurement technique (EMT) is based on the fact that the Schottky junction voltage (also known as built-in voltage) in the case of an FET/HEMT or base–emitter junction voltage under constant current varies linearly with temperature. This technique consists of two steps: determination of junction voltage temperature coefficient (JTC) and calculation of junction temperature rise when the DC power is dissipated. Figure 16.18 shows the simplified test setup for measuring the thermal resistance of FETs. First, the gate is forward biased for a low value of constant gate current. This measures the reference gate–source voltage (V_{gsr}) in low-current mode. Second, the pulsed supply (pulse width 20–50 μs and pulse length $\gg \tau$) is switched momentarily to bias the drain for required DC power dissipation. In this condition, the device is self-biased and the gate is negatively biased with respect to the source. The device channel is at a higher temperature due to DC power dissipation. The gate–source voltage (V_{gs}) is measured again immediately after the device returns to the lower current mode. The delay between the drain supply off and the time of gate–source voltage (V_{gs}) measurement is $1\%\ \tau$. Thus the device under test is subjected to three pulses: low-current mode, device heating mode, and low-current mode.

Figure 16.19 shows the typical variation of gate current versus gate–source voltage for a 150-μm FET at room temperature and two extreme temperatures. The JTC is calculated from a change in gate–source voltage over a specified temperature range at a low value of constant gate current. For this case, the value of JTC is 0.833 mV/°C. Thus over a 100 °C channel temperature change, the gate–source voltage change is 83.3 mV.

The thermal resistance is calculated using

$$R_{th} = \frac{V_{gsr} - V_{gs}}{JTC} \frac{1}{P_{DC} - V_{gsr} I_g} \qquad (16.24)$$

where V_{gsr} and V_{gs} are the gate–source voltage measured at room temperature or constant low gate current I_g; for example, 5 mA and DC power dissipation, P_{DC}.

EXAMPLE 16.6

Consider an FET that has JTC = 0.00083 V/°C, measured when the gate is forward biased at $V_{gsr} = 0.73$ V and $I_g = 5$ mA. Then DC power through the drain is applied at 10 V with an

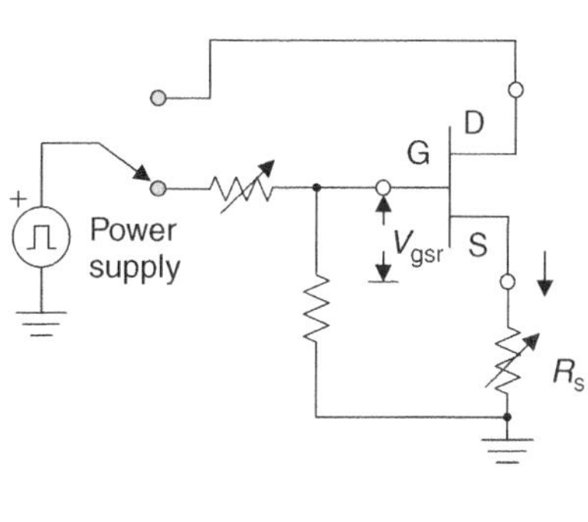

Figure 16.18 Test setup for measuring thermal resistance of FETs.

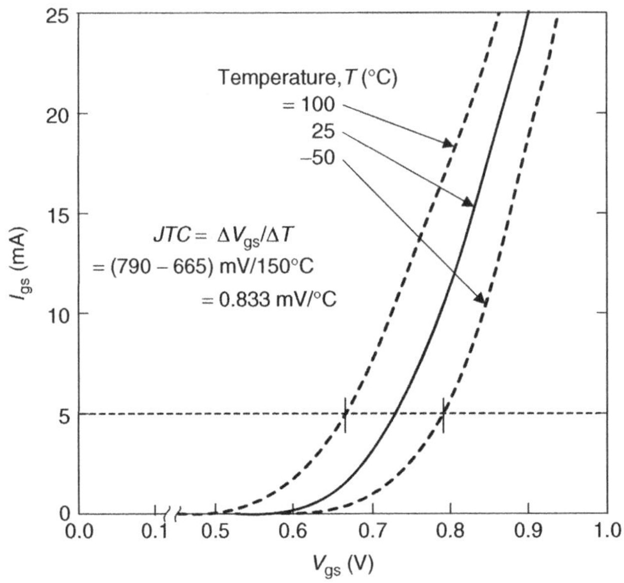

Figure 16.19 Variation of gate current with gate–source voltage at different channel temperatures.

average drain–source current of 1 A; V_{gs} is then measured immediately after the device returns to the lower current mode. The measured value of V_{gs} was 0.65 V. Determine the thermal resistance of the device.

SOLUTION Here, $V_{gsr} - V_{gs} = 0.73 - 0.65 = 0.08\text{V}$.

$$\text{Total power dissipation} = P_{DC} - V_{gsr} \times I_g = 10 - 0.0037\text{W} = 9.9963\text{W}$$

$$R_{th} = (0.73 - 0.65)/(0.00083 \times 9.9963)\,^{\circ}\text{C/W} = 9.64\,^{\circ}\text{C/W}$$ ■

The EMT and liquid crystal method are only suitable for discrete transistors. Thus using these methods it is difficult to determine the thermal resistance of MIC and MMIC power amplifiers designed with several devices in parallel. The IR image technique is commonly used for power amplifiers.

REFERENCES

1. H. F. Cooke, Precise technique finds FET thermal resistance, *Microwaves RF*, pp. 85–87, August 1986. Correction of this paper in *Microwaves RF*, p. 13, February 1987.
2. J. V. Dilorenzo and D. D. Khandewal, *GaAs FET Principles and Technology*, Artech House, Norwood, MA, 1982, pp. 313–347.
3. J. L. B. Walker (Ed.), *High-Power GaAs FET Amplifiers*, Artech House, Norwood, MA, 1993, Chapter chap5.
4. K. Chang (Ed.), *Handbook of RF/Microwave Components and Engineering*, John Wiley & Sons, Hoboken, NJ, 2003, Chapter chap21.

5. G. J. Murphy, D. J. Shippy, and H. L. Luo, *Engineering Analogies*, Iowa State University Press, Ames, IA, 1963.
6. R. C. Levine, Determination of thermal conductance of dielectric-filled strip transmission line from characteristic impedance, *IEEE Trans. Microwave Theory Tech.*, Vol. MTT-15, pp. 645–646, November 1967.
7. I. J. Bahl and K. C. Gupta, Average power-handling capability of microstrip lines, *IEE Microwaves Opt. Acoustics*, Vol. 3, pp. 1–4, January 1979.
8. K. C. Gupta, R. Garg, I. J. Bahl, and P. Bhartia, *Microstrip Lines and Slotlines*, 2nd edition, Artech House, Norwood, MA, 1996.
9. S. B. Cohn, Problems in strip transmission lines, *Proc. IRE PG-MTT*, pp. 119–126, March 1955.
10. R. Mongia, I. Bahl, and P. Bhartia, *RF and Microwave Coupled-Line Circuits*, Artech House, Norwood, MA, 1999, pp. 499–500.
11. *Agilent (Avantek) High-Frequency Transistor Primer, Part III Thermal Properties*, Agilent, Santa Clara, CA.
12. A. D. Kraus and A. Bar-Cohen, *Design and Analysis of Heat Sinks*, John Wiley & Sons, Hoboken, NJ, 1995.
13. *Catalog of Heat Dissipation Components*, EG & G Wakefield Engineering, Wakefield, MA, 1988.
14. R. Soares, (Ed.), *GaAs MESFET Circuit Design*, Artech House, Norwood, MA, 1988, Chapter 6.
15. N. Fukuden et. al., A 4.5GHz 40 watt GaAs FET amplifier, *IEEE MTT-S Int. Microwave Symp. Dig.*, pp. 66–68, 1982.
16. F. N. Sechi, B. S. Perlman, and J. M. Cusack, Computer-controlled infrared microscope for thermal analysis of microwave transistors, *IEEE Int. Microwave Symp. Dig.*, pp. 143–146, 1977.
17. L. Walshak and E. Poole, Thermal resistance measurement by IR scanning, *Microwave J.*, Vol. 20, pp. 62–65, 1977.
18. M. M. Minot, Thermal characteristics of microwave power FETs using nematic liquid crystals, *IEEE Int. Microwave Symp. Dig.*, pp. 495–498, 1986.
19. B. S. Sage, A proposed method for testing thermal resistance of MESFETs, *Microwave Syst. News*, Vol. 7, pp. 66–70, November 1977.
20. *Sage Thermal Resistance Tester Manual*, Sage Enterprise, Inc., Palo Alto, CA.

PROBLEMS

16.1 The thermal resistance of a 2.5-mm device is 30 °C/W. At 10 GHz, the output power (W), gain (dB), and PAE (%) of the device are 2, 9, and 45 for a class-A operation and 2, 8.5, and 55 for class-B operation, respectively. Compare the channel temperatures between these two classes when the ambient temperature is 50 °C.

16.2 A GaAs HPA chip is dissipating 100 W and its thermal resistance is 0.5 °C/W. The chip is mounted on a carrier that is bolted to a heat sink. The carrier's thermal resistance is 0.2 °C/W. Design a heat sink for an ambient temperature of 25 °C when the maximum allowed channel temperature is 175 °C.

16.3 Determine the thermal resistance of a 1-mm gate periphery transistor having 4 fingers with 30-μm gate-to-gate pitch. The thickness of the die is 100 μm. The chip is operating at a channel temperature of 100 °C and device gate length is 1 μm.

16.4 Determine the thermal resistance of a 2.5-mm gate periphery transistor having 10 fingers with 25-μm gate-to-gate pitch. The thickness of the die is 50 μm. The chip is operating at a channel temperature of 150 °C and device gate length is 1 μm.

16.5 A power chip whose size is 2×2 mm^2 is mounted on a carrier using a 25-μm thick AuSn preform. The thermal resistance of the chip is 10 °C/W. The carrier has an area of 1×1 cm^2 and is 1 mm thick. The thermal conductivity values for AuSn and carrier are 1.0 and 1.69 W/cm·°C, respectively. Determine the temperature on the top surface of the chip when 10 W of power is dissipated and the carrier is at 25 °C.

Chapter 17

Stability Analysis

Any amplifier with power gain can be made to oscillate by applying external positive feedback, for example, a high-gain MMIC chip in a plastic or ceramic package with poor isolation. At microwave frequencies unavoidable parasitic effects are often sufficient to cause oscillations if care is not taken in the design and fabrication of the amplifier. Any abrupt change in the DC parameters of the amplifier, output power with no input power, circuits that are very sensitive to their surroundings, and so on are typical indications of unwanted oscillations. Oscillations may occur at frequencies that don't propagate out the amplifier because they are filtered, are blocked by bias capacitors, or are below waveguide cutoff frequencies or at frequencies to which the test equipment is insensitive. It is not unusual for microwave amplifiers to oscillate anywhere between 1 MHz and 40 GHz or higher.

The best single test for oscillations involves using a combination of a broadband spectrum analyzer and sweep oscillator. The absence of both fundamental and intermodulation oscillation signals in the broadband output signal spectrum when the input signal is swept is usually sufficient to assure stable operation. It must be emphasized that any variable change such as impedance levels, load impedance mismatch, supply voltages, input power level, temperature, light intensity, aging, and radiation may quench or create oscillations. Testing often must be done over a broad range of parameters if confidence is to be obtained that the amplifier is stable under those conditions. Fortunately, design techniques can be applied in most cases to build in confidence of stability.

Any positive feedback with proper phase across an active device can be turned into an oscillator. This feedback may be intentional, such as biasing networks between the stages, loops due to devices in parallel, and source inductance, or unintentional, including poor isolation in amplifier housing and poor grounding. Once a stability problem has been identified, it is usually not difficult to correct it in many cases at a slightly lower gain. In most microwave system applications, conditional stability is usually acceptable in the frequency band where component impedances are defined, but unconditional stability may be required out of band where there are usually no system specifications. Ferrite isolators or balanced amplifier configurations (discussed in Chapter 11) are often used to protect amplifiers from unknown impedances. Care must be taken, since stability problems may arise far outside the isolator frequency band where the isolator is neither matched nor unilateral.

In the case of power amplifiers, the stabilization of circuits becomes very critical as compared to small-signal amplifiers. Under large-signal conditions and pulsed

Fundamentals of RF and Microwave Transistor Amplifiers. By Inder J. Bahl

operation, the devices go through large variations in their characteristics. Thus their design requires accurate nonlinear transistor models, suitable CAD tools, and appropriate stability analysis techniques. Since most devices are only conditionally stable, and the systems mandate that the amplifiers must be unconditionally stable, the amplifier's stabilization becomes an integral part of the design.

The oscillations in RF/microwave power amplifiers may be classified into five categories: even mode, odd mode, parametric, spurious parametric, and low frequency. Even-mode stability analysis is adequate for most low-noise and buffer amplifiers using small size transistors. The power amplifier design must be unconditionally stable in terms of even mode and also odd mode, and parametric and low-frequency oscillation conditions must be prevented. In an even-mode operation of a two-port network, the currents flow in and out of the ports, whereas in the odd-mode case there are no currents flowing in and out of the ports but there are internal currents and voltages. These are briefly discussed in the following sections.

17.1 EVEN-MODE OSCILLATIONS

Any transistor, which is not unconditionally stable, can give rise to even-mode oscillations. This occurs when the transistor is connected between the input and output matching networks of an amplifier or connected to biasing networks to perform $I-V$ measurements. As described in several textbooks [1–4], even-mode stability against oscillations can be examined by using two-port S-parameters or an EC model. The parameters S_{12} and S_{21} form a feedback loop that depends on the source and load impedance and may support oscillations. In an ideal amplifier S_{12} would be zero and the amplifier would be unconditionally stable. Stability calculation using an EC model is user friendly because the model parameters such as g_m, C_{gs}, C_{gd}, and L_S can easily be changed to check the circuit stability against these parameters. The stability criteria can be determined in terms of the Rollett condition, K factor, or μ factor expressed in terms of S-parameters. The significance of the stability factor K is that an active device is unconditionally stable for all passive source and load terminations when $K > 1$ or $\mu > 1$. Both linear and nonlinear device models are used to predict the stability conditions of the circuit.

17.1.1 Even-Mode Stability Analysis

Consider a two-port network representing an amplifier as shown in Figure 17.1. If $S_{12} \neq 0$, the input reflection coefficient Γ_{in} with arbitrary Z_L and the output reflection coefficient Γ_{out} with arbitrary Z_S can be expressed as

$$\Gamma_{\text{in}} = S_{11} + \frac{S_{12}S_{21}\Gamma_L}{1 - S_{22}\Gamma_L} \tag{17.1}$$

$$\Gamma_{\text{out}} = S_{22} + \frac{S_{12}S_{21}\Gamma_S}{1 - S_{11}\Gamma_S} \tag{17.2}$$

If the circuit is unconditionally stable, any source or load may be connected to the input or output of the circuit without oscillations. In terms of S-parameters of the device, unconditional stability is assured if the following inequalities are simultaneously

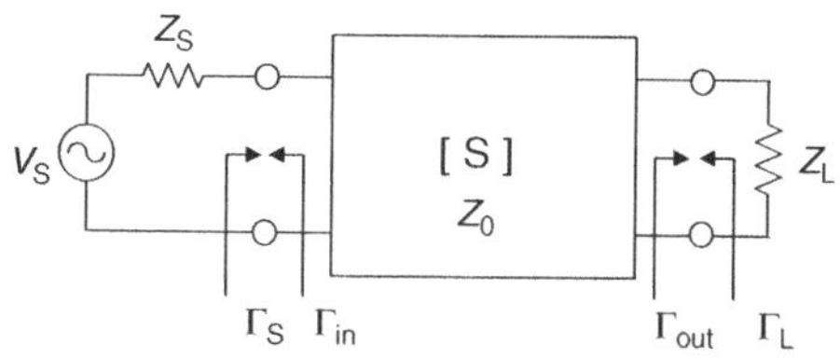

Figure 17.1 Two-port network of a transistor amplifier.

satisfied:

$$|S_{11}| < 1 \quad \text{and} \quad |S_{22}| < 1$$
$$|\Gamma_{in}| < 1 \quad \text{and} \quad |\Gamma_{out}| < 1$$

For $|\Gamma_S| < 1$ and $|\Gamma_L| < 1$, these conditions lead to the following requirement:

$$K = \frac{1 + |S_{11}S_{22} - S_{12}S_{21}|^2 - |S_{11}|^2 - |S_{22}|^2}{2|S_{12}|\,|S_{21}|} > 1 \tag{17.3}$$

The significance of the stability factor K is that a device is unconditionally stable ($K > 1$) for all passive source and load terminations when operated where the small-signal S-parameters are accurate.

If we set $|\Gamma_{in}|$ and $|\Gamma_{out}|$ equal to unity in (17.1) and (17.2), a boundary is established beyond which the device is unstable. Solving for Γ_S and Γ_L, each condition will give a solution of a circle (called stability circle) on a complex reflection plane or on a Smith chart [2]. The stability circle equations are

$$\left|\Gamma_S - \frac{(S_{11} - DS_{22}^*)^*}{|S_{11}|^2 - |D^2|}\right| = \left|\frac{S_{12}S_{21}}{|S_{11}|^2 - |D^2|}\right| \tag{17.4a}$$

$$\left|\Gamma_L - \frac{(S_{22} - DS_{11}^*)^*}{|S_{22}|^2 - |D^2|}\right| = \left|\frac{S_{12}S_{21}}{|S_{22}|^2 - |D^2|}\right| \tag{17.4b}$$

The input and output radius (r) and center (c) are given by

$$\left.\begin{aligned} r_S &= \frac{|S_{12}S_{21}|}{||S_{11}|^2 - |D|^2|} \\ c_S &= \frac{(S_{11} - DS_{22}^*)^*}{|S_{11}|^2 - |D^2|} \end{aligned}\right\} \text{input} \tag{17.5a}$$

$$\left.\begin{aligned} r_L &= \frac{|S_{12}S_{21}|}{||S_{22}|^2 - |D|^2|} \\ c_L &= \frac{(S_{22} - DS_{11}^*)^*}{|S_{22}|^2 - |D^2|} \end{aligned}\right\} \text{output} \tag{17.5b}$$

where

$$D = S_{11}S_{22} - S_{12}S_{21} \tag{17.5c}$$

and subscripts S and L denote source and load, respectively. Here, the origin of the Smith chart is at $\Gamma_{in} = \Gamma_{out} = 0$. In order to plot stability circles, the radius of the Smith chart is considered to be unity. The magnitude of r_S, c_S, r_L, and c_L are calculated with respect to the radius of the Smith chart as a unit, and angles for c_S and c_L are the same as for Γ_{in} and Γ_{out} on the Smith chart. Figure 17.2 shows typical examples of the input plane of unconditionally stable and conditionally stable networks. The shaded area represents the area of the input plane in which instability occurs. If the circuit is potentially unstable ($K < 1$), the source and load impedances should be chosen so they will not fall into the unstable region (e.g., shaded) due to device parameter changes, fabrication variation, and changes in temperature. Under such conditions, the amplifier is said to be conditionally stable and will not oscillate. For an unconditionally stable condition $|c_S| - r_S$ and $|c_L| - r_L$ must be greater than unity. Stability analysis must be carried out from DC (mostly from 50 MHz) to above the frequency where the active devices have power gain. If the circuit is unconditionally stable ($K > 1$), the conditions required to obtain maximum power gain are $\Gamma_{in} = \Gamma_S^*$ and $\Gamma_{out} = \Gamma_L^*$. Solving for a simultaneous conjugate match, the reflection coefficients to be matched are denoted by Γ_{SM} and Γ_{LM} and are given by [1–5]

$$\Gamma_{SM} = \frac{B_1 \pm \sqrt{B_1^2 - 4|C_1|^2}}{2C_1} \tag{17.6}$$

$$\Gamma_{LM} = \frac{B_2 \pm \sqrt{B_2^2 - 4|C_2|^2}}{2C_2} \tag{17.7}$$

$$B_1 = 1 + |S_{11}|^2 - |S_{22}|^2 - |D|^2$$
$$B_2 = 1 + |S_{22}|^2 - |S_{11}|^2 - |D|^2$$

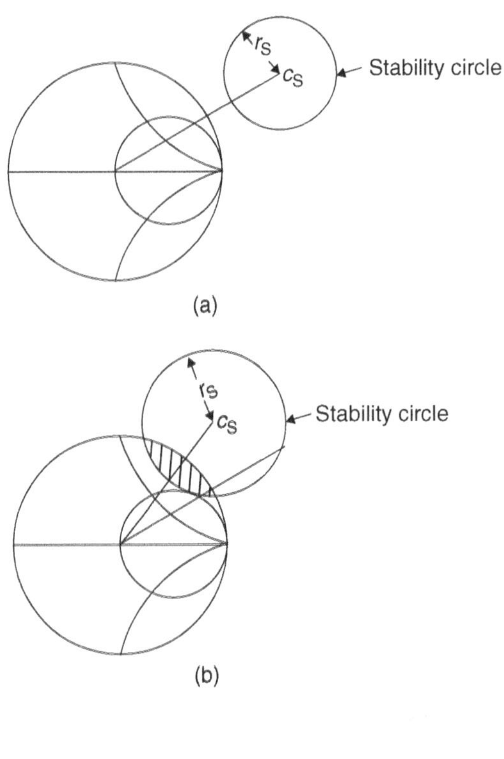

Figure 17.2 Examples of (a) unconditional stability and (b) conditional stability for $|S_{11}| < 1$.

where

$$C_1 = S_{11} - DS_{22}^*$$
$$C_2 = S_{22} - DS_{11}^* \tag{17.8}$$

If $|B_1/2C_1| > 1$ and $B_1 > 0$, the solution with the minus sign produces $|\Gamma_{SM}| < 1$ and the solution with the plus sign produces $|\Gamma_{SM}|>1$. If $|B_1/2C_1| > 1$ with $B_1 < 0$, the solution with the plus sign produces $|\Gamma_{SM}|<1$ and the solution with the minus sign produces $|\Gamma_{SM}| > 1$. Similar considerations apply to Γ_{LM}.

The value of the matched power gain (MG) or maximum available gain (MAG) is

$$MAG = \left|\frac{S_{21}}{S_{12}}\right| (K - \sqrt{K^2 - 1}) \tag{17.9}$$

If the circuit is potentially unstable ($K < 1$), the maximum gain cannot be defined as a unique quantity. It depends on the input and output matching conditions. In this case, the maximum stable gain (MSG) is generally obtained by substituting $K = 1$ in (17.9)

$$MSG = \left|\frac{S_{21}}{S_{12}}\right| \tag{17.10}$$

This means one has to do some stabilization in the device or circuit to achieve $K = 1$. The above described stability definition requires the Rollet condition together with any one of the auxiliary conditions given in (17.8). Edwards and Sinsky [5] have combined the above two tasks into a single stability definition known as the μ factor, given as

$$\mu = \frac{1 - |S_{11}|^2}{|S_{22} - S_{11}^*\Delta| + |S_{21}S_{12}|} > 1 \tag{17.11}$$

The μ factor also measures the relative level of the device's stability. A larger value of μ factor means greater device stability.

Next, we describe examples to study the stability of different devices.

EXAMPLE 17.1

The S-parameters of a low-noise pHEMT are given in Table 5.9a (Chapter 5). Calculate K and μ at 10 and 20 GHz and plot the input stability circle at the frequency where μ is less than unity. Also calculate MSG/MAG at these frequencies.

SOLUTION From (17.5c),

$$|D| = 0.378 \quad \text{at 10 GHz}$$
$$|D| = 0.183 \quad \text{at 20 GHz}$$

From (17.3) and (17.11),

$$K = 0.647, \quad \mu = 0.70 \quad \text{at 10 GHz}$$
$$K = 1.05, \quad \mu = 1.04 \quad \text{at 20 GHz}$$

Thus the device is potentially unstable at 10 GHz and unconditionally stable at 20 GHz.

From (17.5),

$$r_S = 0.865 \quad \text{and} \quad c_S = 1.694\angle 94.87 \quad \text{at10 GHz}$$
$$r_L = 4.814 \quad \text{and} \quad c_L = 5.514\angle 72.92$$
$$r_S = 1.281 \quad \text{and} \quad c_S = 2.311\angle 145.56 \quad \text{at 20 GHz}$$
$$r_L = 3.903 \quad \text{and} \quad c_L = 4.944\angle 71.82$$
$$MSAG = 15.64\ \text{dB} \quad \text{at 10 GHz}$$
$$MSAG = 10.69\ \text{dB} \quad \text{at 20 GHz}$$

Figure 17.3 shows the input stability plot at 10 GHz. The shaded portion depicts the unstable region. The device is stable for a large range of impedance outside the shaded region. ■

EXAMPLE 17.2

Consider a low-noise FET and a low-noise pHEMT whose EC models are given below.

For a 300-μm FET,

$$R_g = 0.5\ \Omega,\ \ R_i = 1.5\ \Omega,\ \ R_s = 1.5\ \Omega,\ \ R_d = 1.5\,\Omega,\ \ R_{ds} = 160\,\Omega,\ \ g_m = 65\,\text{mS},\ \ \tau = 2\ \text{ps}$$
$$C_{gs} = 0.28\,\text{pF},\ \ C_{gd} = 0.05\,\text{pF},\ \ C_{ds} = 0.07\,\text{pF},\ \ L_g = 0.01\,\text{nH},\ \ L_s = 0.005\ \text{nH},\ L_d = 0.01\ \text{nH}$$

For a 200-μm pHEMT,

$$R_g = 0.8\ \Omega,\ \ R_i = 0.5\ \Omega,\ \ R_s = 0.5\ \Omega,\ R_d = 0.5\ \Omega,\ R_{ds} = 200\ \Omega,\ g_m = 75\ \text{mS},\ \tau = 1\ \text{ps}$$
$$C_{gs} = 0.24\ \text{pF},\ \ C_{gd} = 0.03\text{pF},\ \ C_{ds} = 0.04\text{pF},\ L_g = 0.05\text{nH},\ L_s = 0.08\text{nH},\ \ L_d = 0.05\ \text{nH}$$

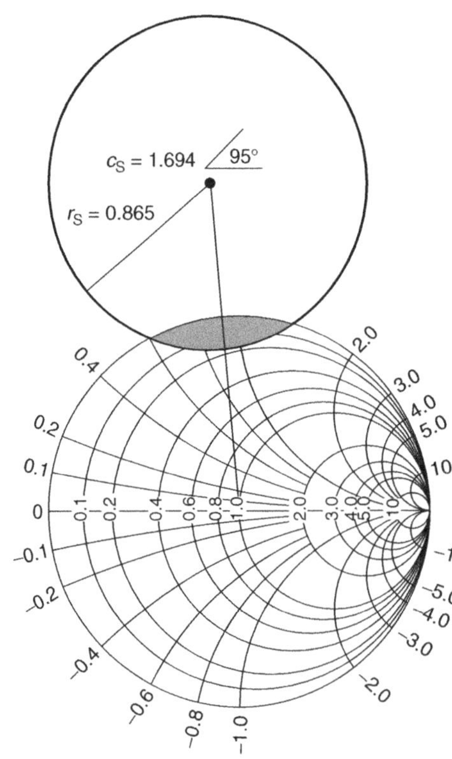

Figure 17.3 Shaded region represents the input impedance range presented to the input of the device for potentially unstable operation.

The pHEMT device also includes bond wire (15–20 mils) inductance. Calculate the μ factor and maximum gain values for these two devices from 2 to 20 GHz with 2-GHz step.

SOLUTION One can convert the EC model into the S-parameters and use equations described previously or any CAD tool to calculate the gain and stability parameters. Here, a commercial CAD tool was used to calculate μ and the maximum stable/available gain values:

Frequency	MESFET		pHEMT	
(GHz)	μ	MSG/MAG (dB)	μ	MSG/MAG (dB)
2	0.09	19.5	0.18	22.7
4	0.18	16.5	0.35	19.7
6	0.26	14.8	0.50	17.9
8	0.35	13.5	0.64	16.7
10	0.43	12.6	0.76	15.6
12	0.50	11.9	0.85	14.8
14	0.57	11.2	0.92	14.0
16	0.63	10.7	0.98	13.3
18	0.69	10.3	1.02	11.8
20	0.75	9.9	1.04	10.7

The pHEMT device has higher gain than the MESFET because of its higher g_m. For each device when $\mu < 1$ and the frequency is doubled, the gain is decreased by about 3 dB. However, when $\mu > 1$, the gain decrease is 4–6 dB. ■

EXAMPLE 17.3

Compare the stability of low-noise transistors used in Example 17.2 with the power pHEMT whose EC model is given in Table 5.10 (#1). The frequency of comparison is 10 GHz. When the values of R_s, L_s, C_{gs}, C_{gd}, and g_m are increased by 50%, 50%, 15%, 25%, and 15%, respectively, determine which parameter significantly affects the maximum available gain and μ factor.

SOLUTION Here again a commercial CAD tool was used to calculate μ and the maximum stable gain values at 10 GHz:

Parameter	Low-Noise MESFET		Low-Noise pHEMT		Power pHEMT	
Value	μ	MSG (dB)	μ	MSG(dB)	μ	MSG(dB)
Nominal	0.43	12.6	0.76	15.6	0.72	17.4
$R_s + 50\%$	0.42	12.3	0.74	15.5	0.69	17.1
$L_s + 50\%$	0.44	12.6	0.88	15.1	0.75	17.4
$C_{gs} + 15\%$	0.45	12.6	0.80	15.6	0.79	17.3
$C_{gd} + 25\%$	0.44	11.7	0.71	14.7	0.70	16.4
$g_m + 15\%$	0.41	13.2	0.76	16.1	0.69	17.9

In all devices the effect of increased C_{gd} on the MSG value is more pronounced. The stability is improved by increasing C_{gs} in the low-noise FET and power pHEMT, and increasing L_s in the low-noise pHEMT, respectively. ■

Next, some examples having even-mode instabilities are described. First consider a self-biased transistor topology, as shown in Figure 17.4a. The device selected is

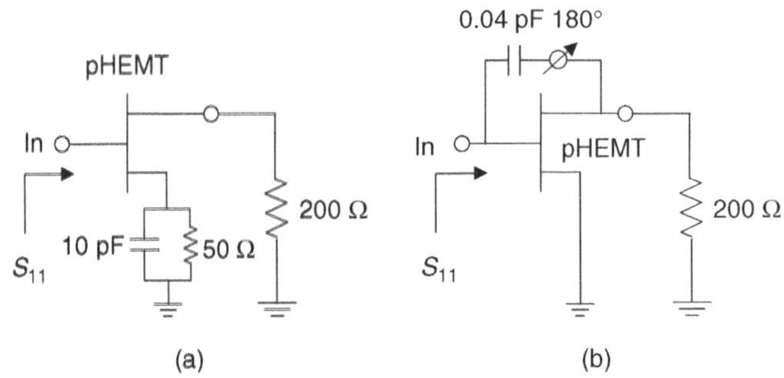

Figure 17.4 (a) Self-biased pHEMT configuration. (b) Effect of a package as a feedback representation.

a pHEMT (Table 5.9a) biased at 2.5 V and 10 mA. Here, the drain–source voltage and gate–source voltage are 2 V and −0.5 V, respectively. The device is operated at 10 GHz. When the device's output is terminated in a 200-Ω load, the simulated K factor and S_{11} are given below and $|S_{22}|$ was less than 1:

Frequency (GHz)	K	$\|S_{11}\|$
1.0	−0.67	1.056
1.5	−0.58	1.062
2.0	−0.48	1.054
2.5	−0.37	1.037
3.0	−0.28	1.015
3.3	−0.22	1.000
3.5	−0.19	0.991
4.0	−0.11	0.962

Thus below 3.3 GHz, the self-bised pHEMT is potentially unstable for 50-Ω input. Here, S_{11} is greater than unity because of the series feedback capacitor. A larger value of the capacitor moves the instability frequency to a lower value.

Figure 17.4b replicates a feedback configuration when a transistor or an amplifier is mounted in a package. Here, about a −28-dB feedback coupling coefficient and a 180° constant phase are chosen. When the device parasitic reactance is ignored, there is a 180° phase difference between the output and input terminals of the transistor. Thus any output signal through the feedback loop reaching the input of the transistor/circuit will have a 360° phase shift and will combine in phase with the input signal. This results in S_{11} greater than unity as given below:

Frequency (GHz)	K	$\|S_{11}\|$
2.0	−0.03	1.015
4.0	−0.05	1.050
6.0	−0.07	1.090
8.0	−0.08	1.108
10.0	−0.08	1.080
12.0	−0.07	1.014
12.4	−0.06	0.999
14.0	−0.02	0.928
16.0	0.05	0.837

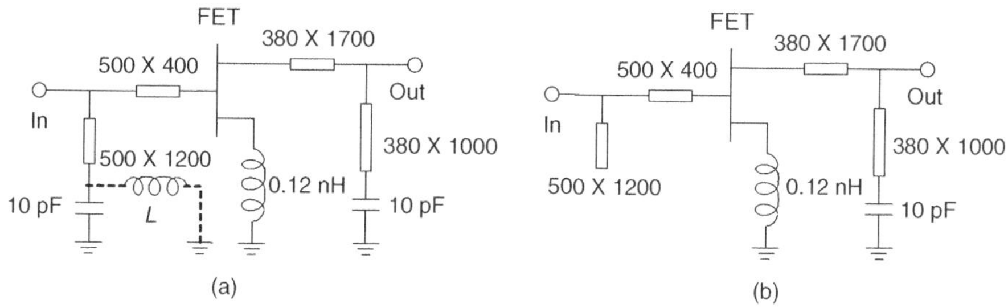

Figure 17.5 (a) A single-stage FET amplifier configuration. (b) Single-stage FET amplifier after 10-pF capacitor resonates out with probe or wire inductance L.

Here, $|S_{22}|$ was also less than 1. Thus below 12.4 GHz, the circuit is potentially unstable.

Another example of a potentially unstable single-stage amplifier is shown in Figure 17.5a. The design parameters are microstrip (width × length) on 15-mil alumina substrate. When a shunt 10-pF capacitor resonates out with probe or wire inductance L, a short-circuited stub becomes an open-circuited stub, as shown in Figure 17.5b, and may give rise to instability. The calculated amplifier performance for Figure 17.5a,b is given below and Figure 17.5b shows potential instability from 20 to 22 GHz.

Frequency (GHz)	K	$\lvert S_{11}\rvert$	$\lvert S_{22}\rvert$	S_{21} (dB)
8	1.09	0.75	0.83	6.1
9	1.15	0.43	0.51	10.4
10	1.28	0.28	0.28	10.2
11	1.31	0.39	0.46	8.1
18	0.68	0.65	0.93	−2.0
19	0.55	0.66	0.99	−2.5
20	0.45	0.77	1.07	−4.1
21	0.38	0.91	1.08	−8.2
22	0.35	0.98	1.02	−15.8
23	0.37	0.99	0.95	−17.6
24	0.40	0.99	0.88	−20.0

17.1.2 Even-Mode Oscillation Suppression Techniques

The earlier two-port stability treatment and examples are applicable only to the sensitivity of the two ports to external impedances, not to internal oscillations. For example, a multistage amplifier might oscillate even though its overall K factor is greater than unity. The stability analysis must be performed on all internal two ports having transistors and on the complete circuit. In a multistage amplifier, internal transistors are terminated by passive matching networks followed by active devices that can give rise to negative resistance. A series of stable two-amplifier stages will be stable, but if additional feedback is introduced, the new circuit must also be analyzed. The circuit shown in Figure 17.6 will be stable if each of the three stages is stable and the two-port combination of stage 2, stage 3, and feedback impedance is stable. The analysis must be performed five times.

The internal feedback (e.g., biasing circuitry) and external feedback (e.g., packages and fixtures) with multiple of 360° phase difference might introduce a

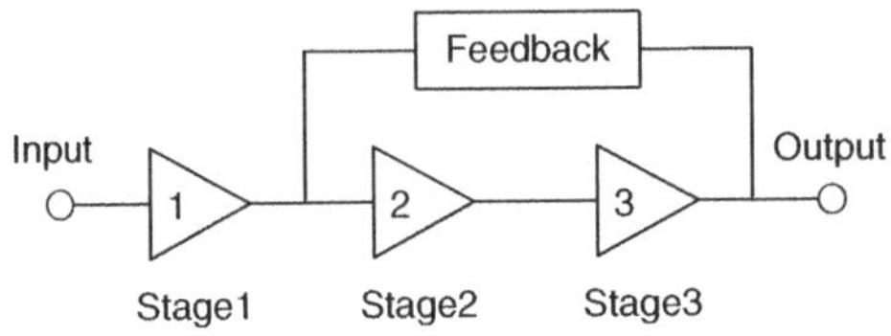

Figure 17.6 Configuration of a three-stage amplifier with feedback.

condition for negative resistance. For a high-gain amplifier to be housed successfully in a package, a commonly used rule of thumb is that the isolation in the package must be at least 15 dB above the maximum gain of the amplifier. For example, in order to test reliably a 30-dB gain amplifier in a package, it requires greater than 45-dB isolation between the input and output ports of the package. If this condition is not met, poor isolation will change the stability factor of the amplifier significantly, no matter how stable the amplifier itself is. The PCB used for testing high-gain amplifiers ($G > 20$ dB) must have a low-inductance paddle ground and the fixture for testing ceramic packages must have a good connection with the bottom of the flange. Commonly, thin conformal shims between the flange and fixture are used. The effect of feedback may also be studied by measuring S-parameters between the input and output terminals of the amplifier placed in a test fixture.

Figure 17.7 shows some of the possible transistor stabilization schemes. Generally, stabilization networks prevent conditionally stable circuits from getting into unstable regions. The stabilization network also lowers the device gain. Table 17.1 provides a comparison of device stabilization schemes for a MESFET and a pHEMT. Tables 5.1 and 5.9 provide the S-parameter and noise-parameter data for these two transistors and the frequency of comparison is 10 GHz. The transistors are potentially unstable. The value of the stabilization parameter was adjusted to obtain $K = 1$. It may be noted from Table 17.1 that the shunt resistor has maximum effect on the noise figure and minimum effect on the gain, while the source inductance has minimum effect on the noise figure and maximum effect on the gain. Therefore a source inductance technique is commonly used for LNA designs. However, at high frequencies the source inductor increases the series feedback and affects the amplifier's stability. As compared to the other two techniques, the shunt resistor and feedback resistor techniques also provide device stabilization below 10 GHz. One can also use a shunt resistor at the drain terminal for circuit stabilization. However, it is not normally used because it increases the DC power requirements and lowers the RF output power.

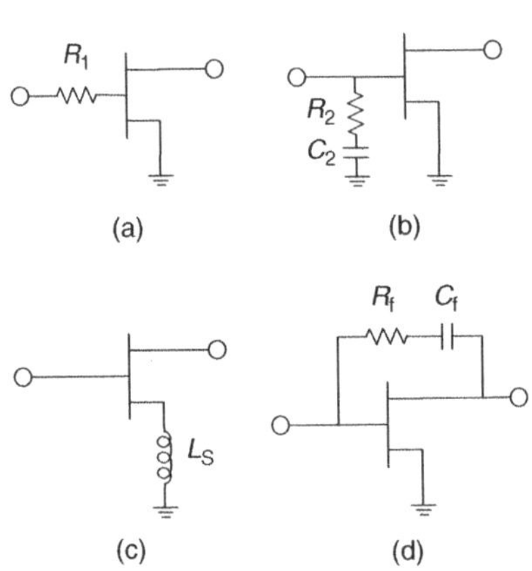

Figure 17.7 Transistor stabilization schemes: (a) series resistor, (b) shunt resistor, (c) source or series feedback inductor, and (d) shunt or parallel feedback.

Table 17.1 Comparison of Device Stabilization Schemes

Parameter	Device[a]	Stabilization Scheme: None	Series Resistor	Shunt Resistor	Source Inductance	Feedback Resistor
K	MESFET	0.377	1.0	1.0	1.0	1.0
	pHEMT	0.647	1.0	1.0	1.0	1.0
MSG (dB)	MESFET	12.7	12.3	12.3	11.8	11.5
	pHEMT	15.8	15.6	15.6	13.0	15.6
NF_{min} (dB)	MESFET	0.81	1.86	4.05	0.79	2.25
	pHEMT	0.39	1.13	1.85	0.38	0.80

[a]MESFET: $R_1 = 10.7\ \Omega$, $R_2 = 97\ \Omega$, $R_f = 485\ \Omega$, $L_s = 0.15$ nH. pHEMT: $R_1 = 9\ \Omega$, $R_2 = 230\ \Omega$, $R_f = 1840\ \Omega$, $L_s = 0.2$ nH.

Accurate bias and temperature-dependent models for transistors, accurate passive component models, and inclusion of bias circuitry and coupling feedback paths in the circuit design simulation can help in predicting even-mode instability. For example, in MMICs or hybrid MICs, a combination of circuit bypass capacitors, bond wire inductance, and external bypass capacitors may result in a parallel resonance. This can create negative resistance inside the circuit and can oscillate at the parallel resonance frequency of the biasing circuitry. Such an oscillation is known as *bias oscillation* and still falls in the category of even-mode oscillation. One of the commonly observed effects of such instability, during testing of an amplifier, is the reduction in small-signal gain from the expected nominal value.

Accurate modeling of bias circuitry along with circuit simulation can predict such instability in terms of the K factor. The circuit can be made stable by changing the capacitor value or wire bond length or by using a resistor in the bias circuitry, as shown in Figure 17.8.

The purpose of the resistor is to damp out the low-frequency oscillations, which occur mostly in the 10–500-MHz range. Generally, the effect of resistors on the RF/microwave performance is negligible. Even-mode oscillations can happen even without applying an RF/microwave signal. For example, an unconditionally stable high-gain amplifier chip placed in a poor isolation package may have microwave frequency oscillations even without an input RF signal. High-gain amplifiers require high isolation or channelized housing with cutoff frequency below the frequency of operation.

For small-signal applications, the standard even-mode ($K > 1$) stability analysis is adequate to avoid microwave even-mode oscillations. However, under large-signal conditions and pulsed operation, it is necessary to use a worst case K factor greater than 1. By designing with S-parameter data for various bias conditions from $V_{ds} = 3$

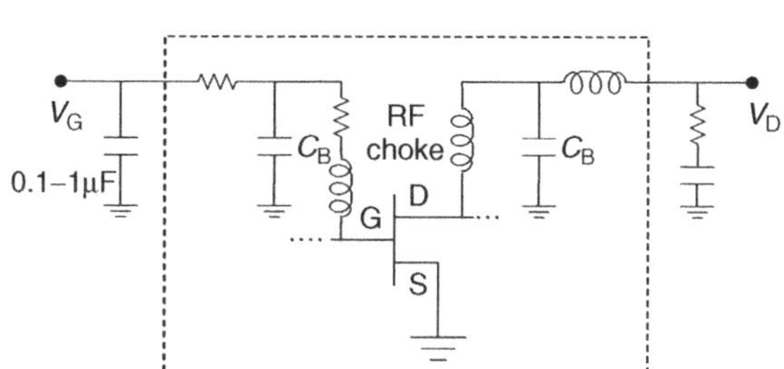

Figure 17.8 An even-mode stabilization scheme.

V and 50% I_{dss} to $V_{ds} = 10$ V and 25% I_{dss}, the envelope that a full cycle of the input signal experiences during the large-signal and pulsed operation is replicated. It was found that for MSAG HPAs biased at 10 V and imposing a $K > 2.0$ condition for $V_{ds} = 10$ V and 25% I_{dss}, small-signal S-parameters are adequate to ensure stable operation under all conditions.

During "on wafer" testing of power amplifiers measured with a network analyzer, a situation may occur where the circuit gain is much lower (10–15 dB) than the expected value. This happens due to improper grounding or low-frequency instabilities outside the frequency range of measurement. This can be verified by connecting the output of the circuit through a coupler to a spectrum analyzer. An effective way to deal with this situation is to connect the DC bias ground and RF ground together with the shortest possible wires/cables and put supply decoupling 100–600-pF capacitors very close to the gate and drain pads.

The biasing networks for Si based amplifiers must be designed carefully to avoid low-frequency oscillations in the range of a few kilohertz to hundreds of megahertz. RF amplifiers are more susceptible to oscillations due to low substrate resistivity, high parasitic capacitance, and high device gain [6]. This can be avoided by using low-frequency resonance-free biasing networks consisting of two sections of RF choke and shunt bypass capacitor. The section connected to the RF circuit is comprised of the low-inductance RF choke and low-value bypass capacitor (20–50 pF) so that their impedance is 5–10 times higher than the impedance presented by the amplifier circuit at that location. This is followed by a high-inductance RF choke, for more isolation between RF and DC, and a high-value bypass capacitor (a combination of 1 nF and 10 μF) for bypassing low-frequency signals from power supply/external circuits.

17.2 ODD-MODE OSCILLATIONS

In power amplifiers a cluster matching network approach is often used, which generally employs an even number of transistor cells. When transistors are placed in parallel and each device has its own partial input and output match, a potential problem with possible odd-mode oscillations can develop. The transistors used in power amplifiers (MMIC or MIC) can have slight variations in I_{ds}, V_P, and g_m or different source or collector inductance, which result in slightly different gains and powers in the transistors. Also, the matching networks used can provide slightly different input and output impedances at the different transistors due to coupling between circuit elements and T or cross junctions. Thus in the parallel transistor configuration, each device as per design does not necessarily see exactly the same match. This slight mismatch creates a condition for odd-mode or differential-mode excitation, due to different RF voltage levels at the transistor drain locations. This leads to negative resistance or odd-mode oscillations. These oscillations are also called parallel-transistor oscillations and may also arise due to finite isolation in the transistors. Consider a two-port amplifier using two transistors combined in parallel employing sections of transmission line as shown in Figure 17.9. In this configuration, there exists a feedback loop between the input and output ports. Due to the high device gain and finite isolation between the input and output of the device, a condition of instability might occur, giving rise to a loop oscillation. The odd-mode oscillations are normally in and near the operating band or much above the operating frequency range of the amplifier. This is because the total length of the loop is close to one wavelength at the operating frequency.

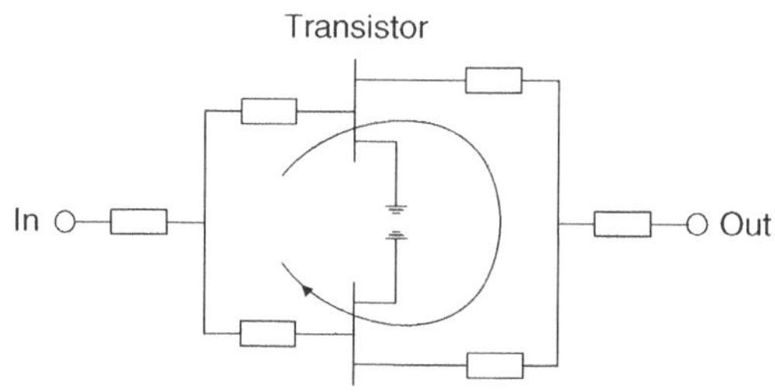

Figure 17.9 An amplifier configuration using two transistors in parallel.

17.2.1 Odd-Mode Stability Analysis

The odd-mode oscillations can generally be predicted by using different small-signal S-parameters measured on a wafer for devices and accurate multiport simulations of the matching networks using EM simulators. Analysis of such oscillations has been described in the literature [7–10]. Mochizuki et al. [11] have analyzed subharmonic (or $f/2$) loop oscillations due to odd-mode excitation. This type of odd-mode oscillation occurs over narrowband and under RF drive conditions near some gain compression region. In this case, the output power suddenly decreases and an $f/2$ oscillation signal appears; when the input power increases further, the power comes back and the oscillations disappear. Such odd-mode oscillations are a strong function of bias voltages and RF input power and are independent of source and load impedance.

The odd-mode stability analysis may be grouped into four categories: odd-mode analysis [7–9], Nyquist stability criterion [12–14], normalized determinant function (NDF) technique [15–17], and global stability analysis [18, 19]. The NDF technique is closely related to the Nyquist stability criterion. The degree of versatility and complexity is lowest for the odd-mode analysis and highest in the case of global stability analysis. The first three techniques are based on linear device models and are discussed next, while for the global stability analysis that is based on a device's nonlinear model, readers are referred to References 18 and 19 for more details.

Odd-Mode Analysis

Consider a single-stage power amplifier shown schematically in Figure 17.10. In this design, four transistors are combined using a cluster matching approach. Planes AA and BB are the symmetry planes for the half and full amplifiers, respectively. R_1 and R_2 are the isolation resistors and their initial values are very large, >1 kΩ. The odd-mode analyses are carried out by treating the symmetry planes as virtual ground or push–pull one at a time. In the case of half of the circuit, when the points of symmetry plane AA, as shown in Figure 17.10, are shorted to ground (forcing a virtual ground), the circuit shown in Figure 17.11 is analyzed for odd-mode or push–pull oscillations. When the input reflection coefficient is calculated over a wide frequency range and its value is greater than 1 with a phase angle of 180° at some frequency, the circuit supports odd-mode oscillations at that frequency [8, 9]. The values of R_1 and R_2 are adjusted so that the input reflection coefficient is less than unity with enough margins reflecting the process variations in devices. The same analysis is repeated at the output by shorting the input, making sure the selections of R_1 and R_2 are correct. Their values do not affect the RF circuit performance as long as the circuit physically and electrically remains symmetrical. The selections of R_1 and R_2 depend on the amplifier design. In MMICs, the physical size of R_2 must be selected carefully to handle the

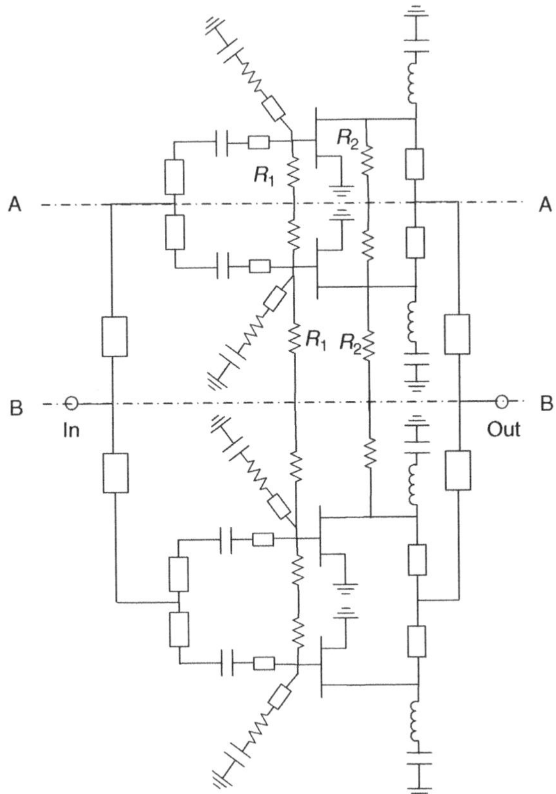

Figure 17.10 A four-FET power amplifier showing symmetry planes for odd-mode analysis.

output power level due to odd-mode excitation or any other imbalance in the output matching circuit that might occur. In the case of FETs/HEMTs, the value of 20 Ω · mm of the device periphery for R_1 and R_2 may be used. A similar analysis is performed for the full circuit. In the case of eight-way combining at the output, one needs to perform odd-mode analysis for quarter, half, and full circuit blocks. The blocks of circuit can also be analyzed for odd mode as a push–pull configuration shown in Figure 17.11b.

The odd-mode analysis discussed above is limited to symmetric structures. However, in power amplifiers when several devices are paralleled, some degree of asymmetry is always introduced. A more general stability analysis technique, such as the Nyquist stability criterion or NDF, is required.

Nyquist Stability Criterion Technique

Consider the general feedback circuit shown in Figure 17.12a, where $A(\omega)$ is the forward or open-loop gain and $F(\omega)$ is the feedback gain. Both $A(\omega)$ and $F(\omega)$ are frequency dependent. In this case the closed-loop gain v_o/v_i is given by

$$\frac{v_o}{v_i} = G_v = \frac{A(\omega)}{1 - A(\omega)F(\omega)} \tag{17.12}$$

where $A(\omega)F(\omega)$ is known as the loop gain. The loop gain, which is a function of ω, is a complex quantity. This can be expressed in real or imaginary quantities or in polar

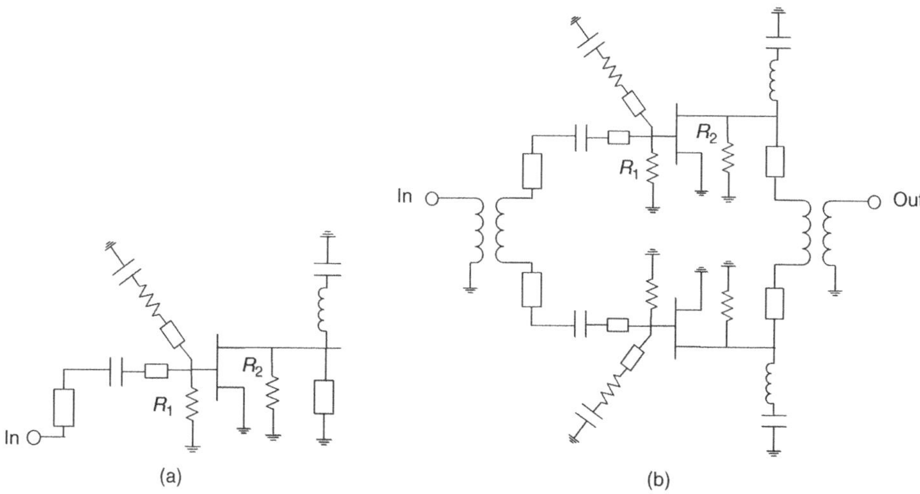

Figure 17.11 Two-FET power amplifier configurations for odd-mode analysis: (a) single-ended and (b) push–pull using baluns.

coordinates. The amplifier will oscillate when

$$1 - A(\omega)F(\omega) = 0 \quad \text{or} \quad |A(\omega)||F(\omega)| > 1$$

and

$$\angle A(\omega) + \angle F(\omega) = 2n\pi, \quad \text{where } n = 0,\ 1,\ 2, \ldots$$

According to the Nyquist stability criterion, for a stable loop gain, the closed-loop system will be unstable when the polar plot of $A(\omega)F(\omega)$ encircles 1 at least once for $-\infty < \omega < \infty$. In other words, the closed-loop gain function has right-half plane poles. This is shown in Figure 17.12b. The stability criterion for any closed-loop system is that the phase shift around the loop must never reach 360° for any frequency at which the gain is unity or greater. If the amplifier stage is a common-source configuration, the signal at the output has an intrinsic 180° phase shift with respect to the input. This dictates that the additional phase shift in the amplifier plus the phase shift of the feedback network must be less than 180° for all frequencies where gain is unity or greater.

Jackson [12] has suggested a practical way to analyze the stability of a microwave power amplifier using commercial CAD tools. First, consider a generic representation of a microwave circuit, where Z_d is the device impedance and Z_r is the impedance of the resonator circuit. Next, let us assume that the circuit is excited by a voltage v_i through an ideal coupler with 0-dB coupling as shown in Figure 17.13. The output voltage v^+ is given by

$$v^+ = \Gamma_d(v_i + v^-) = \Gamma_d v_i + \Gamma_d \Gamma_r v^+ \text{ or } v^+ = \frac{\Gamma_d}{1 - \Gamma_d \Gamma_r} v_i \tag{17.13}$$

where Γ_d, Γ_r are the device and resonator circuit reflection coefficients, and v^- is the reflected voltage from the resonator. There is an analogy between (17.12) and (17.13),

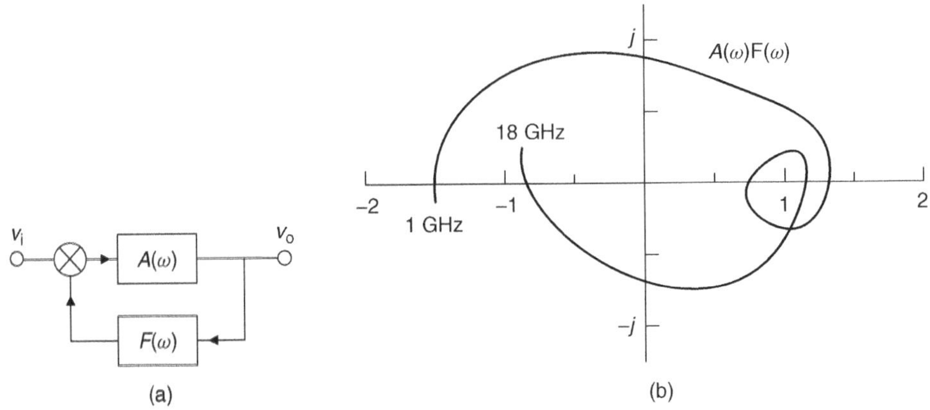

Figure 17.12 (a) Microwave circuit with feedback loop. (b) Plot of $A(\omega)F(\omega)$ versus frequency on a complex plane.

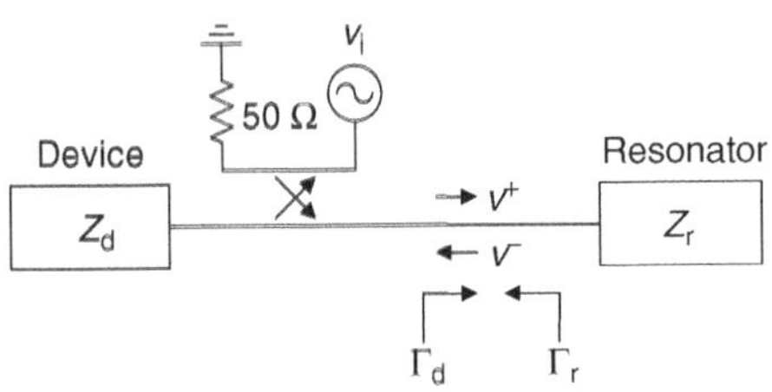

Figure 17.13 Excitation of a microwave circuit with an ideal directional coupler.

and (17.12) represents the closed-loop gain. The circuit in Figure 17.13 is unstable when the polar plot of $\Gamma_d\Gamma_r$ encircles 1 at least once for $-\infty < \omega < \infty$.

The $\Gamma_d\Gamma_r$ product can easily be evaluated by using commercial CAD tools. This can be accomplished by replacing the directional coupler with an ideal circulator, as shown in Figure 17.14. The S-parameters of an ideal circulator are given by

$$[S]_c = \begin{bmatrix} 0 & 1 & 0 \\ 0 & 0 & 1 \\ 1 & 0 & 0 \end{bmatrix} \tag{17.14}$$

As shown in Figure 17.14, S_{11} at port 1 of the circulator is nothing but the loop gain term $\Gamma_d\Gamma_r$. A polar plot versus frequency of this function encircling 1 indicates that the circuit is unstable. A more generic approach of this method is described by Ohtomo [13]. The technique can also be applied to study the instability in very large gate periphery transistors [17, 20] and distributed amplifiers [21].

EXAMPLE 17.4

Design a power amplifier using two transistors connected in parallel. The matching networks are lossless and the center frequency is 10 GHz. Study its even-mode and odd-mode stability. For this study the Toshiba JS8853-AS medium-power GaAs FET in chip form, whose EC model is given below [12], was selected.

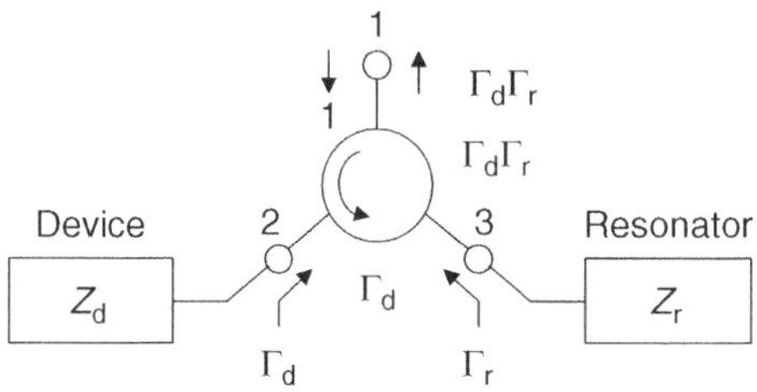

Figure 17.14 Microwave circuit representation of Figure 17.13 with an ideal circulator.

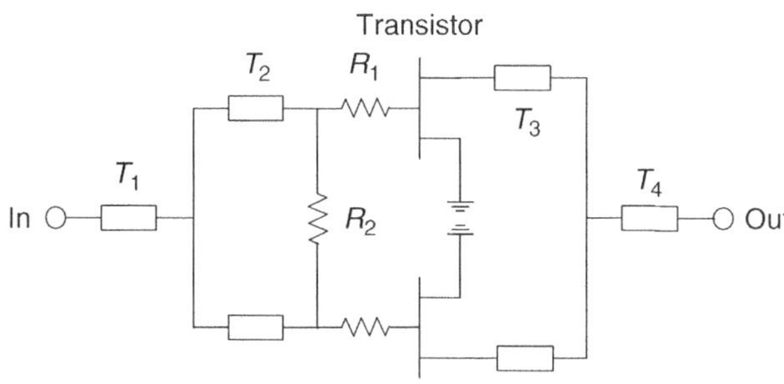

Figure 17.15 An amplifier configuration using two transistors in parallel and stabilization resistors.

$$R_g = 0.62\ \Omega,\ \ R_i = 0.56\ \Omega,\ \ R_s = 0.34\ \Omega,\ \ R_{ds} = 80\ \Omega,\ \ R_d = 0.56\ \Omega$$
$$g_m = 155\ \text{mS},\ \ \tau = 7.2\ \text{ps}$$
$$C_{gs} = 1.4\ \text{pF},\ \ C_{gd} = 0.12\ \text{pF},\ \ C_{ds} = 0.62\ \text{pF}$$
$$L_g = 0.0\ \text{nH},\ \ L_s = 0.01\ \text{nH},\ \ L_d = 0.0\ \text{nH}$$

SOLUTION The amplifier schematic is shown in Figure 17.15 and was optimized using ideal transmission lines. No parasitic or discontinuity effects are included in the design. The design parameters at 10 GHz are given below [13]:

$$T_1 = 8.2\ \Omega, 70.2^\circ;\ \ T_2 = 15.2\ \Omega, 52.5^\circ;\ \ T_3 = 15.0\ \Omega, 72.8^\circ;\ \ T_4 = 11.3\ \Omega, 76.0^\circ$$

The values of R_1 and R_2 are 0 Ω and 100 kΩ respectively. For convenience, this design is referred to as "Design A."

Figure 17.16 shows the schematic used to calculate the open-loop transfer function (OTF) to test the odd-mode stability. The amplifier was simulated over 1–40 GHz; however, limited data was selected that is relevant for this discussion. The calculated performance is given in Table 17.2. At about 7.1 GHz, $|S_{11s}| = 1.08$ and angle = 0 and changes sign. This indicates that the amplifier has odd-mode instability.

Next, we consider the effect of R_1 and R_2 on the amplifier performance and OTF.

1. $R_1 = 1.2\ \Omega$ and $R_2 = 100\ \text{k}\Omega$. In this case $|S_{11s}| < 1$ and the amplifier gain at 10 GHz is reduced from 10.9 dB to 9.2 dB. This also improves the μ factor.
2. $R_1 = 0\ \Omega$ and $R_2 = 194\ \Omega$. In this case $|S_{11s}| < 1$ for all frequencies and the amplifier performance is unaffected. Thus the isolation resistor works more effectively than the series resistor. ∎

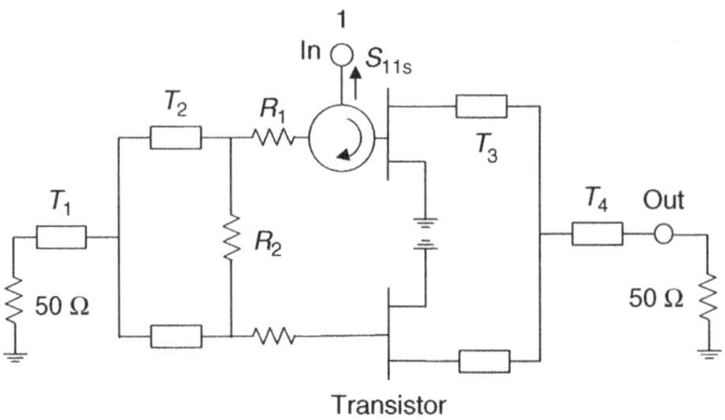

Figure 17.16 Circuit representation of Figure 17.15 with an ideal circulator for odd-mode stability analysis.

Table 17.2 Summary of Design A Amplifier Performance and OTF

Frequency GHz	Input RL (dB)	Gain (dB)	Output RL (dB)	μ	$\|S_{11s}\|$	$\angle S_{11s}$
6.8	0.30	−3.0	0.89	0.92	1.08	6.7
6.9	0.32	−2.8	0.91	0.92	1.08	4.3
7.0	0.33	−2.5	0.94	0.91	1.08	1.7
7.1	0.34	−2.3	0.97	0.91	1.08	−1.1
7.2	0.35	−2.0	1.00	0.91	1.08	−4.2
9.6	6.00	10.6	7.64	0.95	0.68	−0.6
9.8	9.08	11.0	9.69	0.98	0.71	−3.8
10.0	11.35	10.9	10.35	1.01	0.72	−7.0
10.2	10.16	10.3	9.53	1.03	0.73	−10.0
10.4	7.98	9.4	8.47	1.04	0.74	−13.0

EXAMPLE 17.5

Replace the EC model for the Toshiba JS8853-AS medium-power GaAs FET with a pHEMT whose EC model is given in Table 5.10, #1 (Chapter 5) in Design A. Study its even-mode and odd-mode stability.

SOLUTION The simulated amplifier performance and S_{11s} are given in Table 17.3. The amplifier still has odd-mode instability at 8.8 GHz although at higher frequency. This suggests that the odd-mode instability is a strong function of the circuit design rather than the device itself. This will be further discussed in the following.

In Design A the ideal transmission lines are replaced with microstrip sections on a 100-μm thick GaAs substrate and gold conductors are 4.5 μm thick. Microstrip loss is included in the calculations; however, no parasitic or discontinuity effects are included in the design. The amplifier schematic is the same as shown in Figure 17.15. The amplifier design was optimized at 10 GHz using the pHEMT EC model (Table 5.10, #1). The design parameters are given below:

$$T_1 = 12, 3500;\quad T_2 = 90, 2890;\quad T_3 = 12, 1500;\quad T_4 = 190, 1800$$

All dimensions (width, length) are in micrometers (μm). The values of R_1 and R_2 are 1 Ω and 100 kΩ, respectively. This design is referred to as "Design B."

The simulated amplifier (Design B) performance and S_{11s} values are given in Table 17.4. The amplifier still has odd-mode instability at 4 GHz although at lower frequency. Next, the effect of R_1 and R_2 on the amplifier performance and OTF is considered.

Table 17.3 Summary of Design A Amplifier Performance and OTF[a]

Frequency (GHz)	Input RL (dB)	Gain (dB)	Output RL (dB)	μ	$\lvert S_{11s}\rvert$	$\angle S_{11s}$
8.6	0.49	3.8	0.38	0.81	1.11	4.7
8.7	0.53	4.2	0.39	0.80	1.13	2.2
8.8	0.57	4.7	0.40	0.79	1.15	−0.6
8.9	0.62	5.2	0.40	0.78	1.18	−3.7
9.0	0.67	5.8	0.41	0.77	1.21	−7.5
9.6	1.31	9.7	0.51	0.68	0.95	−68.0
9.8	1.77	11.2	0.61	0.65	0.31	−68.3
10.0	2.51	12.8	0.84	0.61	0.33	−5.4
10.2	3.73	14.3	1.33	0.58	0.50	−0.3
10.4	5.54	15.5	2.32	0.56	0.59	−3.6
10.6	7.24	16.1	4.07	0.55	0.65	−7.9
10.8	7.07	15.9	6.52	0.55	0.68	−12.4
11.0	5.73	15.2	9.05	0.57	0.69	−16.8

[a] In this case, the FET is replaced with a pHEMT.

Table 17.4 Summary of Design B Amplifier Performance and OTF

Frequency (GHz)	Input RL (dB)	Gain	Output RL (dB)	μ	$\lvert S_{11s}\rvert$	$\angle S_{11s}$
3.8	0.97	10.5	8.43	0.92	1.22	14.3
3.9	0.94	10.3	8.33	0.92	1.24	6.7
4.0	0.92	10.0	8.25	0.93	1.25	−0.9
4.1	0.90	9.8	8.16	0.93	1.26	−8.5
4.2	0.88	9.6	8.08	0.94	1.26	−16.1
9.6	7.91	14.1	9.34	0.99	0.38	−22.0
9.8	10.70	14.2	9.53	1.01	0.37	−10.8
10.0	12.09	14.1	9.61	1.04	0.36	−0.5
10.2	10.27	13.7	9.58	1.07	0.35	−11.8
10.4	7.84	13.1	9.46	1.10	0.34	−23.3

1. $R_1 = 5.6\ \Omega$ and $R_2 = 100\ \text{k}\Omega$. In this case $|S_{11s}| < 1$ and the amplifier gain at 10 GHz is reduced from 14.1 dB to 11.4 dB. This also improves the μ factor.
2. $R_1 = 1\ \Omega$ and $R_2 = 630\ \Omega$. In this case $|S_{11s}| < 1$.

Next the amplifier Design B was reoptimized at 10 GHz using the pHEMT EC model (Table 5.10, #1) and using a small fixed value of T_3 line length to reduce the loop length. The reoptimized design parameters are given below:

$$T_1 = 40, 4000; \quad T_2 = 190, 3000; \quad T_3 = 12, 300; \quad T_4 = 190, 3350$$

The values of R_1 and R_2 are 1 Ω and 100 kΩ, respectively. This design is designated as "Design C."

The amplifier was simulated over 1–40 GHz. In this case, the amplifier has no odd-mode instability. The amplifier (Design C) performance and S_{11s} values are given in Table 17.5. The amplifier performances for Design B and Design C are similar. This suggests that the odd-mode instability is a strong function of the circuit design.

The amplifier Design B was further investigated by strapping the two drains together by a short high-impedance line (width = 12 μm and length = 200 μm) as shown in Figure 17.17.

Table 17.5 Summary of Design C Amplifier Performance and OTF

Frequency (GHz)	Input RL (dB)	Gain	Output RL (dB)	μ	$\lvert S_{11s}\rvert$	$\angle S_{11s}$
9.0	4.28	13.7	6.22	0.83	0.43	38.4
9.2	5.48	14.1	6.63	0.84	0.42	30.9
9.4	7.03	14.5	7.12	0.85	0.41	23.6
9.6	8.82	14.6	7.66	0.87	0.40	16.3
9.8	10.34	14.6	8.20	0.89	0.39	9.2
10.0	10.64	14.4	8.68	0.92	0.37	2.0
10.2	9.64	14.1	9.05	0.94	0.35	5.2
10.4	8.23	13.6	9.28	0.96	0.33	12.6
10.6	6.96	13.1	9.38	0.98	0.31	20.2
10.8	5.95	12.5	9.38	1.01	0.28	28.3
11.0	5.16	11.9	9.30	1.03	0.26	36.8

In this case each drain sees an open-circuited stub 100 μm long; it has negligible effect on the amplifier performance or it can be absorbed in the design. In this case, the amplifier has no odd-mode instability. This topology is preferred to the one having isolation resistors at the output of transistors because it provides higher combining efficiency. ■

NDF

In a multistage power amplifier using several transistors or a large-periphery transistor using many cells, the stability analysis becomes complex and challenging. An effective way is to use a more general stability analysis technique known as NDF [15–17]. In this method, a composite stability criterion is established that includes all transistor nodes.

A simple representation of a circuit having N active devices for NDF stability calculation is shown in Figure 17.18. Consider a cell with active device 1. Referring to (17.12), the return ratio (RR) is given by

$$RR_1 = -AF = -V_{o1}/V_{e1} \tag{17.15}$$

where V_{o1} is the voltage across the gate capacitor and V_{e1} is the external voltage source for cell 1. In this case the g_m of the sources from 2 to N are set to zero. NDF for

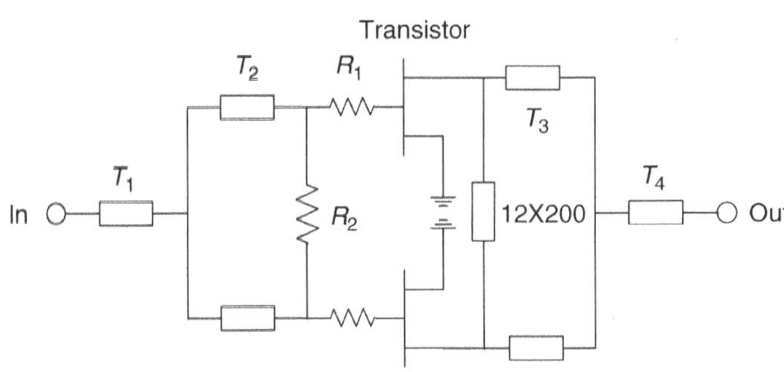

Figure 17.17 An amplifier with modified output match for improved loop instability.

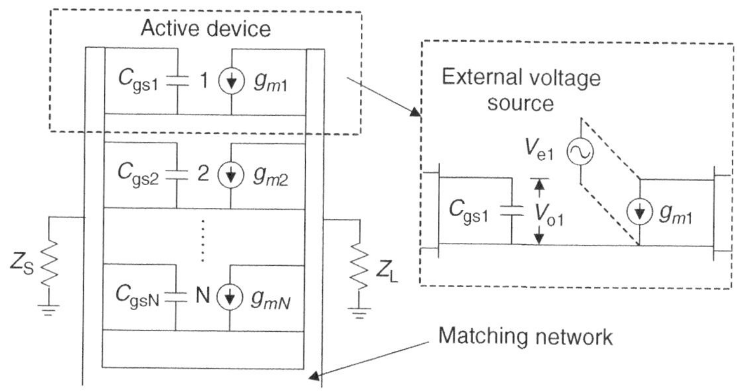

Figure 17.18 A simplified representation of a circuit having N active devices for NDF stability analysis.

source 1 can be calculated from

$$NDF_1 = RR_1 + 1 \tag{17.16}$$

The total NDF for all sources (NDF_t) is given by

$$NDF_t = NDF_1 \times NDF_2 \times \cdots \times NDF_N$$

Generally, the NDF_t response is plotted on a *polar chart*. If it encircles the origin, the N-port circuit is unstable and the cross point on the negative real axis determines the frequency of oscillation [17].

17.2.2 Odd-Mode Oscillation Suppression Techniques

In order to minimize the possibility of odd-mode oscillations, one should use well-matched transistors in terms of magnitude and phase, the same source inductance connected to FETs, and balanced matching networks connected to transistors. Source or emitter imbalance in transistors also leads to odd-mode oscillation. If all conditions described above are hard to meet, one can minimize the odd-mode oscillation possibility either by connecting all gate feeds and all drain feeds by the shortest possible line lengths (or bond wire lengths), or by connecting the required resistors between gate feeds and drain feeds (see Fig. 17.10). The resistors connected between the gate feeds and drain feeds are called isolation resistors. Values of the isolation resistors are determined using odd-mode analysis and their values are generally between 10 and 50 Ω. Often small resistors connected in series in the gate and drain pads are good enough to suppress odd-mode oscillations. The purpose of these resistors is to terminate or damp out these oscillations. Series resistors will also reduce the gain and power of the amplifier. A common technique is to use isolation resistors, which take care of the unbalanced signals. However, isolation resistors also reduce the gain and power of the amplifiers but at lower magnitude. In most cases, the odd-mode oscillation occurs under RF drive conditions.

17.2.3 Instability in Distributed Amplifiers

If the feedback capacitance between the input and output terminals of a transistor is high, this leads to cell/loop instability in distributed amplifiers (DAs). Such instability occurs near the equivalent gate cutoff frequency of a DA (see Chapter 11) due to a large value of feedback capacitor C_{gd} of an FET/HEMT. An oscillation strongly depends on the g_m value of the transistor. This type of instability in DAs generally leads to gain peaking at the high end of the amplifier frequency band and the input/output reflection coefficient becomes greater than unity. The instability becomes more pronounced with an increased value of g_m. If a DA circuit is stable at room temperature, it might oscillate at lower temperatures because of the transistor's increased g_m value.

Instability in DAs can be minimized by selecting the device bias conditions so that the ratio g_m/C_{gd} is largest, by using a cascode configuration, a larger number of cells, and small device size.

17.3 PARAMETRIC OSCILLATIONS

When transistors are driven in a nonlinear region, a parametric subharmonic oscillation occurs due to pumping of a nonlinear reactance by RF voltage at a multiple of the oscillation frequency (which is a subharmonic of the operating or pump frequency). In this case, a subharmonic frequency is injection locked to the fundamental and results in negative resistance at the operating frequency. This negative resistance was originally reported by Manly and Rowe [22, 23].

In BJTs and HBTs, the nonlinear reactance is due to the base–collector capacitance C_{bc}, whereas in MESFETs and pHEMTs, C_{gs} and C_{gd} constitute the pumped nonlinear reactance. In bipolar transistors, the base charge storage has a much longer time constant than the carrier lifetime in unipolar FETs and pHEMTs. This leads to much higher order subharmonic oscillations (on the order of 30) than in FETs and pHEMTs (two or three times lower than operating frequency) [10], as shown in Figure 17.19. The matching networks along with the package parasitics and load termination form an *LC* resonator whose resonance frequency is the subharmonic of the operating frequency. Such resonators support parametric oscillations, which are a strong function of input drive power, bias conditions, frequency of operation, and load termination conditions and can occur at the input or output port of the transistor.

Analysis of parametric oscillations has been reported in the literature [24] and can be modeled using an accurate nonlinear device model along with commercially available nonlinear software. For such applications, the transistor capacitance must be modeled accurately to predict the pumped nonlinear reactance or negative resistance.

Suppression of parametric oscillations in broadband amplifiers poses a serious challenge. However, in narrowband amplifiers this can be achieved either by designing matching networks having sharp roll-off at low frequencies (so that the amplifier has much lower gain at the subharmonic frequencies as compared to the gain at the fundamental frequency) or by using lossy matching and biasing networks. In the latter case, the negative resistance at the device port becomes effectively a positive resistance. Figure 17.20 shows commonly used techniques to prevent parametric oscillations by increasing matching network loss at lower frequencies. Here, a parallel combination of *R* and *C* is used at the input of the transistor. At low frequencies the resistor *R* attenuates the signal, and at the operating frequencies the resistor has little effect on the gain because of the capacitor *C* reactance, which shorts out the resistance.

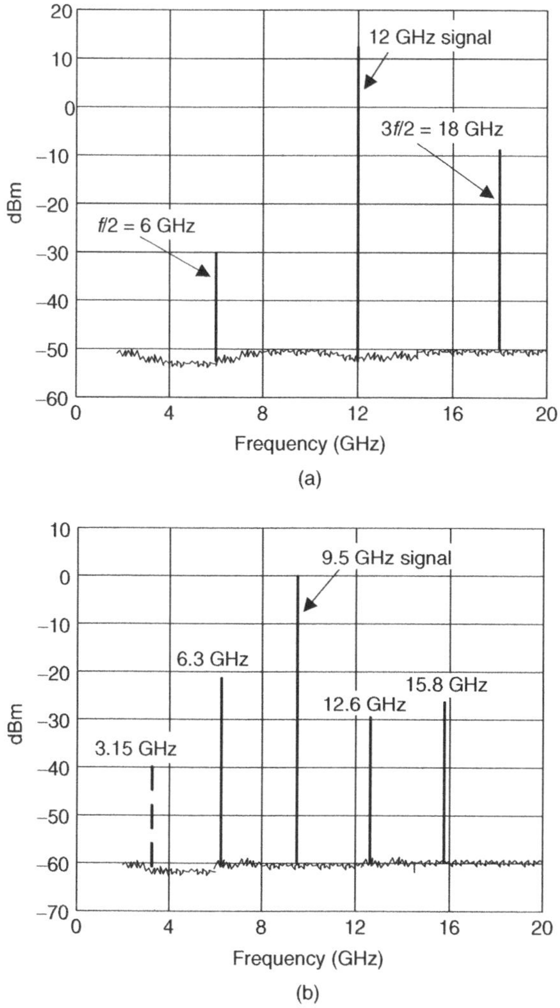

Figure 17.19 Parametric oscillations: (a) at $f/2$ and (b) at $f/3$.

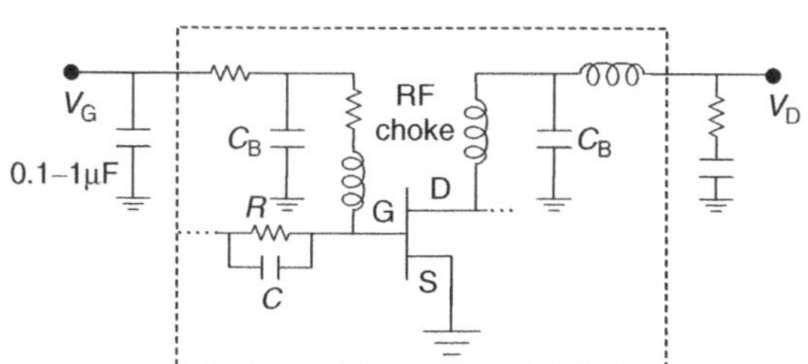

Figure 17.20 Circuit stabilization scheme to prevent parametric oscillations.

17.4 SPURIOUS PARAMETRIC OSCILLATIONS

Spurious parametric oscillations in transistor amplifiers are briefly discussed in References 25 and 26. Spurious parametric oscillations occur as the superposition of a nonlinear amplifier and a linear amplifier bias point determined at the input drive power level. Since an FET's EC model parameters change with drive power level, the

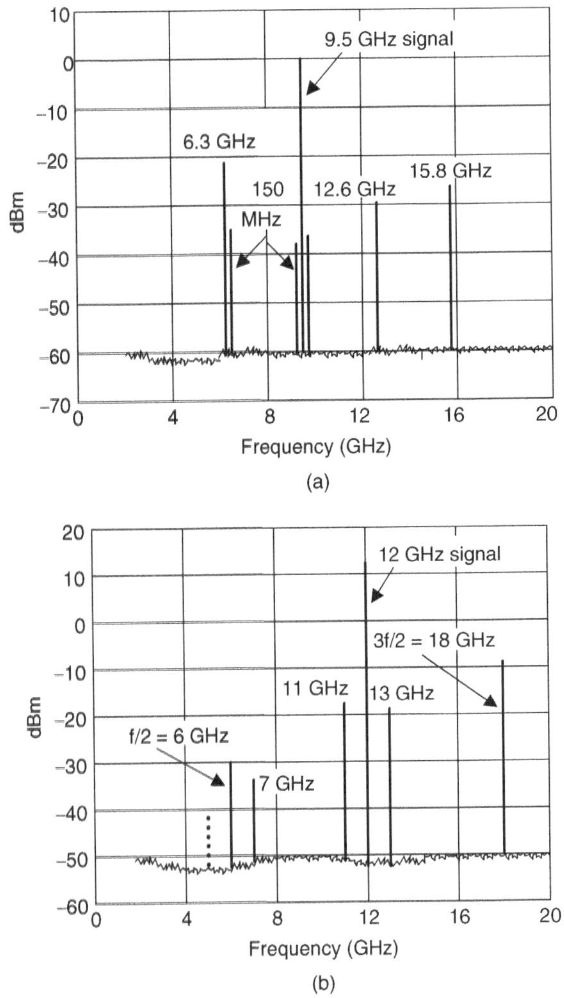

Figure 17.21 (a) Spurious parametric oscillations at (a) $f/3$ and 150-MHz bias and (b) $f/2$ and 7-GHz even-mode/odd-mode.

stability analysis done using small-signal S-parameters does not hold anymore and the amplifier might have even-mode or odd-mode instability at higher drive power levels. If conditions are met to support parametric oscillations due to nonlinear device operation and even-mode or odd-mode oscillations at some other frequency due to the linear operation of the device at that drive power level, the superposition of these oscillations is termed *spurious parametric oscillations*. Techniques used to prevent parametric oscillations also apply here. In addition, one can make sure the design works at other bias conditions where C_{gs} goes down by about 30% and C_{gd} goes up by about 30%. If the IC design has $K > 2$ for each stage at the nominal bias conditions, spurious parametric oscillation should be prevented. Figure 17.21 shows two different types of spurious parametric oscillations.

17.5 LOW-FREQUENCY OSCILLATIONS

The low-frequency oscillations [27] constitute the fifth class, which is attributed to the defects in the transistors, giving rise to complicated nonlinear behavior of the

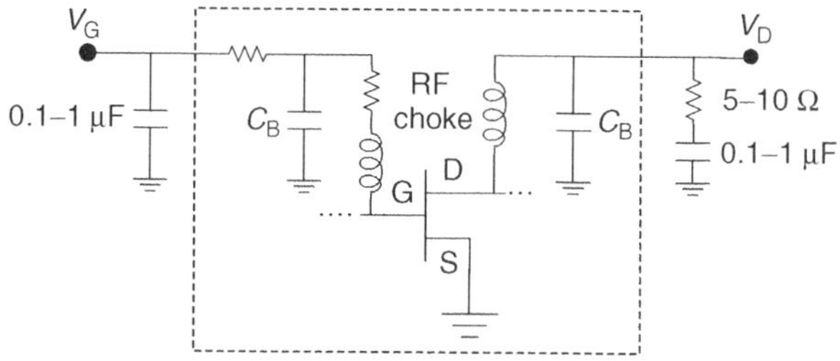

Figure 17.22 Circuit stabilization scheme to prevent low-frequency oscillations.

Table 17.6 Summary of RF and Microwave Oscillations in Amplifiers

Type of Oscillation	Condition of Instability[a]	Analysis	Preventive Measures
Even	DC, SS, LS	Linear and nonlinear	K or $\mu > 1$
Odd	SS, LS	Linear and nonlinear	NDF analysis
Parametric	LS	Nonlinear	Low device input nonlinearity

[a]SS, small signal; LS, large signal.

devices. These oscillations are also known as *chaotic* (*strange*) and cannot be predicted by using the analyses described earlier. Such oscillations are generally observed in high-power transistors/ICs using large-gate-periphery devices and occur in the frequency range of 10 Hz to 50 MHz. Figures 17.20 and 17.22 show effective techniques to prevent low-frequency oscillations. This is mostly done outside the chip. One can also use a resistive corporate biasing network, as described in Section 18.5, to prevent low-frequency oscillations.

In this chapter a comprehensive treatment of theoretical and practical aspects of amplifier stability with several examples has been described. Table 17.6 provides a summary of oscillation conditions, the type of analysis being used to predict the instability in amplifiers, and preventive measures.

REFERENCES

1. T. T. Ha, *Solid-State Microwave Amplifier Design*, John Wiley & Sons, Hoboken, NJ, 1981.
2. G. Gonzalez, *Microwave Transistor Amplifiers Analysis and Design*, Prentice Hall, Englewood Cliffs, NJ, 1984.
3. G. D. Vendelin, A. M. Pavio, and U. O. Rohde, *Microwave Circuit Design Using Linear and Nonlinear Techniques*, John Wiley & Sons, Hoboken, NJ, 1990.
4. K. Chang, I. Bahl, and V. Nair, *RF and Microwave Circuit and Component Design for Wireless Systems*, John Wiley & Sons, Hoboken, NJ, 2002, Chapters 8 and 11.
5. M. E. Edwards and J. H. Sinsky, A new criterion for linear 2-port stability using a single geometrically derived parameter, *IEEE Trans. Microwave Theory Tech.*, Vol. 40, pp. 2303–2311, December 1992.
6. A. Raghavan, N. Srirattana, and J. Laskar, *Modeling and Design Techniques for RF Power Amplifiers*, John Wiley & Sons, Hoboken, NJ, 2008.
7. J. G. Kassakian and D. Lau, An analysis and experimental verification of parasitic oscillations in paralleled power MOSFETs, *IEEE Trans. Electron Devices*, Vol. ED-31, pp. 959–963, July 1984.
8. R. G. Freitag et al., Stability and improved circuit modeling considerations for high power MMIC amplifiers, *IEEE Microwave Millimeter-Wave Monolithic Circuits Symp. Dig*., pp. 125–128, 1988.

9. R. G. Freitag, A unified analysis of MMIC power amplifier stability, *IEEE MTT-S Int. Microwave Symp. Dig.*, pp. 297–300, 1992.
10. S. R. Nelson, Amplifier stability in short course, power amplifiers: from milliwatts to kilowatts ... cool devices with hot performance, in *1998 IEEE GaAs IC Symposium*.
11. M. Mochizuki et al., Nonlinear analysis of $f/2$ loop oscillation of high power amplifiers, *IEEE MTT-S Int. Microwave Symp. Dig.*, pp. 709–712, 1995.
12. R. W. Jackson, Criteria for the onset of oscillation in microwave circuits, *IEEE Trans. Microwave Theory Tech.*, Vol. 40, pp. 566–569, March 1992.
13. M. Ohtomo, Stability analysis and numerical simulation of multi-device amplifiers, *IEEE Trans. Microwave Theory Tech.*, Vol. 41, pp. 983–991, June/July 1993.
14. S. Pisa and M. Zolesi, A method for stability analysis of small-signal microwave amplifiers, *Int. J. RF Microwave Computer-Aided Eng.*, Vol. 8, pp. 293–302, July 1998.
15. A. Platzker, W. Struble, and K. T. Hetzler, Instabilities diagnosis and the role of K in microwave circuits, *IEEE MTT-S Int. Microwave Symp. Dig.*, pp. 1185–1188, 1993.
16. W. Struble and A. Platzker, A rigorous yet simple method for determining stability of linear N-port networks, *IEEE GaAs IC Symp. Dig.*, pp. 251–254, 1993.
17. S. Goto et al., Stability analysis and layout design of internally stabilized multi-finger FET for high power base station amplifiers, *IEEE MTT-S Int. Microwave Symp. Dig.*, pp. 231–234, 2003.
18. S. Mons et al., A unified approach for the linear and nonlinear stability analysis of microwave circuits using commercially available tools, *IEEE Trans. Microwave Theory Tech.*, Vol. 47, pp. 2403–2409, December 1999.
19. A. Suarez and R. Quere, *Global Stability Analysis of Microwave Circuits*, Artech House, Norwood, MA, 2003.
20. F. E. van Vliet and A. de Boer, On the stability of MMICs using transistors with inductive source feedback, in *12th GaAs Symposium Digest*, Amsterdam, 2004, pp. 351–354.
21. P. Gamand, Analysis of the oscillation conditions in distributed amplifiers, *IEEE Trans. Microwave Theory Tech.*, Vol. 37, pp. 637–640, March 1989.
22. J. M. Manley and H. E. Rowe, Some general properties of nonlinear Elements—Part I. General energy relations, pp. 904–913, *Proc. IRE*, July 1956.
23. P. Penfield and R. Rafuse, *Varactor Applications*, MIT Press, Cambridge, MA, 1962.
24. J. F. Imbornone et al., New insight into subharmonic oscillation mode of GaAs power amplifiers under severe output mismatch condition, *IEEE J. Solid State Circuits*, Vol. 32, pp. 1319–1325, September 1997.
25. D. Teeter, A. Platzker, and R. Bourque, A compact network for eliminating parametric oscillations in high power MMIC amplifiers, *IEEE MTT-S Int. Microwave Symp. Dig.*, pp. 967–970, 1999.
26. O. Muller and W. Figel, Stability problems in transistor power amplifiers, *Proc. IEEE*, Vol. 55, pp. 1458–1466, August 1967.
27. D. Miller and M. Bujatti, Mechamisms for low-frequency oscillations in GaAs FETs, *IEEE Trans. Electron Devices*, Vol. ED-34, pp. 1239–1244, June 1987.

PROBLEMS

17.1 The EC model values of a 300-μm GaAs FET are given in Table 5.4. Determine the stability of the device at 15 GHz and identify the element values that give rise to instability.

17.2 The scattering parameters of a GaAs FET are given in Table 5.1, which shows that the FET is potentially unstable at 6 GHz. Draw stability circles at 6 GHz and show how one can stabilize the device without increasing the NF appreciably.

17.3 The scattering parameters for two different transistors are given in Tables 5.1 and 5.9. Calculate the stability factor K in each case and draw input and output stability circles on the Smith chart at 2 GHz. Show which transistor is suitable for a 2-GHz amplifier to obtain 1 dB NF and 15 dB gain.

17.4 The common-source scattering parameters of a GaAs FET are given in Table 5.1. Determine the common-gate and common-drain S-parameters and the stability factor in all the three cases at 2 GHz.

17.5 The scattering parameters of a pHEMT are given in Table 5.9. Calculate the stability of the transistor from 2 to 26 GHz with 4-GHz step. Show the stable region at 10 GHz.

17.6 The S-parameter data for an FET is given in Table 5.1 (Chapter 5) measured in a 50-Ω system.

(a) Calculate the conjugate input and output impedance for maximum gain at 2 GHz. Determine the stability and plot the stability circles on a Smith chart. Calculate the center and radius of input and output stability circles.

(b) If a load of $Z_L = 100\ \Omega$ impedance is connected at the output, determine the input conjugate match impedance and the transducer gain. If $Z_L = 200\ \Omega$, recalculate the above parameters.

(c) Calculate the maximum stable gain of the transistor at 5,10,15, and 20 GHz. Determine the maximum available gain if the conjugate match impedance at the input and output has a 0.5-dB loss.

Chapter 18

Biasing Networks

Biasing networks are important parts of amplifier design. They provide the efficient excitation means for transistors. The biasing parameters can vary with transistor technology; however, the design principles for biasing methods are basically the same. The purpose of this chapter is to include a comprehensive treatment of biasing techniques used from UHF to millimeter-wave frequencies.

There are many biasing schemes for amplifiers used in practice [1–6]. The biasing circuits could be stand-alone or they can be integrated in the matching networks. The reactive effect of biasing network is negligible or absorbed in the matching. In other words, when bias voltage is applied to the device, the RF energy should not leak out through the bias port, otherwise degradation of RF performance including gain, noise figure, output power, PAE, and IP3 takes place. It must isolate the bias voltage applied to various devices. The biasing circuits also provide minimum possible voltage drop and RF loss. Biasing design considerations and schemes primary for FETs/HEMTs are described in this chapter.

18.1 BIASING OF TRANSISTORS

Biasing of transistors comprises two parts: the selection of a bias Q-point for optimum device performance in terms of gain, noise figure, output power, PAE, and linearity, and the biasing networks. Both are critical for the operation of amplifiers, especially dealing with low noise and high power levels.

18.1.1 Transistor Bias Point

One of the challenging jobs in DC biasing circuit design is to select the proper bias or quiescent point (in short Q-point) for the application at hand and maintain constant current over transistor parameter variations due to process and temperature. This in turn is related to the transistor's current–voltage relationship known as $I-V$ characteristics as discussed in Chapters 4 and 5. Figure 18.1a shows the $I-V$ characteristics of an FET. The broken line divides the $I-V$ operation into two regions: linear and saturated. The linear region is resistive in nature and accounts for the signal loss when the device's operation falls in this domain. The saturated region provides gain and power and all transistors for amplification are basically operated in this region. When the gate voltage

Fundamentals of RF and Microwave Transistor Amplifiers. By Inder J. Bahl

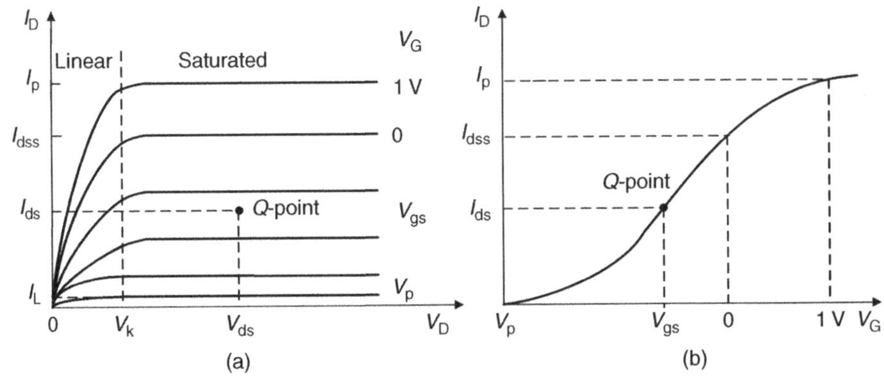

Figure 18.1 Device $I-V$ data and (b) device current versus gate voltage.

is varied from pinch-off V_p to about +1 V (Fig. 18.1), the device current varies from almost zero to I_p. I_p is also known as the open channel or maximum device current I_{max} and occurs at positive gate–source voltage. I_p represents the maximum available drain current in the saturated region and is responsible for obtaining maximum possible device output power.

The most critical device parameter commonly used in biasing the devices is I_{dss}, the saturated current at zero gate bias, or $V_{gs} = 0$ V, or pulsed peak current I_p at a gate bias $V_{gs} = +1.0$ V. The change of I_D with V_G voltage is known as the transconductance defined as

$$g_m = \left.\frac{\Delta I_D}{\Delta V_G}\right|_{\text{operating } V_{ds}} \tag{18.1}$$

Figure 18.1b shows the variation of I_D with V_G at given $V_D = V_{ds}$ voltage, which varies from 3 to 10 V for FETs. Thus over some gate voltage, the current is linear or g_m is constant and above +1 V, and near pinch-off voltage the values of g_m become small. Thus when the devices are operated over constant g_m regions, they result in high gain and linear operation for amplifiers.

The bias selection of an FET/HEMT (V_{ds}, V_{gs}, and I_{ds}) depends on its application. Higher power applications require high V_{ds} and I_{ds} values. The maximum allowed V_{ds} is generally fixed for transistors. The drain voltage is selected as recommended by the manufacturer or foundry and usually set to achieve the maximum output power and best PAE. Much lower than recommended voltage results in lower PAE and poor output power. The drain voltage for standard devices is 2–10 V, whereas for high-voltage devices it can be as high as 50–100 V.

The selection of V_{gs} voltage depends on the class of operation, gain, and PAE requirements, as discussed in Chapter 8. In summary, the factors mostly considered in the selection of bias point are gain, noise figure, output power, PAE, nonlinearity distortion, oscillation suppression, and power dissipation. For low-power and low-noise applications, the devices are biased at low drain–source voltage and current. In general, at lower frequencies the devices are operated at lower currents than at high frequencies, because at higher frequencies the device gain is lower and higher currents result in higher gain values because of higher g_m. Also, low noise figure (NF) applications require low currents, usually $I_{ds} = 15$–25% I_{dss} as NF increases with current. For

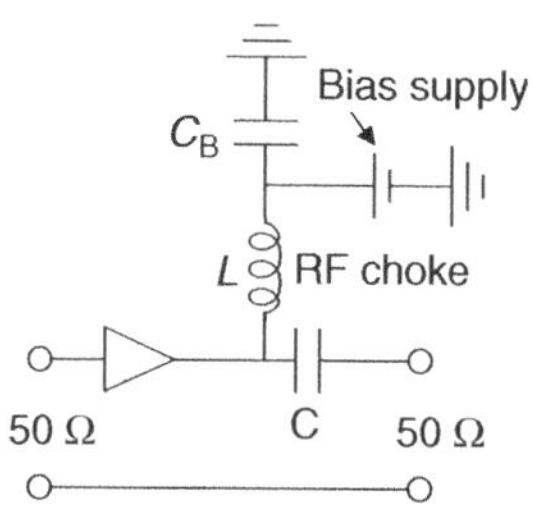

Figure 18.2 A generic biasing configuration.

linear and high-power applications, the devices are operated at about 40–50% I_{dss}, while for a high-efficiency application the Q-point drain current is about 15–35% I_{dss}. At RF frequencies, the drain current is about 10–15% I_{dss}. At high frequencies in the millimeter range, the devices are operated at 50–80% I_{dss} to achieve maximum gain values. For various transistors, the range of gate voltage over which the Q-point current changes $\pm 5\%$ of I_{dss} does not affect significantly the output power and PAE of the device. Thus high-PAE amplifier designs are relatively insensitive to gate bias operating point.

18.1.2 Biasing Schemes

Basically, a biasing network consists of a DC block and an RF choke as shown in Figure 18.2. The purpose of the RF choke is to have very high impedance at the operating frequencies to block the RF from leaking through the biasing network. In this case, the effect of the biasing network is negligible when it presents very high impedance to the matching networks. Chip capacitors in hybrid circuits and MIM capacitors in MMICs are generally used as DC bias blocks. Ideally, the capacitor should present a short circuit at the operating frequencies. The capacitance value C and C_B should be selected such that $\omega C = 2\pi f C$ is very large. Therefore $Z_C = -j(1/\omega C)$ is very small and the signal will pass through the capacitor with little loss or reflection. For example, at 1 GHz, a 100-pF capacitor will give $Z_C = -j1.6\ \Omega$, which is almost a short circuit in a 50-Ω environment. The capacitor value or type is also selected so that there is no parallel resonance in the operating frequency range. This will result in a very high impedance point at the capacitor location that is prone to instabilities in amplifiers, as discussed in the previous chapter. The RF choke blocks the RF signal while acting as a DC short. At low frequencies, generally lumped inductors are used. The inductance value is selected so that its impedance at the operating frequencies is very high. In a 50-Ω environment, a choke impedance in the 500–1000-Ω range is acceptable in most applications. For example, at 1 GHz, a 100-nH inductor will give $Z_L = j628\ \Omega$, which is almost an open circuit in a 50-Ω environment. The choke inductor must have a high Q and the required current-carrying capability. A very large value of $L(dI_D/dt)$ results in a large voltage variation, which may give rise to voltage transients and destroy the devices, especially in high-current and high-voltage power amplifiers. Since at low frequencies the transistors are potentially unstable and the matching networks become open circuited due to low values of DC and RF bypass capacitors, under such conditions the biasing networks provide the required terminations for the amplifier's unconditionally stable operation.

Generally, in a transistor the gate/base current is negligible in comparison to the drain/collector current; the gate/base may be biased through resistors having high values or RF chokes, whereas drain/collectors are biased through RF chokes. Since BJTs

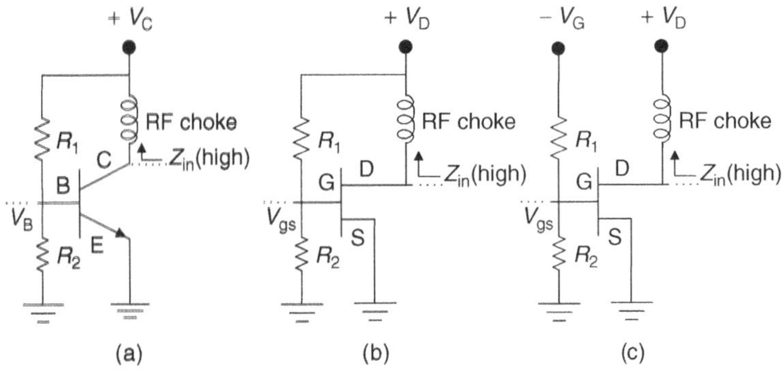

Figure 18.3 Simple DC bias circuits: (a) BJT/HBT, (b) MOSFET/LDMOS, and (c) FET/HEMT.

and HBTs require a positive base–emitter voltage and MOSFET/LDMOS devices require positive gate–source voltage, a single power supply is used. However, FETs and HEMTs use dual power supplies, that is, negative voltage for gates and positive voltage for drains. Figure 18.3 shows simple bias circuits, where a resistive voltage divider is used for base/gate biasing. Here,

$$V_B = \frac{R_2}{R_1 + R_2} V_C \quad \text{and} \quad V_{gs} = \frac{R_2}{R_1 + R_2} V_D \quad \text{or} \quad V_{gs} = \frac{R_2}{R_1 + R_2} V_G \tag{18.2}$$

The values of R_1 and R_2 are selected so that bias current is negligible and does not affect the PAE of an HPA. For E-mode FETs/HEMTs, one needs positive gate voltage, as shown in Figure 18.3b.

Five practical biasing configurations for FETs/HEMTs are given in Table 18.1. The circuits (a), (b), and (e) require two power supplies of opposite polarity. Since in these configurations the source is grounded with minimum possible source inductance, they provide maximum gain. Biasing circuits (c) and (d) with source resistors are widely used for small- and medium-power applications and require only one power supply. As the supply voltage is applied, the gate is simultaneously reverse biased with respect to the source by the series resistor R_S. The value of R_S is selected based on drain–source current I_{ds} and the operating bias. Such a bias scheme is referred to as the self-bias technique and will be discussed later in this chapter.

Gate Bias

A negative voltage is generally applied to the gate terminal for a depletion-mode FET/HEMT device, which is isolated from the RF portion of the circuit. Since gates do not draw appreciable currents, gates can be biased through low-current chokes or $\lambda/4$ transformers or resistors. Resistors are always used in the gates to improve stability as well as to improve isolation between the stages. Since gate current is negligible in many applications, an *LC* type bias filter network is not essential for biasing the gate of FETs. In this case, the gate is biased through an *RC* network, as shown in Figure 18.4a, which provides sufficient isolation between the gate terminal and the bias supply. However, the selection of the right resistance value is important as it affects the amplifier characteristics under saturation. The selection of resistor values depends on the operation of the device as well as the frequency of operation.

Table 18.1 Various FET/HEMT Bias Schemes for Low-Power Applications

Biasing Configuration	Typical Bias Voltages	Amplifier Characteristics	Other Comments
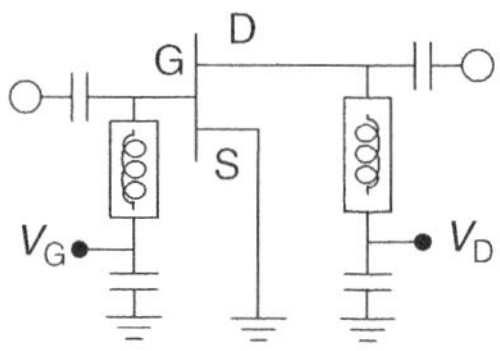 (a)	$V_D = 3$ V $V_G = -1$ V	Low noise, high gain, high power, and high efficiency	Biasing networks are part of matching; insensitive to bias current
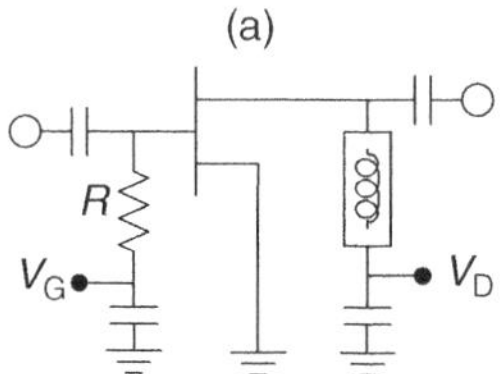 (b)	$V_D = 3$ V $V_G = -1$ V	Moderately low noise, high gain, high power, and high efficiency	Biasing network is part of matching; insensitive to bias current; high value of R provides higher isolation between gate and power supply
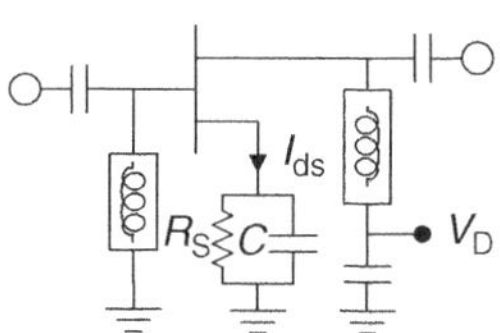 (c)	$V_D = 4$ V $I_{ds}R_S = 1$ V	Moderately low noise, high gain, medium power, and low efficiency	R_S provides automatic transient protection; sensitive to bias current
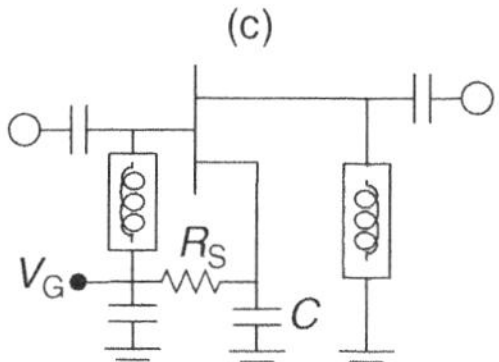 (d)	$V_G = -4$ V $I_{ds}R_S = 1$ V	Moderately low noise, high gain, medium power, and low efficiency	R_S provides automatic transient protection; sensitive to bias current
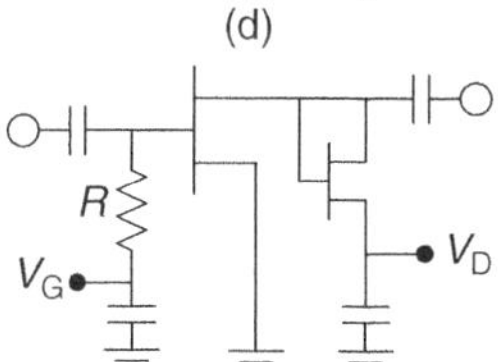 (e)	$V_D = 5$ V $V_G = -1$ V	Moderate noise, high gain, medium power, and low efficiency	Broadband at lower frequencies; sensitive to bias voltage

In linear or small-signal operation, the value of resistor R_1 could be between 500 and 2000 Ω·mm of device periphery, whereas for nonlinear operation, it can be 200 to 500 Ω·mm. The resistance value is based on per mm gate periphery of an FET. For example, using a 200 Ω ·mm design rule, FET sizes of 2 and 5 mm require the gate bias resistance values of 100 Ω and 40 Ω, respectively. Since under hard drive conditions, the gate experiences a much larger value of leakage current as compared to the small-signal conditions, the selection of resistor value becomes critical. Under small-signal conditions the gate current always comes out of the device. The value of I_g is much less than 1 mA/mm. The value of R_1 is selected so that the voltage drop across R_1 is less than about 0.2 V. Under high drive conditions, for negative cycle, the voltage superimposed becomes more negative and the value of I_g increases as shown by curve a in Figure 18.4b. A proper selection of R_1 value makes the gate more positive

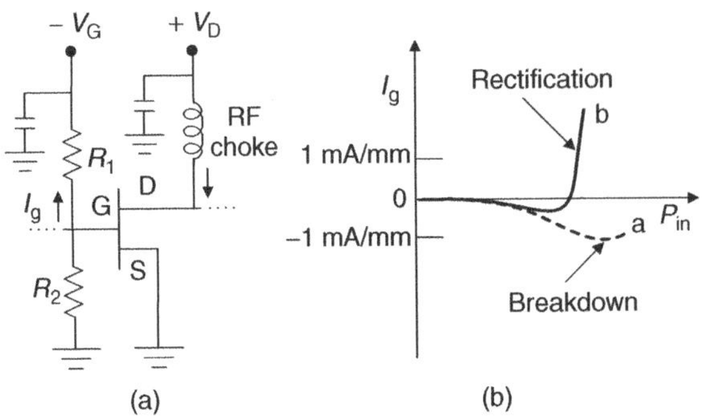

Figure 18.4 A simplified gate bias network and (b) gate current as a function of input power.

and limits I_g to less than 1 mA/mm. In the case of positive cycle, under 1–2-dB compression, the current reverses and starts flowing into the device; that is, the gate starts rectifying and the current starts increasing as shown by curve b in Figure 18.4b. A suitable value of R_1 limits the rectifying current to less than 1 mA/mm.

Gate Current

A gate current on the order of 1–2 mA/mm of gate periphery is normally encountered in FET/HEMT power amplifiers operated in 1–3-dB gain compression for high-efficiency applications. In this case, depending on the load line, the drain voltage and gate voltage swings can extend from the breakdown voltage to rectification, giving rise to breakdown gate current and rectification gate current. Both these currents can add at one frequency or power level and cancel at another frequency and power level. By properly selecting the gate bias or the value of gate bias resistor, the gate current can be minimized.

Under small-signal conditions the gate current flowing through the transistor is negligible. However, if the gate bias is applied through a voltage divider resistor network, the total gate bias current is

$$I_{gt} = I_g + I_b$$

where g and b stand for gate and biasing network, respectively. For the biasing network as shown in Figure 18.4a, I_b is given by

$$I_b = \frac{V_G}{R_1 + R_2}$$

and remains constant. In the case where the gate bias is applied directly to the gate, $I_b = 0$ and $I_{gt} = I_g$.

When the devices are operated at large-signal conditions for high output power and PAE, a significant gate current starts flowing through the device. In the case of FETs, due to reliability requirements, the commonly used limit on the gate current value is 1 mA/mm of gate periphery. However, depending on the input match, output load impedance, and bias voltages, the RF voltage and current swings at the device terminals may exceed this limit. During the positive RF cycle, a large RF voltage at the gate might lead to rectification of the signal, giving rise to positive gate current as shown in Figure 18.4b. On the other hand, during the negative RF cycle, the voltage between the gate and drain terminals may extend into the breakdown region, giving

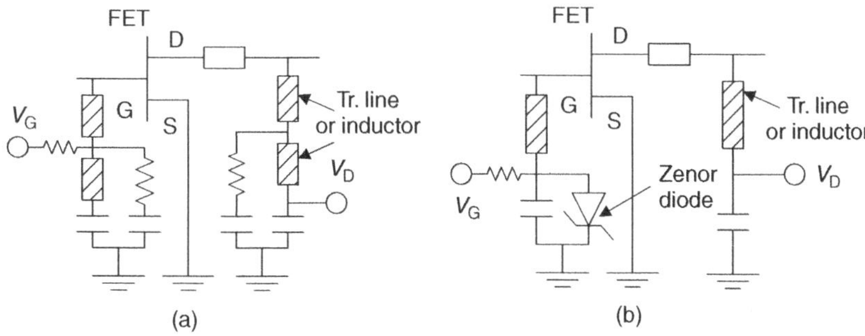

Figure 18.5 Power FET/HEMT biasing schemes.

rise to negative gate current. In a broadband HPA, due to different match impedances and compression points, the gate current may be positive at one frequency and zero or negative at other frequencies; also, it is a function of input power level. The value of gate current and its direction can be controlled by the gate bias voltage, series gate resistor value, and matching networks.

Drain Bias

The drains are usually biased through $\lambda/4$ transformers or RF chokes with minimum resistance value. In both cases, the DC voltage drop and the RF loss are kept to a minimum in order to improve the output power and PAE. However, some RC choking used in the biasing circuitry, as described in the previous chapter, damps out the low-frequency and bias instabilities. Distributed $\lambda/4$ transformers are preferred at higher frequencies due to their high Q values, while inductors are preferred due to their small size at RF or lower microwave frequencies. Figure 18.5 shows typical biasing circuits for FET/HEMT power amplifiers. Sometimes Zenor diodes are used for protection against connecting the supply voltage with the wrong polarity.

In power amplifiers, a proper design of the bias line from the supply to the drain/collector location is very important to realize high-efficiency design. A 30-W HPA with PAE of 30% requires a 10-A current at a supply voltage of 10 V. If the bias line has about 0.1-Ω resistance, it will drop about 1 V and decrease the output power and PAE by about 10%. In order to get negligible voltage drop across the bias line, its resistance must be on the order of 0.01 Ω or less. Applying the bias voltage using multiple bias lines significantly reduces the voltage drop across the bias path.

18.2 BIASING NETWORK DESIGN CONSIDERATIONS

Biasing of devices is an important part of an amplifier design. The design considerations for biasing circuits are higher gain, lower noise figure, high efficiency, oscillation suppression, single source power supply, RF choking, and desirable impedance matching. Biasing circuit design considerations consist of a proper biasing topology and selection of the conductor dimensions to meet the minimum possible resistance and electromigration requirements. These are discussed next.

18.2.1 Microstrip Biasing Circuit

The RF choke at microwave frequencies is generally realized by using a high-impedance λ/4 line, also known as a shunt stub terminated by an RF bypass capacitor C_B as shown in Figure 18.6a. A DC block can be either a capacitor or a 3-dB backward-wave coupler described in Reference 6. For low RF leakage through the biasing network, the impedance ratio of the shunt stub (Z_S) and the through line impedance (Z_0), Z_S/Z_0, must be much greater than unity. The low–high impedance lines serve as a lowpass filter, which prevents the microwave signal leaking into the bias port. In this case, the bandwidth increases when the impedance of the stub increases. For VSWR ≤ 1.05, the bandwidth for $Z_S = 100\ \Omega$ is about 12%. In order to further increase the bandwidth, two sections of quarter-wave long transmission lines are used. If an open circuit is required across the main line for RF signals, a quarter-wave high-impedance line followed by an open circuited quarter-wave low-impedance line are connected. The configuration is shown in Figure 18.6b. Assuming that the characteristic impedance Z_0 of the through line is the same as the system impedance, the normalized admittance with the load ($Y_0 = 1/Z_0$) is

$$y = \frac{Y}{Y_0} = 1 + \frac{Z_0}{Z_{in}} \tag{18.3a}$$

where

$$Z_{in} = jZ_1 \frac{Z_1 \tan\theta_1 \tan\theta_2 - Z_2}{Z_1 \tan\theta_2 + Z_2 \tan\theta_1} \tag{18.3b}$$

Here θ_1, Z_1 and θ_2, Z_2 are the electrical line length, characteristic impedance of the first and second line sections, respectively. Under matched conditions, $Z_{in} = \infty$ and $y = 1$. Maximum bandwidth is obtained when Z_1/Z_2 is large. For example, with $Z_1 = 100\ \Omega$, $Z_2 = 10\ \Omega$, $Z_0 = 50\ \Omega$, and a VSWR $= 1.2$, the maximum bandwidth is about 40%. Commonly, the low-impedance line section is replaced with a radial line section as shown in Figure 18.6c. This arrangement provides better bandwidth than a λ/4 open circuited line section and is smaller in size. The transmission line medium could be either strip line or microstrip. In this biasing scheme, the biasing circuitry becomes an integral part of the amplifier design.

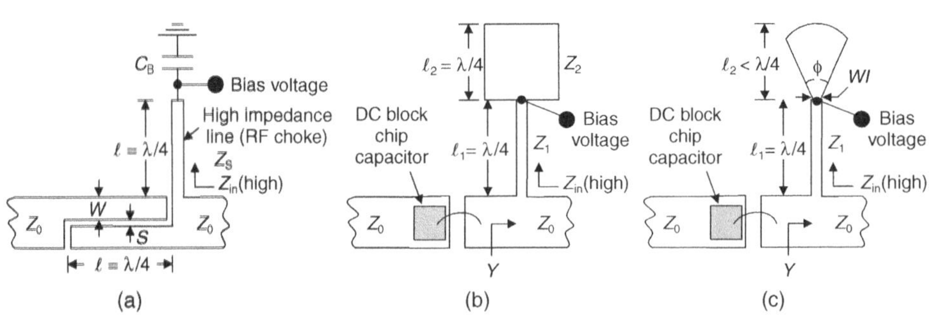

Figure 18.6 Simplified microwave biasing circuits: (a) coupled-line DC block and λ/4 transformer, (b) MIM capacitor DC block and two λ/4 transformers, and (c) MIM capacitor DC block and a λ/4 transformer in series with a radial line stub.

EXAMPLE 18.1

Compare the response of a lumped-element biasing network, Figure 18.2, with a two-section distributed-element biasing network, Figure 18.6b. The design values are $L = 8$ nH, $C_B = C = 10$ pF, $Z_1 = 120\ \Omega$, and $Z_2 = 20\ \Omega$. The center frequency, frequency range, and frequency step are 6 GHz, 4–8 GHz, and 0.5 GHz, respectively. The system impedance is 50 Ω and the biasing networks are assumed to be ideal.

SOLUTION Using a CAD, the calculated S_{11}(dB) and S_{21}(dB) for these two networks are given below:

Frequency	Lumped Element		Distributed Element	
(GHz)	S_{11}(dB)	S_{21}(dB)	S_{11}(dB)	S_{21}(dB)
4.0	−21.2	−0.033	−19.3	−0.052
4.5	−22.3	−0.026	−23.3	−0.020
5.0	−23.2	−0.021	−29.3	−0.005
5.5	−24.1	−0.017	−48.7	−0.000
6.0	−24.9	−0.014	−31.5	−0.003
6.5	−25.6	−0.021	−25.0	−0.014
7.0	−26.2	−0.010	−21.1	−0.034
7.5	−26.8	−0.009	−18.1	−0.067
8.0	−27.4	−0.008	−15.6	−0.122

The simulated S_{11} over the 4–8-GHz frequency range is less than −21.2 dB and −15.6 dB for the lumped-element and distributed-element biasing network, respectively. This demonstrates that the lumped-element technique has a larger bandwidth. ■

EXAMPLE 18.2

Compare the response of two biasing networks shown in Figure 18.6b,c designed on an alumina substrate, $h = 15$ mils (381 μm) and $\varepsilon_r = 9.9$. The transmission medium is microstrip. The design values for Figure 18.6b are $W_1 = 25$ μm and $W_2 = 1500$ μm, and the line lengths are $\lambda/4$ at 6 GHz. In Figure 18.6c, for the microstrip section $W_1 = 25$ μm and length is $\lambda/4$, and for the radial line section $WI = 25$ μm, length is $\lambda/8$, and angle $\phi = 90°$. The frequency range is 4–8 GHz with 0.5-GHz step. The system impedance is 50 Ω and the discontinuity effects may be ignored.

SOLUTION Refer to Table 6.1d. The characteristic impedances for the line widths $W_1 = 25$ μm and $W_2 = 1500$ μm are $Z_1 = 115.3\ \Omega$ and $Z_2 = 21.7\ \Omega$, respectively. The effective dielectric constant values for these two lines are $\varepsilon_{re1} = 5.73$ and $\varepsilon_{re2} = 8.03$, respectively. The free-space wavelength at 6 GHz is 5 cm. The quarter-wave lengths for the microstrip sections are

$$\ell_1 = \frac{5}{4\sqrt{5.73}}\ \text{cm} = 5.22\ \text{mm}$$

$$\ell_2 = \frac{5}{4\sqrt{8.03}}\ \text{cm} = 4.41\ \text{mm}$$

For the radial microstrip section, the effective dielectric constant $\varepsilon_{re2} = 8.03$ is assumed and the calculated $\lambda/8$ line length is 2.205 mm. Using a CAD, the calculated S_{11}(dB) and S_{21}(dB) for these two networks are given below:

Frequency	Two λ/4 Sections		λ/4 and Radial Sections	
(GHz)	S_{11}(dB)	S_{21}(dB)	S_{11}(dB)	S_{21}(dB)
4.0	−18.4	−0.114	−16.4	−0.162
4.5	−22.3	−0.066	−20.1	−0.089
5.0	−28.0	−0.040	−24.8	−0.052
5.5	−44.0	−0.029	−33.3	−0.034
6.0	−31.7	−0.029	−38.5	−0.029
6.5	−24.6	−0.040	−27.1	−0.035
7.0	−20.6	−0.064	−22.3	−0.052
7.5	−17.6	−0.106	−19.0	−0.083
8.0	−14.9	−0.177	−15.4	−0.134

The above data show that both biasing networks have a similar performance response; however, the radial line approach is approximately half in size. ■

18.2.2 Lumped-Element Biasing Circuit

Lumped-element biasing circuits are commonly used at low frequencies and in monolithic circuits. The desirable features of these biasing circuits are compact size, small RF leakage, and ultra-wideband characteristics. A shunt coil inductor, L, also known as an RF choke, is used as a biasing element while a series capacitor C is used to isolate the bias voltage applied to various stages/circuits. Shunt inductor and series capacitor circuits are shown in Figure 18.7 (inset). The normalized admittance of the coil and normalized impedance of the capacitor backed by a 50-Ω matched element

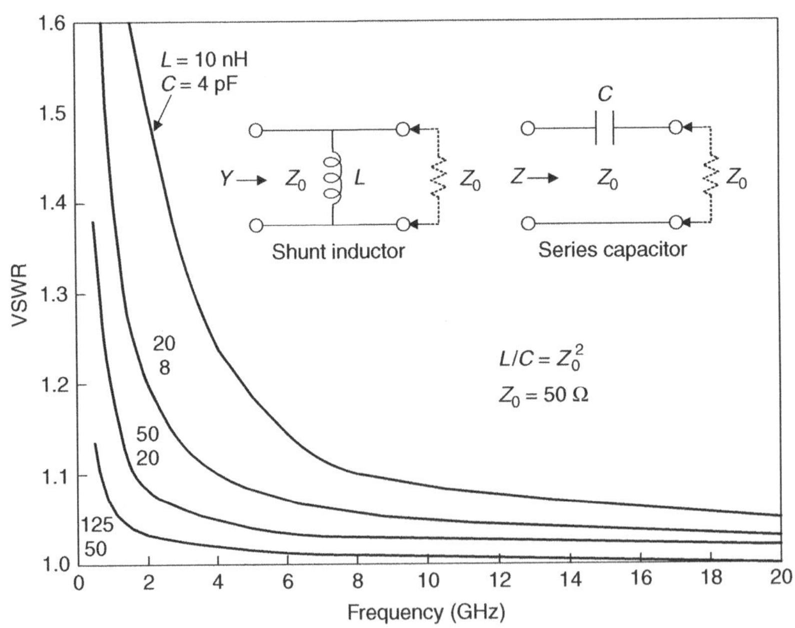

Figure 18.7 Lumped L and C biasing networks and their responses.

are given by

$$y = \frac{Y}{Y_0} = 1 - j\frac{Z_0}{\omega L} \quad (18.4a)$$

$$z = \frac{Z}{Z_0} = 1 - j\frac{1}{\omega C Z_0} \quad (18.4b)$$

If $y = z$, that is, $Z_0^2 = L/C$, both elements will have an equal VSWR at all frequencies. The VSWR in each case may be written

$$VSWR = (1 + |\rho|)/(1 - |\rho|) \quad (18.5)$$

where

$$|\rho| = \left|\frac{1-y}{1+y}\right| = \frac{Z_0/\omega L}{[4 + (Z_0/\omega L)^2]^{1/2}} \quad \text{(for a shunt coil)} \quad (18.6a)$$

$$= \left|\frac{z-1}{z+1}\right| = \frac{1/\omega C Z_0}{[4 + (1/\omega C Z_0)^2]^{1/2}} \quad \text{(for a series capacitor)} \quad (18.6b)$$

The insertion loss (IL) of a reactive discontinuity having a VSWR of S is given by

$$IL = 20 \log\left(\frac{S+1}{2\sqrt{S}}\right) \quad \text{(in dB)} \quad (18.7)$$

Variation of the VSWR corresponding to these elements is shown in Figure 18.7. Higher values of L or C have a lower VSWR.

When both these elements are used simultaneously, the elements should be connected at the same plane. The normalized admittance of the capacitor that is backed by a 50-Ω matched line becomes

$$y_C = \frac{1 + j\frac{1}{\omega C Z_0}}{1 + 1/(\omega C Z_0)^2} \quad (18.8a)$$

This admittance will add to the admittance of the inductor ($y_L = -jZ_0/\omega L$). The total admittance y_t is given by

$$y_t = y_C + y_L \quad (18.8b)$$

When $\omega C Z_0 \gg 1$,

$$y_t = 1 + j\left(\frac{1}{\omega C Z_0} - \frac{Z_0}{\omega L}\right) \quad (18.9)$$

If $Z_0^2 = L/C$,

$$y_t = 1$$

Thus the VSWR will be unity and also independent of frequency as long as L and C are independent of frequency. This network results in ultra-wideband bias circuit

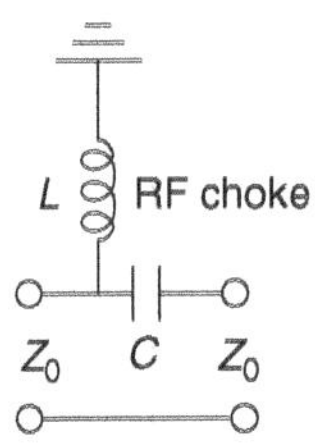

Figure 18.8 An ultra-wideband LC biasing network.

performance. The high-frequency operation of this network is limited by the parasitic reactance associated with lumped inductor and capacitor. The low-frequency operation is limited by the L and C values. With suitably selected biasing network elements, one can design the circuit to work over 30 MHz to 20 GHz.

EXAMPLE 18.3

Determine a relationship for the minimum frequency, $f_{\min}$, of operation for the biasing network in terms of ideal shunt L, series C elements, and system impedance Z_0, shown in Figure 18.8. When the return loss is 20 dB and the system impedance is 50 Ω, calculate the values of shunt inductor L and $f_{\min}$ for the capacitor C values of 2, 5, 10, 15, 20, 50, 100, and 250 pF.

SOLUTION Substituting y_C and y_L in (18.8b), we find

$$y_t = \frac{1}{1 + 1/(\omega C Z_0)^2} + j\frac{1}{\omega C Z_0}\frac{1}{1 + 1/(\omega C Z_0)^2} - j\frac{Z_0}{\omega L} \tag{18.10a}$$

When

$$j\frac{1}{\omega C Z_0}\frac{1}{1 + 1/(\omega C Z_0)^2} - j\frac{Z_0}{\omega L} = 0$$

then

$$L = Z_0^2 C\left[1 + \left(\frac{1}{\omega C Z_0}\right)^2\right] \tag{18.10b}$$

and

$$y_t = \frac{1}{1 + 1/(\omega C Z_0)^2} \tag{18.10c}$$

The magnitude of reflection coefficient ρ is given by

$$\rho = \frac{1 - y_t}{1 + y_t} \quad \text{or} \quad y_t = \frac{1 - \rho}{1 + \rho}$$

$$\frac{1}{y_t} = 1 + \frac{1}{(\omega C Z_0)^2} = \frac{1 + \rho}{1 - \rho} \tag{18.10d}$$

After simplifying, we find

$$\omega = \frac{1}{C Z_0}\sqrt{\frac{1 - \rho}{2\rho}} \tag{18.10e}$$

When the maximum allowed value of ρ is ρ_{max}, the minimum frequency, f_{min}, is given by

$$f_{min} = \frac{1}{2\pi C Z_0}\sqrt{\frac{1-\rho_{max}}{2\rho_{max}}} \tag{18.10f}$$

For $Z_0 = 50\ \Omega$ and return loss = 20 dB or $\rho_{max} = 0.1$, (18.10b) and (18.10f) are reduced to

$$L = 3.056C$$
$$f_{min} = 6.748/C$$

where the units of L, C, and f_{min} are nH, pF, and GHz, respectively. The calculated L and f_{min} values for various C values are given below.

C (pF)	L (nH)	f_{min} (GHz)
2	6.11	3.374
5	15.28	1.350
10	30.56	0.675
15	45.84	0.450
20	61.12	0.337
50	152.80	0.135
100	305.60	0.068
250	764.00	0.027

Thus an LC biasing network operating from 30 MHz to 20 GHz requires an inductor of about 765 nH and a capacitor of 250 pF. ■

18.2.3 High-PAE Biasing Circuit

The high-efficiency power amplifier design requires in-depth analysis of power loss in the matching networks to realize high-PAE performance. In this respect, the low DC loss as well as the low RF loss of the biasing/matching networks become important when these structures have to be optimally designed to transfer maximum power to the load and also provide a proper harmonic termination in the case of a high-PAE power amplifier. One of the primary objectives of the biasing networks is to provide equal power supply with constant voltage to each FET on the chip or ICs in a module in high-power amplifiers when several FETs or chips are combined in order to suppress odd-mode oscillations as well as to have maximum combining efficiency. Since the thickness of microstrip conductors is limited by the process, increased microstrip conductor width and higher conductivity are needed to lower the DC loss. The major power drop in the biasing network is due to a finite resistance of the conductors; the power drop is proportional to I^2R, where R is the resistance of the conductor carrying the current I. Finite conductivity of the conductors gives rise to resistance of the lines. Since the microstrip-like conductors use thin films, the result is high resistance of long lines. If t, W, and L are the conductor's thickness, width, and length, and ρ is the resistivity ($\Omega \cdot$ m) of the thin film or ρ_s is the sheet resistance ($\Omega/\square$), the resistance of the conductor is given by

$$R = \rho\frac{L}{tW} = \rho_s\frac{L}{W}, \quad \text{where} \quad \rho_s = \frac{\rho}{t} \tag{18.11}$$

For given ρ_s of the material, the resistance can easily be calculated from the number of squares of width in a given length. For example, a line of width 50 μm and 1000 μm long (20 squares) has 2.5 times less resistance than a line of width 20 μm and 1000 μm long (50 squares). Thus the key in reducing the resistance is to keep the number of metallization squares as minimum as possible. If a gold conductor line has 20 squares and the sheet resistance $\rho_s = 0.005\ \Omega/\square$, the DC resistance of the line is 0.1 Ω. Copper conductors have about 1.4 times higher conductivity and are fabricated with 3–4 times larger thickness than gold conductors. Thus the sheet resistance ρ_s for copper conductors is on the order of 0.001 $\Omega/\square$ and have much lower DC resistance than gold conductors. For right-angled bends without any chamfering, the effective area is about 0.56 $\square$.

A substrate carrying an HPA is at much higher temperature than a substrate carrying an LNA. Since the conductor resistance also has temperature dependence, its temperature must be considered. The rate of change of resistor value with temperature is known as the temperature coefficient of resistors (TCR) or simply TC and is expressed in percent per degree Celsius or parts per million per degree Celsius (ppm/°C). When the resistance increases with increasing temperature, the TC value has a positive sign; if it decreases, the TC value has a negative sign. The resistor's temperature dependence is given by

$$R_{\text{OT}} = R_{\text{RT}} + TC(T_{\text{OT}} - T_{\text{RT}}) \tag{18.12}$$

where OT and RT denote the operating and room or ambient temperature, respectively. If a resistor has +TC value of 400 ppm/°C, the resistance value will increase with temperature by about 4% at 125 °C; when the ambient temperature is 25 °C, a 100-Ω resistor at room temperature will have a value of about 104 Ω at 125 °C.

18.2.4 Electromigration Current Limits

Electromigration requirements dictate the microstrip and inductor line widths carrying the direct current (DC). In the M/A COM MSAG MMIC process, a very conservative current density value of 2.2×10^5 A/cm^2 as an electromigration limit for 4.5-μm thick gold conductors is used and widely accepted. This translates to a maximum allowed current per unit line width of 10 mA/μm. Consider the example of an HPA that requires 3-A current in the output stage. This warrants a single 300-μm wide microstrip line in the output matching network, or one can use dual bias lines each having 150-μm width. The requirement for such wide, low-impedance shunt lines complicates efforts to shrink the chip area and also gives rise to an imbalance in the output feed lines due to the junction discontinuity effect, impacting significantly the circuit bandwidth. Compact lumped elements are used in the earlier stages of a multistage power amplifier, where power and current densities are less. The current-handling capacity is increased by a factor of 2 when the conductor thickness is doubled.

Conductors used in MICs and PCBs are thicker and wider. In these circuits the electromigration requirements for conductors are not as stringent as in the case of MMICs.

18.3 SELF-BIAS TECHNIQUE

In an FET/HEMT, a self-bias technique (Table 18.1, configurations (c) and (d) configurations is used to realize a desired voltage difference between the gate and source, when only one power supply, generally a positive voltage, is available. In this technique, the bias voltages are applied simultaneously, avoiding the turn-on and turn-off sequence used in the two-supply bias scheme. Another advantage of this scheme lies in the fact that the source resistance provides transient protection to devices. In this case the gate is DC grounded (0 V) and the source is at positive potential; that is, the voltage difference between the gate and the source is equivalent to V_{gs}. This technique is generally used for small-signal amplifier applications. However, one can use a self-bias scheme for low-power (<1 W) driver amplifiers. For higher power levels, where the drain current becomes significant, this technique is not very suitable due to unwanted DC power dissipation in the bias path, which lowers the output power and PAE because of reduced supply voltage across the device terminals.

In a self-bias technique, any increase or decrease in the drain current automatically adjusts the gate bias voltage, to maintain the device current approximately constant. A bypass capacitor is connected across the resistor R_S to realize a very low impedance at the device's source pad to minimize any feedback effect. The value of the capacitor is selected so that its reactance at the operating frequencies is less than 1–2 Ω.

The value of resistor is based on the gate voltage and the drain current given by

$$V_{gs} = -R_S I_{ds} \quad \text{or} \quad V_{gs} = V_p \left[1 - \left(\frac{I_{ds}}{I_{dss}}\right)^p\right] \tag{18.13}$$

where V_p and V_{gs} are negative quantities. A textbook value for the parameter p is 0.5. However, it depends on a fabrication process and for M/A-COM's MSAG FET 5A power process [7] its value is about 0.7.

Let us study the effect of temperature on the self-bias technique. The drain current is a function of gate voltage and temperature:

$$I_{ds} = f(V_{gs}, T) \tag{18.14}$$

Let ΔV_{gs} and ΔT be the changes in the gate voltage and temperature; the change in the drain current ΔI_{ds} is given by

$$\Delta I_{ds} = \frac{\partial I_{ds}}{\partial V_{gs}} \Delta V_{gs} + \frac{\partial I_{ds}}{\partial T} \Delta T \tag{18.15}$$

The change in the gate voltage, due to the change in the drain current, is given by

$$-\Delta V_{gs} = R_S \Delta I_{ds} \tag{18.16}$$

since $\partial I_{ds}/\partial V_{ds} = g_m$, from (18.15) and (18.16), we find

$$\Delta I_{ds} = \frac{1}{[1 + g_m R_S]} \frac{\partial I_{ds}}{\partial T} \Delta T \tag{18.17}$$

Let us define the bias stability coefficient as

$$BS = \frac{\text{change in } I_{ds} \text{ with self-bias circuit}}{\text{change in } I_{ds} \text{ without self-bias circuit}} \tag{18.18a}$$

or

$$BS = \frac{\Delta I_{ds}(R_S)}{\Delta I_{ds}(R_S = 0)} = \frac{1}{1 + g_m R_S} \tag{18.18b}$$

This suggests that the bias stability is improved by $1/[1 + g_m R_S]$ in the self-bias case and is independent of temperature.

In a self-biasing network, the value of capacitor C is selected such that at the lowest operating frequency, $1/\omega C \cong 0.1 R_S$ and $1/\omega C < 3\ \Omega$. A 50-pF by pass capacitor at L-band and a 10-pF bypass capacitor at X-band are usually sufficient for RF grounding and transient protection. This technique lowers the amplifier efficiency due to DC power dissipation in the source resistor and a finite Q of the bypass capacitor. A finite Q of the self-biasing network also increases the noise figure of an LNA by about 0.1–0.2 dB. For power amplifiers where $I_{ds} > 500$ mA, dual polarity supplies are recommended with a sequencing circuit to bias the gate first before applying the drain bias.

EXAMPLE 18.4

Design a self-bias network for a 450-μm gate periphery FET whose EC model is given in Table 5.5 and the Q-point bias current is 90 mA. Determine the required supply voltage and the values of self-biasing network elements operating over 5–10 GHz. Calculate the MSG/MAG and μ factor over the 5–10-GHz frequency range at 0.5-GHz step intervals for the device before and after the self-biasing. Also calculate the bias stability coefficient, BS. Use ideal bias circuit elements.

SOLUTION Figure 18.9 shows the FET configuration before and after the insertion of a self-biasing network. Since the Q-point bias conditions for the device are $V_{ds} = 5$ V, $V_{gs} = -1.2$ V, and $I_{ds} = 90$ mA, the required voltage drop across the self-bias resistor is 1.2 V. In this case, the gate is kept at 0 V. Therefore the supply voltage needed is 6.2 V. The value of R_S is selected such that the voltage across it is 1.2 V; that is,

$$I_{ds} \times R_S = 0.09 \times R_S = 1.2\ \text{V} R_S = 1.2/0.09 = 13.333\ \Omega$$

The value of $BS = 1/(1 + g_m R_S) = 1/(1 + 0.078 \times 13.333) = 0.49$. The value of capacitor C is selected such that at the lowest operating frequency (5 GHz),

$$\frac{1}{\omega C} \cong 0.1 R_S, \quad C = 1/(\omega \times 0.1 R_S) = 23.9\ \text{pF}$$

Using a CAD, the calculated values of the MSG and μ factor are as follows:

Frequency	Dual-Bias		Self-Bias	
(GHz)	MSG (dB)	μ	MSG (dB)	μ
5.0	17.5	0.35	17.2	0.11
5.5	17.1	0.38	16.8	0.16
6.0	16.8	0.41	16.5	0.22
6.5	16.4	0.44	16.2	0.26
7.0	16.1	0.48	15.9	0.31
7.5	15.8	0.51	15.6	0.35
8.0	15.6	0.54	15.3	0.39
8.5	15.3	0.56	15.1	0.43
9.0	15.1	0.59	14.9	0.47
9.5	14.9	0.62	14.7	0.51
10.0	14.7	0.65	14.5	0.54

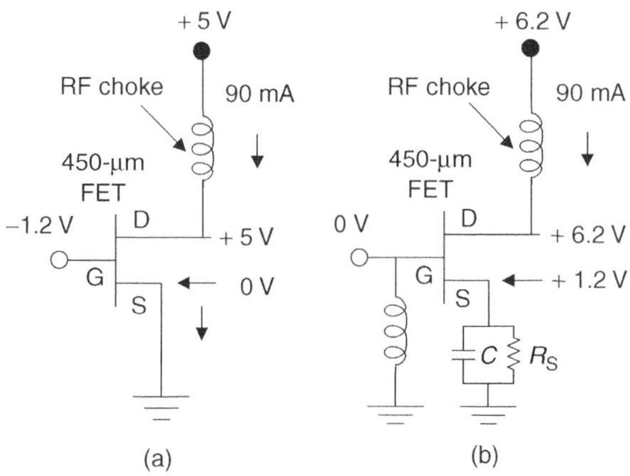

Figure 18.9 Biasing schemes: (a) dual supply and (b) single supply.

It may be noted that the self-bias network has lowered the MSG by about 0.2–0.3 dB due to its finite Q and decreased the value of μ due to the finite series feedback reactance. ■

18.4 BIASING MULTISTAGE AMPLIFIERS

So far, we have described the biasing schemes for single-stage amplifiers. Biasing a multistage amplifier through a single gate power supply and single drain power supply can be very tricky. Biasing gates through a single pad, shown in Figure 18.10a, is not very difficult, as large values of isolation resistors are generally used in the biasing networks. These resistors along with bypass capacitors provide 30–40-dB isolation between the stages. However, when all drain pads are connected together, there is always feedback through the bias lines due to the lack of resistors and not having enough choking. Because of insufficient capacitive bypass and the existence of parallel resonance of the chip capacitors, feedback between stages exists. This feedback results in instability due to other circuitry connected at the pad. In an MMIC design, one can easily minimize the instability by incorporating bias circuitry into the design as shown in Figure 18.10b. The effect of feedback can be minimized by properly selecting the resistors. This figure shows a drain biasing scheme for a four-stage amplifier. No resistor is used in the last-stage drain bias as it draws maximum current. The values

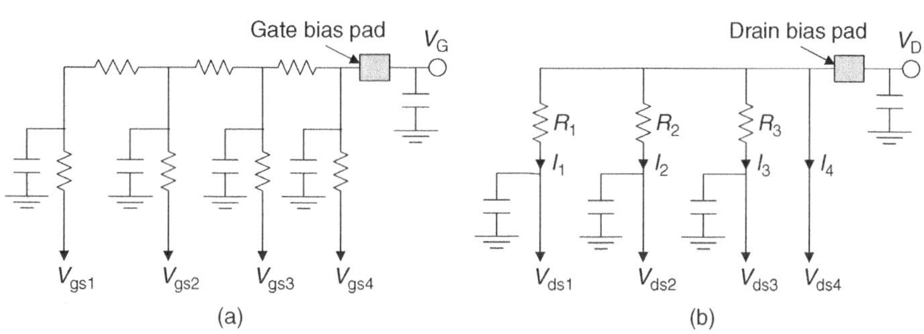

Figure 18.10 Biasing schemes for a 4-stage amplifier: (a) common gate bias and (b) common drain bias.

of resistors are selected so that there is about a 10% of V_D drop across the resistor. In the case of a 2-W driver amplifier operated with a 10-V power supply, the values of R_1, R_2, and R_3 are 20, 10, and 5 Ω for I_1, I_2, and I_3 values of 50, 100, and 200 mA, respectively. Or one can use RF chokes or inductors between stages. Resistors occupy much less space than inductors and usually are preferred to inductors. One should also be extra careful in choosing resistor widths meeting electromigration requirements. In the above example, typical values for MMIC resistor widths on a GaAs substrate are 60, 120, and 240 μm, respectively.

18.5 BIASING CIRCUITRY FOR LOW-FREQUENCY STABILIZATION

In HPAs when a large number of transistors are paralleled using a monolithic or hybrid approach to achieve high power levels, occasionally low-frequency instabilities are observed. These occur due to the very high gain of the devices at these frequencies. These instabilities are suppressed by using proper gate and drain bias networks as discussed in the previous chapter. In this section we describe a gate biasing scheme for low-frequency stabilization.

Consider a single-stage HPA in which four FETs are combined in parallel. Figure 18.11a shows a typical gate biasing scheme while Figure 18.11b shows a gate bias network used for low-frequency stabilization. In the conventional case the gate bias is brought through resistor R_1, whose value is determined by four FETs. However, in Figure 18.11b, the gate bias is brought to each FET through resistor R_1 connected to each FET and its value is determined by a single FET. At lower frequencies, when chokes are virtually shorted and bypass capacitors are open, due to the very high gain of devices (no resistors in series with gate), this can create negative resistance inside the circuit and can oscillate at the parallel resonance of the biasing circuitry. However, a large resistor in series with a gate for each FET/HEMT greatly reduces the condition for negative resistance and stabilizes the circuit. This is further enhanced when the gates of these FETs/HEMTs are isolated from each other by using resistors [8] or capacitors. In this case, the gain of the individual device is much lower than the composite device gain.

18.6 BIASING SEQUENCE

A dual-polarity supply amplifier using FETs or HEMTs requires a particular sequence for the application of bias voltages. This sequencing circuit allows one to bias the gate first before applying the drain bias. This sequencing is required because V_D applied at I_{dss} (drain–source saturation current) may burn the device due to thermal stress. This also avoids possible unstable operation, which might destroy the devices due to transient burnout. To operate the devices safely, first a nominal gate voltage is applied with no drain voltage. Then the drain voltage is turned on and ramped to the desired voltage. Finally, the gate voltage is set for an I_{DQ} value that results in safe thermal operation. Next, input RF power is applied. In power down, the sequence is reversed, where the gate voltage is turned off last. One can also apply both the gate and drain voltages simultaneously by using some sort of time delay network. In this technique, the drain bias network uses a long RC time constant while the gate bias network uses a short RC time constant. This technique is not very suitable for power amplifiers, as a

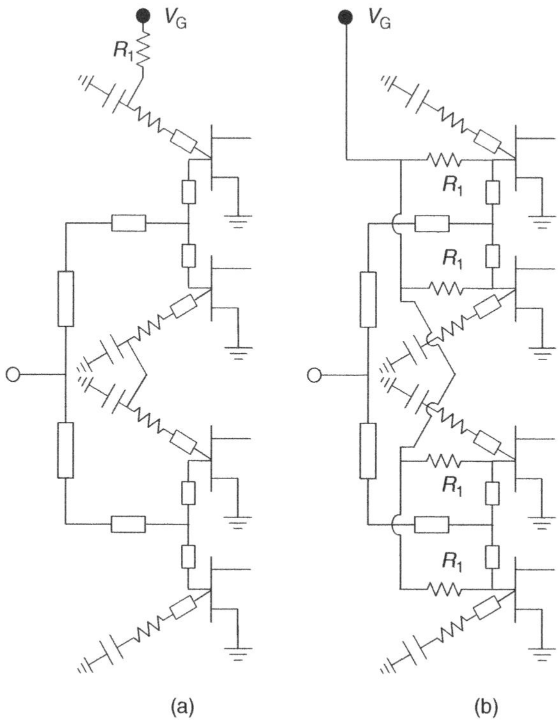

Figure 18.11 Gate biasing schemes for parallel devices: (a) typical series bias and (b) corporate bias for low-frequency stabilization.

lossy drain bias network lowers both output power and PAE, and the pulsed operation is also affected.

REFERENCES

1. G. Gonzalez, *Microwave Transistor Amplifiers*, Prentice Hall, Englewood Cliffs, NJ, 1984, Chapter chap3.
2. J. L. B. Walker (Ed.), *High-Power GaAs FET Amplifiers*, Artech House, Norwood, MA, 1993.
3. K. Chang, I. Bahl, and V. Nair, *RF and Microwave Circuit and Component Design for Wireless Systems*, John Wiley & Sons, Hoboken, NJ, 2002.
4. I. Bahl, *Lumped Elements for RF and Microwave Circuits*, Artech House, Norwood, MA, 2003.
5. I. J. Bahl and P. Bhartia, *Microwave Solid State Circuit Design*, 2nd edition, John Wiley & Sons, Hoboken, NJ, 2003, Chapter chap10.
6. R. Mongia, I. Bahl, P. Bhartia, and J. Hong, *RF and Microwave Coupled-Line Circuits*, 2nd edition, Artech House, Norwood, MA, 2007, Chapter 12.
7. I. J. Bahl, 0.7–2.7 GHz 12-watt power amplifier MMIC developed using MLP technology, *IEEE Trans. Microwave Theory Tech.*, Vol. 55, pp. 222–229, February 2007.
8. S. Goto et al., Stability analysis and layout design of internally stabilized multi-finger FET for high power base station amplifiers, *IEEE MTT-S Int. Microwave Symp. Dig.*, pp. 231–234, 2003.

PROBLEMS

18.1 An RF choke is required for designing a bias-tee operating at 10 GHz in a 50-Ω microstrip configuration. This can be achieved by using a lumped inductor L as shown in Figure 18.2. Find the value of the inductance such that VSWR does not exceed 1.1. Determine the layout of this inductor such that the maximum dimension is less than 50 mils. The minimum line width and spacing that can be conveniently etched are 1 mil and 0.6 mil, respectively, and the film thickness is 0.3 mil.

18.2 The RF choke discussed in Problem 18.1 may also be designed in the form of a microstrip circuit, shown in Figure 18.6b. The maximum and minimum impedances of the microstrip sections are limited to 100 and 20 Ω, respectively. Compare the VSWR performance of this circuit with the lumped circuit in Problem 18.1 at 8.0 and 12.0 GHz. Ignore the effect of variations of λ/λ_0 with frequency.

18.3 An FET is biased using gate and drain voltages of −1 V and 5 V, respectively. The bias current is about 25% of I_{dss}. If I_{dss} is 400 mA and the supply voltage is 6 V, determine the biasing scheme for the self-biased FET configuration. The operating frequency ranges 1–2 GHz. Use a resistive impedance greater than 10 times the capacitive impedance and ideal resistor and capacitor components.

18.4 A self-bias circuit is shown in Figure 18.9b. At 2 GHz using the EC model in Table 5.5 and $Z_L = 100\ \Omega$, determine the range of values for R_S and C for which R_{in} becomes negative. The FET size is 1 mm and the supply voltage is 11 V.

18.5 Discuss the pros and cons of self-biasing techniques for hybrid and MMIC amplifiers.

18.6 Design a biasing network connected in shunt to a 50-Ω microstrip line and working over 6–18 GHz. The desirable return loss looking into the biasing network is 20 dB. Use (a) all lumped elements and (b) a microstrip/radial stub.

Chapter 19

Power Combining

Power combiners are key components in high-power amplifiers. The aim of this chapter is to describe basic power combining principles and planar dividers/combiners. Both device-level and circuit-level power combining techniques are described.

For transistor power devices, with increasing frequency, the power output from a single transistor decreases rapidly. In many applications RF power levels are required that far exceed the capability of any single unit cell device or amplifier. It is therefore desirable to extend the power level of an amplifier by utilizing combining techniques in order to take advantage of the many desirable features of solid state devices, such as small size and weight, reliability, and performance in a broader range of applications.

Although there are fundamental limitations to the power that can be generated from a single transistor, the achievable power levels can be increased significantly by combining a number of devices operating coherently or by accumulating the power from a number of discrete devices. This may be done in one of two ways: by combining power at either the device level or the circuit level. Figure 19.1 shows basic schemes for power combining techniques. Most power combining techniques using matched combiners can provide "graceful degradation" in the case of failure of one or more devices in the combiner.

19.1 DEVICE-LEVEL POWER COMBINING

Device-level combining is accomplished by clustering the devices in a region whose extent is small compared with a wavelength, and is generally limited to the number of devices that can be combined efficiently. Figure 19.1 shows basic schemes used to combine power at the device level and circuit level. Several device cells are paralleled to achieve medium power as a single device. Higher power is obtained by bonding several devices onto a heat sink or a common carrier and connecting them to input and output matching circuitry as shown in Figure 19.2. The entire circuit may be hermetically sealed and used as a single device.

The design of power devices for any purpose (maximum gain, power, efficiency, or linearity) requires careful optimization of electrical and physical design parameters of the structure. Basic device design parameters include minimum parasitic losses, low source inductance, high f_T, high breakdown voltage, matchable input impedance, uniform heat dissipation, low thermal resistance, and minimum loss due to gate resistance and the phase difference between different parts of the device feed structure. The

Fundamentals of RF and Microwave Transistor Amplifiers. By Inder J. Bahl

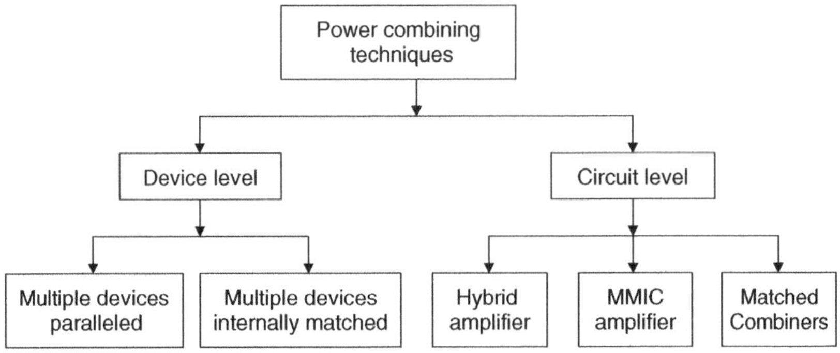

Figure 19.1 Device and circuit power combining schemes.

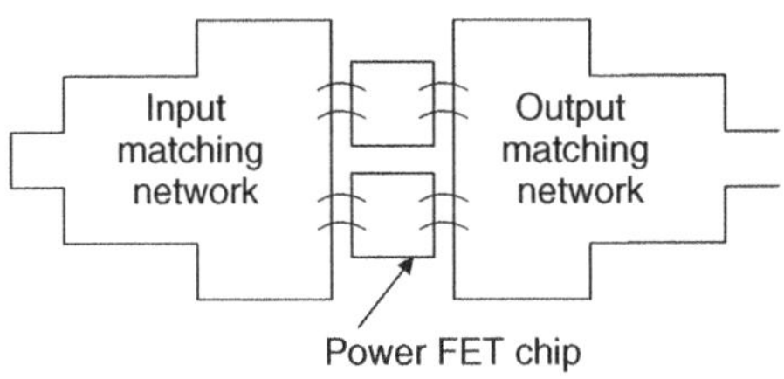

Figure 19.2 Power combining of power FET devices.

FET's/HEMT's output power is proportional to the gate periphery and its input and output impedances are inversely proportional to the gate periphery. Small size power devices also known as cells are normally used in the design of power amplifiers. Single cells are low-power devices; above the cutoff frequency (f_T) the output power decreases rapidly with increasing frequency due to parasitic losses and the transit time. Higher power levels are obtained by combining several cells in parallel on a single die or IC chip. In order to achieve high power levels with high PAE, FET cells or matched FETs must have constant phase and gain. High uniformity of saturated drain–source current (I_{dss}) and f_T across the combined devices is needed to realize uniform phase and gain for the combined unit. When FET cells are combined in phase using reactive techniques on a single MMIC chip, for example, maximum power and PAE are achieved.

As an example, the effect of across-wafer (4-inch diameter) device uniformity of a Ku-band MMIC power amplifier based on MESFET technology was investigated. The effect of device uniformity in terms of I_{dss} and V_p across the wafer on 4-W broadband IC performance is summarized in Table 19.1. The experiment numbers 1, 2, and 3 have the same average value of I_{dss} across the wafers. However, experiment #1 has the best standard deviation in I_{dss} and V_p (i.e., 6 mA/mm and 0.07 V, respectively), while experiment #3 has the worst (i.e., 26 mA/mm and 0.5 V). At 13.5 GHz, the power output and PAE values are 35.9 dBm and 32.6%, and 34.7 dBm and 27.4%, for experiments #1 and #3, respectively.

In order to realize small transistor or MMIC chip size, an interdigital device structure shown in Figure 19.3, which consists of several unit cells, is generally used and optimized for high power gain and low thermal resistance. High-power interdigital structures have typically two problems:

Table 19.1 Effect of Device Uniformity Across the Wafer on 4-W Broadband IC Performance

PCM FET Parameter[a]	Experiment #1	Experiment #2	Experiment #3
Average I_{dss}/mm (mA)	331	328	330
SD I_{dss} (mA)	6 (2%)	12 (4%)	26 (9%)
Average pinch-off voltage (−V)	2.61	2.67	2.61
SD V_p (V)	0.07	0.128	0.5
Number of wafers tested	3	3	1
Number of ICs tested	15	15	5
Average power output (dBm) at 13.5 GHz	35.9	35.4	34.7
Average PAE (%) at 13.5 GHz	32.6	31.0	27.4

[a]PCM, process control monitor; SD, standard deviation.

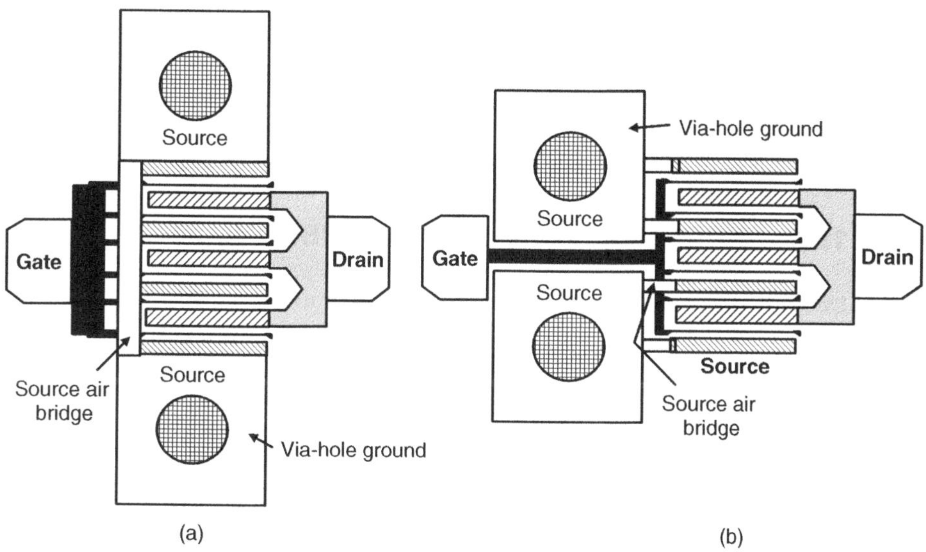

Figure 19.3 FET cell combining using interdigital structure. Source grounding by (a) end vias and (b) offset vias.

1. Interconnection of various cells by crossovers or air bridges, and their source pads grounded through via holes or wire bonds, gives rise to source inductance, which has a significant effect in lowering gain. Gain decreases with increasing number of fingers (due to higher source inductance and increased phase difference between cells) or by lengthening unit gate width (due to increased transmission line loss). Offset via FET configuration, Figure 19.3b, is preferred to end-via FET configuration because it has lower source inductance and also allows use of a larger size cell to realize compact MMIC chips.
2. For maximum gain, all cells must be electrically identical and fed in phase, and each cell should have minimum loss. This mandates minimum possible separation between gates or multiple feed points.

The achievable RF power levels per mm of gate periphery of power FETs/HEMTs operating at 10 V is limited to about 1 W. Thus for high-power applications, very large gate peripheries are required. At S-band, power levels on the order of 100 W for single-chip devices on 30-μm thick GaAs substrate have been demonstrated. In an interdigital structure, for a given substrate thickness, there are basically two design parameters—unit gate width and gate-to-gate spacing—which are to be optimized as a function of frequency. For higher gain, smaller unit gate width and shorter gate-to-gate spacing are required, whereas the low thermal resistance requirements, as discussed in Chapter 16, mandate larger gate-to-gate spacing. If the unit gate width is a significant fraction of a wavelength along the gate transmission line, the RF signal is attenuated due to finite gate resistance. If the gate-to-gate spacing becomes a significant fraction of a wavelength, a portion of the signal gets cancelled due to the phase difference between various paths across the device during combining.

As an example, Figure 19.4 shows the degradation of gain of a 1.25-mm FET as a function of unit gate width for gate-to-gate spacing of 25 μm at 6-, 10-, and 20-GHz frequencies using a single feed. At 20 GHz an optimum unit gate width is about 70 μm. For higher power levels, multiple feeds are usually required. In many applications high power levels are required that far exceed the capability of any single device. High power levels are significantly increased by combining a number of power devices operating coherently in a single circuit using an internally matched technique or in the form of MMICs; these circuits are combined further using thin-film MIC technology for much higher power levels, if required.

19.2 CIRCUIT-LEVEL POWER COMBINING

Device-level combining is generally limited due to the number of transistors that can be combined and effectively matched on a small area. High-power transistors are also matched and combined as an amplifier using the hybrid IC technique. However, monolithic power amplifiers have shown great promise as an alternative to chip combining. These amplifiers have built-in power devices and matching networks. The advantages of

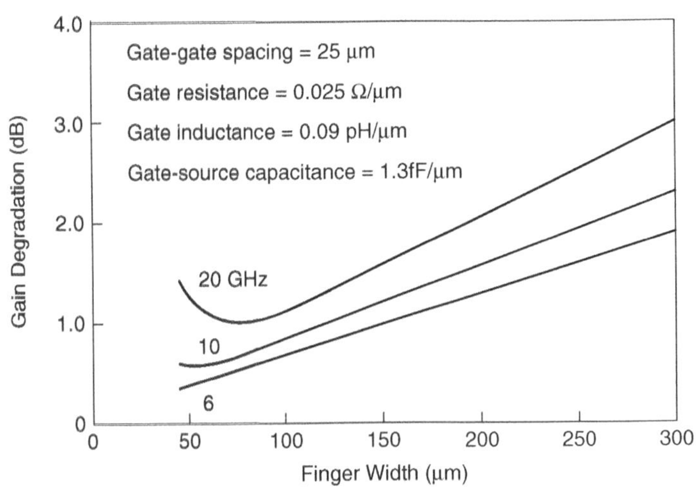

Figure 19.4 Calculated gain degradation as a function of unit finger width for a 1.25-mm gate periphery FET with two end vias.

monolithic power amplifiers are small size, light weight, low cost, and totally matched circuits. Power levels of 50 W and 10 W at C-band and Ku-band, respectively, have been demonstrated by using this technique. Examples of MIC and MMIC HPAs have been described in Chapters 9 through 15.

In this section we describe circuit-level power combining techniques using matched power combiners. Microwave and millimeter-wave power combining techniques have been reviewed in the literature [1–4] and are classified as shown in Figure 19.5. Each combining scheme consists of passive power dividers and combiners, and, in general, they are identical. The desirable characteristics of a combining structure are minimum loss in the matching elements, minimum loss in the combining networks, minimum amplitude and phase imbalance, a good combiner input and output VSWR, even distribution of dissipated heat in the devices, and efficient removal of dissipated heat.

Combining large numbers of transistors on a chip and HPAs using external combiners (due to high loss) become impractical for very high power levels and at high frequencies due to their reduced combining efficiency. Thus spatial (or quasi-optical) power combining (SPC) methods, which have very high combining efficiency, are being used at high microwave and millimeter-wave frequencies to realize very high power modules. The SPC technique has the potential for power combining of solid state devices with very high efficiency with HPA power levels close to TWTs. In this technique, the power signals coming out of HPAs, placed in a guided wave environment, are coherently combined to obtain very high power levels. Since the power combining is done in free space, combining loss is negligible as compared to printed circuits, resulting in high efficiency combining. These techniques are described briefly in Section 19.7.

Resonant Cavity Combiners

In resonant cavity structures, rectangular and cylindrical cavities are used to combine power output from a number of devices. Waveguide dividers/combiners have low loss (= 0.2 dB) and a combining efficiency of 85–90%. These combiners can be used to realize high-power microwave and millimeter-wave sources and proved to be successful

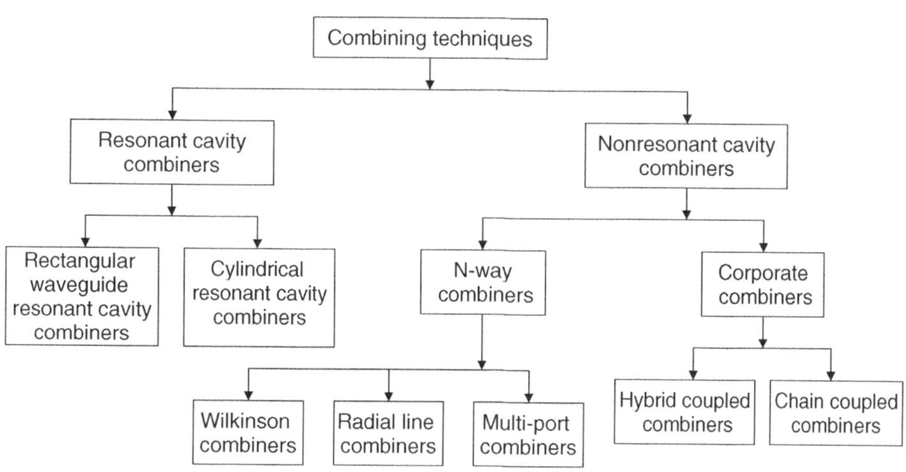

Figure 19.5 Different circuit combining techniques using power combiners.

for narrowband applications up to 220 GHz. For example, a transverse magnetic (TM) mode cavity power combiner [5] has been employed to achieve 80-W power output over 5.9–6.4 GHz.

Nonresonant Combiners

There are two categories of nonresonant circuit-level power combiners: (a) those that combine the output of N devices in a single step (known as N-way combiners) and (b) tree or chain combining structures [1 – 4, 6]. The combining loss at the module level plays a very important role in developing high-PAE power amplifiers. The loss in these networks degrades, for example, in three ways: reduction in output power, reduction in power gain of the circuit, and for a given RF power out, increase in DC power. Several combining techniques are being used, depending on the application at hand. For narrowband and broadband applications, a modified in-phase Wilkinson configuration is frequently used for low loss. This configuration has excellent amplitude and phase balance characteristics to obtain the highest possible combining efficiency. On the other hand, the serial/traveling-wave combiner is compact, with low loss and broadband characteristics. In the case of severe requirements for good input and output match, combiners using Lange couplers, traveling wave combiners, and N-way planar combiners having 90° differential line lengths (also discussed in Chapter 11) are preferred. The combining loss depends on the divider/combiner loss. Power dividers/combiners are described in Section 19.3. Among the various types of combiners, waveguide type combiners have the lowest loss—on the order of 0.05–0.1 dB. Figure 19.6 illustrates the combining efficiency as a function of divider/combiner loss for various values of circuit gain. This figure suggests that it is more efficient to combine high-gain amplifiers.

19.2.1 Graceful Degradation

Consider an N-way matched power combiner, where m amplifiers fail because gates or drains are short circuited. In this case the output voltage at the combiner output port

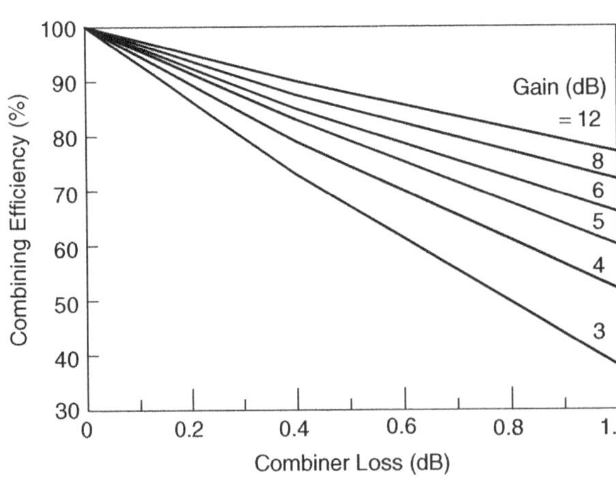

Figure 19.6 Combining efficiency versus combiner loss for various gain values.

is $(N - m)/N$ times the no-failure voltage, and

$$P_o/P_{o,\max} = (1 - m/N)^2 = 10\log[(1 - m/N)^2]\ \text{dB} \qquad (19.1)$$

where P_o is the output power when m amplifiers failed and $P_{o,\max}$ is the maximum output power when there is no amplifier failure. In this case, it has been assumed that the input and output impedance conditions of the working amplifiers are unchanged. Equation (19.1) represents graceful degradation performance and the output power is significantly lower than the sum of the output powers of the remaining amplifiers, if combined reactively, which is given by

$$P_{or}/P_{o,\max} = 1 - m/N = 10\log(1 - m/N)\ \text{dB} \qquad (19.2)$$

where P_{or} is the total power of the remaining amplifiers. The primary difference between the matched and reactive power combining schemes is that in the former case each amplifier unit is isolated, whereas in the latter case all unit amplifiers are connected in parallel.

For example, in a four-way matched combiner when one out of the four combined amplifiers fails completely, the combined output power drops by 2.5 dB, whereas in the reactively combined scheme the total power drops by only 1.25 dB. Similarly, in a three-way matched combiner case, if one-third of the combined amplifiers fail, the combined output power drops by 3.5 dB. This is true in the case of in-phase combining by using an in-phase Wilkinson power divider. However, in the case of traveling-wave combiner, the power loss also depends on the failure condition of the amplifier. Table 19.2 summarizes power loss due to one failure at a time in three-way and four-way traveling-wave combiners. The failed amplifier's input/output might be shorted or opened.

Two possible schemes to improve the graceful degradation performance of combined power amplifiers have been discussed by Saleh [7].

Table 19.2 Power Loss Due to Failure of Power Amplifier Units: One Failure at a Time

Three-Way Traveling-Wave Combiner (Good Match)				
	Power Loss (dB)			
Amplifier Number	Input Short	Input Open	Output Short	Output Open
First	3.3	3.7	3.8	3.3
Second	3.2	3.9	3.2	3.9
Third	3.7	3.3	3.3	3.7
		Average loss is 3.5 dB		
Four-Way Traveling-Wave Combiner (Good Match)				
	Power Loss (dB)			
Amplifier Number	Input Short	Input Open	Output Short	Output Open
First	2.5	2.8	2.4	2.5
Second	2.5	2.6	2.6	2.3
Third	2.5	2.4	2.4	2.5
Fourth	2.5	2.4	2.4	2.6
		Average loss is 2.5 dB		

19.2.2 Power Combining Efficiency

The power combining efficiency as defined in Eq. (3.34), Chapter 3, depends on the imbalance of amplitudes and phases of the signals being combined. The effect of such imbalance on the combining efficiency of HPAs has been discussed by Gupta [8].

Consider an N-way combining scheme using ideal N-way dividers/combiners as shown in Figure 19.7. Assuming perfect isolation between the combining ports, the output voltage can be written

$$V_{\text{out}} = C_1 V_1 + C_2 V_2 + \cdots + C_N V_N \tag{19.3a}$$

where the C's are the voltage transmission coefficients. For an ideal combiner system, the C's are all equal in amplitude and phase with a value of $1/\sqrt{N}$ and all amplifiers have the same phase. Then

$$V_{\text{out}} = \frac{1}{\sqrt{N}}(|V_1| + |V_2| + \cdots + |V_N|) \tag{19.3b}$$

The output power, which is defined by $V_{\text{out}}^2/R_{\text{L}}$, where R_{L} is the terminating impedance, is given by

$$P_{\text{out}} = \frac{1}{N}\left|\sqrt{P_1} + \sqrt{P_2} + \cdots + \sqrt{P_N}\right|^2 \tag{19.4a}$$

If $\theta_1, \theta_2, \ldots$ are the phases associated with each arm, then

$$P_{\text{out}} = \frac{1}{N}\left|\sqrt{P_1}e^{j\theta_1} + \sqrt{P_2}e^{j\theta_2} + \cdots + \sqrt{P_N}e^{j\theta_N}\right|^2 \tag{19.4b}$$

This equation may be used to analyze the effect of imbalance in terms of amplitude and phase: the graceful degradation calculation of the power combining system. Next, several different cases to study the power combining efficiency are discussed.

(a) Let $P_1 = P_2 = P_3 = \cdots = P_N = P_o$ and $\theta_1 = \theta_2 = \theta_3 = \cdots = \theta_N = 0$; then

$$P_{\text{out}} = NP_{\text{o}} \tag{19.5}$$

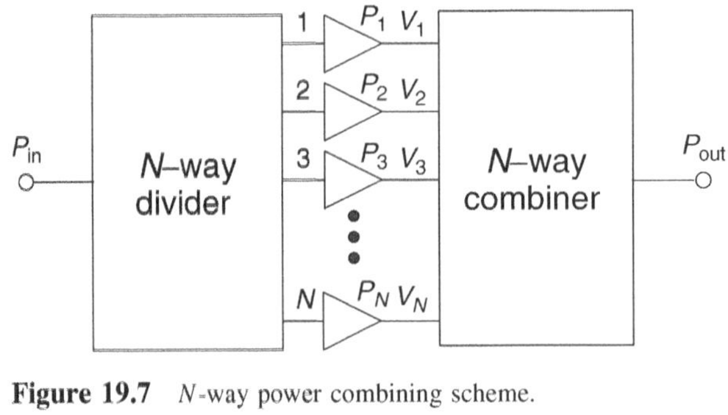

Figure 19.7 N-way power combining scheme.

(b) If $\sqrt{P_2}/\sqrt{P_1} = \Delta_1, \sqrt{P_3}/\sqrt{P_1} = \Delta_2, \ldots, \sqrt{P_N}/\sqrt{P_1} = \Delta_{N-1} < 1$ and $\theta_2 - \theta_1 = \phi_1, \theta_3 - \theta_1 = \phi_2, \ldots, \theta_N - \theta_1 = \phi_{N-1}$, then for $P_1 = P_o$, Eq. (19.4b) becomes

$$P'_{\text{out}} = \frac{P_o}{N}\left|1 + \Delta_1 e^{j\phi_1} + \Delta_2 e^{j\phi_2} + \cdots + \Delta_{N-1} e^{j\phi_{N-1}}\right|^2 \tag{19.6}$$

The power combining efficiency η_c is given by

$$\eta_c = \frac{P'_{\text{out}}}{P_{\text{out}}} = \frac{1}{N^2}\left|1 + \Delta_1 e^{j\phi_1} + \Delta_2 e^{j\phi_2} + \cdots + \Delta_{N-1} e^{j\phi_{N-1}}\right|^2 \tag{19.7}$$

(c) Let all the amplitudes be equal, that is, $\Delta_1 = \Delta_2 = \cdots = \Delta_{N-1} = 1$; then

$$\eta_c = \frac{1}{N^2}\left|1 + e^{j\phi_1} + e^{j\phi_2} + \cdots + e^{j\phi_{N-1}}\right|^2 \tag{19.8}$$

When m amplifiers have phase difference $\phi_i(i = 1, 2, \ldots, m) = \phi$ and the remaining $N - m$ amplifiers have zero phase difference, (19.8) becomes

$$\eta_c = \frac{1}{N^2}\left|(N - m) + me^{j\phi}\right|^2 = \frac{1}{N^2}\{[(N - m) + m\cos\phi]^2 + m^2\sin^2\phi\}$$

or

$$\eta_c = 1 - 2\frac{m}{N}\left(1 - \frac{m}{N}\right)(1 - \cos\phi) \tag{19.9}$$

The value of m for the worst case may be calculated by equating $\partial\eta_c/\partial m = 0$. From (19.9), we have

$$-(1 - \cos\phi)\left(1 - \frac{2m}{N}\right) = 0 \tag{19.10}$$

This leads to $m = N/2$ and the worst-case combining efficiency is given by

$$\eta_c^{\text{worst}} = 1 - \tfrac{1}{2}(1 - \cos\phi) \tag{19.11}$$

(d) Next, only imbalance in the amplifier's signal amplitude, that is, $\phi_i(i = 1, 2, \ldots N - 1) = 0$, is considered. If m amplifiers have $\Delta_i(i = 1, 2, \ldots, m) = \Delta$ and the remaining $N - m$ amplifiers have same amplitude, then

$$\eta_c = \frac{1}{N^2}|N - m + m\Delta|^2 = \left|1 - \frac{m}{N}(1 - \Delta)\right|^2 \tag{19.12a}$$

This represents the case where $N - m$ amplifiers have full power P_o and m amplifiers have a reduced power level Δ. If $N = 4$, $m = 2$, and $\Delta = 0.944$ (-0.5 dB), then

$$\eta_c = 0.945 \text{ or } 94.5\%$$

(e) When m amplifiers fail and other amplifiers have the same amplitude and phase, that is, $\Delta_i(i = 1, 2, \ldots, m) = 0$, $\Delta_{N-m} = 1$, and $\phi_i = 0$, then in this case (19.7) becomes

$$\eta_c = \left|1 - \frac{m}{N}\right|^2 \tag{19.12b}$$

This relation is the same as given by Eq. (19.1). If $N = 4$ and $m = 2$, $\eta_c = 25\%$.

The effect of phase imbalance on the power combining efficiency is much larger than the effect of amplitude imbalance. Worst-case degradation in combining efficiency as a function of maximum allowed phase variations in an N-way combiner is shown in Figure 19.8. It is evident that phase-matched power amplifiers are required for efficient power combining. In high power combining of MMIC power amplifiers, "on-wafer" pulsed large-signal S-parameters are generally measured for bin phase matched ($\pm 10^\circ$) MMIC chips. This maintains the phase combining efficiency at 97%.

An approximate expression for the combining efficiency for amplifiers is given by

$$\eta_c = \left[\frac{1}{2} + \frac{\Delta}{1 + \Delta^2}\cos\theta\right] \tag{19.13}$$

where Δ and θ are the amplitude (voltage ratio) and phase imbalances between the two amplifiers. When two amplifiers have amplitude imbalance of 1.0 dB ($\Delta = 0.8913$) and 20° phase difference ($\cos\theta = 0.94$), the calculated value of η_c is 96.7%.

EXAMPLE 19.1

Consider two power combiners each using four MMIC power amplifier chips. In one case two chips are binned to have 20° phase difference and in the second case it is 30° etermine the combining efficiency for each case.

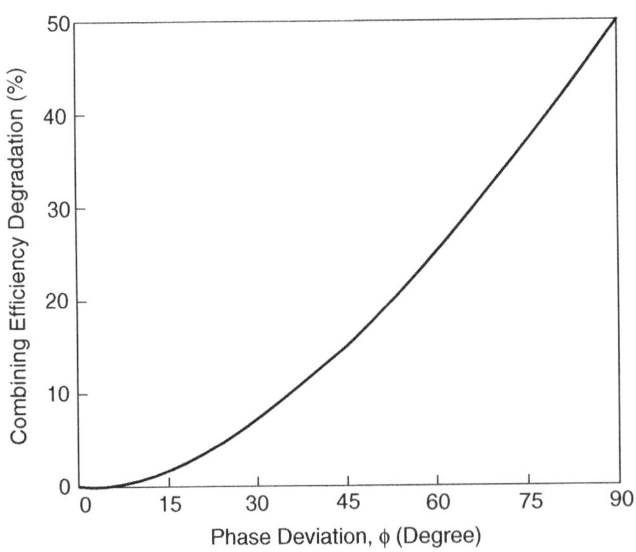

Figure 19.8 Worst-case power combining efficiency as a function of maximum phase difference, that is, $\pm 10^\circ$ is 20°.

SOLUTION In the first case, $N = 4, m = 2$, and $\phi = 20^\circ$. From (19.9) or (19.11), we find

$$\eta_c = 1 - 2\frac{2}{4}\left(1 - \frac{2}{4}\right)(1 - \cos 20^\circ) = 0.97 = 97\%$$

In the second case, $N = 4, m = 2$, and $\phi = 30^\circ$ and the calculated value of η_c is 93.3%. Thus in order to obtain a combining efficiency better than 97%, ϕ should be less than 20° or ±10°. ■

19.3 POWER DIVIDERS, HYBRIDS, AND COUPLERS

Power combiners use several types of power splitters, hybrids, and couplers realized in microstrip or strip line [3, 4] or any other suitable medium. The substrate medium may be plastic or ceramic including LTCC. In this section we describe briefly these components.

19.3.1 Power Dividers

Power dividers/combiners are commonly used for power combining. Most are based on the Wilkinson multiport power splitter.

Wilkinson Power Splitter

A Wilkinson power divider [9, 10], also known as a two-way power splitter, is the most popular type and offers broad bandwidth and equal phase characteristics at each of its output ports. Figure 19.9a shows the schematic diagram of a Wilkinson power divider. The output port isolation is obtained by using a series resistor connected between the output ports. Each of the quarter-wave lines has the characteristic impedance of $\sqrt{2}Z_0$ and the value of the isolation resistor is $2Z_0$, Z_0 being the system impedance.

The bandwidth of a 3-dB power divider was computed [10] using ideal lines, microstrip on GaAs, ideal lumped elements (*LRC*), and MMIC based lumped elements. Table 19.3 compares the bandwidths of these four power divider configurations for two cases of return loss and isolation: 10 and 20 dB. Here, various bandwidth definitions are considered. In some applications input match may be important, while in other cases output match or isolation or both may be critical. For example, in power amplifier combining, the output match and isolation are very important so that the combiner provides the desired 50 Ω to the output of the power amplifiers. With isolation as high as 15–20 dB, due to the expected amplitude and phase imbalance between the amplifiers or even the complete failure of one of the amplifiers, there will be little interaction between amplifiers. In most two-way power dividers/combiners, the bandwidth is limited due to input match and isolation. As expected, networks with higher loss result in larger bandwidths. A Wilkinson power divider based on a microstrip offers a bandwidth of about 35–40%, whereas the output match bandwidth is greater than 130%.

The performance of a Wilkinson coupler can be further improved, depending on the availability of space, by the addition of a $\lambda/4$ transformer in front of the power-division step, or using multisections or a compensation technique [10]. A multisection Wilkinson powerdivider consists of a number of quarter-wave sections with resistive terminations

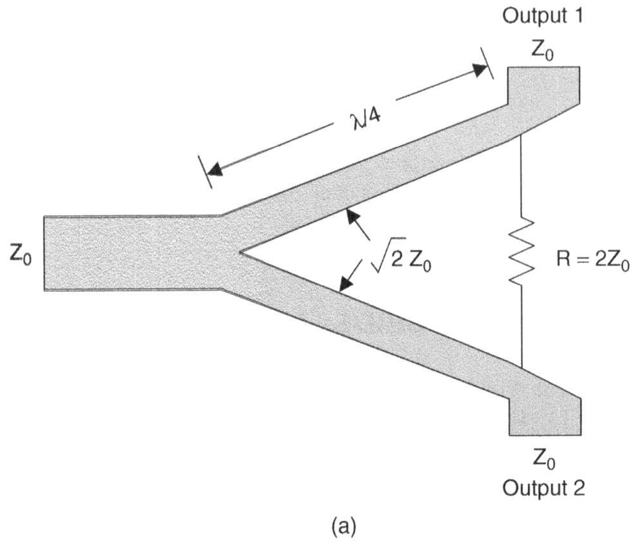

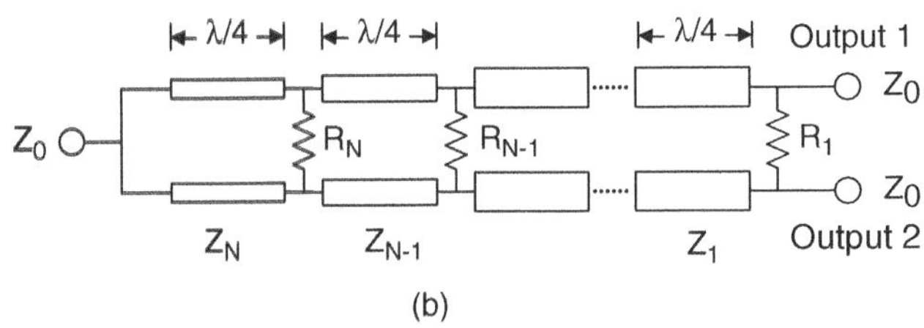

Figure 19.9 (a) Wilkinson power splitter. (b) N-section broadband Wilkinson power splitter.

Table 19.3 Comparison of Percentage Bandwidth for Various Two-Way Dividers/Combiners[a]

	Distributed Element		Lumped Element	
Parameter	Ideal Line	Microstrip	Ideal LRC	MMIC
S_{11} − 10 dB	253	372	102	259
− 20 dB	37	40	17	32.5
S_{22} − 10 dB	505	Not limited	343	551
− 20 dB	132	138	89	94
S_{23} − 10 dB	135	134	92	125
− 20 dB	37	36	23.5	30

[a]Center frequency is 10 GHz. Microstrip or MMIC substrate is GaAs, 75 μm thick, and 4.5-μm thick conductors.

at the end of every section, as shown schematically in Figure 19.9b. Larger bandwidth and greater isolation are obtained when a larger number of sections are used. Design of such couplers is given in References 11 and 12.

Unequal-Power-Split Wilkinson Divider

Figure 19.10 shows the schematic of the unequal-power-split Wilkinson divider [13]. As can be seen from the figure, output impedance transformers are also required, in

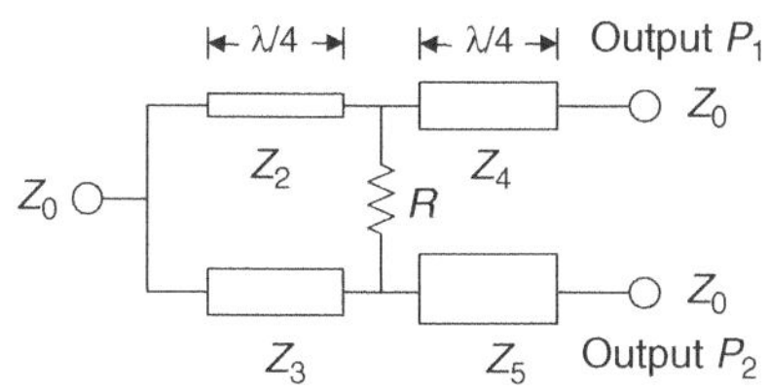

Figure 19.10 Schematic of unequal power-split Wilkinson divider.

contrast to the equal-power-split case. The design equations are as follows:

$$\begin{aligned} K^2 &= \frac{P_2}{P_1}, \quad R = Z_0\frac{1+K^2}{K} \\ Z_4 &= Z_0\sqrt{K}, \quad Z_5 = \frac{Z_0}{\sqrt{K}} \\ Z_2 &= Z_0[K(1+K^2)]^{1/2} \\ Z_3 &= Z_0\left(\frac{1+K^2}{K^3}\right)^{1/2} \end{aligned} \tag{19.14}$$

where P_1 and P_2 are power levels at ports 1 and 2, respectively. These dividers provide two in-phase isolated outputs with a constant arbitrary power division over a wide bandwidth. Design equations for arbitrary power division and termination impedances for a three-port power divider are also available [14].

Serial/Traveling-Wave Structure

Traveling-wave power dividers/combiners have been used for combining three to six power amplifiers because they are compact, have low loss, and are viable well into the millimeter-wave frequency range. They are commonly known as traveling-wave couplers (TWCs). The traveling-wave structures are designed using a cascade of unequal-power-split Wilkinson dividers and an equal-power-split Wilkinson divider each having a different power division ratio. For example, a four-way coupler, as shown in Figure 19.11, uses the power division ratios of 4:1, 3:1, and 2:1. In this divider topology, the signal is continuously divided and can be realized 3-, 4-, 5-, or N-way. In Figure 19.11, each of the transmission line sections has an electrical length of about 90°. The schematic shows the dividers separated by two 90° sections to accommodate relatively wide MIC/MMIC amplifier sizes. Output 4 has three 90° sections (not shown) for the same reason. Additionally, the transmission line impedances were constrained to be no grater than 77 Ω and no less than 20 Ω on a 0.38-mm thick alumina substrate ($\varepsilon_r = 9.9$) due to loss considerations and transverse resonances, respectively.

Compact Wilkinson Divider

The size of the power divider can be reduced by capacitive loading and the schematic of a miniaturized Wilkinson power divider is shown in Figure 19.12. The design equations

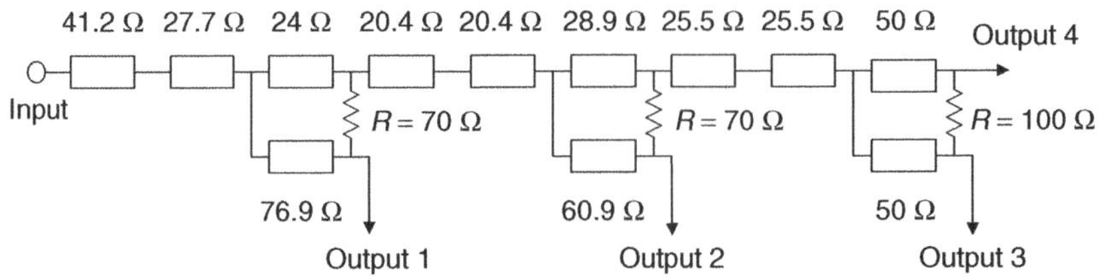

Figure 19.11 Schematic of a traveling-wave combiner. All electrical lengths are 90°.

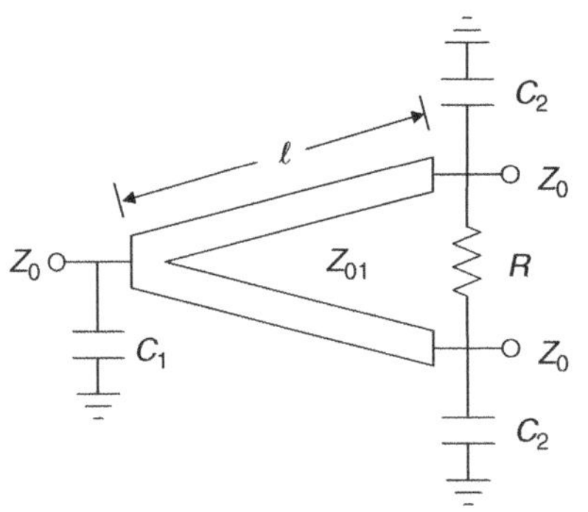

Figure 19.12 Configuration of a reduced-size Wilkinson power splitter.

for this configuration are [15]

$$Z_{01} = \frac{\sqrt{2}Z_0}{\sin(\beta\ell)} \tag{19.15a}$$

$$C_1 = \frac{\cos(\beta\ell)}{\omega_0 Z_0 \sqrt{2}} \tag{19.15b}$$

where $\beta(= 2\pi/\lambda)$ is the propagation constant, Z_0 is the characteristic impedance of the system, and $\omega_0(= 2\pi f_0)$ is the angular frequency at the design frequency f_0. The termination resistance $R = 2Z_0$ and $C_2 = 2C_1$. For example, for $\beta\ell = \lambda/4, \lambda/8$, and $\lambda/12$, the values of C_1 at 10 GHz are 0, 0.16, and 0.195 pF, and the values of Z_{01} are 70.7, 100, and 141.4 Ω, respectively.

19.3.2 90° Hybrids

Hybrids have limited applications in power combining at RF and microwave frequencies because of their large size and narrow bandwidth; however, at millimeter-wave frequencies they are commonly used. The branch-line type of hybrid shown in Figure 19.13 is one of the simplest structures for a 90° hybrid in which the circumference is an odd multiple of λ. The geometry is readily realizable in any transmission medium. Branch-line hybrids have narrow bandwidths on the order of 10%. As shown in Figure 19.13, the two quarter-wavelength long sections spaced one-quarter wavelength apart divide the input signal from port 1 so that no signal appears at port 4. The signals appearing at ports 2 and 3 are equal in amplitude, but out of phase by 90°. The coupling factor is determined by the ratio of the impedance of the shunt and series arms and is optimized to maintain proper match over the

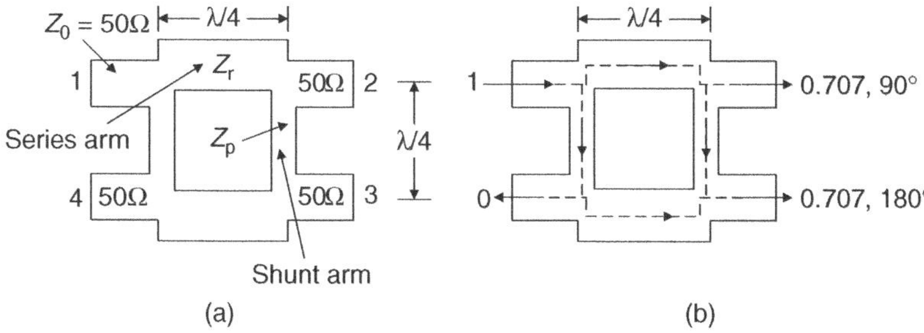

Figure 19.13 Single-section branch-line hybrid: (a) design parameters and (b) voltage split amplitude and phase.

required bandwidth. In terms of Z_r and Z_p, the scattering parameters of a branch-line coupler are given by [16]

$$S_{21} = -j\frac{Z_r}{Z_0}, \quad S_{31} = -\frac{Z_r}{Z_p}, \quad \text{and} \quad S_{41} = 0 \tag{19.16a}$$

For 90° lossless hybrids, the following conditions hold good:

$$|S_{21}|^2 + |S_{31}|^2 = 1 \quad \text{or} \quad \left|\frac{Z_r}{Z_0}\right|^2 + \left|\frac{Z_r}{Z_p}\right|^2 = 1 \tag{19.16b}$$

For 3-dB coupling, the characteristic impedances of the shunt and series arms are Z_0 and $Z_0/\sqrt{2}$, respectively, for optimum performance of the coupler. Z_0 is the characteristic impedance of the input and output ports. For most applications, $Z_0 = 50\ \Omega$; thus shunt and series arm lines have characteristic impedances of 50 and 35.36 Ω, respectively.

19.3.3 Coupled-Line Directional Couplers

Coupled-line directional couplers are frequently used for power combining because of variable coupling and large bandwidth. Among such couplers the Lange coupler [17] is the most popular. When two unshielded transmission lines, as shown in Figure 19.14, are placed in close proximity to each other, a fraction of the power present on the main line is coupled to the secondary line. The power coupled is a function of the physical dimensions of the structure, the frequency of operation, and the direction of propagation of the primary power. In these structures there is a continuous coupling between the electromagnetic fields of the two lines, also known as parasitic coupling. If the coupled lines are of the TEM type (strip lines), the power coupled to port 2 is through a backward wave and the structure is called a backward-wave directional coupler. In such couplers ports 2, 3, and 4 are known as coupled, isolated, and direct ports. The phase differences between ports 1 and 2, and ports 1 and 4 are 0° and 90°, respectively.

Coupled-line structures are based on almost all forms and types of transmission lines/dielectric guides developed to realize microwave and millimeter-wave components. Most popular are strip lines, microstrip lines, coplanar waveguides, image guides,

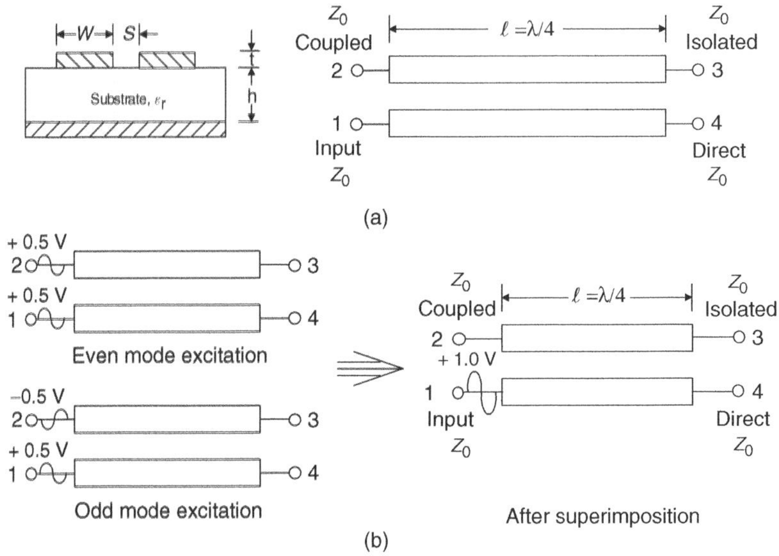

Figure 19.14 TEM coupler: (a) microstrip line coupled lines and (b) even- and odd-mode excitations used for circuit analysis.

and insular and inverted strip guides [16]. In microstrip-like structures (excluding multiconductor structures such as a Lange coupler), tight coupling on the order of about 6 dB over a $\lambda/4$ section is realizable due to practical spacing limitations between the lines. On the other hand, broadside-coupled lines are used extensively to realize tight couplings on the order of 2–3 dB. Coupler examples shown in Figure 19.14 support TEM modes in the case of homogeneous dielectric medium and quasi-TEM modes in the case of nonhomogeneous medium.

In general, the coupled-line structure as shown in Figure 19.14a supports two modes: even and odd. It is the interaction between these modes that induces the coupling between the two transmission lines and the properties of the coupled structures may be described in terms of a suitable linear combination of these even and odd modes (Fig. 19.14b). The even-mode excitation means both microstrip conductors are at the same potential while the odd mode is excited by different potentials. These modes have different characteristic impedances and their values become equal when the separation between the conductors is very large. The velocity of propagation of these two modes is equal when the lines are imbedded in a homogeneous dielectric medium (e.g., strip line). However, for transmission lines such as coupled microstrip lines, the dielectric medium is not homogeneous. A part of the field extends into the air above the substrate and is different for the two modes of coupled lines. Consequently, the effective dielectric constants (and the phase velocities) are not equal for the two modes. The nonsynchronous feature deteriorates the isolation performance of circuits using these types of coupled lines. When the two conductors of a coupled-line pair are identical, we have a symmetrical configuration. The symmetry is very useful for simplifying the analysis and design of such coupled lines. If the two lines do not have the same impedance, the configuration is called asymmetric.

Planar TEM line directional couplers can be either edge coupled or broadside coupled. The design equations for the TEM coupler shown in Figure 19.14 are summarized in the following at the center frequency of the band:

$$\theta_c = \theta_e = \theta_0 = \pi/2, \quad Z_0^2 = Z_{0e} Z_{0o} \tag{19.17a}$$

$$C = -20 \log \left| \frac{Z_{0e} - Z_{0o}}{Z_{0e} + Z_{0o}} \right| \text{ dB} \tag{19.17b}$$

$$Z_{0e} = Z_0 \left(\frac{1 + 10^{-C/20}}{1 - 10^{-C/20}} \right)^{1/2} \tag{19.18a}$$

$$Z_{0o} = Z_0 \left(\frac{1 - 10^{-C/20}}{1 + 10^{-C/20}} \right)^{1/2} \tag{19.18b}$$

where subscripts e and o denote even and odd mode, C is the coupling coefficient expressed in decibels with positive sign, and Z_0 is the terminating or system impedance. To maximize the effective usable bandwidth, it is frequently desirable to overcouple at the design frequency [16], thus permitting a plus and minus tolerance across the frequency range.

For a quasi-TEM line, the condition for input matching is given by

$$Z_0 = \left(\frac{Z_{0e} \sin \theta_e + Z_{0o} \sin \theta_o}{Z_{0e} \sin \theta_o + Z_{0o} \sin \theta_e} \right)^{1/2} \sqrt{Z_{0o} Z_{0e}} \tag{19.19a}$$

and

$$\theta = \tfrac{1}{2}(\theta_e + \theta_o) = \frac{2\pi}{\lambda_0} \frac{\sqrt{\varepsilon_{ree}} + \sqrt{\varepsilon_{reo}}}{2} \ell = 90^\circ \tag{19.19b}$$

where ε_{ree} and ε_{reo} are the effective dielectric constants for the even and odd modes, respectively, and ℓ is the physical length of the coupled section. The expression for the frequency response of the coupling coefficient, $C(\theta)$, is

$$C(\theta) = \frac{jC \sin \theta}{\sqrt{1 - C^2} \cos \theta + j \sin \theta} \tag{19.20}$$

Multiconductor Couplers

The interdigital Lang coupler [17] or multiconductor coupler has always been a popular component in planar circuits because of its broad bandwidth. Figure 19.15 shows a four-finger Lange coupler, although in certain applications the number of elements may be greater than four. The coupler is usually designed using $\lambda/4$ coupled lines for 3-dB coupling and 90° phase difference between output ports 2 and 3. The way the coupler is wired, ports 2 and 3 are coupled and direct, and port 4 is an isolated port. Obviously, the best realization is in the microstrip form.

An interdigital coupler has advantages because of its small size and relatively large separation when compared with the two-coupled-line circuit and has much larger bandwidth when compared with branch-line couplers. However, to connect interdigital fingers, bond wires or straps are required. For given values of the number of fingers, Z_0 (the impedance of various ports), and coupling, the design is obtained in terms of even- and odd-mode impedances of a pair of coupled lines. Once the even- and

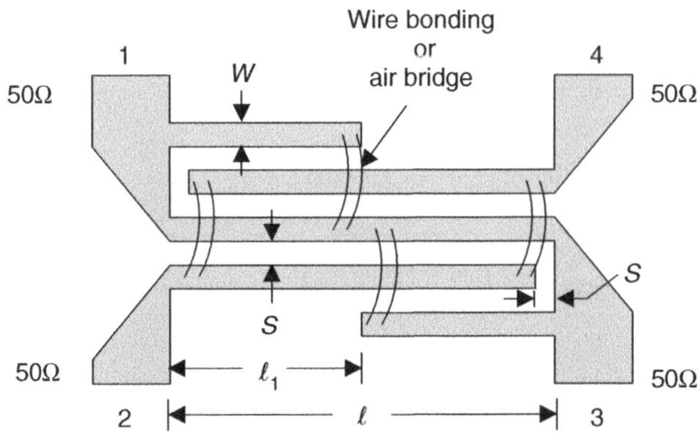

Figure 19.15 Multiconductor Lange coupler.

odd-mode impedances of one pair of coupled lines are known, the width (W) of the lines and the spacing (S) between them can be determined. The design is obtained in terms of even- and odd-mode impedances of one pair of coupled lines because this data is available in the literature for a large class of transmission lines. The working mechanism of the Lange coupler is described in Reference 16.

The design equations for an N-finger (N even) Lange coupler are given by [18]

$$R = \frac{Z_{0o}}{Z_{0e}} \tag{19.21a}$$

$$C = \frac{(N-1)(1-R^2)}{(N-1)(1+R^2)+2R} \tag{19.21b}$$

$$Z = \frac{Z_{0o}}{Z_0} = \frac{\sqrt{R[(N-1)+R][(N-1)R+1]}}{(1+R)} \tag{19.21c}$$

where C is the voltage coupling coefficient between the input and coupled ports at the center frequency of the design and N is the total number of conductors. The impedances Z_{0e} and Z_{0o} denote, respectively, the even- and odd-mode impedances of a pair of coupled lines having the same width and spacing between them as any pair of N-conductor interdigital coupler. It may be worthwhile to remark that the usual relationship $Z_0 = \sqrt{Z_{0e}Z_{0o}}$ is valid in the case of the interdigital coupler only when $N = 2$. For other values of N, this relationship is not satisfied.

The length of the interdigital coupler at the center frequency of design is given by

$$\ell = \frac{\lambda}{4} \tag{19.22}$$

where λ is the average guide wavelength for the even and odd modes.

EXAMPLE 19.2

Consider a 3-dB ($C = 0.707$) Lange coupler having $N = 4$ and $Z_0 = 50\ \Omega$ designed on a 25-mil thick alumina substrate ($\varepsilon_r = 9.9$). Find its dimensions at 10 GHz.

SOLUTION For a given value of voltage coupling factor, C, and number of conductors, N, (19.21b) can be used to find the value of R:

$$0.707 = \frac{(4-1)(1-R^2)}{(4-1)(1+R^2)+2R}$$

or

$$R^2 = 0.2762R - 0.1718 = 0$$

Solving the quadratic equation for the positive root of R, we find

$$R = 0.2985$$

Next, (19.21c) can be used to determine the odd-mode impedance Z_{0o}:

$$Z_{0o} = 50\frac{\sqrt{0.2985 \times 3.2985 \times 1.855}}{1+0.2985} = 52.6\ \Omega$$

Furthermore, the even-mode impedance Z_{0e} can be determined using (19.21a):

$$Z_{0e} = Z_{0o}/R = 52.6/0.2985 = 176.2\ \Omega$$

Using the known values of the even- and odd-mode impedances, the dimensions of the lines and the spacing between them can be determined by using either available nomograms or design equations [16, 19]. Since a Lange coupler is usually realized in microstrip configuration, the normalized dimensions of the coupler are approximately given as $W/h = S/h = 0.08$ [16]. For a 25-mil thick substrate $W = S = 2$ mils. The physical length of the coupled lines is $\lambda/4$ at the operating frequency. At 10 GHz, the coupler length is calculated using (19.19b), and the values of ε_{ree} and ε_{reo} are obtained from Reference 19. The length is

$$\ell = \frac{\lambda}{4} = \frac{\lambda_0}{4(\sqrt{\varepsilon_{ree}} + \sqrt{\varepsilon_{reo}})/2} = \frac{30}{2(\sqrt{6.34}+\sqrt{5.5})}\ \text{mm} = 3.08\ \text{mm} = 121.4\ \text{mils}$$

■

19.4 *N*-WAY COMBINERS

N-way combining structures are simpler than corporate combiners, and they avoid the use of several combining stages, thus making it possible to achieve high-efficiency combining. The structures can be either cavity or nonresonant combining structures. In cavity structures, cylindrical or rectangular cavities are used to combine the power output from a number of devices. These combiner–dividers have low loss ($\cong 0.2$ dB) and a combining efficiency of 85–90%. Note that, in general, combining and dividing circuits are identical.

Many nonresonant N-way combining techniques are also available. Essentially three types of N-way combiners are used for combining large numbers of amplifiers: Wilkinson, radial, and planar. The N-way Wilkinson divider [9], shown in Figure 19.16a, has the advantage of low loss, moderate bandwidth, and good amplitude and phase balance. However, its major disadvantage for power applications is the "floating star-point" isolation resistors. These resistors require a nonplanar crossover configuration, which limits the power-handling capability of the combiner. Fortunately,

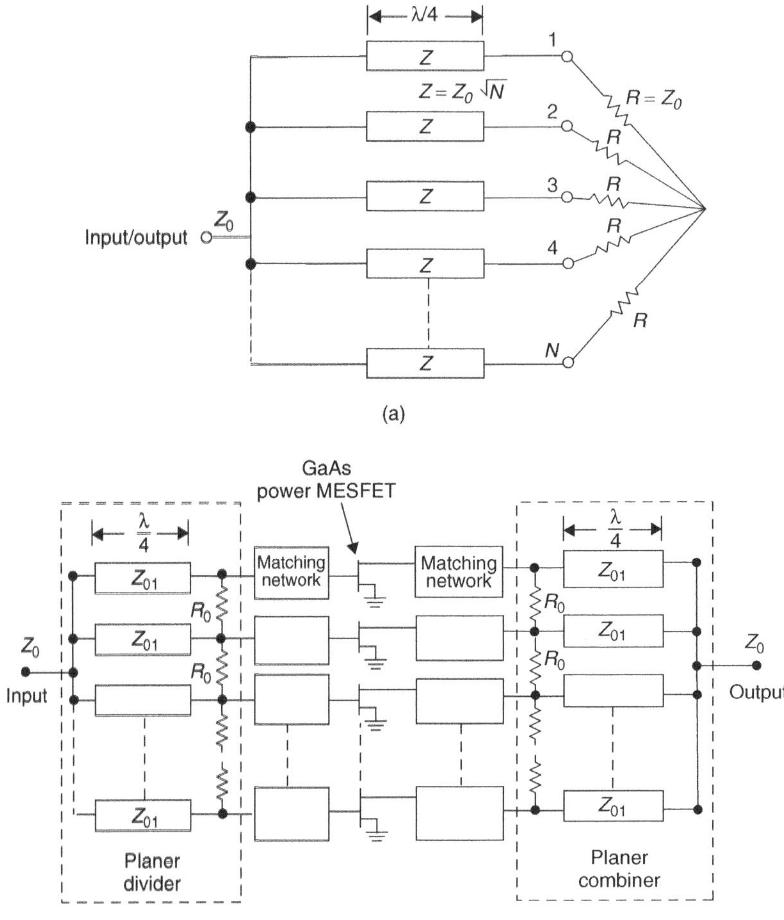

Figure 19.16 (a) Wilkinson N-way divider–combiner. (b) Modified N-way divider–combiner. Isolation resistors suppress possible oscillations.

a simplified version, as shown in Figure 19.16b, can also be used [20]. This particular arrangement has a combining efficiency on the order of 90% and shows much promise for chip combining as well as for MMIC applications. Radial line and planar N-way combiners are shown in Figure 19.17. The radial line [21] combiner has low loss, inherent phase symmetry, and good isolation. Its main disadvantage is that it requires a three-dimensional structure. On the other hand, the planar N-way combiner–divider [22] requires $(N - 1) \times N$ quarter-wave sections for maximum isolation and thus is very large in size. The inherent redundancy in the N-way combiner makes it possible to obtain a graceful degradation characteristic.

19.5 CORPORATE STRUCTURES

A corporate structure (or tree) for combining power from two-way adders is shown in Figure 19.18a or a serial (also known as chain) combiner is shown in Figure 19.19a.

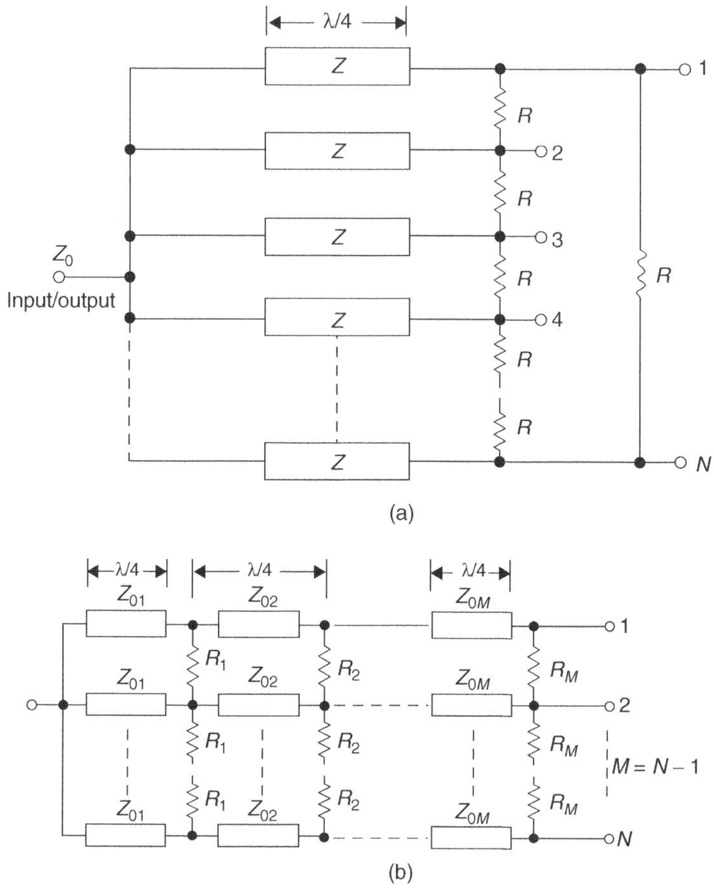

Figure 19.17 N-way divider–combiner: (a) radial line and (b) planar.

The loss in adders limits the combining efficiency. Figure 19.18b illustrates [1] the combining efficiency of the corporate structure versus the number of devices for various loss-per-adder values. The number of devices combined in this way is binary. Examples of two-way adders are directional couplers, hybrids, and the two-way Wilkinson combiner. Among the two-way adders, the Lange coupler is usually preferred because of its good isolation and wideband properties. However, cascading these to obtain high-order combining becomes impractical beyond a four-way combiner due to the relatively high coupler loss (on the order of 0.3–0.4 dB per coupler).

A serial or chain combiner is shown in Figure 19.19a. Here each successive stage of an N-way combiner adds $1/N$ of the power delivered to the output. The number of the stage determines the required coupling coefficients for that stage, as indicated in the figure. One advantage of the chain structure is that another stage can be added by simply connecting the new source to the line after the Nth stage through a coupler with $10 \log(N + 1)$ coupling coefficient. The roles of input and output ports are changed for the divider structure. Losses in the couplers reduce the combining efficiency and bandwidth attainable with this approach. Combining efficiency for the chain structure for each path is shown [1] in Figure 19.19b. The four-way structure on the microstrip can be used to realize a combining efficiency on the order of 90% over an octave or

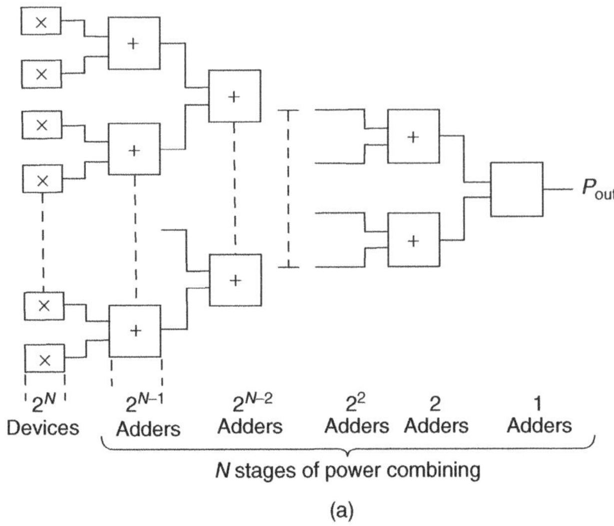

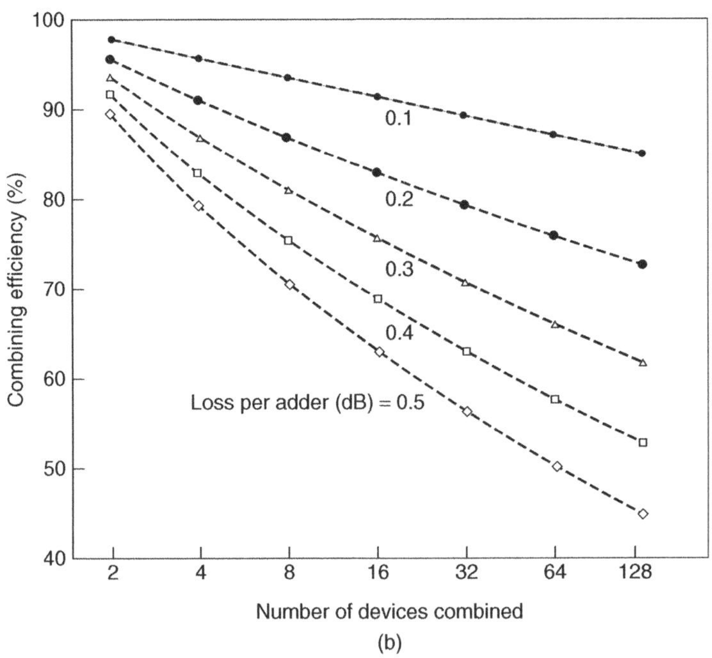

Figure 19.18 (a) Corporate combining structure and (b) combining efficiency for corporate combining structure.

greater bandwidth. The combiner is compact in size and has good input and output VSWR values. However, combining more than five amplifiers becomes impractical due to the realization of the required coupler's dimensions.

Comparing the combing efficiency of various schemes, it is evident that the low-loss combiners are critical for obtaining high power combing efficiency (PCE). For example, if eight HPAs are to be combined and the loss per combiner is 0.2 dB, the PCE values for the N-way, corporate, and serial combining methods are 95%, 88%, and 84%, respectively.

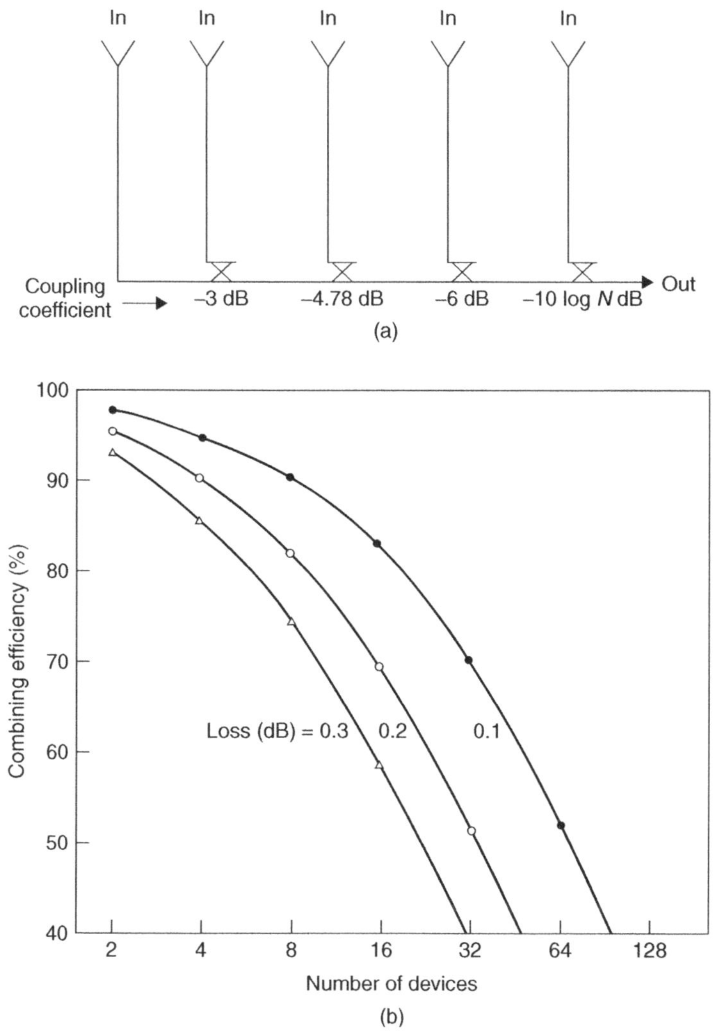

Figure 19.19 (a) Serial combining structure and (b) combining efficiency for serial combining structure. Loss in decibels refers to the loss in each power path in each stage's coupler.

EXAMPLE 19.3

Design a 12-W and 10-dB gain power amplifier at 12 GHz with 40% bandwidth and 30% power added efficiency using monolithic integrated circuit amplifier chips having 50-Ω source and load impedance. The minimum measured performances of these chips are $P_{out} = 5$ W, gain = 12 dB, and PAE = 35% over the desired bandwidth.

SOLUTION In order to obtain a 12-W power output, we need three such chips and two low-loss three-way divider–combiner circuits. The three-way divider–combiner used is of the traveling-wave type similar to that described in Reference 23 and fabricated on a 15-mil thick alumina substrate ($\varepsilon_r = 9.9$). Figure 19.20a shows the assembly of a 12-W HPA using three 5-W MMIC power amplifier chips, and Figure 19.20b shows the typical measured performance. The output power, gain, and PAE values are about 12 W, 11 dB, and 30%, respectively. The measured input return loss was better than 12 dB. ■

(a)

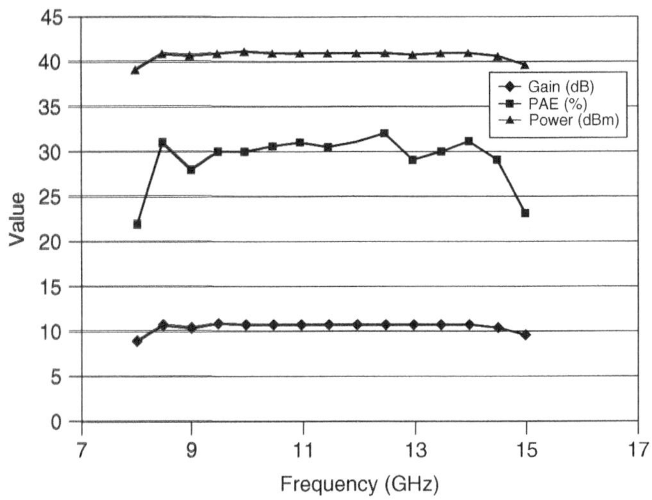

(b)

Figure 19.20 (a) Photograph of 12-W HPA using three 5-W MMIC power amplifiers. (b) Measured gain, power, and PAE of 12-W HPA at $V_{ds} = 10$ V.

EXAMPLE 19.4

Figure 19.21 shows another example of a power combiner to obtain 50-W output power and 37-dB gain at X-band. Here, four 17-W MMIC three-stage power amplifiers were combined in the output stage using a four-way traveling-wave divider–combiner. The two-stage driver amplifier has about 0.5-W output power and 17-dB gain, and the 17-W HPA has about 22-dB gain. Assuming about 1-dB combining loss due to the combiners, the gain of the output stage assembly is about 21 dB. Therefore the input power required is $47 - 21 = 26$ dBm $= 0.4$ W. Thus a 0.5-W driver amplifier has sufficient power to drive the output stage.

19.6 POWER HANDLING OF ISOLATION RESISTORS

In power combiners, it is important to know the power dissipated in the isolation resistors because of coupler or amplifier amplitude and phase imbalance on reflected power. Once the power dissipation is known, in MICs one can select the discrete high-power resistors or design the printed combiner along with the isolation resistors

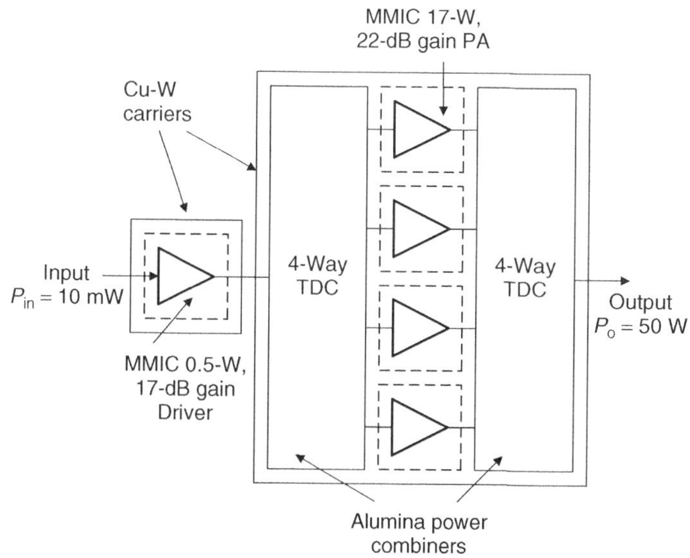

Figure 19.21 Schematic of a 50-W HPA using a driver amplifier MMIC power amplifier and four MMIC HPAs in the output stage.

or high thermal conductivity substrates such as beryllium oxide (BeO) or aluminum nitride (AlN). In MMICs, it is routinely done when several FET cells are combined in parallel. Power handling of isolation resistors is much more severe in combiners using Lange and traveling-wave couplers than in phase combiners such as the Wilkinson or reactive combiners.

For a two-way in-phase power combiner, the power dissipation in an isolation resistor, P_{rdis}, is approximately given by [24]

$$P_{\text{rdis}} = \frac{P_{\text{o}}}{4}(10^{\Delta/20} - 1)^2 \tag{19.23}$$

where Δ is the amplitude imbalance in decibels between the two ports and P_o is the highest power level in watts. For example, if $P_o = 5$ W and $\Delta = 1$ dB, the dissipated power in the resistor is approximately 18.6 mW. This does not include the effects of phase imbalance and reflections. In practice, the effect of phase imbalance and reflections is more severe than the amplitude imbalance.

19.7 SPATIAL POWER COMBINERS

In addition to circuit type combining techniques, quasi-optical or spatial power combining (SPC) techniques are becoming popular [25–32]. In an SPC technique, as shown in Figure 19.22, a power generating source is connected at the input and output with radiating elements. The input signal is coupled to an input antenna, amplified, and radiated and combined in space. In the case of a large beam cross section, many power amplifiers (PAs) are combined in parallel in an array and the combining loss is independent of the number of PAs. An SPC technique has the following advantages over circuit type combining:

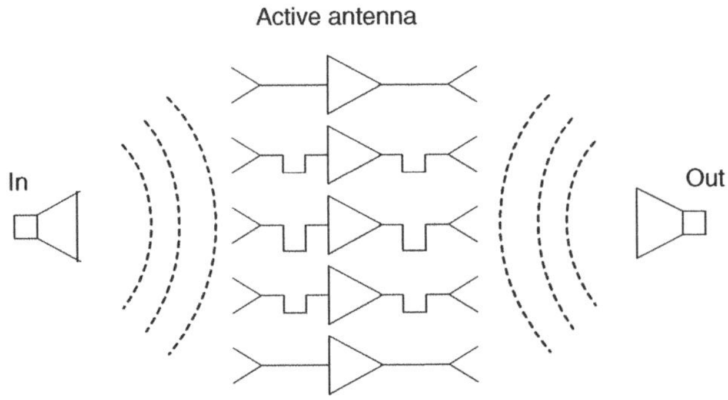

Figure 19.22 Schematic for spatial power combining.

1. Very high combining efficiency for HPAs, on the order of 95–100%, and very low resistive loss due to low-loss Gaussian beam/waveguides are the result.
2. A much larger number of HPAs can be combined to achieve very high output power levels: over 120 W at X-band has been demonstrated.
3. Small footprint, low profile, light weight, and low cost can be achieved.
4. There are built-in heat sinks and antennas.

The spatial power combining architectures include tile and tray approaches. In a tile approach, as shown in Figure 19.23a, the electromagnetic wave propagates normal to the tile surface. A tile is comprised of an array or grid of active antennas. The active antenna element consists of a discrete transistor or an MMIC amplifier and two antennas. This approach has narrow bandwidth and it requires extra care in thermal management. In the tray approach, Figure 19.23b, the electromagnetic wave propagates tangential to the tray surface and it consists of several trays. Each tray again carries several amplifiers along with receiving and radiating elements and has a heat-sinking ground plane. This technique has broadband characteristics.

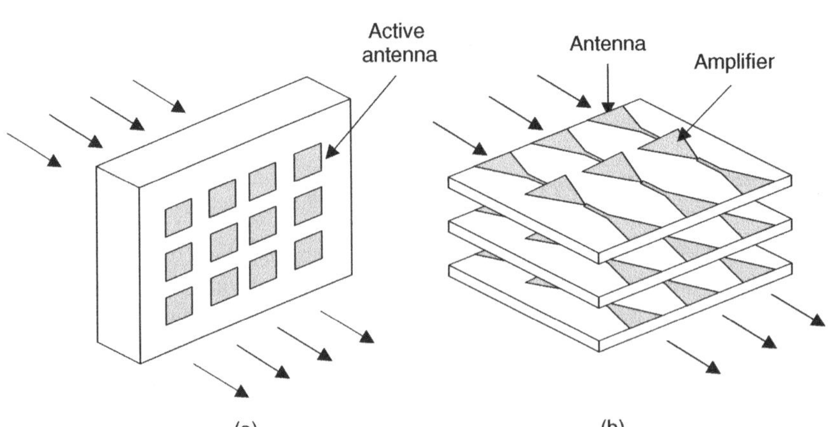

Figure 19.23 Spatial power combining techniques: (a) tile and (b) tray.

Table 19.4 Summary of SPC Power Amplifier Performance

Band	Frequency (GHz)	Output Power (W)	Reference
X	8.1	126	27
Ku	14.5	25	30
Ka	34	25	25
V	61	36	26

Table 19.5 A Comparison of Power Combining Techniques

Combining Technique	Advantages	Disadvantages
N-way W/G cavity	Low loss High efficiency	Nonplanar Complex assembly Narrowband
N-way Wilkinson	Low loss Moderate bandwidth Good isolation High efficiency	Nonplanar Low power
N-way radial line	Low loss Good isolation	Nonplanar Complex assembly
N-way planar	Large bandwidth Good isolation Moderate loss	Large size Low efficiency
Corporate structure	Good isolation Large bandwidth	Impractical beyond four-way due to low efficiency
Serial chain structure	More flexible Octave or greater bandwidth Good efficiency Good isolation Good input and output match	High resolution fabrication required Complex design Impractical beyond five-way combining due to fabrication limitations
Spatial power combining	95–100% combining efficiency Suitable for high power	Complex design Complex assembly

By using SPC techniques, state-of-the-art power levels from X-band to V-band have been demonstrated and several of them are available as products. A summary of SPC output power performance is given in Table 19.4.

19.8 COMPARISON OF POWER COMBINING SCHEMES

A qualitative comparison of various circuit-level power combining techniques is given in Table 19.5.

REFERENCES

1. K. J. Russel, Microwave power combining techniques, *IEEE Trans. Microwave Theory*

Tech., Vol. MTT-27, pp. 472–478, May 1979.
2. K. Chang and C. Sun, Millimeter-wave power-combining techniques, *IEEE Trans. Microwave Theory Tech*., Vol. MTT-31, pp. 91–107, February 1983.
3. I. J. Bahl, Filters, hybrids and couplers, power combiners, and matching networks, in *Handbook of RF/Microwave Components and Engineering* (K. Chang, Ed.), John Wiley & Sons, Hoboken, NJ, 2003.
4. I. J. Bahl and P. Bhartia, *Microwave Solid State Circuit Design*, 2nd edition, John Wiley & Sons, Hoboken, NJ, 2003, Chapter 10.
5. Y. Tokumitsu et al., A 6-GHz 80-W GaAs FET amplifier with a TM = mode capacity power combiner, *IEEE Trans. Microwave Theory Tech*., Vol. MTT-32, pp. 301–308, March 1984.
6. Y. Taniguchi et al., A C-band 25 watt linear power FET, *IEEE Int. Microwave Symp. Dig*., pp. 981–984, 1990.
7. A. A. M. Saleh, Improving the graceful-degradation performance of combined power amplifiers, *IEEE Trans. Microwave Theory Tech*., Vol. MTT-28, pp. 1068–1070, October 1980.
8. M. S. Gupta, Degradation of power combining efficiency due to variability among signal sources, *IEEE Trans. Microwave Theory Tech*., Vol. 40, pp. 1031–1034, May 1992.
9. E. Wilkinson, An *N*-way hybrid power divider, *IEEE Trans. Microwave Theory Tech*., Vol. MTT-8, pp. 116–118, January 1960.
10. I. J. Bahl, Ultra-broadband and compact power dividers/combiners on GaAs substrate, *IEEE Microwave Mag*., Vol. 9, pp. 96–104, April 2008.
11. H. Howe, Jr., *Stripline Ciricuit Design*, Artech House, Norwood, MA, 1974, Chapter 3.
12. C. Q. Li, S. H. Li, and R. G. Basisio, CAD/CAE design of an improved wideband Wilkinson power divider, *Microwave J*., Vol. 27, pp. 125–135, November 1984.
13. L. I. Parad and R. L. Moynihan, Split-tee power divider, *IEEE Trans. Microwave Theory Tech*., Vol. MTT-13, pp. 91–95, January 1965.
14. H.-R. Ahn and I. Wolff, General design equations, small-sized impedance transformers, and their appliclation to small-sized three-port 3-dB power dividers, *IEEE Trans. Microwave Theory Tech*., Vol. 49, pp. 1277–1288, July 2001.
15. M. C. Scardelletti, G. E. Ponchak, and T. M. Weller, Miniaturized Wilkinson power dividers utilizing capacitive loading, *IEEE Microwave Wireless Components Lett*., Vol. 12, pp. 6–8, January 2002.
16. R. Mongia, I. Bahl, P. Bhartia, and J. Hong, *RF and Microwave Coupled-Line Circuits*, 2nd edition, Artech House, Norwood, MA, 2007.
17. J. Lange, Interdigited stripline quadrature hybrid, *IEEE Trans. Microwave Theory Tech*., Vol. MTT-17, pp. 1050–1151, December 1969.
18. A. Presser, Interdigited microstrip coupler design, *IEEE Trans. Microwave Theory Tech*., Vol. MTT-26, pp. 801–805, October 1978.
19. K. C. Gupta, R. Garg, I. Bahl, and P. Bhartia, *Microstrip Lines and Slotlines*, 2nd edition, Artech House, Norwood, MA, 1996.
20. A. A. M. Saleh, Planar electrically symmetric *n*-way hybrid power dividers/combiners, *IEEE Trans. Microwave Theory Tech*., Vol. MTT-28, pp. 555–563, June 1980.
21. J. M. Schellenberg and M. Cohn, A wideband radial power combiner for FET amplifiers, *IEEE Int. Solid State Circuits Conf. Dig*., pp. 164–165, February 1978.
22. N. Nagai, E. Maekawa, and K. Ono, New *n*-way hybrid power dividers, *IEEE Trans. Microwave Theory Tech*., Vol. MTT-25, pp. 1008–1012, December 1977.
23. D. Willems, I. Bahl, and B. Kruger, A quasi-microstrip travelling-wave power divider/combiner for use in high-density packages, *IEEE Microwave Guided Wave Lett*., Vol. 3, pp. 148–149, May 1993.
24. S. R. Nelson, Amplifier stability in short course, power amplifiers: from milliwatts to kilowatts ... cool devices with hot performance, in *1998 IEEE GaAs IC Symposium Workshop Notes*.
25. E. A. Sovero et al., A Ka-band monolithic quasi-optic amplifier, *IEEE MTT-S Int. Microwave Symp. Dig*., pp. 1453–1456, 1998.

26. J. J. Sowers et al., A 36W, V-band, solid-state source, *IEEE MTT-S Int. Microwave Symp. Dig*., pp. 235–238, 1999.
27. N.-S. Cheng et al., A 120-W X-band spatially combined solid-state amplifier, *IEEE Trans. Microwave Theory Tech*., Vol. 47, pp. 2557–2561, December 1999.
28. S. Ortiz et al., A high-power Ka-band quasi-optical amplifier array, *IEEE Trans. Microwave Theory Tech*., Vol. 50, pp. 487–494, February 2002.
29. M. P. DeLisio et al., Power and spectral regrowth performance of 10-W and 16-W Ka-band power amplifiers with single-chip output stages, *IEEE MTT-S Int. Microwave Symp. Dig*., WE3E-5, 2005.
30. Wavestream, A compact efficient 25W Ku-band power amplifier, *Microwave J*., Vol. 49, pp. 144–152, April 2005.
31. M. DeLisio, Spatial power combining techniques, in *IEEE MTT-S International Microwave Symposium Workshop WMH*, 2006.
32. M. P. DeLisio and R. A. York, Quasi-optical and spatial power combining, *IEEE Trans. Microwave Theory Tech*., 50th Anniversary Issue, Vol. 50, pp. 929–936, March 2002.

PROBLEMS

19.1 Assuming that all individual amplifiers in an N-way amplifier are perfectly matched and have the same transmission phase and 100% efficiency of the divider–combiner, prove that the P_{out} of this amplifier is given by

$$P_{out} = \frac{1}{N}\left(\sum_{n=1}^{N}\sqrt{P_n}\right)^2$$

where P_n is the output power of the nth individual amplifier.

19.2 Calculate the efficiency of combining and power output of a four-way chain divider–combiner realized using couplers having 0.1-dB insertion loss. The amplifier has 3-W power output, and the coupling factors are 3, 4.78, and 6 dB.

19.3 Design a single-section coupled-microstrip directional coupler with the following specifications: coupling 10 dB, substrate $\varepsilon_r = 9.0$, substrate thickness = 0.635 mm, system center frequency = 4.0 GHz, impedance = 50 Ω. Neglect dispersion.

19.4 **(a)** Find the impedances of a hybrid ring directional coupler for the following power split ratios:

(1) 0 dB,

(2) 3 dB, and

(3) 9 dB.

(b) For a strip line type coupler, find the corresponding strip width dimensions for a substrate $\varepsilon_r = 3.8$ and ground plane spacing of 2.5 mm.

19.5 Design a 3-dB Lange coupler on a 0.38-mm fused quartz microstrip substrate with isolation and return loss ≥ 25 dB over the 8–12-GHz range. Redesign the circuit for operation over 11–15 GHz.

19.6 Design a four-way symmetric power divider in a microstrip configuration with the following specifications:

Center frequency = 4 GHz

Power in the outermost arms = 20%

Power in the innermost arms = 30%

Input and output impedances = 50 Ω

Microstrip parameters $\varepsilon_r = 9.9$, $h = 0.63$mm, $t = 5$ μm

19.7 Design a 40-W HPA using four 14-W MMIC power amplifiers. The gain and PAE for the MMIC chips are 8 dB and 30%, respectively. The four-way power divider–combiner employs Lange couplers on 15-mil alumina substrate. The Lange coupler's loss is 0.3 dB. The frequency of operation is 5.8 GHz. Determine the Lange coupler's dimensions and the HPA's gain, output power, and PAE by assuming all MMIC chips are matched for amplitude and phase.

19.8 Recalculate the HPA's performance in Problem 19.7 when the MMIC amplifiers have 22-dB gain.

19.9 Recalculate the HPA's performance in Problem 19.7 when the MMIC amplifiers are binned into two groups and there is imbalance between the two groups as follows: **(a)** 0.5 dB lower gain and **(b)** 25° phase difference.

19.10 Design a five-way traveling-way combiner at 5 GHz on a 25-mil thick alumina substrate. Use a quarter-wave transformer at the input of the combiner. Determine the combiner dimensions including isolation resistor values and the combining efficiency. Ignore junction discontinuity and dispersion effects.

19.11 Determine the combining loss of a four-way combiner when the amplifiers have transmission phase differences of 0, $\pi/16$, $2\pi/16$, and $3\pi/16$ between amplifiers. The combiners have been assumed ideal and there is no amplitude difference between amplifiers.

19.12 Show that in a two-way Wilkinson power divider, under matched conditions, the value of the isolation resistor is 2 Z_0, where Z_0 is the terminating impedance at all ports.

19.13 In a coupled microstrip line, show a simplified version of the even- and odd-mode capacitor locations. Also calculate even- and odd-mode capacitance per unit length for a 10-dB coupler on a 25-mil thick alumina substrate.

19.14 Show that, in the Lange coupler, the mutual capacitance between conductors is three times larger than between two lines of similar dimensions.

Chapter 20

Integrated Function Amplifiers

The applications of amplifiers in modern commercial and military systems require cost-effective solutions. A popular technique to achieve product cost goals is to integrate more functions into a single MMIC amplifier chip or into a package or module. For example, a high level of integration at the MMIC chip level reduces the number of chips and interconnects and results in low test and assembly costs, which in turn increases the reliability and reduces the subsystem cost. Some examples of integrated amplifiers described in this chapter include limiter/LNA, transmitter chains with several stages, amplifiers with variable gain and output power, amplifiers with built-in power monitors, temperature compensation for gain, and output mismatch protection. The amplifier packages are discussed in the next chapter.

20.1 INTEGRATED LIMITER/LNA

Due to the fine geometry used in transistors, amplifier circuits are susceptible to damage from high-power spurious EM radiation either from microwave transmitters or the nuclear electromagnetic pulse. Especially LNAs in the front end of microwave systems need high-power protection because these amplifiers can sustain only low input power levels in the range of 10–20-dBm CW or 1–3 μJ of pulsed power. To protect these circuits, a high-power and low-loss limiter with typically less than 0.3–0.5-dB insertion loss is required.

A microwave limiter [1, 2], also known as a receiver protector, allows low input power signals to pass through it while attenuating the high-power signals above its rated threshold power level. There are several different technologies [2] used to realize this component including gaseous or plasma limiters [3, 4], ferrite limiters [5–7], and solid state limiters [8–29]. Gaseous, ferrite, and solid state limiters have input power limiting thresholds ranging from 30 to 90 dBm (megawatt), −25 to 50 dBm (100 W), and 0 to 80 dBm (100 kW), respectively.

To fulfill low-cost and large-volume production requirements, the limiter and LNA functions must be integrable on a single GaAs MMIC chip, forcing the limiter and LNA technologies to be compatible. Good broadband performance of the limiter is also a requirement in order to cover the frequency range of operation. Only solid state diodes and transistors are compatible with monolithic technology. All solid state devices have a limiting function as they have specific maximum output powers. Varactor diodes [2],

Fundamentals of RF and Microwave Transistor Amplifiers. By Inder J. Bahl

p-i-n diodes [8–11, 14, 27], Schottky diodes [15–17, 24], and switching FETs [21] have been used as passive limiting devices. An integrated limiter/LNA was reported recently [28, 29]. The recovery time for 10-W limiters with integrated LNA was on the order of 35 ns [29] as compared to microseconds (μs) for p-i-n diode limiters. Since there is no biasing network required in a passive limiter, the recovery time is expected to be faster than p-i-n diodes.

The discussions in this section deal with the Schottky diode limiters, which are compatible with the FET processes commonly used for low-noise applications. A novel high-power limiter is described that handles greater than 10-W CW power [28] and is integrated with a low-noise amplifier using an MSAG FET process [30]. The design of limiting Schottky diodes and limiter/LNAs with comprehensive measured data are discussed next.

20.1.1 Limiter/LNA Topology

A block diagram of the high-power limiter/LNA MMIC is shown in Figure 20.1. The limiter/LNA design is a balanced configuration in which each single-ended topology is comprised of a two-stage Schottky diode limiter circuit followed by a two-stage LNA. The Lange couplers of the balanced configuration provide a good input and output match at all power levels, minimize LNA noise figure degradation in the presence of an extreme source mismatch (e.g., radiator impedance mismatch at wide scan angles in phased array antennas), and also provide a route for the reflected RF power to an "on-chip" 50-Ω load. The MMIC balanced configuration adds an extra 0.3–0.5 dB to the single-ended amplifier's noise figure.

The two-stage limiter topology is shown in Figure 20.2. The first stage utilizes large Schottky limiting diodes (A) to handle the 37-dBm (half of 10 W due to balanced configuration) input power, and the second stage uses smaller Schottky limiting diodes

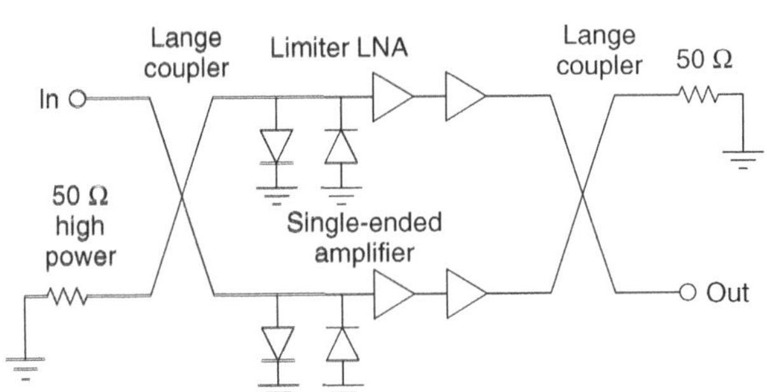

Figure 20.1 Balanced two-stage LNA with limiter.

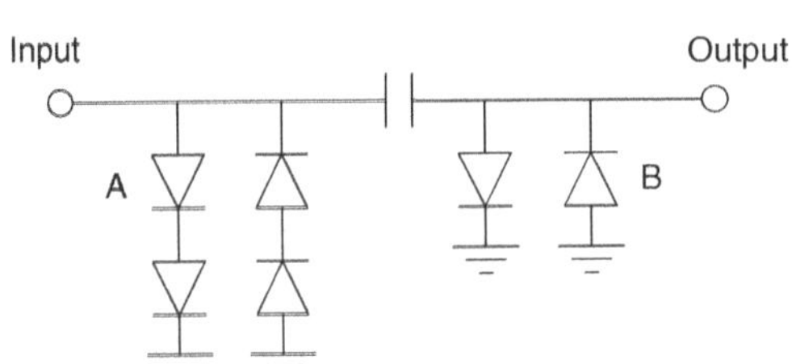

Figure 20.2 Two-stage Schottky limiting diode configuration.

(B), limiting the leakage power below 18–19 dBm. The limiter's first stage employs two 2-diode chains in antiparallel shunt configuration. The limiter's second stage uses two diodes in antiparallel shunt configuration also. A third stage of much smaller Schottky limiting diodes may be used to further limit the leakage power below 14 dBm.

The two-stage LNA configuration has been selected based on a trade-off between the LNA noise figure and the input TOI requirement of the receiver. In the LNA, each stage uses suitable gate periphery FETs with series feedback to provide a simultaneous match for a good VSWR, minimum noise figure, and power handling for the limiter leakage. The capacitive reactance of the limiting diodes is incorporated as a part of the input matching network of the LNA design.

20.1.2 Limiter Requirements

Next, the basic device requirements are briefly described for the limiting circuit to withstand power levels of up to 10-W CW and the design of Schottky diodes.

Figure 20.3a shows a basic front-end limiter using two Schottky diodes. Antiparallel diodes clip both halves of the RF sinusoidal signal. The load R_L represents the LNA's input impedance. Under small-signal conditions, the diodes are open and the impedance presented to the LNA by the diodes is a series combination of C_{j0} and R_S (Fig. 20.3b). Thus the diode capacitance C_{j0} can be absorbed by the input matching network of the LNA using a lowpass structure as shown in Figure 20.3c. As long as the diode's reactance $(1/j\omega C_{j0})$ is much greater than R_S, the noise contribution of R_S to the total NF of the LNA is insignificant.

At high power levels, the diodes turn on and a significant fraction of the incident power (depending on the power level) is reflected back to the source. However, due to R_S and the rectifying action of the ideal diodes, a substantial fraction of the incident power is absorbed into the diodes. This absorbed power increases the channel temperature and may cause damage to the diodes. Thus the diodes must be designed to handle this absorbed power and corresponding high current. If $R_G = R_L$, the power P_a delivered to the diodes is given by

$$P_a = \frac{V_G^2}{4R_G} \tag{20.1}$$

where V_G is the rms voltage across the source (generator) resistor R_G. If I_d is the rms current passing through the diodes, and $R_G = 50\ \Omega$, then $V_G = 50I_d$ and (20.1)

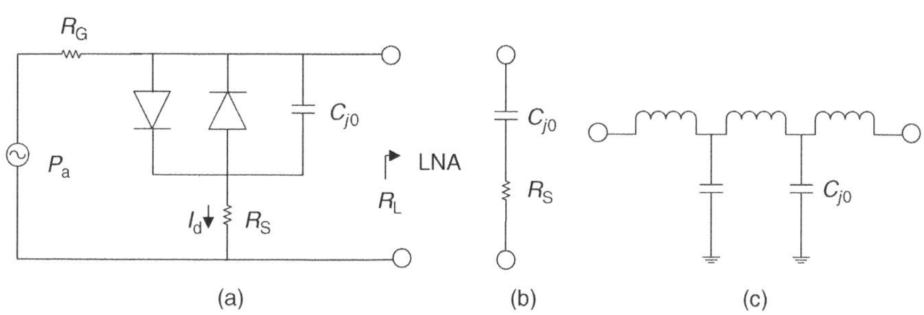

Figure 20.3 (a) Schottky diode power limiter configuration, (b) diode's small-signal lumped-element equivalent circuit, and (c) lowpass equivalent representation of the input match for the LNA.

becomes

$$P_a = \frac{(50I_d)^2}{4 \times 50} = \frac{50}{4} I_d^2$$

or

$$I_d = \frac{\sqrt{2}}{5} \sqrt{P_a} \tag{20.2}$$

The peak current passing through the diodes under this condition will be $I_p = \sqrt{2} I_d$

$$I_p = \tfrac{2}{5} \sqrt{P_a} \tag{20.3}$$

Therefore for 10-W CW operation, the diodes must be able to draw a peak current of 1.26 A without failing. This current will pass through the diode channel as well as its physical structure. In the balanced limiter/LNA design, the peak current passing through the diodes will be halved, or 0.63 A.

The power absorbed by the diodes under large-signal conditions occurs due to the rectification action of the Schottky junction and power dissipated in the series resistor, $I_d^2 R_S$. The time average power over one cycle is approximated by [23]

$$P_D = \frac{2\pi}{\omega} \left[\int_0^{\pi/\omega} V_p I_p \sin(\omega t)\, dt + \int_0^{\pi/\omega} I_p^2 R_S \sin^2(\omega t)\, dt \right] = \frac{1}{\pi} I_p V_p + \frac{1}{4} I_p^2 R_S \tag{20.4}$$

where V_p is about 1.5 V. Integration is carried out over a half-period ($T/2 = \pi/\omega$) because each Schottky diode only conducts over one-half RF cycle.

The increase in diode junction temperature under large-signal conditions is given by

$$\Delta T = P_D R_{th} \tag{20.5}$$

where R_{th} is the thermal resistance of the diode on the GaAs substrate placed on a heat sink. If R_S is 1 Ω, the total power dissipated P_D in a single-ended limiter is about 0.4 W. When R_{th} is about 120 °C/W, the increase in diode junction temperature is calculated to be about 48 °C.

20.1.3 Schottky Diode Design and Limiter Configuration

The design of the Schottky diode and limiter configuration must meet the following requirements:

1. High cutoff frequency $f_c (= 1/2\pi R_S C_{j0})$ in order to have minimum loss/noise contribution.
2. Ability to withstand rectified current during limiting action.
3. Temperature rise less than 80 °C ($T_{ch} = 150$ °C).

For this application, M/A-COM's standard Schottky mixer diode with 2-μm gate length for the anode was used in order to allow higher current and power-handling capability as compared to 1-μm or shorter gate length devices. Figure 20.4 shows the top and cross-sectional views of the Schottky mixer diode. The Schottky diode area and limiter configuration determine its survivable power level. The single-ended limiter consists of two stages. The first stage comprises a set of antiparallel diodes in shunt configuration; each diode set consists of 16 diodes. Each diode has 2×12 μm^2 area. Two of these sets are used in series to increase the voltage-handling capability, which enabled an increase in the heat spreading area by a factor of 4, consequently increasing the current- and power-handling capability without affecting the limiter loss. Typical measured R_S and C_{j0} values for this stage are 1 Ω and 0.4 pF, respectively. The second stage contains another set of antiparallel diodes in shunt configuration. In this case there are 8 diodes and each has 2×8 μm^2 area. Table 20.1 summarizes the typical measured noise parameters of the two-stage limiter shown in Figure 20.5.

Cooke's model as described in Chapter 16 [31] was used to calculate the thermal resistance of the limiting diodes. A thermal resistance of 120 °C/W was calculated for each single-ended limiter stage. Thus if the power dissipated is about 0.5 W

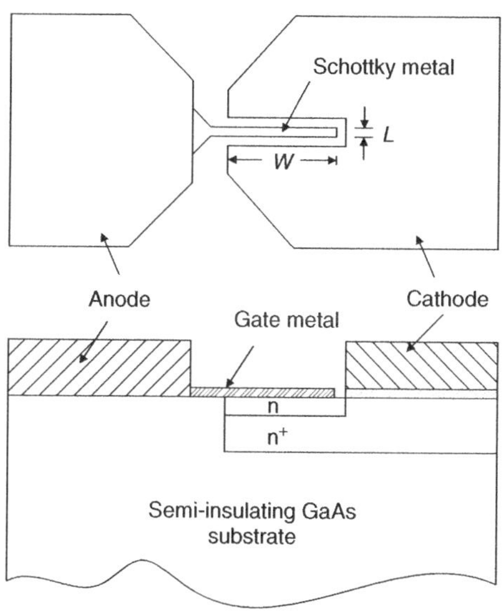

Figure 20.4 Schottky diode's top and cross-sectional views.

Table 20.1 Typical Measured Noise Parameters of the Two-Stage High-Power Limiter

Frequency (GHz)	F_{min} (dB)	Γ_{opt} Magnitude	Γ_{opt} Angle	Normalized R_n
7	0.12	0.19	−30	0.03
8	0.15	0.25	70	0.03
9	0.17	0.32	170	0.01
10	0.20	0.41	−88	0.04
11	0.22	0.49	15	0.09
12	0.25	0.56	121	0.04

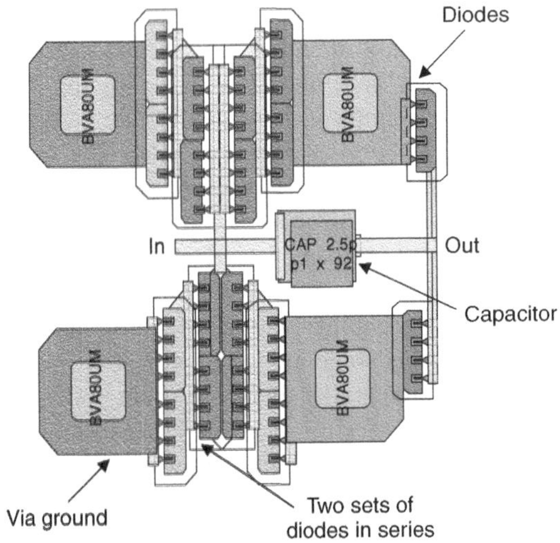

Figure 20.5 Physical layout of the two-stage Schottky diode single-ended limiter configuration.

(approximately 30% higher than the calculated value in the previous section), the rise in the channel temperature is only about 60 °C.

Burnout tests were performed on both diodes and FETs for catastrophic failure. The limiter topology shown in Figure 20.5 was tested up to 10-W CW with no catastrophic failure observed. However, a single-ended matched single-stage LNA circuit using a 300-μm FET could only tolerate up to 1-W CW before catastrophic failure occurred, which translates to 2-W CW power for a 600-μm FET that was used for high-output IP3. Thus the maximum leakage power from the two-stage limiting diode configuration, which is about 20 dBm, is quite safe for 600-μm FETs that were used in the design.

20.1.4 10-W Limiter/LNA Design

The high-power limiter/LNA is a two-stage LNA designed for high IP3, low noise figure, and high-power operation. The single-ended design is based on MSAG 5N low-noise FETs (discussed in Chapter 5) and the matching networks were optimized for minimum noise figure, flat gain, and minimum gain of 15 dB. Both stages use 600-μm gate periphery FETs, each FET employing 8 fingers and a 20-μm gate-to-gate pitch. The 600-μm FET size is selected to achieve the minimum noise figure and effectively handle leakage power from the diodes for high-power operation and higher IP3 performance. The FETs used in each stage have slightly different series feedback inductance values. Typical EC model (Fig. 5.2, Chapter 5) parameters are given below and the measured noise parameters for these two FETs are given in Tables 20.2 and 20.3.

$$R_g = 0.5\ \Omega,\ \ R_i = 1.0\ \Omega,\ \ R_s = 0.8\ \Omega,\ \ R_d = 0.8\ \Omega,\ \ R_{ds} = 110\ \Omega,$$
$$g_m = 120\ \text{mS},\ \ \tau = 2\text{ps}\ \ C_{gs} = 0.55\ \text{pF},\ \ C_{gd} = 0.12\ \text{pF},\ \ C_{ds} = 0.11\ \text{pF}$$
$$L_g = 0.01\ \text{nH},\ \ L_s = L_{ss}, L_d = 0.01\ \text{nH}$$

For the first-stage FET, $L_{ss} = 0.08$ nH and for the second-stage FET, $L_{ss} = 0.01$ nH.

For design purposes, Cooke's model [31], described in Chapter 16, was used to calculate the channel temperature for 600-μm FETs. The thermal resistance for each

Table 20.2 Typical Measured Noise Parameters of the First-Stage FET Biased at 3 V and 25% I_{dss}

Frequency (GHz)	F_{min} (dB)	Γ_{opt} Magnitude	Γ_{opt} Angle	Normalized R_n	G_A (dB)
7	0.61	0.63	86°	0.23	9.7
8	0.70	0.62	97°	0.21	9.1
9	0.79	0.62	106°	0.20	8.6
10	0.88	0.62	115°	0.18	8.2
11	0.97	0.63	126°	0.16	7.8
12	1.06	0.64	131°	0.14	7.5

Table 20.3 Typical Measured Noise Parameters of the Second-Stage FET Biased at 3 V and 25% I_{dss}

Frequency (GHz)	F_{min} (dB)	Γ_{opt} Magnitude	Γ_{opt} Angle	Normalized R_n	G_A (dB)
7	0.62	0.64	85°	0.23	9.9
8	0.71	0.63	96°	0.22	9.3
9	0.81	0.62	105°	0.21	8.7
10	0.90	0.62	114°	0.19	8.3
11	0.99	0.63	122°	0.17	8.0
12	1.08	0.64	129°	0.15	7.7

FET is approximately 210 °C/W. Under full saturation, the power dissipated in each FET is about 0.135 W and the calculated temperature rise of the FET's channel is 29 °C. The S-parameters for 600-μm FETs were also measured using various input power levels from 0 to 27 dBm. The devices were exposed for 10 minutes at each power level and S-parameters were measured. Up to 27-dBm input power levels, no measurable change in the device small-signal performance was observed.

Each FET's gate is biased through a voltage divider network whose effective gate resistance is 1 kΩ and the drain voltage is applied through a 60-Ω resistor. The gate and drain power supply voltages are −5 V and 5 V, so that the gate voltage is about −0.7 V (at 25% I_{dss}) and drain voltage is about 3 V. The circuit design is unconditionally stable. Figure 20.6 shows the schematic of the amplifier and Table 20.4 includes the design parameters.

Lange Couplers

The design parameters of the Lange couplers for this application are as follows:

GaAs Substrate, $\varepsilon_r = 12.9$

Substrate thickness, $h = 125$ μm

Number of fingers, $n = 4$

Line width, $W = 10$ μm

Line spacing, $S = 10$ μm

Coupler length, $\ell = 2200$ μm

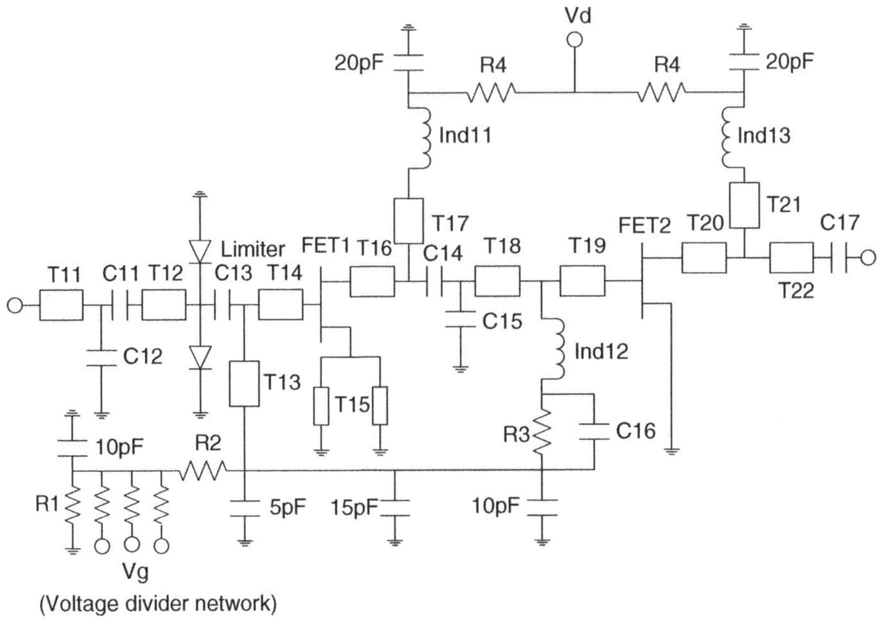

Figure 20.6 Schematic of the single-ended two-stage limiter/LNA.

Table 20.4 Design Parameters of the Two-Stage Single-Ended Limiter/LNA[a]

Transmission Length	MIM Capacitors	Resistors	Inductors
T11: 176 (width = 50)	C11: 0.22	R1: 1160	Ind 11: W = 12, S = 8, Di = 50, n = 3.5
T12: 525	C12: 2.0	R2: 130	
T13: 550	C13: 2.0	R3: 58	Ind12: W = 12, S = 8, Di = 50, n = 2.5
T14: 400	C14: 2.0	R4: 60	
T15: 200	C15: 0.7		Ind13: W = 20, S = 8, Di = 108, n = 1.5
T16: 200	C16: 0.8		
T17: 130	C17: 1.0		
T18: 70			
T19: 250			
T20: 100			
T21: 120			
T22: 500			

[a] Substrate thickness = 125 μm. All transmission lines are 20 μm wide unless specified. All dimensions are in μm, resistors in Ω, capacitors in pF [28]. FET 1, FET 2: 0.4 × 600. Limiter: antiparallel, two stages; first stage 12 μm wide, two 16 cells in series and second stage 8 μm wide, 4 cells. Inductors: W, line width, S, spacing, Di, inner diameter, and n, number.

High-Power 50-Ω Resistor

The 50-Ω termination resistor connected to the input Lange coupler must be capable of handling power levels up to 10 W. The power handling of the resistor is determined by the temperature rise of the resistor film. The parameters that play major roles in the calculation of maximum power handling are (a) thermal conductivity of the substrate material, (b) surface area of the resistor film, (c) thickness of the substrate, (d) ambient temperature, that is, temperature of the medium surrounding the resistor or base plate (heat sink) temperature, and (e) maximum allowed temperature of the resistor film. Normally the base plate temperature is 25 °C and the maximum allowed resistor film temperature is 150 °C; the resistor area calculated using the parallel plate waveguide model [32, 33], also described in Chapter 13, is $615 \times 205\ \mu m^2$. These calculations are based on treating a resistor film as a lossy microstrip line on a 125-μm thick substrate and having a surface resistivity of 150 Ω/□.

Simulated Performance

Both individual stages and the complete single-ended amplifier stage were simulated to be unconditionally stable. The simulated noise figure and gain for the balanced amplifier over the 8–11-GHz frequency range were about 2.7 dB and 15 dB, respectively. The input and output return losses were better than 20 dB.

Circuit Fabrication

The limiter/LNA was fabricated using M/A-COM's ion-implanted planar refractory gate, multifunction self-aligned gate (MSAG) MESFET MMIC process [30], see Fig. 15.2. The MSAG features a full suite of active and passive components fabricated on 4-inch diameter GaAs wafers. The devices were fabricated using the Process 5 low-noise and Schottky mixer diode implants. The process includes Au/Ge/Ni metallization for ohmic contacts, 0.4-μm and 2-μm TiWN Schottky barrier gates, along with thin-film and ion-implanted resistors. The 0.4-μm TiWN gates are covered by a 0.8-μm overlay after planarization. The MSAG TiWN gate is extremely robust (survives 900 °C rapid thermal anneal temperature), which results in an MTTF of 100 years at a channel temperature of 150 °C. A thickness of 2000-Å silicon nitride ($\varepsilon_r = 6.8$) is used for both MIM capacitors and passivation. The air bridges, microstrip lines, and bonding pads are 4.5-μm thick plated gold. The standard MSAG process also uses two layers of polyimide ($\varepsilon_r = 3.2$); interlevel dielectric (3 μm thick), and a scratch protection layer (7μm thick) for mechanical protection of the finished circuitry. Figure 20.7 shows the photograph of the X-band 10-W limiter/LNA.

20.1.5 Test Data and Discussions

The limiter/LNA was tested on wafer using RF probes as well as by assembling die onto gold-plated CuW carriers for RF characterization. CuW material was chosen for good thermal conductivity and thermal expansion match to GaAs and alumina. The carrier 50-Ω input and output microstrip lines were printed on 15-mil thick alumina substrate. The 5-mil thick ICs were die attached using gold–tin (AuSn) solder at 290 °C on a pedestal in order to keep minimum bond wire lengths between the chip pads and the input and output microstrip lines. The estimated base plate temperature

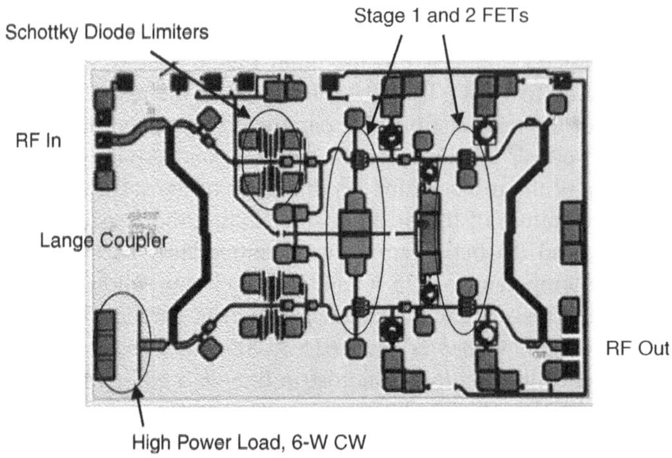

Figure 20.7 X-band 10-W limiter/LNA. Chip size is 14 mm^2. (Photograph courtesy of M/A- COM.)

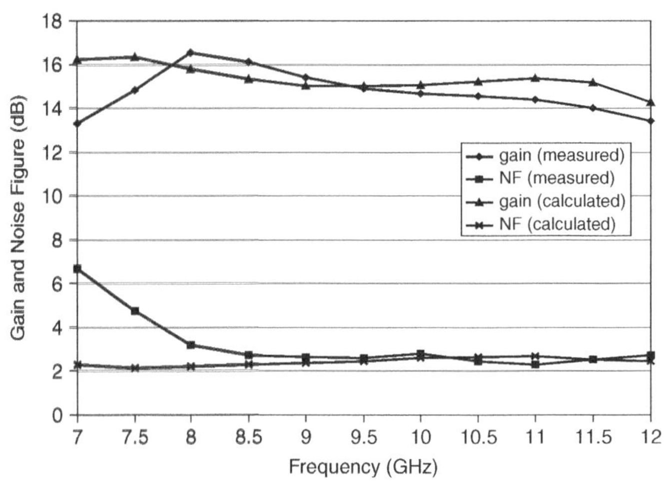

Figure 20.8 Typical measured gain and noise figure of the two-stage limiter/LNA.

for power measurements was about 30 °C. Figures 20.8 and 20.9 show the average data for the limiter/LNA measured on wafer using RF probes for all the chips on a wafer. Figure 20.8 shows the gain and noise figure and Figure 20.9 shows the input return loss and output return loss. For comparison purposes, the simulated gain and NF are also shown in Figure 20.8. Typical measured performance for the limiter/LNA circuit shows gain greater than 14 dB, NF < 2.7 dB, and return loss better than 20 dB over the 8.5–11.5-GHz frequency range. Typical measured output IP3 was about 30 dBm. The measured noise figure and gain for the packaged devices before and after exposure with a CW power level of +40 dBm for 30 minutes were almost identical. The measured CW power handling was greater than 10 W. Another device was exposed to 40-dBm CW power signal for 30 minutes and tested before and after. Then the device was exposed at various increased power levels for 30 seconds and tested before and after. The measured data have shown that no catastrophic failures occurred up to 42.5-dBm (18-W) input power. This concludes the step-by-step design example of an X-band integrated limiter/LNA. The same procedure may be applied to other bands.

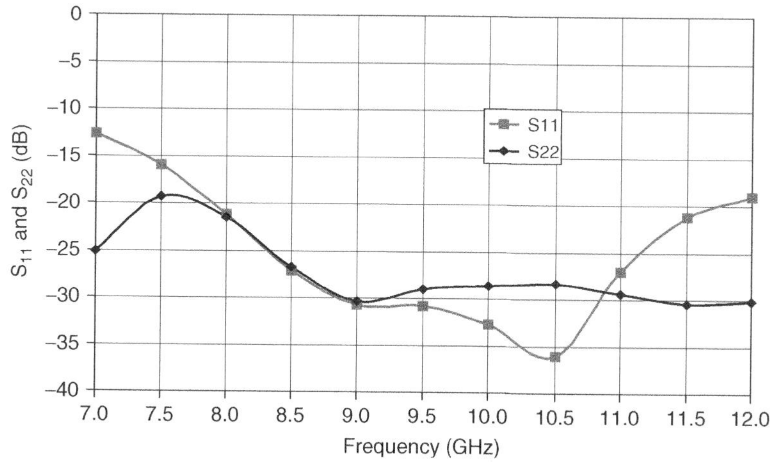

Figure 20.9 Typical measured S_{11} and S_{22} of the two-stage limiter/LNA.

The expected power levels at lower frequencies will be higher and lower at higher frequencies.

20.2 TRANSMITTER CHAIN

In transmitters, the last stage is the most critical element in terms of output power, PAE, thermal, and mechanical designs, whereas the preceding stages known as drivers are relatively less critical. The last stage, commonly known as an HPA, is designed for an optimum PAE, output power, or linearity. If the HPA stage is designed in balanced configuration, it minimizes the interaction effects between the driver and HPA, and between the HPA and antenna. The balanced HPA also behaves relatively more stable than a single-ended version. Generally, in a high-gain transmitter chain,

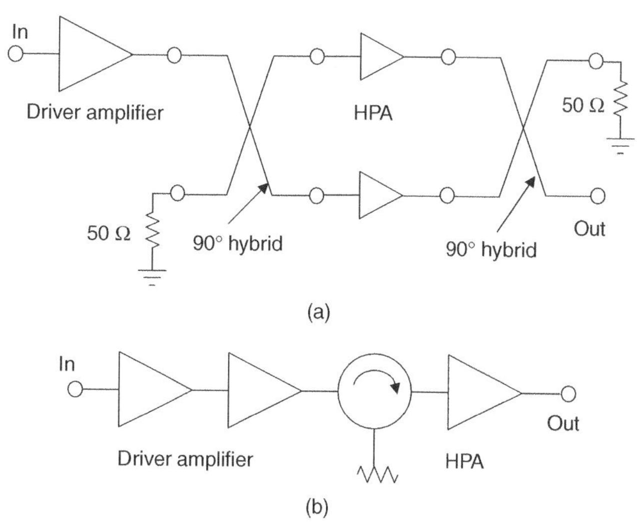

Figure 20.10 Two commonly used transmitter chain configurations: (a) balanced HPA stage and (b) isolator between the driver amplifier and HPA.

isolators or circulators are placed between the amplifier stages at critical positions to minimize the interaction effects. Figure 20.10 shows two commonly used transmitter chain configurations.

The driver stages can be operated between class A and class AB, depending on the requirements. As mentioned in Chapter 8, class-A amplifiers have better linearity and lower AM to PM and harmonic levels as compared to class-B or class-E amplifiers. Thus limiting the distortion in driver amplifiers operating close to class A improves the linearity performance of the transmitter, whereas a driver's class-AB operation improves overall PAE performance. In the former case, the dominant distortion will be in the HPA. Figure 20.11a shows a commonly used transmitter chain configuration with an isolator at the output. Typical RF power, DC power, and gain budget values are also shown. In most applications, the transmitter chain has a variable gain/attenuator stage for gain adjustments for temperature compensation, for equal gain between radiating elements, and fine gain tuning from unit to unit.

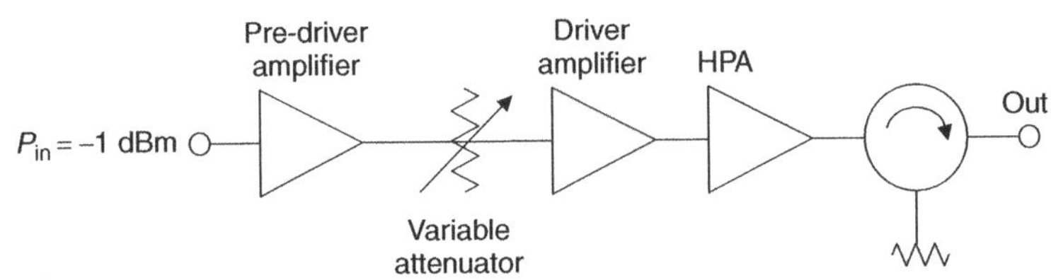

Parameter	Pre-drvr.	Atten.	Driver	HPA	Circulator
Gain, dB	12.0	−2.5	20.0	12.0	−0.5
P_{out}, dBm	11.0	8.5	28.5	40.5	40.0
P_{DC}, W	0.05	0	2.0	30.0	0
Cummulative Gain, dB	12.0	9.5	29.5	41.5	41.0
Cummulative P_{DC}, W	0.05	0.05	2.05	32.05	32.05

(a)

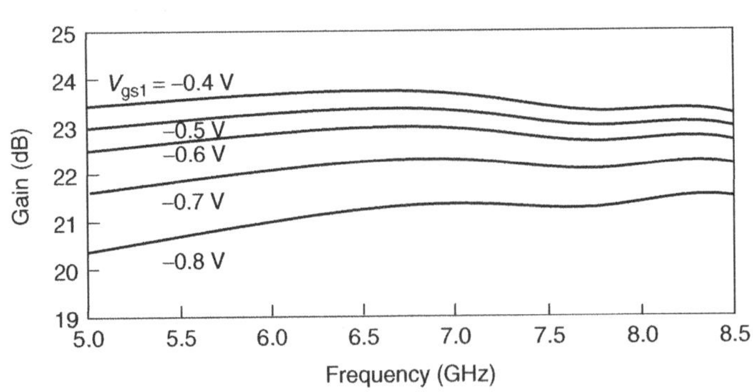

(b)

Figure 20.11 (a) A commonly used transmitter chain configuration. (b) Typical variation of gain of a three-stage amplifier with first-stage gate voltage.

20.2.1 Variable Gain Amplifier

The gain of an amplifier using FETs/HEMTs can be changed by controlling the gate voltage. In general, more positive gate voltage results in higher gain and vice versa, as shown in Figure 20.11b, where the first-stage gate voltage is changed in a three-stage amplifier. However, any change in the gate voltage also affects the input and output impedances of the transistor and will change the input and output match with respect to the nominal bias conditions. It may also change the gain flatness or bandwidth and IP3. Some of these problems in gate-voltage-dependent variable gain amplifiers can be corrected by using balanced configuration or feedback/distributed amplifier topology, discussed in Chapter 11. Since the amplifier gain is not a linear function of gate voltage, external circuitry is generally needed to make gain variation more linear with gate voltage.

The gain of an amplifier may also be varied by inserting a variable attenuator in between the gain stages, as shown in Figure 20.11a. The attenuator, depending on its attenuation value, minimizes the interactions between the stages. However, the insertion of an attenuator between amplifier stages to control the gain of the amplifier chain must be analyzed carefully. In a multistage design having greater than two amplifier stages, the placement of an attenuator affects the noise figure, input and output IP3, output power, and PAE. In the case of a three-stage low-noise and low-power amplifier, the placement of an attenuator after first two stages has much less effect on noise figure than placing it after the first stage. Similarly, in the case of a three-stage power amplifier, the placement of an attenuator after the first two stages has much greater adverse effect on power and PAE than placing it after the first stage. In the high-power amplifier case, one needs an attenuator with a high P_{1dB} level to provide sufficient power to the succeeding stages. Generally, an attenuator with a high P_{1dB} level has higher loss in the reference state, which lowers the PAE of the amplifier. Normally, in an HPA chain, a voltage control feature is inserted in the input side.

For active phased array radar applications, the transmission phase characteristics are extremely important, and it is imperative that the phase remains constant as the gain is varied. Dual-gate FETs may be used to achieve variable gain/attenuation as well as constant phase. They possess two gates: an RF gate and a control gate. The gain/attenuation can be adjusted by varying the bias voltage on the control gate. The RF gate acts in the same manner as the gate on a standard single-gate MESFET. Figure 20.12 shows the schematic and photograph of a variable gain amplifier using a dual-gate FET [34]. Such circuits are usually designed by employing measured S-parameters for the dual-gate FET at various control voltages. Measured amplitude and phase variations versus frequency were obtained with various control voltages. This circuit offers a 10-dB attenuation range with $\pm 3^\circ$ phase change over the 5–6-GHz frequency range. Over this frequency band, transmission amplitude and phase remain constant for any attenuation value. Figure 20.13 shows another configuration of a variable voltage attenuator. Here, control gate voltages V_{gc1} and V_{gc2} are adjusted to get linear attenuation with voltage in the range of 30–40 dB by using the look-up table. Table 20.5 provides typical gate voltage values for a 10-dB attenuation range of an MMIC attenuator.

Variable attenuation can also be realized using switched attenuation sections. These attenuators are usually digitally controlled and consist of a binary combination of 4–6 bits. Broadband digital attenuator bit configurations are shown in Figure 20.14. Small size attenuation bits (0.5 and 1dB) are realized using a single-FET integrated

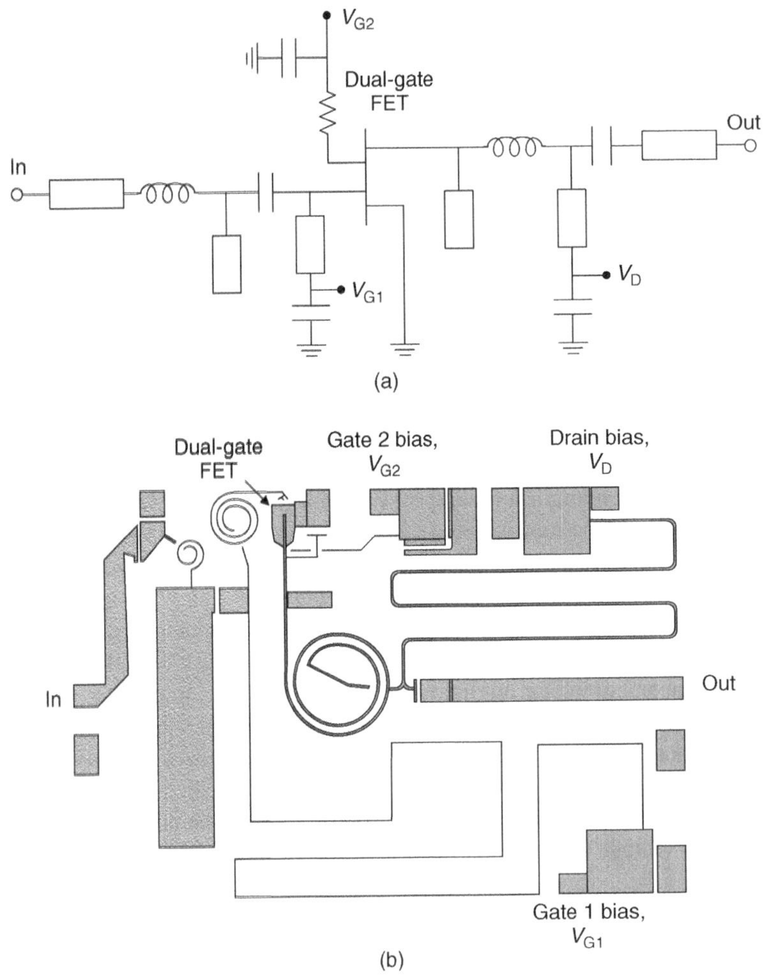

Figure 20.12 (a) Schematic and (b) photograph of a variable gain amplifier using a dual-gate FET.

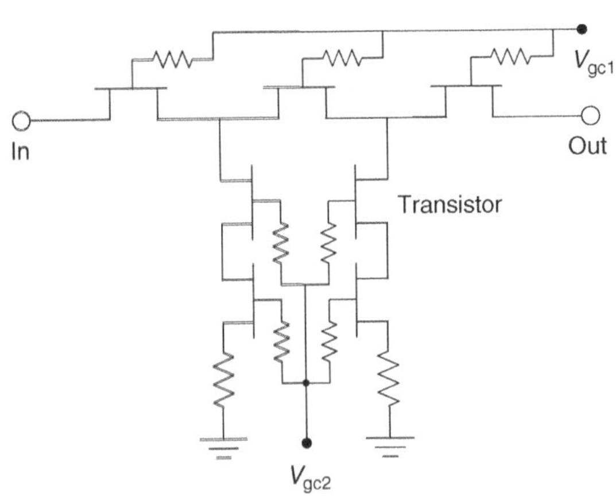

Figure 20.13 Schematic of a variable voltage attenuator using switching devices.

Table 20.5 Gate Control Voltage Values for Various Attenuation Values

Attenuation (dB)	V_{gc1} (V)	V_{gc2} (V)
0	0	−3.74
1	−0.90	−3.60
2	−1.77	−3.50
3	−2.21	−3.45
4	−2.42	−3.36
5	−2.62	−3.26
6	−2.79	−3.20
7	−2.85	−3.15
8	−2.91	−3.10
9	−3.00	−2.90
10	−3.02	−2.73

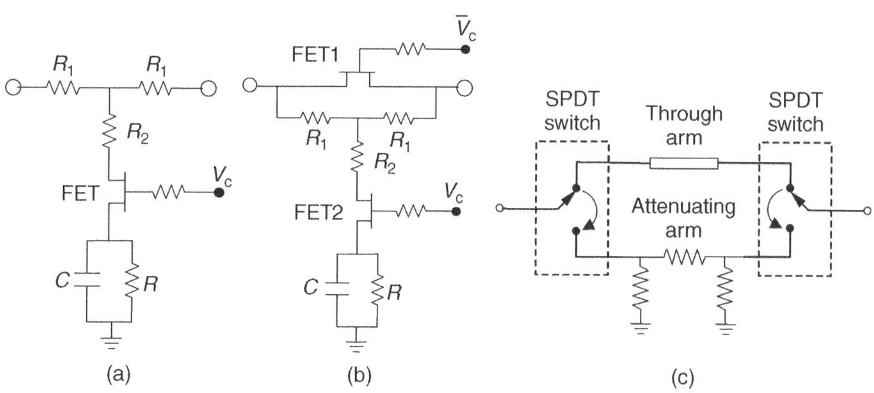

Figure 20.14 Schematics for multi-bit attenuators. (a) single-FET integrated, (b) embedded FET T-network and (c) switchable fixed attenuation bit.

configuration and medium size bits (2 and 4dB) are designed using an embedded FET T- network made up of transistors and resistors in both reference and attenuation paths. Larger bits (8 and 16dB) are generally obtained by switching pi or T-type fixed attenuators with reference to phase compensated through paths using broadband SPDT switches [35].

20.2.2 Variable Power Amplifier

The output power of an amplifier using transistors can be varied by input power level, controlling the drain/collector current and voltage. In a transistor amplifier the DC power is set at the Q-point for optimum output power and PAE. If the input power is lowered significantly so that the transistors are operating well below the saturation point, both output power and PAE are reduced with reducing input power. The output power can also be varied by changing the gate voltage in the FET/HEMT. In this case, to reduce output power the transistor current is reduced by applying more negative voltage. The load presented at the output of the transistor remains the same. At lower input power levels, the prime power is saved and the amplifier becomes efficient over a limited range of output power. The FET/HEMT amplifier's output power is not a strong function of gate voltage.

The most efficient method of changing output power is by varying the drain/collector voltage. In this case, the prime power in the transistor is drastically reduced. The saturated output power is approximately linear with drain voltage below the maximum allowed transistor's supply voltage. The PAE does not change very much with supply voltage. However, any change in the drain voltage also affects the input and output impedance of the transistor and will change the input and output match and the gain with respect to the nominal bias conditions. It may also change the gain flatness or bandwidth. In the variable-power mode, the amplifier is generally designed by using the load line at the highest operating voltage. Figure 20.15 shows the variation of output power and PAE of a three-stage X-band MESFET HPA, as a function of drain voltage. The gate voltage was set at −2 V. The power gain (≅23 dB) was almost constant with drain voltage. The nominal supply voltage for the HPA was 10 V. By lowering the drain voltage from 10 V to 6 V, the output power was reduced approximately from 41 dBm to 37 dBm. Thus the output power slope is 1 dB/V. The PAE was almost constant (40–42%) over 6–10-V operation of the HPA.

Amplifier output power may also be changed by using a variable attenuator at the input. By reducing the input power going into the power amplifier stage/stages, the output can be reduced and the signal becomes linear because the amplifier is operating below saturation. In this case, both the amplifier's match and gain flatness remain unchanged. However, PAE is reduced with output power. Thus realization of variable power with variable drain/collector supply voltage is more efficient.

The output power of a multistage amplifier chain can be changed in steps by switching in and out amplifier stages and using connecting lines as shown in Figure 20.16. This configuration uses three low-loss switches and the power handling of Switch 3 is the highest while for Switch 1 it is the lowest. The P_{1dB} value for Switch 3 is required to be 3–4 dB higher than the output power of the HPA. In this case, the maximum output power occurs when all three switches are in position "A" and the net output

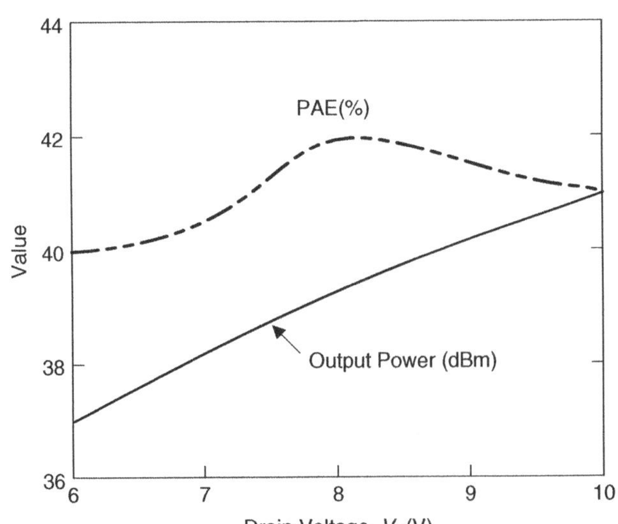

Figure 20.15 Output power and PAE versus drain voltage of a three-stage HPA at 9 GHz.

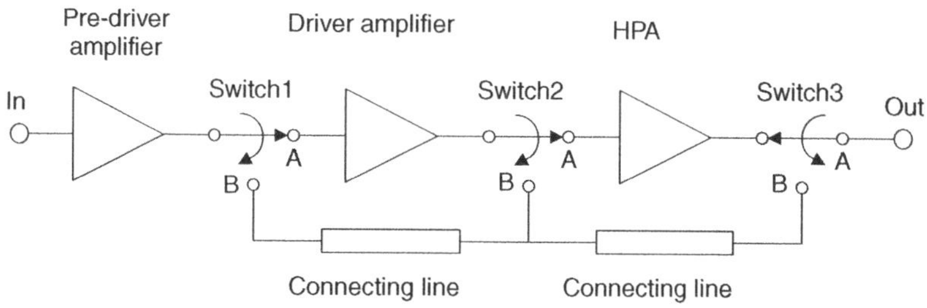

Figure 20.16 Variable output power configuration of an HPA.

power is the HPA's power minus the Switch 3 insertion loss. In this case, the connecting lines are isolated. However, in the other two settings the output power levels are reduced by an aggregate loss of two switches.

20.2.3 Amplifier Temperature Compensation

The transistor amplifier characteristics such as gain, NF, and power are temperature dependent. The gain and output power of an amplifier decrease with temperature, whereas the noise figure increases with temperature. In other words, the higher the temperature, the lower is the gain and the higher is the NF, and vice versa. The temperature coefficients for the transistor amplifier have been described in Chapter 3. In most applications only the amplifier gain is compensated with temperature.

In amplifiers, the gain compensation with temperature (i.e, constant gain with temperature) is obtained by using automatic gain control circuitry. In this case, the temperature is sensed by using a diode and by employing external circuitry; the transistors are rebiased for constant gain. A simple temperature compensation circuit (TCC) for constant gain is described in Reference 36. It is based on the concept that the Schottky diode's threshold voltage changes with temperature. The same concept has been used to monitor channel temperature, as described in Chapter 16.

The gain of an amplifier decreases with the increase of temperature and decrease of the gate voltage (more negative voltage). Therefore the gain variation with temperature may be compensated by properly selecting the gate voltage. In this case, as the temperature increases the gate voltage is increased (less negative voltage) and vice versa. The gain temperature coefficient of an amplifier is about 0.01 dB/°C/stage. The variation of the amplifier gain with gate voltage depends on the Q-point and the transistor used and is not a linear function of gate voltage. In FETs the gain slope is greater near pinch-off. It is close to linear from 10% to 50% I_{dss}. Consider a single-stage amplifier having 10-dB gain and operating over −25 °C to 75 °C. The change in gain with gate voltage is 2.5 dB/V around the Q-point gate voltage of −2.0 V. The expected gain variation over the temperature range is about 1 dB. Thus in this case, we need 0.4-V gate voltage variation for 1-dB gain compensation. The gate voltage needed at −25 °C is −2.2 V and at 75 °C it is −1.8 V.

As discussed in Section 16.6.3, the junction (threshold) voltage temperature coefficient (JTC) for an FET is −0.833 mV/°C. Here the negative sign signifies that the gate–source voltage decreases with increasing temperature. Thus over a 100 °C temperature change, the gate–source voltage change is 83.3 mV and one needs five diodes in series to obtain about 0.4-V change in the gate–source voltage. A simple gain temperature compensation circuit is shown in Figure 20.17a.

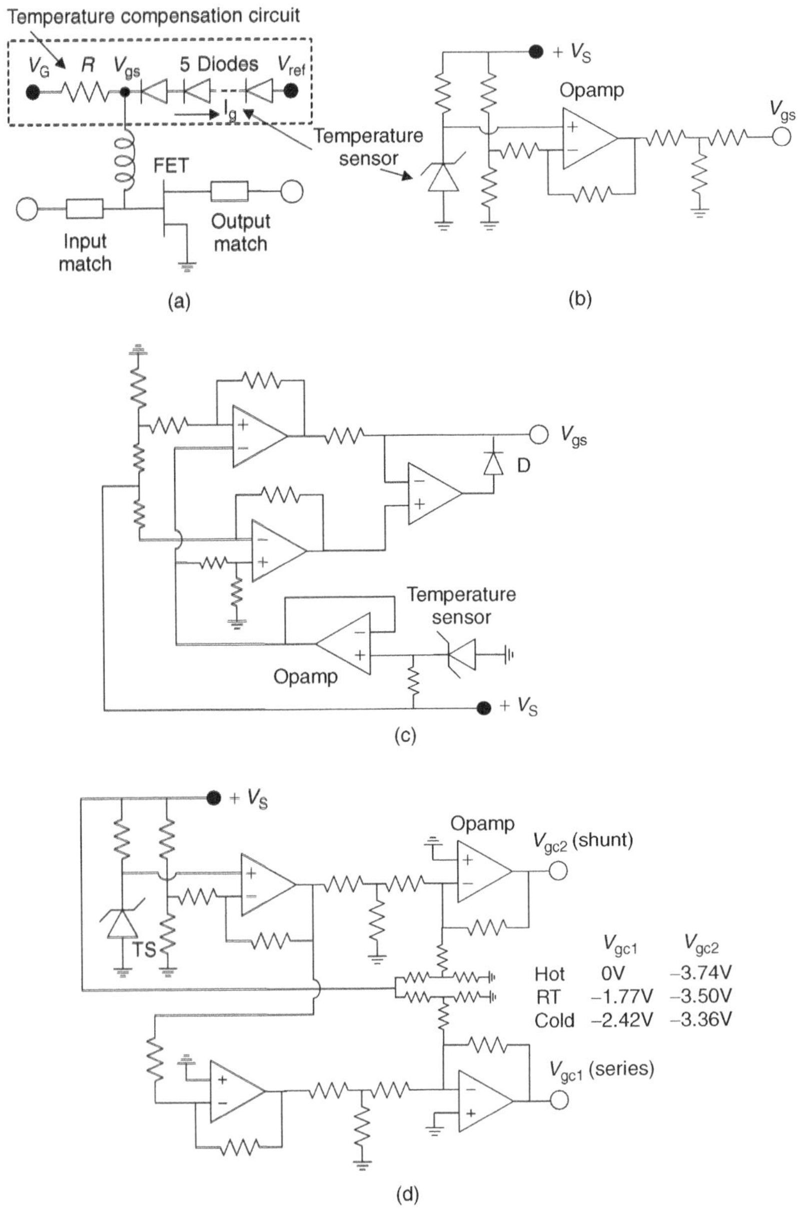

Figure 20.17 Temperature compensation circuits: (a) basic schematic of on-chip diodes, (b) simple configuration using temperature sensor and opamp, (c) schematic providing piecewise linear transfer function, and (d) topology providing dual control voltages for 4-dB attenuation range.

Another temperature compensation circuit using an external temperature sensor is shown in Figure 20.17b. Typical output voltage for such sensors is 10 mV/°C, which is about 12 times greater than for a GaAs Schottky diode. In this TCC, an operational amplifier (usually referred to as an opamp) is used to adjust the gate–source voltage variation applied to the FET. Circuit analysis is straightforward since the TCC works at DC only.

The gain variation with gate voltage varies nonlinearly. In an ideal TCC the temperature is sensed and sets the gate voltage to provide the desired gain. A TCC reported by Goldsmith [37] generates a piecewise linear transfer function between the temperature sensor and the FET gate terminal, and also provides positive control gate voltage needed for the dual-gate FET variable gain amplifier. A simple schematic of the TCC that provides a piecewise linear function between the temperature sensor and the FET gate is shown in Figure 20.17c. The temperature sensor provides a voltage proportional to temperature. The sensed voltage is changed by the opamps to required levels over the desired temperature range at the output. The diode D helps in reversing the gate voltage polarity.

Another TCC used with variable attenuator (Fig. 20.13) is shown in Figure 20.17d. Suppose that a 4-dB gain variation from cold to hot temperature is needed. In this case the TCC resistor setting is used to obtain 2-dB attenuation at room temperature. At a cold temperature the attenuation setting is at 4 dB and at a hot temperature the attenuation setting is at 0 dB. Referring to Table 20.5, the V_{gc1} and V_{gc2} voltages are given in Figure 20.17d.

20.2.4 Power Monitor/Detector

In several microwave applications one needs to monitor or sense the transmitter's output power or control it. Detectors are used frequently as power detectors or power monitors [38–43]. Figure 20.18 shows a simplified schematic of a power detection and control circuitry. A diode based method is the simplest way for power detection. Detectors make use of the nonlinear characteristics of diode devices to generate an output signal containing many frequency components. By proper filtering, the DC component is

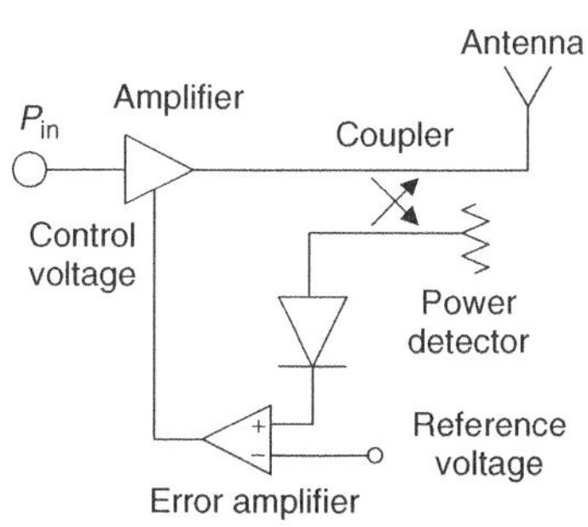

Figure 20.18 Basic schematic of a power control circuit.

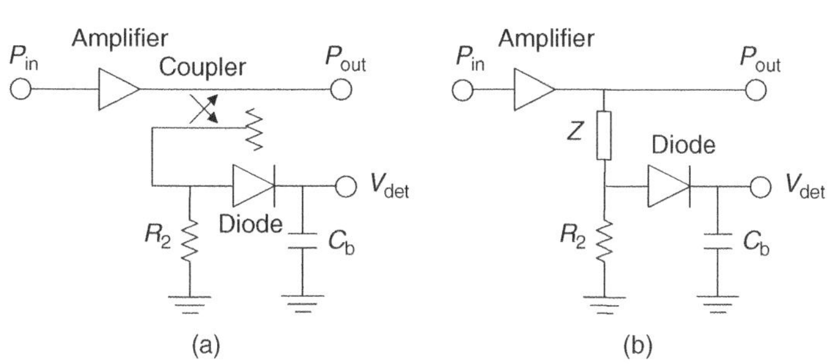

Figure 20.19 Power detector configurations using (a) a loose coupler and (b) high-impedance sampling.

separated and is used as a power monitoring voltage. This works on the principle of a rectifier diode. This device is placed in shunt at the output stage of the amplifier.

Figure 20.19 shows two simplified schematics of a power detection circuit, where the power is sampled by using a 15–20-dB coupler or by using a high-impedance Z network [43]. The value of isolation impedance Z, in a 50-Ω system, is selected such that the power flowing through Z and the diode is negligible and does not load the main circuit. The value of Z is adjusted so that the sampled power is 15–20 dB below the output power P_{out}. The capacitive type sampling has a larger bandwidth potential than resistive or inductive. Next, Schottky diode based power detectors are described to obtain the required detected voltage. In this case, typically a detected voltage of 4V is selected when the sampled power is about 17–18 dBm at the detector interface.

Schottky Diode Configuration

In order to minimize loss, the design of the detector diode requires high cutoff frequency $f_c (= 1/2\pi R_S C_{j0})$ for the Schottky diode. For this application, M/A-COM's standard Schottky mixer diode implant (MDI) available in the MSAG FET process [30, 43] was used. Figure 20.4 shows the top and cross-sectional views of the Schottky diode detector. The Schottky diode dimensions in this case are $0.4 \times 50\ \mu m^2$, which are required to handle an input power level of about 25 dBm. The equivalent circuit of the diode as shown in Figure 20.20 is represented by a parallel combination of R_j and C_{j0} in series with R_S. Typical measured R_S and C_{j0} values are listed in Table 20.6. Two power detector configurations, shown in Figure 20.21, were investigated to obtain small size and wide bandwidth.

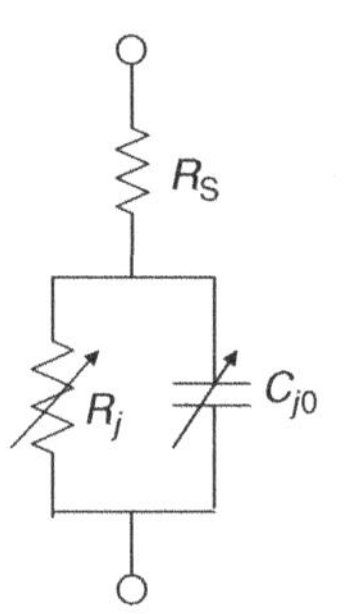

Figure 20.20 Equivalent circuit of a diode.

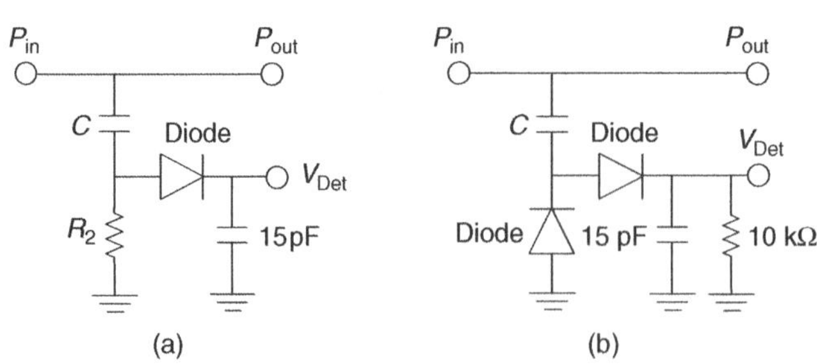

Figure 20.21 Power detector configurations investigated.

Table 20.6 Schottky Diode Parameters

Parameter	Symbol/Formula	Units	Value
Gate length	L	μm	0.4
Unit gate width	W	μm	50
Total periphery	W_g	μm	50
Series resistance	R_S	Ω	6.3
Zero bias junction capacitance	C_{j0}	pF	0.08
Cutoff frequency	f_c	GHz	316

Power Detector Simulated Data

The power detector circuit shown in Figure 20.19b was simulated using a nonlinear diode model. The diode parameters are given in Table 20.6. The power sampling impedance Z can be selected as a thin-film resistor, a coil inductor, or an MIM capacitor. The values of these elements are chosen such that the return loss on the main line remains better than 15 dB. Their values also affect the detected voltage and the bandwidth. Detector simulations predict that the signal sampling through a resistor or an inductor is narrowband, whereas the capacitive sampling results in wideband performance.

Two topologies for the capacitively coupled power detector configurations shown in Figure 20.21 were studied further. In Figure 20.21a the value of R_2 was selected to obtain the largest possible detected voltage and bandwidth. The value of R_2 is about 800 Ω. Simulations showed that the topology in Figure 20.21b requires about 3.5-dB lower input power than the topology in Figure 20.21a to obtain the same detected voltage. Figure 20.22 shows the input power required versus capacitor value for a 4 ± 0.1V power detector (Fig. 20.21b) over 6–18 GHz. The return loss was greater than 15 dB. The detected voltage versus input power for $C = 0.2$ pF is shown in Figure 20.23 at 6 and 18 GHz.

MMIC Power Detectors

Two block diagrams for an MMIC detector, as shown in Figure 20.24, were studied experimentally. Both these circuits were fabricated and tested to determine a broadband frequency response for a 4-V detected voltage for a 2–3-W driver amplifier. The detector circuit shown in Figure 20.24a was studied for use in the configurations shown

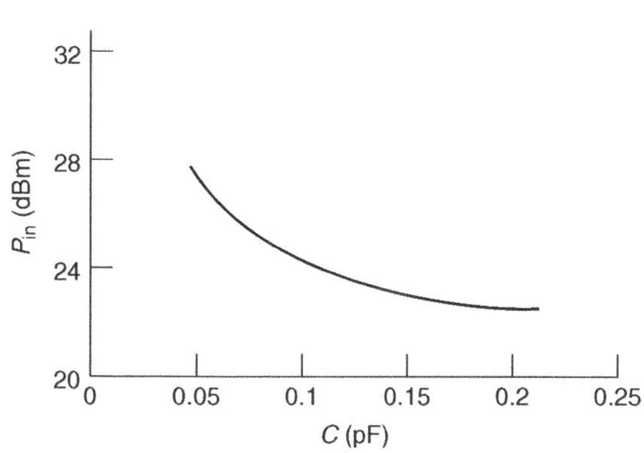

Figure 20.22 A plot of the input power and coupling capacitor for the Figure 20.21b detector configuration to obtain a detected voltage of 4 V.

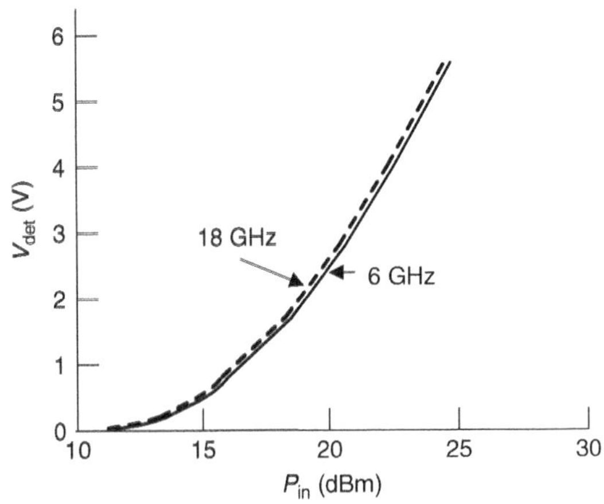

Figure 20.23 A plot of detected voltage versus input power for the Figure 20.21b detector configuration. Capacitor C value is 0.2 pF.

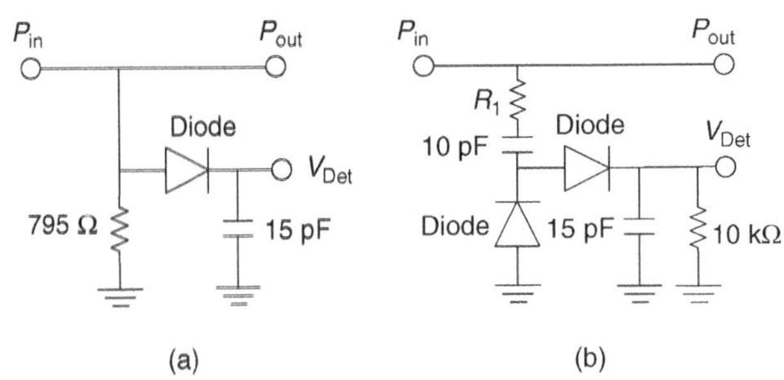

Figure 20.24 MMIC voltage detector configurations investigated.

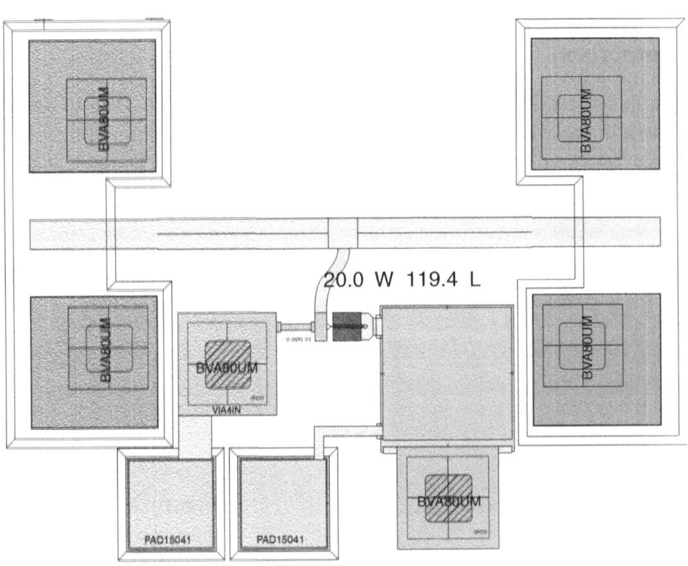

Figure 20.25 Physical layout of the 4-V voltage detector for the Figure 20.24a detector configuration.

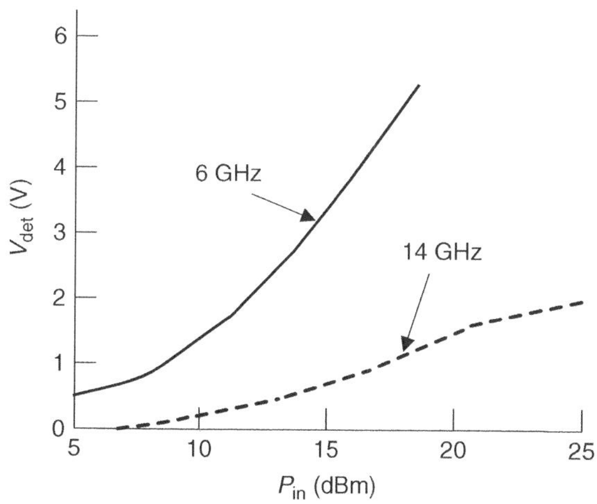

Figure 20.26 Detected voltage versus input power at 6 and 14 GHz for the Figure 20.24b detector configuration. $R_l = 150\ \Omega$.

in Figure 20.19a or 20.19b with a properly selected Z value. The detector circuit shown in Figure 20.24b can be used without a coupler, as shown in Figure 20.19b. At low microwave frequencies (below C-band) the coupler size is large; therefore the topology in Figure 20.19b is preferred. The 10-pF capacitor or the capacitive Z value blocks the RF converted DC voltage flowing back toward the amplifier. Figure 20.25 shows a physical layout for the configuration of Fig. 20.24a.

Test Data and Discussions

The power detector circuits were tested on wafer using RF probes. Figure 20.26 shows the typical measured detector voltage versus input power at 6 and 14 GHz for the circuit shown in Figure 20.24b and $R_l = 150\ \Omega$. Over the 15–25-dB input power range, the detector voltage is linear with input power. In order to get a detector voltage

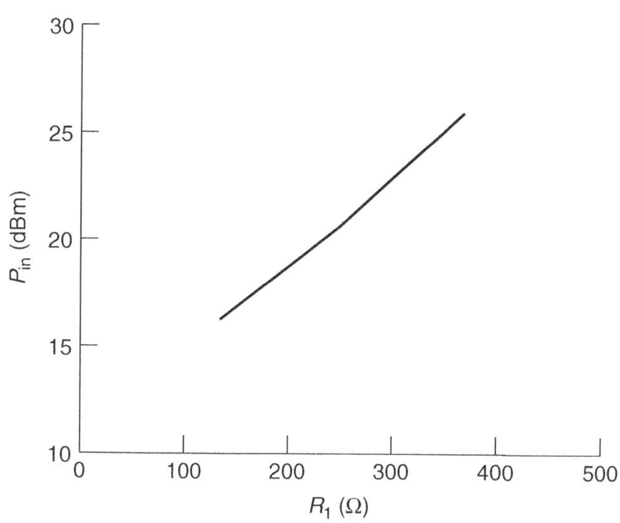

Figure 20.27 A plot of required P_{in} versus R_l value to obtain $V_{det} = 4$ V at 6 GHz for the Figure 20.24b detector configuration.

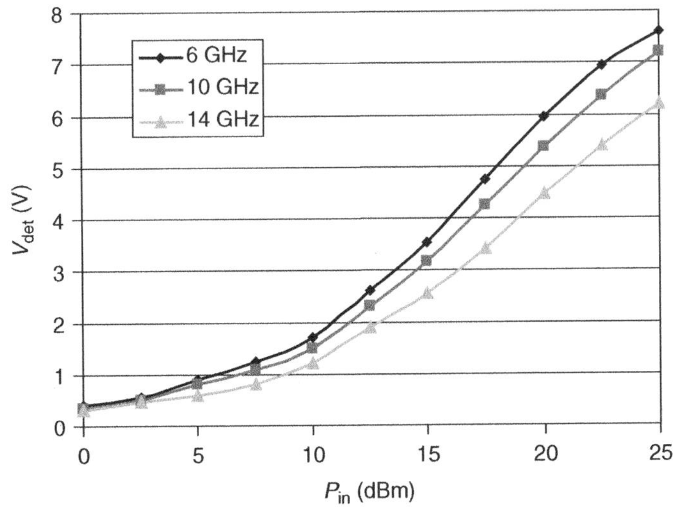

Figure 20.28 Typical measured detector voltage versus input power for the Figure 20.24a detector configuration.

of 4 V, one needs an input power value of about 17 dBm at 6 GHz. At 14 GHz, the maximum voltage detected was about 2 V at 25-dBm input power. Figure 20.27 shows the required input power and R_1 value to obtain a 4-V detected voltage at 6 GHz. The power detector circuit shown in Figures 20.24a and 20.25 were also tested. Figure 20.28 shows the typical measured detector voltage versus input power at 6, 10, and 14 GHz. Over the 10–22-dB input power range, the detector voltage is linear with input power. In order to obtain a detector voltage of 4 V, one needs input power values of about 16 and 19 dBm at 6 and 14 GHz, respectively. Thus one can obtain a detected voltage of 4 ± 0.7 V at 17.5-dBm input power level over the 6–14-GHz frequency range. Preferably, detector curves should not cross each other to avoid confusion in applying corrective voltage to maintain the power level.

20.2.5 Protection Against Load Mismatch

The power amplifiers are designed with 50 Ω as a load at the output. However, if the load presented at the output is quite different from 50 Ω, solid state devices might fail due to the increase in the power dissipation or much higher voltages than the breakdown voltage of the device. There are several techniques being used to protect the solid state devices against large VSWR mismatch at the output. One can use a balanced configuration for the power amplifier, as shown in Figure 20.10, or an isolator at the output, as shown in Figure 20.11a. In both cases, the reflected signal entering the amplifier is terminated into a 50-Ω load. One can also design HPAs for ruggedness against thermal and breakdown voltages. In such cases, the device drain, due to the matching network, does not see a short or open circuit. High-power amplifiers (>10-W output power) have been tested under 10:1 mismatch conditions without any damage.

The HPA can also be designed with a protection circuit like the adaptive bias as a function of output VSWR. In this case, the device does not see large excursions of current and voltage during its operation. Such topology uses external circuitry similar to temperature compensation. One can also use a power detector, as described in the

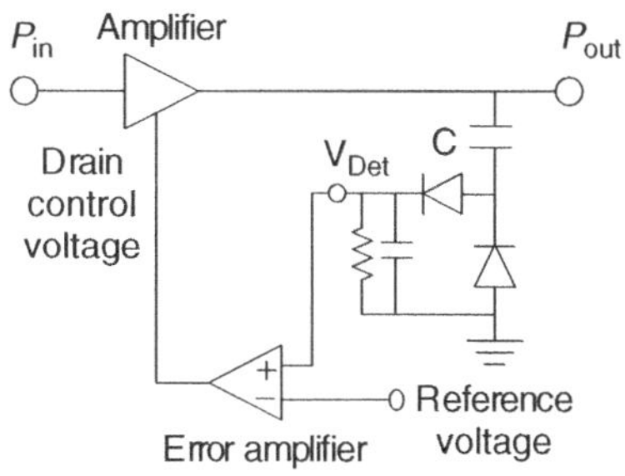

Figure 20.29 Schematic of a load mismatch protection circuit using a power detector.

previous section, with external circuitry to shut off the HPA stage under severe output mismatch conditions. Any severe mismatch will alter the output power and accordingly the detected voltage. The change in detected voltage is fed to the external circuitry, which turns off the amplifier supply drain/collector voltage. Figure 20.29 shows a schematic of a load mismatch protection circuit using a power detector.

20.3 CASCADING OF AMPLIFIERS

For high gain amplification, several amplifiers are cascaded in series. This can be designed as one unit or all amplifiers are matched to 50 Ω and connected in series. If each amplifier is perfectly matched or has a balanced configuration, the total gain of the cascaded unit is the sum of gains (in dB) for each amplifier. When two cascaded amplifiers are not matched to each other, the total gain is the sum of gains (in dB) for each amplifier minus the mismatch loss (in dB) between the amplifiers. If the output of the first amplifier and the input of the second amplifier have return losses better than 10 dB, then the total gain is given by

$$G_t = G_1 + G_2 - ML \tag{20.6}$$

where *ML* is the mismatch loss due to interactions between the stages and its maximum value is 1 dB. Here the subscripts 1 and 2 designate first and second amplifier, respectively.

Accurate values of gain and input and output return losses of a cascaded unit can only be obtained by cascading *S*-parameters for the amplifiers used. In the case of two cascaded amplifiers, when the magnitudes of S_{22}^1 and S_{11}^2 are less than 0.1 (i.e., mismatch interactions are low), the total gain is given approximately by

$$G_t = G_1 + G_2 \tag{20.7}$$

where the superscripts 1 and 2 designate first and second amplifier, respectively.

REFERENCES

1. T. F. Bogart, Jr., Limiters, in *Electrical Engineering Handbook*, 2nd edition (R. C. Dorf, Editor), CRC Press, Boca Raton, FL, 1997, pp. 140–147.
2. R. Kaul, Microwave limiters, in *Wiley Encyclopedia of Electrical and Electronics Engineering* (J. G. Webster, Editor), Vol. 13., John Wiley & Sons, Hoboken, NJ, 1999.

3. H. Goldie and S. Patel, An RF-primed all-halogen gas plasma microwave high-power receiver protector, *IEEE Trans. Microwave Theory Tech.*, Vol. MTT-30, pp. 2177–2183, December 1982.
4. S. D. Patel et al., Microstrip plasma limiter, *IEEE MTT-S Int. Microwave Symp. Dig.*, pp. 879–882, 1989.
5. B. Lax and K. J. Button, *Microwave Ferrites and Ferrimagnetics*, McGraw-Hill, New York, 1962.
6. J. L. Carter and J. W. McGowan, X-band ferrite-varactor limiter, *IEEE Trans. Microwave Theory Tech.*, Vol. MTT-17, pp. 231–232, April 1969.
7. J. D. Adam and S. N. Stitzer, Frequency selective limiters for high dynamic range microwave receivers, *IEEE Trans. Microwave Theory Tech.*, Vol. 41, pp. 2227–2231, December 1983.
8. D. Leenov, The silicon PIN diode as a microwave radar protector at megawatt levels, *IEEE Trans. Electron Devices*, Vol. ED-11, pp. 53–61, February 1964.
9. N. J. Brown, Design concepts for high-power PIN diode limiting, *IEEE Trans. Microwave Theory Tech.*, Vol. MTT-15, pp. 732–742, December 1967.
10. J. F. White, *Semiconductor Control*, Artech House, Norwood, MA, 1977, Chapter 7.
11. R. V. Garver, *Microwave Diode Control Devices*, Artech House, Norwood, MA, 1978, Chapter 9.
12. R. J. Hamilton, Jr., Wide band dual-gate GaAs FET output limiters, *IEEE MTT-S Dig.*, pp. 256–258, 1979.
13. R. J. Hamilton, Jr. et al., Devices vital to IFM limiting amplifiers, *Microwave Syst.-News*, Vol. 9, pp. 59–66, September 1979.
14. S. D. Patel and H. Goldie, A 100 kW solid-state coaxial limiter for L-band, *Microwave J.*, Vol. 24, pp. 61–65, December 1981.
15. E. J. Crescenzi, Jr. et al., Wide-band limiting amplifiers with low second harmonic distortion, utilizing GaAs MMIC limiters, *IEEE MTT-S Int. Microwave Symp. Dig.*, pp. 328–331, 1985.
16. I. J. Bahl, E. L. Griffin, and E. R. Schineller, GaAs monolithic limiting amplifiers, *IEEE GaAs IC Symp. Dig.*, pp. 71–73, 1986.
17. I. J. Bahl, E. L. Griffin, and E. R. Schineller, Monolithic limiting amplifiers for EW systems, *Microwave J.*, Vol. 30, pp. 205–209, September 1987.
18. P. Sahjani and E. Higham, PIN diode limiters handle high-power input signals, *Microwaves RF*, Vol. 29, pp. 195–199, April 1990.
19. J. McAdoo et al., A terminal protection device for gallium arsenide MMIC technology, *GOMAC*, pp. 441–445, 1991.
20. J. Ozaki et al., C-band GaAs MMIC limiting power amplifier with small insertion phase variation, *IEEE MTT-S Int. Microwave Symp. Dig.*, pp. 331–334, 1991.
21. C. F. Vasile, FET adaptive limiter with high current FET detector, US Patent No. 5,157,289, 1992.
22. T. Parra et al., X-band low phase distortion MMIC power limiter, *IEEE Trans. Microwave Theory Tech.*, Vol. 41, pp. 876–879, May 1993.
23. R. A. LaRue et al., Broadband microwave and millimeter-wave 10 watt CW power limiters, *GOMAC*, pp. 360–363, 1994.
24. P. Huang et al., A 9–16GHz monolithic HEMT low noise amplifier with embedded limiters, *IEEE Microwave Millimeter-wave Monolithic Circuits Symp. Dig.*, pp. 185–186, 1995.
25. C. Trantanella, M. Pollman, and M. Shifrin, An investigation of GaAs MMIC high power limiters for circuit protection, *IEEE MTT-S Int. Microwave Symp. Dig.*, pp. 535–538, 1997.
26. R. Kaul et al., MMIC-compatible terminal protection device, *IEEE MTT-S Int. Microwave Symp. Dig.*, pp. 1679–1682, 1998.
27. D. G. Smith et al., Designing high-power limiter circuits with GaAs PIN diodes, *IEEE MTT-S Int. Microwave Symp. Dig.*, pp. 329–332, 1999.
28. I. J. Bahl, 10W CW broadband balanced limiter/LNA fabricated using MSAG MESFET process, *Int. J. RF Microwave Computer-Aided Eng.* Vol. 13, pp. 118–127, March 2003.

29. J. Looney, D. Conway, and I. Bahl, An examination of recovery time of an integrated limiter/LNA, *IEEE Microwave Mag.*, Vol. 5, pp. 83–86, March 2004.
30. I. J. Bahl et al., Multifunction SAG process for high-yield, low-cost GaAs microwave integrated circuits, *IEEE Trans. Microwave Theory Tech.*, Vol. MTT-38, pp. 1175–1182, September 1990.
31. H. F. Cooke, Precise technique finds FET thermal resistance, *Microwaves RF*, Vol. 25, pp. 85–87, August 1986. Correction of this paper in *Microwaves RF*, p. 13, February 1987.
32. I. J. Bahl and K. C. Gupta, Average power-handling capability of microstrip lines, *IEE Microwaves Opt. Acoustics*, Vol. 3, pp. 1–4, January 1979.
33. K. C. Gupta, R. Garg, I. J. Bahl, and P. Bhartia, *Microstrip Lines and Slotlines*, 2nd edition, Artech House, Norwood, MA, 1996.
34. D. Fisher and I. Bahl, *Gallium Arsenide IC Applications Handbook*, Academic Press, San Diego, CA, 1995.
35. K. Chang, I. Bahl, and V. Nair, *RF and Microwave Circuit and Component Design for Wireless Systems*, John Wiley & Sons, Hoboken, NJ, 2002, Chapter 7.
36. K. Yamauchi et al., X-band MMIC power amplifier with an on-chip temperature compensation, *IEEE Trans. Microwave Theory Tech.*, Vol. MTT-49, pp. 2501–2506, December 2001.
37. C. L. Goldsmith, Simple circuit flattens gain over temperature, *Microwaves RF*, Vol. 28, pp. 159–165, May 1989.
38. A. K. Sharma et al., A high performance transceiver chipset for millimeter-wave commercial digital communication systems, *IEEE Microwave Millimeter-Wave Monolithic Circuits Symp. Dig.*, pp. 45–48, 1995.
39. I. Bahl and P. Bhartia, *Microwave Solid State Circuit Design*, 2nd edition, John Wiley & Sons, Hoboken, NJ, 2003.
40. K. A. Lee et al., A InGaP/GaAs HBT WLAN power amplifier with power detector, *12th GAAS Symp. Dig.*, pp. 539–542, 2004.
41. Y. Lu, I. Gresham, and A. Jenkins, A closed-loop pulsed power control circuit for UWB 24GHz automotive radar transmitter, *IEEE MTT-S Int. Microwave Symp. Dig.*, pp. 1672–1675, 2006.
42. T. Zhang et al., Bipolar microwave RMS power detectors, *IEEE J. Solid State Circuit*, Vol. 41, pp. 2188–2192, September 2006.
43. I. Bahl, Broadband power detectors, *IEEE Microwave Mag.*, Vol. 8, pp. 82–86, June 2007.

PROBLEMS

20.1 Design a temperature compensated single-stage amplifier with constant gain over 0–80 °C. Assume 10 dB gain and the change in gain with gate voltage is 2 dB/V.

20.2 In a balanced limiter/LNA, a single-stage amplifier has 12-dB gain. Calculate the NF of the balanced configuration when each Lange coupler and input and output match have 0.5-dB loss and NF_{min} of the LNA transistor is 0.3 dB.

20.3 Design a driver amplifier with 10-dB gain variation. The output power is 30 dBm at P_{1dB} and the amplifier has three stages. The output power for the attenuator is 25 dBm at P_{1dB}.

20.4 Describe the critical design requirements of a power transmitter with 40-dB gain, 100-W output power and 30% PAE. Both input and output VSWR values are 1.5:1.

20.5 Describe the critical design requirements of a receiver chain with 2-dB noise figure and 30-dB gain. Both input and output VSWR values are 1.5:1.

20.6 Design a mismatch protection circuit at the output of a 50-W HPA at 6 GHz. The mismatch is 6:1.

Chapter 21

Amplifier Packages

RF packages are widely used for both commercial and military applications including wireless communication, sensing, and radar. For high-volume commercial applications, amplifiers are generally housed in low-cost plastic packages, whereas military applications often use semicustom/custom packages designed for performance, reliability, and low- to medium-volume manufacturing. RF packaging technologies are advancing rapidly in terms of modeling, frequency bandwidth, and cost. Some of these packages are usable up to 40 GHz. This topic of RF and microwave packaging has been treated in several books [1–6], book and handbook chapters [7–12], and many other publications [13–17]. This chapter provides an overview of RF and microwave packages being used for amplifiers and their assembly techniques.

21.1 AMPLIFIER PACKAGING OVERVIEW

Selection of a suitable package and assembly technique plays an important role in the performance, cost, and reliability of amplifier products. When amplifier circuits are packaged, the effect of the package assembly technique and package environment must be kept to a minimum. Minimizing package complexity is important for minimizing package cost. Minimizing the number of dielectric layers and the overall size assists dramatically in the improvement of production yields and lowering costs. However, a trade-off exists between the simplicity and functional features of these packages.

Amplifier packaging can be performed at three levels, as shown in Figure 21.1. MMIC amplifiers can be mounted in individual packages; MIC amplifiers can be fabricated with support circuitry and placed in a housing, commonly referred to as a first-level (amplifier-to-package) package. The ICs can also be packaged at the subsystem level, referred to as second-level (package-to-board or housing) assembly. The most commonly used first-level package assembly consists of die bonding and wire bonding. Other techniques include flip-chip assembly and tape-automatic bonding (TAB) [6]. The second-level (package-to-board or housing) assembly techniques can be surface mount or through-hole. The surface mounted packages include chip-scale package

Fundamentals of RF and Microwave Transistor Amplifiers. By Inder J. Bahl

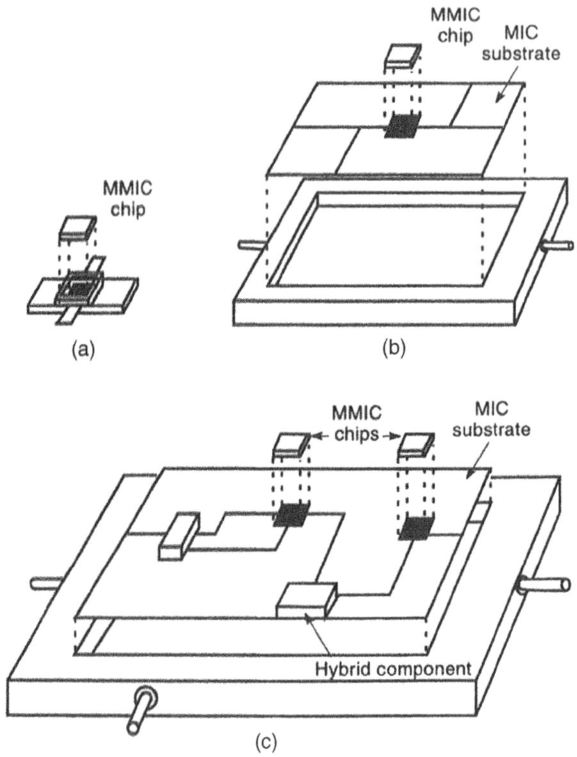

Figure 21.1 Three MMIC packaging levels: (a) MMIC in package, (b) MMIC with support circuitry, and (c) MMICs with hybrid and support circuitry.

Table 21.1 Advantages and Concerns for Single-Chip and Multichip Packaging

	Single Chip	Multichip
Advantages	Each device tested Good yield as module Device-to-device isolation Hermeticity for individual device Volume manufacturability	Size reduction Light weight Assembly cost reduction/cycle time reduction Performance enhancement Reliability enhancement Power reduction Known good die testing and yield
Concerns	Electrical performance at next level assembly module level	Rework Use of via for RF line DC/RF shielding

(CSP), printed circuit board (PCB), and ball-grid array (BGA). The major challenges in first-level and second-level package assembly technologies are to control the effect of assembly on the amplifier's performance. The most critical parameters are wire lengths, lead frame lengths, and grounding effects leading to variations in amplifier performance from unit to unit, and maintaining an adequate thermal path to keep chip temperatures low. However, with modern automated assembly techniques, the production yield has been drastically improved by minimizing the variability in the wire bond lengths. Table 21.1 compares single-chip and multichip packaging techniques [11].

The packaging requirements depend on the application at hand. For example, in wireless communications applications below 18 GHz, GaAs MMICs are mounted into surface mount plastic packages in order to achieve low cost goals. For applications with relatively low-power operation, thermally they are acceptable. For high-frequency, high-performance, and high-power applications, metal base ceramic packages are often required, as they have low thermal resistance, good hermetic properties, high-power capability, and good reliability.

21.1.1 Brief History

The packaging of transistors in hermitically sealed metal cans and in plastic packages appeared in the late 1950s and early 1960s. In the 1970s, high-performance microwave transistors were housed in hermetic ceramic packages and were available for both low-noise and medium-power applications. The enormous developent of transistor and RF IC/MMIC packages started in the 1970s and 1980s, respectively. To date, there are numerous commercially available ceramic and plastic packages capable of providing adequate performance with GaAs MMICs and Si ICs over the entire 0.1–20-GHz frequency range.

For RF and microwave amplifier applications, the initial package concepts were transistor outline (TO), dual-in-line package (DIP), transistor 90-mil ceramic package, and metal/ceramic flatpack. TO can packages were introduced in the 1950s and are available in different lead versions including TO-5, -8, and -10. They have up to 14 pins. Requirements for greater than 14 leads called for DIPs or flat ceramic packages in the 1960s. DIPs are available in both ceramic and plastic. The 14-pin DIP is a rectangular flat package with two rows of pins/leads on its two sides. After attaching the die to the lead frame and wirebonding the interconnects, the structure is molded in plastic. DIPs are low-cost packages and are the commonly used package type for low-frequency and low-power applications. The 90-mil metalized ceramic package was introduced in the 1970s and was developed for small-signal FETs. Examples of TO, 90-mil transistor, DIP, and metal flatpack packages are shown in Figure 21.2.

Numerous types of ceramic packages were developed during the late 1970s and 1980s primarily for transistors and MMICs, as shown in Figure 21.3. The most popular ceramic packages were for power transistors as well as for internally matched power amplifiers. Packages were devloped both with and without leads. Today, the most popular ceramic package is with leads and was developed for medium-power MMIC amplifier applications having moderate volumes such as VSAT and point-to-point radio applications. A photograph of this type of package with lid is shown in Figure 21.4.

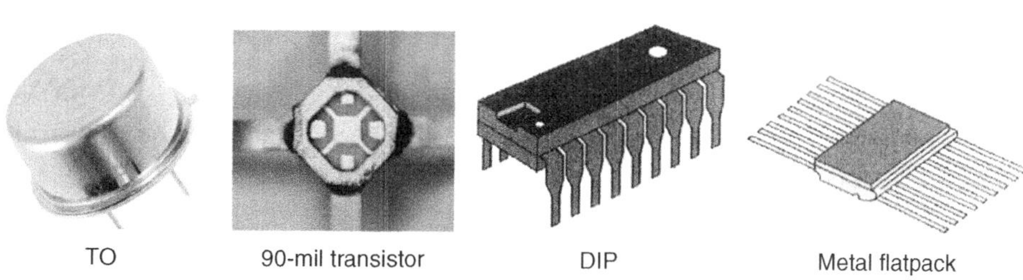

Figure 21.2 Transistor and IC packages.

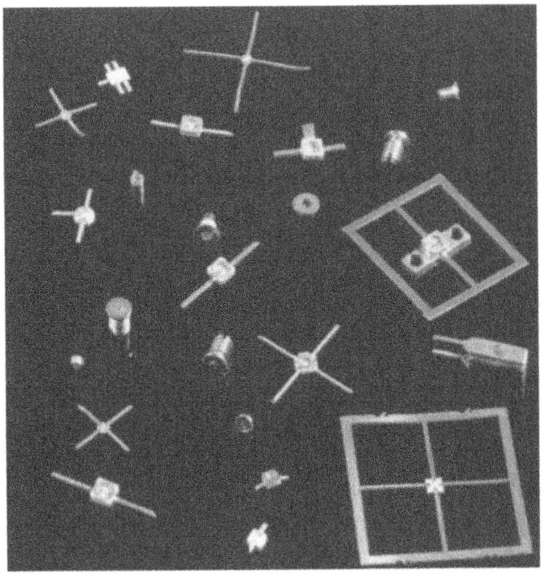

Figure 21.3 Examples of ceramic transistor and IC packages.

Figure 21.4 Medium-power ceramic IC package for MMIC amplifiers.

Examples of high-power transistor ceramic packages are shown in Figure 21.5. The packages are capable of delivering in the range of 100-W RF output power. High-performance medium-power (less than 5 W) ceramic packages in large quantity are available in the \$2–3 range.

The work on plastic packaging continues to make them more versatile, to extend their frequency range to higher frequencies, to allow them to handle more power, and to make them less expensive. These developments led to surface mount plastic packages such as the small outline transistor (SOT) and the small outline integrated circuit (SOIC) packages shown in Figure 21.6. SOIC packages have 8–16 pins and work reasonably well up to 2 GHz. In order to improve the RF performance and power dissipation for power ICs, custom-fused lead frames with low-signal lead parasitic reactance and reduced-ground bond inductance are being used in custom plastic packages. Later, the thin, small outline package (TSOP) was introduced due to constraints on package thickness. Plastic packaging, which includes both package and assembly costs, often costs less than \$0.25 per package.

Figure 21.5 Examples of high-power transistor ceramic packages.

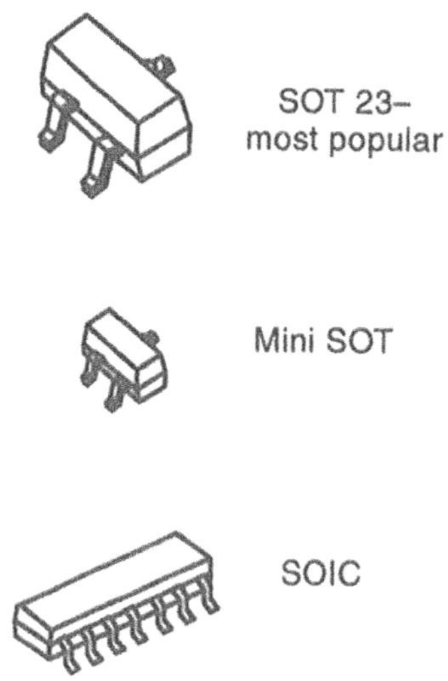

Figure 21.6 Plastic packages for RF and microwave applications.

One of the first successful demonstrations of a microwave packaging application was the direct broadcast satellite (DBS) receiver using a TO can package [15]. This product was one of the first to be sold in the millions. In the late 1980s, the MMIC based DBS receiver was available in a TO can and later available in ceramic metal packages, plastic dual-in-line packages, and small outline integrated circuit packages.

Low-cost and low-inductance requirements in the semiconductor industry were the driving force behind the enormous development of the high-performance leadless molded plastic packages. These packages are surface mount leadless packages (SMLPs) also known as power quad flat no lead (PQFN) or simply QFN. These packages typically have leads on all four sides, but newer very small outlines are two sided. PQFN packages come in a variety of sizes and lead frame configurations. Pitch, the distance between the leads, varies from 0.3 to 1 mm. The bonding pad size is 12 mils by 12 mils. Figure 21.7 shows examples of plastic IC packages including SOIC and PQFN.

Figure 21.7 Examples of plastic IC packages including SOIC and PQFN.

Table 21.2 Salient Features for Different Materials Used in Single-Chip RF Packages

Metal Wall	Ceramic Wall	Glass Ceramic	Plastic
Broadband	Broadband	Low loss	Low cost
Low loss	Lower cost than metal wall	Suitable for single-chip and MCM	Suitable for high-volume applications
Antiresonat structure	Low loss	Surface mount	Surface mount
Excellent shielding	Good shielding	Suitable for high-volume applications	
Excellent isolation	Good isolation	Excellent design flexibility	Adequate isolation

21.1.2 Types of Packages

Single-chip packages can be grouped based on package material. These are metal wall, ceramic wall, glass ceramic, and plastic packages. Amplifiers are packaged in all of these materials depending on frequency, dissipated power, cost, and so on. Their major characteristics are compared in Table 21.2. For amplifier applications, basic package types are ceramic and plastic.

Ceramic Packages

There are many types of ceramic packages offered by manufacturers in either off-the-shelf or custom outlines. A ceramic package uses a ceramic material as the base material between leads for high isolation and low loss. In ceramic packages, the amplifier die is usually mounted such that the IC is in an air cavity with a metal or ceramic lid on top. Also, the power amplifier die is soldered to a metal base for the best heat transfer from the package. In power packages, the metal base or flange is then directly attached to a heat sink. Ceramic packages can be manufactured in such a way that the product will be either hermetic or nonhermetic, depending on the environmental requirements. Hermetic seal can be one of the key advantages of

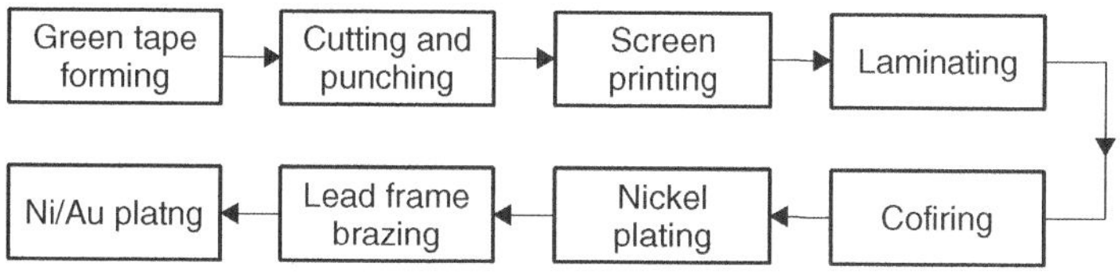

Figure 21.8 Flow process for manufacturing ceramic packages.

Figure 21.9 Air cavity ceramic packages showing open, lid, and backside package views.

the ceramic based package, especially for challenging environmental requirements. Hermetic seal adds considerable cost to the amplifier product.

There are several methods being used for manufacturing ceramic packages. These methods are similar to hybrid circuit fabrication techniques including thin film, thick film, LTCC, and HTCC. Figure 21.8 shows a typical flow process for the manufacturing of ceramic packages. Major suppliers of ceramic packages include Kyocera and NTK. Ceramic packages are shipped in plastic waffle packages. Figure 21.9 shows an air cavity surface mount low-profile ceramic package. The ceramic packages are surface mounted on PCB or soldered into modules.

Plastic Packages

In contrast to a ceramic package, the amplifier die in a plastic package is encapsulated with a plastic molding compound so that no air cavity or lid is involved. The molding compound can affect the amplifier frequency response, especially for frequencies above 3 GHz. High-frequency designs should be simulated with the loading of the plastic compound on the matching networks. Leads in the plastic package have lower isolation (40 dB versus 60 dB) because of the material properties of the plastic. In addition, plastic packages are nonhermetic and are sensitive to moisture. Packages are rated (moisture sensitivity levels) by the amount of moisture the plastic allows to penetrate the die and possibly cause moisture induced stress. A moisture sensitivity level (MSL) of 1, for example, means that the plastic part does not need to be baked prior to a solder reflow operation. An MSL of 2 or higher means that some precaution needs to be taken to protect the plastic amplifier product against moisture damage.

Even though plastic packages have reduced isolation and hermetic performance, they are superior in price to ceramic devices for high-volume applications. Because plastic injection molds can handle large batches of material, the plastic package amplifier product will be lower in cost. Also, innovations in package and molding compound design have allowed plastic packages to be used for high-gain and high-frequency amplifier products.

In the plastic package, the lead frame (LF) is the central supporting structure to which the die is attached. The lead frame is stamped from a thin sheet of metal. The metal sheet is usually of Kovar for low-power application and copper/copper alloy for power packages. The sheet thickness is about 8–10 mils. The LF carries the die throughout the assembly process. Plastic packages are mostly modified versions of standard lead frame designs including the molding material. Diced wafers are supplied to the plastic packaging manufacturer and the manufacturer performs all the steps including pick-and-place, epoxy dispense, wire bonding, molding, marking, and sawing packages. For power amplifier products, thermally as well as electrically conductive epoxy is required. An example of thermally conductive epoxy is Abeltherm's 2600 series with a conductivity of 20 W/m·K (AuSn solder $\sim$ 60 W/m·K). Plastic packages are shipped in plastic tubes, cans, and tapes, and reels. Major suppliers of plastic packaging include Amkor, Carsem, Unisem, Asat, and Azimuth. The plastic packages are usually surface mounted on PCB in their final configuration.

21.2 MATERIALS FOR PACKAGES

Next, we discuss several types of materials used in the packages.

21.2.1 Ceramics

Ceramic package substrate materials can be composed of alumina (Al_2O_3), glass (SiO), glass alumina (LTCC and HTCC), aluminum nitride (AlN), boron nitride (BN), beryllium oxide (BeO), or silicon carbide (SiC). Among these, alumina is the most popular. Ceramic materials have very high melting points and chemical stability because of their strong bonds. Ceramic materials have an excellent combination of electrical, mechanical, thermal, and dimensional properties. The physical properties of these materials are given in Chapter 14.

21.2.2 Polymers

Organic polymers are used in the molding of plastic packages. Some of these materials are also described in Chapter 14. Commonly used polymers are thermoplastic and thermoset. Thermoplastic polymers have high molecule weight and can be melted and reformed any number of times. Below their glass transition temperatures, they are stiff. Commonly used thermoplastic polymers are polystyrene, fluorocarbon, polyethylene, and silicon polyimide. The dielectric constant for these materials ranges from 2 to 3.

Thermoset polymers have higher glass transition temperatures and are more rigid than the thermoplastic materials. Commonly used thermoset materials are benzocyclobutene (BCB), polyimide, epoxies, polyester, polyurethane, alkyds, and phenolics. The dielectric constants for these materials range from 2.5 to 10 and are much larger than for thermoplastic polymers.

21.2.3 Metals

Kovar is the most popular lead frame or pin material for leaded ceramic packages. Kovar is an alloy with a composition of 53% Fe–17% Co–20% Ni. It has good

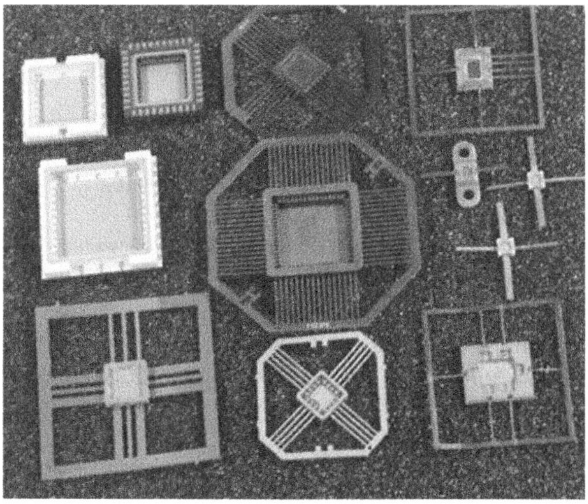

Figure 21.10 Several MMIC ceramic packages.

thermal expansion match with alumina, Si, GaAs, and sealing glass but poor thermal conductivity. Due to the poor thermal conductivity, the use of Kovar is limited to package leads and to low-power applications as a package base/flange. The base or flange material for high-power applications is generally a composite metal such as CuW or CuMoCu. The lead frame (LF) material most widely used for low-power plastic packages is Alloy 42 (42% Ni–58% Fe). However, for medium-power applications copper clad or copper alloy is used as the LF material.

21.3 CERAMIC PACKAGE DESIGN

Many of the packaging considerations for MMICs are similar to those for hybrid MICs. Most ceramic/metal packages should meet the environmental requirements of MIL-S-19500 and test requirements of MIL-STD-750/883. The package must pass rigorous tests of hermetic properties; thermal and mechanical shock; moisture resistance; resistance to salt atmosphere, vibration, and acceleration; and solderability. In order to minimize the effect of the package on MMIC performance, electrical, mechanical, and thermal modeling of packages must be performed. Several MMIC ceramic package styles are shown in Figure 21.10.

The most important electrical characteristics of microwave packages are low insertion loss, high return loss and isolation, and no cavity or feed through resonance over the operating frequency range. When a chip or chip set is placed in the cavity of a microwave package, there should be minimum degradation in the chip's performance. Generally, this cannot be accomplished without accurate electrical and electromagnetic modeling of the critical package elements. Microwave design must be applied to three parts of the package: RF feed through, cavity, and DC bias lines. Of the three, the design of the RF feed through is the most critical in determining the performance of packaged MMIC chips. Salient features of MMIC package design are discussed next.

21.3.1 Design of RF Feedthrough

The basic requirement for a feed through is to match the MMIC to a 50-Ω transmission line because no tuning of the circuit is desirable after packaging [16, 17]. The design of the feed through should be such that when MMIC chips are packaged, their performance is unaltered. The feed through should satisfy the following requirements:

- Low insertion loss per feed through (<0.01 dB/GHz)
- High return loss per feed through (>20 dB) over the operating frequency band.

In order to meet the above-mentioned requirements, the feed through structure should have the following properties:

1. Planar, such as microstrip or stripline.
2. Good match for the electric and magnetic fields at and between all interfaces.
3. Minimum number of transitions in transmission media or line widths as well as in dielectric interfaces in order to have minimum discontinuity effects in the signal path.
4. Minimum number of discontinuities in the ground path.
5. Minimum feed through length but long enough to reduce interaction between the discontinuities.
6. Minimum ground inductance.

Figure 21.11 shows a ceramic package with a feed through. Two commonly used conductor configurations for feed throughs are shown in Figure 21.12. The step-in-width and tapered conductors are used to minimize the effect of capacitive loading due to the ceramic/glass wall.

It is desirable to make the line lengths, connecting the chips to the leads, as small as possible to avoid large phase angles in the input and output reflection coefficients

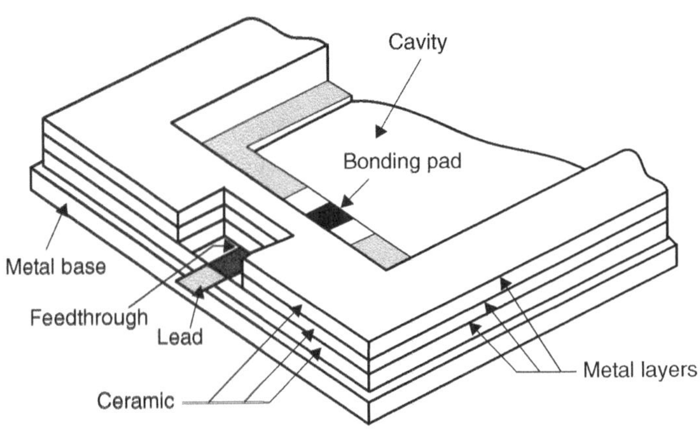

Figure 21.11 A ceramic package with feed through.

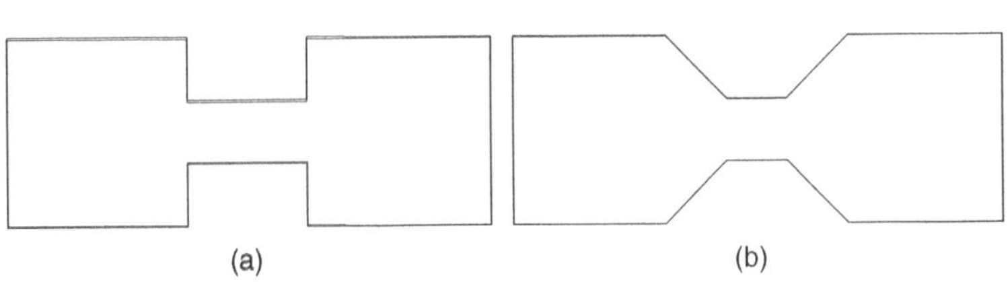

Figure 21.12 Two commonly used conductor configurations under the ceramic/glass wall.

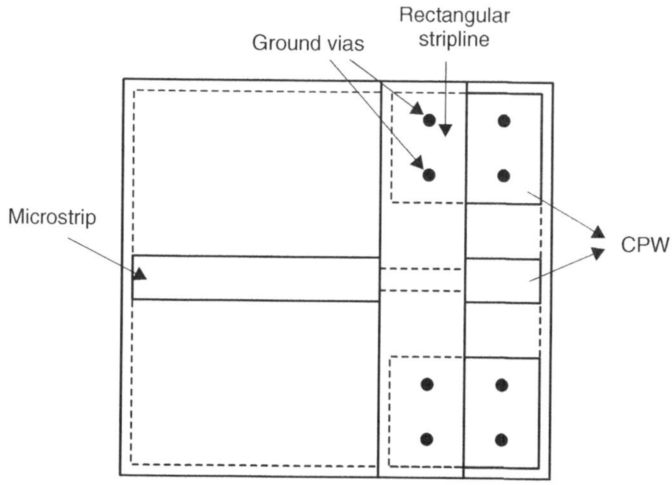

Figure 21.13 Feed through using microstrip–rectangular stripline–CPW media.

as the operating frequency is increased. Such large phase angle differences lead to an inability to match the packaged device over wide bandwidths, which can cause significant degradation in the circuit performance. The best solution to this problem is to maintain a 50-Ω characteristic impedance of the feed structure from the chip position all the way up to the lead position with minimum discontinuities.

The feed structure shown in Figure 21.13 consists of a microstrip section, a rectangular stripline section (due to the package wall), and a CPW section. The microstrip section is inside the package cavity. These transmission media have discontinuities at the interfaces due to changes in dielectric media as well as in the transmission mode. These transitions degrade package performance even though the feed through may have a 50-Ω characteristic impedance throughout. These effects must be accurately measured, modeled, and corrected. The use of a rectangular stripline in the package wall gives rise to slightly higher insertion loss (about 10% more), although the overall performance improves at higher frequencies due to improved RF grounding. Figure 21.14 shows a typical measured performance of a ceramic feed through. Up to 15 GHz, the measured insertion loss is less than 0.3 dB and the return loss is better than 20 dB.

21.3.2 Cavity Design

The basic requirements for a cavity are to house the MMIC chip without degrading its performance and accommodate the I/Os and DC bias lines as well as other chip compnents such as capacitors, inductors, and resistors as required in the package. The design of the cavity should satisfy the following requirements:

- Metallic wall ring
- Nonmetalic or grounded lid
- No propagation of waveguide modes
- No cavity resonances

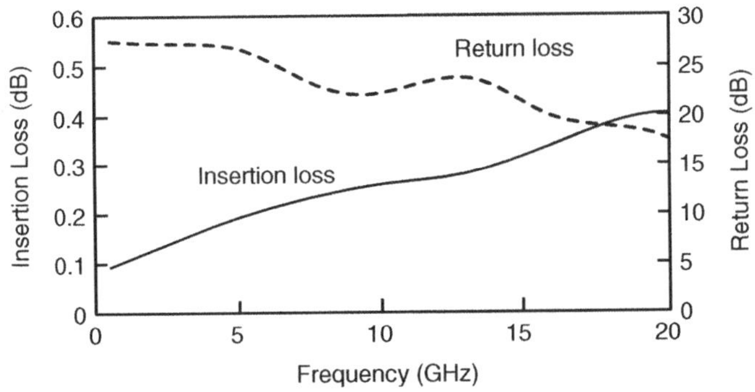

Figure 21.14 Typical measured performance of a ceramic feed through.

A grounded lid and/or grounded metal wall ring helps in improving isolation between the RF ports and RF and DC leads. This also minimizes the interaction between the MMIC and package environments. The cavity of the MMIC package must be designed so that the cutoff frequency for the dominant TE_{10} waveguide mode is beyond the maximum operating frequency. If this condition is not met, there is a high probability that the energy launched to the package, which is carried by the TEM mode, may couple to the TE_{10} mode or any other mode supported by the cavity. At higher frequencies this mode conversion may influence the insertion loss in the package more than the dissipated loss or the impedance mismatch loss in the feed through. A cavity size that does not support waveguide modes also improves lead-to-lead isolation and prevents spurious mode propagation.

The package wall height is selected to provide adequate thermal impedance and to facilitate the lid sealing operation, while not affecting the IC performance. It is generally less than the cavity width. The package wall ring is brazed to the carrier, which is grounded to the package body. Finally, the lid is soldered or epoxied to the wall ring.

Waveguide Modes

As shown in Figure 21.15, the package cavity can be treated as an H-plane dielectric slab loaded structure, where the electromagnetic wave is propagating in the z-direction. The dominant waveguide mode in the cavity can be either quasi-TE_{10} mode or hybrid LSM_{11} (longitudinal-section magnetic) mode, depending on the ε_r value and guide dimensions. The dielectric-filled waveguide cutoff frequency (f_{cd}) normalized with respect to the TE_{10} mode cutoff frequency ($f_c = c/2W$) for an empty rectangular waveguide is given by

$$\frac{f_{cd}}{f_c} = \sqrt{1 - \frac{h}{H}\frac{\varepsilon_r - 1}{\varepsilon_r}} \tag{21.1}$$

where h, H, and ε_r are the substrate thickness, cover height, and substrate dielectric constant, respectively.

In order to suppress these modes, W/H should be less than 0.5 and h/H should be less than 0.1. Consider a ceramic package design working up to 18 GHz. For $\varepsilon_r = 9.7$ and $h/H = 0.1$, f_{cd}/f_c from (21.1) for the lowest mode (LSM_{11}) is 0.955. For a cutoff

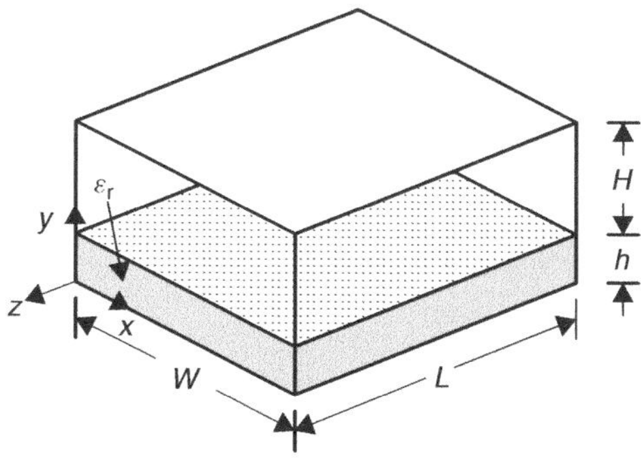

Figure 21.15 Dielectric loaded cavity configuration.

frequency f_{cd} of 18 GHz, this gives $f_c = 18.85$ GHz, or $W(= c/2f_c)$ should be less than 8 mm.

Resonant Frequency

For waveguide dimensions $H < W < L$, the lowest-order resonant mode is LSM_{111} or TE_{101}. The resonant frequency for $H < W < L$ is given by

$$f_0 = \frac{15}{W}\left(\sqrt{1 - \frac{h}{H}\frac{\varepsilon_r - 1}{\varepsilon_r}}\right)\sqrt{1 + \left(\frac{W}{L}\right)^2} \tag{21.2}$$

where W is in centimeters and f_0 is in gigahertz. Thus the lowest resonant frequency occurs when L is very large as compared to W and is given by the cutoff frequency for the dominant mode. Resonance can also occur when the largest dimension equals half the guide wavelength. For $W/L = 0.5$, $W = 0.5$ cm, $\varepsilon_r = 12.9$, $h/H = 0.1$, the lowest resonant frequency $f_0 = 31.96$ GHz.

21.3.3 Bias Lines

Bias lines and input/outputs must be designed in a manner such that they do not degrade the insertion loss and return loss performance due to coupling, crosstalk, or resonances. Unused pads, stubs, or bias lines must be grounded in order to increase the resonant frequency and maximize the isolation between the input and output ports. Unterminated or unisolated bias leads may lower the resonant frequency by a factor of 2. Ground connections between the bias lines may help to increase the resonant frequency and increase isolation. In the design of bias lines, RF ground and DC ground must be sufficiently separated to prevent undesired coupling, especially for very-high-gain applications.

21.3.4 Ceramic Package Construction

The selection of the substrate material and thickness for ceramic packages depends on the electrical performance requirements, cost, and frequency range of interest. The substrate thickness is selected to match its height with MMIC thickness; otherwise, a pedestal for mounting MMIC chips is required because these chips are typically about

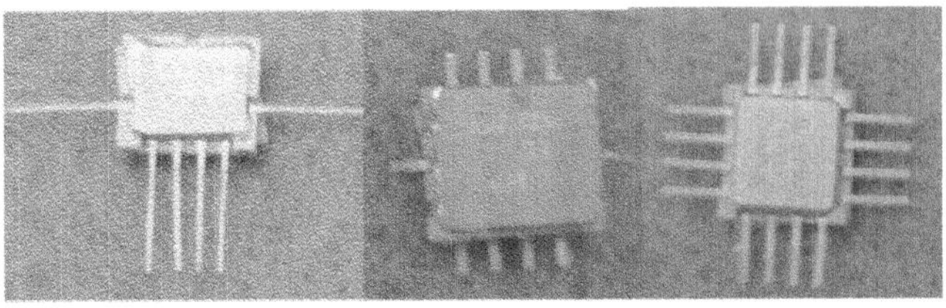

(a) (b) (c)

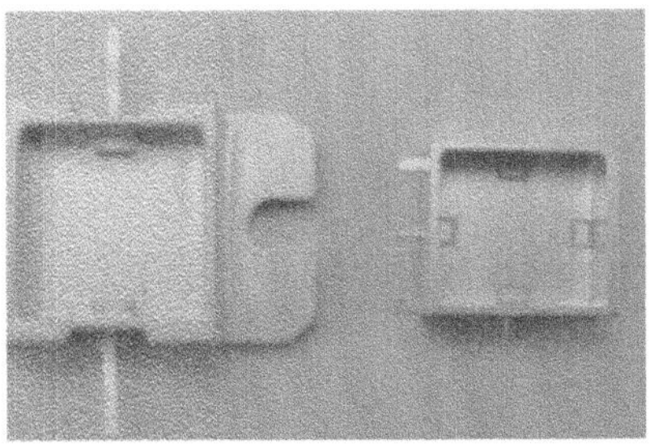

(d) (e)

Figure 21.16 Examples of leaded low-cost ceramic packages: (a) 6-lead, (b) 10-lead, (c) 16-lead, (d) transistor, and (e) MMIC.

3–5 mils thick. Microwave packages generally use 10–20-mil thick alumina substrates, whereas millimeter-wave packages use 4–5-mil thick quartz. A low dielectric constant is generally preferred because it makes the package interconnects electrically insensitive and tolerant to microstrip dimensions and broadband frequency ranges and results in a high yield. The microstrip width and thickness determine the characteristic impedance and the DC resistance, whereas the spacing between the two conductors on the same plane controls the crosstalk because of coupling. Generally, sufficient space between the MMIC, the package walls, and the lid is provided in order to minimize any interactions. The effect of the package lid on the MMIC characteristics is kept to a minimum by keeping the lid above the MMIC surface about five times the package substrate thickness. In the ceramic package design, the effect of the type of lid becomes critical in terms of amplifier stability. In high-gain applications, the use of a ceramic or plastic lid is adequate, whereas the use of a metal lid might need some absorber material to minimize the feedback between the output and input leads. Figure 21.16 shows several types of ceramic packages.

Many styles of ceramic (alumina, Al_2O_3), beryllium oxide (BeO), and aluminum nitride (AlN) packages with metal bases (copper, copper–tungsten, or copper molybdenum) are available for MMIC power amplifiers. Their cost depends on package size, frequency of operation, metal used, and volume. Some of these packages can be used up to 40 GHz. Packages working up to 20 GHz can be obtained for less than \$3 in large volume. In small quantities, they cost between \$20 and \$50, not including nonrecurring engineering (NRE) tooling cost. Typically, the measured dissipative loss per RF feed is less than 0.3 dB at 20 GHz. These packages provide much higher frequency of operation, low lead frame inductance, very low ground connection inductance, and much lower thermal resistance than the plastic packages. Ceramic-type packages are well suited for high-frequency, small-signal, and medium-power MMICs and high-power

Figure 21.17 Test fixture for ceramic packaged driver amplifiers. Bypass capacitors are 0.1-μF ceramic chips.

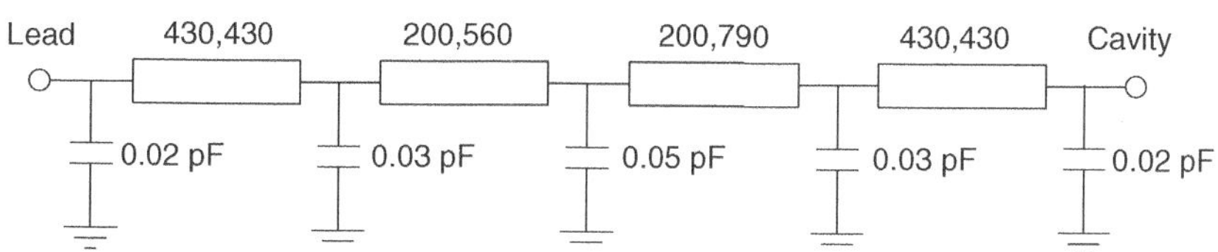

Figure 21.18 An equivalent circuit model for a ceramic package feed through.

transistors. Figure 21.17 shows the test board for a commonly employed ceramic package used for driver amplifiers.

21.3.5 Ceramic Package Model

The equivalent circuit (EC) model for M/A COM's standard ceramic package feed through was developed by using *S*-parameter measurements up to 20 GHz. The measurements were performed with the help of RF probes and suitable TRL deembedding standards designed for the ceramic package. Figure 21.18 shows the distributed EC model for the feed through. The microstrip dimensions (width and length) are in micrometers (μm). The substrate ($\varepsilon_r = 9.5$) is 10 mils thick. Shunt capacitors account for the additional step-in width and different transmission media discontinuities. No interaction effects between leads located on opposite sides were included.

21.4 PLASTIC PACKAGE DESIGN

The design of a plastic package basically consists of lead frame and plastic polymer, and both are available as standard items from plastic packaging manufacturers. Several different types of lead frame materials such as nickel–iron and copper-based alloys are being used. Their selection for a particular application depends on factors such as cost, performance, and ease of fabrication. The lead frame can be modified by paying an additional tooling cost to the manufacturer. Basic design of a lead frame includes its material, package thickness, cavity size, number of total leads, and lead-to-lead pitch. For a given package size the only package design variable is the number of leads.

The lead frame (LF) is the backbone of a plastic package and its mechanical design depends on the application at hand. The desirable features for the LF materials are good strength, good thermal expansion match with Si and GaAs materials, and high thermal conductivity. The LF material may be grouped into three categories: nickel–iron, copper clad strip, and copper-based alloys. Kovar and Alloy 42 (42% Ni–58% Fe) are the most widely used LF materials for low-power applications due to their low thermal conductivity. Copper-based LF materials have very high thermal conductivity and are used for power amplifiers. Copper-clad LF materials were developed to match mechanical properties of Alloy 42 while retaining copper's high thermal conductivity. Cladded material is fabricated by rolling copper foil onto stainless steel. Copper alloys are obtained by mixing other metals into copper to obtain alloy properties suitable for plastic packages. The frames are either chemically etched or mechanically stamped from metal sheets. Typical sheet thickness is from 8 to 10 mils. The portion of the lead frame that is to be wire bonded is silver plated. The dielectric constant and loss tangent values of the organic molding compound are about 3.7 and 0.01, respectively.

21.4.1 Plastic Packages

There are several versions of high-performance PQFN packages available. Some of them are shown in Figure 21.19a and listed in Table 21.3 with preferred GaAs or Si die size for each package. The cavity size is slightly larger than the die size. Large cavity size is frequently used for multistage medium-power amplifiers.

Small outline transistor (SOT), small outline integrated circuit (SOIC), and plastic quad flatpack no leads (PQFN) plastic packages are commonly used. The SOIC packages work reasonably well up to 2 GHz. The measured dissipative loss in an SOIC 8-lead package is on the order of 0.2 dB at 2 GHz. Plastic packages such as 4-mm PQFN-16LD can be used up to 18 GHz, and the measured loss in a PQFN package is on the order of 0.2 dB at 18 GHz.

Packaged amplifiers are generally tested by mounting them on a printed circuit board (PCB). The materials for the PCB are described in Chapter 14. Figure 21.19b shows a PCB used for the testing of PQFN packages.

21.4.2 Plastic Package Model

The equivalent circuit (EC) model for PQFN packages were developed by characterizing them using an EM simulator. Figure 21.20 shows an EC model for the 4-mm PQFN 16-lead package. Here, four input ports represent four leads on one side of the lead frame. The EC model also includes coupling/crosstalk between the leads. No interaction effects between leads located on different sides were included.

In the case of plastic packaging, capacitive loading due to the molding material occurs and affects the transistor and transmission line characteristics. The effects are more pronounced at higher frequencies and must be considered in the amplifier design

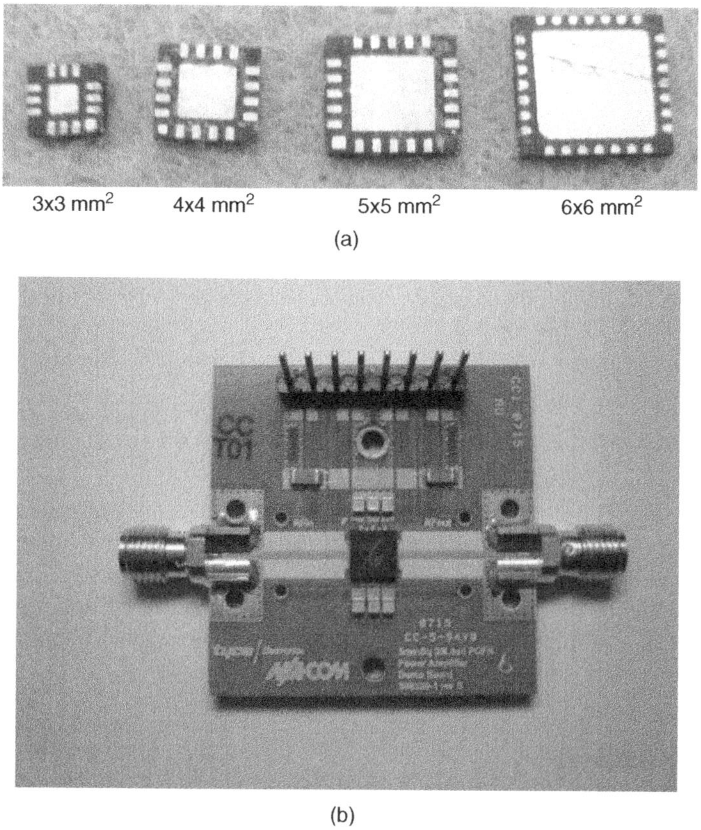

Figure 21.19 (a) High-performance PQFN plastic packages. (b) PCB used for testing of PQFN packages.

Table 21.3 Examples of PQFN Packages with Maximum Die Size

Package Style	Max Die Size (mm^2)
4-mm PQFN-20LD	2.15 × 2.15
4-mm PQFN-24LD	2.45 × 2.45
5-mm PQFN-20LD	3.15 × 3.15
5-mm PQFN-28LD	3.15 × 3.15
6-mm PQFN-28LD	4.45 × 4.45

phase. For transistors covered with the molding material, device models are developed using the measured data, while for transmission lines (microstrip and CPW) covered with dielectric overlay the device models are EM simulated.

21.5 PACKAGE ASSEMBLY

After selecting a suitable package, the next critical step is die attach and wire bonding [6, 9]. In an amplifier assembly, the first step is to attach chip components onto carriers, pedestals, package cavities, and substrates or substrates onto carriers, pedestals, package

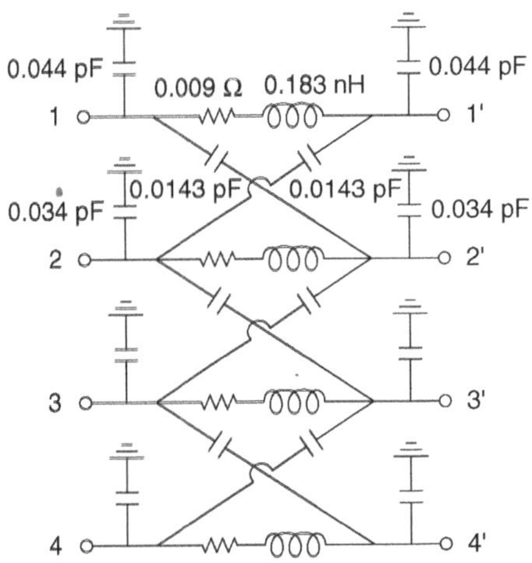

Figure 21.20 An equivalent circuit model for a 4-mm PQFN 16 lead package.

cavities, and so on. The chip attachment technique is called the *die attach process*. The important considerations for die attach are to have low thermal resistance and strong mechanical bond. In the case of hybrid assemblies, both die and surface mount packaged die including semiconductor chips, capacitors, inductors, and resistors are used. The die form minimizes size, weight, effect of parasitic reactance, and die-to-die propagation delays.

21.5.1 Die Attach

There are two methods used for die attachment: epoxy die attach and eutectic die attach.

Epoxy Die Attach

The epoxy die attach process is commonly used for mounting passive components and low- and medium-power devices. The epoxies are cured at relatively low temperatures, are easy to work, are applied using automatic dispensers, and are military and space qualified. Epoxy is available in two types: silver (Ag) epoxy and gold (Au) epoxy. Ag epoxy is commonly used as it is less expensive than Au epoxy. Epoxy die attach is a simple process consisting of epoxy preparation, applying controlled epoxy quantity under the chip area, and placing and pushing the chip with tweezers (in the case of a manual process) in the middle of the epoxy. All these steps are done at room temperature. Next, the assembly is placed in a curing oven. The oven temperature and the curing time are recommended by the epoxy manufacturer and usually are 150 °C and 60 minutes. Since the thermal resistance of epoxy is higher than eutectic die attach materials (see Chapter 16), the epoxy die attach technique is normally used for low-power applications or where dissipated power is lower. For example, the epoxy process is used exclusively for all low-noise amplifiers and plastic package assemblies.

Although some military and space systems do not allow the use of conductive epoxies for many amplifiers, they can be satisfactory for all but the highest power applications if kept thin. If the bonding is well done, the chip should break during the die shear testing before it is removed from its mounting. Weak bonds are usually caused by insufficient cleanliness, inadequate epoxy curing, or inadequate heating. Ag epoxy should not be used where it will come into contact with lead–tin solders or high-tin solders. There can be an anodic reaction that can cause failure of the die bond.

Eutectic Die Attach

Eutectic die attach is performed using a heated stage. The commonly used solder material is gold–tin (AuSn) in the ratio 80:20 with chips that have backside gold metallization. The metallization of 1-μm gold is good enough for eutectic soldering. For higher temperature operation, a gold–germanium (AuGe) eutectic solder can also be used. The chips are thoroughly cleaned in a trichloroethylene solution to remove all grease and dirt and dried with methanol.

Eutectic solder material is available in preform and usually its size is slightly smaller than the chip size. The work stage or chuck is preheated. The temperature of the heating stage should be set such that the temperature of the bonding area quickly rises to within 50–75 °C of the melting point of the solder preform. A 1-mil thick preform is generally used. At the final solder step, a jet flow of heated forming gas or nitrogen, which has a gas temperature of about 100 °C above the solder melting point, is used. The solder should melt in a few seconds after applying the jet of air. A preform is placed where the die is to be soldered and it promptly melts. The die is placed on the melted preform with care and scrubbed back and forth. The carrier is removed from the heated stage and allowed to cool. The solder time is generally less than 5 minutes. If the chuck temperature or solder time or both are substantially increased, the die attach process might degrade the semiconductor chip performance.

There are several methods used to ensure good solder die attach including die shear, X-ray, and sonoscan. Among these, the nondestructive techniques such as the sonoscan are preferred to determine good versus bad solder die attach. This technique is widely used for high-power amplifiers.

21.5.2 Die Wire Bonding

After the die attach process, discrete lumped elements and semiconductor devices (transistors or MMICs or both) are interconnected with each other or connected to package substrate pads or to leads using wire bond, flip-chip, and tape-automated techniques, as illustrated in Figure 21.21. These are briefly discussed next.

Wire Bonding

The process of connecting a die using wires is known as wire bonding. In wire bonding, two similar metals are "bonded" together under the influence of pressure and temperature at well below their melting points. Both the wire and pad are gold. If the wire is

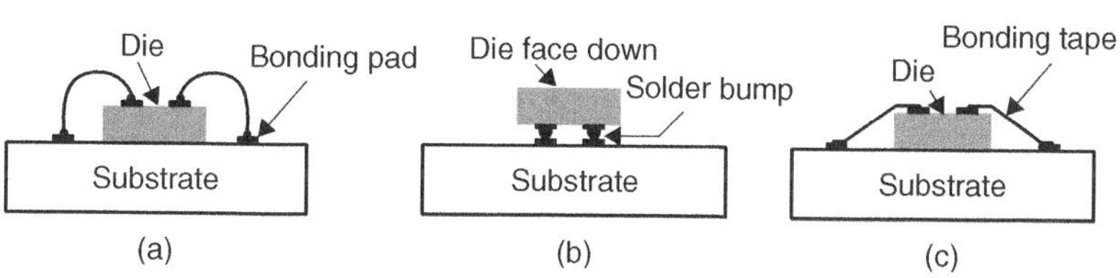

Figure 21.21 Die-to-substrate pad bonding techniques: (a) wire, (b) flip-chip, and (c) tape-automated bonding.

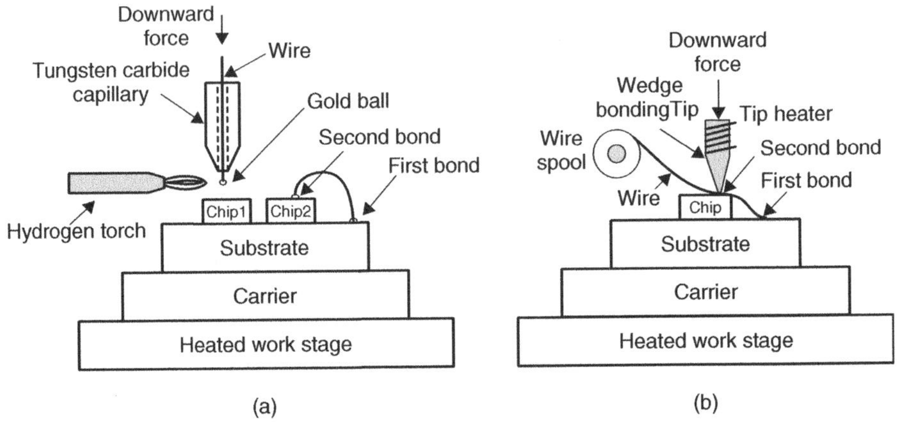

Figure 21.22 Wire bonding techniques: (a) ball and (b) wedge.

made from gold, the pad to be attached has to be made of gold. The wires used are in the rang of 0.5–2.0 mil diameter. This technique is also known as thermal compression bonding. Two methods for thermal compression wire bonding are used: ball bonding and wedge bonding. Wedge bonding is also performed by using ultrasonic techniques. Ball and wedge wire bonding techniques are illustrated in Figure 21.22.

Ball Bonding In the ball bonding technique, a small flame from a hydrogen torch or an electric arc melts the very end of the wire to form a small ball. Using a tungsten carbide tip (called capillary), the ball is pushed onto the bonding pad. At the same time the capillary is instantaneously electrically heated and pressed against the ball contact, making a mushroom shaped ball. This technique does not require a heated work station. However, the bonding pad size is larger and the bond wire length is longer due to the longer loop.

Wedge The wedge bonding technique uses a narrow and pointed tungsten carbide bonding tool to press a gold wire into the gold bonding pad. The tip of the bond tool is 2×2 mils^2 in area. Both the bonding wedge tip and work stage are heated and a precisely controlled downward force is used to facilitate the bonding of the wire onto the pad. Figure 21.22b shows the graphical steps for the bonding process. A combination of tool force, tip temperature, and work stage temperature, known as the bonding schedule, allows reliable wire bonding. Since the wedge bonds are narrow they do not need a pad size very much larger than the wire diameter. Also, wires are placed in the plane of the pad; wire bond lengths are short and have low inductance. At millimeter-wave frequencies, wedge bonding is preferred to ball bonding due to the low pad and bond wire parasitic reactance effects. Both wire and ribbons are used in the bonding process. Ribbons are mostly used for low-inductance and high-current applications. The ribbon thickness and width are in the range of 0.5–1.0 mil and 5–10 mil, respectively.

It is difficult to give definite parameters for the force, time, and temperature for an optimum bonding schedule. Different wire, bonding surface, or semiconductor die characteristics require different bonding conditions. In general, the bonding parameters should be adjusted to maximize reproducibility at a high bond test.

For DC and RF connections, either ball or wedge bonds can be used. For the best microwave performance above 10 GHz, wedge bonds of shortest length employed on the RF interconnects is preferred over ball bonds. The failure of transistors and monolithic and discrete capacitors due to voltage stress is often observed during assembly. Bonding equipment should be checked regularly for sources of surge voltage and should be properly grounded at all times. All test handling equipment should be grounded to minimize the possibility of static discharge.

Ultrasonic Wedge Bonding In this technique, the bond wire is pressed against the bonding pad and both are subjected to an ultrasonic vibration, enabling the formation of the desired bond between them by diffusion. In a wedge bonder, ultrasonic vibration energy deforms the wire and induces diffusion to make a strong bond between the wire and pad. This technique is widely used where low process temperatures and connections between two dissimilar metal compositions are required. It uses both aluminum and gold wires. Using this technique, one can make wire bond between the same metal and different metal types. In plastic packages, gold wires are attached to silver-plated copper alloy leads.

Flip-Chip Bonding

In flip-chip bonding (FCB), the die is mounted upside down onto a package, carrier, module, or PCB. The die has via solder bumps at predetermined locations and bonded to the pads. Since this technique does not use wire bonds for connections, it has very small parasitic reactances. The FCB technology has high-performance characteristics and is suitable for multichip packaging technology. The downside of FCB for amplifiers is that the heat flow path must now be through the solder bumps, which can be a challenge. A die attached to a metal slug or heat spreader will have the lowest thermal resistance.

Tape-Automated Bonding

Tape-automated bonding (TAB) is used primarily for large silicone ICs having large I/O requirements. In this technique, a TAB tape having repeated wide copper interconnection patterns formed lithographically [6] from a metal laminate is used. The IC pads are aligned to the interconnection stripe and joined by thermocompression bonding. All bonds are made simultaneously and it is called gang bonding. Since this technique is suitable for large dies having a large number of pads, it is not used for RF and microwave dies having a small number of pad connections.

Wire Bond Model

The electrical models for a single wire and multiple wires have been described in Chapter 6. As a first-order approximation, the lead frame parasitic reactance may be absorbed in the wire bond inductance to realize the transmission line characteristic impedance of 50 Ω.

21.5.3 Assembly of Ceramic Packages

As an example, a ceramic package flow diagram is illustrated in Figure 21.23. The first step is to apply silver epoxy (low-power devices) or place solder preform in the cavity

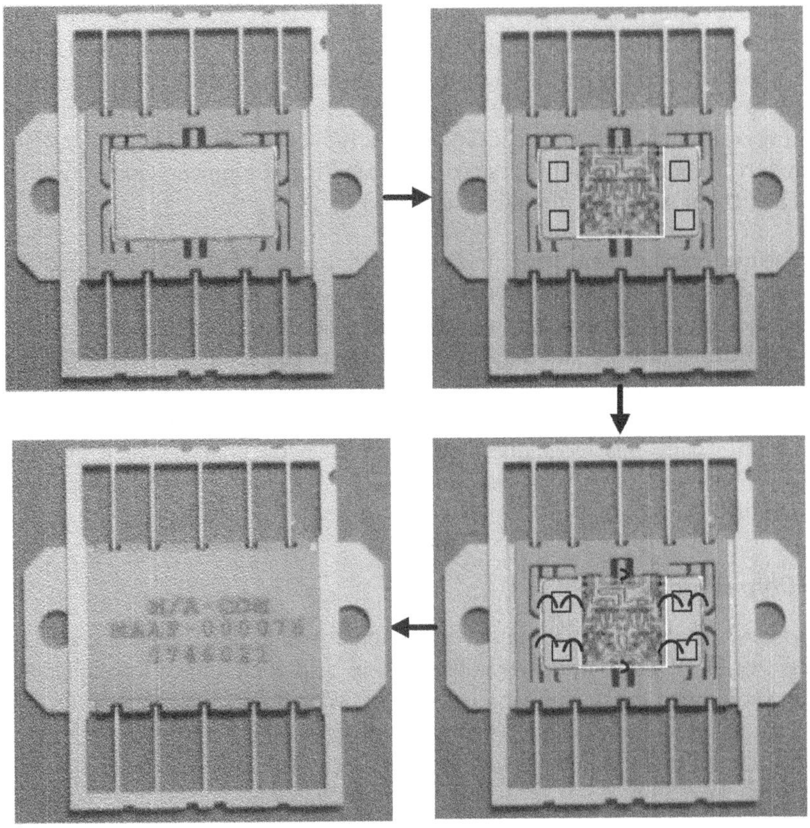

Figure 21.23 Flow diagram for the MMIC housed in a ceramic package.

of the package (power devices). In the latter case, the package is placed on a hot plate. The next step is to place the semiconductor die that is to be packaged. In moderate- and high-volume applications, this step is usually done by an automatic pick-and-place machine. After this, chip capacitors with values of 100–200 pF are attached with silver epoxy. This is followed by wire bonding. Next, the lid is attached to the package wall. Finally, the product is marked and the lead frame is trimmed or cut off. The marking is usually done by laser scribing.

21.5.4 Assembly of Plastic Packages

As an example, a plastic packaging flow diagram is illustrated in Figure 21.24. The ICs are packaged using pick-and-place techniques. The first step is to apply silver epoxy. The second step is to place the semiconductor die that is to be packaged. This is followed by wire bonding and molding. Steps such as pick-and-place, epoxy dispensing, wire bonding, and molding are performed by using automatic machines and robots for high reproducibility of performance and low cost. The next step is to mark the product and finally to saw or punchout the lead frame to separate each plastic package. The marking is done using either laser scribing or inking. Figure 21.25 shows a medium-power GaAs MMIC amplified attached to a lead frame.

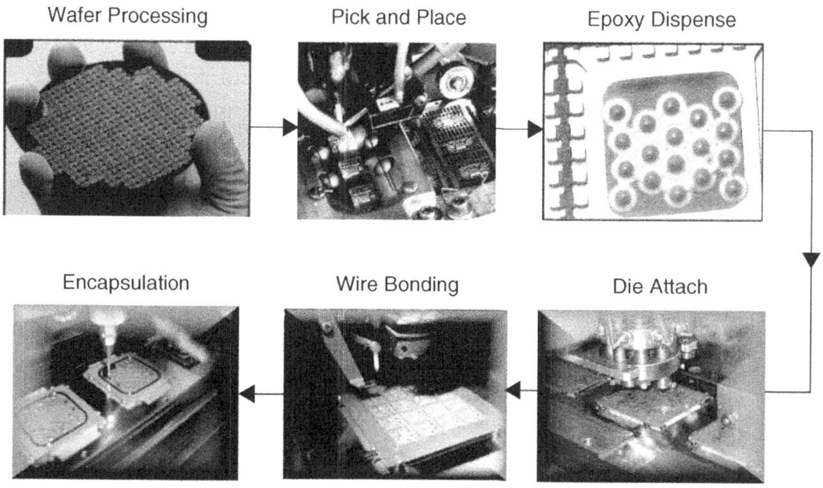

Figure 21.24 Flow diagram for the MMIC housed in a plastic package.

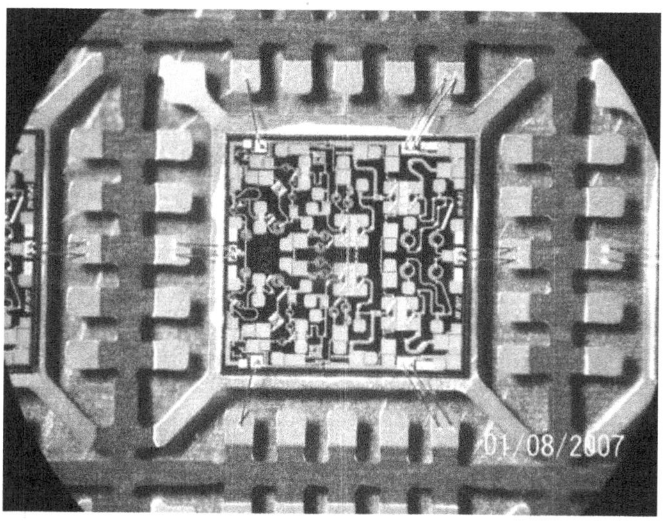

Figure 21.25 Illustration of a GaAs medium-power amplifier die bonded to the lead frame.

21.5.5 Hermetic Sealing and Encapsulation

Many commercial and military applications require hermetically sealed packages. A ceramic package and a metal housing are hermetically sealed to protect the unit from moisture and environmental hazards. In this technique, after mounting the die and wire bonding, a lid, made from ceramic, glass, or metal, is attached to the top of the package housing. The lid is coated with solder material along its boarder and then thermally attached. Lids are also attached using brazing, glass sealing, and welding. Only ceramic/glass packages and metal housings are considered practical for hermetic seals.

In some applications the package encapsulation is achieved by glop-top, molding, and cavity fill techniques. In these methods, the die and wires are covered with a polymer material. Package encapsulation is less complex and provides limited protection from environmental hazards. Due to the improved chemical purity of encapsulation materials, the reliability of plastic encapsulated circuits has been enhanced over the last decade [6].

21.6 THERMAL CONSIDERATIONS

Thermal modeling of packages must be performed in addition to mechanical and electrical designs. Thermal modeling of packages becomes very important when the packages are used with power FETs and power ICs. Since thermal effects are frequency independent, thermal modeling techniques used for low-frequency packages can also be used for microwave packages. The basics of the thermal design of amplifiers are treated in Chapter 16.

The thermal design of the layer stack from the amplifier die to the heat sink is critical in high-power amplifier products. By properly designing the boundaries between the GaAs chip and the heat sink for good thermal path, heat generated in the active devices can be removed efficiently. A thinner GaAs substrate, a void-free and reliable die attach, high thermal conductivity of the base plate, and a coefficient of thermal expansion similar to that of GaAs and alumina are the bases for good thermal design. CuW material has an excellent thermal conductivity (248 W/m·°C) and its thermal expansion coefficient ($5.7 \times 10^{-6} \Delta\ell/\ell$) is very close to GaAs ($5.6 \times 10^{-6} \Delta\ell/\ell$) and alumina ($7.1 \times 10^{-6} \Delta\ell/\ell$). Finally, the MIC and MMIC packaged amplifier products must be correctly soldered or epoxied to the high-conductivity base plate.

Transistors and amplifiers produced in large volume are housed in air-cavity ceramic or overmolded plastic packages. These packages protect the devices from the external environment. Packages for high-voltage transistors and internally matched FETs used for wireless infrastructure or base stations, radar, and other high-power applications generate large heat fluxes and the flange is designed to handle the dissipated heat. Achieving a uniform die attachment, at least under the active area of the device, is very critical for thermal management of dissipated power and a high-conductivity flange is the main path for efficient heat flux removal. Since voids in the die attach area have very low thermal conductivity, they either significantly degrade or damage transistors. The quality of the die attach can be examined using a scanning acoustic microscope, also known as a sonoscanner. In this method ultrasonic energy is applied at the back of the flange. Since the ultrasound signal will not pass through voids, they are detected using acoustic imaging.

In high-power amplifiers (HPAs) and high-voltage HPAs (HVHPAs), where heat removal is of prime importance in packages and assemblies, thermal management becomes the predominant issue. The current heat spreaders are comprised of BeO, AlN, CuW, CuMo, CuMoCu, and SiC and their thermal conductivity values range between 150 and 350 W/m·K. In GaAs and Si based transistor amplifiers, the heat flux is in the range of 100–300 W/cm^2. However, in HVHPAs the heat flux is much higher than 300 W/cm^2. These heat flux values are an order of magnitude higher than a high-power microprocessor's heat flux level. To handle very high heat flux values, one needs diamond-like materials or composite materials as heat spreaders with thermal conductivity values of over 500 W/m·K.

21.7 CAD TOOLS FOR PACKAGES

Successful development of packaged RF and microwave amplifiers requires seamless integrated CAD tools, that is, design tools that can simultaneously optimize the circuit along with the package used for housing the circuit. Typically, two approaches are used for such integrated solution: (a) circuit/network simulators for circuits and multiport equivalent circuits for packages; and (b)circuit/network simulators for circuits and EM field simulators for packages [11]. In the first approach, a package is represented by using a multiport network/equivalent circuit (EC) model as described in Section 21.4.2. The EC model is based on the physical nature of a package, including the lead frame, isolation between ports, and grounding and shielding effects. The EC model parameters are extracted by using extensive EM field simulations characterizing the package. The EC model for a package is then used with RF simulators to design amplifiers, including the package effects. The EC models for packages may be replaced with artificial neural network (ANN) models trained by EM simulation of the package. The EC or ANN models for commonly used packages may be developed to a level suitable for RF circuit designers to use with commercial CAD tools.

In the second method, circuit/network simulators for circuits and EM field simulators for packages are integrated to use interactively. There were some attempts to interlink several different simulators to optimize amplifier circuits in different packages; however, their usage is limited to an analysis level due to extensive simulation time requirements. CAD tools for RF packages are continuously being developed, including their integration with RF CAD tools. In addition to integrated electrical design tools for packaged amplifiers, thermal and mechanical designs need to be integrated as well to develop a comprehensive CAD tool. As a first-order approximation, the lead frame and wire parasitic reactances may be absorbed in the design of amplifiers.

21.8 POWER AMPLIFIER MODULES

A power amplifier module consists of several amplifier circuits and components and is built to specified requirements. In a set of modules, each module has identical dimensions, electrical performance, mechanical structure, and thermal properties. In large product quantities, modules are assembled quickly and have low production cost. Figure 21.26 shows an example of a 50-W module consisting of six MMIC amplifier chips. The module has about 50-dB gain. Figure 21.27 shows another module consisting

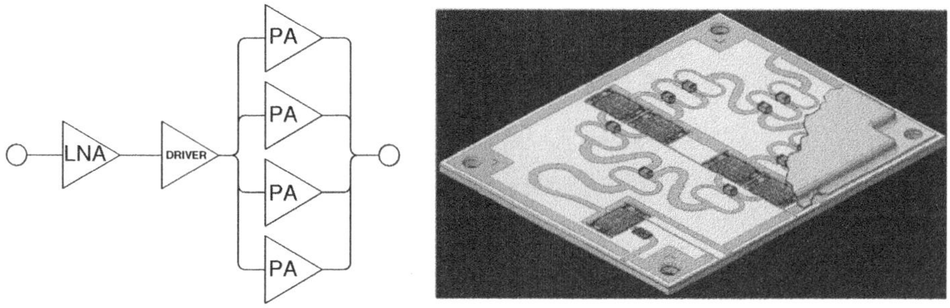

Figure 21.26 High-power module configuration.

Figure 21.27 High-power module configuration using two LDMOS transistors.

of two high-power LDMOS packaged transistors combined using a Wilkinson power combiner.

Surface mount technology is mostly used in the system-in-package (SIP) modules. The SIP technology is commonly used to integrate multiple functions in a very small package in a low-cost manner. Active devices, including transistors, diodes, and ICs, and passives are integrated together using epoxy or solder in a single package. Figure 21.28 shows a photograph of an SIP module.

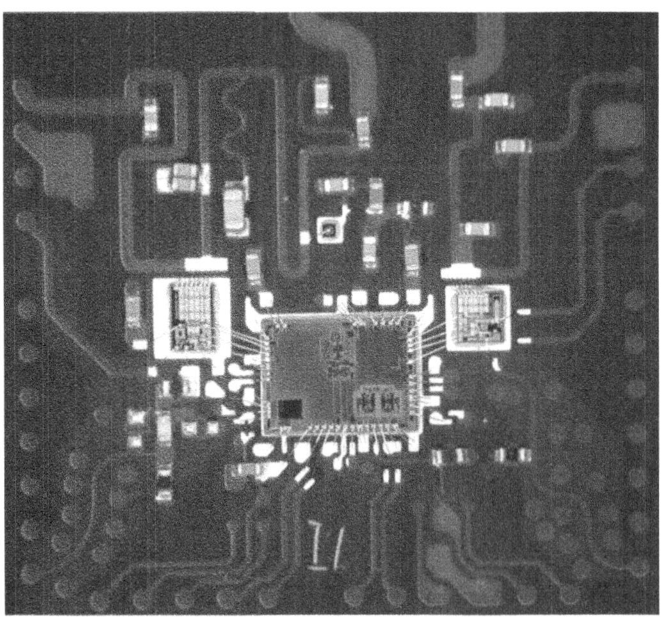

Figure 21.28 SIP module configuration.

REFERENCES

1. R. R. Tummala and E. J. Rayaszewski (Eds.), *Microelectronic Packaging Handbook*, Van Nostrand Reinhold, New York, 1989.
2. L. T. Manzione, *Plastic Packing of Microelectronic Devices*, Van Nostrand Reinhold, New York, 1990.
3. J. E. Sergent and C. A. Harper (Eds.), *Hybrid Microelectronics Handbook*, McGraw-Hill, New York, 1995.
4. P. E. Garrou and I. Turlik, *Multichip Module Technology Handbook*, McGraw-Hill, New York, 1998.
5. T. K. Gupta, *Handbook of Thick- and Thin-Film Hybrid Microelectronics*, John Wiley & Sons, Hoboken, NJ, 2003.
6. R. K. Ulrich and W. D. Brown (Eds.), *Advanced Electronic Packaging*, 2nd edition, John Wiley & Sons, Hoboken, NJ, 2006.
7. N. G. Einspruch and W. R. Wissemen (Eds.), *VLSI Electronics Microstructure Science, Volume 11, GaAs Microelectronics*, Academic Press, New York, 1985, Chapter 8.
8. R. Goyal (Ed.), *Monolithic Microwave Integrated Circuits: Technology and Design*, Artech House, Norwood, MA, 1989, Chapter 10.
9. A. Sweet, *MIC and MMIC Amplifier and Oscillator Circuit Design*, Artech House, Norwood, MA, 1990.
10. M. Golio (Ed.), *RF and Microwave Handbook*, CRC Press, Boca Raton, FL, 2000, Section sec6.10.
11. I. J. Bahl, *Lumped Elements for RF and Microwave Circuits*, Artech House, Norwood, MA, 2003, Chapter chap13.
12. Y. C. Lee et al., Packaging RF devices and modules, in *Encyclopedia RF and Microwave Engineering*, K. Chang (Ed.), Vol. 4, John Wiley & Sons, Hoboken, NJ, 2005, pp. 3590–3614.
13. S. Pinel et al., RF/wireless packaging, in *Encyclopedia RF and Microwave Engineering*, K. Chang (Ed.), Vol. 5, John Wiley & Sons, Hoboken, NJ, 2005, pp. 4516–4537.
14. K. Lim et al., RF-system-on-package (SOP) for wireless communications, *IEEE Microwave Mag.*, Vol. 3, pp. 88–99, March 2002.
15. C. A. Tavernier et al., High performance multilayered high temperature cofired ceramic for wide band packaging, *IEEE MTT-S Int. Microwave Symp. Dig.*, pp. 2277–2280, 2003.
16. D. Fisher and I. Bahl, *Gallium Arsenide IC Applications Handbook*, Academic Press, San Diego, 1995, Chapter 10.
17. I. J. Bahl and E. L. Griffin, Semiconductor Chip Housing, US Patent # 4701573, October 20, 1987.
18. I. J. Bahl, Microwave Feedthrough Apparatus, US Patent # 5428327, June 27, 1995.

PROBLEMS

21.1 An MMIC is to be packaged in plastic packaging. List the critical design requirements for the package. Ground connections are through vias.

21.2 Design a ceramic cavity package for 30-GHz operation for GaAs MMICs.

21.3 Describe the pros and cons of ceramic and plastic packages. Frequency of operation, power handling, and cost need to be addressed.

21.4 Determine the lead frame model for plastic and ceramic packages. As a first-order approximation, assume 50% lead frame lengths (inside cavity) made of microstrip and the other 50% leads are in air. Leads are 10 mils wide and 30 mils long. The other parameters are ceramic, $\varepsilon_r = 9.9$, and $h = 10$ mils; and plastic, $\varepsilon_r = 3.5$ and $h = 5$ mils.

21.5 Determine the lead frame model when two wires of 1-mil diameter, 5-mil spacing, and 20-mil long are also included inside the cavity in Problem 21.4.

21.6 The S-parameters of an amplifier are given below. The amplifier is placed in a plastic package whose lead frame model is given in Figure 21.20. Calculate the composite S-parameters of the packaged amplifier when a feedback between the input and output package leads represented by a capacitor of $C = 0.01$ pF is included. Also calculate the stability factor before and after packaging. Ignore bond wire effect.

S-parameters of the amplifier are:

Frequency (GHz)	$\lvert S_{11}\rvert$	$\angle S_{11}$	$\lvert S_{21}\rvert$	$\angle S_{21}$	$\lvert S_{12}\rvert$	$\angle S_{12}$	$\lvert S_{22}\rvert$	$\angle S_{22}$
13.0	0.223	−72.00	13.39	120.86	0.0032	−37.80	0.126	−7.64
13.5	0.258	−79.91	13.13	88.65	0.0035	−64.66	0.159	−14.72
14.0	0.272	−89.08	12.96	56.95	0.0039	−89.48	0.182	−24.17

21.7 Repeat Problem 21.6 with $C = 0.02$ pF.

21.8 Compare qualitatively ceramic and plastic packages for 4–5-W power amplifier applications. Assume 30% efficiency.

Chapter 22

Transistor and Amplifier Measurements

RF and microwave measurements [1–30] can be classified into two categories: one used for the characterization of transistors and the other for amplifier evaluation. Although the basic test setups and procedures are similar, transistor characterization test setups may need more components such as tuners and bias tees, while amplifiers may not need bias tees and use simpler calibration procedures. This chapter deals with several test methods used for transistor and amplifier measurements. *Many test instrument companies provide application notes and test procedures to perform accurate measurements. Readers are advised to refer to manufactures' manuals and instructions for updated information.*

22.1 TRANSISTOR MEASUREMENTS

Transistors are characterized for small-signal S-parameters, small-signal and large-signal equivalent circuit (EC) models, noise models, and nonlinear source-pull and load-pull data. Next, we briefly describe the transistor measurements on which EC model parameter values are based: DC current–voltage ($I-V$) and S-parameter data, noise parameter data, and source-pull and load-pull data. On-wafer $I-V$ and RF measurements using high-frequency probes provide accurate, quick, nondestructive, and repeatable results up to millimeter-wave frequencies. An overview of device measurements is given Figure 22.1.

In transistors, often one terminal is used as the input and the other as the output and the third is grounded. In FETs/HEMTs, normally the source is grounded and this configuration is known as common-source. A common-source configuration has maximum gain, while common-gate and common-drain configurations have low gain but can be matched to 50 Ω over wide bandwidths without using any external matching networks as previously discussed in Chapter 11.

Fundamentals of RF and Microwave Transistor Amplifiers. By Inder J. Bahl

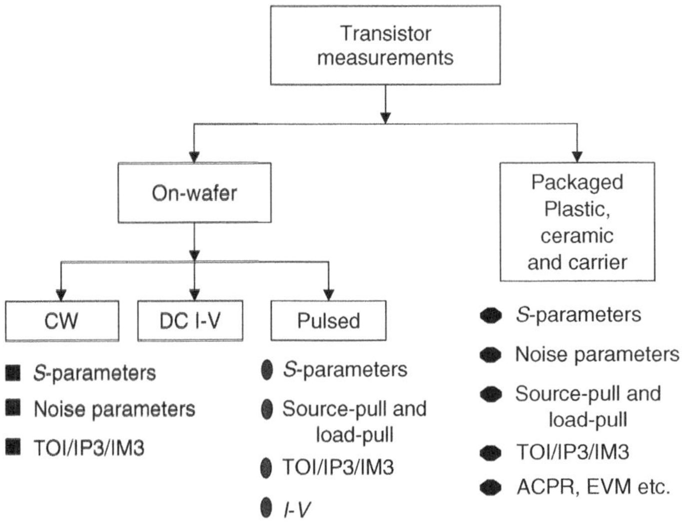

Figure 22.1 An overview of transistor measurements.

22.1.1 *I–V* Measurements

For linear operation, transistors are characterized at an operating bias or quiescent point known as the *Q*-point. In this case, the device $I-V$ data is generally not required. However, under large-signal conditions, the devices go through large voltage and current swings. Such voltage and current excursions could reach voltage breakdown and open-channel current-limiting values. Thus it is very important to measure DC $I-V$ data for devices to assess the voltage and current capabilities. Such data is measured between drain–source or collector–emitter electrodes with gate voltage or base voltage/current as a control.

A set of $I-V$ measurements is an integral part of bias-dependent and nonlinear device models. This involves the measurements of parasitic resistances of the device and the current–voltage relationship. Many devices have dispersion characteristics; that is, current is a function of frequency. For such devices, the measured DC g_m and calculated RF output power values based on DC $I-V$ data are higher than the measured RF g_m and RF output power, respectively. Depending on the device type, it has been observed experimentally that pulsed I_{peak} is lower than DC I_{peak}. This is due to trapping phenomena in devices and happens at very low frequencies. This anomaly is minimized by taking $I-V$ data using an RF technique or pulsed rather than DC data. Thus in order to assess accurately the device power capability, it is important to use a pulsed technique (pulse width of 1 μs or less) to obtain $I-V$ data for the transistors under test.

The DC or pulsed $I-V$ data can be measured using a home-grown test setup or commercially available equipment. The pulse width is on the order of 1 μs and RF frequency is on the order of 1 MHz. Figure 22.2 shows measured DC and pulsed $I-V$ data for a 625-μm FET. When thermal effects are ignored, the measured pulsed current for a power FET/pHEMT is always lower than the DC values. The difference between the DC and pulsed peak current could be as large as 30%. When the devices have no dispersion effects, that is, DC g_m is the same as the RF g_m, due to lower

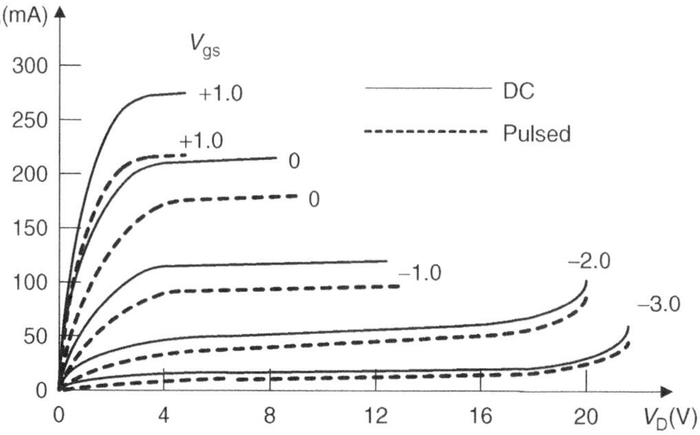

Figure 22.2 Measured DC and pulsed $I-V$ data for a 625-μm FET showing current dispersion.

thermal heating, the pulsed current value is higher than the DC current value at higher drain/collector voltages.

22.1.2 *S*-Parameter Measurements

S-parameters are measured using a vector network analyzer (VNA). A VNA is a very complex system, which basically consists of a synthesized source, dual reflectometer test set, a multichannel receiver with downconverters and IF detectors, A/D converters, a data processing unit, and a screen. The system makes very broadband measurements with hundreds of points with fine resolution. The DUT is connected to the dual reflectometer test set through probes/connectors and cables. Both small-signal CW and pulsed power VNAs are available.

VNA measurements are affected by drifts (due to temperature and time related changes), random uncertainties (system noise and connector repeatability), and systematic errors (component imperfections and mismatch error due to couplers, connectors, and cables). Systematic errors are repeatable and can be removed by using vector calibration procedures as supported by VNA vendors. Modern calibration procedures use several types of calibration standards applicable to various transmission media (e.g., coaxial, waveguide, CPW, and microstrip).

The RF performance of devices is determined using small-signal, two-port *S*-parameter measurements, although large-signal *S*-parameter measurements are also performed but are not commonly used. The *S*-parameter measurements are made in a 50-Ω system using a VNA. Basic *S*-parameter measurements are small signal and consist of measuring incident, reflected, and transmitted waves with both magnitude and phase. The VNA measures the amplitude and phase of ingoing and outgoing signals at each port and determines the ratios compared to a reference signal as well as to a reference plane. Computing the ratios of these quantities and postprocessing the data at each frequency yields the *S*-parameters. The complete process takes about 1 ms per frequency. Since there exist lines/cables, connectors/launchers, and so on between the device under test (DUT) and the VNA's reference planes, deembedding standards are used to accurately extract the *S*-parameters for the DUT at its reference planes. The deembedding procedure is an integral part of the VNA measurements. Since *S*-parameter measurements are based on two-port vector calibration and a

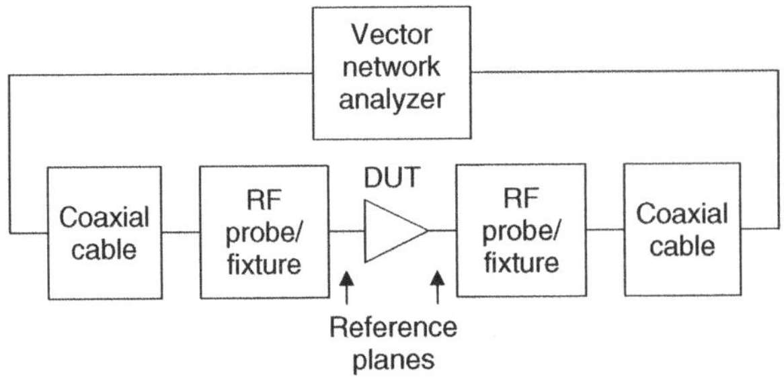

Figure 22.3 A basic S-parameter measurement setup.

deembedding procedure, their accuracy and reproducibility are much better than any other measurement method based on scalar calibration. Figure 22.3 shows a basic S-parameter measurement test setup.

Modern VNAs support one- and two-port S-parameters in a 50-Ω system with single connection or probe touch down. The EC model parameters are extracted by converting S-parameters into Z- and Y-parameters. The device S-parameter measurements ($V_{ds} = 0$), referred to as cold measurements, are also used to accurately extract the extrinsic device EC model parameter values.

Calibration Methods

Choosing the proper calibration method for on-wafer measurements is critical for obtaining accurate measurements. The calibration, in turn, is only as good as the calibration standards and lack of on-wafer calibration standards gives rise to errors in the measured results. Basically, there are two important elements of calibrating on-wafer: calibration standards (including calibration verification devices) and calibration methods/procedures.

The first step in the calibration process is to create standards and calibration verification devices. The standards are used to establish an unchanging reference point from which all measurements will be based. The verification devices are secondary standards that are used to verify (cross-check) the accuracy of the calibration.

The calibration process includes the methods and the procedures that are implemented to remove systematic errors and to maximize measurement repeatability. Some of the "procedures" that the user needs to implement to assure a good calibration process would include the following items: instrument maintenance, connector use and care, equipment cleaning, temperature and humidity control, operator training, step-by-step instructions, probe placement, probe alignment, and probe pressure. Using "markers" near the probe launching position on the calibration standards helps in making repeatable measurements. Without strict adherence to good calibration and measurement practices, accurate and repeatable measurements are not possible.

Typically, a network analyzer is used to vectorially remove measurement system errors. This is accomplished by measuring the calibration standards. After measuring the calibration standards, a 12-term error model may be calculated and stored in the network analyzer. These error terms quantify errors over the calibrated frequency range.

Several of today's popular calibration methods include short-open-load-thru (SOLT), thru-reflect-line (TRL), and line-reflect-match (LRM). The majority of users employ the SOLT method because on-wafer "standards" are commercially available. These commercially available on-wafer calibration standards are fabricated on a sapphire substrate, known as an impedance standard substrate (ISS), which includes shorts, transmission lines, and loads. The ISS also includes devices that are used to verify the calibration's accuracy such as opens, shorts, loads, inductors, and capacitors. The accuracy of these standards over a given frequency range can be evaluated in terms of precision and repeatability. In MICs and MMICs, the calibration standards can be placed on the same substrate on which devices/components are fabricated.

SOLT Calibration

The SOLT calibration technique is widely used with coaxial standards as well as with on-wafer standards. However, it has shown marginal utility in IC media, especially above 15 GHz, because of the unavailability of well-characterized open and short standards. The ISS "open" (probe lifted above the wafer) exhibits fringing capacitance that must be determined for each probe while the short has inductance that also must be measured and corrected. Frequency dispersion in these standards makes calibration less precise. The parasitic effects associated with each standard change their expected characteristics and can cause errors in the calibration data. These parasitics must be determined in order to minimize the measurement error. Figure 22.4 shows "on-wafer" a set of microstrip calibration standards for the SOLT method.

TRL

The "thru-reflect-line" calibration technique requires a minimal number of physical standards: a thru line, one or more delay lines, and two equal-length reflects of high unknown reflection coefficient. Thus the only standards for this method are equal impedance line lengths. In this method, the reference planes can be set relative to the thru line or reflects. Figure 22.5 shows a microstrip calibration standards for the TRL method as applied to on-wafer calibration. Reproducible and low parasitic reactance via holes are required in this case. The vertical lines denote the calibration reference planes. Thus the TRL calibration technique is accurate and has simple standards that can easily be placed on the same substrate on which the component/ICs are fabricated—therefore

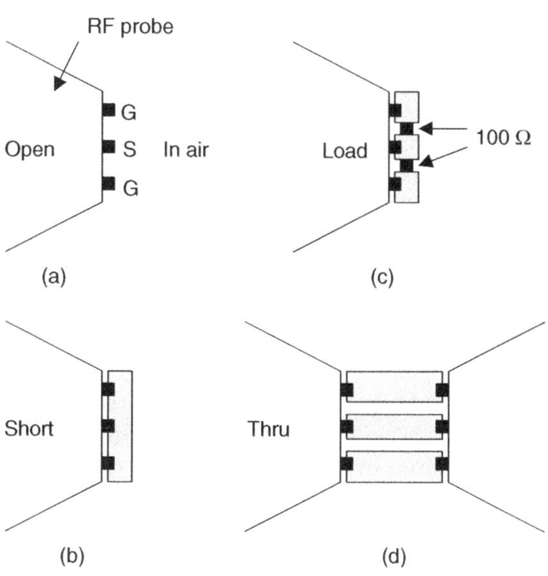

Figure 22.4 On-wafer calibration standards for the SOLT method: (a) open, (b) short, (c) load, and (d) thru line.

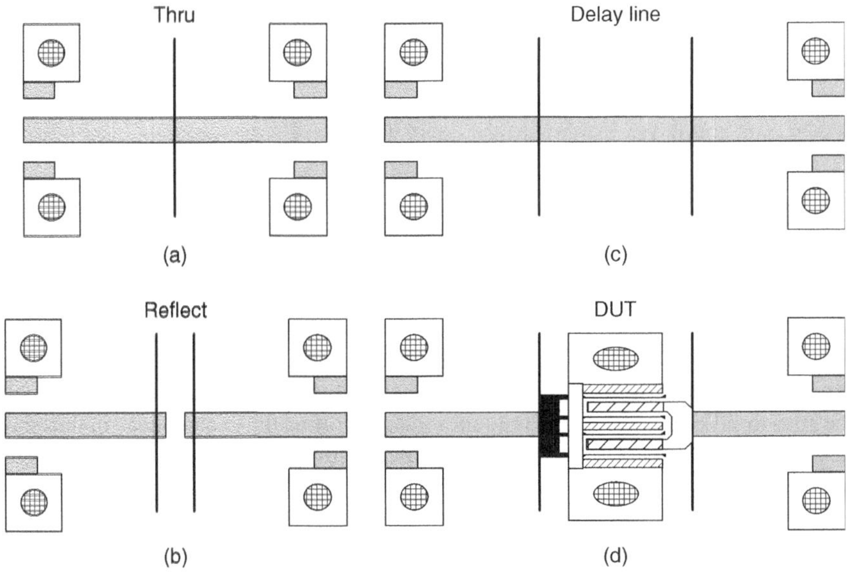

Figure 22.5 TRL calibration standards and measurements: (a) thru, (b) open as reflect, (c) delay line, and (d) device under test embedded in TRL standards. Grounds are connected through vias to the backside.

ensuring a common transmission medium. This calibration technique has better flexibility in locating the reference planes and also minimizes radiative crosstalk effects between the two probes because they are not as close as those used in the SOLT calibration technique for thru measurements.

A TRL calibration requires a thru line, a highly reflective impedance at each port, and one or more delay lines (the number of delay lines required is determined by the bandwidth of the calibration). The resulting reference planes can be located at the center of the thru line. The length of the thru line should be as short as possible but long enough to avoid interaction between the probes. This ensures that the effects due to the excitation of higher order modes caused by the CPW to microstrip transition are included in the error terms. The length of the delay lines and the number of delay lines required depend on the frequency band. Two (three) different delay line lengths are needed to calibrate the system from 0.5 to 18 GHz (0.5–40 GHz), but measurements are made over the full range in one sweep. To calibrate at lower frequencies, a relatively long line length is required. Generally, the electrical length must be greater than about 20° at the lowest frequency and less than 160° at the highest frequency of operation. The electrical line lengths approaching integral multiples of π will result in ill-conditioned error terms.

LRM Calibration

The "line-reflect-match" calibration technique combines some of the advantages of SOLT and TRL without their disadvantages. In the TRL method, line standards have limited bandwidths and can only be used over an 8:1 frequency range. At low frequencies, line standards become inconveniently long. The LRM calibration method is similar to the TRL method except that the delay line standards are replaced with broadband matched loads, which can be realized along with other standards on the same

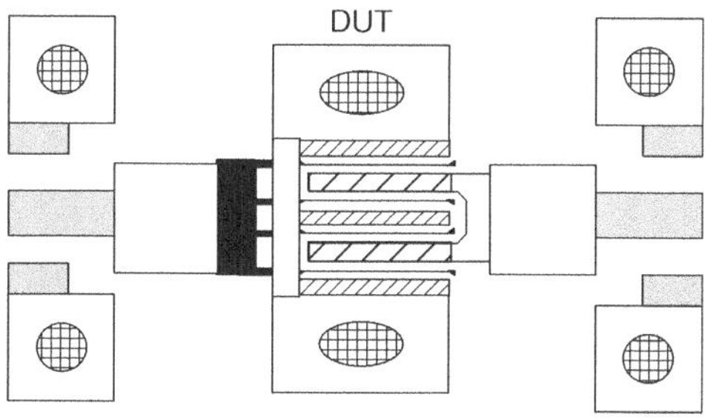

Figure 22.6 Device under test embedded in SOLT configuration.

substrate. Thus there are three steps to the LRM calibration method: first, connect a thru line; second, connect a pair of identical reflective impedances; and third, connect a pair of matched loads. The improved accuracy of the LRM calibration over the SOLT and TRL techniques has been demonstrated over 1.5–40 GHz.

Figures 22.5d and 22.6 show an FET in a series configuration to be tested for S-parameters using the TRL/LRM and SOLT calibration deembedding techniques, respectively. The device structure shown in Figure 22.6 is suitable for hybrid amplifier implementation, while in Figure 22.5d it is suitable for IC amplifier implementation. In Figure 22.6 there are gate and drain pads for wire bonds. An overview of calibration standards is given in References 4 and 5.

22.1.3 Noise Parameter Measurements

There are several methods used to characterize transistors for noise modeling. A commonly employed measurement based noise model for transistors, supported by commercial CAD tools, uses noise parameters as described in Chapters 4 and 5. The noise model consists of noise parameters (F_{min}, Γ_{opt}, and R_{n}) and S-parameters. A noise parameter extraction method is similar to source-pull and load-pull techniques, described in detail in the following section. A basic "on-wafer" noise parameter extraction test setup is shown in Figure 22.7. It consists of input and output electronic tuner modules (ETMs). An ETM is comprised of a solid state tuner and a low-loss SPDT

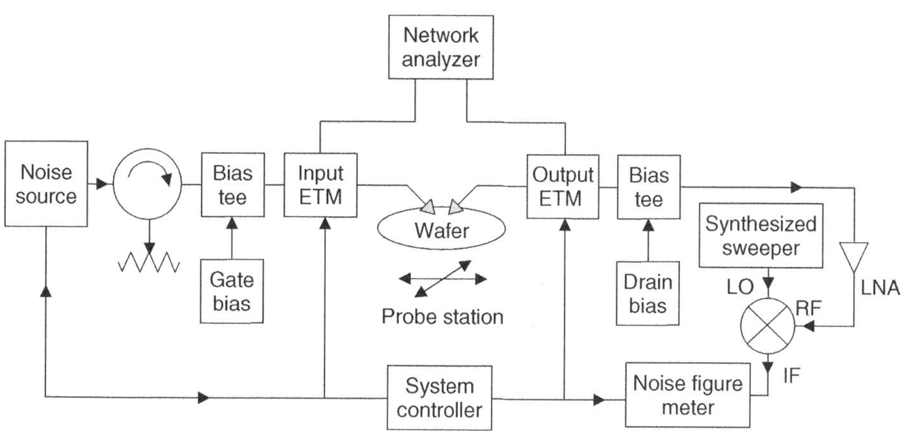

Figure 22.7 A basic "on-wafer" noise parameter extraction test setup.

switch. By using the SOLT or TRL deembedding technique, the impedance reference planes are accurately located at the input and output terminal points of the device under test. The input is matched to obtain F_{min} and constant noise figure contours (Γ_{opt} and R_n), and the output is matched for complex conjugate impedance. The S-parameters are measured in 50 Ω.

22.1.4 Source-Pull and Load-Pull Measurements

The source-pull and load-pull measurement techniques are well known for the characterization of transistors to determine source and load conditions to achieve an optimum solution in terms of output power, PAE, IP3, and/or ACPR. These measurements, however, require a bench loaded with expensive laboratory equipment including signal synthesizers, power meters, spectrum analyzer, network analyzer, tuners with drivers, circulator, variable attenuator, and couplers. In this section, the source-pull and load-pull measurements are briefly discussed.

Source-Pull and Load-Pull for Power and PAE

In the design of high-performance power amplifiers, one requires an accurate nonlinear device model or accurately measured source-pull and load-pull data. This data allows the design of the optimal matching networks for power amplifiers to achieve maximum power and PAE performance simultaneously by knowing the right source and load condition for fundamental as well as harmonic frequencies. These measurements can be made using either on-wafer RF probes or packaged devices.

There are many methods [16–26] used for obtaining source-pull and load-pull data. They are the classical manual techniques, an automatic method using passive and solid state tuners, the active load-pull method, the load-pull method that provides simultaneously optimized harmonic termination, intermodulation distortion, or ACPR. A basic source-pull and load-pull setup is shown in Figure 22.8. The input power is selected based on the estimated associated power gain value. The input tuner is set for maximum gain and good input match. The output tuner is adjusted to obtain maximum output power or PAE or both. Then the tuners are disconnected to measure the source impedance and load impedance with the help of a network analyzer. This procedure is repeated for various source and load settings at several frequencies in the frequency range of interest. In most cases, these measurements are performed in a fully automated fashion. A typical automatic load-pull test setup with fixed input tuner is shown in Figure 22.9. To make source-pull and load-pull measurements in a

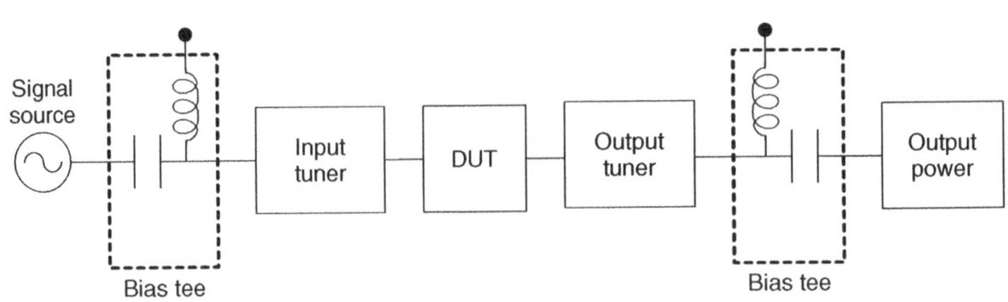

Figure 22.8 A basic source-pull and load-pull test setup.

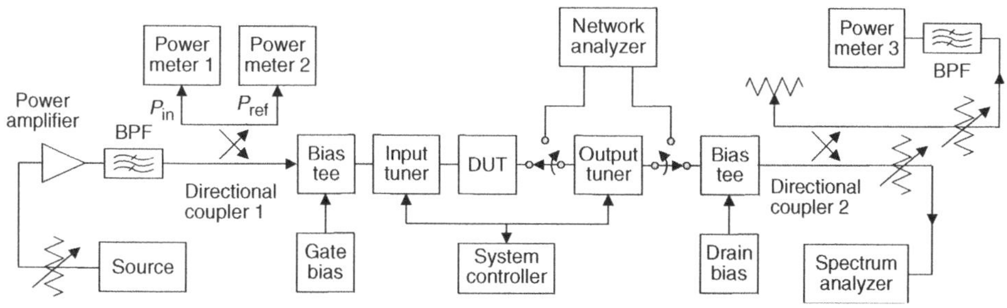

Figure 22.9 Typical load-pull test setup with fixed input tuner.

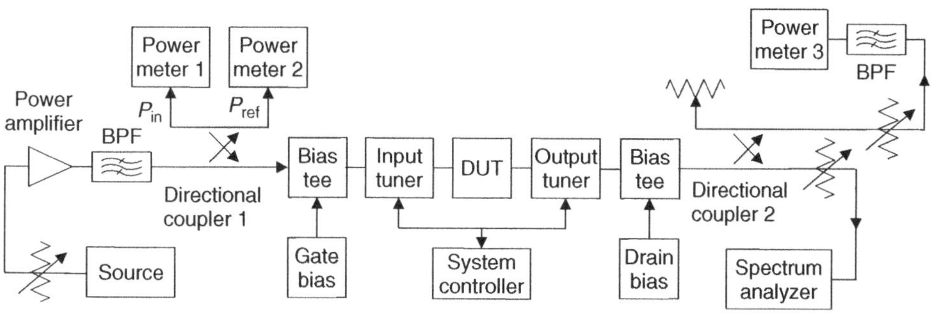

Figure 22.10 Typical source-pull and load-pull test setup.

fully automated way, the test setup shown in Figure 22.10 is used. The input power and the input match are fixed; the load-pull data on the Smith chart depicting various constant-power contours or constant-PAE contours or both, as described in Chapter 5, are generated to help design the output matching network for the power amplifier circuit. Similarly, the output match is fixed; the source-pull contours on the Smith chart depicting input match, gain, power, and PAE are generated.

Spectrum analyzers (SAs) are used to measure power levels of harmonics, IM products, and oscillations. They are an integral part of any setup using transistors in the nonlinear mode. The accuracy of power levels measured using SAs is not as good as by using VNAs or power meters.

Large transistors have low impedances, so their accurate source-pull and load-pull characterization is difficult due to the limited impedance range available from mechanical tuners. The loss between the transistor terminals and the tuners limits the realizable reflection coefficient at the transistor terminals in the range of 0.8–0.9. In such cases, a prematch technique for transistors using $\lambda/4$ transformers on-chip or off-chip is preferred. Figure 22.11 shows a schematic of the prematch FET configuration. In this method, the transistor's impedances are transformed to measurable values, and source-pull and load-pull data are taken using tuners. Using deembedding techniques, the measured source-pull and load-pull data are transformed at the device terminals. The prematch technique has a limited bandwidth because of the $\lambda/4$ transformers. One can also use a suitable transistor size so that its projected source and load impedances are well within the measurable range of tuners. Then the measured source and load impedances are scaled for other device sizes, as described in Chapter 5.

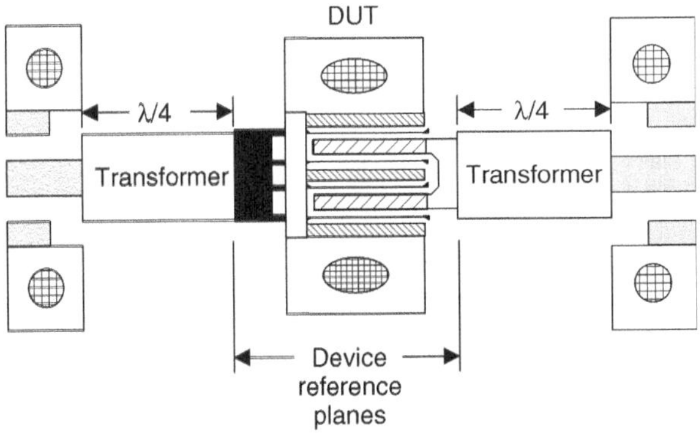

Figure 22.11 Schematic of a prematch technique using λ/4 transformers.

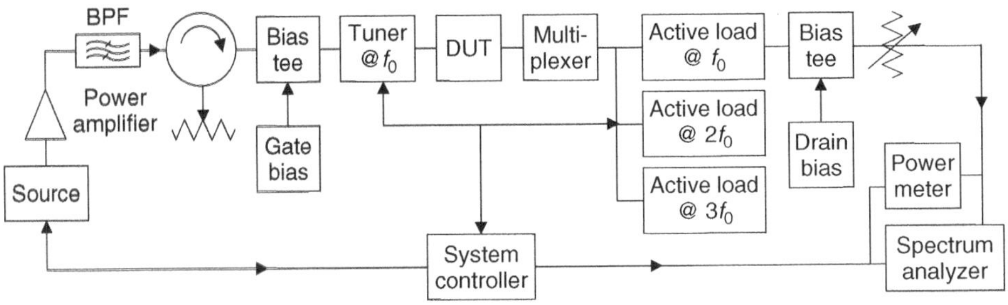

Figure 22.12 Typical active source-pull and load-pull with harmonic tuning capability test setup. Output load-pull includes second and third harmonics.

The load-pull techniques using mechanical turners described above have one major shortcoming in terms of losses in the tuners and between the device under test and the output tuner, due to the test fixture, the connector (or RF probe), and the connecting cables. This loss limits the realization of larger reflection coefficients required for higher power devices or higher frequencies. This problem is alleviated using an active load-pull technique. Major advantages of the active load-pull technique are achieving a reflection coefficient of up to unity at the device terminal as well as harmonically tuning the load for fundamental, second-, and third-harmonic frequencies. A typical active source-pull and load-pull with harmonic tuning capability test setup is shown in Figure 22.12.

Source-Pull and Load-Pull for IP3

Many systems require improved IM3/IM5 intermodulation distortion performance and ACPR performance in addition to output power and PAE. Several load-pull techniques now can characterize transistors for these parameters. Thus one can determine optimum source and load conditions for active devices based on the specific application. A typical source-pull and load-pull test setup for IP3 characterization is shown in Figure 22.13. This setup needs two synthesized sources for two tone measurements. The two carrier frequencies are generally 5–10 MHz apart. Other test procedures are similar to those used for power and PAE characterization. ACPR is characterized using a digitally modulated carrier as described in the amplifier measurements.

For large device size or high-voltage operated devices, the tuners are unable to realize correct source and load impedance values due to the finite loss in the cables,

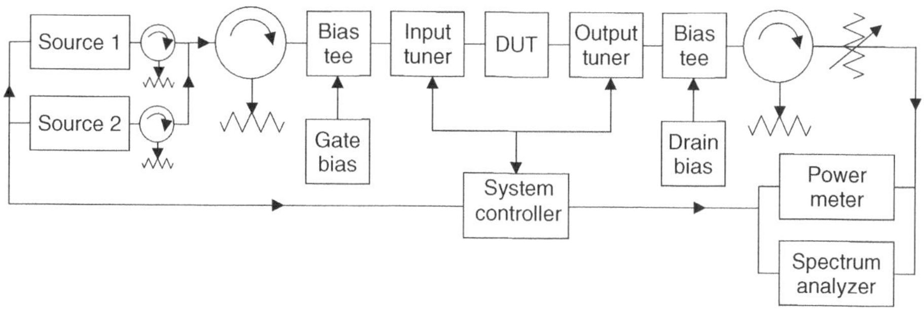

Figure 22.13 Typical source-pull and load-pull test setup for IP3 measurement.

connectors, and RF probes. In such cases, the devices are prematched using λ/4 transformers to move the device impedance closer to the center of the Smith chart. By proper deembedding techniques, the device source and load impedances are extracted.

22.2 AMPLIFIER MEASUREMENTS

Most test procedures described for transistors are also valid for amplifiers and vice versa. In amplifier characterization, the tuners are either removed or set for 50 Ω looking into the tuners. A comprehensive overview of amplifier testing is given in Figure 22.14.

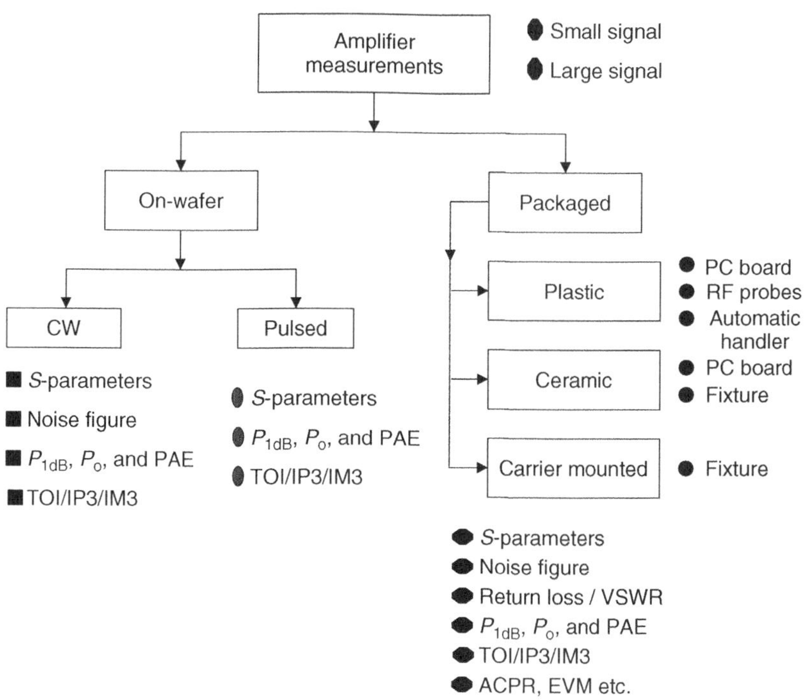

Figure 22.14 Overview of amplifier measurements.

22.2.1 Measurements Using RF Probes

Most small-signal MMIC amplifiers and plastic packaged driver amplifiers are characterized by using RF probes. Pulsed power measurements are performed to test power amplifiers on wafer for screening before being mounted onto carriers or into packages. These include *S*-parameters (gain, isolation, input and output return losses), IP3, NF, and phase noise. The RF probes are normally calibrated using SOLT standards available on commercial sapphire substrates. To check the validity of measurements, gold standards or benchmark circuits are generally used. The MMIC amplifiers are tested on wafer (Fig. 21.15a), while plastic packaged amplifiers are loaded into a graphite tray (usually can hold about 100 circuits) and are automatically tested (Fig. 21.15b).

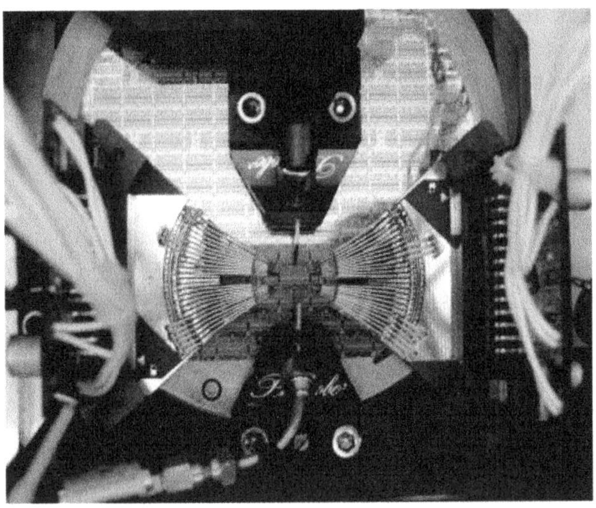

(a)

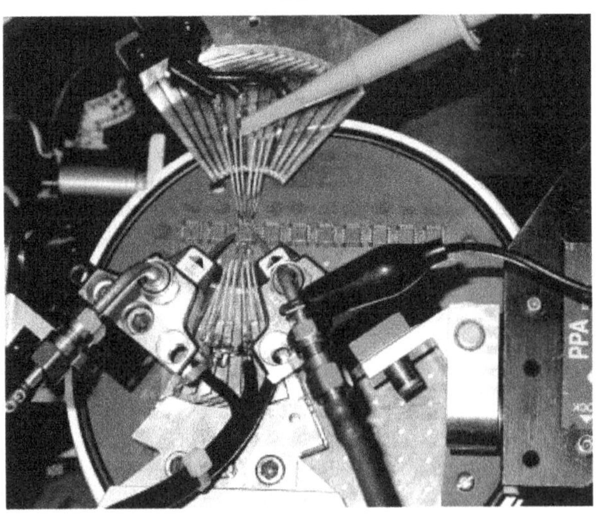

(b)

Figure 22.15 Basic *S*-parameters measurement setup: (a) on-wafer HPA pulsed power measurement and (b) testing plastic packaged driver amplifier with RF probes.

22.2.2 Driver Amplifier and HPA Test

Plastic packaged devices are also tested in tubes using automatic handlers. Prototypes are tested by mounting them on PCBs. Ceramic packaged amplifiers are characterized using fixtures. In this case, the output power at the fundamental and harmonic frequencies are measured as a function of input power. The fundamental output power is usually measured using a power meter (CW or pulsed), while the harmonics are measured by employing a spectrum analyzer (SA). Also, the SA is used to monitor for oscillations during power testing. In testing power amplifiers, it is very important to keep the base plate temperature to within specified limits. During the test, an HPA is mounted on either a heat sink or a cold/hot plate to maintain the base plate temperature as per specification.

Figure 22.16a shows a test setup for P_{out} versus P_{in} measurements. In this setup, harmonic levels are measured with a spectrum analyzer. The amplifiers are biased using on-chip biasing networks or external bias tees (not shown). The input power is varied to obtain linear as well compressed gain measurements as per requirements. The power amplifier is used to obtain correct power levels at the input of the DUT. The

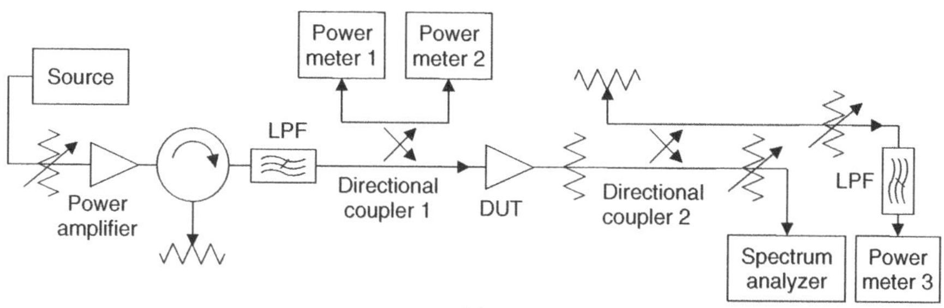

(a)

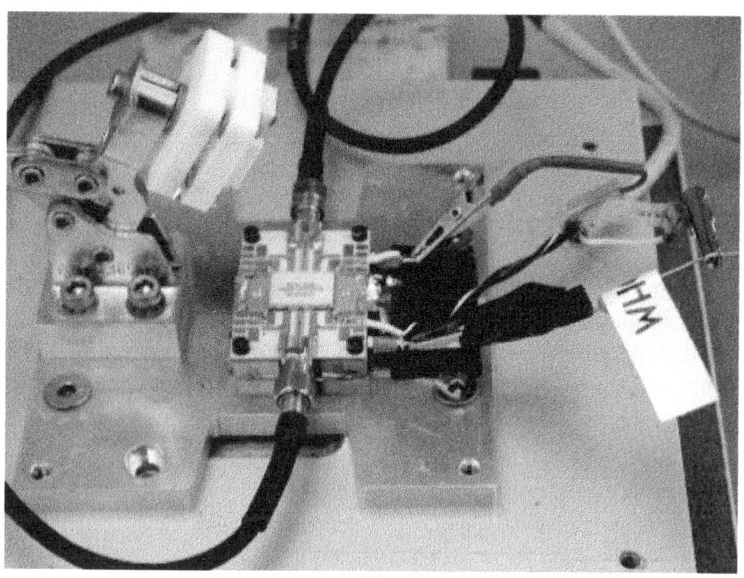

(b)

Figure 22.16 Typical P_{out} versus P_{in} measurement setup: (a) schematic and (b) photograph.

LPFs are used to filter out the harmonics. The system is calibrated with respect to the input and output terminals of the DUT. Power meters 1, 2, and 3 measure the reflected power, incident power, and output power, respectively, with respect to the input and output ports of the DUT. At each input power level, the gate and drain currents are also measured. From P_{out} versus P_{in} data and drain current at each input power level, PAE and $P_{1\ dB}$ are calculated. Figure 22.16b shows the photograph of the ceramic packaged HPA test setup.

Since this setup measures scalar quantities, most components used are narrowband and are selected based on the frequency of operation. The scalar measurement accuracy/reproducibility is better when the components and the complete test station are matched to 50 Ω (return loss better than 20 dB). The smaller the interactions between the components due to mismatch, the better will be the power measurement accuracy.

22.2.3 Large-Signal Output VSWR

Input return loss of an amplifier under large-signal conditions is measured using pulsed power S-parameters or the setup shown in Figure 22.16. However, large-signal measurement of output return loss of a power amplifier is not trivial. By using a VNA, mostly small-signal return loss measurements are made. A simple method for measuring the large-signal (at full power) return loss is the slotted line technique of standing waves [3, 27]. In this technique, standing-wave patterns of electric field along the slotted coaxial line or waveguide are measured using a standing-wave ratio (SWR) meter or power meter. The electric field is detected by the probe, which has negligible perturbation on the standing waves.

Figure 22.17 shows a simplified slotted line SWR measurement test setup. The probe is moved along the line to obtain a maximum power meter reading (i.e., p_{max}). Then the probe is again moved along the line, in either direction, to obtain a minimum power meter reading (i.e., p_{min}). The VSWR, reflection coefficient, and output return loss are calculated as follows:

$$VSWR = \sqrt{\frac{p_{max}}{p_{min}}}, \quad \Gamma = \frac{VSWR - 1}{VSWR + 1}, \quad RL = -20 \log \Gamma \tag{22.1}$$

22.2.4 Noise Figure Measurements

The test setup shown in Figure 22.7 may also be used to measure the noise figure (NF) of amplifiers by setting the ETM to 50 Ω. In NF measurements, the system is calibrated using an excess noise ratio (ENR) source. Two simplified NF measurement test setups are shown in Figure 22.18. Below 1.5 GHz, the NF of the amplifier under

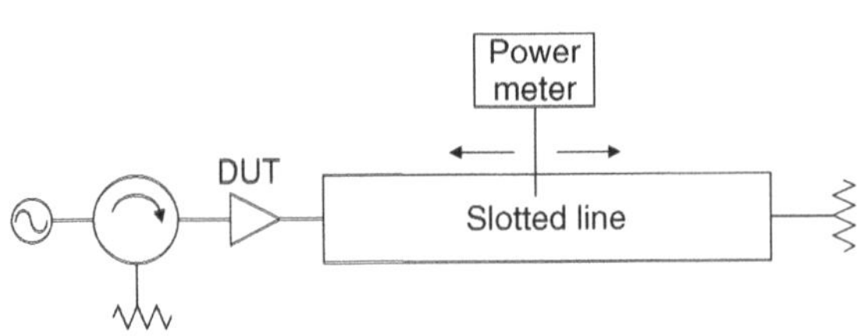

Figure 22.17 Simplified slotted line SWR measurement setup.

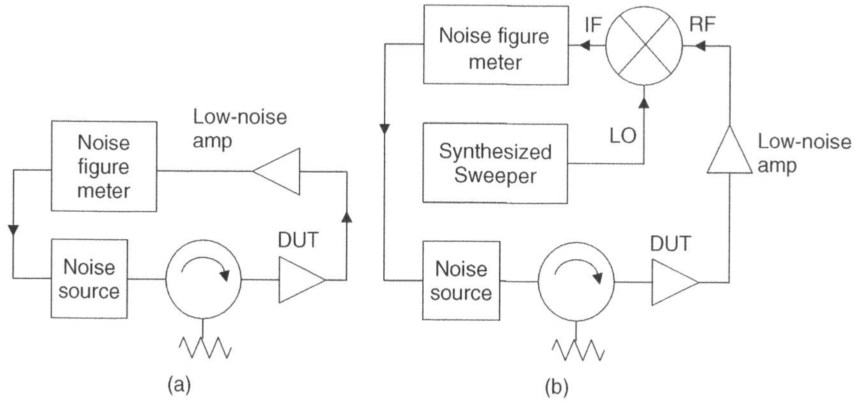

Figure 22.18 Simplified NF measurement setups: (a) below 1.5 GHz and (b) above 1.5 GHz.

test can be measured directly with a noise figure meter, whereas above 1.5 GHz, external mixing is required. The intermediate frequency (IF) is usually about 70 MHz. With these test setups, NF and gain of the device under test are measured. Since these are scalar measurements, by maintaining constant temperature and good return loss in the system, we can get repeatable and accurate measurements.

22.3 DISTORTION MEASUREMENTS

There are several techniques to measure distortion in amplifiers [3, 10, 28]:

- AM–AM and AM–PM, single-tone signal
- TOI or IP3/IM3, two-tone signals
- C/Iratio $= 2(\text{IP3} - P_\text{f})$, two-tone signals
- ACPR, digitally modulated signals
- $NPR = (C + I)/I$, multitone signals
- EVM, digitally modulated signals

These are discussed next.

22.3.1 AM–AM and AM–PM

The simplest method to measure distortion in an amplifier is to measure AM–AM and AM–PM. AM–AM can be measured as described in the previous section. To measure AM–PM, one can use a VNA. This also allows one to measure AM–AM. A simple test setup shown in Figure 22.19 may be used to measure these quantities. In this case, the values of attenuator pads are selected to maintain safe power levels for the VNA.

From the measured magnitude and phase of S_{21} versus input power at the reference planes, the values of AM–AM and AM–PM at the required output power are calculated. For example, in a two-stage 6-W HPA, the magnitude and phase of S_{21} at 5.5 GHz vary with input power as follows:

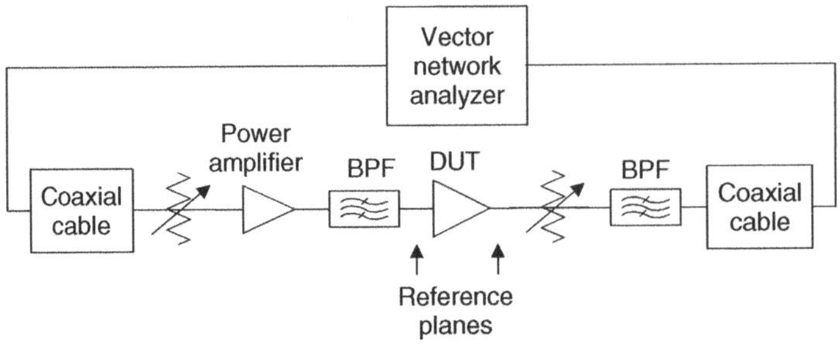

Figure 22.19 AM–AM and AM–PM measurement setup.

Input Power (dBm)	$\|S_{21}\|$ (dBm)	$\angle S_{21}$	Gain (dB)
15	35.0	50.0°	20.0
16	36.0	50.2°	20.0
17	36.8	50.8°	19.8
18	37.5	53.0°	19.5
19	38.0	55.9°	19.0
20	38.3	58.9°	18.3
21	38.3	62.0°	17.3

The output power at the P_{1dB} point is 38 dBm (6 W) and AM–PM at P_{1dB} is $55.9 - 53.0 = 2.9\,^\circ$/dB, which stays almost the same at higher gain/power compression points.

22.3.2 IP3/IM3 Measurement

A commonly used test setup to measure IP3/IM3 is shown in Figure 22.20. In this measurement a spectrum analyzer and two frequency synthesizers are used. The signal levels of the two fundamental tones, which are separated by 10 MHz in this case, are adjusted to equal magnitude by attenuators placed after the frequency sources. The bandwidth of the spectrum analyzer is selected so that one can read fundamental tones as well as closely spaced intermodulation products. The output IP2 and IP3 are then

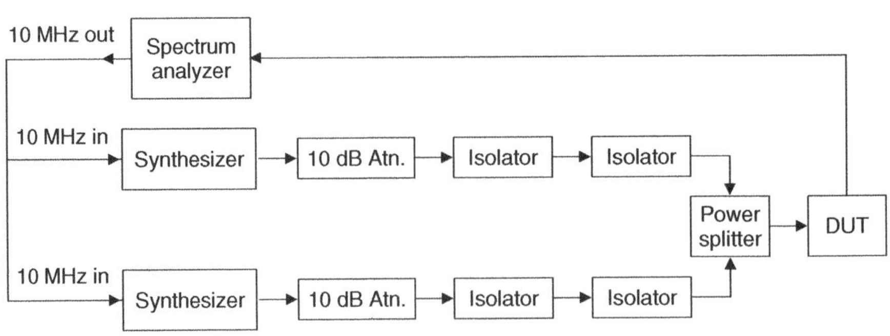

Figure 22.20 An IP3 measurement setup.

calculated using the following relationships:

$$IP2 = 2P_f - IM2 \tag{22.2}$$

and

$$IP3 = 0.5(3P_f - IM3) \tag{22.3}$$

where P_f, $IM2$, and $IM3$ are the power levels (dBm) of each of the two fundamental tones and in each of the second and third intermodulation products, respectively.

In this measurement three precautions are generally observed. First, isolators are placed between the DUT and the attenuators to minimize the interaction effects in order to keep the intermodulation component levels minimum at the input of the DUT. Second, the attenuator range in the spectrum analyzer is adjusted so that it operates in the linear region at the maximum power output from the DUT. Third, the power supply has the required filtering and no spurious signal should cause interaction between the supply and the DUT.

In the measurement of IP3, extra care is needed in the design of bias tees or biasing networks to keep the drain bias and gate bias voltage ripples to a minimum. In order to measure the C/I ratio better than 40 dBc, the gate bias voltage ripple and drain bias voltage ripple must be below 1% and 2%, respectively. The asymmetry in measured IM products, varying their levels with different frequency changes and elevating their levels at lower input power levels, may be due to transistor memory effects. The biasing networks (not shown) also must have high RF loss (30–40 dB) at frequencies up to 50 MHz to obtain reliable IP3 data.

22.3.3 ACPR Measurement

To perform adjacent channel power ratio (ACPR) measurements, two methods are used [28]: the resolution bandwidth (RBW) and integration bandwidth (IBW). Both these techniques differ in speed of measurement and in sensitivity to spectrum ripple. Figure 22.21 shows the measurement setup for these two methods.

In the RBW method, the RBW filter's bandwidth is set equal to 30 kHz and the video bandwidth (VBW) filter is set approximately 3 times the RBW filter's bandwidth, that is, 90 kHz to minimize the averaging response error. Then the ACPR is calculated by measuring the power spectral density (PSD) using a time-domain sweep of the

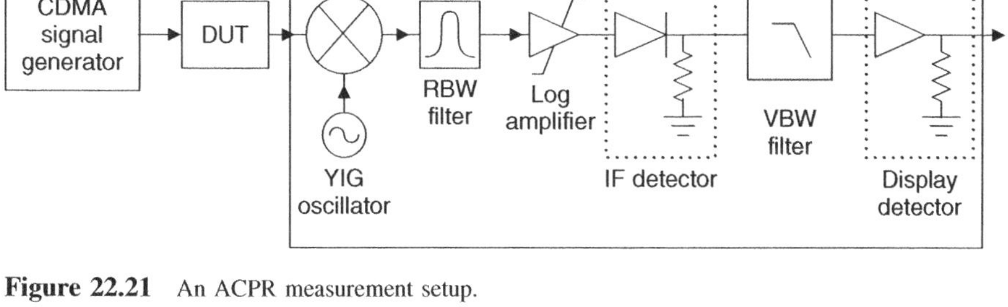

Figure 22.21 An ACPR measurement setup.

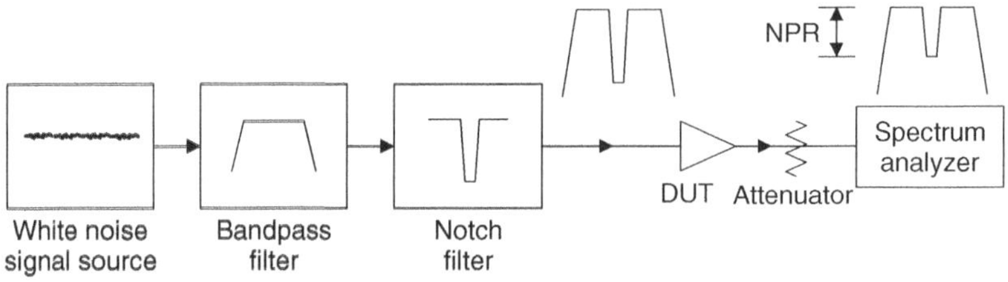

Figure 22.22 NPR measurement setup.

signal over specified bandwidths for the main as well as the offset channels as given in Eq. (3.30). For example, for CDMA power amplifiers, the two offset channels are 885 kHz and 1.98 MHz and 30 kHz is the measurement bandwidth. In the IBW method, the swept spectrum is used to calculate the PSD. Here, the RBW filter is set narrow in comparison to the desired measurement bandwidth. The results for ACPR measured by both methods are in close agreement.

22.3.4 NPR Measurement

The linearity of amplifiers is also evaluated by measuring the noise power ratio (NPR) using the test setup shown in Figure 22.22. In this technique the total distortion contributed by all adjacent channels is measured in the middle of the operating band. NPR measurement is analogus to the two-tone intermodulation method but represents multiple carriers passing through a nonlinear amplifier and measures carrier-to-interference level. As an example, in this measurement, a 25-MHz wide noise signal is generated at a carrier frequency using a white noise source and a bandpass filter. The bandpass filter is selected to limit the signal bandwidth that is equal to the bandwidth of the device under test. This signal then passes through a notch (bandstop) filter with a bandwidth of one channel equal to 30 kHz [5]. The rejection band of the bandstop filter is generally slightly larger than the passband of the DUT. In this case, the notch filter resonators need very high Q and in most cases 50 dB of measurement range is adequate. Due to intermodulation distortion, the noise floor of the channel is raised and the ratio of carrier-to-noise level is termed NPR. When measured in decibels (dB), it is the difference in carrier and noise power levels and is expressed in dBc.

22.3.5 EVM Measurement

In the EVM measurement a required digitally modulated test signal is generated. An arbitrary waveform generator (AWG) is used to generate I- and Q-baseband envelopes. Commercially available software are used to generate envelope waveforms. The I- and Q-baseband envelopes are modulated with a carrier frequency using an RF modulator. The EVM measurement is done by using a vector signal analyzer (VSA), which is basically a receiver. The VSA can handle a variety of modulation formats and frequencies. Figure 22.23 shows a typical EVM measurement test setup.

22.4 PHASE NOISE MEASUREMENT

Modern applications including radars and satellite communications require low phase noise in LNAs and power amplifiers to extend the system's operating range, precision,

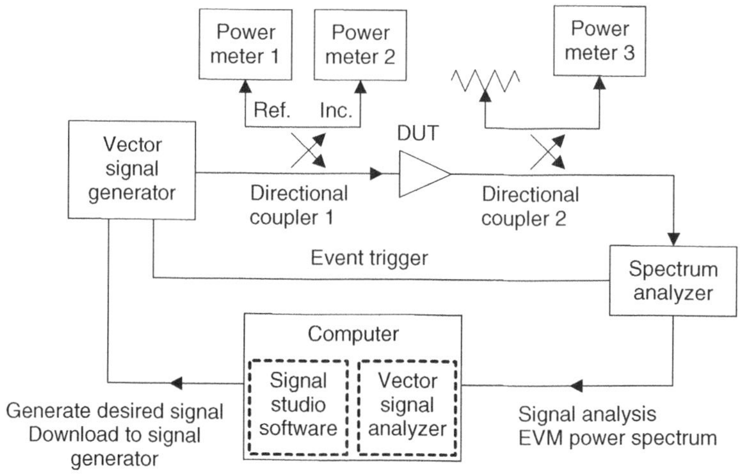

Figure 22.23 Typical EVM measurement test setup.

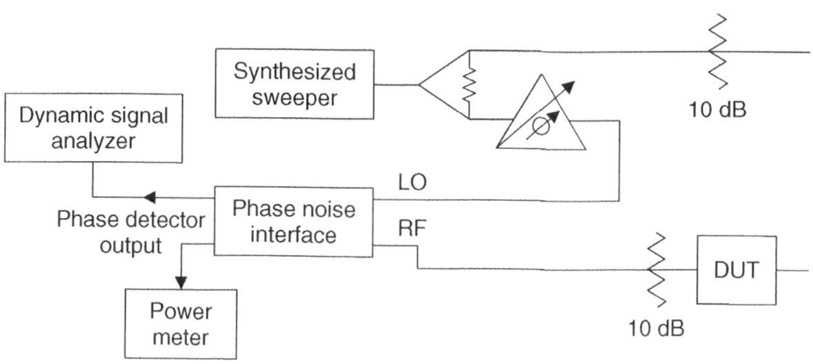

Figure 22.24 Typical phase noise measurement test setup.

and sensitivity. LNAs are designed for low noise figure, and predriver amplifiers in transmitters are often designed for low noise figure to maintain the transmitter noise figure at a low level. Apart from these design practices, no other design approaches have been reported for low phase noise in amplifiers. Basically, the phase noise in amplifiers depends largely on the transistor used. It has been observed that a larger FET/HEMT size in an LNA results in lower phase noise, and a smaller device aspect ratio in a multistage power amplifier provides lower phase noise.

There are several methods of making phase noise measurements, each with its own set of advantages and disadvantages. An overview of some of the common methods is given in Reference 7. Among these, the phase detector method using two sources yields the best overall sensitivity. Figure 22.24 shows a typical phase noise measurement test setup using Agilent instruments. The setup is calibrated by replacing the DUT with a low-loss through line and removing both attenuator pads. The source power is adjusted so that the measured power levels at the LO and RF ports are close to the maximum ratings on the equipment. In this case, the system's phase noise is in the range of −155 to −160 dBc at 1-kHz off center frequency.

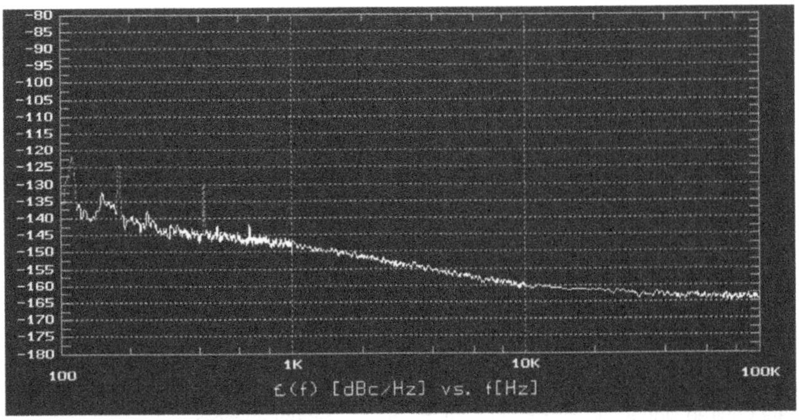

(a)

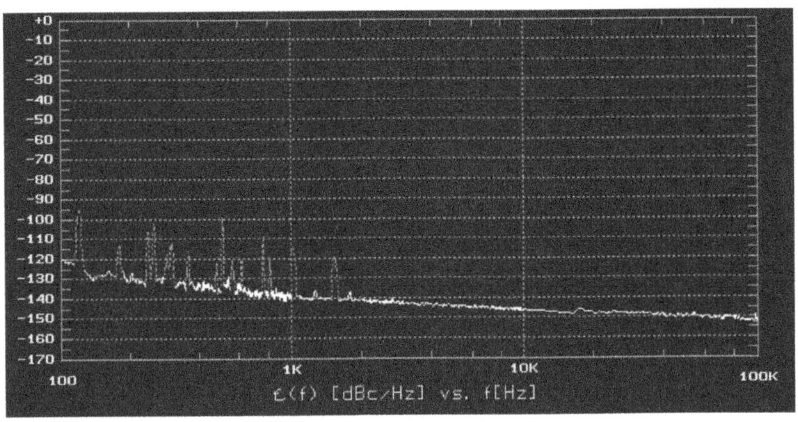

(b)

Figure 22.25 Typical measured phase noise of (a) a limiter/LNA and (b) an HPA.

Then the DUT is connected and the test sequence recommended by the equipment manufacturer is followed. Typical phase noise measured for a limiter/LNA described in Section 20.1 and for a 10-W HPA are shown in Figure 22.25. For more accurate and spike-free phase noise response, it is recommended that the measurements be performed in a screened/EM shielded room or in an environment with no other power radiating sources nearby.

22.5 RECOVERY TIME MEASUREMENT

High-power limiter/LNAs, described in the previous chapter, are widely used in commercial and military microwave systems, and one needs to measure their small-signal gain recovery time after the removal of the high-power signal. The circuit selected for these measurements was the commercially available M/A-COM limiter/LNA MMIC [29] and has been described in Section 20.1. This limiter/LNA has an operating bandwidth of 8.5–12 GHz, a nominal gain of 15 dB, an NF< 3 dB, and an input TOI of 13 dBm. Operating bias voltages and currents are 5 V, 130 mA nominal and −5 V, 4 mA nominal for the drain and gate, respectively.

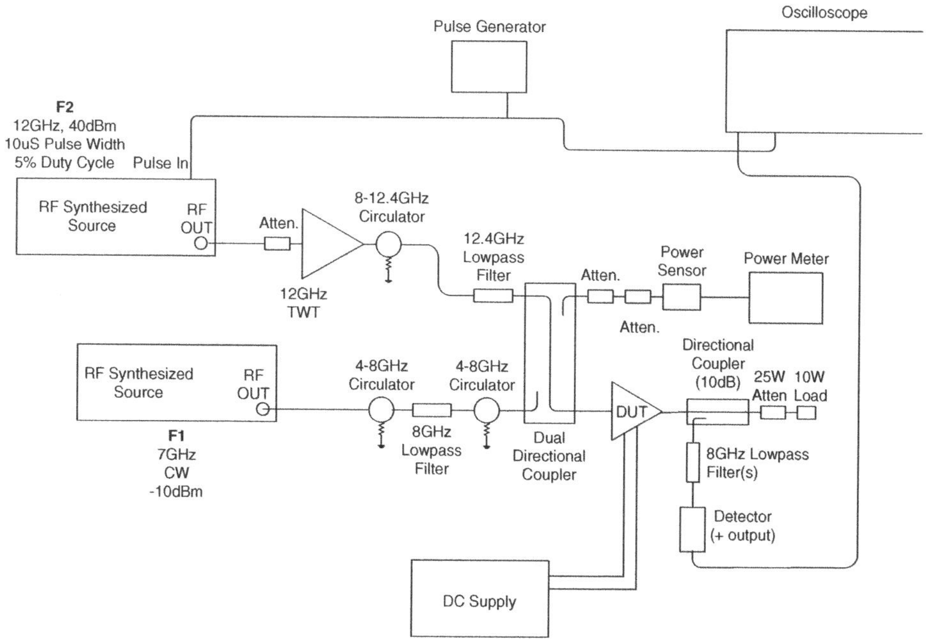

Figure 22.26 Limiter/LNA recovery time measurement test setup.

The limiter recovery time was measured by pulsing the input RF signal from a small-signal level to a high-power state (limiter active), then using a detector to measure the response time as the RF level drops back to normal operating levels [29]. Due to equipment limitations, a method to pulse the RF from a small but detectable signal level to a large-signal level was not readily available. Consequently, a system to measure recovery time using two RF tones was developed. By injecting a CW small-signal level at the low end of the DUT's band of operation (F1) and a pulsed high-power signal at the upper end (F2), it was possible to measure the small-signal recovery time by separating the two signals at the output of the device using typical components found in most test labs. The frequencies of the two signals were chosen to allow the use of existing lowpass filters to remove the F2 signal so only the response of the F1 signal would be measured. For this test, a dual directional coupler was used to combine the signals. The F1 signal was injected through the reflected power port of the coupler and the pulsed F2 signal was injected at the thru path. This uses the reverse isolation of the coupler to separate the small-signal F1 tone from the pulsed high-power F2 tone. For this device, F1 was a 7-GHz, −10-dBm CW signal and F2 was a 12-GHz, 40-dBm pulsed signal at 10-μs pulse width at a 5% duty cycle and they were separated using 8-GHz lowpass filters. At the output of the device, a directional coupler was used to sample the combined signals and lowpass filters were used to filter out the pulsed F2 tone. The recovery time measurement was then made using a positive output voltage detector connected to a high-frequency oscilloscope. Refer to Figure 22.26 for the test setup block diagram.

The graph in Figure 22.27 represents the RF levels of the two frequency tones versus time. When the high-power F2 signal is on ("high"), the Schottky diodes of the limiter are effectively short-circuited to ground. The CW F1 signal is subsequently attenuated and goes low.

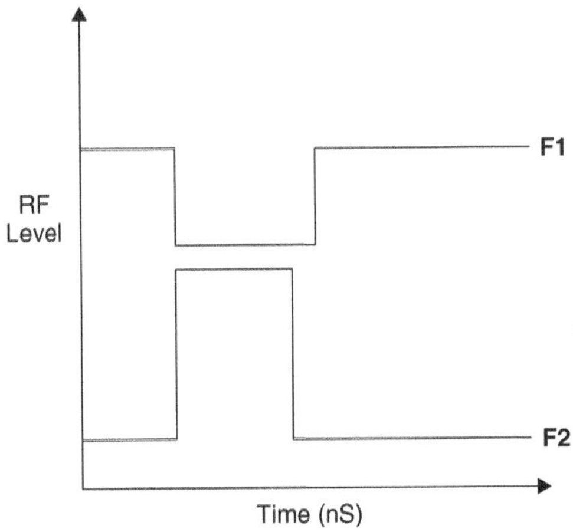

Figure 22.27 Representational plot of RF response times between F1 (CW, small-signal) and F2 (pulsed, large-signal) tones.

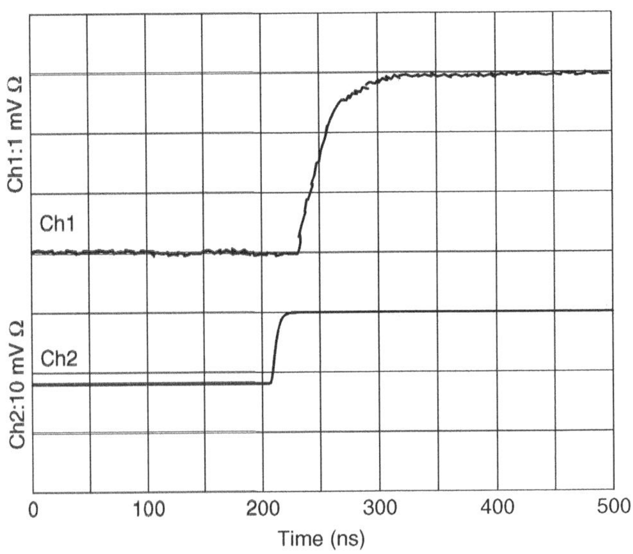

Figure 22.28 LNA small-signal recovery time plot of F1 tone (Ch1) under high-power pulsed stimulus (F2 tone, $P_{out} \sim 10$ W, Ch2, negative detector).

Figure 22.28 shows a typical RF recovery time plot from the oscilloscope using the internal "rise time" measurement capability to determine the 10% to 90% rise time (in this example 36.6 ns). By using a negative detector to look at the F2 pulse (at incident port of dual directional coupler), one could also measure the delay between the two signals. This plot was taken at the full rated power of the limiter, 10 W CW. A limiter recovery time acceptance limit of <100 ns could easily be achieved based on this data. Under small-signal conditions, the 10% to 90% rise time was less than 5 ns.

A brief description of transistor and amplifier measurements has been included in this chapter. In addition to basic measurement setups, experienced test designers use many other modifications in the setups and tricks to make accurate RF and microwave measurements. Readers are referred to the listed references for further details.

REFERENCES

1. Agilent, Santa Rosa, CA.
2. *Automated Tuner System*, Maury Microwave Corp., Ontario, CA. Focus Microwaves Inc., Pointe Claire, Quebec, Canada.
3. K. Chang (Ed.), *Handbook of Microwave and Optical Components*, John Wiley & Sons, Hoboken, NJ, 1989, Vol. 1, Chapter 9.
4. I. Bahl, G. Lewis, and J. Jorgenson, Automatic testing of MMIC Wafers, *Int. J. Microwave Millimeter-Wave Computer-Aided Eng.*, Vol. 1, pp. 77–89, January 1991.
5. M. Golio (Ed.), *RF and Microwave Handbook*, CRC Press, Boca Raton, FL, 2000.
6. K. Chang, I. Bahl, and V. Nair, *RF and Microwave Circuit and Component Design for Wireless Systems*, John Wiley & Sons, Hoboken, NJ, 2002.
7. I. Bahl and P. Bhartia, *Microwave Solid State Circuit Design*, 2nd edition, John Wiley & Sons, Hoboken, NJ, 2003, Chapter 9.
8. I. J. Bahl, 10W CW broadband balanced limiter/LNA fabricated using MSAG MESFET process, *Int. J. RF Microwave Computer-Aided Eng.*, Vol. 13, pp. 118–127, March 2003.
9. J. C. Pedro, Intermodulation, in *Encyclopedia RF and Microwave Engineering*, K. Chang (Ed.), Vol. 3, John Wiley & Sons, Hoboken, NJ, 2005, pp. 2189–2215.
10. M. T. Abuelmaatti, Intermodulation Measurement, in *Encyclopedia RF and Microwave Engineering*, K. Chang (Ed.), Vol. 3, John Wiley & Sons, Hoboken, NJ, 2005, pp. 2215–2231.
11. A. Ferrero, V. Teppati, and U. Pisani, Microwave measurements, in *Encyclopedia RF and Microwave Engineering*, K. Chang (Ed.), Vol. 3, John Wiley & Sons, Hoboken, NJ, 2005, pp. 2802–2812.
12. H.-C. Chang, Phase noise and measurements, in *Encyclopedia RF and Microwave Engineering*, K. Chang (Ed.), Vol. 4, John Wiley & Sons, Hoboken, NJ, 2005, pp. 3802–3809.
13. D. Vye, Measuring S-parameters: the first 50 years, *Microwave J.*, Vol. 51, pp. 34–52, March 2008.
14. T. G. Ruttan et al., Multiport VNA measurements, *IEEE Microwave Mag.*, Vol. 9, pp. 56–69, June 2008.
15. A. Rumiantsev and N. Ridler, VNA calibration, *IEEE Microwave Mag.*, Vol. 9, pp. 86–99, June 2008.
16. J. M. Cusak, S. M. Perlow, and B. S. Pelaman, Automatic load contour mapping for microwave power transistors, *IEEE Trans. Microwave Theory Tech.*, Vol. MTT-22, pp. 1146–1152, December 1974.
17. Y. Takayama, A new load-pull characterization method for microwave power transistors, *IEEE MTT-S Int. Microwave Symp. Dig.*, pp. 218–220, 1976.
18. R. B. Stancliff and D. D. Poulin, Harmonic load-pull, *IEEE MTT-S Int. Microwave Symp. Dig.*, pp. 185–187, 1979.
19. G. P. Bava et al., Active load techniques for load-pull characterization at microwave frequencies, *Electron. Lett.*, Vol. 18, pp. 178–180, 1982.
20. R. S. Tucker and P. D. Bradley, Computer-aided error correction of large-signal load-pull measurements, *IEEE Trans. Microwave Theory Tech.*, Vol. MTT-32, pp. 296–300, March 1984.
21. R. Actis and R. A. McMorran, Millimeter load pull measurements, *Appl. Microwave*, Vol. 1, pp. 91–102, November/December 1989.
22. B. Hughes, A. Ferrero, and A. Cognata, Accurate on-wafer power and harmonic measurements of MM-wave amplifiers and devices, *IEEE MTT-S Int. Microwave Symp. Dig.*, pp. 1019–1022, 1992.
23. M. Demmler, B. Hughes, and A. Cognata, A 0.5–50GHz on-wafer, intermodulation, load-pull and power measurement system, *IEEE MTT-S Int. Microwave Symp. Dig.*, pp. 1041–1044, 1995.

24. F. M. Ghannouchi, G. Zhao, and F. Beauregard, Simultaneous load-pull of intermodulation and output power under two-tone excitation of accurate SSPA's design, *IEEE Trans. Microwave Theory Tech.*, Vol. 42, pp. 929–934, June 1994.
25. P. Berni et al, An experimental study of the effects of harmonic loading on microwave MESFET oscillators and amplifiers, *IEEE Trans. Microwave Theory Tech.*, Vol. 42, pp. 963–950, June 1994.
26. J. M. Nebus et al., Improvement of the active load-pull technique for the optimization of high power communication SSPAs, *Int. J. Microwave Millimeter-Wave Computer-Aided Eng.*, Vol. 5, pp. 169–160, July 1995.
27. I. J. Bahl and K. C. Gupta, Measurement of parameters of an artificial dielectric using a partially filled parallel plate waveguide, *Int. J. Electronics*, Vol. 28, pp. 173–177, February 1970.
28. M. Slovick, Measuring ACPR in CDMA amplifiers, *Microwave J.*, Vol. 41, pp. 74–80, December 1998.
29. J. Looney, D. Conway, and I. Bahl, An examination of recovery time of an integrated limiter/LNA, *IEEE Microwave Mag.*, Vol. 5, pp. 83–86, March 2004.
30. P. Hulbert, Use pulsed $I-V$ testing to characterize RF devices, *Microwaves RF*, Vol. 47, pp. 56–65, August 2008.

PROBLEMS

22.1 Transistor S-parameters were measured in common-source configuration and given in Table 5.1a. Determine the S-parameters over 2–4 GHz with 0.5-GHz step for common-gate and common-drain configurations. Calculate MAG and input RL and MAG and output RL, respectively. Determine their sizes for a 50-Ω match at 2 GHz.

22.2 What are the basic requirements for the test setup shown in Figures 22.18b and 22.16 for accurate NF and power measurements.

22.3 Describe the importance of filters, attenuator pads, and circulators in scalar measurement setups.

22.4 Describe the basic equations based on transmission line theory to measure VSWR.

22.5 Describe the critical test equipment requirements for IP3 and phase noise measurements.

22.6 A four-port coupler is to be tested using a VNA. Describe the test procedure.

22.7 Convert the S-parameters in Problem 22.1 into a 25-Ω system and show that the value of MAG does not change.

22.8 List the various critical requirements for accurate IP3 measurements.

22.9 Describe the advantages and disadvantages of SOLT, LRM, and TRL deembedding techniques for on wafer S-parameter measurements from RF to millimeter-wave frequencies.

Appendix A

Physical Constants and Other Data

Permittivity of vacuum, $\varepsilon_0 = 8.854 \times 10^{-12} \cong (1/36\pi) \times 10^{-9}$ F/m

Permeability of vacuum, $\mu_0 = 4\pi \times 10^{-7}$ H/m

Impedance of free space, $\eta_0 = 376.7 \cong 120\pi\,\Omega$

Velocity of light, $c = 2.998 \times 10^8$ m/s

Charge of electron, $e = 1.602 \times 10^{-19}$ C

Mass of electron, $m = 9.107 \times 10^{-31}$ kg

Mass of proton, $M = 1.67 \times 10^{-27}$ kg

Boltzmann's constant, $k = 1.380 \times 10^{-23}$ J/K

Planck's constant, $h = 6.547 \times 10^{-34}$ J·s

10^7 erg = 1 J

1 joule = 0.6285×10^{19} eV

1 electron volt = energy gained by an electron in accelerating through a potential of 1 V

Conductivity of copper, $\sigma = 5.8 \times 10^7$ S/m

Conductivity of gold, $\sigma = 4.1 \times 10^7$ S/m

$j = \sqrt{-1}$

$\pi = 3.1416$

Fundamentals of RF and Microwave Transistor Amplifiers. By Inder J. Bahl

Appendix B

Units and Symbols

B.1 SI UNITS AND THEIR SYMBOLS

In 1960, the International System of Units was established as a result of a long series of international discussions. This modernized metric system, called SI, from the French name, Le Systeme International d'Unités, is now, as a general world trend, to replace all former systems of measurement, including former versions of the metric system.

In the SI system, four physical quantities are classified as fundamental: *length, mass*, *time*, and *charge*. For practical purposes, *temperature* is included here as a basic unit. In Table B.1 the first five are basic quantities and the rest are derived quantities; that is, their dimensions can be expressed as a combination of the first five.

B.2 METRIC PREFIXES

The nomenclature in this decimal structure is derived from a system of prefixes, which are attached to units of all sorts. For example, the prefix "kilo" means 1000, hence kilometer, kilogram, and kilowatt mean 1000 meters, 1000 grams, and 1000 watts, respectively. Most of our everyday experiences with metric units will involve some of the prefixes listed in Table B.2.

Fundamentals of RF and Microwave Transistor Amplifiers. By Inder J. Bahl

Table B.1 SI Units and Their Symbols

Quantity	SI Unit	SI Symbol	Dimensions
Length	meter	m	Basic
Mass	kilogram	kg	Basic
Charge	coulomb	C	Basic
Time	second	s	Basic
Temperature	kelvin	K	Basic
Frequency	hertz	Hz	1/s
Energy	joule	J	$kg \cdot m^2/s^2$
Force	newton	N	$kg \cdot m/s^2$
Power	watt	W	J/s
Pressure	pascal	Pa	N/m^2
Electric current	ampere	A	C/s
Electric potential (voltage)	volt	V	J/C
Electric field	volts/meter	V/m	J/C/m
Resistance	ohm	Ω	V/A
Resistivity	ohm·meter	Ω·m	V·m/A
Conductance	siemen	S	A/V
Capacitance	farad	F	C/V
Permittivity	farad/meter	F/m	F/m
Magnetic field	ampere/meter	A/m	A/m
Inductance	henry	H	V · s/A
Permeability	henry/meter	H/m	H/m

Table B.2 SI Prefixes

	Prefix	Symbol	Factor by Which the Unit is Multiplied	
	exa	E	10^{18}	
	peta	P	10^{15}	
	tera	T	10^{12}	
→	giga	G	10^{9}	All uppercase
→	mega	M	10^{6}	↕
→	kilo	k	10^{3}	All lowercase
	hecto	h	10^{2}	
	deca	da	10^{1}	
			10^{0}	
	deci	d	10^{-1}	
→	centi	c	10^{-2}	
→	milli	m	10^{-3}	
→	micro	μ	10^{-6}	
→	nano	n	10^{-9}	
→	pico	p	10^{-12}	
→	femto	f	10^{-15}	
	atto	a	10^{-18}	

Appendix C

Frequency Band Designations

Table C.1 Radio frequency and Radar Band Frequency Designations

GLF	30 Hz–300 Hz	Extremely low frequency
VF	300 Hz–3 kHz	Voice frequency
VLF	3 kHz–30 kHz	Very low frequency
LF	30 kHz–300 kHz	Low frequency
MF	300 kHz–3 MHz	Medium frequency
HF	3 MHz–30 MHz	High frequency
VHF	30 MHz–300 MHz	Very high frequency
UHF	300 MHz–3 GHz	Ultrahigh frequency
SHF	3 GHz–30 GHz	Superhigh frequency
EHF	30 GHz–300 GHz	Extremely high frequency

Table C.2 Lettered Radar Band Designations

Band Designator	Frequency Range	
VHF	30–300 MHz	
UHF	300–1000 MHz	
L	1000–2000 MHz	μ wave
S	2000–4000 MHz	μ wave
C	4000–8000 MHz	μ wave
X	8000–12,500 MHz	μ wave
Ku	12.5–18 GHz	μ wave
K	18–26.5 GHz	μ wave
Ka	26.5–40 GHz	μ wave
Millimeter	>40 GHz	

Appendix D

Decibel Units (dB)

The illustration of a parameter having orders of magnitude on a linear chart is impractical. Also, the manual addition of two quantities is easier than multiplication. The decibel unit (dB) allows one to present amplifier gain and power quantities in a simple way.

D.1 POWER RATIO TO dB

The Decibel (dB)

The decibel is a logarithmic unit of power ratio, although it is commonly also used for current ratio and voltage ratio. If the input power P_i and the output power P_o of a network are expressed in the same units, then the network insertion gain or loss is

$$G = 10 \log \frac{P_o}{P_i} \text{dB} \quad \text{or} \quad G = -10 \log \frac{P_i}{P_o} \text{dB}$$

For example, if $P_o = 10$ W and $P_{in} = 2.5$ W (like testing a single-stage HPA under compression),

$$G = 10 \log \left(\frac{10 \text{ W}}{2.5 \text{ W}} \right) = 10 \log 4 = 6 \text{ dB}$$

If $P_o = 2$ W and $P_{in} = 10\text{mW} = 0.01\text{W}$ like testing a driver for small-signal gain),

$$\begin{aligned} G &= 10 \log \left(\frac{2 \text{ W}}{0.01 \text{ W}} \right) = 10 \log 200 \text{ dB} \\ &= 10(\log 2 + \log 100) = 10(0.3 + 2) = 23 \text{ dB} \end{aligned}$$

The decibel unit (dB) is used for characterizing or measuring gain, noise figure, return loss, insertion loss, conversion loss, and so on.

Fundamentals of RF and Microwave Transistor Amplifiers. By Inder J. Bahl

dB to Power Ratio

$$G = 10\ \log(P_o/P_i)$$
$$P_o/P_i = \text{antilog}(G/10) = 10^{G/10}$$

For example, for $G = 5$ dB,

$$P_o/P_i = 10^{5/10} = 10^{0.5} = \sqrt{10} \cong 3.16$$

Table D.1 provides the relationships between power ratio and the decibel unit for commonly used power ratio values.

The dBm and dBW

The absolute power levels are expressed in dBm, which is defined as the power level P in reference to 1 mW; that is,

$$P(\text{dBm}) = 10\ \log \frac{P(\text{mW})}{1\text{mW}}$$

If the reference power level is 1 W, then the power ratio is P(dBW).

If the signal is expressed in dB, with respect to the carrier level it is expressed in dBc. For example,

$$P = 1\ \text{mW} = 10 \log \frac{1\text{mW}}{1\text{mW}} = 0\ \text{dBm}$$
$$P = 5\ \text{mW} = 10 \log \frac{5\text{mW}}{1\text{mW}} = 7\ \text{dBm}$$
$$P = 10\ \text{W} = 10 \log \frac{10 \times 10^3\text{mW}}{1\text{mW}} = 40\ \text{dBm}$$
$$P = 10\ \text{W} = 10 \log \frac{10\text{W}}{1\text{W}} = 10\ \text{dBW}$$

Table D.1 Relationships Between Ratio, R, and Decibel Unit (dB)

Ratio, R	log R	dB	Ratio, R	log R	dB
1	0	0	10	1	10
2	0.3	3	100	2	20
3	0.48	4.8	10^3	3	30
4	0.60	6	10^4	4	40
5	0.70	7	10^5	5	50
6	0.78	7.8	0.1	−1	−10
7	0.85	8.5	0.01	−2	−20
8	0.90	9.0	0.001	−3	−30
9	0.95	9.5	10^{-4}	−4	−40
			10^{-5}	−5	−50

The conversions of typical power units to mW are given as follows:

$$1\text{ MW} = 10^9\text{ mW} \quad 1\text{kW} = 10^6\text{ mW}$$
$$1\text{ W} = 10^3\text{ mW} \quad 1\mu\text{W} = 10^{-3}\text{ mW}$$
$$1\text{ nW} = 10^{-6}\text{ mW} \quad 1\text{pW} = 10^{-9}\text{ mW}$$

More Examples

(a) $3\ mW = 5$ dBm
(b) 50 mW = 17 dBm
(c) 60 W = 6×10^4 mW = 48 dBm
(d) 4 kW = 4×10^6 mW = 66 dBm
(e) 2.5 MW = 2.5×10^9 mW = 94 dBm
(f) 5 μW = 5×10^{-3} mW = −23 dBm
(g) 8 nW = 8×10^{-6} mW = −51 dBm
(h) 20 pW = 2×10^{-8} mW = −77 dBm

$$P_c = \text{carrier level} = 10\text{ dBm}$$
$$P_s = \text{spurious level} = -20\text{ dBm}$$
$$P_c/P_s = 10\text{ dBm} - (-20\text{ dBm}) = 30\text{ dBc}$$
$$\text{Or } P_s/P_c = -20\text{ dBm} - 10\text{ dBm} = -30\text{ dBc} \leftarrow \text{more accurate, reference is carrier}$$

Input reflection coefficient $S_{11} = 0.1$

$$\text{Return loss} = 10\ \log \frac{1}{|S_{11}|^2} = 20\text{ dB}$$

Always with positive sign

Input reflection = $20\ \log |S_{11}| = -20$ dB

Always with negative sign

Table D.2 provides decibel operations and their meanings.

dBm to Power

$$P(\text{dBm}) = 10\ \log \frac{P(\text{mW})}{1\text{mW}}$$
$$P(\text{mW}) = \text{antilog}[(P\text{dBm})/10]$$
$$= 10^{P(\text{dBm})/10}$$

Table D.2 Decibel Operations and Their Meanings

Operation	Resulting Unit	Physical Meaning	Allowed?
1. dB + dB	dB	Product of two numbers	Yes
2. dB − dB	dB	Comparing two numbers	Yes
3. dBm + dBm	XX	Multiplying two powers	No
4. dBm − dBm	dB or dBc	Comparing two powers	Yes
5. dBm + dB	dBm	Power amplification	Yes
6. dBm − dB	dBm	Power attenuation	Yes

For example,

$$P(\text{dBm}) = 23\ \text{dBm}$$
$$P(\text{mW}) = 10^{23/10} = 10^{2.3} = 200\ \text{mW}$$

D.2 VOLTAGE RATIO

So far, only power ratios and power amplification have been described. However, in low-frequency electronics, voltage amplification (i.e., voltage gain) is an important factor. Although voltage gain can also be expressed as a ratio, this particular ratio cannot be immediately converted to decibels. Remember that decibels originated in power comparisons. It is possible in many cases, however, to express powers in terms of the associated voltages. Since power = (voltage)^2/resistance, the following can be done. Since

$$\text{Power A} = V_A^2/R_A$$

and

$$\text{Power B} = V_B^2/R_B$$

then

$$\begin{aligned} \text{dB} &= 10 \times \log(P_A/P_B) \\ &= 10 \times \log \frac{V_A^2 \cdot R_B}{V_B^2 \cdot R_A} \end{aligned}$$

Normally, the two resistances R_A and R_B are chosen to be equal for purposes of comparison. This means that the power generated by two separate voltages, V_A and V_B, can be compared when applied across resistors of equal values (or across one standard resistor).

$$\begin{aligned} \text{dB} &= 10 \times \log(V_A^2/V_B^2) \\ &= 20 \times \log(V_A/V_B) \end{aligned}$$

D.3 CURRENT RATIO

Similarly, current gain can also be expressed in decibels as

$$\text{dB} = 20 \log(I_A/I_B)$$

Appendix E

Mathematical Relationships

1. Quadratic equation

$$ax^2 + bx + c = 0$$
$$x = \frac{-b \pm \sqrt{b^2 - 4ac}}{2a}$$

2. Properties of logarithms

$$\log_a x = N, x = a^N$$
$$e = 2.71828$$
$$\log_e x = \ln x = 2.30259 \log_{10} x$$
$$\log_{10} x = 0.43429 \log_e x$$
$$\log_a b = 1/\log_b a$$

3. Trigonometric identities

$$\sin^2 x + \cos^2 x = 1$$
$$\tan x = \sin x/\cos x$$
$$\sin x = (e^{jx} - e^{-jx})/2j$$
$$\cos x = (e^{jx} + e^{-jx})/2$$
$$\sin(x \pm y) = \sin x \cos y \pm \cos x \sin y$$
$$\cos(x \pm y) = \cos x \cos y \mp \sin x \sin y$$
$$\sin 2x = 2 \sin x \cos x$$
$$\cos 2x = \cos^2 x - \sin^2 x$$
$$\sin x \cos y = 1/2[\sin(x + y) + \sin(x - y)]$$
$$\sin x \sin y = 1/2[\cos(x - y) - \cos(x + y)]$$
$$\cos x \cos y = 1/2[\cos(x + y) + \cos(x - y)]$$

Fundamentals of RF and Microwave Transistor Amplifiers. By Inder J. Bahl

4. Hyperbolic functions

$$\sinh x = (e^{x} - e^{-x})/2$$
$$\cosh x = (e^{x} + e^{-x})/2$$
$$\tanh x = \sinh x/\cosh x$$

5. Circle equation

$$(x - a)^2 + (y - b)^2 = r^2$$

The circle has a center at (a, b) and its radius is given by r.

Appendix F

Smith Chart

A Smith chart is a graphical aid used for solving impedance matching problems involving the use of Eq. (7.15) for Z_{in}. Writing $Z(z)/Z_0 = R + jX$, where Z_0 is the normalizing impedance and $\rho(z) = u + jv$, we have

$$R + jX = \frac{1 + u + jv}{1 - u - jv} \tag{F.1}$$

Equating real and imaginary parts on the two sides of (F.1), we find

$$\left(u - \frac{R}{1+R}\right)^2 + v^2 = \frac{1}{(1+R)^2} \tag{F.2}$$

$$(u-1)^2 + \left(v - \frac{1}{X}\right)^2 = \frac{1}{X^2} \tag{F.3}$$

Equations (F.2) and (F.3) represent two families of circles in the complex $\rho(= u + jv)$ plane. The first set, given by (F.2), corresponds to circles for various values of parameter R. These circles have their centers at $(R/(1+R), 0)$ and their radii are given by $1/(1+R)$. These are called *constant-resistance circles*. Equation (F.3), on the other hand, represents *constant-reactance circles* with centers at $(1, 1/X)$ and radii given by $1/X$.

From (7.15) it may be noted that for any positive Z_L, the maximum value of $|\rho|$ is unity. Thus the area of interest in the complex (u, v) plane lies inside a circle with radius $|\rho|$ equal to unity. A graphical representation of this area with constant-R and constant-X circles, as given by (F.2) and (F.3) shown therein, is known as a *Smith chart*. The layout of the Smith chart is shown in Figure F.1. Commonly used charts have more constant-R and constant-X circles than shown here. Also, constant-$|\rho|$ circles are normally not printed on the chart. Value of R and X as shown are normalized with respect to Z_0.

A few more observations about the Smith chart are in order. The point $R = 1$, $X = 0$ corresponds to $u = 0$, $v = 0$, the origin of the ρ plane, that is, the center of the Smith chart. The distance from the origin to any other point gives the magnitude of ρ and the angle between the $X = 0$ line, and the line connecting the point to the origin gives the phase angle of ρ. When one moves along the transmission line from the load end toward the generator, the ρ vector moves clockwise. The information is noted on

Fundamentals of RF and Microwave Transistor Amplifiers. By Inder J. Bahl

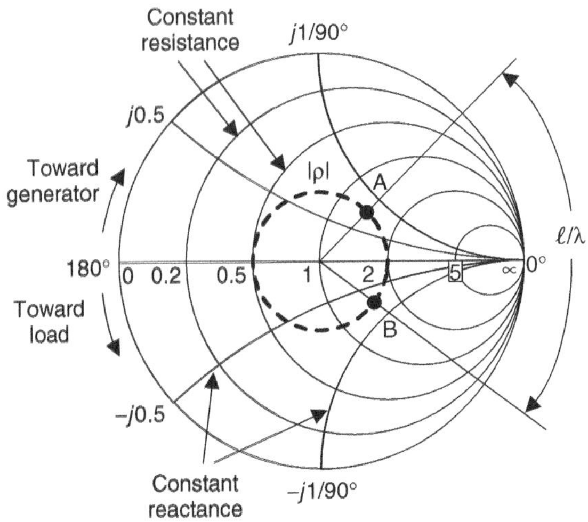

Figure F.1 Smith chart.

the Smith chart as "toward generator." Also, the angles $2\beta\ell(=4\pi\ell/\lambda)$ are marked in terms of ℓ/λ. The chart covers a range of ℓ/λ equal to 0.5, which corresponds to $2\beta\ell = 2\pi$. For $\beta\ell$ greater than π, $\rho(z)$ and $Z(z)$ repeat themselves.

In order to calculate Z_{in} for a given Z_L (i.e., to solve Eq. (7.15)), one uses the fact that as one moves along a uniform section of a transmission line, $|\rho|$ remains constant. Z_{in} is obtained as follows: (a) locate Z_L/Z_0 on the Smith chart, say, point A in Figure F.1; (b) draw a constant-$|\rho|$ circle passing through A; (c) move on this circle clockwise (toward generator) by the required distance ℓ/λ from point A to point B; and (d) read R and X at point B to obtain Z_{in}/Z_0. Finally, R and X are multiplied by Z_0 to obtain their absolute values.

Appendix G

Graphical Symbols

Table G.1 Summary of Graphical Symbols for Microwave Components, Devices, and Circuits

One-Port Components

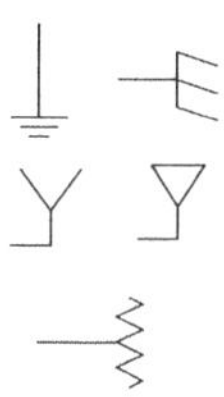

- Short-circuit/ground
- Antenna
- Matched load

Two-Port Transmission Lines

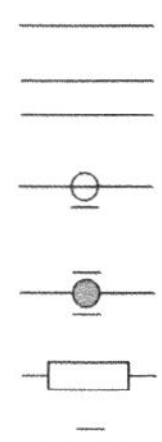

- Transmission line
- Coaxial line
- Stripline
- Microstrip

Two-Port Devices

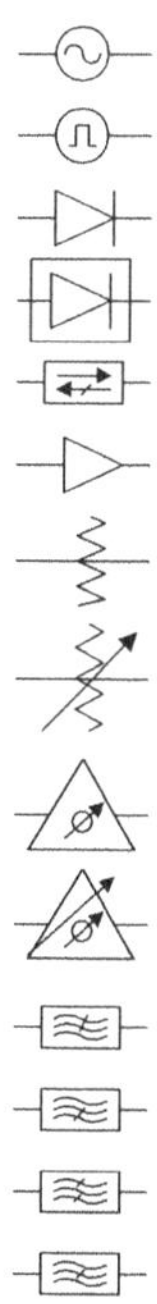

- CW generator
- Pulse generator
- Diode
- Detector
- Isolator
- Amplifier
- Fixed attenuator
- Variable attenuator
- Fixed phaseshifter
- Variable phase shifter
- Lowpass filter
- Highpass filter
- Bandpass filter
- Bandstop filter

Multiport Components

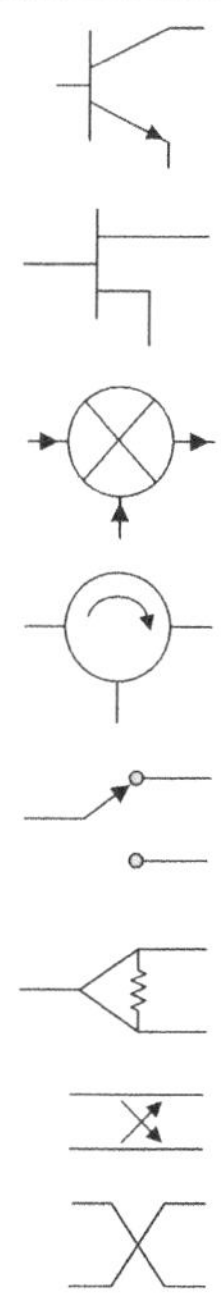

- BJT
- MESFET
- Mixer
- Circulator
- Switch
- Power divider
- Directional coupler

Fundamentals of RF and Microwave Transistor Amplifiers. By Inder J. Bahl

Appendix H

Acronyms and Abbreviations

ACL	Adjacent channel leakage
ACPR	Adjacent channel power ratio
AGC	Automatic gain control
AM	Amplitude modulation
AMPS	Advanced mobile phone service
ANN	Artificial neural network
APHC	Average power-handling capability
BER	Bit error rate
BGA	Ball-grid array
BiCMOS	Bipolar complementary metal oxide semiconductor
BJT	Bipolar junction transistor
BOC	Back of the chip
BPA	Balanced power amplifier
BPSK	Biphase shift keying
BW	Bandwidth
BV	Breakdown voltage
CAD	Computer-aided design
CD, CG, CS	Common drain, common gate, common source
CDMA	Code division multiple access
CMOS	Complementary metal oxide semiconductor
CPW	Coplanar waveguide
CSP	Chip-scale package
CTE	Coefficient of thermal expansion
CW	Continuous wave
DA	Distributed amplifier
DBS	Direct broadcast satellite
DC	Direct current
DECT	Digital European cordless telecommunication
DIP	Dual-in-line package
DL	Dissipative loss
DPD	Digital predistortion
DR	Dynamic range
DUT	Device under test

Fundamentals of RF and Microwave Transistor Amplifiers. By Inder J. Bahl

EC	Equivalent circuit
EER	Envelope elimination and restoration
EM	Electromagnetic
EMT	Electrical measurement technique
ENR	Excess noise ratio
ESR	Effective series resistance
ETM	Electronic tuner module
EVM	Error vector magnitude
EW	Electronic warfare
FCC	Federal Communications Commission
FET	Field effect transistor
FM	Frequency modulation
FP	Field plate
FSK	Frequency shift keying
GCN	Gain compensation network
GMSK	Gaussian-filtered minimum shift keying
GPS	Global positioning system/satellite
GSM	(Groupe Special Mobile) global system for mobile communications
HBT	Heterojunction bipolar transistor
HCA	Harmonic control amplifier
HEMT	High electron mobility transistor
HIT	Harmonic injection technique
HIPERLAN	High-performance radio LAN
HPA	High-power amplifier
HRA	Harmonic reaction amplifier
HTCC	High-temperature cofired ceramic
HV	High voltage
I and Q	In-phase and quadrature phase
IC	Integrated circuit
IF	Intermediate frequency
IL	Insertion loss
IM	Intermodulation
IM3, IM5	Third- and fifth-order intermodulation products
IMD	Intermodulation distortion
IMPATT	Impact ionization avalanche transit time diode
IP	Intermodulation product
IP2 and IP3	Second- and third-order intermodulation intercepts
IR	Infrared
ISS	Impedance standard substrate
ISM	Industrial, scientific and medical
JTC	Junction voltage temperature coefficient
JTRS	Joint tactical radio system
LAN	Local area network
LDMOS	Laterally diffused metal oxide semiconductor
LE	Lumped element
LF	Lead frame
LINC	Linear amplification with nonlinear control
LLM	Low-loss match

LNA	Low-noise amplifier
LO	Local oscillator
LPF	Lowpass filter
LRM	Line, reflect, and match
LSM	Longitudinal-section magnetic
LTCC	Low-temperature cofired ceramic
MAG	Maximum available gain
MCP	Multichip package
MESFET	Metal semiconductor field effect transistor
MIC	Microwave integrated circuit
MIM	Metalinsulatormetal
MIS	Metal insulator semiconductor
MMIC	Monolithic microwave integrated circuit
ML	Mismatch loss
MN	Matching network
MODFET	Modulation doped field effect transistor
MOSFET	Metal oxide semiconductor field effect transistor
MSAG	Multifunction self-aligned gate
MSG	Maximum stable gain
MTTF	Mean time to failure
NDF	Normalized determinant function
NF	Noise figure
NFmin	Minimum noise figure
NPR	Noise power ratio
NRE	Nonrecurring engineering
OFDM	Orthogonal frequency division multiplex
OTF	Open-loop transfer function
PA	Power amplifier
PAA	Phased array antenna
PAE	Power added efficiency
PAR	Phased array radar, peak-to-average ratio
PCB	Printed circuit board
PCM	Pulse code modulation
PCN	Personal communication network
PCS	Personal communication service
PD	Predistortion, power density
pHEMT	Pseudomorphic high electron mobility transistor
PM	Phase modulation
PPR	Point-to-point radio
PQFN	Power quad flat no lead
QPSK	Quadrature phase shift keying
QAM	Quadrature amplitude modulation
RF	Radiofrequency
RFIC	Radiofrequency integrated circuit
RL	Return loss
RMS	Root mean square
RR	Return ratio

SA	Spectrum analyzer
SAG	Self-aligned gate
SIP	System in package
SMLP	Surface mount leadless package
SOIC	Small outline integrated circuit
SOLT	Short, open, load, and through
SONET	Synchronous optical network
SOT	Small outline transistor
SPC	Spatial power combiner
SRF	Self-resonance frequency
SWR	Standing-wave ratio
TAB	Tape automatic bonding
TCR	Temperature coefficient of resistance
TE, TM	Transverse electric, transverse magnetic
TEM	Transverse electromagnetic
TDD	Time division duplex
TDMA	Time division multiple access
TO	Transistor outline
TOI	Third-order intercept
TRL	Through, reflect, and line
T/R	Transmit/receive
TSOP	Thin small-outline package
TSSOP	Thin-shrink small-outline package
TWC	Traveling-wave coupler
TWTA	Traveling-wave tube amplifier
UHF	Ultra high frequency
UWB	Ultra wideband
VHF	Very high frequency
VHV	Very high voltage
VLSI	Very large-scale integration
VNA	Vector network analyzer
VOD	Voice on demand
VOIP	Voice over internet protocol
VSAT	Very small aperture terminal
VSWR	Voltage standing-wave ratio
WAN	Wide area network
WBG	Wide bandgap
WiMAX	Worldwide interoperability for microwave access
WLAN	Wireless local area network
WLL	Wireless local loop

Appendix I

List Of Symbols

Symbols and acronyms used in the text are defined below. Because of the large number of quantities to be represented and the undesirability of using alphabets other than English and Greek, it has been necessary to use the same symbols to represent different quantities at different places. In each instance, the symbol has been defined upon introduction to avoid misinterpretation of its meaning.

A	Area
AuSn	Gold–tin
B	Signal bandwidth, Susceptance
c	Velocity of electromagnetic waves in free space
C	Capacitance, coupling coefficient, specific heat of semiconductors
C_1, C_t	Capacitances
CuW	Copper–tungsten
e	Electron charge
E	Electric field, error vector
f, f_ℓ, f_h, f_0	Frequencies
f_c	Cutoff frequency
f_{max}	Maximum frequency, frequency at unity power gain
f_r	Resonant frequency
f_T	Frequency at unity current gain
F	Noise figure ratio
F_{min}	Minimum noise figure
g_m	device transconductance
G	Conductance, gain
GaAs	Gallium arsenide
GaN	Gallium nitride
G_A, G_T, G_{max}	Gains
h	Substrate thickness
HP_2, HP_3	Second- and third-harmonic power levels
InP	Indium phosphide
k	Wave number, Boltzmann constant
k_0	Free-space wave number
K	Rollett stability factor, thermal conductivity
K_{th}, K_{GaAs}	Thermal conductivities
l, l_1, l_2	Lengths of transmission lines
L	Length of the microstrip, inductance

M	Magnetic field, mutual inductance
n	Impedance transformation ratio, number of fingers
N	Noise power, number of amplifiers
P	Power
P_{1dB}	Power at 1-dB gain compression
P_D, P_{diss}	power dissipated
P_{DC}	DC power
P_{in}	Power input
P_o, P_{out}	Power output
P_T	Total power
Q, Q_C, Q_L	Quality factor
Q_T	Total quality factor
R	Resistance
R_{in}	Real part of input impedance
R_n	Equivalent noise resistance
R_s	Surface resistivity
R_S, R_G	Source or generator resistance
R_{th}	Thermal resistance
S	VSWR, scattering parameters
S	Siemen unit
SiC	Silicon carbide
SiGe	Silicon germanium
t	Strip thickness, time
$\tan d$	Loss tangent of dielectric material
T, T_a, T_{ch}	Temperatures
V, V_0, V_G	Voltages
W	Strip width of microstrip line transistor unit width
X, X_L, X_S	Reactances
X_{in}	Imaginary part of input impedance
Y	Admittances
Y_0	Characteristic admittance
Y_{in}	Input admittance
Z	Impedance
Z_0, Z_{0m}, Z_m	Characteristic impedances
Z_{in}	Input impedance
Z_L	Load impedance
Z_S	Source impedance
α	Attenuation constant, current gain
α_c	α due to conductor loss
α_d	α due to dielectric loss
α_L	$\alpha_c + \alpha_d$
α_T	Total attenuation constant
β, β_0	Phase constants, current gains
γ	Propagation constant
Γ, Γ_L, Γ_m, Γ_S	Reflection coefficients
δ	Skin depth
∇	Del operator
Δf	Bandwidth
ΔT	Temperature difference
ε	Permittivity
ε_0	Free-space permittivity
ε_r	Relative permittivity/dielectric constant
ε_{re}	Effective ε_r

ε_{ree}	Effective ε_r for even mode
ε_{reo}	Effective ε_r for odd mode
η	Efficiency, impedance
η_0	Free-space impedance ($= 120\pi$ ohms)
η_c	Combining efficiency
η_D	Drain efficiency
θ	Angle, electrical length
λ	Wavelength in microstrip
λ_0	Free-space wavelength
μ	Permeability, stability factor
μ_r	Relative permeability
μ_0	Free-space permeability
ρ	Charge density, reflection coefficient, resistivity, density
σ	Conductivity
τ	Transit time, thermal time constant
ϕ	Phase angle
ω	Radial frequency
ω_0	Center radial frequency
Ω	Ohm unit

Appendix J

Multiple Access and Modulation Techniques

The frequency reuse in cellular systems necessitated the development of several multiple access techniques for allowing users to share the same frequency band. The two simplest techniques are frequency division multiple access (FDMA), which assigns each user within a geographical region a specific frequency channel; and time division multiple access (TDMA), which assigns several users to a common frequency band, but transmit on a rotating basis during specifically assigned time slots. Thus, TDMA has higher capacity than FDMA. Another technique is code division multiple access (CDMA), which allows many users to transmit at the same time within the same frequency band. Each user is assigned a unique "code" or "signature" sequence within the CDMA system. This code sequence allows the transmitter to generate a signal, which may be uniquely recognized by the intended receiver. The CDMA has higher capacity than TDMA.

To send the required signal consisting of voice, data and video the carrier is modulated by the signal to be transmitted. There are several modulation schemes in use. These are:

GMSK: Gaussian minimum shift keying

BPSK: binary phase shift keying

QPSK: quadrature phase shift keying

OFDM: orthogonal frequency division multiplexing

GFSK: Gaussian filtered frequency shift keying

Pi/4 QPSK: $\pi/4$ quadrature phase shift keying

Pi/4 DQPSK: $\pi/4$ differentially encoded quadrature phase shift keying

MPSK: multiple phase shift keying

Fundamentals of RF and Microwave Transistor Amplifiers. By Inder J. Bahl

Index

Note: Following a page number, "t" denotes a table; "f" denotes a figure.

Fundamentals of RF and Microwave Transistor Amplifiers. By Inder J. Bahl

Printed and bound by CPI Group (UK) Ltd, Croydon, CR0 4YY
16/09/2026
14963415-0007